CZECHOSLOVAK ACADEMY OF SCIENCES

AF761715

CZECHOSLOVAK ACADEMY OF SCIENCES

Scientific Editor: *Prof. Dr. Bohumír Rosický, DrSc.*

Foreign Language Editors: *F. G. A. M. Smit, A. M. Wright*

Graphic Design by *Vladimír Kalenský*

PROCEEDINGS OF THE 3rd INTERNATIONAL CONGRESS OF ACAROLOGY

held in Prague
August 31 — September 6, 1971

Edited by
Milan Daniel and Bohumír Rosický

Springer-Science+Business Media, B.V.
1973

Additional material to this book can be downloaded from http://extras.springer.com

ISBN 978-94-010-2711-3 ISBN 978-94-010-2709-0 (eBook)
DOI 10.1007/978-94-010-2709-0

© Springer Science+Business Media Dordrecht 1973
Originally published by ACADEMIA, Publishing House of the Czechoslovak Academy of Sciences 1973
Softcover reprint of the hardcover 1st edition 1973

CONTENTS

Symposium on Scanning Electron Microscopy in Acarology

Section 1

Soil Mites

Section 2

Marine and Freshwater Mites

Section 3

Plant Mites

Section 4

Taxonomy of Higher Categories, Phylogeny and Zoogeography

Section 5

Mites of Stored Products

Section 6

Parasitic Mites, Faunistics and Taxonomy (excluding ticks)

Section 7

Ticks, Faunistics and Taxonomy

Section 8

Mites and Ticks of Medical and Veterinary Importance

Section 9 A

Mites and Ticks in Natural Foci of Tick-borne Encephalitis

Section 9 B

Mites and Ticks in Natural Foci of Other Diseases than TBE

Section 10

Biological, Chemical and Physical Control

Section 11

Physiology, Genetics and Behaviour

Section 12

Ecology of Mites and Ticks

PREFACE

The Third International Congress of Acarology (Prague, August 31 – September 6, 1971) was held under the auspices of the Presidium of the Czechoslovak Academy of Sciences which had accepted the proposal of the Committee of the International Congress of Acarology to organize this meeting of acarologists in Czechoslovakia and subsequently requested the Institute of Parasitology of the Czechoslovak Academy of Sciences to set about this task. The members of the Organizing Committee of the Congress express their gratitude to the Presidium of the Czechoslovak Academy of Sciences, especially to its President Academician J. Kožešník who personally attended the Congress.

The Congress sessions were attended by more than 200 specialists from 32 countries and by World Health Organization representatives. In twelve sections and one symposium 153 papers were read, all of which are included in this volume. The papers are published in extenso, with the exception of a few communications presented by authors who decided to publish them in professional journals. In such cases abstracts of the papers are included in the volume for the sake of completeness.

The sessions were held in congress languages and simultaneous translation was provided into English and Russian. All papers in this volume are printed in English and this necessitated a language revision of the major part of the manuscript. Thanks to the kind efforts of Mr. F.G.A.M. Smit and Mrs. A.M. Wright of the British Museum (Natural History), the Zoological Museum, Tring, this arduous task was accomplished at short notice. Invaluable assistance in the translation of some papers and in the preparation of manuscripts for the press was rendered by Mrs. E. Bělyjová, Mrs. M. Dašková and Mrs. J. Wagnerová of the Institute of Parasitology, Czechoslovak Academy of Sciences. Also invaluable were the efforts exerted by Drs V. Černý, F. Dusbábek and E. Honzáková, the staff members of the same Institute, who assisted with the correction of proofs. To all of them the Editors express their deep gratitude. Sincere thanks are also due to the members of ACADEMIA, the Publishing House of the Czechoslovak Academy of Sciences, especially to Dr. M. Cílková and Mr. J. Kudrnovský, for their cooperation and understanding enabling this volume to be published in the shortest possible time.

The Editors

Committee of the International Congress of Acarology

(acting from 1967 to 1971)

N.G. Bregetova, U.S.S.R.
G.O. Evans, U.K.
G. Mathys, France
P.W. Murphy, U.K.
E. Piffl, Austria
M. van de Vrie, Holland
T.A. Woolley, U.S.A.

Executive Committee of the International Congress of Acarology

(acting from September 1971)

M. Daniel, Czechoslovakia, Past–President[1]
F.J. Radovsky, U.S.A., Secretary [2]
D.A. Griffiths, U.K., Treasurer[2]
C. Athias–Henriot, France[2]
Yu. S. Balashov, U.S.S.R.[2]
M. Costa, Israel[2]
A. Filipponi, Italy[2]
H. Hoogstraal, U.S.A.[2]
E. Piffl, Austria[3]
M. van de Vrie, Holland[3]

[1] Acting in his place as Past-President at the request of Prof. B. Rosický
[2] Elected September 1971
[3] Elected July 1967

Officers of the Congress

President

B. Rosický

Vice-Presidents

N.G. Bregetova, U.S.S.R.
G.W. Wharton, U.S.A.

Sectional Chairmen

C. Athias-Henriot, France
C. Bader, Switzerland
Yu.S. Balashov, U.S.S.R.
J. Boczek, Poland
N.G. Bregetova, U.S.S.R.
G.O. Evans, U.K.
D.A. Griffiths, U.K.
C.A. Hall, U.S.A.
H. Hoogstraal, U.S.A.
W. Knülle, West Berlin
V.V. Kucheruk, U.S.S.R.
B. Rosický, Czechoslovakia
V.I. Volgin, U.S.S.R.
G.W. Wharton, U.S.A.

Organizing Committee in the Host Country of the Third International Congress of Acarology

Chairman

B. Rosický

Vice-Chairman

M. Daniel

Secretary

K. Samšiňák

Committee Members

J. Ašmera
V. Černý
F. Dusbábek
E. Honzáková
J. Kolman
J. Kramář
J. Kratochvíl
M. Kunst
M. Mrciak
J. Nosek
M. Přívora
J. Řeháček
J. Schmid
J. Vaněk
P. Verner

Honorary Members of the International Congress of Acarology

The following distinguished acarologists were elected by the Third International Congress of Acarology as Honorary Members for life, in recognition of their valuable contributions to Acarology.

E. Browning, U.K.
F. Grandjean, France
Taiji Imamura, Japan
I.I. Sokolov, U.S.S.R.

OPENING ADDRESS

J. KOŽEŠNÍK
President of the Czechoslovak Academy of Sciences, Prague

Ladies and Gentlemen,

speaking on behalf of the Czechoslovak Academy of Sciences and its Presidium allow me to welcome all participants of the Third International Congress of Acarology in Prague.

The Institute of Parasitology of the Czechoslovak Academy of Sciences has been requested by the Presidium to organize this international assembly of scientists as that Institute employs the largest group of our acarologists.

Acarology as a science, dealing with the vast group of mites and ticks, is well qualified to contribute an important share to biological knowledge and to that of natural laws in the animal kingdom and to the problems associated with the advance in science and technology. The results of acarological research, be it the study of soil fauna, of mites causing damage to stored products and plants or the research of the role of ticks and mites as vector of diseases of man and domestic animals, directly help to increase the efficiency of plant and animal production as well as to protect the health of man and domestic animals.

These results have been widely acknowledged by workers in agriculture, industry, human and veterinary medicine and have been making the public aware of the possibilities for protecting their health and food products.

The Czechoslovak Academy of Sciences recognises the importance of acarology. Some of its institutes, namely the Institute of Parasitology, the Institute of Entomology and the Institute of Virology have made considerable advances in this scientific field. The fact that the Czechoslovak Academy of Sciences has been asked to organize the Third International Congress of Acarology is evidence of the esteem enjoyed by Czechoslovak acarologists and their work abroad. Especially in tick research the Czechoslovak school holds a prominent place.

During this Congress all main problems of acarology will be discussed. Classification of mites, soil mites, mites infesting plants and stored products, water mites, mites and ticks parasitizing man and animals, their distribution and role in epidemiology will be some of the main items discussed here during the next few days.

I have noted with pleasure that the problems of natural focality of diseases, a theory developed by the former foreign member of our Academy, the late Academician E. N. Pavlovsky, have also been included in the programme.

I am convinced that the Congress participants will contribute to further development of acarology by their papers and fruitful discussion. Indeed, I believe that this Congress, organized in the capital of Socialist Czechoslovakia, will be another milestone in the development of acarology, a scientific discipline which holds great promise yet for both the theory of the development of organisms and the practical side, especially in plantgrowing, the food industry, and in human and veterinary medicine.

I wish your Congress the best of success and wish you all a very pleasant stay in our capital and in our Socialist Republic whose people will follow this event with great interest.

OPENING ADDRESS

B. ROSICKÝ

Vice-President of the Czechoslovak Academy of Sciences, Prague
President of the 3rd International Congress of Acarology

Ladies and Gentlemen, dear Colleagues,

on behalf of the Organizing Committee of the Third International Congress of Acarology I have the honour and pleasure in opening this Congress and in welcoming all participants from abroad as well as from Czechoslovakia.

I cordially welcome here the President of the Czechoslovak Academy of Sciences, Academician J. Kožešník and the members of the Committee of the International Congress of Acarology.

I would also like to welcome Professor G. Anastos who has been delegated by the World Health Organization to discuss the WHO project concerning the research of tick biology.

The Institute of Parasitology has organized this Congress in close cooperation with the Committee of the International Congress of Acarology.

As a result of a short preliminary meeting of the International Committee before the opening of this Congress it is proposed that its Vice-Presidents be Dr N. G. Bregetova (USSR) and Dr G. W. Wharton (USA).

In 12 sections and 1 symposium our main task at the Congress will be to deal with the present problems of acarological research. The comparison of results obtained by different laboratories, the exchange of views on submitted materials, analyses and hypotheses may benefit future trends of acarology and may solve various open problems satisfactorily.

The Organizing Committee has tried to do its best to fulfil all the wishes of our honoured guests and I do hope that our sessions here will not only broaden and deepen the co-operation of acarologists in the scientific field but will also help to establish good personal contacts.

Speaking on behalf of the Organizing Committee and all staff members of the Institute of Parasitology and other acarological research centres in the Czechoslovak Socialist Republic I do hope that the friendly exchange of views will contribute to further advances in the development of acarology.

ACAROLOGY AND ITS PRACTICAL IMPORTANCE

Introduction

BOHUMÍR ROSICKÝ
Institute of Parasitology, Czechoslovak Academy of Sciences, Prague

The Third International Congress of Acarology is resuming the work of the previous meetings of acarologists in Fort Collins in 1963 and in Nottingham in 1967. Acarology is a young science which has, according to Bregetova (1966), lagged behind entomology by about 150 years. However, during these last few years acarology has been rapidly making up for lost time. As a science which should develop progressively and prove that it is greatly contributing not only to theory but to practice as well, it should necessarily throw light upon the scope and importance of its object. Thus it was not by chance that at the First Congress G. W. Wharton in his presidential address recapitulated the development of acarology. Up to 1963 a total of 17,500 species of mites and ticks had been recorded. The number of species then undescribed was assessed to be from half a million to one million.

At the Second Congress its President T. E. Hughes outlined the new trends which should be followed in acarological research.

After consulting my fellow acarologists as to what should be the subject of my introductory paper to the Third Congress, I arrived at the conclusion that a survey of the practical importance of acarology would be quite suitable. A new scientific field carries weight only if its results have practical value. The development of science is usually subsidized, provided that the sum invested will be reflected in national economy.

The Acari constitute an animal group distributed from Polar regions to highest altitudes and with its very different ecology it has a prospect of playing significant role in national economy. The mites form a very large and important component of fauna whose adverse economic effects have been overlooked until recently.

A notable feature which is of great practical consequence in mites, is their size. In many groups, especially in parasitic or phytophagous ones, they are tiny organisms which can be compared with spores of some microscopic fungi, with some protozoans, yeast-plants or even with large bacteria. Their size ranges from 100 to 500 μm. This fact characterizes their possible influence upon animals and plants and points to specific possibilities of distribution unknown in other animals.

Another peculiarity of mites, especially of some parasitic and phytophagous species, is their reproductive ability. For example, it is not exceptional to find up to 70,000 specimens of the species *Dermanyssus gallinae* or *D. hirundinis* in a single nest of some bird species. The data on the number of eggs laid by a single female tick (according to species up to 17,000 eggs may be produced) are common knowledge. Fiebiger (1923) reports Gerlach's assessment of fertility of scab mites of the genera *Psoroptes* and *Chorioptes*. If one female lays 15 eggs only, in 15 days there will be 10 females and 5 males. In 3 months there will be one and a half million mites! No wonder that for example on a scabby chamois 600-1,000 scab mites of the genus *Sarcoptes* have been found per 1 cm^2 of skin. Up to several tens of thousands of mites of the family Acaridae have been found per 1 kg of stored products. In cases of heavy infestations, from 1,000 to 3,000 specimens of the species *Tyrolichus casei* may be found per 1 cm^2 of cheese rind. They destroy the cheese surface and the resulting powder contains large numbers of faeces, egg membranes, dead mites and exuviae. In 1 g samples of house-dust collected in Japan Oshima Shiro (1970) found an average number of 1,045 mites belonging to 40 species.

Being one of the foremost animal groups as regards numerosity, mites and ticks play an important role in biogeocenoses (ecosystems) of separate biomes of the Earth.

From tundra soils, the vegetation of temperate zone and from the Tropics to the Antarctic region they form an important component of the edaphic fauna. Their numbers and varied ecological relationships - from saprophagia to parasitism - condition the communities which markedly influence the processes in soil. During recent years the studies on soil mites have been intensified. Apart from clearly theoretical aspects it has been ascertained that 1. mites play a decisive role in humification; studies have shown that the forms of practical importance are mostly common and widespread species with a large number of specimens; 2. mites living in soil may be vectors or intermediate hosts of different pathogenous agents (from viruses to helminths).

Soil mites are typical representatives of a soil microfauna, inhabiting soil pores and other minute soil spaces. In contrast to the macrofauna they are incapable of active penetration of the soil. They inhabit natural soil spaces filled with air (primarily in upper layers). Together with the springtails (Collembola) they are the most important representatives of the aerobiontic microfauna. In grassland (meadows, pastures etc.) of Central Europe the number of mites is said to be between 17,000 and 270,000 per 1 m^2 (average number 100,000). In fields their numbers range from 10,000 to 90,000 per 1 m^2 (average number 35,000) (Tischler, 1965). Their biomass in meadow soils is reported to be 1 - 2 g m^2. In forest habitats their representation is about 2,5 times that of grassland and more than 20 times that of field habitats (Gatilova, 1964). The soil mites are mainly phytophagous, fungivorous and saprophagous forms, predators and parasites. Of exceptional importance are the oribatid mites which participate markedly in the circulation of substances in soil. They triturate mechanically substances, which would otherwise be inaccessible to microbes, into increasingly smaller particles and accumulate them in their faeces. For example, according to Margowsky and Prusinkiewicz (1955) in a riverbank meadow 78,400 mites/m^2 processed 107.6 cm^3 of organic substances per year. The mites of different groups, together with other tiny arthropods, control the growth and development of soil microflora and take part in its dissemination. In contrast to other microfaunal components they are more resistent to dessication. The soil microfauna (with its decomposers) compensates for the annual energy turnover through respiration. The species composition of small decomposers (Nematoda, Enchytraeidae, Acari and Collembola) varies as regards the dominant or insignificant species, but the metabolic activities remain surprisingly the same (Edwards et al., 1970). Oribatid mites are especially abundant in humus rich in moulds and fungi. Predacious mites occur in all types of the soil. The numbers of mites in soil depend on the season of the year, on anthropogenic interference with the soil (manuring, tilling) etc.

The above data show the significance of soil mites in processing the organic matter into humus, for the life of the soil as an organic component of ecosystems in which they help to maintain an equilibrium and to increase the fertility of soils.

Several soil mites may be vectors of various agents pathogenous to man (e.g. chiggers of the family Trombiculidae whose larvae transmit rickettsia *R. orientalis*, responsible for scrub typhus, are typical microcavernicolous inhabitants of the soil penetrating the deeper soil layers. The soil also serves as a shelter for ticks during their life span. Among typical oribatid mites some species are utilized as intermediate hosts by cestodes of sheep and other ruminants, hares, rabbits, etc.

In the ecosystem (biogeocenosis) mites are present along with all other biotic components, from plants to animals. As pointed out above they take part in processes in the soil and undoubtedly contribute to its organic character. They affect the root system of plants, attacking also the plants themselves. They are present in the nests of animals or on their cover or in their body cavities.

Diverse ecological relationships among mites may be utilized in biological control and in integrated control of mites themselves. Predatory mites attack not only phytophagous mites, but various species of insect pests as well (e.g. some aphids, bugs of the family Reduviidae). There are a number of examples in literature showing that mites of the family Phytoseiidae and Ascidae control populations of plant pests of the family Tetranychidae. Mites damaging stored products are destroyed by other mites of the family Cheyletidae. However, their usage in biological control is rather problematic. Very effective in the control of Tetranychid mites are predators killing adult females. They decrease the pest population much more efficiently than those predators which feed on eggs or immature stages incapable of reproduction. Also among obligatory or facultative parasites attacking some insects are species which could be used in biological control of mites. The use of mites in biological or integrated control of some pests is an example of positive exploitation of mites for human gain.

As plant pests, mites have adapted themselves to most diverse ecological conditions. There are species which just live on plants and on the other hand others which feed on plants. Mites of the family Tetranychidae, for example, pierce the surface layers of leaves and buds, feeding on the contents of cells and excreting substances into the host plant which are harmful not only locally but are responsible for a general withering of the plant. Sometimes even the fruits are affected. Tetranychids are not greatly specialized to one particular host plant and often live on many species of plants. Well-known are the species which infest a number of plant species while new and harmful species are still being discovered. Tetranychids may cause damage in large areas. For example, the species *Oligonychus ununguis* attacked more than 800,000 acres of fir stands in Montana and Idaho in 1957 (Krantz, 1970).

Economically important damage is caused by these plant pests in green houses which offer an exceptionally favourable environment for some species.

Other species penetrate the plant tissues or live inside the plants, causing their deformations (the so-called leaf curl) or new formations, specially galls (gall-forming mites). There is no crop in the temperate, subtropic or tropic zone of the Earth which would not be attacked by mites. The harmful effects of mites on plants have not been thoroughly elucidated and in some instances they have been apparently underestimated.

When sprays of synthetic contact insecticides began to be used, the selection of mites (e.g. Tetranychids) which remain unaffected by insecticides and continue to multiply freely without being endangered by their natural enemies, has become an important phytopathological problem. Apart from this the preservation of predatory mite species must be taken into account as they are very important in the control of pests but are distroyed by spraying. A considerable section of the chemical industry producing pesticides also prepares substances against resistant mite species. The growing resistance in mites and ticks to acaricides in use poses another problem. New and more efficacious substances must therefore be looked for. Thus emerges a new national economic task associated with mites, a group completely different from insects.

The plant mites, however, are not only direct pests or parasites. Like parasites of vertebrates they may play a role as vectors of plant diseases. They disseminate, mostly mechanically, mycotic agents causing decay. Recently the problem of their role in the dissemination of plant viruses has come to the foreground. Five strains of viruses are transmitted by mites (Shykhuis, 1963).

There is scarcely any other animal group which can become adapted to synanthropic life as mites do. Besides distinct synanthropic forms, omitting the obligatory or facultative parasites, there are quite a few groups in which species may be found which adapt themselves spontaneously to living conditions in structures built by man throughout the world (hemisynanthropic types). Mites spontaneously colonize human dwellings and outhouses, being attracted by a possibility of feeding on food stored by man (commensals), by the presence of domestic animals and their activities (coprophagous and saprophagous species), or just for sheltering. These relationships may be very diverse. Synanthropic mites are thus spontaneous members of human settlements (like other synanthropic animals). The mites not only freely penetrate accessible temporary or permanent human dwellings and adjoining agricultural buildings, but also the most modern houses built from new building materials.

In future a further development of some species in new building materials may be anticipated and therefore the present technology of some organic materials should be adapted so as to prevent any mite development. Only in this way calamities arising in dwellings where insufficiently processed materials have been used, may be avoided. Samšiňák (1958) describes a case in which pressed shells of sunflower seeds were used for producing insulating material (the so-called sopalite) which is laid under parquet floors. Remnants of oil preserved in the pressed shells supported mite populations even after the pressing and other processing of the material. In this way new flats incorporating sopalite insulating material were overrun with mites. One of the species found was responsible for causing dermatitis in inhabitants of the flats, mainly children.

The synanthropization of mites is associated with their incidence in house dust which accumulates in certain places in dwellings. The inhaled house dust containing remnants of dead mites primarily of the genus *Dermatophagoides* (here again the size phenomenon in this group becomes evident) causes asthmatic attacks and allergic reactions of different types, especially allergic rhinitis (Spieksma, Spieksma-Boezeman, 1967). Mites of the family Pyroglyphidae, mainly of the genus *Dermatophagoides*, have often been found in house dust, even in newly finished modern flats built in climatically humid regions. The species involved are *D. pteronyssinus* and *D. farinae* and some others, e.g. *Malayoglyphus intermedius* in South East Asia (Fain et al., 1969). Fain (1966, 1967) presumes that their food consists of dandruff or skin scales. The mites were also found in the dust of mattresses (Maunsell et al., 1968). In households where persons suffered from asthma, large numbers of mites were found in the dust. As many as 94 % of persons allergic to house dust showed a skin reaction to extract from the mite *D. pteronyssinus*. During other investigations 61 samples of dust were collected in houses and all proved to contain house dust mites, up to 160 specimens per 0.1 g of dust (Larson, 1969). Species of the genus *Dermatophagoides* are common in birds'nests, where they feed on detritus, and occasionally they have been found in the feathers of birds. Likewise, some other genera and species of the family Pyroglyphidae have been found in nests of birds and mammals or in stored products. These findings indicate sources from which the mites have spread to man.

Acarologists began to study this phenomenon most intensively in order to elucidate the origin of various types of allergy in man caused by house dust mites and thus to provide pre-requisites for their control. Perhaps it will not be an exaggerated assertion if I say that many allergies may, after detailed examination, be found to be associated with the presence of house dust mites in older as well as modern human settlements.

The phenomenon of synanthropization is also met with in mites infesting stored products of organic origin. Mites living originally in small numbers on seeds, plants, decaying vegetation, in nests of mammals etc. are gradually introduced to all types of stored products due to man's economic activities (food storage). Mites attack either stored unprocessed products (grain, vegetables, fruits, meat, textile plants, hay etc.) or prepared products (flour and flour products, cheese, wine, dried fruit, etc.). They are primarily mite species of the family Acaridae.

Acarid mites are important pests because they multiply rapidly, spread readily from place to place while their hypopus survives many years in the stage of latency.

Therefore they easily become a source of new contaminations, resisting all standard insecticides. When they first occur they escape detection. By their activities they harm the quality of stored products as they destroy and contaminate them with their excrements and secretions, exuviae and with dead individuals making them unfit for consumption and thus dangerous to health of man as well as domestic animals. There is no group of arthropods which could be so closely associated with human dwellings and agricultural buildings as the family Acaridae.

Thus a new scientific field, concerned with mites as pests of stored products, has originated (Hughes, 1961). Infestation of stored products by mites poses a great problem all over the world. An assessment of damage caused on global scale has not yet been made. As regards Czechoslovakia, according to Verner and Pulpán (1965) the damage caused in stored cereals by the species *Acarus siro* and *Glycyphagus destructor* has been estimated to be about 8 million Czechoslovak crowns (equivalent to £ 400,000). While inspecting 36 store houses of cheese in Czechoslovakia Pulpán and Verner (1965) ascertained that in hard cheeses a layer up to 4 cm deep was damaged. The mere presence of acarid mites poses a serious hygienic problem

Mites are also important pests of cultivated fungi, especially mushrooms. In the United States and Canada species of the genera *Linopodes* (family Eupodidae) and *Tyrophagus* (family Acaridae) occur in large numbers in commercial mushroom houses where they often destroy the whole crop of mushrooms (David, 1944).

Acarid mites infesting stored products may often escape attention and thus the food infested by them (especially grain, flour etc.) becomes toxic to man and domestic animals. The mites consumed together with food cause intestinal and urinary afflictions in animals, in more serious cases even abortion and paralysis, e.g. in mares and saws (Speransky, 1971).

Mites consumed with food are of no parasitological importance. Recent studies showed that excrements of mammals always contained dead mites from food or forage. The toxic effects in man and domestic animals caused by grain, flour and other food products polluted by mites have often been described in the literature. The mite activities in some food products or forage may also lead to their rapid contamination with dangerous fungi. Mites developing in mouldy grain become wrapped in spores of fungi which cover their bodies or are excreted with their faeces. Aucamp (1960) ascertained that ground nuts may become contaminated by the fungus *Aspergillus flavus* due to activities of mites which thus contribute to the contamination of ground nuts with aflatoxin.

The body of acarid mites is covered with setae and fine hairs which may cause dermatitides, conjunctivitis, damage to the respiratory tract, colic in the digestive organs etc. of persons handling food products infested with mites (in agriculture, food industry etc.). Well known, for example, is a dermatitis (copra itch) caused by mites in workers handling copra. Laarman (1952) reports that all workers unloading copra in the Rotterdam docks suffered from itchy rash on neck and arms, caused by the mite *Cosmoglyphus laarmani.* The skin reaction became apparent as late as 24-72 hours after contact with the mites. The symptoms disappeared as soon as the contact with the contaminated substance ceased. Laarman did not find any traces of mite feeding.

The role of mites parasitizing animals is of no less importance. From multipods and insects, to Echinodermata and mammals there is no animal group which is not parasitized by mites. From the economic point of view of greatest interest are parasites of man, domestic and other production animals. Especially abundant among mites are ectoparasites and parasites living in the skin of vertebrates.

All developmental stages of somatic parasites (mites of the genera *Sarcoptes*, *Psoroptes*, *Chorioptes*, *Otodectes*, *Knemidocoptes* etc.) live on their hosts, no stage lives off its host in an outer environment. To the somatic parasites belong species which can cause most extensive damage to the human body (scabies) and great economic losses as well (mange).

The modern age also introduces new epidemiological aspects on the dissemination of scabies. Forms are now more often encountered which do not reveal the classical clinical picture known from the literature.Subclinical forms arise being influenced by regular washing with soap and use of underwear and clothes made from artificial fibres. In the tropics, scabies in its moderate clinical form, still belongs to the most frequent skin afflictions (Clarke, 1970).

Considerable damage is caused by sarcoptic, psoroptid and other mites of game animals. A classical case of sarcoptic mange caused by *Sarcoptes scabiei* varietas *caprae* or varietas *rupicaprae* in chamois *(Rupicapra rupicapra)* in the Austrian Alps is still well remembered. Due to the presence of goats infected with sarcoptic mange and grazed in habitats of chamois, large numbers of chamois perished on many mountain ranges. Another case of sarcoptic mange in wild boars kept in a forest reservation was described by Fiebiger (1923).

Many afflictions caused by mites living as ectoparasites or endoparasites were described for animals kept in zoological gardens. Animals kept in an environment which, in spite of the best zoo conditions, does not correspond with their normal living requirements, become predisposed to mite infestations.

To the widespread somatic ectoparasites belong the members of the family Demodicidae. The incidence of *Demodex folliculorum* in man is estimated by some authors to be as high as 100 %. In dogs considerable damage is caused by *D. canis*, especially in breeds with fine hair and kept indoors as pets. The percentage of demodicosis in dogs cannot be assessed because the disease is not being systematically registered. Planned diagnostics indicates that demodicosis as a clinical disease is considerably spread and may be significant in dog breeding, because it is a congenital disease. The significance of demodicosis lies in the fact that the whole litter (or 70 - 80 %) is usually contaminated, even repeatedly from a clinically healthy mother. *Demodex canis* was demonstrated in 52.9 % of examined samples of normal skin of dogs. Literature data indicate that the number of cases of demodicosis has lately increased. During the last year 14 dogs were treated for demodicosis at the Institute of Parasitology. The dogs belonged to 8 breeds and represented valuable individuals assigned for further breeding.

Parazitation of cattle by somatic mites *(Sarcoptes*, *Chorioptes*, *Psoroptes*, *De-*

modex etc.) results in damaged hides and considerable losses in the quality of hides used in the leather industry. In 1960 the damage in hides produced in Czechoslovakia was estimated to be 20 - 30 million Czechoslovak crowns.

The next group of mites belongs to the nidicolous ectoparasites. They are gamasoid mites (Dermanyssidae, Laelapidae etc.), burrow ticks of the family Ixodidae, and the family Argasidae. They cause damage by direct feeding, but a number of species is known as vectors transmitting diseases to man and animals. Mites of the family Dermanyssidae have been recorded as hemisynanthropic species, mainly *Dermanyssus gallinae* and *D. hirundinis* which under certain ecological conditions penetrate human dwellings from birds' nests built on houses and thus represent a serious nuisance, even in cities. In the tropics a similar role is played by the species *Ornithonyssus bursa*. The same mites often endanger poultry breeding.

The last and very numerous group constitute external ectoparasites, attacking their hosts in nature (ticks - Ixodidae, Trombiculidae, as well as some Gamasoidea). Chiggers (Trombiculidae) often cause persistent dermatitides. For example, Červa and Daniel (1964) recorded about 3,000 cases of dermatitides caused by the chigger *Neotrombicula autumnalis*, on the basis of data collected by health centres in Central Bohemia in 1960. The cases occurred primarily in the suburbs of Prague and in the adjoining villages.

Ticks are not only known as vectors of dangerous diseases of man and animals, but they are also a nuisance to cattle, annoying and disturbing the animals by blood sucking. The damage caused by ticks in animal production is of the following types:

1. Damage caused by direct tick feeding on cattle. According to our investigations in untended pastures in Slovakia, the blood loss per head may be as high as 1 litre in one season. This calculation is quite realistic. We based it on the weighing of engorged females of *Dermacentor marginatus* and ascertained that the weight of a fully engorged female may reach 0.8 g. An additional loss of blood in cattle should be taken account of due to the digestion of blood by ticks. The literature data indicate the loss of blood to be 3 - 5 litres per head. Ticks sucking in "clusters" cause additional bleeding from wounds cut by cluster feeding. The local blood coagulation is weakened by the effects of secretions of the salivary glands of ticks. In the sites of cluster feeding, the blood dries up and forms blood cakes in the animals' coat or among its tail hairs. A high infestation may cause anaemia in cattle. Such a constant weakening of cattle, especially in a spring following an unfavourable winter, causes a decrease of weight (at least by half in comparison with healthy heads), a decrease in yield of milk during lactation, a retardation in growth of young cattle and a reduction in the performance of draught animals.

The decrease in milk yield in infested cows constitutes about 25 - 40 % in comparison with uninfested cows. The blood-loss due to tick feeding is associated with constant annoyance of cattle by the attached ticks, which cause dermatitides, accompanied by itching, skin inflammation, swelling of the sites of attack and even their ulceration. There are also toxic effects of the excretions of the salivary glands which may induce general symptoms of poisoning in infested animals. This envenomination has adverse effects on the condition of domestic animals throughout the period when the ticks are feeding on cattle.

2. The second group includes damage inflicted to cattle by ticks as vectors of various diseases. Foremost are cattle diseases caused by parasitic protozoa of the class Piroplasmasida.

3. In some regions of the world ticks induce the so-called tick paralysis. It affects mostly lambs, calves, etc. and is known in Australia, North America and other regions. It may also affect man.

4. Further economic losses are inflicted to animal hides by attached ticks. At the site where the tick is sucking a perforation and infiltrate occurs, thus devaluing the hides. Ticks feeding in clusters (e.g. the genus *Dermacentor)* cause ulceration of skin resulting in extensive skin damage.

5. The damaged skin permits the entry of other infectious germs which may cause local or general afflictions of domestic animals and which are disseminated mechanically by flies, through urine, faeces etc. This predisposition promotes infection with some fungi, leptospirae, Brucella or it may be the point of entry for myiasis--causing fly larvae.

6. Ticks threaten the health of pastured domestic animals as well as of persons coming into contact with tick-infested animals or their products (milk, hides, wool). Epidemics of tick-borne encephalitis disseminated by drinking unboiled goat milk, the spread of Q-fever, tularaemia, etc. are common knowledge.

The importance of ticks and chiggers, however, is stressed by their role as vectors. Mites and especially ticks are known as vectors of arboviruses, rickettsiae, bedsoniae, bacteria, spirochetae, protozoa, parasitic worms, etc.

An important milestone in the history of the importance of mites to human health were the events in South East Asia during the Second World War, connected with the high incidence of tsutsugamushi. Philip (1948) reports that 4 days of activities of one British division in the Ceylon jungle resulted in 756 soldiers becoming ill and hospitalized with tsutsugamushi transmitted by chiggers (Trombiculidae).

A great many causative agents transmitted by mites and ticks induce diseases which are localized in nature in the so-called natural foci of diseases (Pavlovsky, 1939). The mites in natural foci play a double role - primarily that of vectors with epidemiological importance, i.e. they transmit infections directly to man. For example, the tick *Ixodes ricinus* transmits the virus of tick-borne encephalitis of the Central-European type to man. Besides, they play a role as maintainance hosts of pathogenous agents in the biocenosis of natural foci. They propagate the infection among reservoir animals. In this propagation may take part also species which are normally unimportant in the epidemiological process proper and which do not attack man or domestic animals. Many species of gamasoid mites and burrow ticks are rightly considered to be possible vectors, as evidenced in publications by various acarologists (see, for example, surveys of Bregetova, 1966, Zemskaya, 1962 etc.). The spatial structure of natural foci is extremely closely associated with characteristic features of the ecology of vectors (and of the reservoir animals as well).

Modern scientific methods indicate the growing importance of the role of ticks in the spread of some arboviruses. At present more than 50 arboviruses transmitted by ticks are known. Group B includes a complex of arboviruses causing tick-borne encephalitis, one of the most important infections of the central nervous system in man and widespread in nontropical Eurasia. This group also includes causative agents of the Omsk haemorrhagic fever and Kyasanur forest disease. More than 40 arboviruses which do not belong to the B group, have been isolated from ticks (families Ixodidae and Argasidae). They are causative agents of well-known diseases such as Colorado tick fever, Crimean haemorrhagic fever and Kemerovo tick fever. New arboviruses are being isolated from ticks. The elucidation of their relationship to man or domestic animals is the object of urgent research throughout the world.

The activities of man associated with modern forms of recreation involves man in a closer contact with mites and ticks (families Ixodidae, Argasidae, Sarcoptidae, Trombiculidae, Dermanyssidae, Acaridae etc.). Large numbers of domestic pets kept under urban and rural conditions may be introducing mites into homes and may thus be responsible for illnesses afflicting their owners or other members of the family.

Of the endoparasites of vertebrates especially the species *Cytodites nudus*, causing peritonitis or enteritis in the body cavities of hens, should be taken into account; next in importance are parasites living in the lungs of monkeys (*Pneumonyssus*), in those of rodents (Pneumocoptidae), in nasal cavities of seals and walruses (Halarachnidae) and of birds (Rhinonyssidae), etc. Research on these parasitic groups results in new discoveries.

From an economic point of view a special group constitutes parasites and predators destroying bees. Apart from classical acarinosis, there is a number of predators which deserve attention, especially the species *Varroa jacobsoni*.

Fish are also parasitized by mites. In Czechoslovakia, e.g., *Hydrozetes lacustris* parasitises the skin and gills of fish. It mainly attacks fry and is also a pest of spawn (Dyk, 1961).

A special chapter on the importance of mites is the problem of their passive dissemination over great distances Their size plays an extraordinary role again. Their spread by ship, railway, road or air transport has been repeatedly proved. It takes place continuously as a result of their size. Transportation of fruits, vegetable, grain flour, etc. is a suitable means for their distribution, especially in the case of members of the family Acaridae. Other movements of hosts play also a role in forms which are parasitic (ticks). Of epidemiological and epizootological importance is the transportation of ticks and other mites over long distances by migratory birds. This problem has recently attracted great attention.

In this survey I could only briefly outline the economic and medical importance of mites and their role in nature. The significance of mites in national economy on a global scale will have to be critically studied in detail. All facts hitherto known about mites and ticks make us ponder on their role in the economy of man and emphasize our responsibility towards human society.

References

Aucamp, J.L.: The role of mite vectors in the development of Aflatoxin in groundnuts. J. Stored Rex. Prod. Res. 5 : 245 - 249, 1969.

Bregetova, N.G.: Economical importance of mesostigmatic mites in medicine, veterinary and agriculture. In: Zagadnienia akarologii. PWRL, Warszawa, pp. 51 - 76, 1966.

Červa, L., Daniel, M.: K častosti výskytu trombikulosy ve středních Čechách. Čs. parasitol. XI: 71-76, 1964. (In Czech.)

Clarke, G.H.V.: Diseases of the skin. In: Jelliffe: Diseases of childern in the subtropics and tropics. Edward Arnold, London, pp. 604-638, 1970

Davis, A.C.: The mushroom mite/*Tyrophagus lintneri* (Osborn)/ as a pest of cultivated mushrooms. U.S.D.A. Tech. Bull. 879: 26, 1944.

Edwards, C.A., Reichle, D.E., Crossley, D.A.: The role of soil invertebrates in turnover of organic matter and nutrients. In: Reichle (ed): Analysis of temperate forest Ecosystems. Springer, Berlin, Heidelberg, N.York, pp. 147-172, 1970.

Evans, G.O.: Proceedings of the 2nd International Congress of Acarology. Akadémiai Kiado, Budapest, pp. 652, 1969.

Fain, A.: Nouvelle description de *Dermatophagoides pteronyssinus* (Trouessart, 1897) importance de cet acarien en pathologie humaine (Psoroptidae: Sarcoptiformes). Acarologia 8: 302-327, 1966.

Fain, A.: Le genre *Dermatophagoides* Bogdanov 1864, son importance dans les allergies respiratoires et cutanées chez l' homme (Psoroptidae: Sarcoptiformes) Acarologia 9: 179-225, 1967.

Fain, A., Cunnington, A.M., Spieksma, F.Th.M.: *Malayoglyphus intermedius* n.g., n. sp., a new mite from house dust in Singapore and Djakarta (Pyroglyphidae: Sarcoptiformes). Acarologia 11: 121-126, 1969.

Fiebeger, J.: Die tierischen Parasiten der Haus-und Nutztiere, sowie des Menschen. Braumüller Wien u. Leipzig, pp. 442, 1923.

Gatilova, F.G.: On the fauna of Oribatei in the forest-steppe of the Volga highland. In: Pochvennaya fauna SrednegoPovolzhya.Nauka, Moskva, pp. 120—132, 1964.

Hughes, A.M.: The Mites of Stored Food. Ministry Agr. Fish. Food. Tech. Bull. 9, London 287, 1961.

Krantz, G.W.: A manual of acarology. O.S.U. Book Stores, Corvallis, Oregon (U.S.A.) pp. 335, 1970.

Laarman, J.J.: Copra itch in Netherland. Documenta de Medicina Geographica et Tropica 4, 3: 268-272, 1952.

Larson, D.G. et al.: Preliminary studies on *Dermatophagoides farinae* Hughes and house dust allergy. J. Med. Ent. 6 : 295—299, 1969.

Margowski, Z., Prusinkiewicz, Z.: The influence of ecological factors on the occurrence and vital activities of the small soil fauna, Poznanskie Towarz. przycaciol nauk 26: 1-31, 1955. (In Polish.)

Maunsell. K. et al.: Mites and house-dust allergy in bronchial asthma. The Lancet 15 : 1267 — 1270, 1968.

Oshima Shiro: Studies on the mite fauna of the house-dust of Japan and Taiwan with special reference to house-dust allergy. Eisei dobutsu. Jap. J. Sanit. Zool. 21: 1-17, 1970.

Pavlovsky, E.N.: (On the natural foci of infectious and parasitic diseases.) Vestn. Akad. Nauk SSSR 10: 98-108, 1939. (In Russian.)

Pavlovsky, E.N.: Natural nidality of transmissible diseases with special reference to the landscape epidemiology of zooanthroponoses. Univ. of Illinois Press, Urbana and London; 261, 1966.

Philip. C.B.: Tsutsugamushi disease(Scrub Typhus) in World War II. J. Parasitol. 34 : 169 — 191, 1948.

Rosický, B., Kratochvíl, J.: Synantropie savců a úloha synantropních a exoantropních savců v přírodních ohniskách nákaz. Čsl. biol. 2: 283-295, 1953. (In Russian.)

Samšiňák, K.: Roztoči ze slunečnicových semen. Čsl. parasitol. V /1/: 185-198, 1958.

Slykhuis, J.L.: Mite transmission of plant viruses. Advances in Acarology 1: 326-340, 1963.

Speransky, V.V.: Acaridoses in man caused by tirogliphoid mites.(Obzor literatury).Med. Parazitol. (Moskva) 49: 488-491, 1971.

Spieksma, F. Th. M., Spieksma-Boezeman M. I. A.: The mite fauna of house dust with particular reference to the house dust mite *Dermatophagoides pteronyssinus* (Trouessart, 1897) (Psoroptidae: Sarcoptiformes). Acarologia 9: 226-241, 1967.

Tischler, W.: Agrarökologie. VEB G. Fischer, Jena, pp. 499, 1965.

Verner, P.H., Pulpán, J.: Control of tyroglyphoid mites in stored grain by the predatory mite *Cheyletus eruditus* (Schrank) Can. J. Zool. 43: 417–432, 1965.

Verner, P.H., Pulpán, J.: Roztoči škodící v mlékárenském průmyslu na tvrdých sýrech a jejich hubení.Sb. Věda a výzkum v potr. prům. XVI. 95-114, 1965.

Zemskaya, A.A.: Gamasoid mites. In:Perenoschiky vozbuditeley prirodnoochagovych bolezney. Medgiz. Moskva, pp. 291-323, 1962. (In Russian.)

SYMPOSIUM ON SCANNING ELECTRON MICROSCOPY IN ACAROLOGY

Chairman

D. A. Griffiths (U.K.)

Assistant Chairmen

J. Lom and J. Vávra (Czechoslovakia)

SENSORY RECEPTORS IN TICKS AND MITES[1]

R. C. AXTELL, R. F. FOELIX, L. B. COONS and M. A. ROSHDY[2]
Department of Entomology, North Carolina State University, Raleigh, N. C. USA

Sensory receptors in Acari are poorly known although extensive research has been reported on other Arthropoda, particularly Insecta. Our investigations of the setae on the first tarsus of the hard ticks, *Amblyomma americanum* (L.) and *Haemaphysalis (Alloceraea) inermis* Birula revealed the presence of chemo- and mechanoreceptors, judging from ultrastructural characteristics. A consistent pattern of distribution of tarsal sensilla in the region of Haller's organ was observed, by scanning electron microscopy (SEM), in the different species and life stages. Presumably chemoreceptive sensilla, resembling one type in the ticks, were found on the first tarsus of the mesostigmatid mite, *Macrocheles muscaedomesticae* (Scopoli).

A. americanum

Setal Pattern. The distribution of setae in the vicinity of Haller's organ on the dorsal aspect of the first tarsi is shown in Fig. 1. Both sexes of the adults (Fig. 2a) have the same number and distribution of setae as the nymphs which have been studied in detail (Foelix and Axtell,1971 and in press). The larval stage (Fig. 2b) differs in the anterior pit setae and the medial setae. The paired distal setae differ in diameter and length; the lateral (away from the body) d_2 is smaller than the mesad d_1. The distal setae were called the anterior bristle group by Lees (1948), M_3 by Zolotarev and Sinitsyna (1965) and front hair tuft (vorderes Haarbüschel) by Schulze (1941).

There are 7 anterior pit setae (terminology of Lees,1948, but called X_2 by Zolotarev and Sinitsyna,1965) in the nymph and adult (Fig. 3a). For convenience, a_1 designates the lateral (behind d_2) seta and a_2 designates the seta on the opposite side of the group and behind d_1. Between a_1 and a_2, but closer to a_2, are the remaining 5 setae. The larva has only 5 setae in the anterior pit (Fig. 3b); missing

---

[1] This research was supported by the Office of Naval Research and PHS Training Grant ES00069. Specimens of *Haemaphysalis inermis* were provided by Dr. Harry Hoogstraal, NAMRU–3, Cairo, UAR.

[2] Visiting Professor, Permanent address: Department of Zoology, Ain Shams University and NAMRU–3, Cairo, UAR.

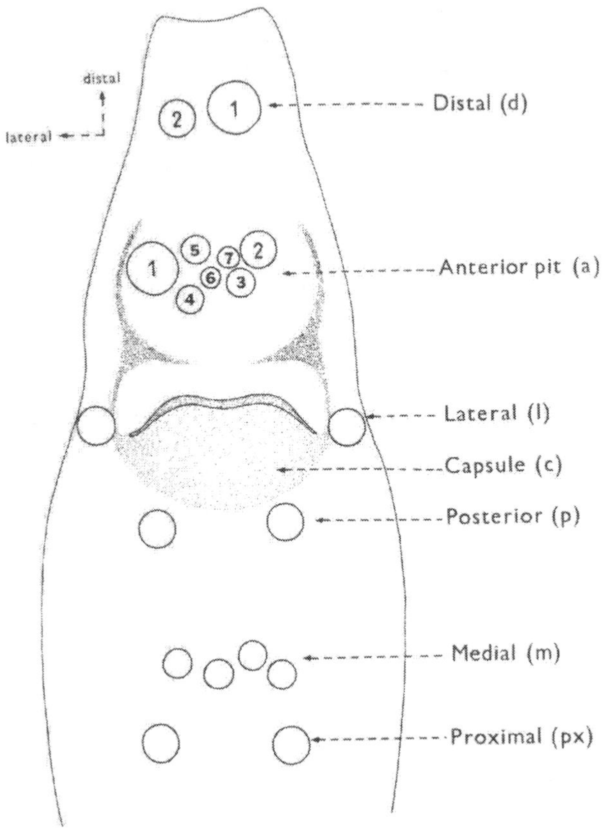

Fig. 1. Location and designations of Haller's organ (capsule and anterior pit) and associated setae on the dorsal surface of the first tarsi of *A. americanum* (nymph and adult).

are a_2 and either a_3 of a_4 (which is not certain).

There are 7 capsule sensilla (X_1 of Zolotarev and Sinitsyna, 1965) in the nymph (Foelix and Axtell, in press) while the adult has 8 according to Bruce (1971) but he used only SEM and a cuticular spine could easily be mistaken for an innervated seta (= sensillum). We have not examined the capsule sensilla in the adult and larva.

On each side of the capsule is a "lateral" seta and behind the capsule are paired "posterior" setae in all life stages. Proximal to the posterior setae is a group of 4 "medial" setae in the adult and nymph (called "4-group" by Foelix and Axtell, 1971). There are only 2 medial setae in the larva. Proximal to the medials are the paired "proximal" setae in all life stages. Collectively the medial and posterior setae were called "posterior bristle group" by Lees (1948) and rear hair tuft (hinteres Haarbüschel) by Schulze (1941). The X_6 of Zolotarev and Sinitsyna (1965) apparently refers to the medials in *Hyalomma asiaticum* if one assumes the posterior setae are absent.

Ultrastructure. The fine structure of these setae has been investigated in the nymph (Foelix and Axtell,1971 and in press). Presumably the same types of ultrastructure would be found in the larva and adult. The most distinctive setae are those with large cuticular pores (1200 Å diam, and observable in SEM, Fig. 4) containing a "plug" (Fig. 5). Such a pore plug has not been reported in other arthropods. The setae of this type are d_1, a_1, a_2 and the capsule sensilla which have thinner walls than the others. Setae of a second type (a_3 and a_5) have small pores (100-200 Å diam.) in longitudinal surface grooves and a spoke-wheel appearance of the cuticle shaft in cross-section. The third type of setae have apparently a single opening at the tip. This was determined for the proximal setae and may be true for a_4, a_6 and a_7. A fourth type includes only the medial setae which have small slit openings to the exterior from internal "vase-shaped" channels.

All of these setae are innervated by varying numbers of dendrites (2-9) which branch when there are multiple pores and are unbranched in the presence of a single opening at the tip. Morphologically they correspond to known chemoreceptors in insects. The entrance of fine silver granules through the cuticular pores has been demonstrated in some of the sensilla (Foelix, in press).

In addition to the sensillae described above, there are several (d_2, l, p) with thick cuticule, several unbranched dendrites in the shaft (Fig. 6) and 2 characteristic mechanoreceptive dendrites at the base. No pore openings were found but presumably there is an opening at the tip. These sensilla are apparently combined chemo- and mechano- receptors.

The dendritic ciliary structure of all sensilla differs from other invertebrate receptors in commonly having 11 double-tubules without central elements (11 + 0 instead of 9 + 0 or 9 + 2).

H. inermis

The setal pattern in *H. inermis* (Fig. 7a, b) is the same as described for *A. americanum*. The adults (both sexes) have the same number and distribution of setae as the nymphs. The distal setae differ in the same manner; d_2 smaller than d_1. There are 4 medial setae although they are positioned slightly different than in *A. americanum*. The anterior pit contains 7 setae with a_1 and a_2 easily recognized by size, shape and position (Fig. 8a). There are 2 small pointed setae (a_6 and a_7); a_3, a_4 and a_5, could be designated on the basis of their positions.

The larva of *H. inermis* differs from the other stages in the same manner as in *A. americanum*. There are 5 anterior pit setae (Fig. 8b) with a_1 easily recognized and a_2 missing. Pores were easily observed on a_1 by SEM. The small pointed setae are a_6 and a_7, but the other 2 setae cannot be reliably designated.

The fine structure of these setae has not been determined. However, their locations and appearance when observed by SEM suggest they correspond to those of *A. americanum*.

M. muscaedomesticae

The same types of tarsal sensilla in ticks may exist in other Acari. The tip of the first tarsus of the mesostigmatid mite *M. muscaedomesticae* (Fig. 9a) has 8 blunt-tipped setae (Coons,1970) which have surface protuberances after treatment in ethanol but not after killing in hot water. These setae may be olfactory receptors; olfaction has been demonstrated for the apical setae of the first tarsus (Farish and Axtell,1966, Jalil and Rodriguez,1970). Presumably the protuberances were from pores in the wall of the setae which are probably of the type shown in cross-section in Fig. 9b (notice the 5 dendrites). This resembles closely the medial sensilla in *A. americanum.* Thick-walls with (Fig. 9c) and without (Fig. 9d) pores were observed in other sections of apical sensilla.

These observations on ticks and a mite demonstrate the need for additional research on the receptors in Acari. Morphological observations will provide the basis for electrophysiological and behavioral studies which are necessary for elucidation of the functions of the various types of sensilla. Observations on *H. inermis* and preliminary studies on *Dermacentor* and *Ixodes* suggest considerable uniformity in setal types and distribution in the region of Haller's organ in hard ticks. Further comparative studies using *A. americanum* as the "model" are in progress. Similar studies are in progress on soft ticks using *Argas (Persicargas) arboreus* as a "model."

References

Bruce, W.A.: Posterior capsule of Haller's organ in the lone star tick, *Amblyomma americanum* (Acari: Ixodidae). Florida Entomologist 54: 65-72, 1971.

Coons, L. B.: Fine structure of selected organ systems of the mite *Macrocheles muscaedomesticae* (Acarina: Mesostigmata: Macrochelidae). Doct. thesis, N. C. State Univ., Raleigh, N. C., 163 p. 1970.

Farish, D.J., Axtell,R.C.: Sensory functions of the palps and first tarsi of *Macrocheles muscaedomesticae* (Acarina: Macrochelidae), a predator of the house fly. Ann. Entomol. Soc. Am. 59 : 165–70, 1966.

Foelix, R.F.: Permeability of tarsal sensilla in the tick *Amblyomma americanum* (L.)(Acarina, Ixodidae), Tissue & Cell. (In press.)

Foelix, R.F., Axtell, R.C.: Fine structures of tarsal sensilla in the tick, *Amblyomma americanum* (L.). Z. Zellforsch. 14: 22-37, 1971.

Foelix, R.F., Axtell, R.C.: Ultrastructure of Haller's organ in the tick *Amblyomma americanum* (L.). Z. Zellforsch. (In press.)

Jalil, J., Rodriguez,J.G.: Studies of behavior of *Macrocheles muscaedomesticae* (Acarina: Macrochelidae) with emphasis on its attraction to the house fly. Ann. Entomol. Soc. Am. 63: 738-46, 1970.

Lees, A. D.: The sensory physiology of the sheep tick, *Ixodes ricinus* L. J. Exptl. Biol. 25: 145-207, 1948.

Schulze, P.: Das Geruchsorgan der Zecken. Untersuchungen über die Abwandlungen eines Sinnesorgans und seine stammesgeschichtliche Bedeutung. Z. Morph. Ökol. Tiere 37: 491-564, 1941.

Zolotarev, E. Kh., Sinitsyna, Ye.Ye.: Chemoreceptive organs on the fore legs of ixodid ticks. Vestn. Mosk. Univ. 20: 17-25, 1965. (In Russian).

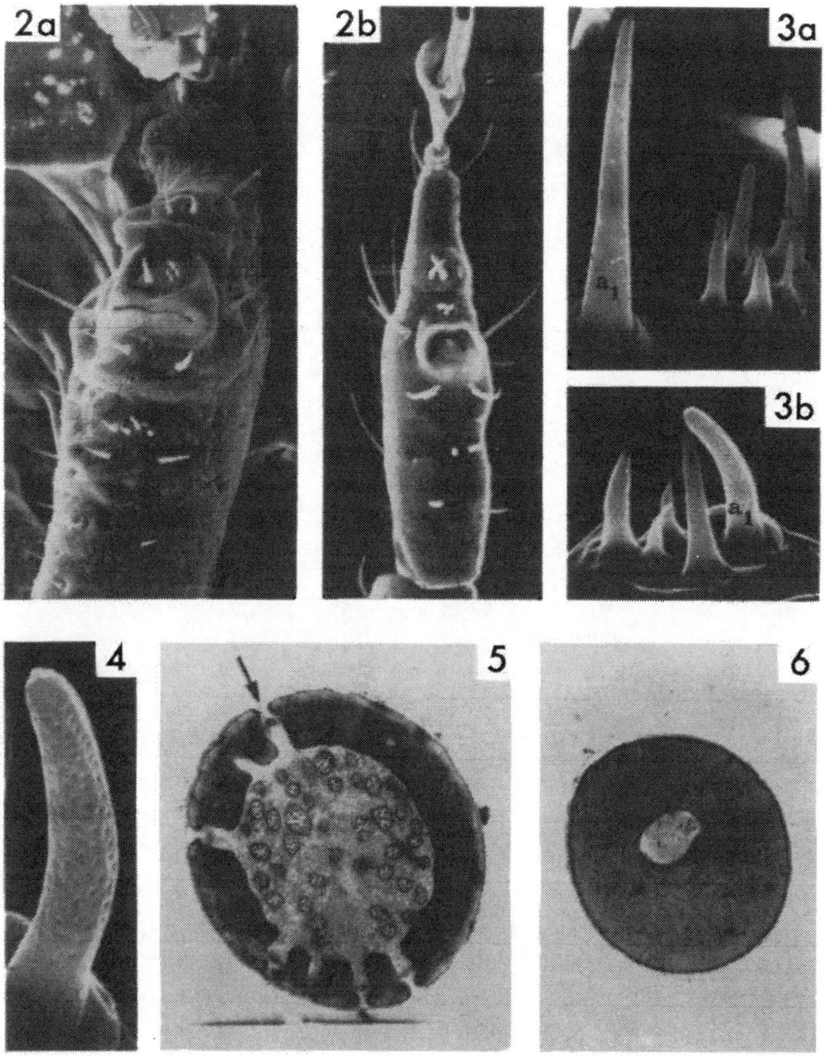

Plate I.

Fig. 2. First tarsus of *A. americanum,* dorsal aspect showing Haller's organ and associated setae: a) Adult female, 178 ×; b) Larva, 216 ×.

Fig. 3. Sensilla of anterior pit of Haller's organ of *A. americanum:* a) Adult (7 sensilla), 1200 ×; b) Larva (5 sensilla), 1800 ×.

Fig. 4. Surface pits indicating pores in the sensillum wall. (a_1 of *A. americanum* larva). 4800 ×.

Fig. 5. Cross-section of a sensillum (a_1 of *A. americanum* nymph) showing pores, pore-plug (arrow) and dendritic branches in the lumen. 13320 ×.

Fig. 6. Cross-section of a sensillum (d_2 of *A. americanum* nymph) which does not show pores, but a single opening at the tip is presumed. 16512 ×.

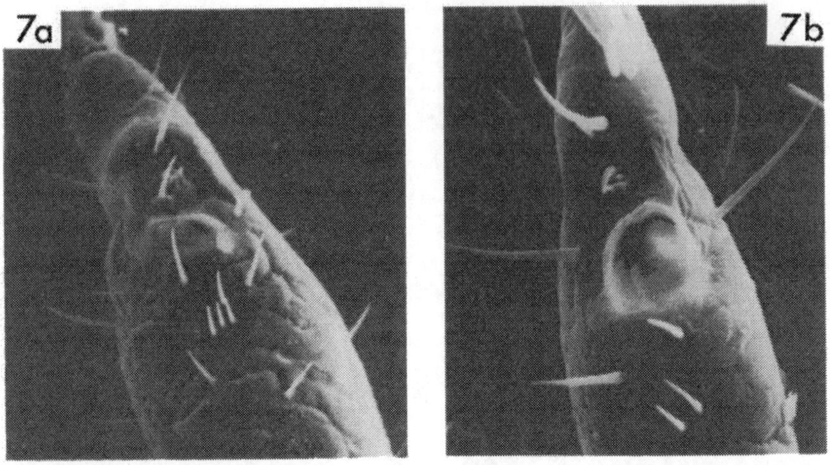

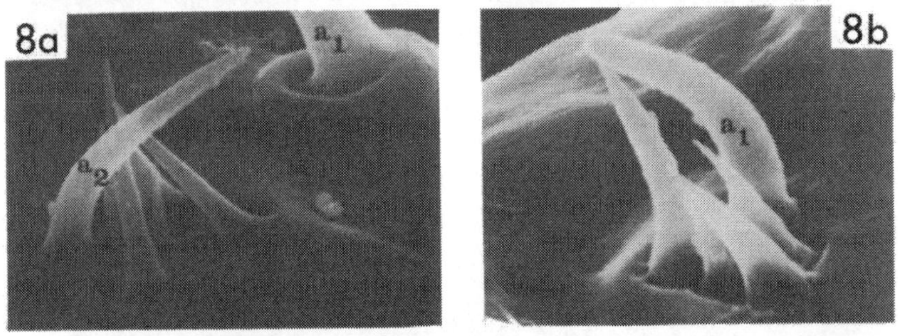

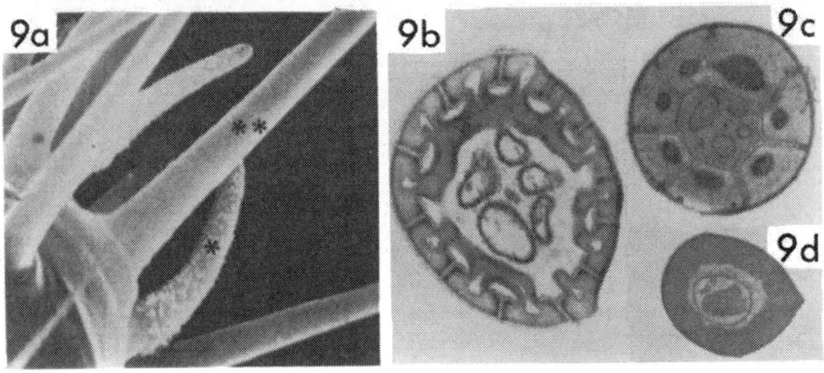

Plate II.

Fig. 7. First tarsi of *H. inermis*, dorsal aspect showing Haller's organ and associated setae: a) Adult female, 157 ×; b) Larva, 490 ×.

Fig. 8. Sensilla of anterior pit of Haller's organ of *H. inermis:* a) Adult (7 sensilla), 1470 ×; b) Larva (5 sensilla), 3185 ×.

Fig. 9. Setae at the tip of the first tarsus of the mesostigmatid mite, *M. muscaedomesticae:* a) Blunt pegs with protuberances (*) and long pointed setae (**), 1715 ×; b)—d) Cross sections of tarsal setae.

SPECIMEN PREPARATION FOR, AND THE ADVANTAGES OF, EXAMINING ACARINE MORPHOLOGY WITH THE SCANNING ELECTRON MICROSCOPE

D. A. GRIFFITHS
Pest Infestation Control Laboratory, Ministry of Agriculture, Fisheries and Food, Slough, England

Specimen Preparation

The results obtained from the scanning electron microscope (SEM) are in the form of micrographs. Like its counterpart, the transmission electron microscope (TEM), their quality depends as much on the correct preparation of the specimen as it does upon the technical competence of the instrument.

Preparation techniques for the TEM are well established, but because the SEM is comparatively new and relatively few are available to biologists, methods of preparing specimens vary considerably and tend to reflect individual tastes. Insufficient time has elapsed for the development or exchange of ideas required to establish competent techniques. My experience in the preparation of acarine material is presented here as an initial contribution in this field.

Fortunately, specimen preparation for the SEM is simpler and takes less time than that for the TEM. It generally involves three stages; the specimen must be dried, fixed on the aluminium stub (which plugs into the specimen holding stage of the instrument) and finally rendered conductive by coating with a fine layer of an evaporated metal.

It is possible to examine live mites without drying or applying a conductive coat, but this is a technique better reserved for special purposes rather than one to be used in order to obtain scanning electron micrographs. For the latter the best results are achieved with freeze-dried specimens obtained directly from living material or, freeze or air-dried specimens taken from spirit material.

Soft bodied mites including all the Astigmata and most of the Prostigmata need to be freeze-dried. Hard bodied mites, freshly killed in alcohol, can be exempted from the freeze-drying part of the treatment, but freeze-drying is preferable.

Tissue driers incorporating a freezing platform (-50 to -60°C) and a vacuum system (10^{-4} torr) are satisfactory. The mites are plunged for not more than thirty seconds into freezing isopentane, maintained at its melting point of -158°C, by a surrounding bath of liquid nitrogen, before transfer to the drier.

Small, 1 cm x 1 cm diameter copper cylinders with a base of brass gauze (80 meshes per cm) are used to hold the sample. The containers are made by covering the end of a section of copper tubing 1 cm high x 1 cm diameter with a circle of gauze 2 cm in diameter. This is inserted, gauze end first, into a similar section of tubing the diameter of which is just big enough to provide a firm fit which ensures that the gauze base is tightly stretched. Some 200 or so astigmatid mites can be accommodated in one cylinder. As each cylinder is removed from the isopentane, it should be placed momentarily onto an absorptive paper tissue to drain off excess isopentane.

Material is usually left in the drier for 24 hours. Astigmatid mites which have dried successfully are easily recognised since the body retains its natural shape but is opaque and starkly white. Sclerotized mites tend to become less shiny and paler in colour. A phenomenon commonly encountered is that in a batch of material (within the same cylinder) perhaps half the specimens will emerge from the treatment completely undried. Although dead, their bodies are fully inflated, shiny, exactly as in living specimens. This effect is more common in soft bodied mites, especially ectoparasitic Astigmata. When it happens, it is worthwhile removing the dried specimens and then soak the unaffected specimens, still in their container, in 5 per cent alcohol for 12 hours. After which re-submit them for freeze-drying. There is no obvious explanation why two specimens, of the same sex, side by side in the container should react so differently to the treatment.

Certain groups (one example is the Euptyctima or armadillo mites) contract their limbs and gnathosoma when plunged into the freezing isopentane. Therefore, when this is likely to occur, specimens should be relaxed by placing them in a warm solution of propylene phenoxytol over-night. Wash off with distilled water and proceed with the freeze-drying.

Heavily sclerotised mites stored in alcohol need only be removed from the preservative and air-dried prior to coating. Soft bodied mites, preserved in this manner, usually become dehydrated and show some signs of collapse. They need to be brought slowly up through the alcohols into a final solution of 5 per cent strength for twelve hours and then freeze-dried. Providing the softer tissues have not deteriorated during storage, satisfactory micrographs can be obtained.

Before coating, the specimens must be attached to the flat surface of the small, circular aluminium specimen stub. If the specimen is unique, a small spot of gum chloral may be used as an adhesive since this is easily dissolved in water and the specimen retrieved. When specimens are reasonably plentiful the following method, carried out under a low-power microscope, is advised. Cut a square centimeter from a roll of double-sided adhesive Sellotape and attach firmly to the surface of the stub. With a surgical scalpel score the square into alternate thick and thin strips to provide 3 or 4 thin strips spread evenly across the stub. They should be about as wide as the width of the mites in the sample. Remove the upper protective coating from the double-sided tape and, using the point of the scalpel, lift up and remove the thick strips. This leaves the stub covered by 3 or 4 parallel strips of tape presenting a sticky surface upwards. When micrographs of whole specimens are required, or if the specimens are small (e.g. Tarsonemidae), the mites need only be pressed lightly against the sticky surface to provide sufficient adhesion. If only part of the specimen is to be examined, i.e. a particular structure, the mites should be arranged on the sticky strips to expose the structure to best advantage and then fixed more firmly into position by applying a minute blob of quick drying, conductive liquid, running from that part of the specimen which is touching the tape, across the tape, to the metal surface. The liquid is made up of one part of Durofix adhesive[1] to three parts of Silver Dag[2] diluted with amyl acetate. Very small quantities can be applied to the required site using a fine mounted entomological pin which has been sharpened by dipping the point into nitric acid. However, care is required in order not to swamp the specimen with liquid. Depending on size, three to five mites can be placed on one strip so that one stub can hold up to 20 specimens.

Since biological specimens are non-conductive, it is essential to provide a low resistance leakage path between the surface of the specimen and the stub. This is achieved by depositing a thin layer of evaporated metal over both. A layer of pure gold between 300 and 500 Å units thick, applied in a vacuum of 10^{-4} torr, generally proves suitable.

Surface contaminants such as detritus and natural secretions of an oily or waxy nature can effectively obscure surface detail as well as interfere with the coating process. The removal of contaminants has proved difficult and the general rule is to search for and pick out the cleanest specimens in the sample. Techniques which have given some measure of success are outlined below.

Free living Astigmata taken from culture are usually very clean. The dipping in freezing isopentane seems to be sufficient to dislodge particles of food adhering to their bodies and generally no further treatment is required. The mouthparts of astigmatid ectoparasites are often covered with blood and sebaceous secretions. This can sometimes be removed by soaking the specimens in 20 per cent solution of Decon-75[3]. Some success with removing particles from hard bodied mites has been achieved by immersing the specimens in a 0.5 per cent mixture of Duron[4] in iso-butyl alcohol and briefly placing the vial in an ultrasonic bath. The Duron, which is an anti-static agent, releases the particles which are then shaken free by the ultrasonic waves. After removal, the specimens are washed in iso-butyl alcohol and freeze-dried or left to dry in air. A number of solvents have been used in an attempt to de-wax oribatid mites. The best results have been obtained from briefly washing the specimens in trichlorethylene.

Live mites, lightly attached to thin strips of adhesive tape, are able to withstand the 10^{-4} torr vacuum and bombardment by an electron beam for which the accelerating voltage has been reduced to 1 or 2 kV. I have subjected a number of species of soft-bodied Astigmata to a maximum of 15 minutes in the specimen chambers. During this time a proportion have been scanned by the beam and micrographs have been obtained. Some of the latter collapse and die but others survive providing ex-

---

[1] Manufactured by Rawlplug Co Ltd., Kingston-upon-Thames, England.

[2] Colloidal dispersion in silver; manufactured by Acheson Colloids Company, Prince Rock, Plymouth, England.

[3] A non-foaming surface active agent. Decon Labs Ltd., Brighton, England.

[4] Manufactured by: J.C. Thompson & Co. (Duron) Ltd., Bradford.

posure is reduced to a minimum. Those not exposed to the beam usually survive. When removed from the chamber, all survivors remain inert for about ten minutes. On transfer to a rearing cell they resume what appears to be a normal life. The females lay viable eggs.

Micrographs obtained from live mites are poor in quality since the undried, uncoated material becomes "charged up". Also, resolution drops off as the accelerating voltage (normally 10 to 20 kV for mites) is reduced. However, this technique may be important since, eventually, it should be possible to adapt the instrument for microsurgery, either with conventional micro-probes inserted into the specimen chamber or, more probably, using a method based upon beam-etching, a technique which is now under development. Ideas for some of the techniques described came originally from Mr. B. Martin and Mr. C. Ogden, British Museum (Natural History) to whom I am most grateful.

Instrument Performance

Although in recent years the versatility of the optical microscope has been enormously extended by the development of techniques such as interference-, phase-contrast- and fluorescent-microscopy, the instrument is fundamentally limited by the properties of light. The wave-length of light limits attainable resolution to about 5000 Ångström units[5] and effective magnification to about 1000 times. Moreover, the depth of fields is very small and for examination under high magnification, specimens generally have to be flattened, sectioned or otherwise distorted.

The scanning electron microscope eliminates these disadvantages. With this instrument the maximum resolution varies between 300 and 150 Ångström units. Magnifications of up to 100,000 times can be obtained, although for most work with mites magnifications above 20,000 times are rarely required. Far more important is the tremendous depth of field. This is at least 300 times greater than an optical microscope working at similar magnification.

Acarine examples

Figures 1 to 6, a comparative series on the mouthparts of free-living and parasitic Astigmata, have been chosen to illustrate some specific advantages to acarology which this instrument can bestow. The definition and three dimensional appearance of structures in situ are important not only to taxonomists but as an aid to the study of function in the Acari.

Glycyphagus destructor (Fig. 1) represents a free-living Astigmata in which the palps are segmented, freely mobile, bearing finger-like sensory organs at their tips.

[5] 1 Ångström unit = 10^{-10}m.

The malae, or lower lips are massive and closely attenuate the chelicerae. In life, the chelicerae, which bear narrow scissor-like chelae, move alternatively, piston-like between the lips. The dorsal internal portion of each lip is extended into a blade-like process which acts as a scraper against each side of the teeth of each chelicera as it is withdrawn. This structure shows up distinctly in Fig. 5 which is of *Tyrophagus putrescentiae*, a free-living member of the Acaridae. The structure, although obviously with the same function as the rutella present in some Cryptostigmata is not homologous. In this figure, the palp sensory organs (different in shape) are clearly visible.

Fungi probably form an important part of the diet of both *Glycyphagus* and *Tyrophagus*. In contrast, *Dermatophagoides farinae*, Fig. 2, is associated with humans and appears to feed upon skin detritus. The palps, although segmented, do not appear to have much mobility, nor well developed sensory organs at their tips. The lower lips are vestigial. The whole of the mouth is dominated by the massive chelicerae, admirable adapted to crush tough skin particles. In Fig. 3, the mouthparts of a feather mite (Analgesidae) show adaptations which are difficult to interpret. The palps appear to have taken over the function carried out by the lower lips in *Glycyphagus* and the lips have developed into cup-like lobes. These may have a special function in relation to feathers.

Figures 4 and 6 represent an extreme case of adaptation. The mite, *Psoroptes cuniculi* lives inside the ears of domestic rabbits where they scrape off the dermal layers and feed upon the blood and other secreations which exude from the wounds. The palps, Figs. 4 and 6, in contrast to those of *Tyrophagus* have fused laterally and their tips have become adapted for scraping. The lower lips appear to be perfectly structured to act as a suction syphon to take up fluid.

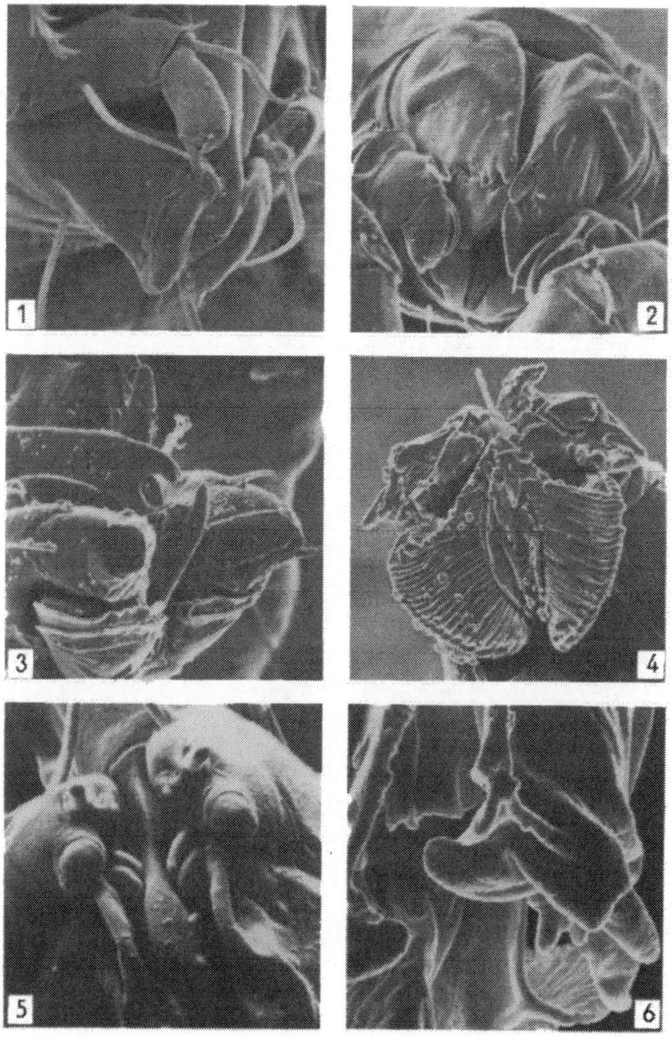

Plate I.

Figs. 1 to 6. **A comparative study of the mouthparts of some free-living and ectoparasitic** Astigmata.

Fig. 1. ***Glycyphagus domesticus*** **(De Geer). 1160×.**

Fig. 2. ***Dermatophagoides farinae*** **Hughes. 580×.**

Fig. 3. **Analgesidae — feather mite unidentified. 1450×.**

Fig. 4. ***Psoroptes cuniculi*** **(Delafond). 1624×.**

Fig. 5. ***Tyrophagus putrescentiae*** **(Schrank). Palp tarsi bearing sensory organs, lower lips bearing flattened scraping processes internally, which are adpressed on each side of the cheliceral teeth. 3248×.**

Fig. 6. ***Psoroptes cuniculi.*** **Dorsum of modified palp tip, with suctorial lower lip showing beneath. 3248×.**

ULTRASTRUCTURE OF THE EGG ENVELOPES OF A UROPODID MITE

D.R. KOLATA, K.J. NORSTOG and C.J. ROHDE, Jr.
Northern Illinois University, Dekalb, Illinois 60115, U.S.A.

During the course of a continuing investigation relating to the effects of environmental factors upon hatching in the eggs of a mite prey and its mite predator, an examination of the microscopic and ultrastructure of the membranes surrounding the developing embryo of the predator was undertaken. The purpose was to determine the nature of these membranes and the changes in them during the period of incubation.

This report is based upon eggs in late vitellogenesis prior to oviposition and within a maximum of 24 hours incubation following oviposition.

Methods

Gravid females of *Fuscuropoda marginata* (C.L. Koch, 1839) were held securely to a piece of heavy paper by spraying them with a light coat of rubber cement. The dorsal shield was picked off using an insect pin, the eggs removed and placed directly into a cold (4°C) solution of 6% glutaraldehyde in 0.1M sodium cacodylate buffer (pH 7.2) for 6-8 hours and then rinsed in 0.1M sodium cacodylate buffer for four 1-hour periods. Postfixation in 1% osmium tetroxide in 0.1M sodium cacodylate buffer took 6-8 hours and was followed by four 5-minute rinses in cold 0.1M sodium cacodylate buffer. Eggs which had been oviposited were treated similarly.

Fixed eggs were dehydrated in a cold (4°C) acetone series of 10% concentration increments at 5-minute intervals. Three changes of 100% acetone spaced at 20-minute intervals allowed for good penetration of the resin. Eggs were transferred singly, with a small amount of acetone, to a Beem capsule for infiltration with the embedding mixture.

The embedding stock mixture consisted of Epon 812 (62 ml). Araldite 502 (50 ml) and dibutyl phthalate (8 ml) (Mollenauer, H. H., personal communication). When used, 4 ml of the stock mixture were combined with DDSA (dodecenyl succinic anhydride) (5 ml) and 13 drops (approximately 0.6 ml) of DMP - 30 (2, 4, 6 tridimethylamino methyl phenol). Good penetration of the resin was accomplished by putting the eggs in 40, 60, 80 and 100 % of the mixture in 4-hour intervals, under vacuum, with acetone as the solvent. Before the 100% mixture was added, each egg was removed and placed in a clean Beem capsule to assure the purity of the polymerized medium. Polymerization was carried out in an oven at 47°C for 12 hours followed by 6 hours at 60°C.

Thin sections (600 - 1,000A) were cut with glass knives and mounted, unsupported, on 400 mesh grids. The specimens were stained with 1% aqueous uranyl acetate for 15 minutes and lead citrate (Reynolds, 1963) for 20 minutes. Due to the small size of the egg it was often difficult to orient it in the Beem capsule making it impossible to determine exactly where a section was made. Further, the egg contains much yolk, a problem which leads to the crumbling of many sections.

Specimens were examined with JEM - T7 and Hitachi Hu - 125E electron microscopes at 60 and 125 Kv respectively. Micrographs were taken on DuPont Cronar film and printed on Agfa Brovira paper (grades 4 and 5).

Results and Discussion

As seen in the egg prior to oviposition, the following were observed as shown by diagram (Figure 1): (1) an outermost, highly vesicular layer of tissue, the follicle wall; (2) two thin layers of granular material, the wax layer; (3) a layer of degenerating follicle cells, the theca, which remains with the egg at oviposition; (4) a dense layer of uniform thickness, the chorion; (5) a space occupied by an amorphous, flocculent material; (6) a unit membrane, the vitelline membrane; and (7) the ooplasm bounded by a unit membrane, the plasmalemma.

The actual appearance as seen from the micrograph (Plate I) reveals the outer layer, or follicle wall (1) as 3 - 4 μm thick and containing vesicles and large, irregular vacuoles. Pinocytotic activity, clearly seen at higher magnifications, was apparent along both the inner and outer faces of this layer. The direction of vesicular movement was toward the two thin layers of granular material (2). This inward movement of the vesicles suggests a possible role in the deposition of a superficial wax layer on the surface of the egg. Lees and Beament (1948) reported a superficial wax layer secreted on the surface of the egg in ticks and that similar layers of wax are present below the chorion in the red spider mite. The two layers, here considered collectively as the wax layer (2), were 0.1 μm thick and rested upon the layer of degenerating follicle cells. This cellular layer (3) was approximately 1 μm thick and contained Golgi complexes formed of flat cisternae with accumulations of opaque secretory droplets. Numerous lysosomes were present, suggesting participation in cell digestion. Some of these cells contained disrupted membranes, flattened vesicles and residual bodies and were in advanced stages of atrophy. It appears that these cells degenerate and form a covering, or theca, and with the wax layer remains attached to the chorion at oviposition (see also Figure 3).

Referring to Plate I, it is noted that the theca is separated by a basement membrane 150 A thick (see also insert in Plate II) from the chorion (4). The chorion appears as a granular layer, approximately 1.5 μm thick, composed of a narrow outer zone of greater electron density and an inner, thicker zone which is less electron dense. These regions are more clearly resolved in Plate II. They may be compared with the exochorion and endochorion in insects (Wigglesworth, 1953). The chorion and theca become progressively dense and impermeable to reagents through the incubation period of approximately 7 days. Examination of the developing embryos requires the removal of the chorion and theca (Rohde, 1964).

Below the chorion there is a region (5) approximately 5 μm thick and containing flocculent material which may be interpreted as having been precipitated by fixation, perhaps within a fluid or semi-fluid matrix. Unit membranes are present at the outer and inner boundaries of this space. The inner or vitelline membrane (6) is about 70 Å thick.

The ooplasm (7) contained an abundance of lipid vacuoles averaging from 2-3 μm in diameter and were surrounded by glycogen granules (Figure 1). These granules occurred singly or in clusters and rosettes. Mitochondria and Golgi complexes were found scattered among the large, abundant yolk spherules. Multivesicular bodies were often seen at the periphery of the ooplasm. According to Smith (1968), these structures, as seen in insects, probably represent a type of lysosome containing hydrolytic enzymes and appear to be concerned with the digestion of materials brought into the cell by pinocytosis. In view of the fact that pinocytotic activity takes place along the surface of the ooplasm, these vesicles probably serve the same function in *Fuscuropoda*.

In addition to the preceding, the changes occurring in the mite egg upon and immediately following oviposition (see Plate II) are as follows. The material associated with the egg follicle (see Plate I) is missing. The wax layer (2) immediately external to the theca appears to be more closely appressed to its surface. The theca (3) retains its cellular components but there is evidence of cellular degeneration in the form of disrupted membranes. Numerous lysosomes are present and are interpreted as relating to the degenerative processes in the theca. The chorion (4) appears to be somewhat more electron dense than in the pre-oviposition egg. The space (5) beneath the chorion has diminished dramatically in thickness (from about 5 μm to 0.2 μm) and the vitelline membrane (6) consequently is in close proximity to the chorion. The inner membrane of the theca appears to be a basement membrane (3a).

It was impossible to determine whether the egg coverings are formed by the follicle cells and oocyts as in insects (Slifer, 1937; Hopkins and King, 1966; Richards, 1969) or by the oocyte alone as in the tick, *Ornithodorus moubata* (Aeschlimann and Hecker, 1967).

Acknowledgements. We wish to thank Miss Dian Molsen of the Electron Microscopy Laboratory and Miss Jane Glaser of the Graphics Laboratory for their assistance.

References

Aeschlimann, A., Hecker, H.: Observations préliminaires sur l'ultrastructure de l'ovocyte en dévelopement chez *Ornithodorus moubata*, Murray (Ixodoidea: Argasidae). Acta Trop.: 24: 225-243, 1967.

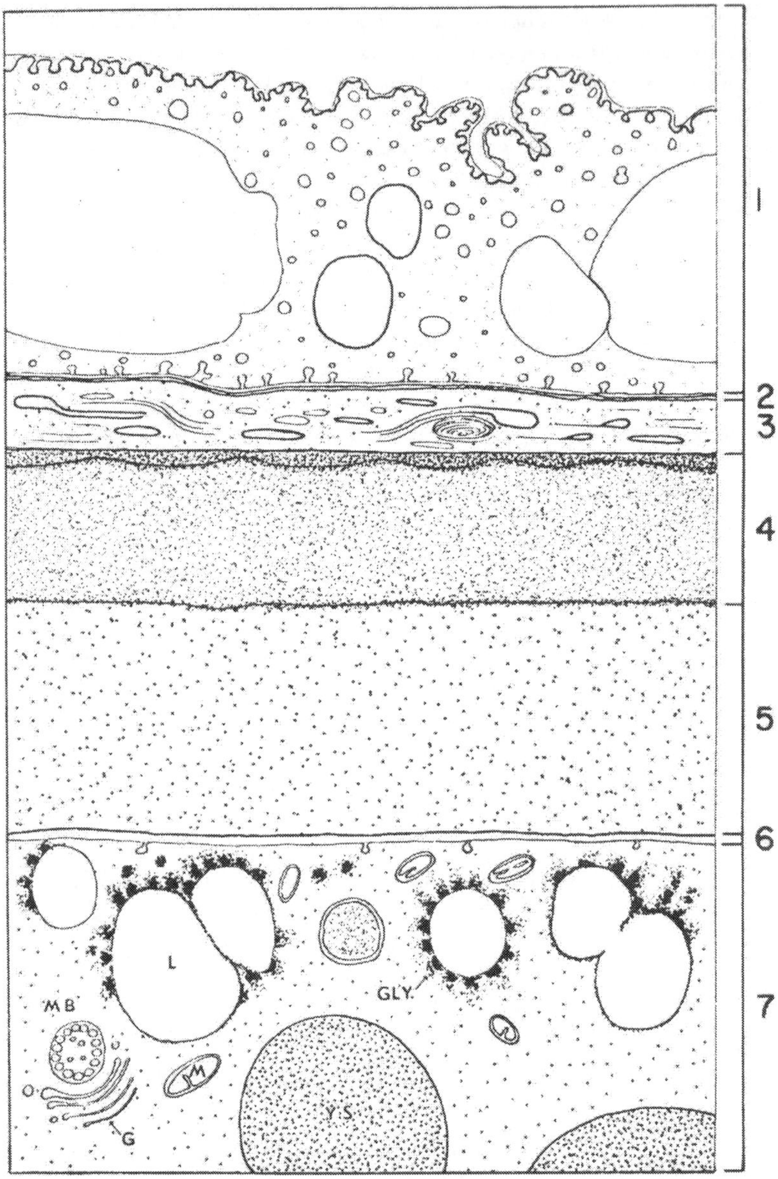

Fig. 1. Diagram of the principal layers surrounding the ooplasm of egg prior to oviposition: (1) wall of egg follicle, the outer and inner surfaces revealing much pinocytotic activity; (2) two thin layers of granular material, together interpreted as a wax layer; (3) a layer of degenerating follicle cells, the theca; (4) the chorion, a dense layer of uniform thickness showing increasing density at its point of contact with the theca; (5) a fluid-filled space; (6) a unit membrane, the vitelline membrane; (7) the ooplasm bounded by a unit membrane, the plasmalemma.

Hopkins, C.R., King, P. E.: An electronmicroscopical and histological study of the oocyte periphery in *Bombus terrestris* during vitellogenesis. J. Cell Sci. 1 : 201-216, 1966.

Lees, A.D., Beament, J.W.L.: An egg-waxing organ in ticks. Quart. J. Microscop. Sci. 89: 291-292, 1948.

Reynolds, E. S.: The use of lead citrate at high pH as an electron-opaque stain in electron microscopy. J. Cell. Biol. 17: 208-212, 1963.

Richards, J.G.: The structure and formation of the egg membranes in *Nasonia vitripennis*. J. Microscop. 89: 43-53, 1969.

Rohde, C. J.: Some techniques in the preparation of stained whole mounts and serial sections of mite embryos and adults. Acarologia. Supplement. 1st International Congress of Acarology. Fort Collins, Colorado (1963), pp. 208-214, 1964.

Slifer, E.H.: Egg membranes in *Melanoplus* (Orthoptera). Quart. J. Microscop. Sci. 79: 493-506, 1937.

Smith, D.S.: Insect Cells, Their Structure and Function. Oliver and Boyd, Edinburgh, 372 pp., 1968.

Wigglesworth, V. B.: Principles of Insect Physiology. Methuen & Company, Ltd. London, 434 pp., 1953.

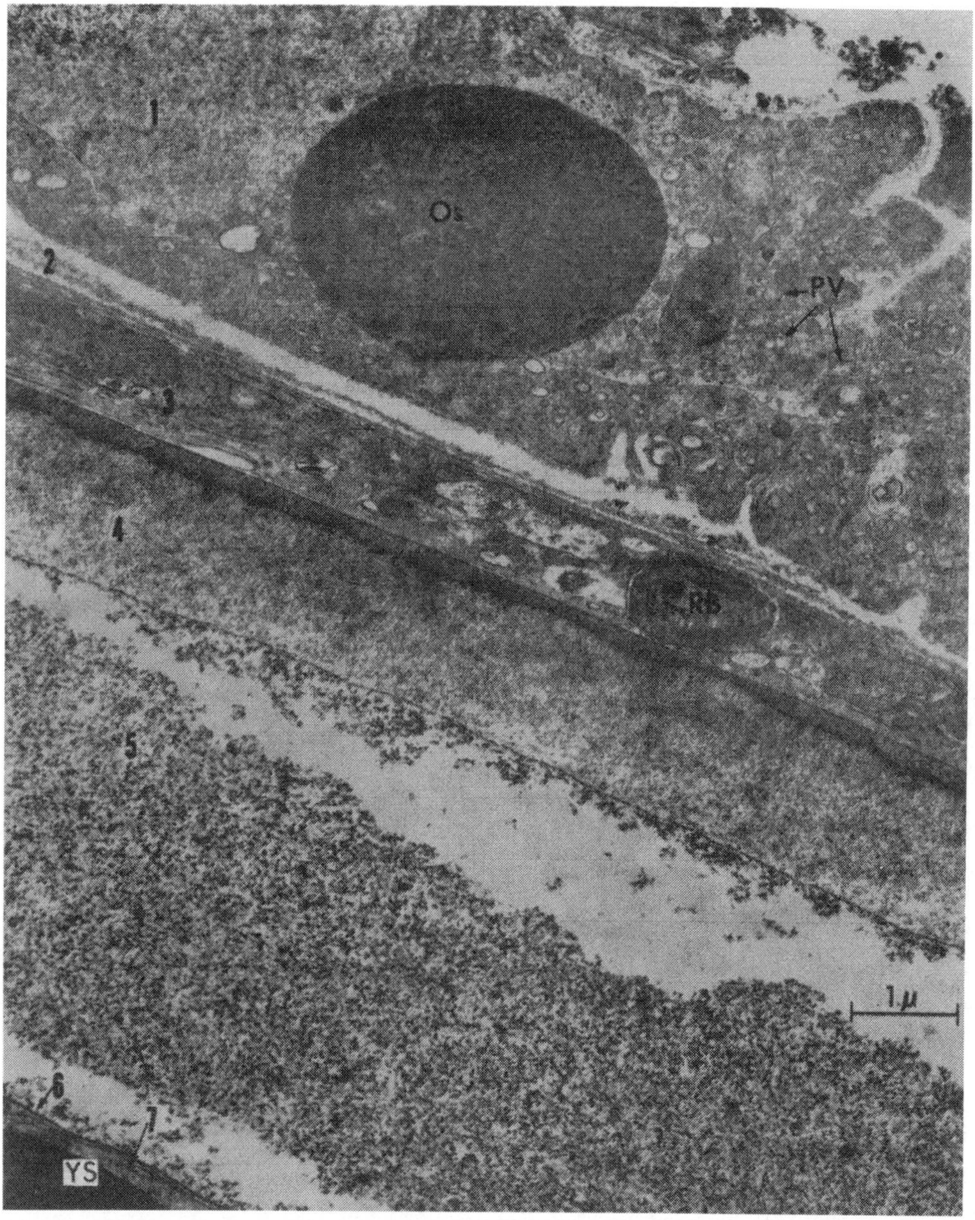

Plate I.

Electronmicrograph from an egg prior to oviposition (15120 ×): (1) follicle wall seen with pinocytotic vesicles (PV) and a large non-membrane-bounded osmiophilic body (Os) which is interpreted as lipid in nature; (2) wax layer; (3) theca, with folded membranes visible and residual body (RB); (4) chorion; (5) space in which flocculent material is evident; (6) vitelline membrane; (7) ooplasm with a portion of dense yolk spherule (YS) visible.

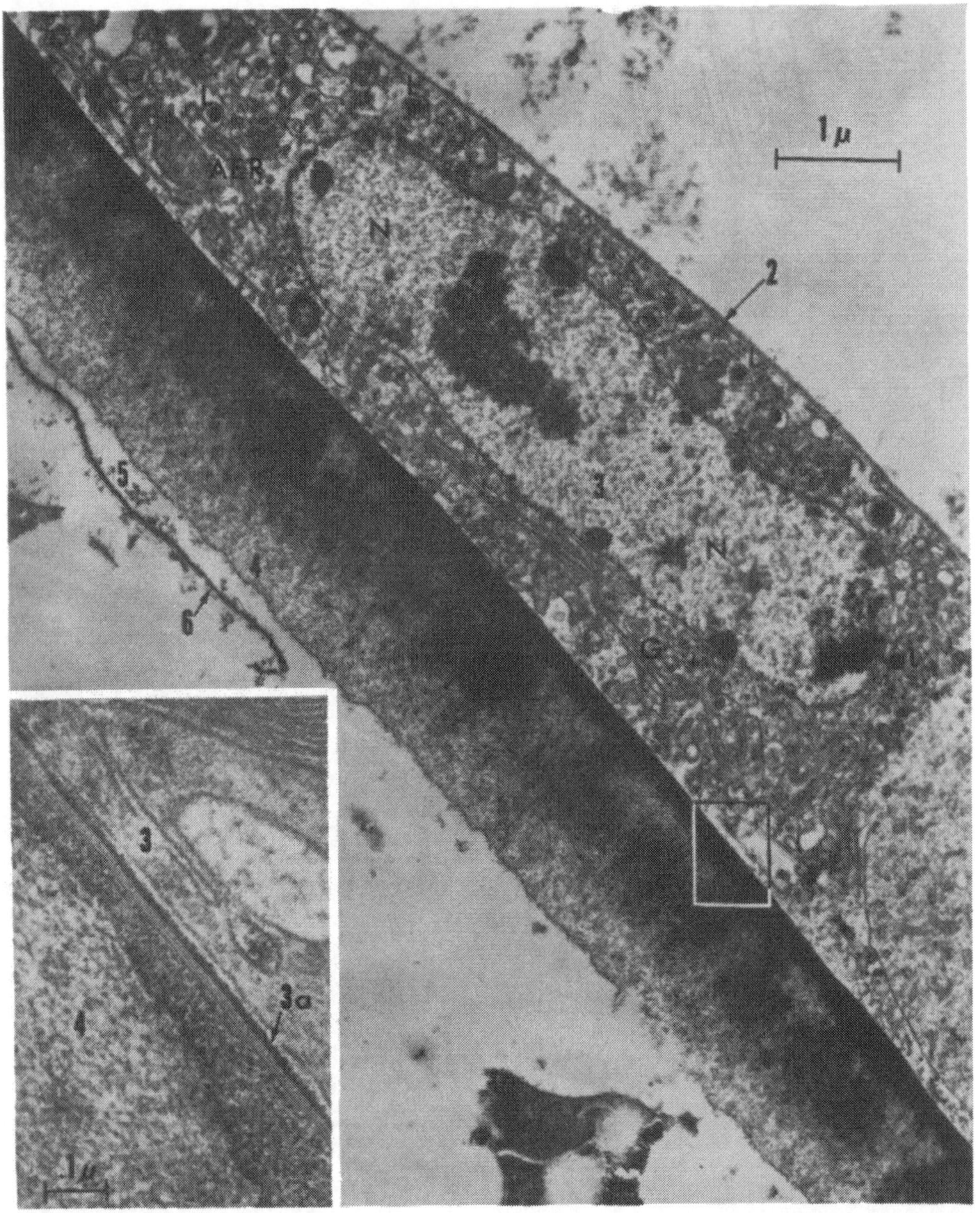

Plate II.

Electronmicrograph of egg less than 24 hours old following oviposition (17206×): (1) the follicle wall is absent; (2) wax layer; (3) theca, with degenerating cell but showing intact nucleus (N), agranular endoplasmic reticulum (AER), Golgi complex (G) and numerous lysosomes (L). Note chorion (4), the reduction in space size (5) and proximity of vitelline membrane (6) to chorion.

Insert: Appearance of theca (3) and chorion (4) at higher magnification (85500×). Note intermediate layer or basement membrane (3a).

SCANNING ELECTRON MICROSCOPE INVESTIGATIONS ON TICKS, WITH SPECIAL REGARD TO THE SENSILLAE, HALLER'S ORGAN AND THE APICAL PALPAL SEGMENT

W. SIXL, E. DENGG and H. WALTINGER

Institute of Zoology of the University Graz and Central Laboratory for Electron microscopy, Graz, Austria

The ticks used for investigation were collected in Czechoslovakia and Austria. The *Ornithodoros papillipes* and *Rhipicephalus sanguineus* originated from laboratory cultures. My thanks are due to Dr Černý, Institute of Parasitology of the Czechoslovak Academy of Sciences, Prague and Dr Nosek, Institute of Virology, Slovak Academy of Sciences, Bratislava, for some tick specimens.

The ticks were killed in alcohol and dried between filter papers. When specimens were heavily contaminated with debris, this was removed by means of a forceps. After this procedure, single specimens were exposed to osmium tetroxyde vapour for 48 hours. Finally, they were treated with carbon dioxide and gold in apparatus EPA 100 (Leybold-Heraeus) developed in Graz.

At present we have compared the following species: *Ixodes canisuga, I. hexagonus, I. redikorzevi, I. ricinus; Haemaphysalis concinna, H. inermis, H. punctata; Dermacentor marginatus, D. reticulatus; Ornithodoros papillipes; Argas reflexus.*

The following important structures were investigated: the areae porosae, spiracles, hypostome, chelicerae, apical palpal segment, Haller's organ and the sensillae of the apical palpal segment.

Schulze (1941) pointed out that it is possible to identify tick species on the basis of comparative morphology of Haller's organ. In the same work, Schulze reported that the pit of Haller's organ contained two types of sensory setae.

Arthur (1956) published the first detailed description of this organ as it appears in *I. hexagonus* and other tick species. He considered that the pit in *I. hexagonus* possessed 4 types of sensillae. In our investigations using the scanning electron microscope, females of three different Ixodes species were compared. All there species have the same number of sensillae and, respectively, these appear to be morphologically identical. However, distinct specific differences were observed in the structure of the capsula of Haller's organ.

Compared with the previous species great differences have been found in the morphology of the sensillae of the genera *Dermacentor, Rhipicephalus* and *Ornithodoros.* The palpus and the apical palpal segment have, in different genera, a characteristic shape and size and bear a constant number of sensillae. There are no secondary sexual differences between the sensillae of males and females. Thus, they represent generic markers which can be used for systematic classification.

In addition, studies of the apical portion of the hypostome (corona) in *Haemaphysalis concinna, H. inermis* and *H. punctata* indicate that distinct morphological differences exist at specific level. Finally, it has been possible to determine in detail the fine structure of the areae porosae, the hypostome and the chelicerae.

The function of the various structures and, in particular that of the sensillae, must eventually be determined by experimental behavioural studies. But, meanwhile the scanning electron microscope opens up new criteria for both taxonomic purposes and for morphological investigations of surface ultrastructure. Once the fine structure and the function of organs such as sensillae is understood, it may be possible to characterize the behaviour of ticks under natural conditions. In turn, this will undoubtedly contribute to the intelligent control of ticks which occur in natural foci of diseases.

References

Arthur, D.R.: The morphology of the British Prostriata, with particular reference to *Ixodes hexagonus* Leach. III. Parasitology 46, 261-307, 1956.

Schulze, P.: Das Geruchsorgan der Zecken. Untersuchungen über die Abwandlungen eines Sinnesorganes und seine stammesgeschichtliche Bedeutung. Z. morph. ökol. Tiere, 37, 491-564, 1941.

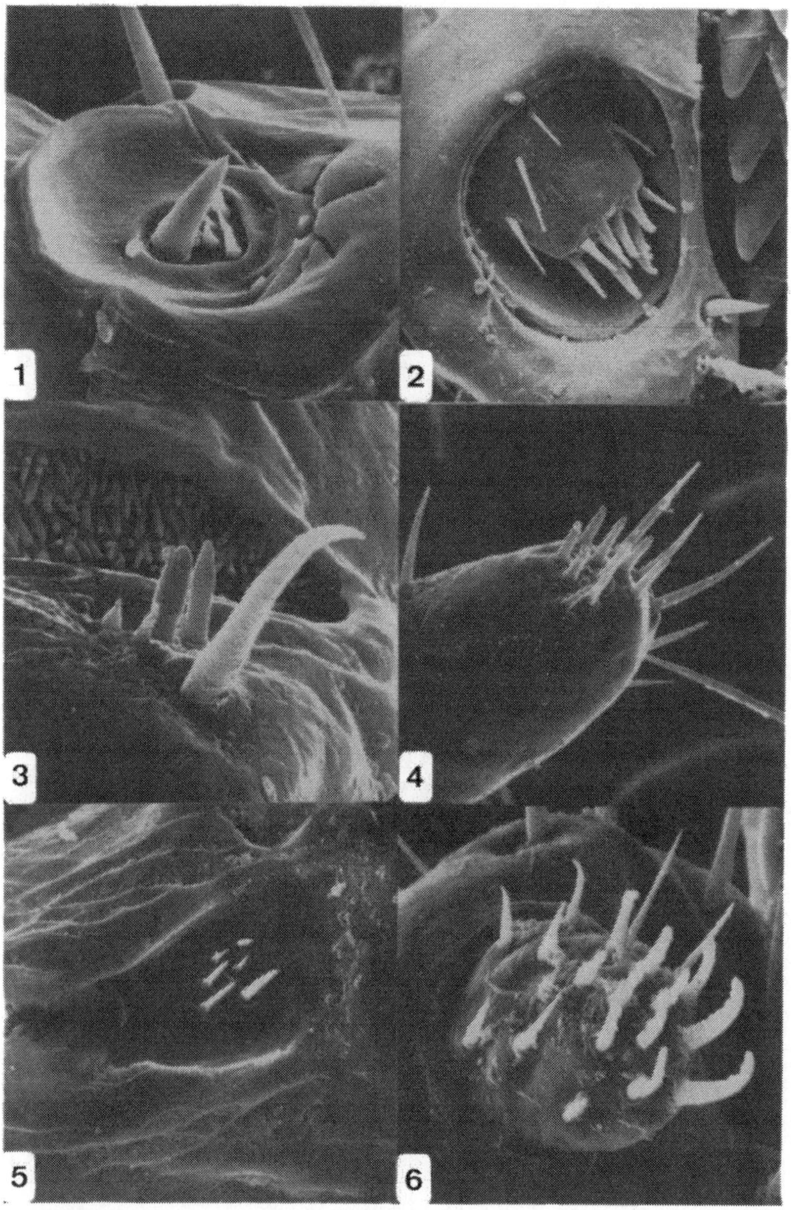

Plate I.

Fig. 1. *Rhipicephalus sanguineus;* Haller's organ (743 ×).

Fig. 2. *Rhipicephalus sanguineus;* palpal tarsus (352 ×).

Fig. 3. *Ornithodoros papillipes;* Haller's organ (1045 ×).

Fig. 4. *Ornithodoros papillipes;* palpal tarsus (512 ×).

Fig. 5. *Ixodes canisuga;* Haller's organ (578 ×).

Fig. 6. *Ixodes canisuga;* palpal tarsus (715 ×).

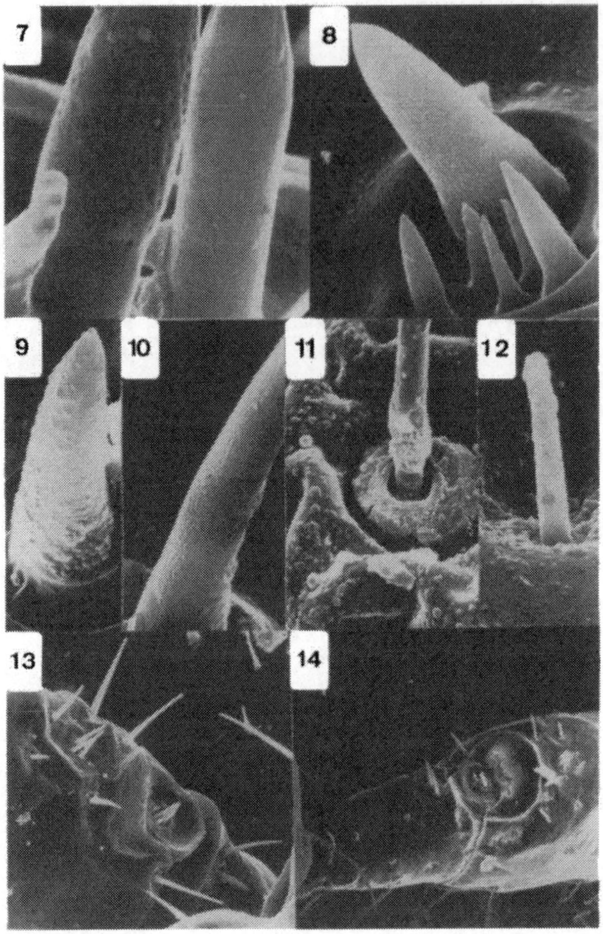

Plate II.

Fig. 7. Part of the sensillae-group of Haller's organ of *Ornithodoros papillipes* (female). Original magnification 4085 ×.

Fig. 8. Group of sensillae of Haller's organ of *Rhipicephalus sanguineus* (male). 2021 ×.

Fig. 9. Secretion on the great sensilla of Haller's organ of *Rhipicephalus sanguineus* (female). 2064 ×.

Fig. 10. The graet sensilla in Haller's organ of *Haemaphysalis inermis* (male). 2150 ×.

Fig. 11. Sensilla (trichobothriumlike) from the ventral view of *Ornithodoros papillipes* (female). 860 ×.

Fig. 12. Hair on the shield of *Dermacentor marginatus* (female). 860 ×.

Fig. 13. Haller's organ and the surrounding hairs and sensillae of *Rhipicephalus sanguineus* (male). 204 ×.

Fig. 14. Haller's organ and the surrounding sensillae of *Dermacentor marginatus* (male). 101 ×.

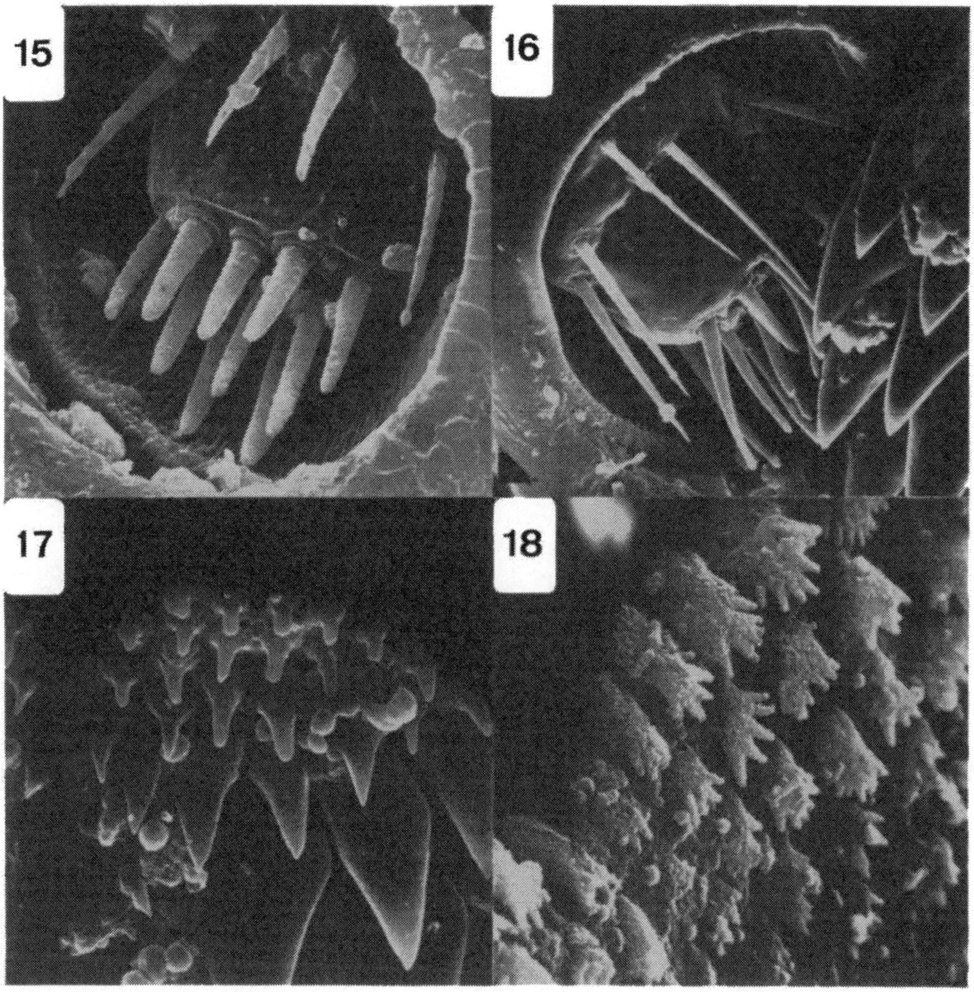

Plate III.

Fig. 15. Palptarsus (apical part of palpus) of ***Dermacentor marginatus*** (male). 680×.

Fig. 16. Palptarsus of ***Haemaphysalis inermis*** (female). 714×.

Fig. 17. Apical part of hypostom (corona) of ***Haemaphysalis inermis*** (female). 1428×.

Fig. 18. Apical part of hypostom (corona) of ***Haemaphysalis concinna*** (female). 1360×.

Proceedings of the 3rd International Congress of Acarology, Prague, 1971

SCANNING ELECTRON MICROSCOPY OF ORIBATID MITES

T.A. WOOLLEY

Department of Zoology, Colorado State University, Fort Collins, Colorado, U.S.A.

Even though the scanning electron microscope has been used to good advantage in studies of a number of acarines, relatively few micrographs have been published of oribatid mites. Sasa, et al. (1971) used the instrument for taxonomy of mites and mosquitoes. Griffiths (1970) showed parts of *Acarus siro.* By SEM Bernini (1968-69) illustrated some features of *Tropacarus brevipilus, Oribatula travei* and *Zygoribatula exarata.* Woolley and Yamamoto (1970), and Woolley (1970) showed some morphology of *Oppia coloradensis* Woolley, and *Tenuiala kurti* Woolley and Higgins, respectively. Brody (1970) described the development of integument for *O. coloradensis* by TEM and SEM. The author has made other studies of oribatids by SEM, some results of which are described and figured in this paper.

The SEM is an instrument of exciting potential in acarology. In oribatids it can be used to determine details of 1) general morphology not visible in the same magnifications nor with the same clarity as in light microscopy because of the 3-dimensional micrographs (Figs. 1B, 2D, 3B, C, F, 4D, F); 2) composition of integument (Fig. 1B, C-F; Fig. 3 B-F; Fig. 4, B-D); and 3) taxonomy, chaetotaxy of body, legs, etc. (Fig. 1A; 2D, E).

Most specimens in this study were Au-coated, but observations were also made of living oribatids. Such observations give insights into movement and behaviour although motion of a living mite may cause poor micrographs by deflection of the electron beam.

Oppia coloradensis has a very torose integument adjacent to the exobothridial hair, pedotecta I & II and the foveae of legs II and III (Fig. 1 A-D). The stellate knobs (Fig. 1 E) are most distinctive. Similar tuberculation is noticeable in *Liacarus cidarus* Woolley (Fig. 2 F) and is distinctive for Liacaroidea as noted in previous research (Woolley, 1968) but reemphasized by SEM. The inner surface of the bothridium is also replete with more elongated tubercles (Fig. 1 F).

By SEM *Hafenrefferia gilvipes* (Koch) shows structural details not visible by light microscopy (Fig. 2 A-E). The toothed anterior margin of the bothridium seems to be distinctive and differs from this organ in *Tenuiala kurti* and *T. nuda.* The dorsal setation of tarsus I, distinctive for species in Tenuialidae, (Woolley and Higgins, 1953), is reaffirmed by SEM for *T. kurti* (Woolley, 1970) and *H. gilvipes* (Fig. 2 D). The use of SEM for differentiation of other chaetotaxic features is an easy prospect for the future.

The integument of oribatids is an open subject for study by SEM, and is morphologically striking in *Gymnodamaeus chalazionus* Woolley (Fig. 3 A - F). The perforated integumental tubercles are different on the dorsal and ventral surfaces. Dome-shaped knobs are found in the bothridium, on the dorsal surface, the legs and leg hairs (Fig. 3 C). Thimble-shaped tubercles cover the ventral surface (Fig. 3 E, F). Fat and wax-dissolving solvents do not affect the structures of these integumental tubercles. Further investigations by SEM are in progress that will determine not only the variations of integument in various oribatids, but the possible composition of surface structures and cerotegument as well.

Spiny integumental placoids are distinctive in *Eremaeus* sp. (Fig. 4 A - D). Although these are located on the general surface, some of the placoids can be displaced (Fig. 4 D). Investigations are continuing on the formation and composition of these structures.

Examples of *Eremaeus* sp. were studied to observe the ovipositor. In one specimen an emerging egg was seen at the distal tip of the ovipositor (Fig. 4 E - F).

Although both TEM and SEM were not used for this paper, it is becoming increasingly evident that the combination of these two techniques (Brody and Wharton, 1970) provides the greatest amount of useful information in studies of this kind. In the absence of TEM techniques, however, SEM will still yield many useful and important facts regarding its 1. application to taxonomy, through clarification of such aspects as chaetotaxy; 2. morphology, structural composition and interpretive function; 3 the ever-present application of this technique to teaching through the strikingly aesthetic as well as informative micrographs or videotapes that result.

Acknowledgements. The writer acknowledges the assistance of T. Yamamato, Jeolco, Burlingame, California, and the use of a JSM-U3 instrument, and Howard Heck, Marathon Oil Company, Littleton, Colorado, and the use of a JSM-2 instrument in obtaining the micrographs.

References

Bernini, F.: Notulae oribatologicae I. Redia 51 : 329 - 375, 1968 - 69.

Brody, A. R.: Observations on the fine structure of the developing cuticle of a soil mite *(Oppia coloradensis)*. Acarologia 12(2) : 421 - 431, 1970.

Brody, A. R., Wharton, G. W.: *Dermatophagoides farinae:* Ultra-structure of Lateral Opisthosoma Dermal Glands. Trans. Am. Micr. Soc. 89(4) : 499 - 513, 1970.

Griffiths, D.A.: A further systematic study of the genus *Acarus* L. with a key to the species. Bull. Brit. Mus. Nat. Hist. 19(2) : 85 - 118, 1970.

Sasa, M., et al.: The use of scanning electron microscopy in morphology and taxonomy of some mites and mosquitoes. Japanese J. Exptl. Med. 41(2) : 135 - 138, 1971.

Woolley, T.A.: Some observations on external anatomy of oribatid mites by the scanning electron microscope. BioScience 20 (23) : 1253 - 1257, 1970.

Woolley, T.A.: The systematics of the Liacaroidea, (In press.)

Woolley, T.A., Higgins, H. G.: A review of the family Tenuialidae with a description of a new species from Colorado and Utah. Bull. Chicago Acad. Sci. 10(4) : 45 - 60, 1955.

Woolley, T.A., Yamamoto, T.: Close up on a mite. Nature 226 - 1005, 1970.

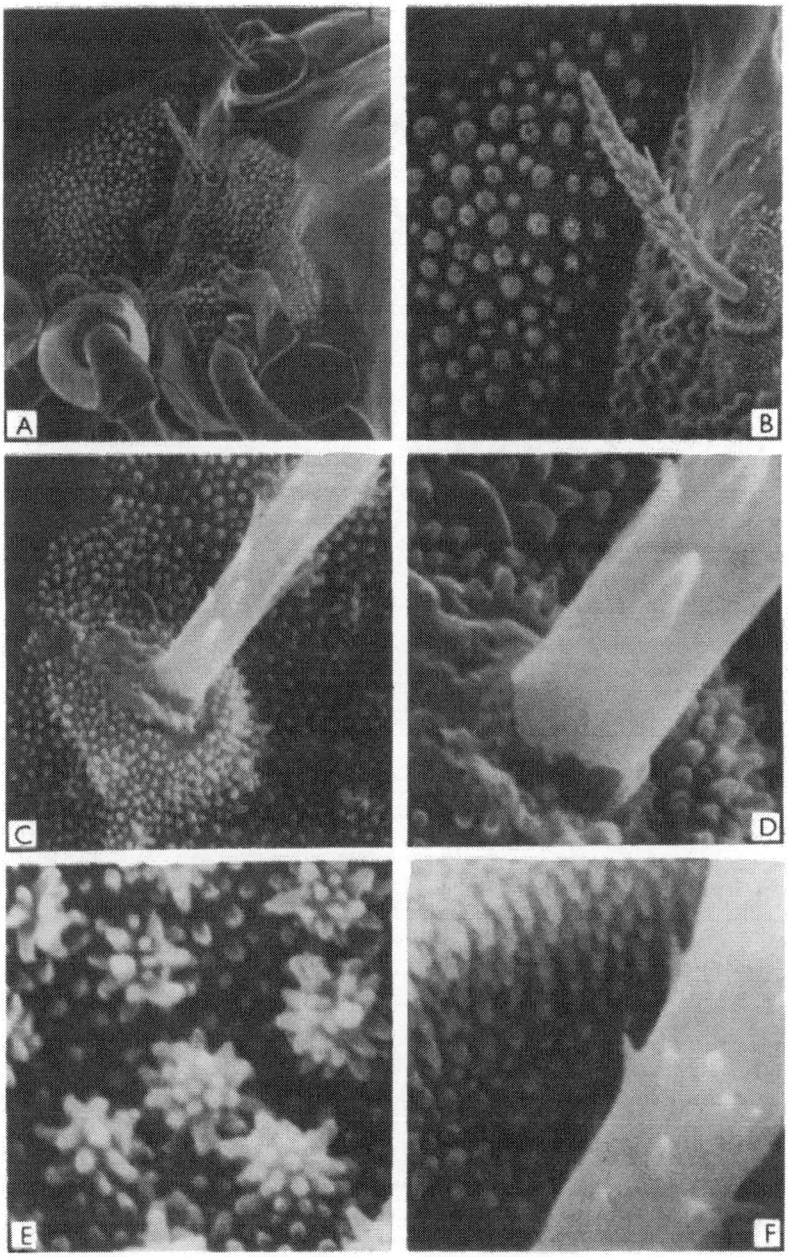

Fig. 1. *Oppia coloradensis* Woolley: A) trichobothrium, exobothridial hair, integument above legs, 580 ×; B) exobothridial hair and stellate tubercles, 2900 ×; C) enlarged view of exobothridial hair, 5800 ×; D) same enlarged, 17400 ×; E) stellate tubercles, 17400 ×; F) tuberculate inner surface of bothridium with pedicel of sensillum partly showing, 17400 ×.

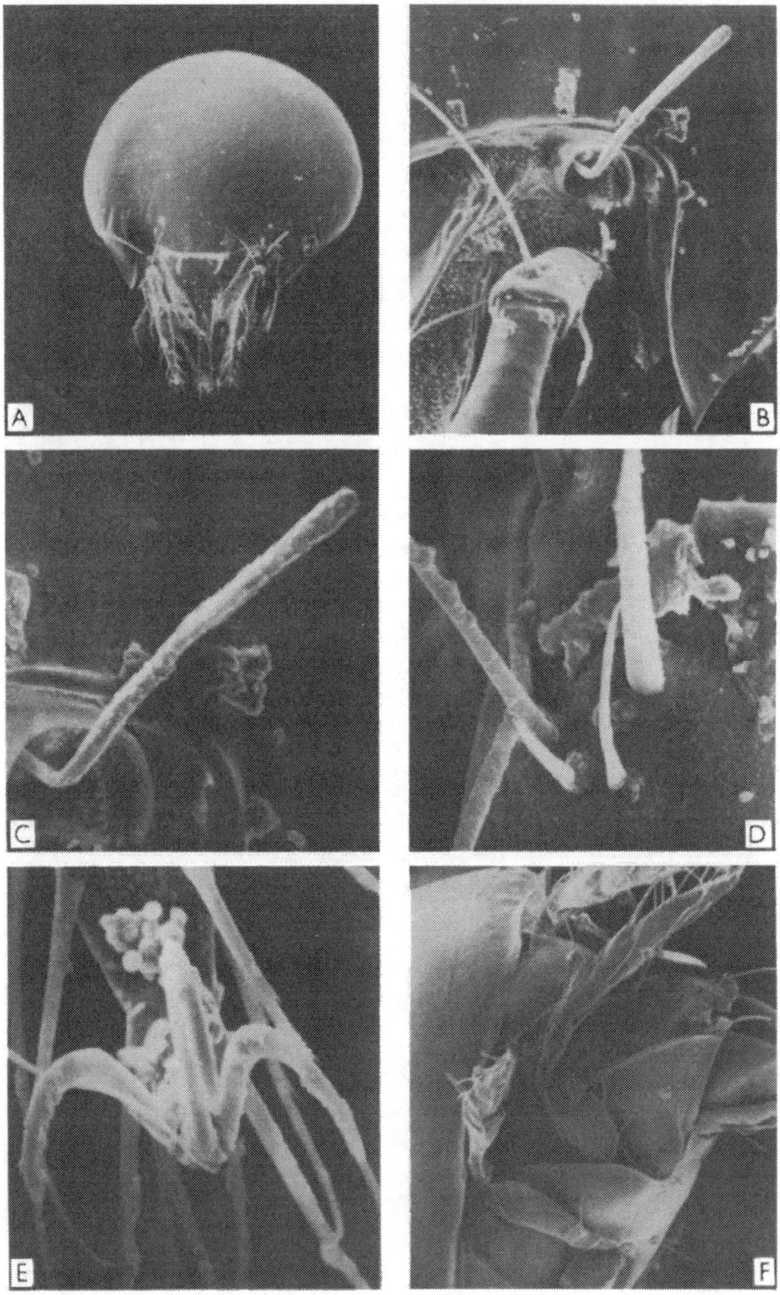

Fig. 2. *Hafenrefferia gilvipes* (**Koch**): **A**) anterodorsal view, 58 × ; **B**) left humeral process, trichobothrium, leg and lamellar margin, 377 × ; C) enlarged view of trichobothrium, 696 × ; D) Solenidial setae of tarsus I, 1740 × ; E) tarsal claws, 1450 × ; F) *Liacarus cidarus* Woolley, prodorsum and legs, 116 × .

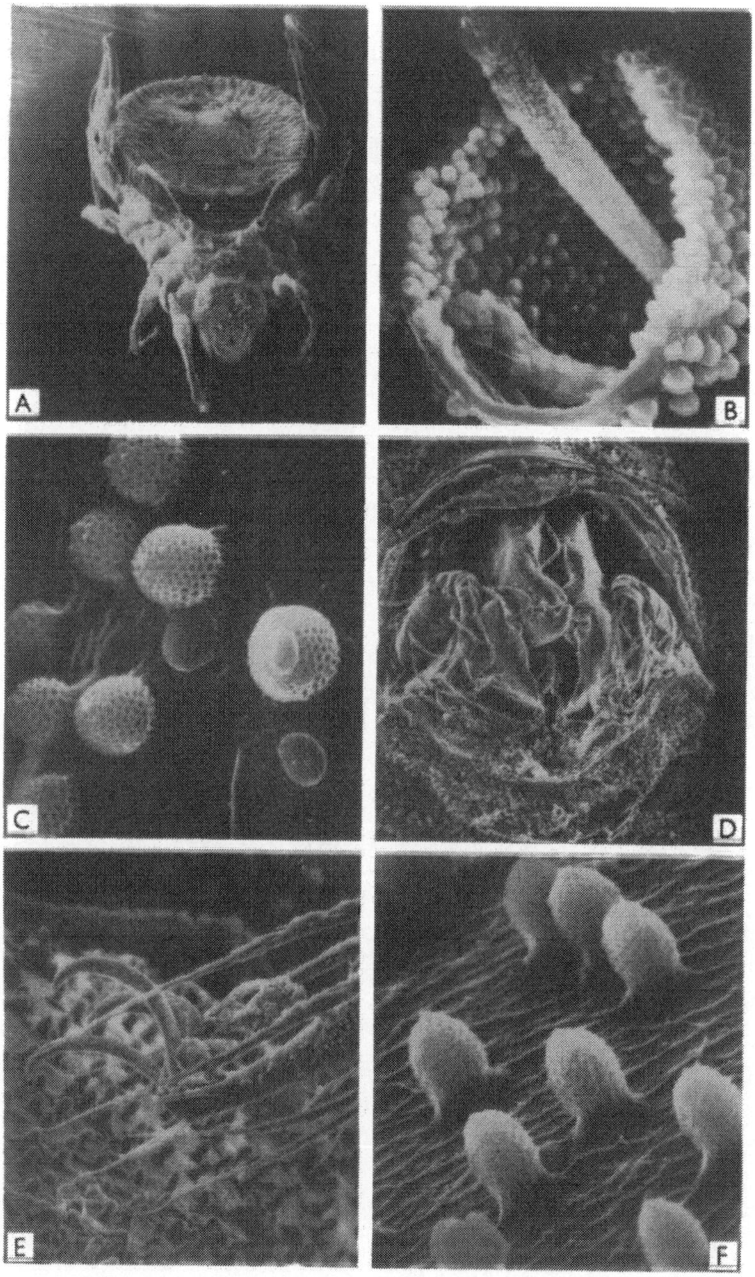

Fig. 3. *Gymnodamaeus chalazionus* **Woolley: A) antero-dorsal view,** 57 × ; **B) trichobothrium** showing surface with dome-like tubercles, 1710 × ; C) **enlarged view of tubercles,** 5700 × ; D) chelicerae, rutella, palps and gnathosomal integument, 399 × ; E) tarsal claws and ventral integument, 1425 × ; F) enlarged view of thimble-shaped ventral tubercles, 5700 × .

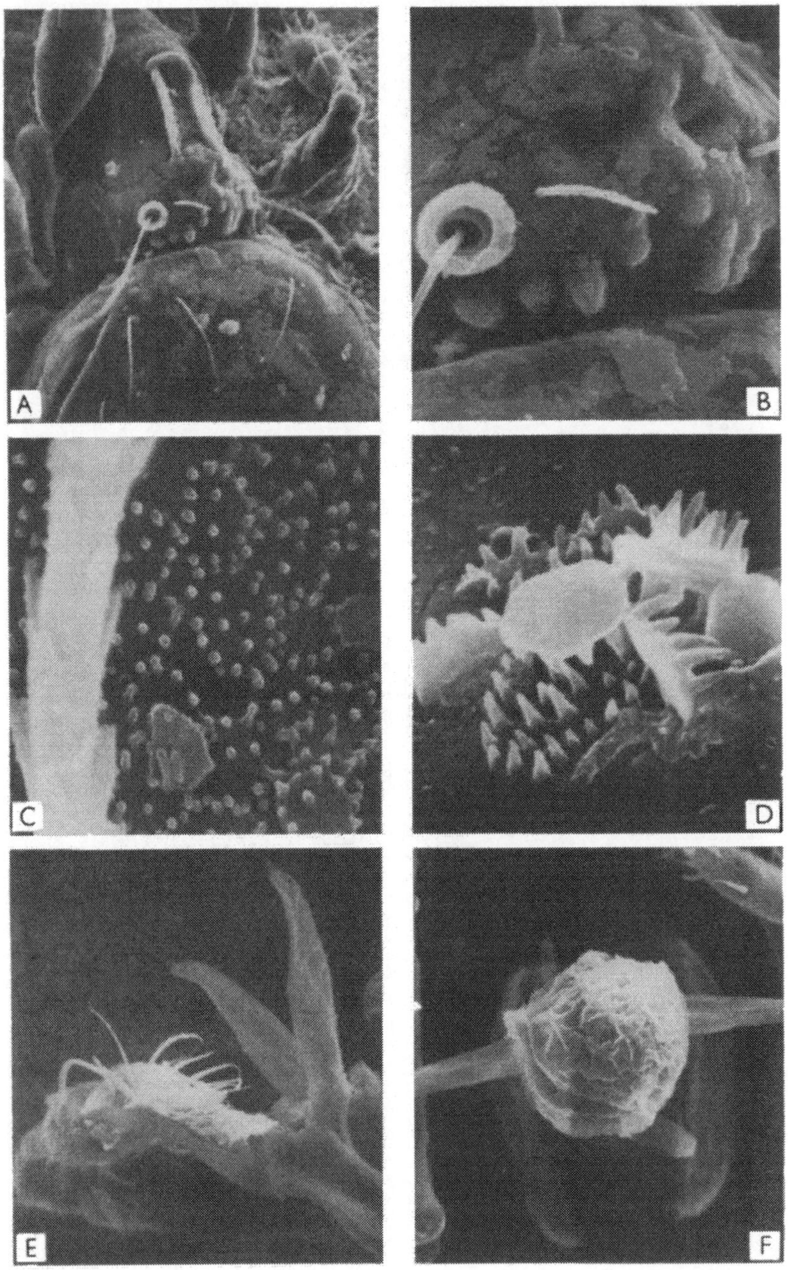

Fig. 4. *Eremaeus* sp.: A) prodorsum, 177×; B) enlarged view of trichobothrium, prodorsum and posterior knobs, integumental placoids, 5900×; D) placoids in place and displaced, 7670×; E) ovipositor and tarsus of leg, 295×; F) ovipositor with egg partially extruded, genital plates in background, 295×.

Section I

SOIL MITES

Chairman
C. Athias–Henriot (France)

Assistant Chairman
M. Kunst (Czechoslovakia)

SOIL MITES (ORIBATIDS) CLIMBING TREES

JUN-ICHI AOKI
Department of Zoology, National Science Museum, Tokyo, Japan

In general conception, habitats of oribatid mites are soils including mosses, litter, rotten woods and other plant detritus on the soil surface. As Travé (1963) and a few other acarologists took notice, however, some species of oribatid mites are often found on living trees and they all have been considered as species wandering between soils and trees. Thus, we have had no doubt that a tree fauna of oribatid mites, if exists, is quite dependent upon the soil fauna under the tree in question. Having investigated the Japanese oribatid fauna in soils as well as on trees, I came to the conclusion, however, that this conception seems to be wrong.

During the period of 1968 to 1970, investigations on tree arthropod fauna at 4 localities in Japan (Fig. 1) were carried out by Prof. Zenpei Yamashita of Mie University. Several kinds of trees selected in each locality were smoked by using 3 cans of the fumigant, Varsan Bomb No. 1 (600g in total weight, containing 150g lindane and 30g DDVP). All the arthropods dropped from above on white sheets settled under the tree were collected with a battery-operated insect sucker. Among the tree arthropods thus collected, the oribatid mites were sent to me for identification and counting.

1. Systematic Position of the Oribatid Families Found on Trees

The oribatid mites collected by fumigation comprised of about 40 species (15 identified species) belonging to 20 families as shown in Table 1. Generally speaking, there is no tendency that the families found on trees are restricted in certain groups of Oribatei. Most of the superfamilies of Oribatei superiores contain families dwelling on trees except Hermannielloidea and Microzetoidea. On the other hand, only Nothroidea have such tree forms in Oribatei inferiores and the lower family groups such as Palaeacaroidea, Parhypochthonoidea, Hypochthonoidea, Phthiracaroidea and Perlohmannoidea were not collected from living trees.

2. Three Life-forms of Oribatei

As to vertical stratification, the oribatid mites are divided ecologically into the following 3 groups (Fig. 2):

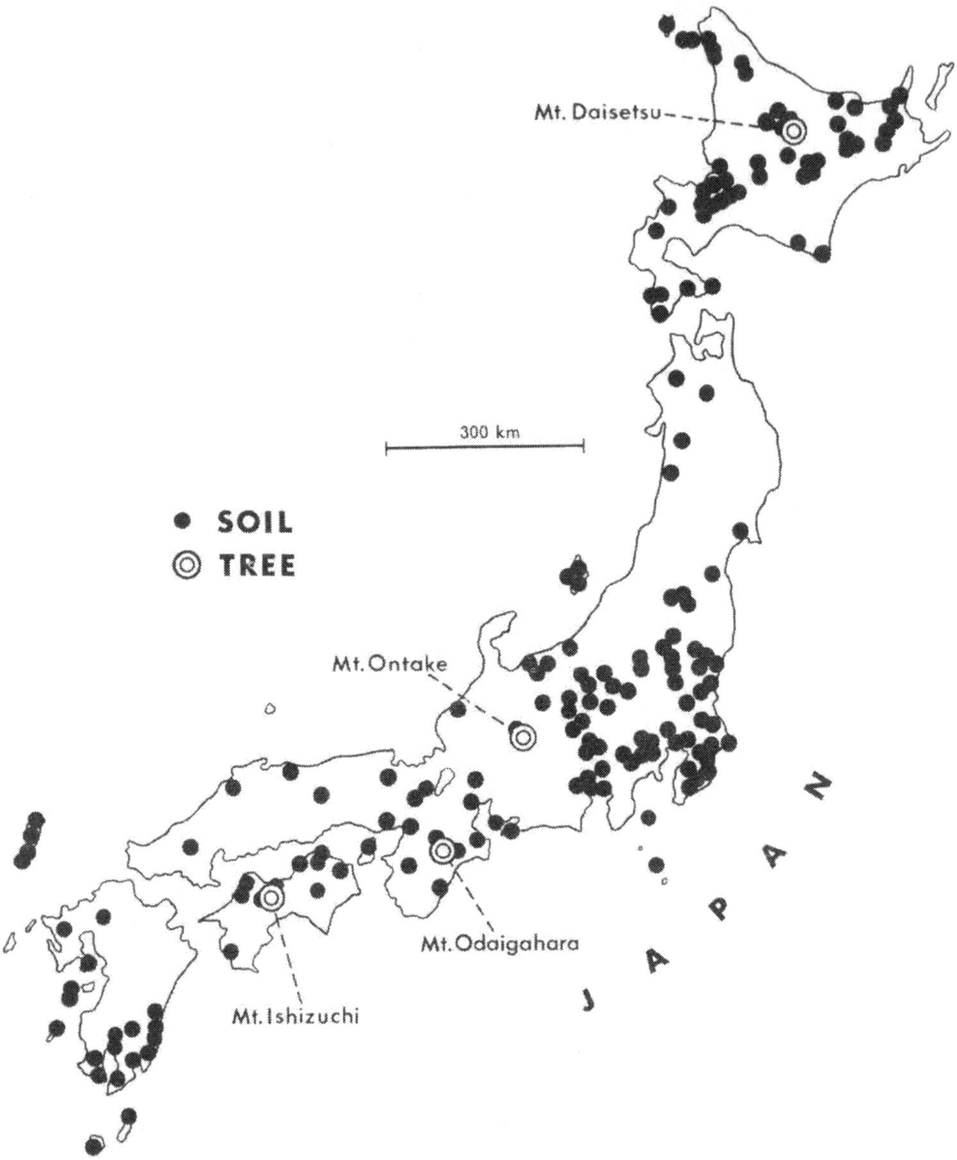

Fig. 1. Sampling points of litter and soil material for extracting and trees for smoking.

Table 1. List of oribatid families found in Japan

	I. ORIBATEI INFERIORES
	PALAEACAROIDEA
	Palaeacaridae
	PARHYPOCHTHONOIDEA
	Parhypochthoniidae
	HYPOCHTHONOIDEA
	Hypochthoniidae
	Eniochthoniidae
	Brachychthoniidae
	Cosmochthoniidae
	MESOPLOPHOROIDEA
	Mesoplophoridae
	PHTHIRACAROIDEA
	Phthiracaridae
	Oribotritiidae
	Euphthiracaridae
	PERLOHMANNOIDEA
	Perlohmanniidae
	Epilohmanniidae
	Eulohmanniidae
	Lohmanniidae
	NOTHROIDEA
	Nothridae
■	Camisiidae
■	Trhypochthoniidae
	Malaconothridae
	Nanhermanniidae
■	Hermanniidae
	II. ORIBATEI SUPERIORES
	A. PYCNONOTICAE
	HERMANNIELLOIDEA
	Hermanniellidae
	LIODOIDEA
■	Liodidae
	Plateremaeidae
	Gymnodamaeidae
	CEPHEOIDEA
■	Cepheidae
	MICROZETOIDEA
	Microzetidae
	ZETORCHESTOIDEA
	Zetorchestidae
	Gustaviidae
	EREMAEOIDEA
■	Eremaeidae
○	Megeremaeidae
	Amerobelbidae
	Eremobelbidae
	LIACAROIDEA
○	Liacaridae
○■	Metrioppiidae
	Astegistidae
	Xenillidae
	Tenuialidae
	CARABODOIDEA
○■	Carabodidae
■	Tectocepheidae
	Nippobodidae
	OPPIOIDEA
	Oppiidae
	Autognetidae
	Suctobelbidae
○	Otocepheidae
	Tokunocepheidae
	AMERONOTHROIDEA
○	Cymbaeremaeidae
	B. PORONOTICAE
	PELOPOIDEA
■	Pelopidae
	ORIBATELLOIDEA
○	Achipteriidae
	Oribatellidae
	CERATOZETOIDEA
○■	Ceratozetidae
○■	Mycobatidae
	Chamobatidae
	Mochlozetidae
	GALUMNOIDEA
○■	Galumnidae
■	Parakalummidae
	ORIBATULOIDEA
○	Oribatulidae
	Haplozetidae

■: Family containing wandering species. ○: Family containing arboreal species.

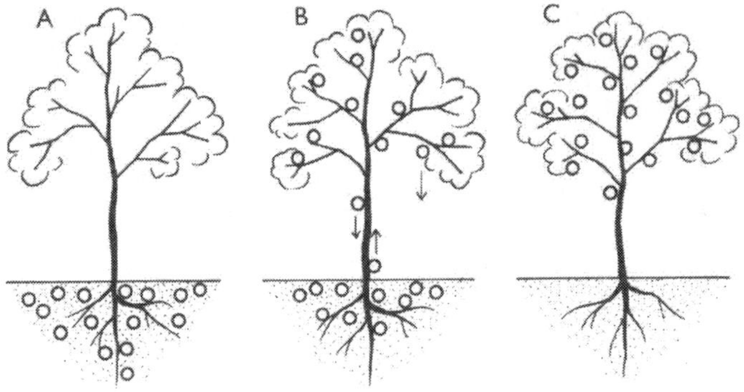

Fig. 2. Oribatid mites divided ecologically into three life-forms. (A) Soil species. (B) Wandering species. (C) Arboreal species.

(1) Soil forms. —Oribatid mites of this group live restrictedly in soil environment in a wide sense and, for the time being, have not been found on living trees. The lower superfamilies in Oribatei inferiores, except Nothroidea, may be true soil forms in this sense. In Nothroidea, it is interesting to note that Nothridae have not been found on trees, although their close relatives, Camisiidae,live abundantly on trees In Oribatei superiores, none of the members of the families Oppiidae and Suctobelbidae have been collected from trees, though they are most abundant in soils. It is, however, insecure to decide true soil species or families at the present moment, because our investigations are restricted to summer season, to small number of localities and to only several tree species. The soil forms in Japan comprise by the present time 149 species belonging to 53 families (Only Cymbaeremaeidae and Megeremaeidae do not contain soil form).

(2) Wandering forms. —The mites of this group enlarge their soil habitat to living trees. They are collected from living trees as well as from soil and litter. Thus, they wander between soil and tree, being called here "the wandering forms or species". The following 12 species belonging to 10 families are considered the mites of this life-form: *Camisia segnis* (Hermann), *C. spinifer* (C.L. Koch), *Trhypochthonius japonicus* Aoki, *Hermannia gibba* (C.L. Koch), *Cepheus latus* (C.L. Koch), *Eremaeus tenuisetiger* Aoki, *Ceratoppia bipilis* (Hermann), *Carabodes bellus* Aoki, *Eupterotegaeus armatus* Aoki, *Tectocepheus velatus* (Michael), *Eupelops acromios* (Hermann), and *Protokalumma parvisetigerum* Aoki. Future increase of knowledge will add many more species to this category.

(3) Arboreal forms. — The oribatid mites collected by smoking trees at 4 localities in Japan consisted, qualitatively as well as quantitatively, mostly of new species which have never been found by 15 years soil survey in Japan. This fact shows that these species are true arboreal forms. The determined species among them are the following 9 species belonging to 8 families: *Megeremaeus expansus* Aoki et Fujikawa, *Dendrozetes caudatus* Aoki, *Liacarus breviclavatus* Aoki, *L. clavatus* Fujikawa et Aoki, *Carabodes breviclava* Aoki, *Fissicepheus amabilis* Aoki, *Scapheremaeus yamashitai* Aoki, *Achipteria curta* Aoki and *Incabates major* Aoki.

3. Comparison of the Fauna among Strata, Tree Species and Localities

In Mt. Ishizuchi area (one of the 4 survey area), similarity index of species composition (Ja = Jaccard's index) was calculated each between tree-tree, tree-soil or soil-soil of different kinds. As shown in Fig. 3, the similarities are fairly high among trees as well as among soils regardless of tree species, while the indexes show low values in any combination of trees and soils. The another test to compare the tree fauna of different 4 localities (Mt. Ishizuchi, Mt. Odaigahara, Mr. Ontake and Mt. Daisetsu, enough remote from one another) showed that they were unexpectedly similar to one another Ja being 18 - 57 which are distinctly higher than Ja among trees and soils in the same locality (Mt. Ishizuchi). These facts indicate that trees have their own rich oribatid fauna which are mostly independent of soil fauna.

	39	43	26	F	3	8	5	5
		52	22	A	3	7	8	6
			17	B	4	4	3	3
				T	8	13	9	5
F	A	B	T		t	b	a	f
				t		50	23	55
				b			86	42
				a				63
				f				

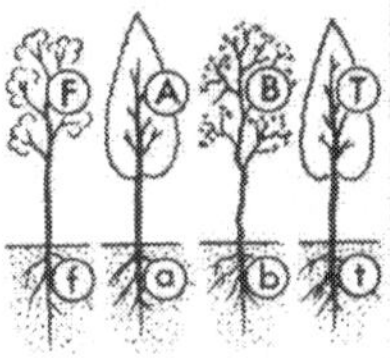

Fig. 3. Similarity index (Jaccard's index) among the species compositions of oribatid mites collected from 4 species of trees and soils beneath them in Mr.Ishizuchi area. F/f: *Fagus crenata*, A/a: *Abies veitchi*, B/b: *Betula ermani*, T/t: *Tsuga sieboldi*.

4. A Special Morphological Feature of Arboreal Oribatid Mites

It is well known that the sensilli of oribatid mites are quite diverse in shape, showing a great variety of modifications (simple seta-like, whip-like, bacilliform, spindle-shaped, clavate, barbed, feathered, forked, etc.). Examination of arboreal oribatids revealed, however, an interesting fact that their sensilli are similar in shape to one another regardless of systematic positions and have a tendency to become short and clavate (Fig. 4). In the genus *Liacarus*, for example, both the arboreal species (*Liacarus breviclavatus* and *L. clavatus*) have short, strongly clavate sensilli, while the soil species (*L. yesoensis, L. acutidens, L. orthogonios, L. coracinus, L. contiguus, L. nitens, L. flammeus, L. gammatus* and *L. bacillatus*) have lanceolate, spindleshaped or bacilliform sensilli. This tendency is well observed in every family which contain arboreal species. Many of the wandering species have also short and clavate sensilli, though they sometimes have sensilli of other shapes, showing an intermediate condition between arboreal and soil species (Fig. 4). Such a modification of sensilli seems to be one of the adaptation of tree-dwelling oribatids, but its meaning is quite unknown at the present moment.

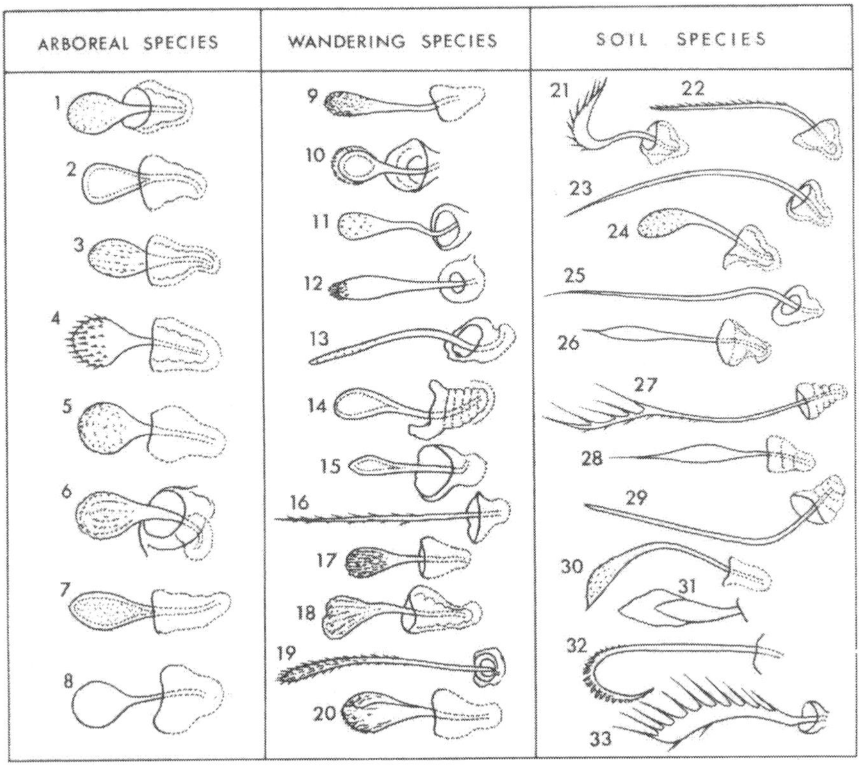

Fig. 4. Variation in shape of sensilla of the arboreal (left), the wandering (center) and soil (right) species of oribatid mites. (1) *Dendrozetes caudatus*, (2) *Liacarus breviclavatus*, (3) *Liacarus clavatus*, (4) *Carabodes breviclava*, (5) *Fissicepheus amabilis*, (6) *Scapheremaeus yamashitai*, (7) *Achipteria curta*, (8) *Incabates major*, (9) *Eupelops acromios*, (10) *Camisia segnis*, (11) *Protokalumma parvisetigerum*, (12) *Trhypochthonius japonicus*, (13) *Hermannia gibba*, (14) *Cepheus latus*, (15) *Camisia spinifer*, (16) *Ceratoppia bipilis*, (17) *Tectocepheus velatus*, (18) *Eupterotegaeus armatus*, (19) *Eremaeus tenuisetiger*, (20) *Carabodes bellus*, (21) *Oppia nova*, (22) *Protoribates* sp. (23) *Eremobelba japonica*, (24) *Scheloribates latipes*, (25) *Nothrus silvestris*, (26) *Gustavia microcephala*, (27) *Parhypochthonius aphidinus*, (28) *Liacarus contiguus*, (29) *Nemacepheus dentatus*, (30) *Pergalumna duplicata nipponica*, (31) *Phthiracarus japonicus*, (32) *Machadocepheus kanekoi*, (33) *Cryptacarus hirsutus*.

References

Travé, J.: Écologie et biologie des Oribates (Acariens) saxicoles et arboricoles. Vie et Milieu Suppl. 14: 267 pp., 1963.

ULTRASTRUCTURAL OBSERVATIONS ON THE ALIMENTARY CANAL IN ORIBATIDAE (ACARI)

F. BERNINI
Institute of Zoology of the University, Siena, Italy

So far, the alimentary canal in Acari, in Oribatidae in particular, has been studied almost exclusively by means of the light microscope. However, in spite of the difficulty in dissecting so small animals and laborious manipulation the use of the electron microscope affords an essential aid towards our better understanding of the histological details of the intestinal cells, by documenting their peculiarities at an ultrastructural level. The present study was undertaken to this purpose.

The intestine in Oribatidae is of a sarcoptiform type (Baker and Wharton, 1958). Despite some divergences, all authors who have investigated this field (Michael, 1883, Berlese, 1897, Warren, 1947, Tarras-Wahlberg, 1960, Woodring and Cook, 1962, Hoebel-Mävers 1967, Tarman, 1968, Bäumler, 1969) have reached fairly precise conclusions about the morphology of the gut.

In the present study four species were examined, differing in their systematic position and alimentary "habitus". *Phthiracarus sp.* and *Steganacarus anomalus* are macrophytofagous, *Xenillus tegeocranus* is a "non-specialized", and *Scheloribates laevigatus* a microphytofagous species.

In all mentioned species, allowance being made for proportion differences between the various segments, the gut may be schematically represented as in Fig. 1.

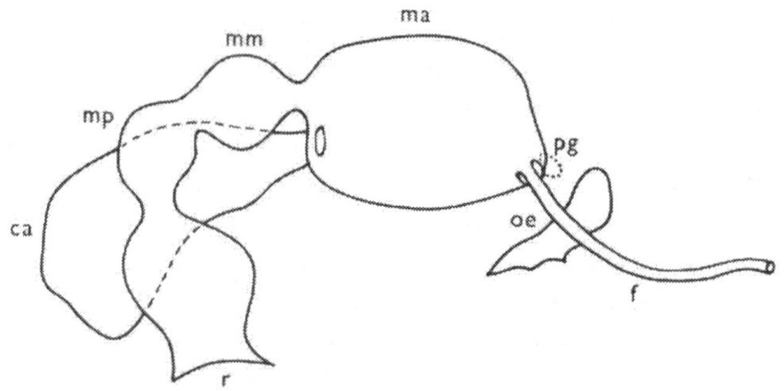

Fig. 1. Schematical drawing of the whole alimentary canal of the Oribatid mites. f – pharynx, oe – oesophagus, pg – proventricular glands, ma – anterior portion of the mesenteron, mm – middle portion of the mesenteron, mp – posterior portion of the mesenteron, ca – coeca, r – rectum.

It is consistently formed by a pharynx, an oesophagus and a mesenteron in which three portions may be identified, namely, the anterior mesenteron bearing caeca, the mid-mesenteron and hind-mesenteron; finally, the posterior end (hind-gut) forms the rectum.

Results

The pharynx and oesophagus, making up the fore-gut and the rectum, or hind-gut, have the same ectodermal origin and morphology. Hence, they may be described together. In these intestinal tracts (Pl. I, 1) the epithelium consists of very low cells with a flat nucleus, poor in cytoplasmic organelles. The cuticle investing these epithelia, though differing in thickness, shows an identical triple-layered structure. The most osmiophilic outer layer corresponds to cuticulin, the intermediate one to the tanned fraction of the procuticle, and the thicker third one may be identified with the endocuticle. In addition, in the rectum a peculiarity stands out which allows it to be distinguished from the other intestinal segments. Its cells are crammed with glycogen, as shown by Thiéry's method (1967). In this respect two hypotheses may be tentatively framed. According to the first one, the rectum would make up a sort of cistern like that formed by the parenchima in *T. farinae*, as reported by Hughes (1950). In this case there is a passage from the hemocoel into the cells. According to the second hypothesis the rectum would be actively involved in the absorption of carbohydrates from the intestinal lumen immediately resulting in glycogen synthesis by its own cells. The epithelium of the anterior mesenteron, which closely follows the fore-gut, consists of columnar cells. Their outer border is furnished with short microvilli next to which a number of mitochondria are seen to aggregate. These cells possess two types of cytoplasmic inclusions. One type, less frequent, is made up of lysosome-like osmiophilic granules possibly containing digestive enzymes. The other type, very common, consists of urospherites (Pl. I, 2). The latter, like what is known to take place among insects and other Acari (Wright and Newell, 1964) might be regarded as guanine. The cells become progressively and almost completely crowded with them, while the cytoplasm undergoes extreme reduction. At this stage the cell is dead, and through the breakdown of its membrane, its contents are released into the intestinal lumen. Although the cells devoid of these urospherites were never encountered in the anterior-mesenteron, it was possible to trace their genesis in the coecal cells. They arise within a vacuole, by virtue of a progressive condensation of substances of external origin. Occasionally, in this tract, cells with a polyploid nucleus and extensive ergastoplasmic whorls are also found. The epithelium is coated by a thick basement membrane and two muscle layers, a transverse inner one and a ribbon-like longitudinal layer. In the mesointestinal coeca, the epithelium though much lower than the preceding one, is mostly composed of cells similar to those described above. However, this cell type is associated with another one characterized by the cells being almost entirely occupied by a large transparent vacuole.

The anterior mesenteron is consistently followed by the mid-mesenteron (colon, after Hughes; small intestine, after Berlese) of variable length according to species. The epithelial cells, (Pl. II, 1) bearing microvilli, are fairly small, with cytoplasmic organelles showing no peculiar features. These cells possess, as a rule, two types of inclusions, namely, transparent vacuoles displaced towards the cell base, and apical lysosome bodies. In these cells, the most outstanding peculiarity is exhibited by the basal boundary, which is consistently richly indented. In some sites, deep invaginations of the basement membrane and even the passage of drops into the hemocoel are found. In *S. laevigatus* another cell type was found, consisting of cells with secretion drops possibly of mucopolysaccharide nature (Pl. II, 2). In this segment a thick basement membrane and two muscle layers are found, the innermost of these, arranged in ribbons, penetrates the basal infoldings.

Following a constriction of the intestinal lumen the last mesenteron segment (Hughes's postcolon, Berlese's colon) is found. Generally, (Pl. I, 3) the epithelial cells are tall and possess a strikingly high microvillous border. In all species examined, these cells are very rich in inclusions. All cytoplasmic organelles are also well developed. This epithelium also reveals a high activity represented by its indented basal boundary and images of pinocytosis.

Discussion

The present investigation has brought to light some structures which will be of help in the study of the functional physiology of the gut from these animals.

The occurrence of urospherites in the anterior mesenteron and caeca supports the claim of their excretory function, (besides that of secretion and absorption!) (Berlese, 1897; Warren, 1947). This device is similar to that reported in other arthropods (Gouranton, 1968) both when devoid of specific excretory apparatuses as Collemboles (Dallai, 1966) and Pseudoscorpions (Baccetti and Lazzeroni, 1967), and when provided with them, as *Anystis sp.* (Wright and Newell, 1964). The concretions found by Berlese (1897) in the mesenteron were believed by him to be the result of a chemical reaction. The bolus would be highly acidic, and the pH of the intestinal lumen would be maintained within bearable limits through the neutralization of the absorbed uric acid by means of alcali resulting in urate precipitation.

Among the structures revealed by the electron microscope, of significance are the basal invaginations indicating a massive absorption of fluids from the hemocoel by the cells of the mid-mesenteron. This phenomenon is suggestive of a secretion of different nature, hence of a different function of this tract.

As to the peritrophic membrane, no structure similar to it would be observed. However, considering that Hughes (1950) reported the occurrence of the cells engaged in the secretion of such membrane in the colon (mid-mesenteron), the secretion found by us in the same segment, might be indirect evidence of its presence.

In brief, from the above findings the following conclusions may be drawn:

a) the presence of most highly specialized mesenteron tracts equates the intestine of the species studies here with that of the other Arachnida.

b) The absence of any specialization at the proctodeal level and the cell plasticity both point to primitivity.
c) As to the problem of water absorption, we presume that the last mesenteron tract is engaged in fluid absorption, being provided by the generic devices involved in it.
In no segment does the gut attain such specialization as is known to occur in the insect rectum.
d) Finally, appreciable divergencies in the digestive apparatus between the four species, differing in their alimentary habits, do not exist.

References

Baccetti, B., Lazzeroni, G.: Primi reperti ultrastrutturali sul canale alimentare di uno Pseudoscorpione. Redia, L. pp. 351-63, Tav. II, 1967.

Baker, E.W., Wharton, G.W.: An introduction to Acarology. New York, 1952.

Berlese, A.: Ricerche sugli organi e sulla funzione della digestione negli Acari. Riv. Pat.Veg. 5, pp. 129-195, pls. 8, 9, 1897.

Bäumler, W.: Zur Morphologie, Biologie und Ökologie von *Hermannia gibba* (Acarina Oribatei) unter Berucksichtigung einiger Begleitarten. Z. angew. Entomol. 66 (3 und 4), pp. 257-77 und pp. 337-62, 1970.

Dallai, R.: L'ultrastruttura dell'intestino di *Orchesella villosa* (Geoffroy) (Insecta, Collembola). Ann. Ist. Mus. Zool. Univ. Napoli 17 (5), pp. 1-18, 1966.

Gouranton, J.: Composition, structure, et mode de formation des concrétions minérales dans l'intestin moyen des Homoptères Cercopides. J. Cell Biol. 37 (2), pp. 316-28, 1968.

Haarløv, N: On the alimentary system of *Nothrus palustris*. Abs. Paps., XIIIth. Int. Cong. Entomol., Moscow, p. 96., 1968.

Hoebel-Mävers, M.: Funktionsanatomische Untersuchungen am Verdaungstrakt der Hornmilben (Oribatei). Diss. Thesis, Braunschweig, pp. 1-46, 1967.

Hughes, T. E.: The physiology of the alimentary canal of *Tyroglyphus farinae*. Quart. J. Micr. Sci. 91 (I), pp, 45-61, 1950.

Michael, A.D.: British Oribatidae. Ray Society, London, vol. I., 1883.

Tarman, K.: Anatomy, Histology of Oribatid gut and their digestion. Biol. Vest., Ljubljana 16, pp. 67-76, 1968.

Tarras-Wahlberg, N.: Taxonomical, Anatomical and Biological Comments on *Diapterobates humeralis* (Hermann 1804) (Oribatei Acari). Including a Description of the Tritonymph. Kungl. Fysiog. Sall. Lund. Forh. 30 (12), pp. 91–101, 1960.

Thiery, J.P.: Mise en evidence des polysaccharides sur coupes fines en microscopie électronique.. J. Micr. 6, pp. 987-1018, 1967.

Warren, E.: On the genitalsystem and gut of the Oribatid Mites, *Cepheus tegeocranus* (Herm.), and the reaction of these organs to a rayfungus parasite. Ann. Natal Mus. II, pp. 1-36, 1947.

Woodring, J.P., Cook, E.F.: The Internal Anatomy, Reproductive Physiology and Molting Process of *Ceratozetes cisalpinus* (Acarina Oribatei). Ann. Entomol. Soc. Am. 55, pp. 164–181, 1962.

Wright, K.A., Newell, I. M.: Some Observations on the Fine Structure of the Midgut of the Mite *Anystis* sp. Ann. Entomol. Soc. Am. 57, pp. 684-93, 1964.

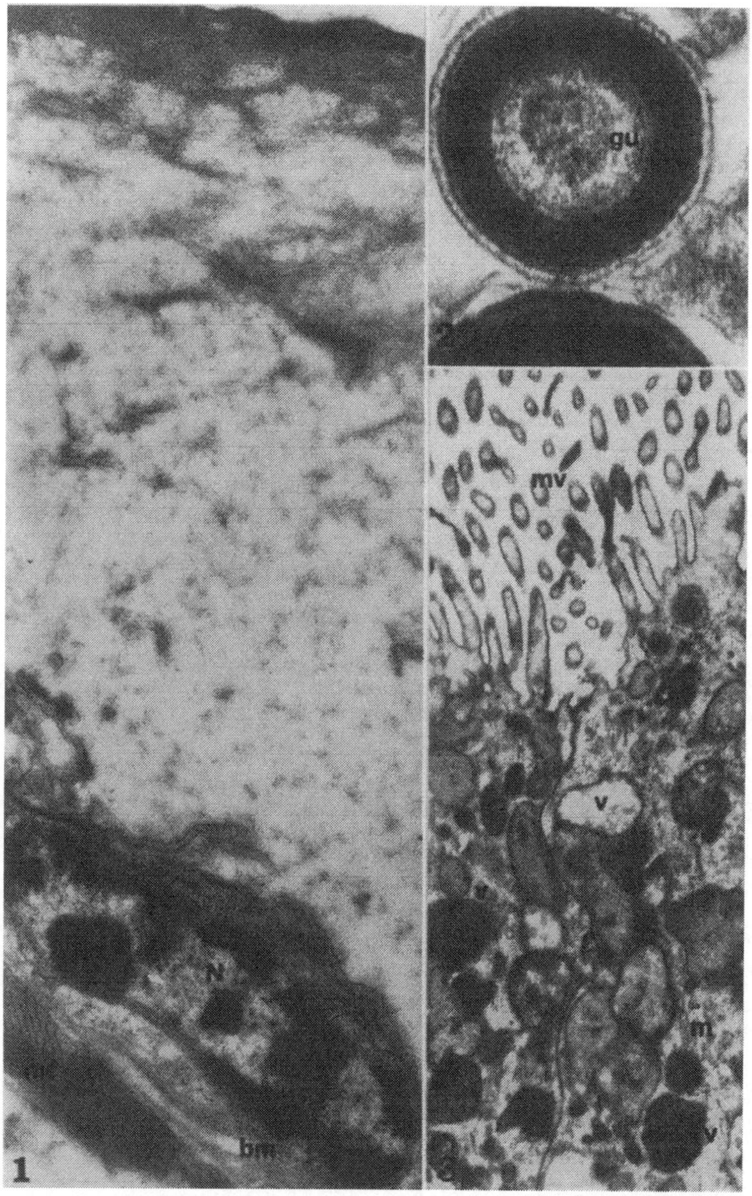

Plate I.

1. Cross section of epithelial cell and muscle tunic (mf) of the pharynx from *Xenillus tegeocranus*. Note the triple-layered intima coating the epithelium. N: nucleous and bm: basement membrane. 25500×.
2. A portion of the cytoplasm of a cell from the anterior mesenteron in *Xenillus tegeocranus*, containing urospherites (gu). 25500×.
3. Apical portion of a hind-mesenteron cell in *Phthiracarus sp.*, containing abundant vacuoles (v) and furnished with an extremely tall microvillous border (mv). m: mitochondria. 15300×.

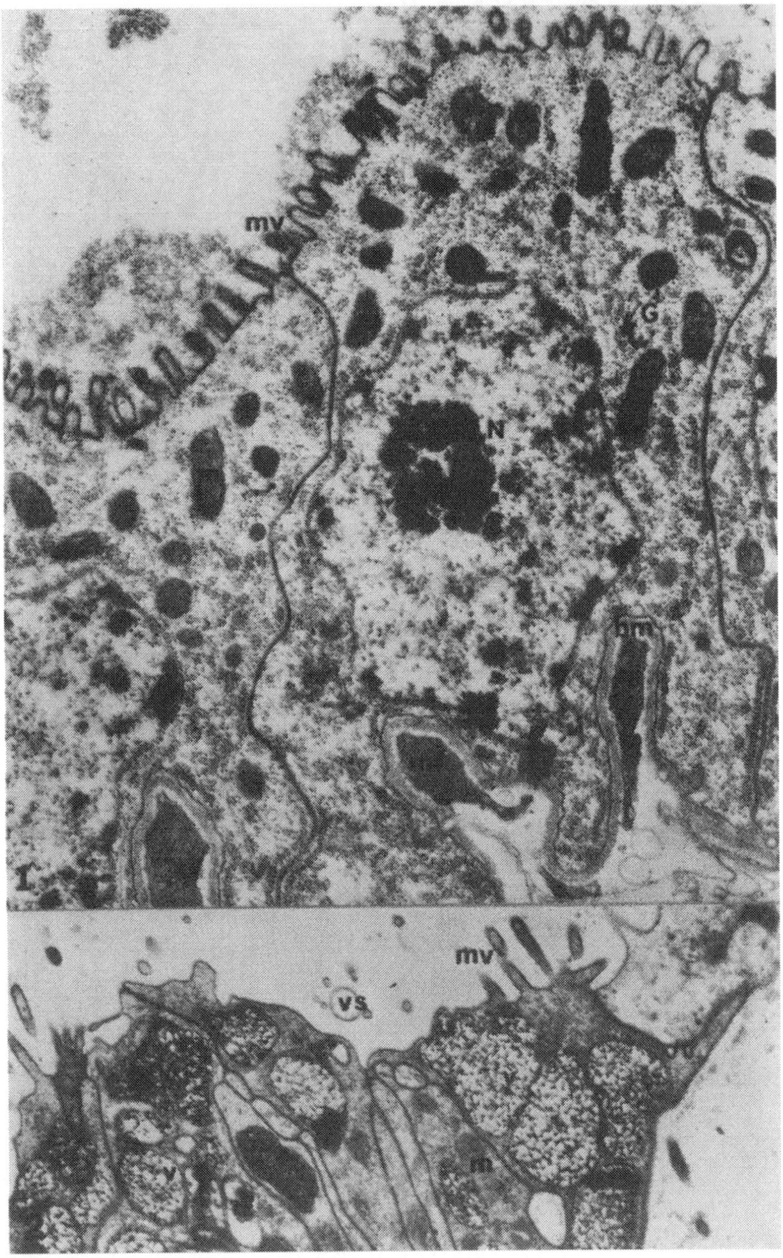

Plate II.

1. Cross section of a cell from the mid-mesenteron in *Scheloribates laevigatus:* the mitochondria (m) clusters at the cell apices, the microvillous border (mv) and the Golgi's complex (G) may be seen. 16380 ×.
2. Mid-mesenteron apical cell in *Scheloribates laevigatus,* containing secretory vacuoles (v) possibly of polysaccharide nature. vs: vesicle. 16380 ×.

DISTRIBUTION OF ORIBATIDS NEAR BARROW, ALASKA[1]

K. K. BOHNSACK
Department of Zoology, San Diego State College, San Diego, U.S.A.

Six biotopes representing differences in vegetation, soil, drainage and microelevation were selected for intensive sampling of the tundra turf during the summers of 1962 and 1963 in the vicinity of Barrow, Alaska. The Barrow region lies in the Coastal Plain Province of the Alaskan Arctic slope approximately 71° 18' north latitude. About 2/3 of the area consists of frost features such as high-center and low-center polygons. A few old raised beaches occurred in the study area. There is generally poor drainage because of the low topography and the impermeable permafrost. The normal annual precipitation is about 11 cm and the mean annual temperatures for the region as reported by the U.S. Weather Station at Barrow for 1931-60 was -12.4° C, with a record high of 25.5° C and a record low of -48.9° C. The maximum summer soil thaw in the study areas varied from 28-52 cm.

A circular sod sample measured 43 cm^2 to a depth of 2.5 cm and 1,006 samples comprising a combined surface area of 4.3 m^2 were taken during the two summers. From these samples 441,743 arthropods were removed with Tullgren funnels. The Acarina composed 34 % of the arthropods collected and the oribatids comprised almost 72 % of the mites and about 1/2 of them were immatures.

Ten species (10 genera) were recorded. Hurd (1958) reported 15 oribatid species and Hammer (1955) 10 species from the Barrow region. Eight of Hurd's species and five of Hammer's occurred in my samples (Table 1). The mean number of oribatids per sample each summer was 106 or approximately 24,300/m^2. Because the immature mites could not be determined with certainty, the discussion which follows deals primarily with the adult mites.

The same four species, *Oppia clavigera, Trichoribates lucens, Liochthonius sellnicki* and *Hermannia subglabra* were the most abundant and accounted for 80 % of the adult oribatids each summer. *Mycobates consimilis* and *Fuscozetes sellnicki* were the rarest species. In the temperate zone ecologists have found the species structure of a community to consistently contain a comparatively few species that are common (represented by a large number of individuals) and a comparatively large number of species that are rare at any one time or place. Crossley and Bohnsack (1960) found in a pine forest in Tennessee that one oribatid species composed 41.2 % of the total

[1] Sponsored by the Arctic Institute of North America under contractural arrangement with the United States Office of Naval Research.

Table 1. A comparison of the oribatid species reported from the Barrow region by different collectors

Species	Hurd	Bohnsack	Hammer
Camisiidae			
Camisia horrida (Hermann)	X	X	X
Brachychthoniidae			
Liochthonius sellnicki (Thor)	X	X	
Hermanniidae			
Hermannia subglabra Berlese	X	X	
Hermannia reticulata Thorell			X
Ameronothridae			
Ameronothrus lineatus brevipes Willmann	X		X
Gymnodamaeidae			
Gymnodamaeus gildersleeveae Hammer	X		
Metrioppiidae			
Ceratoppia quadridentata (Haller)	X	X	X
Ceratoppia sphaerica (L. Koch)	X		X
Pryoppia lanceolata Hammer	X	X	X
Oppiidae			
Oppia clavigera Hammer	X	X	
Oppia maritima (Willmann)			X
Ceratozetidae			
Diapterobates notatus (Thorell)	X	X	X
Trichoribates lucens (L. Koch)	X	X	X
Trichoribates polaris Hammer	X		X
Melanozetes meridianus Sellnick	X		
Fuscozetes sellnicki Hammer		X	
Mycobatidae			
Mycobates aff. conitus Hammer	X		
Mycobates consimilis Hammer		X	
Gen. et. sp.	X		
Total: 19 species	15	10	10

collection, and that the next highest species in abundance amounted to only 8 % of the total mites collected. This kind of species structure was not the case for the oribatid mites in the tundra. Rather here it consisted of a number of comparatively common species and a similar number of rare species, at least 1/2 of the species occured in 50 % of the samples.

A spectrum of dry to moist habitats was sampled. More mites occurred in the drier tundra areas, providing vegetation was present, than in the wet tundra. For both summers, in most areas there was a general increase in the mean numbers of mites per sample with the advancing summer days, and a drop in the population in early September.

All oribatid species occurred in each of the collecting areas, although some were absent one summer in an area and present in the other summer. The second most abundant mite, *Trichoribates lucens*, as well as *Liochthonius sellnicki*, the third most abundant mite, had the highest consistent frequencies in all areas.

When the species frequencies for the different areas and summers were compared, no similar ranking by species was seen. One must conclude that distributional differences occurred both within areas, between areas and between summers. In order to determine whether the differences between areas indicated different oribatid faunas, Sorenson's Quotient of Similarity (Sorenson, 1948) was used. The calculations suggested that the oribatid assemblage in the different tundra biotopes which were sampled represented a single fauna despite variations in abundance and frequencies of individual species.

The mean species weight was determined by weighing 10-95 mites with a quartz helix balance after they had been dried in an oven for 24 hours at 60° C. The weights of the mites ranged from 0.9 to 34.4 micrograms. The mean biomass for all oribatids (adults and immatures) was 230 mg/m^2. The adults comprised 78 % of the biomass in 1962 and 72 % in 1963. Two species, *Hermannia subglabra* and *Trichoribates lucens*, accounted for over 2/3 of the biomass of the mites in both years. Four species composed over 90% of the adult biomass each summer. *Oppia clavigera* which was one of the most abundant mites (first in abundance in 1962 and second in 1963) made up less than 2 % of the biomass.

MacFadyen (1963) calculated the biomass for oribatid mites in grassland as 2,000 mg/m^2 and 400 mg/m^2 for an upland Juncus moor. Engelmann (1961) found an oribatid biomass of 53.9 mg/m^2 for grassland in Michigan. The Arctic oribatid biomass of 230 mg/m^2 determined in this study is considerably lower than MacFadyen's calculations for grassland in England and only a little more than half of his calculations for the moor biomass, but it is fourfold greater than Engelmann's reported for grassland in Michigan.

In summary, the oribatids of the Arctic tundra in Alaska represent a depauperate species population; nevertheless they compose a large percentage of the mites and represent a considerable biomass. Furthermore, even though the species vary in frequencies and abundance in different biotopes, the oribatids represented a single faunal assemblage.

References

Crossley, D.A., Bohnsack, K.K.: Long-term ecological study in the Oak Ridge area. III. The oribatid mite fauna in pine litter. Ecol. 41 : 785–790, 1960.

Engelmann, M.D. The role of soil arthropods in energetics of an old field community. Ecol. Monographs 31: 221-238, 1961.

Hammer, M.: Alaskan Oribatids. Acta Arctica, Fasc. VII: 1-36, 1955.

Hurd, P.D. Jr.: Analysis of soil invertebrate samples from Barrow, Alaska. Final Report, Arctic Institute of North America 24 pp., mimeo., 1958.

MacFadyen, A.: The contribution of the microfauna to total soil metabolism, pp. 3-17. In J. Doeksen and J. Van Der Drift (eds.) Soil Organisms. Proceedings of the Colloquium on Soil Fauna, Soil Microflora and their Relationships. North Holland Publ. Co., Amsterdam, 1963.

Sorenson, I.: A method of establishing groups of equal amplitude in plant sociology based on similarity of species content and its application to analyses of vegetations on Danish commons. Biol. Skr. Danske Vidensk. Selsk. 5 (4): 1-34, 1948.

THE SYSTEMATICS AND GEOGRAPHICAL DISTRIBUTION OF MITES OF THE SUPERFAMILY BELBOIDEA DUB., 1958 (ACARIFORMES; ORIBATEI)

E. M. BULANOVA-ZAKHVATKINA
Faculty of Biology and Soil Sciences, Moscow State University,
Moscow, USSR

The present superfamily embraces a large group of oribatids recognized in the literature by the two families Damaeidae Berl. and Belbidae Willm.

Based upon the morphological similarity of adult mites and peculiarities of the ontogeny we singled out three groups of family rank: Damaeidae, Belbidae and Belbodamaeidae B. - Z. Each group is distinguished by the common appearance of its members and specific features of development (the development time of separate phases, chaetotaxy of the body and extremities, availability of exuviae). To diagnose the above taxons there were used: the structure and character of the proterosomae sclerotization and notogaster, and the form and relative measurements of certain segments of limbs. The characteristics of taxons within the families are based on morphological features such as the structure of extremities and chaetotaxy of their particular limb segments.

For the diagnosis of some species the following morphological features were mainly used: the form and dislocation of the parastigmal apophysis, presence or absence and number of proterosomal, hysterosomal and ventral tubercles as well as the form of gastronotic bristles, their distribution and relative measurements.

At present as many as 152 species of belboid mites are known, with over 90 species in the USSR fauna. These are distributed as follows: Damaeidae - *Damaeus* Koch. - 1 species; *Paradamaeus* B. - Z. - 1 species; *Hypodamaeus* B. - Z. - 27, including 18 known in the USSR; *Spatiodamaeus* B. - Z. - 11 (5 in the USSR); *Epidamaeus* B. - Z. - 38 (24 in the USSR). Belbidae: *Belba* Heyden - 28 (15 in the USSR); *Metabelba* Grandj. - 24 (10 in the USSR); *Allobelba* Kunst - 1; *Metabelbella* B. - Z. - 6 (5 in the USSR); *Neobelba* B. - Z. - 1; *Subbelba* B. - Z. - 2; *Parabelbella* B. - Z. - 3. Belbodamaeidae: *Belbodamaeus* B. - Z. - 3; *Porobelba* Grandj. - 1; *Damaeobelba* Selln. - 1; *Hungarobelba* Balogh - 2.

Nearly all the species of Belboidea known were recorded in the Northern hemisphere (precisely in Holarctica) with the exception of: *Metabelba obtusus* Hammer, 1966 (New Zealand) and *M. flagellata* Balogh and Mahunka, 1969 (South America).

The distribution of separate genera of Belboidea in the world fauna is tabulated in the Table 1. with data showing how many species of each genus are known for the given territory. The greatest numbers of species have been recorded in Eurasia.

Table 1. The distribution of individual genera of Belboidea in the world fauna

Regions / Genus and number of species	USSR																					
	European part					Asian part																
	Northern	North-Western	Central part	Crimea North Caucasus	Transcaucasus	Ural West Siberia	Tuva Transbaikalian	Altai East Siberian	Kamchatka	Sakhalin Kuril Islands	Middle Asia	Scandinavia	Great Britain	West Europe	Greenland	North America Canada	Alaska	Japan	South China	Viet-Nam	South America	New Zealand
Damaeidae																						
Damaeus		1												1								
Paradamaeus		1										1		1								
Hypodamaeus		2	2	6					1	2	3			10					2			
Spatiodamaeus	1	3		1		1		1		1		2	3	5		2						
Epidamaeus	6	3	4	1	3	3	4	2	1	3	4	2		7	1	5	4	2	1			
Belbidae																						
Belba	1	5	2	3	1	2	3	3		4	5	1	1	14					1			
Metabelba	1	4	3	5	3						1	1	1	15				1		1	1	1
Allobelba														1								
Metabelbella	1		1	3	1									1								
Neobelba		1												1								
Subbelba		1				1				1				1								
Parabelbella						1	1	1			2											
Belbodamaeidae																						
Belbodamaeus	1		1		1	1	1	1														
Porobelba	1	1	1	1				1	1	1		1	1	1		1						
Damaeobelba										1				1								
Hungarobelba		1		1						1				1								

The analysis of findings on the geographical distribution demonstrates a certain relationship of the group as a whole to specific ecological conditions. This becomes perfectly clear after research on the USSR fauna in which the group under study is represented in taiga and mixed forest areas by the greatest number of species.

The genus *Epidamaeus* is extensively distributed in the USSR fauna except in the Crimea only.

Along with its typical transasian species the genus *Belba (B. verrucosa* B. - Z., *B. rossica* B. - Z.) contains many endemics of the Crimean and Caucasus damp subtropics *(B. dubinini* B. - Z., *B. limasetosa* B. - Z.) and Primorye *(B. sellnicki* B. -Z.), as well as the semideserts and deserts of the Middle Asia *(B. meridionalis* B. - Z., *B. submeridionalis* B. - Z., *B. tenuisetosa* B. - Z., and *Belba* sp. undet).

10 species of the genus *Metabelba* have been known so far in the USSR fauna. The species of the genus *Metabelbella* are of particular interest due to their morphology and the nature of their distribution. Their most northern species *M. insperata* B. - Z. was found at Dixon while the most southern one *M. zachvatkini* B. - Z. in Lenkoran (the Talysh Mountain Ridge).

The species of the genus *Parabelbella* are typical transasian species as are the species of the genus *Belbodamaeus*. The European species *Damaeobelba minutissima* Sell. was found in this country in the Far East only. The only species of the genus *Porobelba - P. spinosa* (Sell.) is well known in the USSR fauna. The observed small deviation in the morphology of individual specimens from different regions do not give groups for creating new species and subspecies.

The present state of our knowledge of this group of mites does not allow us at present to make a full review of ranges for particular species.

References

Aoki, J.: Results of the Speleological Survey in South Korea 1966. V. Bull. Nat. Sci. Mus., 9 (4), 1966.

Balogh, J., Mahunka, S.: The Zoological Results of the Hungarian Soil Zoological Expeditions to South America. 11. Opusc. Zool. Budapest, 9 (1), 1969.

Bulanova-Zachvatkina, E.M.: Oribatid mite fauna of the USSR and their distribution. In: Oribatei and their role in the process of the soil formation. Vilnius 1970 (In Russian).

Hammer, M.: Some Oribatids from Kodiak Island near Alaska. Acta Arctica, fasc. 14, 1967.

THE PHENOLOGY OF *PLATYNOTHRUS PELTIFER* (C.L.K.)

D. J. L. HARDING
The Polytechnic, Wolverhampton, Staffs., England

Certain aspects of the life cycle of this parthenogenetic oribatid have been studied in the laboratory and in two woodland sites.

Scotland Plantation, Nottinghamshire, 1962-4

Fig. 1 shows seasonal fluctuations in the abundance of *Platynothrus peltifer* in the forest floor of this broadleaved site (*Quercus* sp., *Fagus silvatica* and *Sambucus nigra;* see Harding, 1969). Between April and October, when a particular immature instar decreased from a peak density the succeeding instar began to increase from zero. Changes in the density of nymphs at the beginning of 1963 involved overwintering populations, and did not include zeros. The decrease in tritonymphs between August and September was accompanied by an increase in the number of very pale, lightly sclerotized adults, which made up 50 per cent of the September adult sample. These particular young adults could not have given rise to larvae until the following year, since gravid adults (which are always well sclerotized) were absent between October and March, having been seen first on 29th April, 1963. It appeared that most larvae were derived from eggs laid during the summer by adults which had overwintered, possibly supplemented by contributions from mites which became adult during the early summer.

The Ercall, Shropshire, 1968-71

A clearing on this wooded hillside has a dense field layer of *Pteridium aquilinum,* and is surrounded by *Quercus* sp., *Fagus silvatica* and *Sorbus aucuparia;* the forest floor includes L, F and a thin H layer, and has a pH of *ca.* 3.4.

Field Cultures

Mites were buried in the forest floor of the Ercall in plastic containers with a plaster of Paris floor and lined with glass-fibre filter paper. Food included *Pleurococcus,* decomposing oak and beech leaves, and microbially conditioned cellophane

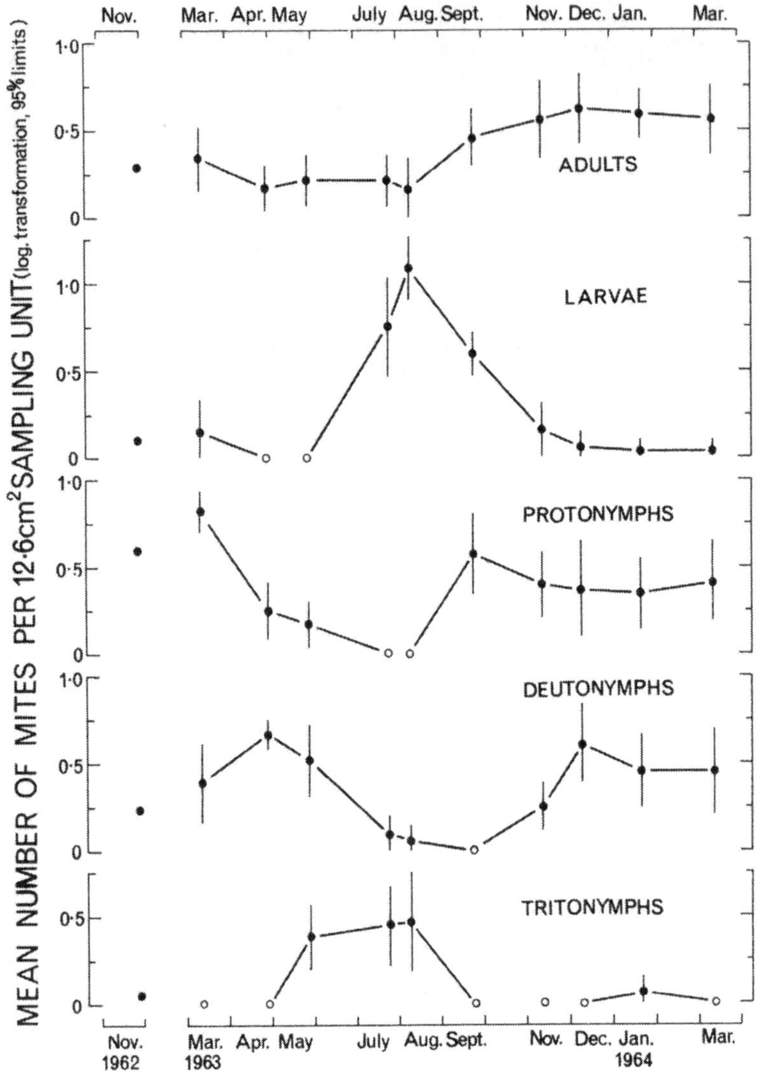

Fig. 1. Population fluctuations of *Platynothrus peltifer* in Scotland Plantation 1962–4 (10 sampling units per sample).

Mites which became adult between May and early August usually darkened within one or two weeks of their final moult, and began to lay eggs after a further 2 - 4 weeks; hatching occurred 3 - 6 weeks later. Adults which emerged from tritonymphs from mid-August onwards failed to produce eggs until the following summer; these eggs were first seen in mid-May, with larvae appearing at the beginning of July. The latest recorded oviposition was in mid-September.

Sampling of the Ercall forest-floor

Provisional results of sampling over a 4-year period indicate that immature peaks tend to be less discrete than in Scotland Plantation. Tritonymphs are present throughout the year, but are more abundant during the late summer; very young adults were recorded mainly between June and November, usually representing less than 10 per cent of an adult sample.

The most striking feature of the Ercall samples is the regularity with which gravid adults and their larval offspring first appear at specific times of the year (see Fig. 2; Table 1 includes corresponding data from other sites); on the Ercall these times are late April-early May for gravid adults, and mid-July for larvae. The date when gravid adults were last seen on the Ercall was more variable, but was generally in September.

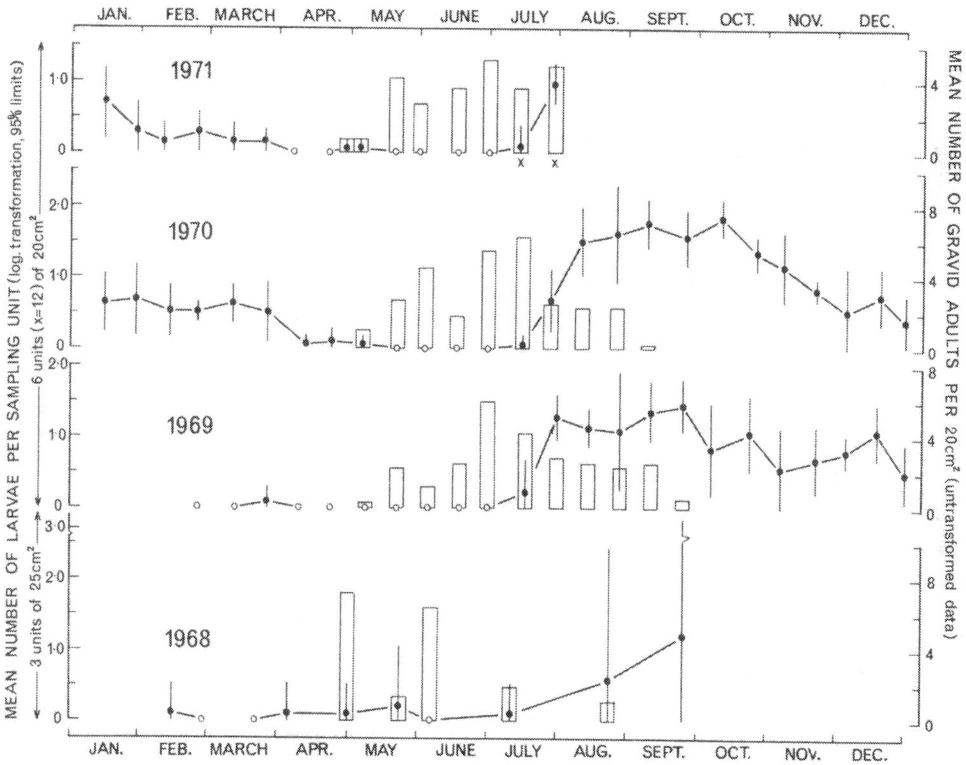

Fig. 2. Abundance of gravid adults (histograms) and of larvae (circles) of *Platynothrus peltifer* on the Ercall, 1968–71.

Table 1. Earliest occurrences of gravid *Platynothrus peltifer* and of larvae

Site	Gravid adults		Summer larvae	
	First recorded	Previous sample	First recorded	Previous sample
Ercall, Shropshire, 1968	May 2	April 4	July 11	June 6
" " 1969	April 30	April 23	July 17	July 2
" " 1970	May 6	April 22	July 15	July 1
" " 1971	April 28	April 20	July 13	July 7
Scotland Plantation, Notts., 1963	April 29	March 13	July 23	May 27
Domleo's Spinney, Notts., 1971	April 22	April 7		
Rothamsted, Herts., 1971	April 29	April 15		
Henley-on-Thames, Oxon., 1971	April 17	March 14		
Furzebrook, Dorset, 1971	April 13	March 23		

Discussion

The restriction of egg production by *P. peltifer* to the period April-September contrasts with the situation in *Rhysotritia duplicata* (Grandjean) in which gravid adults occur throughout the year (see Harding, 1969), but may be somewhat similar to the June-January occurrence of gravid adults of *Nothrus silvestris* Nicolet, observed by Webb (1970). Lebrun (1971) recorded high proportions of gravid mites in adult populations of *N. palustris* (C.L.K.) at all times of the year, but overwintering adults failed to lay eggs until May, when temperatures exceeded 10° C. Lebrun has suggested that there are two phenological cycles in this species, two larval peaks corresponding to the main emergence times for adults in July and October.

The extrinsic factors which are most likely to control the markedly regular, seasonal occurrence of gravid adults of *P. peltifer* are daylength and temperature. Paris and Pitelka (1962) correlated breeding activity in the isopod *Armadillidium vulgare* with daylength, while Lebrun (1970) has shown that the light regime has a marked effect on the duration of the life cycle of *Nothrus palustris* in culture; this species, like *P. peltifer*, inhabits the upper layers of the forest floor. However, in *P. peltifer* daylength could hardly be involved in the case of field cultures, since these were in lightproof containers. Furthermore, temperature has been shown to influence the timing of egg production in the laboratory and in the field. Adults collected in January and cultured in the dark at constant temperatures of 10°, 15° and 20° C laid eggs in March and April, earlier than eggs were seen in natural populations. Adults from artificially heated grassland plots, kindly supplied by Mr. Ian Haines of Rothamsted Experimental Station, contained eggs on 18th March, 6 weeks earlier than in control plots. In this grassland, as also in other sites including the Ercall, the proportion of adults which were gravid was very low at the beginning of the egg-bearing season; this might be related to horizontal and vertical distribution, with associated temperature differences, or to intrinsic differences between individuals.

The regular occurrence of the first larvae in mid-July, 10 weeks after the first gravid adults, suggests that the duration of this interval may be independent of temperature variation, at least within the range experienced in 1968-71.

There is no evidence for overwintering by eggs of this species, which would be expected to result in a second larval peak, in the spring. Protonymphs and deutonymphs, however, each tend to have two density peaks, one before and one after the winter; these presumably correspond to different oviposition dates. Most of the early larvae probably develop as far as deutonymphs before winter, whereas late developers overwinter as protonymphs or larvae, moulting in the spring to produce peaks in the succeeding instars. Geographical and annual variation in the proportions of immatures originating from eggs laid at different times of the year could account for the different forms of population fluctuation, including the discreteness of the peaks and the presence of zeros. Finally, it should be noted that tritonymphs do not have two density peaks, but that at least in Scotland Plantation there was a high plateau during the summer, based on the moulting of mites which became deutonymphs before the winter, as well as of those which overwintered as protonymphs. The timing of the final moult to the adult stage determines whether an individual will produce eggs in its first summer as an adult, or in the following year, or, conceivably, in both years.

References

Harding, D. J. L.: Seasonal changes in the abundance of Cryptostigmata in the forest floor of an oak woodland. In Proc. 2nd int. Congr. Acarology (ed. G. O. Evans), 31-5, 1969.

Lebrun, Ph.: Écologie et biologie de *Nothrus palustris* (C. L. Koch 1839) - Acarien-Oribate. III. Acarologia 12, 193-207, 1970. IV. Acarologia 12. 827-48, 1971.

Paris, O.H., Pitelka, F.A.: Population characteristics of the terrestrial isopod *Armadillidium vulgare* in California grassland. Ecology 43, 229-48, 1962.

Webb, N. R.: Population metabolism of *Nothrus silvestris* Nicolet (Acari). Oikos 21. 155-9, 1970.

AN ORDINATION OF ORIBATID MITE FAUNA UNDER DIFFERING FOREST COVERS

J. R. HILL
Department of Entomology, MacDonald Campus of McGill University,
Ste Anne de Bellevue, Quebec, Canada

This paper outlines part of the results obtained from a comparison of the oribatid mite fauna collected from forest soil samples at 12 sites (number 1-12) under six different forest cover types within the Morgan Arboretum near Montreal, Quebec, Canada. The forest cover types occur in two vegetative catenas: the Beech-Hemlock *(Fagus-Tsuga)* catena with acid sandy soil from which 2 sites each of Hemlock, Beech, Beech-Red Maple, and Grey Birch-Red Maple *(Betula - Acer)* stands were chosen; and the Sugar Maple-Basswood *(Acer - Tilia)* catena with weakly acid clay soil from which 2 sites each of Aspen *(Populus)* and White Ash *(Fraxinus)* stands were chosen. The experiment was designed to attempt to determine whether the fauna is more closely related with the forest cover type or with other environmental variables. Sites were chosen varying greatly in the relative distance between sites within the relatively small area of 150 hectares.

Methods

Pairs of samples 3.33 cm in diameter were taken 6-10 cm deep according to the depth of the humus layers. The fauna was collected in an extractor which combined some features of both Murphy's (1958, 1962) split funnel extractor and Kempson's extractor (Kempson et al., 1963). The cores were first sub-divided into subcores, each 2 cm deep, before extraction; specimens were collected over a picric acid solution.

The specimens were cleared in lacto-phenol solution, washed in distilled water, and either permanently mounted in Berlese fluid and ringed with Gurr's Glyceel or temporarily mounted in lactic acid in a cavity slide (10 per slide) for identification.

Observations

In total 37,194 oribatids belonging to 154 species were identified (many still undescribed), although 15 were represented by only a single individual each. When numbers on the basis of 2 cores 3.3 cm in diameter were converted to populations per square metre, they compared well with other workers' results, the highest obtained in one Beech-Red Maple habitat being equivalent to 1,475,000/ m^2. There were generally much lower populations in the Aspen and White Ash habitats, where the soil was a clay loam of the mull type. The maximum populations obtained in the two White Ash habitats were equivalent to only 34,000 and 46,000/ m^2.

Analysis

Initially Sorensen's Quotient of Similarity, using only presence or absence data, was used to attempt to group the sites, but results were inconclusive, although when only the dominant species, which were considered to be those making up 2 % or more of the population with a constancy of 50 % or more, were used, there was a greater spread in the values obtained. This occurred because the majority of the species are widely distributed, though in lower numbers in the less favourable sites.

From this observation it was evident that a more sophisticated analysis, taking into account the numbers of each species present, was necessary to separate dissimilar sites.

Essentially the problem is to classify or ordinate the habitats on the basis of a large number of species, taking into account their populations, i.e., on the basis of multiple variables. Factor Analysis has been developed for this sort of problem where we have a non-manipulative experiment (i.e., no control) with each individual having many variables. It is an exploratory technique for simplifying a complex set of data where the controlling ecological factors are partly or entirely unkown (Greig-Smith, 1964, p. 172). Advantages for ecological research are that it is unbiased making no prior assumptions, it does not require a randomized sampling program, and it is not necessary to have a normal or even known frequency distribution. However, since no prior assumptions are made, there are difficulties in interpretation. It is useful for a first hypothesis. It has been used in plant ecology to classify or ordinate plant associations (see Goodall, 1954; Orloci, 1966 et al.), but, to the knowledge of the writer, has not been used in animal ecology. Recently it has been used for classification of geological strata in oil exploration (see Parks, 1966, et al.). The procedure involves the following:

1. The information is set up in a data matrix tabulating entities (in this case sample cores) against attributes or variables (in this case numbers of each species found in each core).
2. From the data matrix is calculated a correlation matrix using in this case a log $n\ (x + 1)$ transformation and eliminating the sparse variables to render the data more coherent.
3. Thereafter Cluster Analyses with print-out dendrograms and principal component analyses were carried out after using the Varimax method of analytical rotation to eliminate the effect of correlated variables. It must be remembered that an analysis of this nature is an approximation of the information which could be derived from the estimated true data matrix, i.e. it screens out errors and discontinuities,

but inevitably results in a loss in information. It is necessary to choose an analysis which as closely as possible normalizes as well as simplifies the information and also retains as much of the variability (information) as possible.

Of the analyses used, a *Q*-mode (comparison of samples on basis of species present) using 90 variables (species) with log $n(x + 1)$ transformation which by various manipulations retained 82 % of the variability when 9 factors were extracted, best represented the data. Other analyses were used as a check.

Results and Discussion

A rather drasticly condensed summary of the Varimax Factor matrix from this analysis is presented in Table 1. The factors extracted may be described as hypothetical variables which are condensations of the values of the numbers of each species in each sample. The values indicated represent the range of the coefficients of the 8 cores taken at each site.

Factor 1. The two Hemlock sites come out on the same factor but since the coefficients do not overlap they are nevertheless not very closely related, though they appear more similar to each other than to any other habitat.

Factor 2. No. 3 White Ash site is closely related to 4 of the 8 cores from No. 12 White Ash site but not to the other 4 evidently because 2 different microhabitats occur in the No. 12 samples.

Factor 3. The 2 Grey Birch-Red Maple and the 2 Beech-Red Maple sites come out more strongly on the same factor than on any other although only No. 1 Grey Birch-Red Maple and No. 9 Beech-Red Maple sites have strong similarity. No. 10 Grey Birch-Red Maple and No. 7 Beech-Red Maple sites do not have a high degree of similarity either to each other or to sites No. 1 and No. 9. They also express some of their variability separately at lower levels on Factors 8 and 9.

Factor 4. No. 11 Aspen site showed no similarity to any other site.

Factor 5. No. 8 Aspen site also was dissimilar to any other habitat.

Factor 6. In No. 12 White Ash site 4 cores expressed some of their variability on this factor but a somewhat larger proportion came out on Factor 2, the other 4 cores were closely related but not strongly related to the first 4 cores.

Factor 7. The 2 beech habitats are very closely related.

Factor 8. Not clear.

Factor 9. Also not very reliable but probably expressed some of dissimilarity of site No. 10 as composed with any other site.

In Figure 1 the 9-factor Cluster Analysis, when compared at about the 0,2 level with the Varimax Factor Matrix, shows a very similar result. Nearly the same amount of information was retained in this analysis.

Figure 2, showing an 8-factor Cluster Analysis, which accounts for 80.31 % of the variability, was similar but somewhat less accurate. Elimination of the last

Table 1. Summary of Varimax Factor Matrix 90 Variables log $n(x+1)$

	Site \ Factor	1	2	3	4	5	6	7	8	9
Beech-Hemlock Catena	1. Grey Birch-Red Maple			.77–.88						
	10. " " " "			.43–.66						.36–.57
	7. Beech-Red Maple			.63–.77					.32–.44	
	9. " " "			.80–.88						
	2. Beech	.33–.48		.36–.55				.60–.70		
	4. Beech	.36–.56		.17–.41				.57–.69		
	5. Hemlock	.55–.72							.43–.53	
	6. "	.79–.91								
Sugar Maple Basswood Catena	8. Aspen					*.67–.81				
	11. "				.67–.89					
	3. White Ash		.63–.85							
	12. " " part		.53–.71				.41–.65			
	12. " " part		.06–.26				.81–.92			
	Variance	16.01	6.52	22.61	7.34	5.66	6.00	10.60	4.33	3.27
	Cumulative Variance	16.01	22.53	45.14	52.48	58.14	64.14	74.73	79.06	82.33 %

* Except 2 cores with values of .27 and .42 indicating different microenvironments.

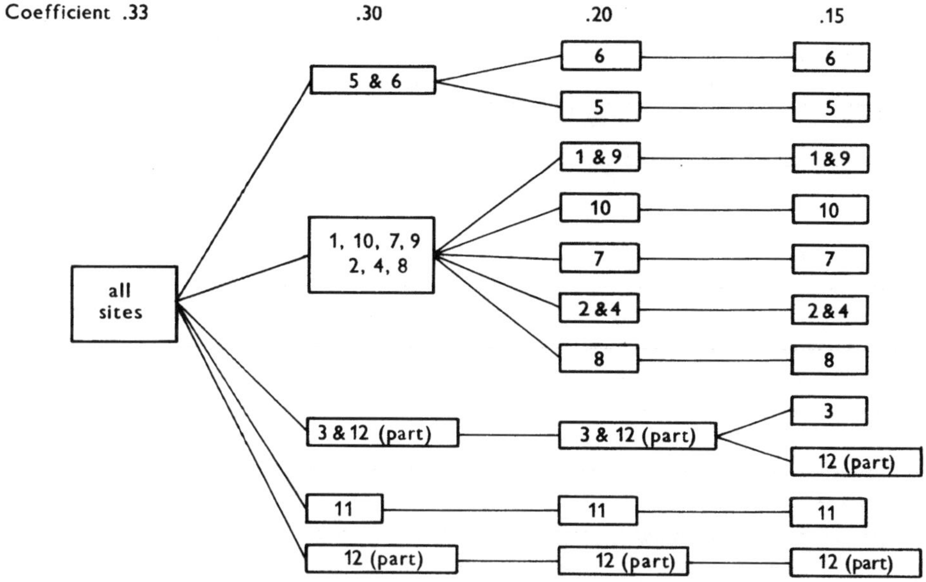

Fig. 1. Summary of 9–factor cluster analysis.

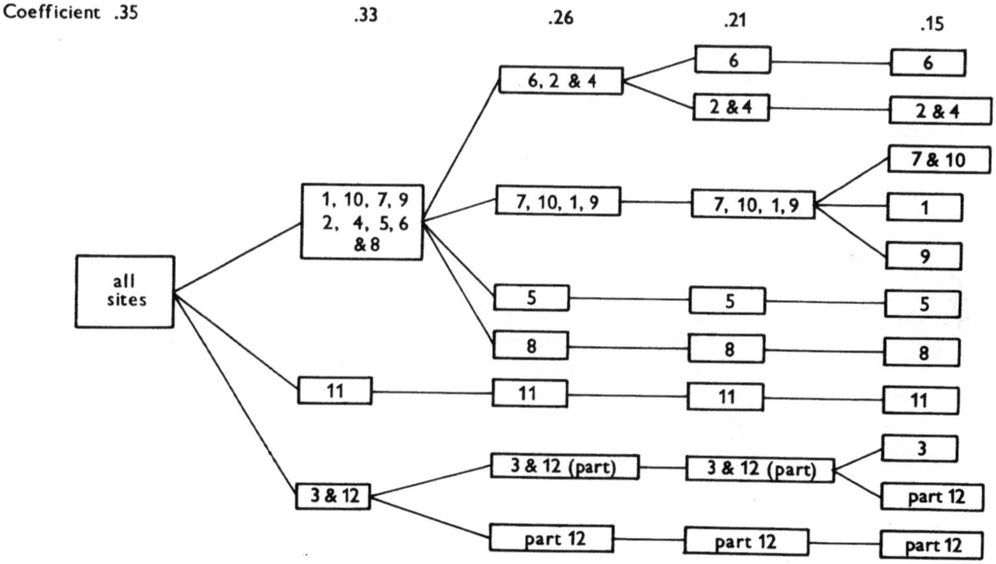

Fig. 2. Summary of 8–factor cluster analysis.

factor results in slightly greater loadings on the factors retained as well as loss of information which may obscure dissimilarities (e.g. No. 6 Hemlock site is shown as being more similar to an No. 2 and 4 Beech sites than to No. 5 Hemlock). Ten-factor analysis did not significantly increase the reliability of results over 9-factor analysis.

Conclusions

1. In general a fair degree of similarity between the faunas of the same or similar forest stands has been shown.
2. The low degree of similarity between the two White Ash and dissimilarity of Aspen sites was possibly due to the small areas of these stands within other stands and possibly to their relative unfavourability for oribatids and hence the occurrence of more highly contagious populations.
3. The forest cover appears to express quite a strong influence on the fauna, but other environmental variables which await analysis apparently have a strong influence in some cases.

References

Goodall, D. W.: Objective Methods for the Classification of Vegetation III. An Essay in the use of Factor Analysis Australian J. Bot. 2: 304-24, 1954.

Greig-Smith, P.: Quantitative Plant Ecology. 2nd ed. Butterworths. London. 256 pp., 1964.

Kempson, D., Lloyd, M. and Ghelardi, R.: A new Extractor for Woodland Litter. Pedobiologia 3, (1): 1-21, 1963.

Murphy, P. W.: The Quantitative Study of Soil Meiofauna I. The Effect of Sample Treatment on Extraction Efficiency, with a Modified Funnel Extractor. Entomol. Exptl. Appl. 1: 94-108, 1958.

Murphy, P. W.: The Split-funnel Extractor - a Modified Tullgren Funnel. In:Progress in Soil Zoology. P. W. Murphy (ed.) pp. 174-8, 1962.

Orloci, L.: Geometric Models in Ecology. The Theory and Application of Some Ordination Methods. J. Ecol. 54: 193-216, 1966.

Parks, J. M.: Cluster Analysis applied to Multivariate Geological Data. J. Geol. 74 (5): pt. 2, pp. 703-15, 1966.

THE EVOLUTIONARY ECOLOGY TRENDS AND TEMPO OF EVOLUTION IN PALEARCTIC ORIBATEI

D.A. KRIVOLUTSKY
Institute of Animal Morphology, Acad. Sci. USSR, Moscow, USSR

There are 4 main directions of ecological specialisation (the main life-forms) in soil dwelling Oribatei in the Palearctic from one initial primitive type: 1) Euedaphon (*Epilohmannia, Simkinia,* etc.), 2) Hemiedaphon (*Hypochthonius, Oribatula, Scheloribates* etc.), 3) Epiedaphon (*Galumna, Trichoribates, Damaeus* etc.), 4) the dwellers of minute soil cavities (*Oppia, Suctobelba, Microzetes* etc.). The number of species and intraspecific variability varies widely in different life-forms of *Oribatei* and often does not depend on population density and species area. There are relatively small numbers of euedaphic species; a greater diversity and variability is shown in epiedaphic forms with weakly sclerotized integuments. The greatest diversity and intraspecific variability occur in the inhabitants of minute soil cavities. The tempo of oribatid mite evolution, which differs with the life-form is associated with the environmental variability, ability of species to endure unfavourable factors and capacity for dispersal within the natural limits. The greatest tempo of evolution in the dwellers of minute soil cavities can be illustrated by the number of endemic genera in different life-forms of palearctic oribatei (Table 1.) and in the oribatoid mite fauna in the regions of the USSR with brown forest soils (Figs. 1, 2.).

Table 1. The number of endemic genera in the different life-forms of Oribatoid mite fauna in the Palearctic

Life-forms	The total number of genera	The number of endemic genera	%
The dwellers of minute soil cavities	37	15	41
Epiedaphic forms	148	48	32
Hemiedaphic forms	36	3	8
Euedaphic forms	10	3	30

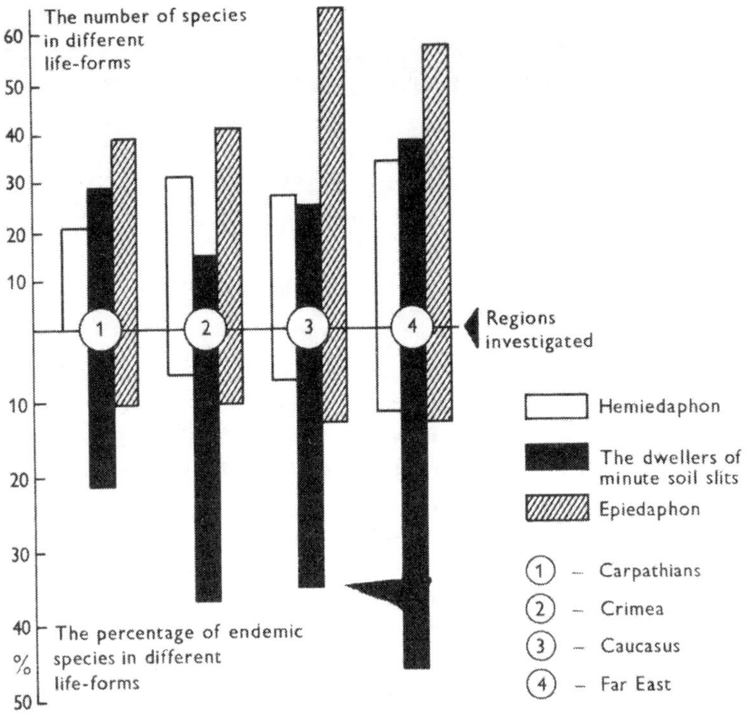

Fig. 1. The oribatoid mite species specificity in different life-forms in the brown forest soil type in the USSR.

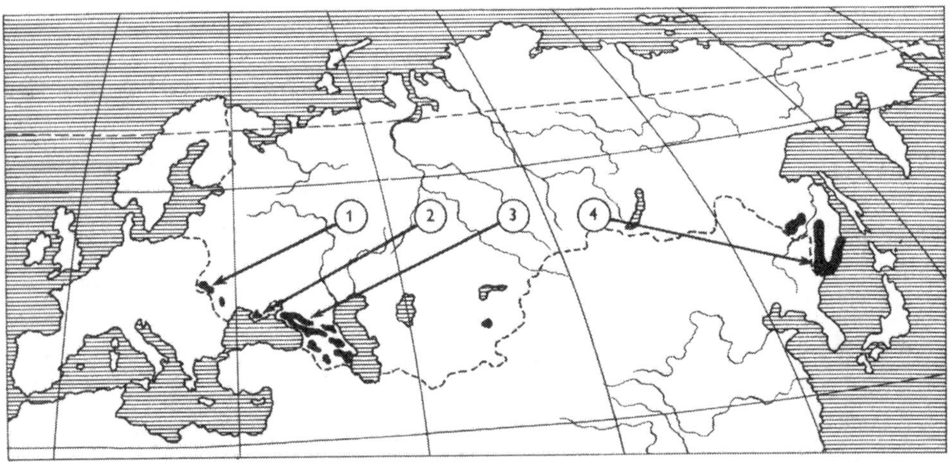

Fig. 2. The distribution of the brown forest soil type in the USSR. I – Carpathians, 2 – Crimea, 3 – Caucasus, 4 – Far East.

It is interesting, that fossil oribatid mites in the upper Jurassic period are representatives of recent families Cymbaeremaeidae (genus *Jureremus*), Trhypochthoniidae (genus *Palaechthonius*) and recent genera *Cultroribula* and *Achipteria* (Plate 1). Thus, the evolution of oribatoid mites is a very slow process and the tempo of evolution in these mites is one of the slowest in the terrestrial animals.

References

Krivolutsky, D. A.: Ecological specialization and production of forms in oribatoid mites. Zool. J., 47, 6: 820-827, 1968. (In Russian).
Krivolutsky, D. A., Krasilov, V.A.: The fossil jurassic oribatoid mites. (In litt.).

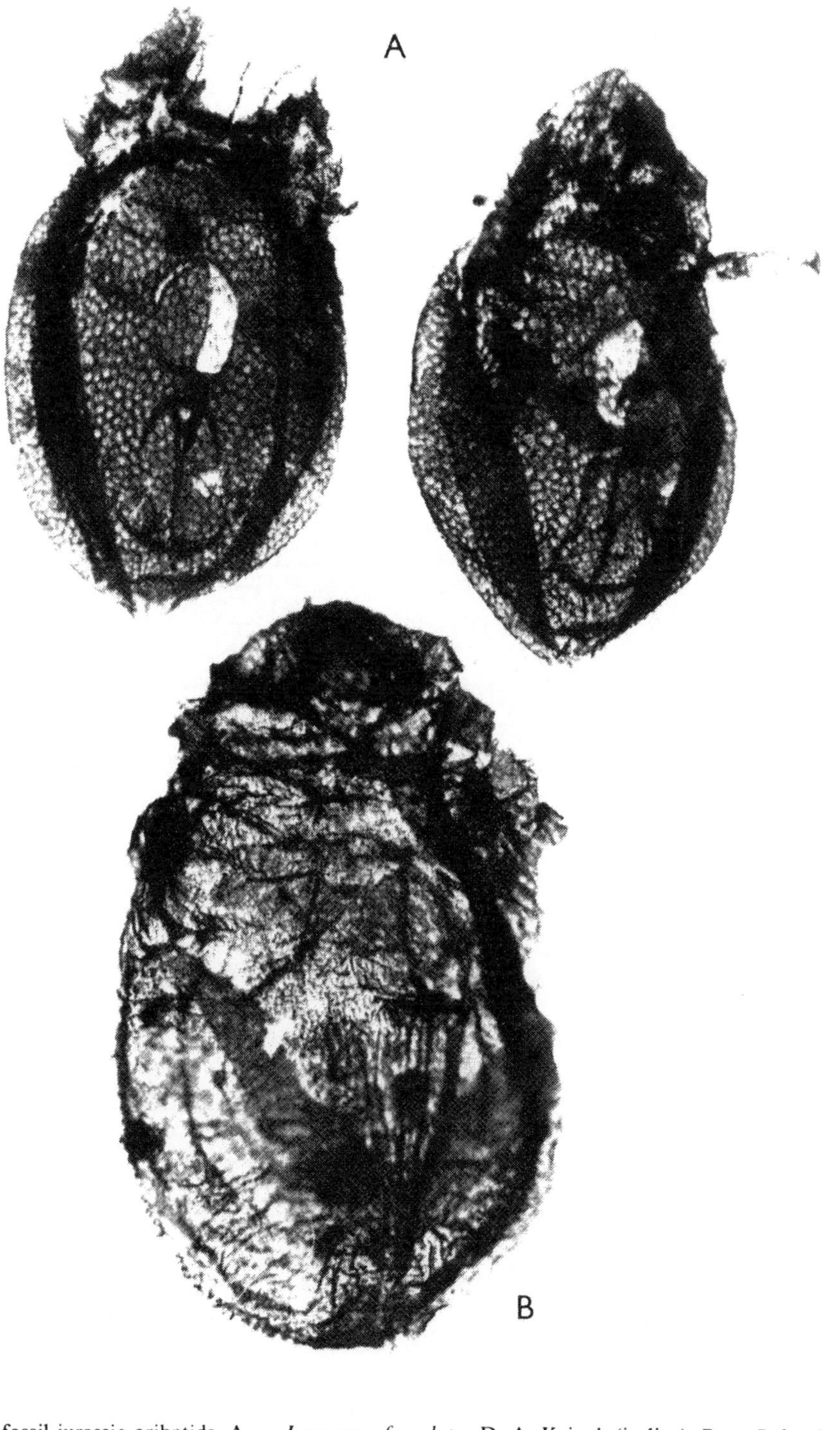

Plate I. Some fossil jurassic oribatids. A — *Jureremus foveolatus* D. A. Krivol. (in litt.), B — *Palaechthonius krasilovi* D. A. Krivol. (in litt.).

THE ORIBATID MITE FAUNA OF OAK FORESTS OF THE TRANS-CARPATHIANS (EAST CARPATHIANS) AND OTHER AREAS OF THE EUROPEAN PART OF THE USSR

G. F. KURCHEVA
A. N. Severtsov Institute of Evolutionary Morphology and Ecology of Animals
Academy of Sciences of the USSR, Moscow, USSR

Oak forests are widely distributed in the European part of the USSR, mainly in the southern part of the forest zone and the forest-steppe as well as in the steppe zone and mountain regions (the Carpathians, the Crimea, the Caucasus). In these forests varied soils are found.

A comparative study was carried out of the oribatid mite fauna of oak forests of the Transcarpathian Tissa lowland and that of dubravas of 16 large areas of the above mentioned zones and regions (more than 125 oak forests were studied). A good deal of material collected recently was used (Bashkirova, 1953; Gatilova, Krivolutsky, 1968; Gordeeva, 1970, et al.).

All the plain forests studied are represented by *Quercus robur* and the mountain forests - by *Q. petraea*, *Q. pubescens* (the Crimea) and *Q. ibericus* (the Caucasus).

As a whole the fauna of the oak forests of the European part of the USSR amounts to more than 278 species of oribatid mites. In the plain oak forests of the Transcarpathians (soddy podzolic soils) 65 species of oribatid mites are found. 20 of these species inhabit more than a half of the oak forests studied (*Hypodamaeus riparius*, *Oppia minus*, *Quadroppia quadricarinata*, *Suctobelba hammeri*, *Cultroribula trifurcata v. rotundata*, *Carabodes areolatus*, *Oribatula tibialis*, *Chamobates voigsti*, *Ch. lapidarius* etc.). The majority of these species are widely distributed (*Oppia nova*, *Tectocepheus velatus*, *Scheloribates laevigatus*, *Liebstadia similis*, *Ceratozetes gracilis*, *Achipteria coleoptrata*, *Parachipteria punctata*, *Euphthiracarus loricatus*, *Steganacarus striculus* et al.). *Hypochthonius rufulus* and *Xenillus tegeocranus* as well as *Suctobelba perforata* and *Zetorchestes micronychus*, rarely found in other forests, are typical of plain broad-leaved (mainly) and coniferous broad-leaved forests of the European part of the USSR.

8 species of oribatid mites which inhabit the plain oak forests of the Transcarpathians (*Neonothrus humicola*, *Metabelba propexus*, *Belbodamaeus tuberculatus*, *Liacarus gracilis*, *Lepidozetes singularis*, *Tectoribates connexus*, *Eupelops aurites*, *Phthiracarus annonymum*) are not found in other dubravas. 5 species, besides the Transcarpathians, are recorded only in the oak forests of the Crimea - *Ctenobelba pectinigera* the Caucasus - *Tropacarus pulcherrimus*, the western part of the forest-steppe zone - *Nanhermannia nana* and *Oppia maritima*, the central part of the cher-

nozem forest-steppe zone - *Oppia translamellata*. The other species of mites typical of the Transcarpathian dubravas are more or less found in other oak forests of the European part of the USSR. *Xenillus latus, Poroliodes farinosus* and *Tropacarus carinatus*, besides being found in the Crimea and the Caucasus, are recorded in dubravas of 2-5 different plain areas. A number of species are found in 3-5 plain regions but they are not found in the Crimea and Caucasus (*Nothrus palustris, Platynothrus peltifer, Suctobelba forsslundi, Eremaeus hepaticus, Chamobates cuspidatus, Eupelops planicornis*). Such species as *Nothrus silvestris, Metabelba pulverulenta, Suctobelba acutidens* and *Protoribates capucinus* which are often recorded in other plain oak forests and *Oppia nitens, Ceratoppia bipilis* are as yet recorded only in the Crimea. But such species as *Hermanniella dolosa, Oppia fallax v. obsoleta, Suctobelba trigona, Carabodes labyrinthicus, Minunthozetes pseudofusiger, Phthiracarus ligneus*) are recorded in the Caucasus dubravas but they are not found in the Crimea.

The species *Oppia ornata* (Kurcheva, 1970) widely distributed in the Transcarpathians and mostly inhabiting these coniferous forests is found also in appreciable quantity in Ukrainian dubravas. To the east, this species is found in the oak forest (few individuals) only in the Volga forest-steppe region (Chuvash Republic) - in a sandy soil of the oak forest mixed with spruce. On the other hand, in this forest the species *Euphthiracarus cribrarius* is recorded which is found also only in the dubrava on chernozem (central forest-steppe). In the Transcarpathians this species is found only in leaf-bearing forests (beech and mainly oak forests)'

The oribatid mite fauna of the Transcarpathian plain dubravas is quite different from that of other forests of the European part of the USSR. However the oribatid mite groups of the latter are not very similar either.

References

Bashkirova, E. Ya.: The oribatid mite fauna in the wind forest strip region of the northern part of the steppe zone. Zool. J., 32, 1114-1125, 1953. (In Russian.)

Gatilova, F. G., Krivolutsky, D. A.: The oribatid mite fauna (Oribatei) of dubravas of the European part of the USSR. In: Materials on fauna and ecology of soil-inhabiting invertebrates, Kazan', Kazan' State University, p. 98-116, 1968. (In Russian.)

Gordeeva, E. V.: Oribatid mites in the Crimea soils. In: Oribatid mites (Oribatei), their role in soil-forming processes. Vilnius, Acadamy of Sciences of Lithuanian SSR, p. 119-129, 1970. (In Russian.)

Kurcheva, G. F.: Oribatid mites of the Transcarpathians. In: Oribatid mites (Oribatei), their role in soil-forming processes. Vilnius, Academy of Sciences of Lithuanian SSR, p. 73-79, 1970. (In Russian.)

THE BIONOMICS OF *HUMEROBATES ROSTROLAMELLATUS* GRANDJEAN (CRYPTOSTIGMATA–CERATOZETIDAE) ON FRUIT TREES

P. W. MURPHY and A. N. BALLA
School of Agriculture, University of Nottingham, Sutton Bonington, Loughborough, England

Humerobates rostrolamellatus Grandjean, a member of the family Ceratozetidae, is a medium-sized heavily sclerotized oribatid mite some 750-800μ long and having well developed pteromorphae. It has a widespread occurrence, and has been recorded from a range of habitats including heathland and woodland soil. It is considered to be a somewhat xerophilous species; it is particularly prevalent on the bark of unsprayed fruit trees of various kinds including apple and plum, and is also found on deciduous trees (Travé, 1963). It rarely occurs on the leaves. It is found occasionally in dwelling houses in such numbers as to constitute a nuisance (Scott, 1958). It is believed to feed on algae and other lower plants and has been successfully cultured on *Pleurococcus* sp. (Green, 1961) in the laboratory.

In previous work at the School of Agriculture on the bionomics of this species Martin (1960) showed that the adult mites moved down from the branches towards the trunk in autumn. Jalil (1963) studied its life cycle in the laboratory, and in field sampling he found that the immature postembryonic stages were present on the younger wood throughout the year. However his method of recovery of these stages was inadequate to provide reliable quantitative data and no eggs were found. The objects of the present investigation were to study quantitatively the seasonal movements of the adults, and seasonal occurrence and location of the eggs and juvenile stages on mature apple trees in a grass orchard. This paper refers to some of the more striking features of the study; a more detailed account of these findings will be published elsewhere.

Experimental Procedure

The experimental site is an old unsprayed grass orchard at the School of Agriculture having an area of about two acres. The trees are mainly apple with some plums and are for the most part 60-70 years old. Most of the sampling was carried out on two half-standard apple trees with trunks 40-50 cm in height from ground level to the point where the first branches arise, and having a diameter of about 30 cm at 25 cm from ground level.

For sampling purposes each tree was divided into five regions: the lower and upper parts of the trunk, the proximal (inner) and distal (outer) halves of the branches and the young shoots towards the tips of the branches. Three sampling procedures were used. In the first – referred to as visual sampling – each sampling unit consisted of a 4-cm^2 area of bark selected at random within each of the trunk and branch regions. Each area was examined for adult mites using an illuminated magnifying lens, and the number present was recorded. On each sampling occasion eight sampling units were examined from each of four regions (shoots excluded) of the two trees. The branch stations were duplicated to provide 16 sampling units for each branch region making a total of 48 units per tree.

The second series of samples consisted of the shoots removed from the extremities of the branches. On each sampling occasion four shoots per tree were taken at random each from a separate branch. The shoot lengths had a range of 6-44 cm. The leaves were removed from the shoots which were then cut into short lengths, and the mites and eggs present were recovered by means of a washing technique based in part on that of Morgan and Downing (in Morgan, Chant, Anderson and Ayre, 1955). The shoots were soaked in an aqueous solution of sodium hypochlorite (5 per cent concentration) and the liquid was gently heated. The liquid was then reduced to a convenient volume by filtration, examined under a stereomicroscope, and the number of each stage on each shoot recorded.

The third series of samples consisted of 50 live adult mites taken directly from a plum and two apple trees reserved for this purpose and not used for the visual or shoot samplings. These were mounted and examined under the microscope to determine the sex, the number of eggs when present, and the presence or absence of food in the gut. The mites were collected from the trunk, branches or shoots, the region selected being as far as could be ascertained the one with the maximum density of mites at the time of sampling.

The three sampling procedures were repeated at approximately fortnightly intervals between December, 1962 and March, 1964 but because of the work involved all three could not be maintained throughout the whole of this period nor was it possible to carry them out on the same day. The visual samplings and collecting of adult mites were made on 26 and 25 occasions respectively whilst the shoots were sampled on 17 occasions. In the visual samplings there were many empty sampling units, and as was to be expected there was an increase in sample variance with increasing mean. The regressions of log variance on log mean suggested that the rate of increase was greater for eggs and juveniles than for adults For this reason the counts (x) were transformed using fractional-power transformation (z) as follows:

visual sampling for adults: $Z = x^{0.2}$;
shoot sampling for adults: $Z = x^{0.3}$;
shoot sampling for eggs and juveniles: $Z = x^{0.1}$.

Results

The overall means (untransformed data) of the visual samplings of the two trees were 0.97 and 0.77 adults respectively. Analysis of variance of the transformed data indicated that these densities differed significantly ($P<0.001$). The seasonal movements, however, were not dissimilar and for present purposes the data for the two trees have been combined to provide mean values for the trunk and inner and outer regions of the branches for each occasion (Figure 1). In a similar analysis

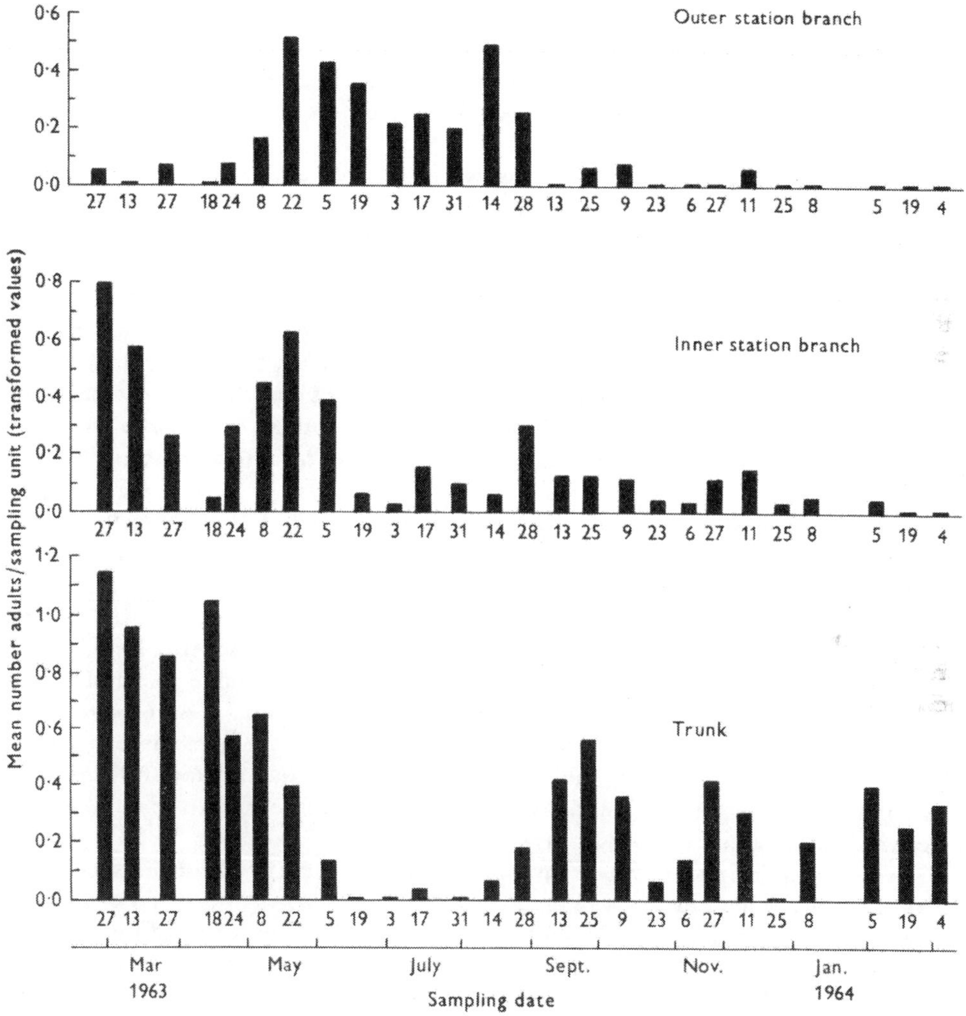

Fig. 1. Mean number (fractional-power transformation) of adults per 4 cm² of bark on trunks and lower surface of branches of two half-standard apple trees in a grass orchard between February, 1963 and March, 1964 (16 sampling units/region/occasion).

of the shoot data for each stage of the life cycle neither trees nor shoots differed significantly except for deutonymphs where the shoot densities were significantly different ($P < 0.001$). These data have also been combined to provide mean values for the two trees on each occasion (Fig. 2).

In February-March, 1963 when the visual samplings commenced few adults were present on the outer region of the branches (Figure 1). From the end of April to the beginning of May adult numbers increased in both branch regions, and by the latter part of June few mites could be found on the trunks. It is likely that a decline in adult numbers occurred in the outer branch region in June, and evidence from the shoot samples of a marked increase in numbers on the younger wood in mid to late June suggest a continuing outward movement of adults. There is evidence from the shoot and visual samples of a movement of adults in the reverse direction commencing in the latter part of August with adult density on the trunk reaching its highest point towards the end of September.

The adult collections provide information on the sex ratio, incidence of feeding and occurrence of gravid females. Overall the females exceeded males in the ratio of 1.6 : 1.0 and on individual occasions the ratio rarely fell much below 1 : 1. In December, 1962 when sampling commenced the proportion of adults with gut content was less than 5 per cent. There was a sharp increase between the end of January and 11th February, and in March a high proportion of individuals had ingested material in their gut. This situation continued although with fluctuations some of which were large until the beginning of August from which time there was a steady decline until December, 1963 when sampling ended. The results for presence of eggs in females are somewhat similar to those for gut content except that the increase in proportion of gravid females is somewhat later commencing after mid-April and reaching 100 per cent in the latter part of May. With the exception of two sampling occasions it remained at this level until August. During this period the mean number of eggs per egg-bearing female increased to a peak (10.7) at the beginning of June, then declined but rose to a second peak (10.2) in mid-July. In the latter sample only half the females had eggs present. The 5 per cent fiducial limits of the mean of this sample were almost double those of the previous peak presumably due to a mixed population consisting in part of overwintering adults with low, and current-season adults with high numbers of eggs. At the end of August to the beginning of September there was a sharp fall in the proportion of gravid females, and this continued until October when less than 20 per cent of the females contained eggs.

Eggs were found on the shoots towards the end of June at about the time when adults were becoming numerous (Fig. 2), and were present until the end of August. Larvae were first evident in appreciable numbers in mid-July followed by protonymphs in mid-August. Deutonymphs appeared in the latter part of September. It should be noted that few tritonymphs were recovered from the shoots and that none was present after mid-July.

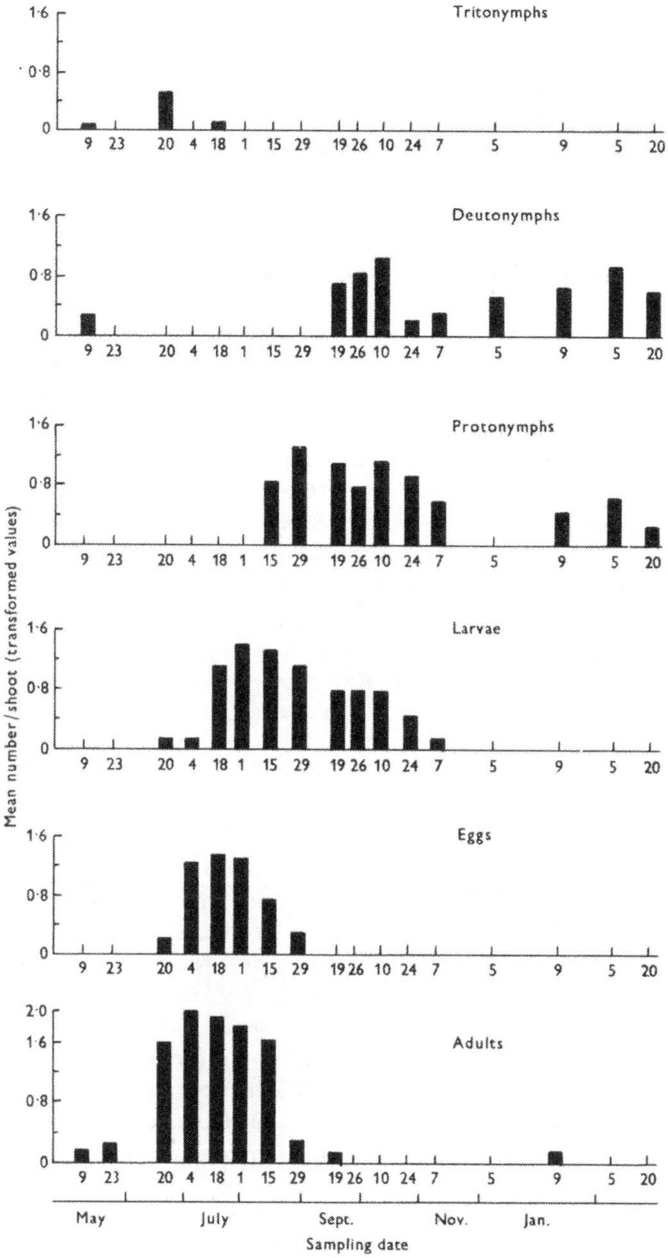

Fig. 2. Mean number (fractional-power transformation) of adults, eggs and juveniles per shoot from two mature apple trees between May, 1963 and February, 1964 (4 shoots/tree/occasion).

Discussion

From the data presented we would suggest the following sequence of events for the adults of *H. rostrolamellatus.* The mites on the lower parts of the tree, which are presumed to be overwintering adults, commence feeding in late winter and this is followed by a movement up the tree in mid-April at a time when the proportion of females with eggs is increasing sharply. The upward movement continues until the adults reach the younger wood, and it is possible that in the case of females oviposition does not take place until this movement has been completed. It is suggested that after oviposition there is less feeding among these adults and a movement downwards to the base of the tree commences. It is quite possible that at least some of these mites reach the soil surface beneath the trees but this aspect has not been investigated in this study. (Table 1.)

During the winter period the juveniles, present mainly as protonymphs and deutonymphs, occur towards the extremities of the branches although it is possible that the deutonymphs, at least, move inwards from the peripheral shoots. In May and June lightly sclerotized adults were observed on the shoots, and it is suggested that emergence of adults from the overwintering juveniles reached a peak by the beginning of July when a relatively large number of these adults were found. It is believed that the second rise in number of eggs per egg-bearing female occurring in mid-July refers particularly to these recently emerged adults. If this is correct then the larvae stem from two sources: overwintering adults, and current-season adults emerging in early summer after overwintering as juveniles. The virtual absence of adults on the outer region of the branches during winter would suggest that those that become adult in the current season also move down the tree – in the case of females presumably after oviposition of their eggs, and the second build up in density on the trunk occurring in November (Figure 1) may refer to these mites. It seems likely that some of the fluctuations in the proportion of mites with gut content and proportion of females with eggs may be due to the presence of a mixed population consisting of overwintering adults moving downwards and current-season adults moving upwards. Whether adults of both sexes are similar in respect to movement and feeding are aspects that require further study although it should be noted that appreciable numbers of males were recovered from the shoots. A tentative timetable for these biological events is given in Table 1. It should be stressed that the phenological picture provided refers to a single year's sampling, and with the added limitation that sampling occurred at fortnightly rather than weekly intervals which might have been more appropriate.

Some of the events recorded are quite precisely timed. Both Martin (1960) in 1959 and Green (1961) in 1960 observed a downward movement of adults at the end of August and beginning of September on trees in the same orchard. Jalil's (1963) data, which refer to apple trees about three miles away, sampled in 1961-62, also show a fall in numbers on the shoots in the latter part of August, 1962, and a sharp rise in June-July 1961 and 1962. In addition the times at which egg-bearing

Table 1. A tentative timetable for the bionomics of the overwintering adults and immature stages of *Humerobates rostrolamellatus* Grandjean on fruit trees in a grass orchard (Interrogation marks indicate uncertainty in the timing or observation listed)

Date	Overwintering adults	Overwintering immature stages
Early February	Feeding commences	
Mid-April	Movement up tree commences. Gravid females present	
? Early June	Peak in number of eggs/ egg-bearing female	
Mid to late June	Adults present in large numbers on terminal shoots	Tritonymphs still present
Early July	Eggs numerous on terminal shoots	Current-season adults numerous
? Mid-July	Oviposition drawing to a close	Peak in number of eggs/ egg-bearing female
Mid to late July	Larvae appear	
? End July	Reduction in numbers feeding	Oviposition commences
Mid-August	Movement down tree commences	
Mid to late August		? Reduction in numbers feeding
Early September		Majority of eggs hatched
Late September	Adults numerous on trunk	
Mid-October	? Movement of deutonymphs from shoots commences	
? November		? Current-season adults appear on trunk

females and larvae were first observed are in close agreement with those reported by Harding (1971) for the years 1968 – 71 for *Platynothrus peltifer* (C. L. Koch), an oribatid mite inhabiting leaf litter. It may be that the phenological precision evident in certain of the data applies particularly to the overwintering adults and the early postembryonic life of their offspring, and that for the remainder the timing of biological events such as duration of the postembryonic stages is more variable. It is possible that this latter variation has led to difficulties when populations from different habitats and different years are compared.

The genetic and ecological implications of an arrangement whereby the overwintering adults and their progeny have a degree of reproductive discreteness and temporal and spatial separation are of some interest. If the hypothesis suggested i correct then the young adults reproducing for the first time may be reproductively isolated from their parents, and it is only when they reproduce a second time that they have an opportunity of coming in contact with their forbears. Presumably the chance of this happening is lessened because of mortality of the older mites. The progeny of the two generations, on the other hand, live side by side although there may be some separation of these if the early deutonymphs occur on the older wood or tend to move back from the shoots.

References

Green, M.B.: Studies of the Distribution of *Humerobates rostrolamellatus* Grandjean on Apple Trees. Biol. Sci. Dissertation, University of Nottingham, 102 pp., 1961.

Harding, D. J. L.: The phenology of *Platynothrus peltifer* (C. L. K.) Proc. 3rd Int. Congr. Acarology, Prague 1971.

Jalil, M.: Morphological and Biological Studies of Three Oribatid Mites with Particular Reference to their Life Cycles. M. Sci. Thesis, University of Nottingham, 2 vol. 287 + 196 pp., 1963.

Martin, T. J.: *Humerobates rostrolamellatus* Grandjean an Oribatid Mite Inhabiting Fruit Trees. Biol. Sci. Dissertation, University of Nottingham, 122 pp., 1960.

Morgan, C. V. G., Chant, D. A., Anderson, N.H. and Ayre, G. L.: Methods for estimating orchard mite populations, especially with the mite brushing machine. Can. Entomol. 87. 189-200. 1955.

Scott, H.: *Humerobates rostrolamellatus* Grandjean, an oribatid mite, invading a house. Entomol. Mon. Mag. 94, 69–70, 1958.

Trave, J.: Écologie et biologie des Oribates (Acariens) saxicoles et arboricoles. Vie et Milieu Suppl. No. 14, 267 pp., 1963.

AUTECOLOGICAL-ZOOGEOGRAPHICAL ANALYSIS OF MOSS MITES (ACARI, ORIBATEI) BASED ON THE FAUNA IN THE ENVIRONS OF POZNAŃ

A. RAJSKI
Faculty of Zoology, W. S. R., Szczecin, Poland

The areas of about 180 species were analysed.*

1. Twenty-one groups of species have been distinguished, covering different geographical areas. For each species a definition of the area and its characteristic features has been given: the deliniations of the areas across Poland, the disjunctions and the typical refugial territories. The particular species have also been determined in ecological respect. The definitions of the areas are elaborated versions of the method of Meusel et al. (1965).

The investigated fauna proved to be little specific. Only 35 % of the species cover European areas, while species covering larger areas than the European ones predominate (65 %). The Central European species constitute a small share (7,35 %). Among the species whose areas are directed towards the environs of Poznań, southern (10.16 %) and western (7.35 %) species prevail over northern (4.52 %), eastern and southeastern ones (2.82 % each).

2. An object of the analysis was also the dependence of the moss mite fauna in the environs of Poznań on the humidity of the macroclimate. It turned out that most species (69.49 %) are relatively independent of the humidity of the macroclimate and occur in the oceanic, suboceanic and subcontinental zones. Moreover 12.43 % of species have also been found in territories of the continental climate. To the species, confined only to two zones, (a) the oceanic and suboceanic ones, (b) the suboceanic and subcontinental ones, belong 6.78 % each. Finally, only 4.52 % constitute species that occur in the typical suboceanic zone (i.e. the tested territory).

3. On the grounds of the structure of present-day areas (disjunctions and refugial territories), and to some extent on the basis of palaeontological evidence (moss mites of Baltic amber), an attempt has been made to establish the oldest fragments of the areas, and thus to denote the minimal absolute age of the species examined. The basis for these considerations are the following assertions and facts: moss mites (and also certain other animals) generally occur in connection with a specific type of vegetation, usually representing a higher phytosociological unit (ordo etc.), or a determined floristic formation; in accordance with the results hitherto obtained,

* Full text in Acta zool. crac., Krakow, 1970, 15 (3) : 161–258 pp.

the passive transference of moss mites on an intercontinental scale is of little probability; characteristic for animal species, and especially for moss mites, is the stability of their ecological requirements during the whole period of their existence; polytopic formation of species is rare particularly in the animal world and does not play an important role among living organisms; an analysis of recent areas of species of plants or animals does not authorize to conclude that the history of their formation would reach into a past beyond the Palaeocene; it is assumed that the surface of continents and oceans was shaped in accordance with Wegener's theory of continental drift. The areas of all species examined have been analysed according to the palaeographical reconstructions and the above mentioned assertions and facts.

There is some evidence concerning the Palaeocene derivation of fragments of areas of 20 species, i.e. 11.30 % of the oribatid fauna investigated. Other facts permitted to ascertain that the oldest parts of the areas of 77 species, i.e. 43.50 % of the fauna investigated, were formed in the Neocene. In 47 cases, the areas investigated came partly into existence in the Quarternary. However, with only 20 species, i.e. 11.30 % of the fauna investigated, the areas have exclusively Quarternary features.

In the case of 60 species (33.90 %) it was not possible to establish the characteristic features of the areas that could be linked with a definite historical era.

4. Relatively satisfactory knowledge of the development of the geophysical environment and vegetation in the environs of Poznań after the retreat of the Scandinavian Ice Sheet as well as about the ecological requirements of moss mites, both with relation to micro- and macroclimate, permitted to formulate a hypothesis on the development of the moss mite fauna in the environs of Poznań in the course of the late Glacial Epoch and in the Holocene. It concerns only those species that nowadays occur in the territory investigated and can very briefly be presented as follows:

Under the conditions of the Oldest Dryas, in the environs of Poznań only those species of moss mites could occur which nowadays are represented in the Arctic region, tolerating the continental climate and able to live in a territory that is not covered by a woody vegetation, in particular by forests. Species occurring among moss and lichens on hard substrate would come into this category. Eleven species (6.2 %) of the fauna investigated comply with these requirements. All of them have holarctic or larger areas.

Within the Bölling period, in the territory under consideration, some species that had not been adapted to extreme continental climates, probably settled down and these are occurring now in the bioclimatic Arctic zone. In ecological respect they are forms of open territories or independent of vegetation, with different humidity requirements. Of such species there are 28 in the fauna investigated, i.e. 15.82 %. All of them occupy very large areas. These species have disappeared from the Wielkopolsko-Kujawska Lowland, or at least, have been considerably limited, in the Older Dryas when the climate has deteriorated.

A real turning-point in the development of the moss mite fauna in the environs of Poznan has probably occurred in the Allerød period when conditions were being created for the settling of forest forms, peat-inhabitants, heliophilous forms and those connected with the habitat of mosses and lichens on a hard substrate, nowadays reaching in the north only as far as the bioclimatic boreal zone. Together with forms that reached and settled down here in previous periods, the fauna of moss mites at that time probably numbered already about 120 species, i.e. 68.93 % of the present fauna. A characteristic feature of that period was the appearance of a great number of temperate holarctic and palaearctic species and a certain number of European species except the South European ones.

The second regression in the development of the fauna investigated occurred in the Younger Dryas, in connection with the repeated deterioration of the climate. Probably 37 species vanished, mainly forest ones of European and temperate holarctic areas. They may have returned to the territory investigated already in the pre-Boreal period.

In the Boreal period, thermophilous species of Southeast European origin as well as European (except West European) and European-Manchurian mesohygres that are typical for deciduous forests, might have already occurred in the Wielkopolsko-Kujawska Lowland. The fauna examined was enriched at that time by probably 30 species, i.e. 17 % of the present fauna and perhaps already numbered 152 species, i.e. 86 % of the known ones found here today. The remaining 14 % of species, chiefly West and Central European ones, have probably appeared in the environs of Poznań in the Atlantic period, being mostly forest meso- and polyhygres as well as other forms nowadays connected with the suboceanic climate.

References

Meusel, H., Jäger Weinert, E.: Vergleichende Chorologie der Zentral europäischen Flora. Jena I ÷ 583 pp., II - 258 pp., 1965.

BIOLOGICAL AND ECOLOGICAL OBSERVATIONS ON ORIBATID MITES

G.M. SHEREEF
Faculty of Biology and Soil Science, Moscow University, Moscow, USSR

Methods

Oribatid mites were first reared successfully by Michael (1884) who used small glass cells made by fixing glass rings to microscope slides. Grandjean (1950) and Hartenstein (1962) cultured the mites in Stender dishes on a filter paper substrate, while Lipovsky et al. (1957) were the first to use a substrate consisting of a mixture of plaster of Paris and charcoal.

In the present investigation oribatid mites were kept alive in Petri dishes on a substrate consisting of a mixture of plaster of Paris and charcoal mixed in the ratio 9:1. Various fungi were fed to the mites by introducing sections of agar with adherent fungi from pure cultures. For more detailed observations, eggs from the Petri dish cultures were transferred with a fine brush to a plaster of Paris charcoal substrate in small glass cells made by fixing a glass ring (10 x 5 mm) to a microscope slide. The culture cells and dishes were kept in a moisture-saturated atmosphere at 25° C.

For an investigation of the suitability of various fungi as a diet for *Belba meridionalis* B-Z, culture cells 20 mm in diameter and 10 mm deep were used. Three-hundred tritonymphs of this species were isolated from the Petri dish cultures. After 12 days these moulted to become adults; after a further 4 days, with the onset of spermatophore production, males and females could be differentiated. Seven adults (2 males and 5 females) were placed in each cell, and a small portion of fungus (without agar) was then introduced, each cell receiving a different species of fungus. These cells were kept in a saturated atmosphere at 25° C, and examined daily to assess the amount of feeding and faecal-pallet production which had occurred. Fresh fungi were supplied as needed. The numbers of eggs, larvae, nymphs and adults in each cell were noted after 104 days.

Results

Life-cycle observations

Belba meridionalis B-Z and *Spatiodamaeus subverticillipes* B-Z were successfully cultured on *Penicillium martensii*, and *Palaeacarus kamenskii* A. Zach., *Eremobelba geographica* Berl. and *Granuloppia* sp. on *Aspergillus flavus*. Data on the life-cycles of these species are presented in Fig. 1. It will be noted that the duration of the life-cycle of *P. kamenskii* is considerably longer than that of the other species, and that in case of all species in which males are produced, the duration of the male life-cycle is shorter.

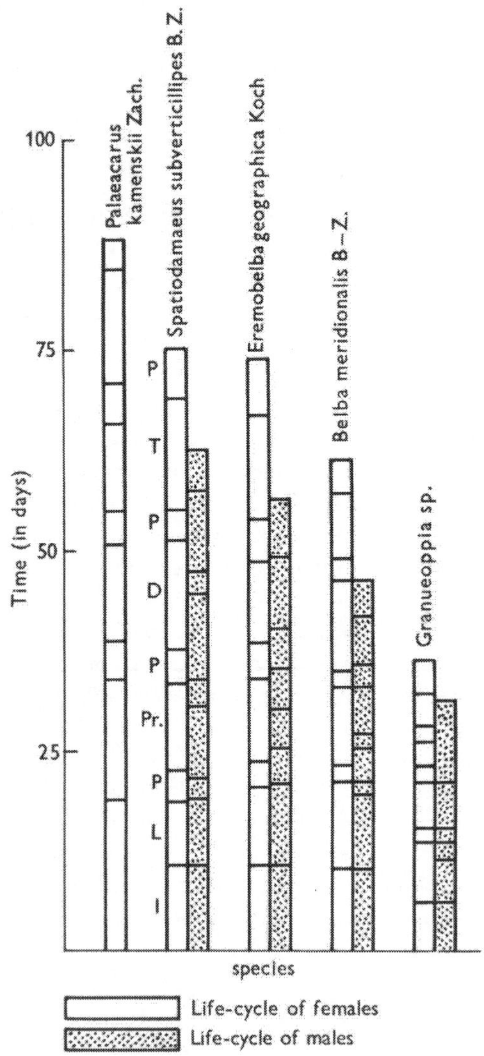

Fig. 1. Life-cycles of 5 species of oribatids: (I) Incubation period, (P) Pre-protonymph, (P) Pre-deutonymph, (P) Pre-tritonymph, (P) Pre-adult, (L) Larva, (Pr) Protonymph, (D) Deutonymph, (T) Tritonymph.

B. meridionalis, S. subverticillipes and *Granuloppia* sp. could produce spermatophores. Parthenogenesis was definitely established for *P. kamenskii. E. geographica* was not seen to produce spermatophores even though observations were carried out for more than a year. Parthenogenetic reproduction was not observed in this species.

Feeding experiments with *Belba meridionalis*

The results of the comparative study with 12 species of fungi are shown in Fig. 2. It will be noted that the mites reproduced on all 12 fungi, but, as assessed by numbers of individuals of all stages present in the cells after 104 days, the most

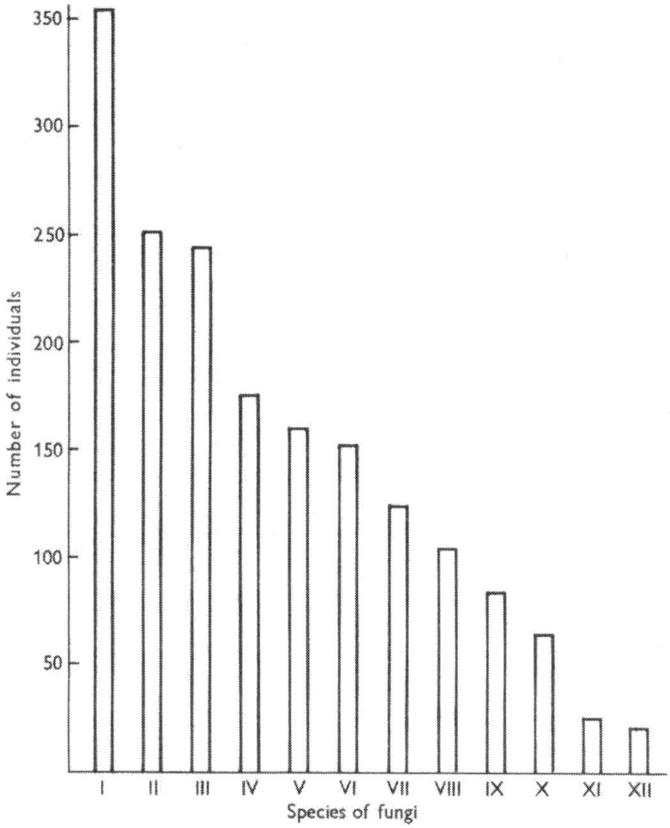

Fig. 2. Reproduction of *Belba meridionalis* (in 104 days) on 12 species of fungi: I – *Trichoderma* sp., II – *Penicillium martensii*, III – *Fusarium* sp., IV – *P. cyclopium*, V – *P lilacnium*, VI – *P. spinulosum*, VII – *P. brevi-compactum*, VIII – *F. solani*, IX – *Cladosporium* sp., X – *F. oxysporium*, XI – *P. viridicatum*, XII – *Mucor ramannianus*.

favourable diet was *Trichoderma* sp. The least favourable were *P. viridicatum* and *Mucor ramannianus*. It is interesting to find that a fungus most suitable for one species of mite is not necessarily the most suitable one for others. Thus, in earlier observations (Shereef, 1970) it was found that *M. ramannianus* was the third most suitable, and *Cladosporium* sp. the least favourable diet for *Epidamaeus kamaensis* Sell. Again Sinha and Mills (1968), working with 10 species of *Penicillium*, concluded that *P. cyclopium* was the most, and *P. impilicatum* the least favourable diet for the astigmatid mites *Acarus siro* L. and *Tyrophagus putrescentiae* Oudms.

Spermatophore production

Woodring and Cook (1962) reported that the males of *Ceratozetes cisalpinus* Berl. and *Scheloribates laevigatus* Koch started to produce spermatophores about 10 days after emergence. The maximum rate of spermatophore production by one male was about 20 in five hours.

In the present study the average number of days between the emergence of the males and the onset of spermatophore production for the various species was found to be as follows: *B. meridionalis*, *E. plumosus* and *E. kamaensis*, 3; *B. rossica*, 7 and *Granuloppia* sp., 13. The influence of the presence of females on spermatophore production varies. In the case of the 16 species observed, only *S. subverticillipes* failed to produce spermatophores in the absence of females. Males of this species kept in isolation for 9 months failed to produce a single spermatophore, but when brought into the presence of females they deposited spermatophores after 9 days.

The number of spermatophores produced appears to vary considerably from species to species. Thus, when ten "fresh" males (i,e,, males which had just started to produce spermatophores) were isolated in glass cells the number of spermatophores per male produced in one month was found to be: *Oppia concolor* Koch, 1058; *E. apophysus*, 149; *E. kamaensis*, 136 and *Porobelba spinosa* Sell., 74. The highest number produced by a single male of *B. meridionalis* during the whole adult period (15 months) was 3071.

The morphology of the spermatophores

The spermatophore in species belonging to the families Damaeidae, Belbidae and Belbodamaeidae, was found to be very similar in structure to that of *Damaeus quadribastatus* (as described by Cancela da Fonseca,1969). In these families, and indeed in a number of other families, the pedicel commonly terminates in an ampulla, but this ampulla was not found in *Peloribates pilosus* Mar. Hamm., *Carabodes sp.* nor in *Diapterobates numerosus* Sell. (fig. 3). In *Oppia* sp., *O. concolor*, *Hermaniella granulata and D. numerosus*, the pedicels are straight, but in the other species studied that were curved near the tip. In *Oppia* sp., *O. concolor* and *Granuloppia* sp.

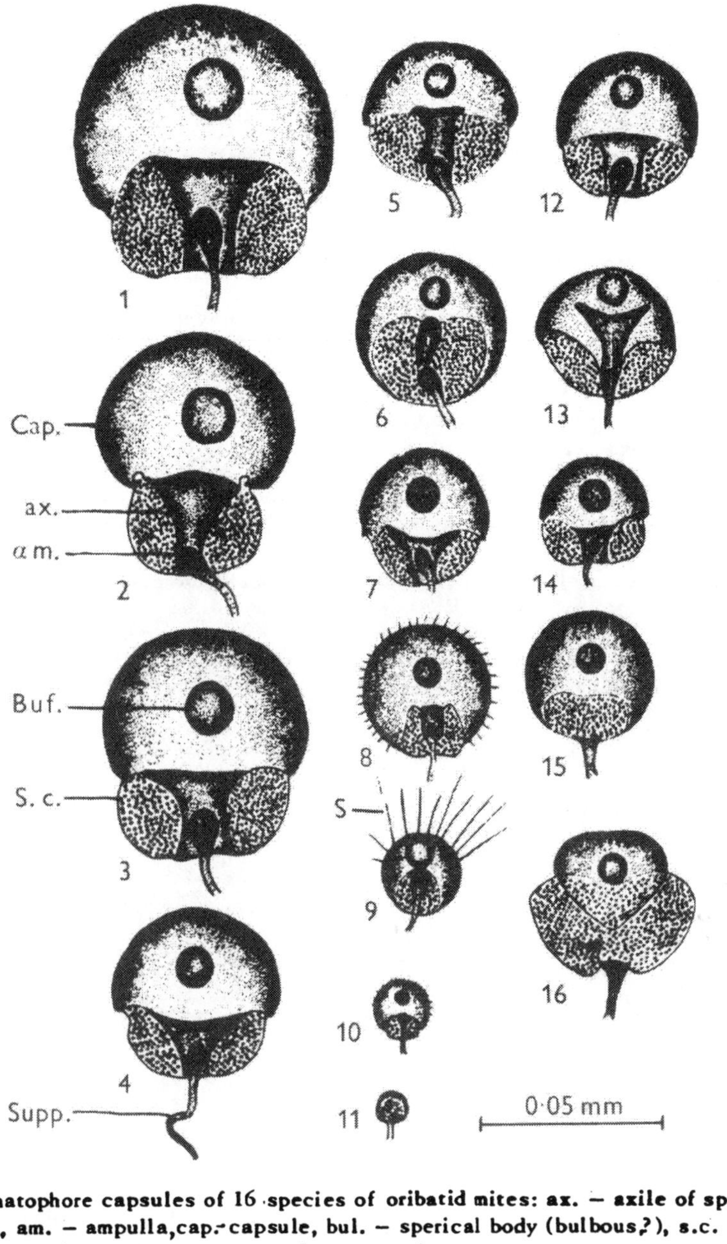

Fig. 3. Spermatophore capsules of 16 species of oribatid mites: ax. – axile of spermatophore, am. – ampulla, cap.- capsule, bul. – sperical body (bulbous,?), s.c. – sperm sac., s.-setules. supp. – support of spermatophore.
1 – *Epidamaeus plumosus*, 2 – *Spatiodamaeus subverticillipes* 3 – *Epidamaeus* sp., 4 – *Belba meridionalis*, 5 – *E. grandjeani*, 6 – *Hermanniella granulata*, 7 – *E. kamaensis*, 8 – *Oppia concolor*, 9 – *Granuloppia* sp., 10 – *Oppia* sp., 11 – *Carabodes* sp., 12 – *B. rossica*, 13 – *Metabelba rara*, 14 – *Porobelba spinosa*, 15 – *Peloribates pilosus*, 16 – *Diapterobates numerosus*.

the tip of the spherical capsule was seen to be covered with minute setules, but in the other species investigated the tips of the capsules were smooth and sticky.

References

Cancela da Fonseca, J.: Le spermatophore de *Damaeus quadribastatus* Märkel et Mayer (Acarien Oribate). Proc. 2nd. Int. Congress Acarology, Sutton Bonington 1967, 227–232, 1969.

Grandjean, F.: Observations éthologiques sur *Camisia segnis* et *Platynothrus peltifer*. Bull. Mus. His. Nat. Paris, (2) 22 : 224–231, 1950.

Hartenstein, R.: Soil Oribatei 1 – Feeding specifity among forest soil oribatids. Ann. Entomol. Soc. Am. 55 : 202–205, 1962.

Lipovsky, L.J., Byers, G.W., and Kardos, E.N.: Spermatophores - the mode of insemination of chiggers. J. Parasitol. 43 : 256–262, 1957.

Michael, A.D.: British Oribatidae Vol. I, London. Ray Society, 1884.

Shereef, G.M.: Observations on the feeding, reproduction and faeces obtained from oribatids fed on different species of *Penicillium* and *Aspergillus*. IVe Coll. Int. Zool. Sol. 1970, Dijon, France, 1970.

Sinha, R.N., and Mills, J.T.: Feeding and reproduction of the grain mite and the mushroom mite on some species of *Penicillium*. J. Econ. Entomol. 61 : 1548–1552, 1968.

Woodring, J.P., and Cook, E.F.: The biology of *Ceratozetes cisalpinus* Berl., *Scheloribates laevigatus* Koch and *Oppia neerlandica* Oudms. (Oribatei) with a description of all stages. Acarologia 4 : 101–137, 1962.

THE POSTEMBRYONIC DEVELOPMENT OF PELOPIDAE

L.G. SITNIKOVA
Zoological Institute of the Academy of Sciences, Leningrad, USSR

The members of the family Pelopidae are very difficult to idetify, mainly because of their great individual variability. Adult mites show considerable variation in body size and even in the shape of hairs, colour, ceratogument thickness and in many other structures. As in immature stages Pelopidae, like all higher oribatids, differ drastically from the adults, it is difficult to determine what species they belong to. About a hundred species have so far been described but immature stages are known of only 6 - 10 of them.

A reassessment of the family and an elucidation of the taxonomical importance of the characteristics called for the examination of the origin of certain imaginal structures in their ontogenesis. Many oribatologists hold the view that in higher oribatids imaginal structures such as lamellae, translamella, tutorium, pedotectum and a few others arise suddenly, during the final moults so that there are few, if any external morphological relationships between adult and immature stages. It therefore seemed of particular importance for the systematics of Pelopidae to identify the structures peculiar for the species through the entire postembryonic development.

To achieve that goal we cultivated *Eupelops torulosus* under laboratory conditions; this species is a typical representative of the family and the most common. No less than 10 specimens of each stage were examined. Immature individuals underwent total examination while only isolated parts of adults were studied.

In the immature stages *E. torulosus* has a peculiar lamellar extension of the proterosoma, fixed by ridges (rostral tectum). Each bothridia carries a dorsal ridge that continues towards the rostral part of proterosoma and runs along it. Another ridge originates ventro-basically on the first pair of legs. Both ridges are linked with the base of the rostral hair by lamellar extensions. As the individual develops, the lamellar regions and the ridges expand and become thicker and curved, while in the adult individual they are responsible for the formation of tutorium and of the genal part. The dorsal ridge of the rostral tectum develops into the main ridge of the adult tutorium (Fig. 1).

Both in the immature and adult stages the rostral transverse ridge carries three lobes: two convex lateral lobes and a concave median one. The shape of the ridge remains essentially unaltered in the course of development and it is only the size of the ridge that alters. Triangular lamellar extensions originate on the lateral lobes of an adult. Thus, the shape of the transverse ridge and of the translamella remain stable during ontogenesis.

The medial longitudinal ridges of the immature stages retain the same location in the adults being overlain dorsally by laminar extensions that merge with lamellar ones. The laminar extensions form the second layer of the proterosoma coverings.

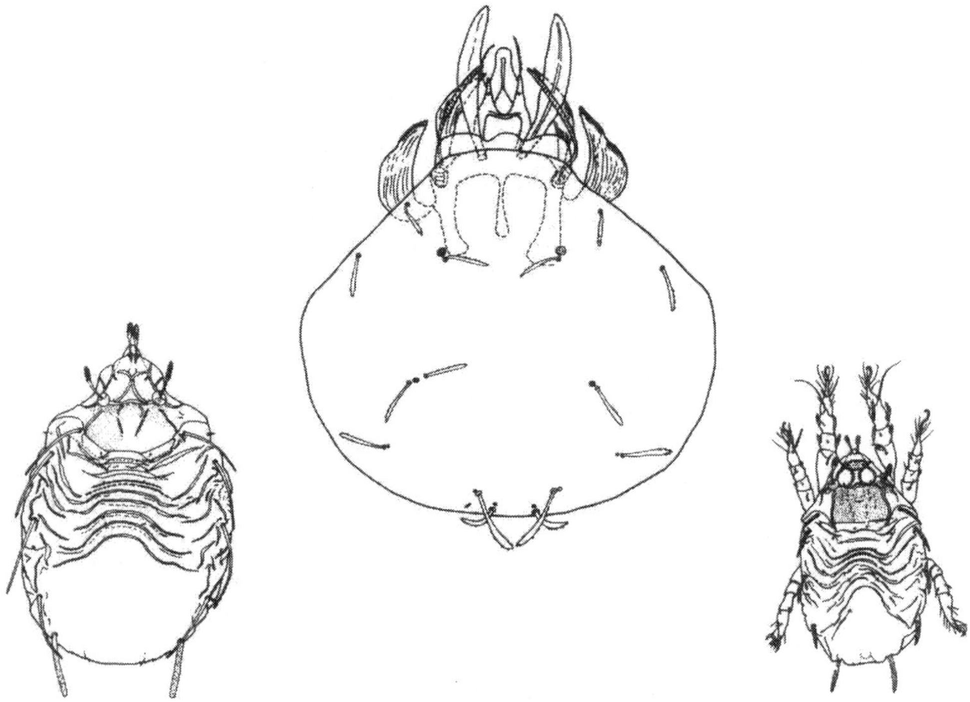

Fig. 1. The dorsal side of larva, tritonymph and adult of *Eupelops torulosus*.

The medial ridges produce a funnel-shaped structure in adult mites. The origin of this structure puzzled Grandjean (1934). However, the above funnel-shaped structure is an important feature in the identification of all developmental stages as well as of the adults as the shape of this structure remains essentially unchanged in all stages. This feature is all the more important in the case of *E. acromios*. In the immature stages the basal proterosoma part, the only lying below the bothridial ridge, is strongly developed. The rugose hysterosoma of larva and nymphs extends towards and overlaps the proterosoma, whereas in the adult the entire basal proterosoma part is covered by a notogastral plate. The proportions of proterosoma and hysterosoma vary considerably with the onset of imaginal moults. The proportions are: larval proterosoma ~41 % of the total length, tritonymphal proterosoma 36 %. adult proterosoma - as little as 25 %. Similar proportions apply to the entire family. The basal part of the proterosoma is transformed into thin transparent membranes which form a stable yet mobile link between proterosoma and the notogastal shield.

A study of the specific features of the dorsal hair growth is of great importance for the identification of developmental stages of the Pelopidae. In immature stages the serrate, spatulate rostral hairs develop into narrow lanceolate, flat, fluffy hairs. A maximal growth of rostral hairs occurs in protonymph and in adult; the hairs are 1.5 times the size of those in the preceding stage. In addition to the modification in shape of the Pelopidal rostral hairs, they are subject to considerable displacements during imaginal moults. The interlamellar setae in immature stages are hair-like and scarcely discernible (2-13 µm), in the adult they are more than 10 times as large (145 µm), becoming broadly spatulate and covering the entire medial part of the proterosoma. In the genus *Peloptulus* the interlamellar hairs show but a slight increase in size and they retain their shape. Trichobothria that are pinlike, with a very fluffy head in larva and nymphs, become thin spindle-shaped, slightly serrated longitudinally troughout the head. Thus, the shape and the size of all proterosomal hairs in the adult individual (except exbothridial and interlamellar hairs in *Peloptulus)* are very markedly changed:

All notogastral hairs can be subdivided into two groups according to their shape and growth patterns. Falling into the first group are the hairs of the l row, c_2,K_1 ; into the second- the hairs of the d row and K_2 (PN_1). The hairs of the l row in the larva are spindle-shaped, and strongly serrated. In the protonymph these hairs seem elongated, being more than twice as long as broad, rod-like and sparsely serrated. They are slightly enlarged in deuto- and tritonymphs, in adults constituting 1/2 - 3/4 of the tritonymphal hairs. The hairs c_2 of the immature stages are similar in shape to those of l row, but their growth is somewhat retarded, though with the maximal change in length occurring in the deutonymph, not in the protonymph. Similar modifications affect hairs K_1. d-row and K_2 (RN_1) hairs which are very small (4-13 µm) in immature stages but are 6-9 times larger in adults reaching the size of all the other notogastral hairs.

As can be seen from the above, the shape of the proterosoma, trichobothria and medial notogastral hairs does not change. Changes, if any, are concerned with the size of the hairs. The shape and the size of the lateral notogastral hairs are considerably altered in protonymph, c_2 - in deutonymph. Adults show marked variations in the shape and size of rostral and interlamellar hairs and trichobothria (except *Peloptulus).* The shape and the size of all notogastral hairs change strongly those of the lateral hairs to a smaller extent than those of the tritonymphal, whereas the medial hairs change in shape and increase several times in size. It should be stressed that the growth patterns of lateral hairs and trichobothria are the same in principle, though not simultaneous, i.e. the hair of the head of the trichobothria extends rapidly between the indents resulting in a clearcut rod-like shape.

An analysis of the material available of immature stages and adults of the Pelopidae leads us to the following conclusions:

1. The shape and arrangement of the rostral ridge, and likewise of the translamella, remain the same in adults throught their development.

2. The shape and arrangement of the medial longitudinal ridges and of the funnel-shaped structure are also same in adults in all stages.

3. In immature stages the species characteristics are manifest in trichobothria, interlamellar and marginal hairs.

4. In adults the species characteristics are manifest in the rostral slot, trichobothria, the lamellar keels and the arrangement of notogastral hairs.

THE OCCURRENCE OF *ORIBATULA TADROSI* (POPP), (ACARINA, ORIBATULIDAE) ON PEAR TREES IN THE U.A.R.

M. S. TADROS
Department of Plant Protection, Faculty of Agriculture,
Kafr El-Sheikh, Alexandria University, Alexandria, U.A.R.

Introduction

In 1961 the author carried out an ecological investigation on the European red spider mite, *Panonychus ulmi* (Koch). During the course of that study he noticed the fluctuations of *Oribatula tadrosi* (Popp) found on foliage and twigs. In the winter of 1962/1963 the present investigation was carried out and concerns:

1. A survey of the different species of oribatids found on pear twigs in a certain locality in Egypt.
2. A study of the natural population density of the most prevalent species occurring on tree twigs, i.e. *O. tadrosi* (Popp.).
3. The fluctuations of both nymphs and adults of the prevalent species found on trees.

Methods and techniques

Few studies have been made on soil oribatids in Egypt. Tadros (1966) listed twelve species which have been collected from the Giza region. The author obtained the moss mites from soil, foliage and twigs. Others obtained them from various materials: Sengbush (1957) got them from green algae of leaf litter and branches, Kunst (1962) separated oribatids from dry parts of shore meadows, Aoki (1966) collected the mites from circulionid beetles while Woolley (1966) and Seyd (1968) separated them from lichens and moss.

The present observations were carried out in a pear orchard of 3.5 acres, in the middle of the Nile Delta at Abou-Sir Village, Garbia Province (about 100 km west of the eastern branch of the river Nile).

Samples were taken over a period of two complete years, during the leafless season of pear trees, starting in January 1962 and ending in April 1963 by the same methods as described by Tadros (1971).

Results and discussions

I. Monthly means

The figures in Tables I-II indicate that the *O. tadrosi* population in 1963 was smaller than that of 1962. This fact may be due to either control measures during the 1963 season in the orchard or to the washing effect of rainfall during that season. It was also obvious that March was the month of a high population, January of a low. This phenomenon may be due to the opening buds and new foliage supplying an excess area for mites to develop.

Table I. Monthly means of *O. tadrosi* (Popp) at 3 levels in all directions in 1962 and 1963

Levels	January				February				March			
	1962	1963	Total	Mean	1962	1963	Total	Mean	1962	1963	Total	Mean
Lower	2.00	2.00	4.00	2.00	7.00	0.00	7.00	3.50	279.00	66.00	345.00	172.00
Middle	1.00	3.00	4.00	2.00	10.00	2.00	12.00	6.00	70.00	109.00	179.00	89.50
Upper	1.00	0.00	1.00	0.50	50.00	0.00	50.00	25.00	29.00	42.00	71.00	35.50
Total	4.00	5.00	9.00	4.50	67.00	2.00	69.00	34.00	378.00	217.00	595.00	297.50

Table II. Seasonal means of *O. tadrosi* (Popp) in four directions in all levels in 1962 and 1963

Directions	1962	1963	Total	Mean
East	63.00	14.00	77.00	38.50
West	70.00	70.00	140.00	70.00
North	239.00	25.00	264.00	132.00
South	77.00	115.00	192.00	96.00
Total	449.00	224.00	673.00	336.00

II. Directional Means

As shown in Table II the means of the two seasons indicate that sites facing North were most favoured by mites while those facing East, least. This is probably due to the fact that most mites prefer moist conditions owing to their original existence in soil.

III. Level Means

Table III shows that lower regions of the trees were the most preferred by the mite. This fact may be due to the favourable degree of R.H. found near the soil because of evaporating water, or to the shelter offered there. The figures also indicate that the mites disappeared from tree tops due to the harmful effects of direct sunlight, incorrect temperature and rain fall.

Table III. Monthly means of *O. tadrosi* (Popp) in three levels in 1962 – 1963

Year	Levels			Total
	Lower	Middle	Upper	
1962	288.00	81.00	80.00	449.00
1963	68.00	114.00	42.00	224.00
Total	356.00	195.00	122.00	673.00
Mean	178.50	97.50	0.00	336.50

IV. The mite dispersion on tree foliage

In April 1963 newly emerging mites were found on trees. In that month the mite population was composed of immature stages and adults, as shown in Table IV. It is obvious that adult fluctuations on leaves were nil in that month.

On the other hand the immature stages were more dominant on leaves. This phenomenon may be due to the new spiny leaves having a low proportion of parenchym and this may not be attractive for adult mites. Moss mites may also immigrate from earth to trees when they are extremely numerous in sòil at that time of year as recorded by Tadros et al. (1965) under wheat.

Table IV. Monthly means of the moving stages (nymphs and adults) of *O. tadrosi* on twigs and leaves of pear trees, in 3 levels in April 1963

Levels	Cuttings			Leaves		
	Nymphs	Adults	Total	Nymphs	Adults	Total
Lower	1.00	0.00	1.00	1.00	0.00	1.00
Middle	1.00	0.00	1.00	3.00	0.00	3.00
Upper	1.00	2.00	3.00	7.00	0.00	7.00
Total	3.00	2.00	5.00	11.00	0.00	11.00

V. The occurrences of oribatids on both leaves and soil

It was found that *O. tadrosi* was the dominant species on tree foliage, some scattered individuals of *Protoribates* spp. or *Epilohmannia cylindrica* being encountered.

On examining soil samples, eleven species were recorded in four groups arranged in descending numerosity as follows:

1. *Oribatula tadrosi* (Popp), (Oribatulidae).
2. *Protoribates* spp. (Haplozetidae), *Epilohmannia cylindrica* (Berlese) (Epilohmanniidae), *Lohmannia paradoxa* (Haller) (Lohmanniidae), *Eulohmannia* sp. (Eulohmanniidae).
3. *Plakoribates multicuspedus* (Popp) (Oribatellidae), *Oppia sticta* (Popp) (Oppiidae).
4. *Pseudotritia ardua* (Rathke) (Phthiracaridae), *Passalozetes africanus* (Grandjean) (Carabodidae), *Microzetes auxiliaris* (Grandjean) (Microzetidae), *Allogalumna exigua* (Popp) (Galumnidae).

As found by Tadros (1965) and as indicated by the species listed, it is obvious that species that live in a disturbed layer of the soil could migrate to tree foliage and buds in search of a new biotope and this would probably be due to lack of humidity in the disturbed layer, as recorded by Madge (1961) and the author (1965).

References

Aoki, Jun-ichi: Epizoic symbiasis: An oribatid mite *Symbioribates papuensis*, representing a new family, from cryptogamic plants growing on backs of papuan weevils (Acari: Cryptostigmata). Pacific Ins. 8: 281-289, 1966.

Kunst, M.: *Anoribatella* n. g., a new genus of oribatid mites from central Europe. Acta Univers. Carolinae, Biol., No. 1: 89-98, 1962.

Madge, D.S.: The behaviour of free living mites as affected by humidity. Dept. Entomol. Roth. Exptl. State., Harp. Herts: 108, 1961.

Sengbush, H.G.: Checklist of oribatid mites in the vicinity of mountain lake biological station, Virginia, (Acarina: Oribatei). Univ. J." Sci.,8: 120-134, 1957.
Seyed, E.L.: Studies on the moss mites of *Snowdonia*, (Acari: Oribatei) 1 Moel Hebog. Entomologist, Feb. 1968: 37-41, 1968.
Tadros, M.S.: The occurrence of *O. tadrosi* (Popp). (Acarina: Oribatulidae) on pear trees in U.A.R. Bull. Soc. Entomol. Egypte. (In press).
Tadros, M.S., Wafa, A.K. El-Kifl, A.H.: Ecological studies on oribatids in Giza region. Bull. Soc. Entomol. Egypte, XLIX: 1-37, 1965.
Woolley, T.A.: A new genus and species of oribatid mite from Colorado (Acarina: Oribatei). J. Kansas Entomol. Soc. 40: 32-37, 1967.

THE CENOSIS OF ORIBATID MITES AS AN AID IN BIO-INDICATION (ACARINA, ORIBATOIDEA)

J. VANĚK
Institute of Landscape Ecology, Czechoslovak Academy of Sciences, Prague, Czechoslovakia

Lately we have been witnessing serious endeavours to elucidate the mechanism of the anthropic influences as well as efforts to determine the characteristics and scope of changes caused by them in ecosystems. In order to obtain evidence of the biologic effectiveness of anthropogeneous alternations bio-indicating organisms are used. Of course, the sensitiveness with which they indicate modifications varies in accordance with the kind of influence and species of the respective organism, and the generalisation of the results obtained by the use of bio-indicating species is often of problematic value only.

To testify in a more generally valid way the biologic effectiveness of the anthropic influences in nature, a cenological bio-indication can be used. In such a case the whole cenosis, which forms an organic and functional ingredient of the whole ecosystem, a collection of species possessing an unequal sensitivity towards the influence under review, is being measured and expressed by their numerical characteristics. The alternations of the cenosis' characteristics measured and expressed in this way express also besides the qualitative and quantitative changes the alternations of the cenosis' structure as well, and give even the possibility of deriving changes of its function. They possess therefore a more conclusive validity of a more general nature than the modifications by which the bio-indicating species studied react as far as anthropic influence is concerned.

In order to demonstrate the biologic effectiveness of the anthropogeneous changes in soil biosystems we have chosen a representative, systematically homogeneous, partial biocenosis of the soil mites of the Oribatoidea group and we ascertained the modification of its qualitative, quantitative and structural characteristics as a consequence of some specific anthropic influences.

The objects of interest were the alterations of the respective cenoses caused by:
1. the effects of industrial emissions,
2, the advance of an underground fire in a coal seam,
3. the application of certain insecticides.

The above influences are of three different kinds. It was of much interest to race the respective differences of the reactions of the Oribatei cenosis, and to find ut in which respect they will be similar or identical in character.

The qualitative properties of the cenosis were expressed by three numerical characteristics, the quantitative by five, the structural by six and in addition, another six comparative numerical criteria were used. The above mentioned 20 indices were calculated both for the cenosis of the oribatid mites, which were influenced by the anthropogeneous effects studied and for the cenosis of the control areas.

In the case of industrial emissions, the problems were investigated in 12 different localities of which six were affected and six were selected for the sake of comparison. By sample analysis 32,615 mites were obtained. The basis of cenologic analyses of oribatid mites was formed by 21,793 mites representing 132 species.

In areas affected by the emmissions the following changes of cenotic characteristics were observed: species quantity reduction by 41.6 % .; species density reduction by 37.5 % ; Oribatei specimens density reduction by 69.5 % ; Oribatei dominance reduction by 16.22 % ; decline in the qualitative homogeneity by 24.7 p. c.; drop in quantitative homogeneity by 63.0 % ; and, finally, extension of minimum areas by 174.33 % .

It was possible to prove that industrial emissions produce chronic poisoning in the cenosis of oribatid mites which manifests itself in the following ways:

1. In case of permanent exposure to harmful substances of low concentration: reduction in the number of species, increase in the number of specimens of the remaining species, limitation of their dispersion and extension of the whole cenosis minimum area.

2. Whenever harmful substances in greater concentration are permanently active or when low concentrations of poisonous emission components are affecting the cenosis, significant reduction in the number of species, as well as of specimens, a definite decrease of the values of the cenosis structural characteristics, particulary the great increase of the cenosis minimum area can be observed. The structural characteristics retain a trend of permanent or regressive development.

3. Whenever great concentrations of harmful substances or poisonous components of emissions act in the form of a single impulse only, the cenosis reacts by a vehement decrease in all characteristics mentioned under point 2, with a very pronounced increase of the minimum area. The extent of these changes can be seen as catastrophic. After the residual effects of the emission impact disappear, progressive development trends of the quantitative as well as of the structural characteristics appear, but the restoration of the original values of all the characteristics of oribatid mites cenosis is a long-term affair lasting several decades.

Industrial emissions create effects of a chronic nature.

As regards the unrepeated, one-time influences, we observed them both in cases of underground fires in a coal seam (a short-term influence), as well as after the application of pesticides (single unrepeated influence). After the phase of the underground fire has passed, it became possible to record the successive stages within the space of 0-45 years on a transect placed into the axis of its advance, to determine the successive trends and to determine the attained regeneration degree of the Oribatei cenosis.

In the case of the above mentioned fire the problems were studied in one locality divided into five zones. Their condition was recorded: before the passing of the fire front (checking), the fire front itself, the primordial colonization of area damaged by fire, the advanced colonization of the area damaged by fire, and also areas that remained untouched for up to 45 years, in order to restore their cenosis of oribatid mites. The basic material consisted of 40 samples, from which 4 201 animals of the mezoedaphon category were obtained; among them were 2570 oribatid mites, belonging to 69 species, and their cenoses were subjected to a detailed analysis.

In the zone of the fire front a total destruction of the bio-indicating cenosis was ascertained. Of the original numbers only 10 % of the species remained, merely 19.7 %, of the primary density of specimens, the dominance of the oribatid mites decreased by 72.7 %, the qualitative homogeneity dropped by 58.7 %, the average number of specimens of one species diminished by 17.7 %, whereas the cenosis minimum area increased by more than 492.5 %.

After the fire front had passed and its acute manifestations had subsided in the zone of the initial colonisation, a cenosis predetermined on the one hand by a nearly complete soil sterilization and on the other hand by fire symptoms still reverberating e.g. by higher soil temperatures, decreased humidity, etc., came into being. In this particular case the characteristics indicate a community which is ecologically specific, poor in species comprising large numbers of specimens. The cenosis minimum area is much larger than in the case of the control cenosis. The structural characteristics indicate a community which is unstable, unstructured and inadequately formed and one which shows higher values only with regard to the density of specimens per volumetric unit.

In spite of the fact, that up to 45 years elapsed since their damage, neither the cenosis of the advanced colonisation zone, nor zones of the regenerated cenosis, attain in one single case the values of the characteristics of the control cenosis, native to the zone before the fire front passed.

When investigating the effects of the application of pesticides, identical or very similar processes and phenomena were observed as when research into the fire effects was carried out. Especially the progress of colonization and cenosis regeneration was similar. However, both the start and the intensity of the course they ran were determined by the time required for the inactivation of the effective substances of the pesticides and the duration of residual biologic effects of components caused by the desintegration of the pesticides. Naturally, the respective period differed in accordance with local conditions and the kind of pesticide. In any case, even within the first phase of colonization of treated plots, a quantitative expansion of a few species took place, the consequences of the pesticides applied were always evident in the characteristics of the quality and structure of cenosis during the whole period of observation after their application, i.e. for 3-5 years.

In conclusion it can be stated, that the analyses of the cenosis of oribatid mites, carried out to assess the biologic effects of the anthropic influences in soils, have always yielded relatively clear and convincing results permitting to make a representative judgement of the diversity and stability of the whole soil biosystems, as well as changes in their characteristics due to human interference. When evaluating the biological consequences of anthropogeneous changes in the soil, different procedures can be applied: on the one hand the classic methods of physics and chemistry to demonstrate the physical as well as chemical aspects of such changes, on the other hand bio-indications by means of species to render evident the specific effectiveness of the respective changes. Finally, representative zoocenologic (or phytocenologic) bio-indications can be used to prove the biologic effects of such changes, availing oneself, e.g., of the partial biocenosis of oribatid mites in the original biosystem of the locality under investigation.

By changing their characteristics the bio-indicating cenoses indicate the mixture of direct, as well as undirect, biotic and abiotic changes caused by human intervention. They are not encumbered by any one specific sensitivity, and indications of disturbances obtained by an analysis of their characteristics carry more weight when evaluating the consequences of anthropic influences, and are valid in a more general way, than other information obtained by measurements of a chemical or physical nature, or by a specific bio-indication.

References

Balogh, J.: Lebensgemeinschaften der Landtiere. Budapest - Berlin 1958.

Balogh, J.: Grundzüge der Zoozönologie. Budapest. 1954.

Palissa, A.: Bodenzoologie in Wissenschaft, Naturhaushalt und Wirtschaft. Berlin 1964.

Pelz, E.: Rauchschädendiagnose. Archiv F. Forstwesen 8: 750-759, 1959.

Sobotka, A.: Vliv průmyslových exhalátů na půdní živěnu smrkových porostů Krušných hor. Lesnický časopis 10 (XXXVII) 11: 987–1002, 1964.

Tischler, W.: Synökologie der Landtiere. Stuttgart, 1959.

Vaněk, J.: Použitelnost půdní zoologie pro lesnickou typologii. Sborník lesnické fakulty v Praze II: 25-40, 1959.

Vaněk, J.: Změny mezoedafonu školek po použití HCH proti ponravám chroustů. Lesnictví: 337-346, 1959.

Vaněk, J.: Zoocenologická indikace biologického stavu půd. Sborník VLÚ při VŠZL 8: 253-267, 1966.

Vaněk, J.: Industrieexhalate und Moosmilbengemeinschaften in Nordböhmen. In: Progress in Soil Biology; Braunschweig: 331-339, 1967.

Vaněk, J.: Exhaláty a společenstva pancířníků v severočeské oblasti kouřových škod. In: Imise a lesní cenosy; Sborník ÚTOKu, ČSAV: 69-94, 1968.

Vaněk, J.: K otázce biologické rovnováhy, biologické hodnoty a biologické stability krajiny. Acta Ecol. Nat. Reg. - Terplán, Praha, 1-2: 13-16, 1970.

SOME ASPECTS OF THE ENERGETICS OF SOIL MITES

J. A. WALLWORK
Department of Zoology, Westfield College, University of London, England

Introduction

Numerous investigations (see Cummins & Wuycheck,1971) have shown that calorific values for biological materials of all kinds fall within a range of about 3000--9500 cals/g, and that there is considerable variation between the estimates for different groups of organisms. Since fatty tissue has an appreciably higher calorific value than protein or carbohydrate, it may be possible to use calorific values as an index of the nutritional state of an animal or population in the field (Slobodkin & Richman,1961). This report presents and discusses calorific data for seven common species of soil mite of a beech forest floor, namely *Damaeus onustus, Steganacarus magnus, Platynothrus peltifer, Hermannia scabra* (Cryptostigmata), *Pergamasus* sp., *Parasitus* sp. and a species of Macrochelidae (Mesostigmata).

Methods

Samples of mites were collected regularly from beech litter over an 18-month period. The individuals collected on each occasion were dried at 60° C for three days prior to being burned in a Phillipson microbomb calorimeter (Phillipson,1964).

Results

Average values for each species, obtained by pooling individual calorific estimates, and expressed as cals/g ash free body weight, are given in Table 1. These results provide several interesting features:

1. Firstly, there appears to be a tendency for carnivores to have a higher calorific content than detritivores. Thus, the cryptostigmatids *D. onustus, S. magnus, P. peltifer* and *H. scabra* which function, ecologically, as primary consumers, all have calorific values below 5000 cals/g, whereas two of the three carnivorous mesostigmatids have values in excess of 5000 cals/g.
2. There also appears to be an inverse relationship between calorific value and ash content (Fig. 1). The very high ash content of *S. magnus* is striking, and reflects the fact that large deposits of $CaCO_3$ occur in the cuticle of this species. However, ther

Table 1. Calorific data for 7 species of soil mite

Species	NO. OF RUNS	CALS/G ASH FREE (MEAN ± S.E.)	% ASH (MEAN)
Damaeus onustus Ads.	21	4907 ± 112	4.7
Juvs.	3	4313 ± 68	11.0
Steganacarus magnus	16	3193 ± 126	46.5
Platynothrus peltifer	1	3538	14.0
Hermannia scabra	1	4660	25.0
Pergamasus sp.	20	5269 ± 131	4.9
Parasitus sp.	2	5698 ± 226	4.0
Macrochelidae sp.	5	4873 ± 164	4.0

is no obvious explanation for why the very low calorific value should be related to this characteristic, although it is possible that large amounts of inorganic material in the exoskeleton may interfere with the complete combustion of the pellet. This possibility is now being studied.

3. Another interesting feature is that there appears to be a direct relationship between calorific content and metabolic rate. This has not been completely documented, but the trend can be expected since we know that predatory forms have higher respiration rates than detritivores (Webb,1970), and it is evident from (1) above that carnivores tend to have higher calorific values than detritivores.

4. Following on from (2) and (3) above, it is to be expected that there will be an inverse relationship between ash content and metabolic rate. This work suggests as much, but more results are needed.

Discussion

The ecological significance of the first of these findings is fairly clear. It is accepted that populations at the base of ecological pyramids have a lower free energy of formation (calorific value) than the upper trophic levels. In this work, fungal material on which cryptostigmatid mites mainly feed was found to have an average calorific content of 2665 cals/g. The cryptostigmatid species studied provide an average calorific value (from Table 1) of 4104 cals/g. The average value for carnivorous Mesostigmata (also from Table 1) is 5242 cals/g. Thus, in moving from the base of the trophic pyramid to the higher levels, there is a pro-

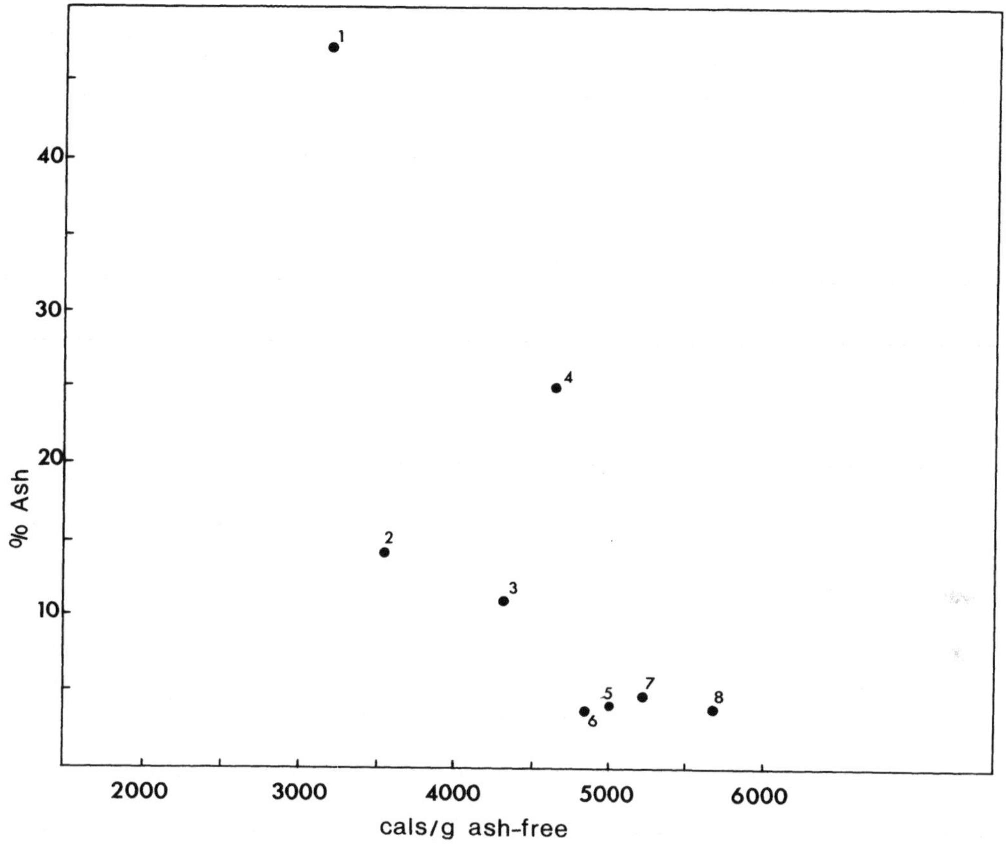

Fig. 1. The relationship between calorific content (cals/g ash free) and % ash for various species of soil mites (1 = *Steganacarus magnus*; 2 = *Platynothrus peltifer*; 3 = *Damaeus onustus* juvs; 4 = *Hermannia scabra*; 5 = *Damaeus onustus* ads; 6 = Macrochelidae sp.; 7 = *Pergamasus* sp.; 8 = *Parasitus* sp.)

gressive increase in calorific value.

Primary consumers, such as Cryptostigmata, are then an important intermediate stage in the upgrading of energy from the low level found in plant materials to the high level found in carnivores. There are strong energy constraints on the trophic efficiencies of these primary consumers since, in order to move energy from the primary producer level up the ecological pyramid, much of the energy they assimilate must be expended in metabolic activities designed to raise the remainder to a higher level. Indeed, it is known that primary consumers are poor "converters", and about 90 % of the energy assimilated may be dissipated in respiration. Clearly, the energy step from primary producers to primary consumers cannot be too great, otherwise the upgrading process would be more than the metabolism of the consumers could cope with.

In order to interpret the relationships between calorific value, ash content and metabolic rate, we need to draw on information about the biology of the species concerned. In the Cryptostigmata studied, the heavy exoskeleton makes up most of the dry weight, and this is augmented by deposits of $CaCO_3$ in *S. magnus*, as already noted. The heavy body armour of such mites implies that they are rather sedentary animals, and this is confirmed by observations on live material. It may be noted also that Cryptostigmata such as *D. onustus* and *S. magnus* are active and apparently feeding all the year round in the field. As stated earlier, these mites act as intermediate stages in upgrading energy because of their relatively low calorific contents. It is likely that they can tolerate such low calorific values because they are rather sedentary, and because they can obtain a continuous supply of energy throughout the year. They do not need food stores in the body and, thus, can exist on a low metabolic rate. Hence, they will tend to be rather inactive animals requiring heavy body armour to protect against predation.

The exoskeleton of carnivorous species is lightly armoured to allow greater mobility, metabolic rates are generally higher, as befits a more active life, and since prey may not be always available, there may be a need to store energy in the body (hence the higher calorific values).

It should be possible to test these arguments. Thus, if the Cryptostigmata are receiving a continuous supply of energy throughout the year, and are utilizing, rather than storing, as they receive it, we would expect very little seasonal fluctuation in calorific value in field populations. On the other hand, carnivorous Mesostigmata may be expected to show greater changes in calorific content, notably when prey is scarce and food reserves have to be utilized. In Figs 2 and 3 are plotted monthly data for the detritivirous *D. onustus* and the predatory *Pergamasus*, which show that, in both, there is an increase in calorific content during late winter and spring which is coincident, at least in *D. onustus*, with the production of eggs. A subsequent decline during mid-summer is more pronounced in *Pergamasus* than in *D. onustus*. During this particular summer, leaf litter in the sampling locality became extremely dry, and it is likely that microarthropod populations in general experienced some degree of food stress. That the carnivorous *Pergamasus* reacted more strongly to this situation than did *D. onustus*, may be a reflection of the greater ability of the former to convert assimilated material into food reserves and to mobilize these when required.

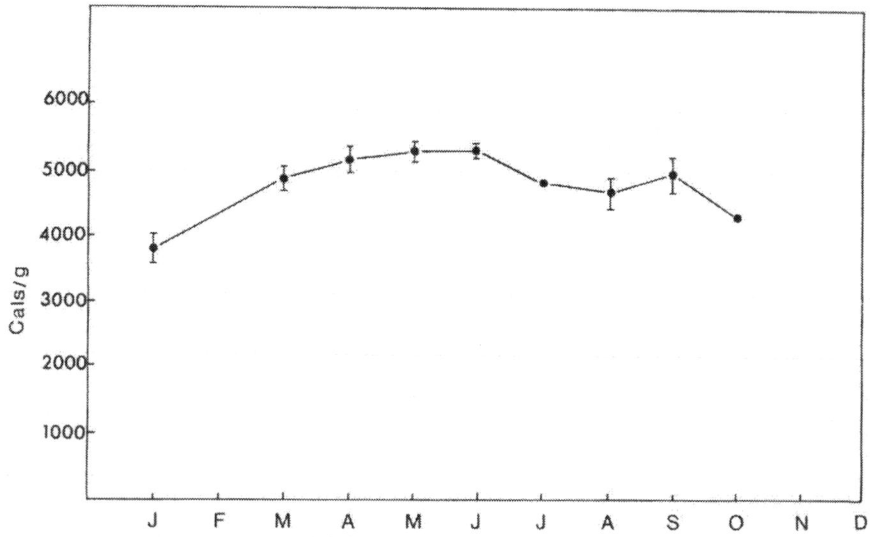

Fig. 2. Monthly variation in calorific content of adult *D. onustus* populations taken from the field.

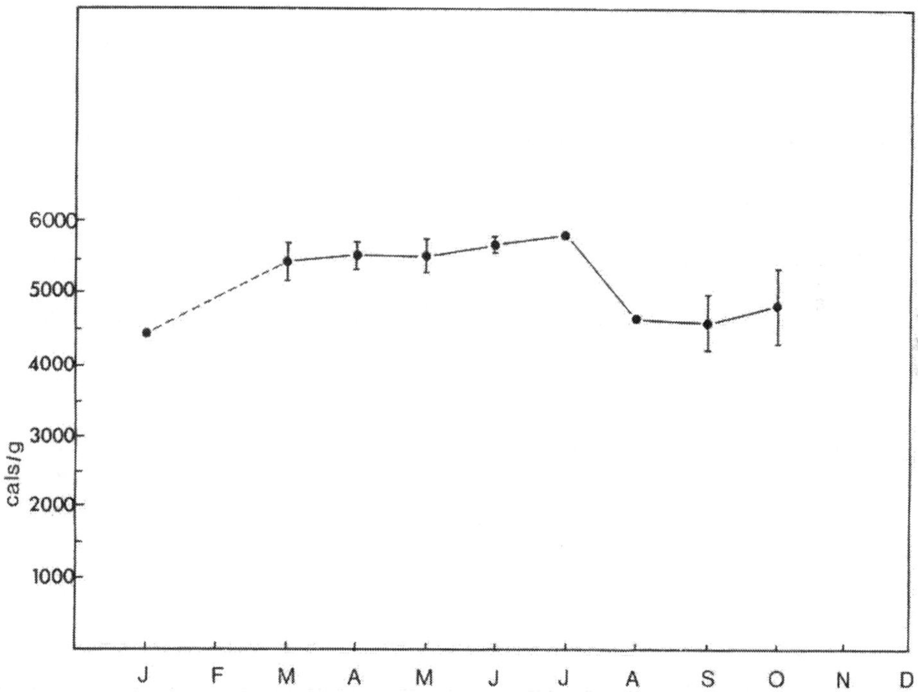

Fig. 3. Monthly variation in calorific content of *Pergamasus* sp. populations taken from the field.

References

Cummins, K.W., Wuycheck, J.C.: Calorific equivalents for investigations in ecological energetics. Mitt. Int. Ver. Limnol. 18: 158 pp., 1971.

Phillipson, J.: A miniature bomb calorimeter for small biological samples. Oikos 15: 130-39, 1964.

Slobodkin, L.B., Richman, S.: Calories per gram in species of animals. Nature 191: 299, 1961.

Webb, N.R.: Oxygen consumption and population metabolism in some mesostigmatid mites. Pedobiologia 10: 447-56, 1970.

Proceedings of the 3rd International Congress of Acarology, Prague, 1971

THE STRUCTURE OF AN ADULT POPULATION OF *STEGANACARUS MAGNUS* (NIC.) (CRYPTOSTIGMATA)

N. R. WEBB and G. W. ELMES
Furzebrook Research Station, Wareham, Dorset, England

Introduction

This paper reports the results of a study on the structure of an adult population of the litter dwelling Phthiracarid mite *Steganacarus magnus* (Nic.) and forms part of a more detailed population study being undertaken on this species, the results of which will be published elsewhere.

The population, under study, occurred in a sandy heathland soil in Southern England. Monthly samples consisting of large cores 50 cm in diameter and 10 cm in depth were collected for fourteen months commencing in November 1968, and were heat extracted by conventional Tullgren techniques. The mites collected were examined and assigned as either freshly moulted or mature, depending on the degree of darkening of the exoskeleton. (When freshly moulted from the tritonymphal stage, the exoskeleton is pale and, subsequently, darkens). The mites were then dissected in order to determine the number of eggs and prelarvae (deutovia) present in the reproductive tracts of the females.

Results and discussion

Figure 1a indicates the percentage of the population which has newly moulted from the tritonymphal stage and possesses a pale coloured exoskeleton. From this figure it can be seen that there is continual recruitment to the adult population throughout the entire year. The curve of recruitment reaches a peak in the months of April and May and has a minimum during September. The peak in recruitment is subsequently reflected in the numbers of eggs and prelarvae recorded (Figure 2a).

About a month later in June the peak number of eggs is recorded from the reproductive tracts of the females and by this month the exoskeleton has darkened. The time that this apparently takes in the field corresponds closely with the time observed from laboratory cultures.

A peak in the number of prelarvae occurs in July, about a month after peak egg production. Secondary peaks of egg and prelarva production occur again in August and September respectively and are again separated by an interval of about one month, suggesting that this is the time required for development from egg to prelarva.

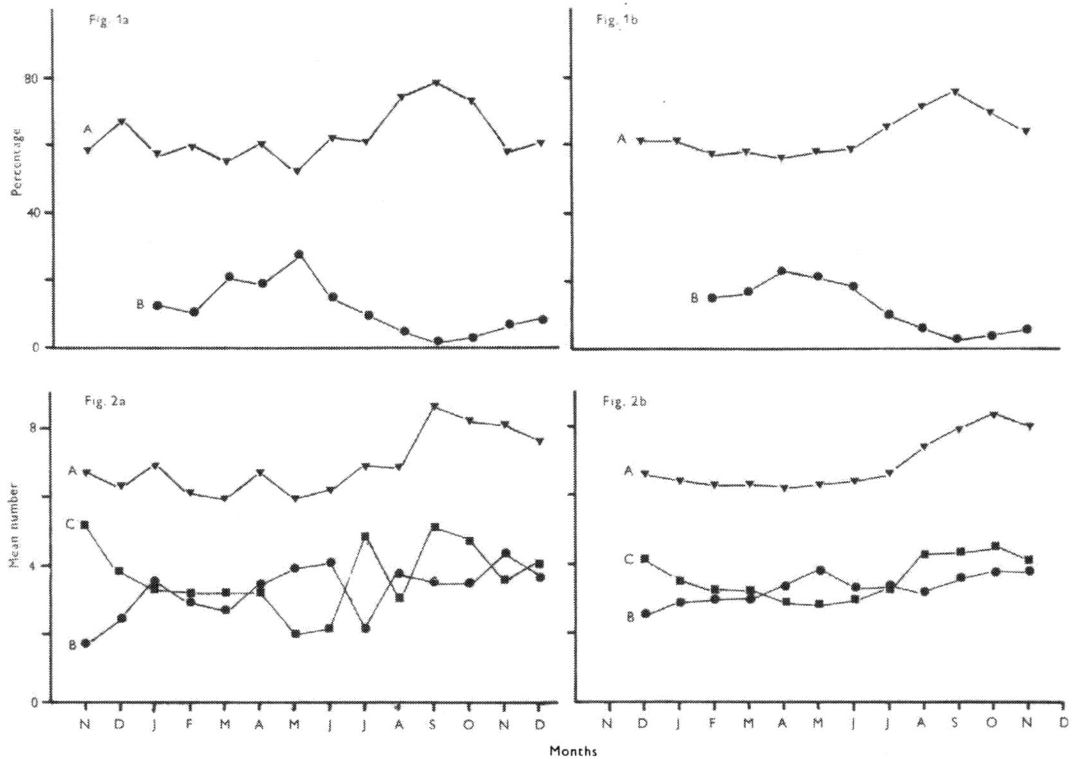

Fig. 1 a. The number of mature females expressed as a percentage of the population of mature individuals (A) and the number of immature adult individuals expressed as a percentage of the total adult population (B) of *Steganacarus magnus.*

Fig. 1 b. The three way moving average of the percentages plotted in Figure 1 a.

Fig. 2 a. The mean number of eggs and prelarvae per female (A), the mean number of eggs per female (B) and the mean number of prelarvae per female (C) found throughout the year for *Steganacarus magnus.*

Fig. 2 b. The three way moving averages of the figures plotted in Figure 2 a

Thereafter egg and prelarva production declines to a winter minimum.

The seasonal trends in the figures are shown more clearly if a three way moving average is taken (Figures 1b and 2b). These figures indicate that egg production builds up to a peak in May and that the prelarval production is greatest from July until the autumn. This autumnal peak corresponds closely with the peak in the percentage of females in the total population, (Figure 1b) and the minimum recruitment of new adult individuals.

Throughout the year about 60 % of the adult population consists of females and, although this proportion remains fairly constant, it approaches 80 % in September. At this time the population consists mostly of females and, consequently, egg and prelarva numbers are highest. In addition, recruitment to the population is least. It is not known whether the longevity of the males is similar to that of the females. From analysis of the weight distributions of the immature adult individuals (those which have freshly moulted) their sex can be ascertained (Elmes and Webb,1972 in press). From this analysis it would appear that recruitment to the male population, during the spring, reaches a peak approximately a month after peak recruitment to the female population, although it must be added that some recruitment to the populations of both sexes takes place throughout the year.

The developmental history beyond the prelarval stage remains unknown at present. In particular it would be interesting to know how long the prelarvae remain within the reproductive tracts of the female. It will be noticed from Figure 2a that during August there is a decline in the number of prelarvae which may indicate that they are "laid" at this time, but, alternatively, this decline could be interpreted as arising from recruitment, to the female population, of new individuals. If the prelarvae are not "laid" they may be retained within the body of the female, to be released and to complete their development upon her death.

References

Elmes, G.W., Webb, N.R.: An interpretation of morphometric and gravimetric differences shown by adult *Steganacarus magnus* (Nic.) (Acari: Cryptostigmata). J. Zool. London 172. (In press.)

Section II

MARINE AND FRESHWATER MITES

Chairman

C. Bader (Switzerland)

Assistant Chairman

L. Halík (Czechoslovakia)

THE DORSAL SHIELDS OF THE THYASIDAE

C. BADER
Museum of Natural History, Basel, Switzerland

The earliest specialists in watermites, e.g. Koenike, Piersig and Wolcott, noticed the dorsal shields of the Thyasidae, but they were unable to make use of these sclerotized plates for taxonomic purposes. By studying these shields and their associated glands, Lundblad was the first to prove that there are features associated with these plates which distinguish the different genera of the family. In 1927 Lundblad published his famous scheme dealing with both plates and glands and this has been accepted by all workers on Hydracarina. Lundblad found the following organs:

1. Plates	-	frontale (unpaired, pigment present or absent)
	2	frontalia (prae-, postfrontalia)
	2	okularia (prae-, postokularia)
	5	dorsocentralia
	4	dorsolateralia
2. Glands	7	dorsoglandularia
	4	lateroglandularia

The ciphers 2, 4, 5 and 7 appear in an irregular succession. According to the morphology of the Chelicerata we should expect 2 and 4 only. Lundblad himself realized this, and though later on he tried to make some unessential changes, he was not convinced about it. Motas and coworkers proposed another explanation, although retaining their scepticism.

After studying an immense number of watermites and the abundant literature on the subject, I finally succeeded in erecting a revised scheme with the expected numbers 2 and 4. This scheme for dorsal shields can be applied not only to the Thyasidae, but to other families as well. The Feltriidae, e.g., have the same distribution of plates and glands while the Protziidae and Lebertiidae, lacking the plates (with the exception of the postokularia), also demonstrate a similar arrangement of all their glands.

A detailed study of the genus *Thyas* convinced me that Lundblad was mistaken about the anterior plates. He found five dorsocentralia but in reality there are only four. Lundblad's dorsocentrale No. 1 is in fact the postfrontale, hence only the expected four dorsocentralia remain. The position of the postokulare is important and always decisive. This small plate furnished with a fine hair is in all cases situated between the prae- and the postfrontale. It is liable to fuse with one of these shields or with both. Due to this fact only four dorsocentralia can be observed in all Thyasidae. (Fig. 1.)

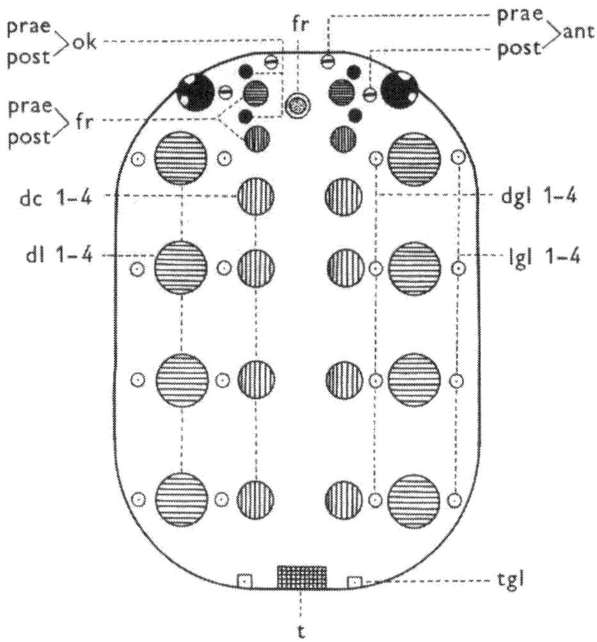

Fig. 1.:Dorsal Shields and Glands of the Thyasidae.

1. Plates: fr = Frontale (unpaired), and = Antenniformia (Prae-, Post-Antenniforme), okc= Okularia (Prae, Postokulare), dc_{1-4} = Dorsocentralia 1–4, dl_{1-4} = Dorsolateralia 1–4, t = Terminale;

2. Glands: dgl_{1-4} = Dorsoglandularia 1–4, lgl_{1-4} = Lateroglandularia 1–4, tgl = Terminoglandulare.

According to Lundblad there are seven dorsoglandularia. I too have found seven glands, but they should be arranged in a different order. Lundblad designates the first pair as "Antenniforme", thus reducing the number to six. Dorsoglandulare No. 2, however, is the second antenniforme, situated in all Thyasidae in the anterior part of the body, close to the frontalia and okularia, both belonging to the "Cephalon" of the Chelicerata body. One should therefore distinguish between the Postantenniforme and the Praeantenniforme, so that there remain apparently only five dorsoglandularia. Dorsoglandulare No. 7 in the rear of the body has not the same structure as the remaining 4 glands and it can be associated with an unpaired shield in the posterior part of the body. For this "abdominal plate", occasionally found in some genera of Thyasidae, *e.g.*, *Panisus*, I propose the new term Terminale, and the associated gland Terminoglandulare. This results in a revised scheme (see also figure):

The dorsal shields

1. Plates – frontale (unpaired)			– terminale (unpaired)
	2 frontalia (prae, post)	4 dorsocentralia	
	2 okularia (prae, post)	4 dorsolateralia	
2. Glands			
	2 antenniformia (prae, post)	4 dorsoglandularia	1 terminoglandulare
		4 lateroglandularia	
chelicerata	cephalon	thorax	abdomen

In the soft skinned species of the genus *Thyas* all the plates are small and separated, the other genera of Thyasidae exhibit different ways of fusion and decrease of the platelets. In *Panisus*, e.g., the unpaired frontale fuses with the two frontalia and postokulare, thus forming uniform frontal shield between the eye-capsules. For the Feltriidae Cook named this shield the "anteromedial plate". I adopt this name for the Thyasidae as well. The dorsocentralia and the dorsolateralia of other genera can fuse in different ways and finally the dorsum may be covered by a large dorsal shield. The Ctenothyasidae possess such a uniform plate, including all the above mentioned plates and glands.

The ancestors of the freshwater mites are terrestrial Trombidiform mites. At the moment the opinion is prevailing that the soft skinned Thyasidae with the small plates are "primitive". A progressive development should then lead to the armoured forms of the family. The genus *Thyas* is said to have immigrated into freshwater only recently. In this case, however, we ought to find in the terrestrial group of the Trombidiformes similar soft-skinned representatives. Yet they do not exist!

My recent investigations in the Swiss National Park have produced some interesting results. *Panisus michaeli* does not prefer the very sources of Alpine rivers but the upper courses immediately below the source. This region (Quellbach) is furnished with watermosses and is a few meters long. It is this region where the specimens of *Panisus michaeli* live, but they leave the water for several hours, time and time again. They are in reality inhabitants of the "moisture-biotope" (Biotop des Feuchten). *Panisus* is well armoured like some terrestrial forms, e.g. *Tarsotomus*, which lives in wet places too. In contradiction with the prevailing opinion I suggest that the immigration into freshwater took place via the mosses of the upper courses of rivulets or streams and not via the calm places of the sources.

From the upper courses of rivers the strongly armoured mites penetrate either the sources or the fast running parts of the rivers. The dorsoventral muscles are fastened onto the dorsal (and ventral) shields. The stronger the water current, the stronger the musculature becomes, and the bigger the plates must be. In the calm spring waters, on the contrary, the mites need not attach to the substrate, the musculature is reduced and the plates degenerate. In this case we cannot consider *Thyas* to be a primitive genus as it was adapted to life in water earlier than *Panisus* etc. The latter with its larger plates is the primitive form, only recently adapted to an aquatic environment. From *Panisus* two progressive paths can be traced, the first leading to the soft skinned inhabitants of sources, the second to the strongly armoured mites of fast running waters.

Parallel to this evolution runs a transformation of the genital area. Most of the terrestrial mites of Trombidiformes have three pairs of genital suckers, protected by valves; the same applies to the Thyasidae. The third pair is more or less stalked. During colonization of the sources the valves were reduced, and the number of stalked suckers increased. These features appear not only in the Thyasidae but also in the Protziidae: *Partnunia*, e.g., has small valves, *Protzia* has none at all, while both have many stalked suckers. But the genus *Protziella* has the same anteromedial plate as in the genus *Vietsia*, a member of the Thyasidae. The near-relationship between Thyasidae and Protziidae is therefore evident. Both have in common: 1) epimera arranged in four typical groups of two, all of them separated from one another by skin areas; 2) dorsodistal portion of the fourth palpal segment extending beyond the insertion of the fifth segment, forming a chelate palp (scherenförmig); 3) walking legs with well developed claws and without swimming hairs; 4) mid-gut characterized by many diverticulae.

Viets united the Protziidae with the Eylaidae and Limnocharidae to form the superfamily Limnocharae. All of these have genital suckers fastened in the soft skin of the body, they are more or less stalked. As the organization of *Eylais* or *Limnochares* is so different from that of *Protzia*, it seems impossible to unite these genera.

Certain members of Thyasidae have settled in the backwaters of the rivers, these too have a reduced dorsoventral musculature and the plates are missing excep for the anteromedial plate. This is characteristic of Hydryphantidae. The Hydrodromidae have a similar organization, yet in addition they possess swimming hairs on their legs.

According to our discussion the superfamily of Hydryphantae Viets, 1931, consists of

1. Thyasidae,
2. Ctenothyasidae,
3. Hydryphantidae,
4. Hydrodromidae and in addition
5. Protziidae and so on.

References

Bader, C.: Die Wassermilben des Schweizerischen Nationalparks. Ergebn. wissenschaftl. Untersuch. Schweiz. Nation. Park. (In press.)

Cook, D.R.: Studies on the Thyasinae of North America. Am. Midl. Naturalist, V 62: 402–428, 1959.

Lundblad, O.: Die Hydracarinen Schwedens. Zool. Bidrag, Uppsala, V 11: 185–540, 1927.

Lundblad, O.: Zur Kenntnis von *Lundbladia petrophila* (Michael) und den verschiedenen Entwicklunsrichtungen bei Thyasinen. Zool. Bidrag, Uppsala, V 14: 219–252, 1933.

Motas, C., Tanasachi, J., Orghidan, T.: Sur un Hydracarien nouveau de la sous-famille des Thyasinae (*Placothyopsis fibulata* n.g., n. sp.) et sur l'homologie des formations cuticulaires dans ce group. Ark. Zool., V 11: 167–185, 1957.

Viets, K.: Die Milben des Süsswassers und des Meeres. Jena, VEB Gustav Fischer Verlag. 1956.

THE GROUPS OF WATER MITES (HYDRACHNELLAE) IN PONDS AFTER PEAT-DIGGING

H. BAZAN – STRZELECKA
University of Lodź, Lodź, Poland

The purpose of this work was to analyze the population of water mites in peat-ponds, the seasonal and 24-hour variations in the formation of groups having been taken into consideration.

The study comprised three ponds selected from the peatbog at Konopnica, Wieluń district. They differed in the size, depth, and the species of plants growing in them. In each pond three areas were marked out from each of which a series of samples was collected. Water mite specimens were taken during the vegetative seasons 1961 - 1962. The samples were collected with about a fortnight's interval between each series. Moreover, in pond II observations of three 24-hours cycles were made. In each 24-hour cycle specimens were taken 5 times with an 8-hour interval between each capture.

A total of 7680 specimens of water mites was collected and determined. The occurrence of 28 species was recorded, most of which have already been noted in another paper on water mites in peat-ponds (Bazan - Strzelecka, 1963). Among the species which were not noted in that paper the following ones were now found: *L. connata, L. fulgida, F. musculus, U. figuralis, N. vernalis, N. deltoides, T. barbigera, P. coccinea, M. orbiculata, A. crassicaudatus.*

An attempt was made to investigate whether there were different groups of water mites inhabiting the ponds examined. 21 species of water mites were analyzed, species occurring only occasionally being disregarded. In the material examined three groups of species were differentiated:

Group A comprised five species of water mites - which makes 23.5 % of the total number of species studied taken in pond I only: *P. lutescens*, which outnumbered all the others found in this locality, *F. musculus* found in a somewhat smaller number of specimens, and *N. limosa, U. figuralis, P. nodata.*

Group B comprised six species-28.6 per cent characteristic of pond II only: *M. orbiculata* which, like *P. lutescens* in pond I, outnumbered all the other species found in this locality, and *L. connata, L. fulgida, T. barbigera, P. coccinea* and *A. crassicaudatus.*

Group C comprised ten species - 47.9 per cent of water mites occurring in all three ponds examined. Here belonged the ones of more or less equal numerosity: *U. crassipes, H. despiciens* and *A. globator;* next came the species caught in great numbers in pond II but only in a small number in the remaining ponds: *B. ver-*

sicolor and *H. globosa;* finally the species caught in various numbers in all the ponds examined: *L. undulata, P. conglobata, A. tricuspidator, A. bicuspidator* and *L. aquatica.*

The number of specimens and the number of species in the groups of water mites inhabiting the ponds examined show visible variations during the vegetative season. The course of these variations emphasized distinctness or similarity between the particular ponds. A similar course of variation in the numbers of individuals was recorded in ponds I and II; in the latter the number of individuals was much greater. The maxima of the number of species did not correspond with the maxima of the number of individuals. Most species were found in pond II. These variations followed a different course in pond III, a shift of the climax by a month being recorded. The maximum of the number of species agreed with the maximum of the number of individuals.

The analysis of the process of dominance also indicated the dissimilarity of particular ponds. In pond I *H. despiciens* dominated first, then *P. lutenscens* followed by disappearance of dominance in September. In pond II *M. orbiculata* dominated. From the middle of July the picture of dominance was more confused; *P. coccinea* slightly dominated over *L. undulata,* which began to dominate from 18 August 1961 and continued to dominate till the end of the period studied – with an exception when on 2 Sept. 1961 *A. globator* dominated slightly. In pond III the dominants were *U. crassipes* and *A. globator;* afterwards the number of individuals was too small to observe the occurrence of dominance.

It can be generally stated that there are great differences between the groups of water mites inhabiting the ponds examined. These are manifest in the number of species, in the number of individuals within the species and in the dominance of certain species in the groups found in the ponds.

The analysis of the 24-hour variations in the groups of water mites covered three 24-hour cycles. During the captures of the mentioned cycles made by day the atmospheric conditions were nearing fine or rather fine – whereas conditions were varied during the night. In July the night was windless and there was very bright moon light; on 2 Aug. it was very dark and it was raining; on 17 Aug. the night was rather bright and windless.

The number of water mites taken at night was as a rule greater than the number of those taken during the day. In ponds I and II this regularity was recorded in all the captures of the first 24-hour cycle, whereas in pond III the number of individuals taken during the day in the cycle under discussion equalled the number of those taken at night. In the second 24-hour cycle the number of individuals taken during the day in ponds I and II nearly equalled the number of those taken at night. In pond III the number of individuals taken during the day was only slightly greater than that taken at night. Finally, in the third cycle the number of individuals taken at night from all three ponds examined was markedly greater.

As a rule the number of particular species taken at night was greater than that taken during the day, the index was greater than one. Especially high indices were shown in the first cycle by *H. despiciens* – a nearly eightfold growth in number; in the second cycle by *U. crassipes* a sixfold growth in number; in the third cycle by *A. globator* a fivefold growth in number.

However no variations of dominance in 24-hour cycle were recorded. A similar occurrence of differences in the number of plankton crustaceans taken during the day and at night in the Charzykowo lake was observed by Patalas (1954) and in the Tajty lake by Rybak (1960). The authors interpreted this phenomenon in a similar way, assuming that crustaceans are more active during the day, being affected by light and hence possibly actively avoiding capturing apparatus.

As the members of particular families of water mites display considerable differentiation in swimming skill, active escape from the dipper is at least for most of species impossible. It applies first of all to the inefficient swimming forms. The differences between the day and the night captures may therefore supposed to be due to different locations in the biotope at the different times of the 24-hour cycle. It is possible that during the day water mites mostly inhabit plants or bury themselves in the mud while at night they stay in the free water zone and hence the number of water mites taken at night increases. Of course, this interpretation requires verification by further investigation and study.

References

Bazan-Strzelecka, H.: Attempts at an analysis of groupings of water mites in an ox-bow lake and a pond after peat-digging. Ekol. Pol., 9: 521-530, 1963.

Patalas, K.: Ilościowe badania nad zmianami w rozmieszczeniu skorupiaków planktonowych w j. Charzykowo. Arch. Hydrobiol. 2: 73-89, 1954.

Rybak, J.: Rozmieszczenie skorupiaków plankt. w litoralu i pelagialu. Ekol. Pol. 8: 38-56, 1960.

A NEW METHOD FOR TRAPPING WATER-MITES IN THE BENTHOS OF A LAKE

J. C. CONROY
Department of Biology, University of Winnipeg, Winnipeg, Manitoba, Canada

In trying to assess the interspecific relationships set up between water-mites in Marion Lake, British Columbia, several trapping methods were used. Of these methods available none were found to be adequate for the problem under investigation and a new method of sampling the benthic water-mites was devised.

Description of Trap

The trap consisted of a base formed by a 61 cm square frame over which was nailed wire mesh with a central circle, 35.7 cm in diameter, removed. The base supports an inverted plastic funnel which is held 10 cm above the mud-water interface. The funnels were designed as insect emergence traps by Hamilton (1965) who had modified an earlier suggested design of Morgan et al. (1963). The funnels consisted of "transparent vinyl plastic, 0.020 inches thick, which had a base diameter of 35.7 cm, and which covered an area of 0.1 square meters. All seams were heat sealed. Eight ounce jars were used as collecting vessels. They were attached to the funnels by means of bakelite lids, 5.9 cm in diameter, each of which had a hole, 4.5 cm in diameter, cut in it. A glass jar, with the same outside diameter as that of the bakelite lids, was forced through each funnel after the smaller ends of the funnel had been immersed in boiling water. A lip was made on each of the smaller funnels by placing the water-heated plastic on a flat surface and pressing the mouth of the emergence jars over each funnel. This was best accomplished by rotating the jars as pressure was applied. A #32 stainless steel hose clamp was used to attach the bakelite lid to the funnel" (Hamilton 1965). The funnel was then secured to the base by heavy-gauge wire. (See Figs. 1, 2.)

In operation the trap was placed on the lake bottom and its position marked by a numbered styrofoam float, which was anchored in position by a heavy nylon line. Sufficient line was used to ensure that fluctuations in the lake level will not cause the trap to float away. The collecting jar was filled with water. To collect a sample the trap was raised to just below the surface of the water, inverted while still below the water to ensure no loss of water from the collecting jar, and then raised into the boat. Excess water, still in the large plastic funnel was decanted off before the collecting jar was unscrewed, numbered and capped. A new jar filled with water,

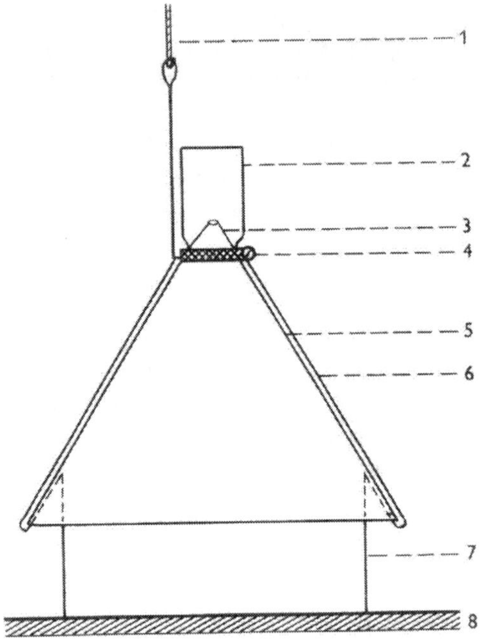

Fig. 1. 1 – line to float; 2 – 8oz jar; 3 – small cone; 4 – 32 clamp; 5 – large cone; 6 – supporting wire attached to clamp; 7 – supporting wire to (8) base

with a small cone in situ, was then put in its place. The trap was then returned to the lake bottom. The length of time the trap was kept on the lake bottom depends on the type of lake being studied.

In operation, it was found that algae tended to accumulate on the perspex funnels. It is extremely important to keep the perspex free of algal growth as this will soon render it opaque, and thus impair the trap's efficiency. It is for this reason that regular cleaning of the cones be carried out, to reduce the opportunity for attachment and growth of algae.

Results

Tests run comparing the bottom traps, herein described, with various other methods of trapping water-mites in the benthos of a lake showed that the new traps caught more mites/trap or mites/square meter than did any other method used. Pieczynski traps (Pieczynski, 1961) yielded 4.1 - 32.6 mites/square meter from Marion Lake, while the new traps during the same time period yielded 25.0 - 679.0 mites/square meter.

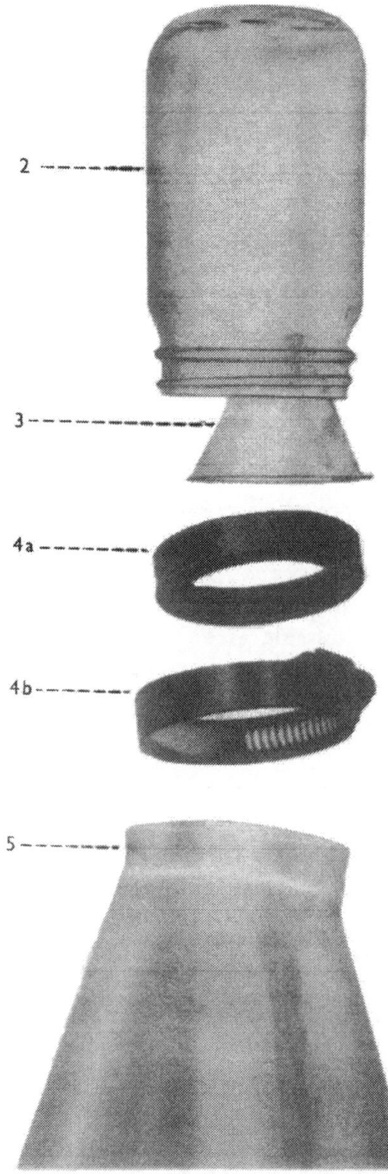

Fig. 2. 2 – 8oz; jar; 3 – small cone; 4a – lid; 4b – clamp; 5 – large cone

Eleven species of water-mite were recorded using Pieczynski traps, while twenty species were recorded from the new traps during the same period. Two of the species recorded from the Pieczynski traps were not taken in the new traps. One of these, however, *Piona interrupta*, was recorded many times in the new traps later on in the survey, while the other, *Lebertia martesensis*, was only recorded on the one occasion throughout the three years sampling of the lake .

A test was run to see if water-mites were either entering or leaving the traps as they were raised or lowered to the lake bottom. In fiftyeight consecutive lowerings and raisings of the traps no mites were recorded as entering the traps. Because the traps sink fairly slowly to the bottom, it is my opinion that the mites are able to swim away in time and so don't get trapped during this phase of the operation. In checking to see if mites were lost as the trap was raised to the surface, twenty-five mites were placed in the collecting jar which was then attached in place to the trap. A muslin cloth was tied across the mouth of the large funnel to prevent any mites entering the trap and the trap was lowered to the bottom, where it remained for a minute. The trap was then raised to the surface and the number of mites remaining in the collecting jar was checked. In one hundred repetitions of this test, only one mite left the collecting jar. It was concluded that very few mites are lost during the raising operation.

Tests were also run on the traps to check the hypothesis that traps changed every day catch the same number of water-mites as do traps changed ever 2, 3 or 4 days. No significant differences were noted and it was concluded that there was no reason to suppose that the effectiveness of the traps drop off with the length of time in the water (see Table 1). Similar results were noted by Pieczynski for his traps (Pieczynski, 1965). Using the formula given by Pieczynski (op. cit.) for the calculation of the standard error, values for the new traps of 7.82, 2.98, and 11.09 were obtained.

Discussion

Three different variations of the Pieczynski traps were tested (i.e. with funnels of 15 cm diameter; same with formalin; and with funnels of 25 cm diameter) but none of them were particularly successful when compared to the new traps, which caught not only more mites/square meter but also caught more species. Because of this discrepancy in the numbers of species caught by the two trap-types (11 in the Pieczynski, and 20 in the new traps), I concluded that the Pieczynski traps did not offer an equal opportunity to be captured to all species of water-mite. In comparing the numbers of mites caught by the new traps with the numbers reported by Pieczynski (1965), and in comparing the standard error values for the two trap-types, Table 2 was obtained. The discrepancy in catch sizes may explain the lower standard error values for the new traps.

Because mobility plays such an important part in trapping mites (using either trapping method) it is obvious that setting the new traps at 10 cm above the mud-water interface may appear to bias the catches with respect to the more active and abundant species. Tests were made comparing the traps set at 10 cm with those set at 5 cm. No statistical difference was noted between these two heights either in numbers of mites caught/trap, in the relative abundance of the different species, or in order of species dominance. Dip-net sweeps and cylinder samples gave slightly

Table 1. Changes in the numbers of water-mite individuals and species, and dominance relations (in %), in relation to length of time traps remained in the lake

Exposure time (days)	1	+ 1	+ 1	+ 1	4
# of traps	10	10	10	10	40
Avr. no./trap per day	21.7 23.1 58.5	21.4 23.3 58.9	21.6 22.7 63.0	21.7 22.8 62.9	21.75 (1) 23.4 (2) 61.75 (3)
Total # species	9 11 9	10 9 10	12 8 12	15 8 16	17 (1) 12 (2) 19 (3)
Dominant mite species Test 1	A: 76.5 B: 12.4 C: 5.1 D: 3.2	A: 77.1 B: 12.6 C: 5.1 D: 2.8	A: 66.7 B: 16.2 C: 8.3 D: 8.3	A: 66.4 B: 16.6 C: 7.8 C: 7.8	A: 71.4 B: 14.7 C: 6.7 D: 6.1
Test 2	A: 60.2 D: 16.9 B: 15.6 C: 3.9	A: 60.1 D: 16.7 B: 15.0 C: 3.9	A: 40.1 B: 32.6 D: 16.7 C: 6.2	A: 40.4 B: 32.9 D: 17.1 C: 6.7	A: 49.7 B: 23.9 D: 16.9 C: 5.3
Test 3	A: 86.3 D: 2.7 B: 2.2 G: 2.1	A: 87.3 D: 2.7 B: 2.0 G: 2.0	A: 83.0 B: 7.3 E: 2.5 F: 1.7	A: 83.1 B: 7.3 E: 2.7 F: 1.7	A: 83.6 B: 4.7 E: 1.9 F: 1.7

A: *Unionicola crassipes;* B: *Piona carnea;* C: *Arrenurus cascadensis;* D: *Piona constricta;* E: *Unionicola gracilipalpis;* F: *Arrenurus pseudocylindratus;* G: *Piona debilis.*

different results with *Piona carnea* and *P. constricta* being the dominant mites as opposed to *Unionicola crassipes, P. carnea,* and *P. constricta,* the dominant forms from the bottom traps. Pieczynski (1969) points out that *U. crassipes* is a more active form at night and hence will be more prevalent in sweeps taken at night than in the day. A 24-hour series run continuously for three days in August 1969 showed that in Marion Lake *U. crassipes* was active at night, and less active during the day, while both *P. carnea* and *P. constricta* were more active during the daylight hours. Net sweeps made at night showed *U. crassipes* as dominant and *P. carnea* and *P. constricta* as being subdominants. Pieczynski (1969) reports *U. crassipes* as being mainly a bottom dweller (98 % of specimens being recorded from the bot-

Table 2. Standard error values comparison between the Pieczynski and the Conroy trap

Author	PIECZYNSKI**						CONROY		
Area	Medium			Large					
	78.5 sq.cm			176.7 sq.cm			1000 sq.cm		
Indicator	SEV	Avr/trap	Avr/sq.m	SEV	Avr/trap	Avr/sq.m	SEV	Avr/trap	Avr/sq.m
Test 1	10	5.8	738.9	33	9.0	509.3	7.82	38.52	385.2
Test 2	32	12.0	1528.8	80	26.1	1477.3	2.98	16.26	162.6
Test 3	14	5.0	636.9	22	9.6	543.6	11.09	66.58	665.8

** Data from Pieczynski, 1965.
SEV – Standard error value.

tom). This is in complete contrast to the Marion Lake data which showed *U. crassipes* as being the dominant water-mite in plankton samples (90 %, 8712 individuals) and also as being the dominant form in the bottom traps (62.6 %, 25,433 individuals).

In considering the efficiency of bottom traps, one should refer to the observations of Morgan et al.(1963) who noted two potential losses in efficiency - the possibility of swamping of the emergency-type cone traps (not applicable here) and also the shading effect of the trap itself on ascending mites. They point out however that it is probably only those animals that ascend near the edges of the trap that can take successful avoiding action. It is also noted by these workers that the larger the area of the trap, the smaller this edge-effect relative to the size of catch. Because the trap diameter of 35.7 cm was the largest possible from vinyl plastic sheets available, it was considered to meet the edge-effect requirement as best as possible.

References

Hamilton, A.L.: An analysis of a fresh-water benthic community with special reference to the Chironomidae. Ph. D. Thesis, The University of British Columbia, Vancouver, B.C., 1965.

Morgan, N.C., Hall, W. B., Waddell, A.B.: A comparison of the catches of emerging aquatic insects in floating box and submerged funnel traps. J. Anim. Ecol. 32: 203--219, 1963.

Pieczynski, E.: The trap method of capturing water mites (Hydracarina). Ekol. Pol., B, 7: 111-115, 1961.

Pieczynski, E.: Effect of the size of traps and their exposure time on the catches of aquatic invertebrates, chiefly water mites (Hydracarina). Bull. Acad. Pol. Sci., Cl. II, 13: 577-581, 1965.

Pieczynski, E.: The trap method for ecological studies on water mites (Hydracarina) in lakes. Proc. 2nd Int. Congr. Acarology, 1967: 103-106, 1969.

THE WATER-MITES (HYDRACHNELLAE, ACARI) OF A STONY STREAM

T. GLEDHILL
Freshwater Biological Association
The River Laboratory, East Stoke, Wareham, Dorset, England

During 1959 and 1960 the author collected water-mites from Ford Wood Beck, a small stony stream in the English Lake District which has been the subject of several studies (Macan, 1957 a, b and 1958; Mackereth, 1957 and 1960).

Twenty-eight species from 17 genera belonging to 12 families have been identified. Data on the species found and their distribution throughout the stream are presented.

The stream and sampling stations

The stream and most of the sampling stations (Fig. 1) have been fully described by Macan (1957) but the three stations A, B & C have been added by the writer. Briefly, Ford Wood Beck may be described as a small, stony, moorland stream flowing into Blelham Tarn which is situated on the north west side of Windermere. The stream is less than 2km long and nowhere more than 2m wide. The depth varies from a few centimetres to 60cm or more after heavy rain and the stream bed is stony throughout most of its length. The lower reach of the stream flows through a valley.

Station 1 is near the mouth of the stream, is about 2m wide and has stones of varying sizes overlying gravel and sand.

Station A is an area where the stream flows over a rocky outcrop colonized by *Fontinalis* and *Eurbyncium rusciforme* Milde.

Station B is a somewhat narrow reach of the stream in the lower part of the upper basin of the valley. The bottom consists of small stones overlying gravel. *Sparganium ramosum* Huds. subsp. *neglectum* (Beeby) Schinz & Thell is able to grow here and *Callitriche* has become established in one or two places where some silting occurs.

Station 2 is situated towards the upper end of the valley and is similar to Station 1.

Station 3 lies at the foot of the steep part of the stream where cascades alternate with pools full of stones and moss-covered boulders.

Station 4 is a narrow and shallow reach just below the source where although the gradient is still fairly steep, the boulder clay has not been eroded as it has at Station 3.

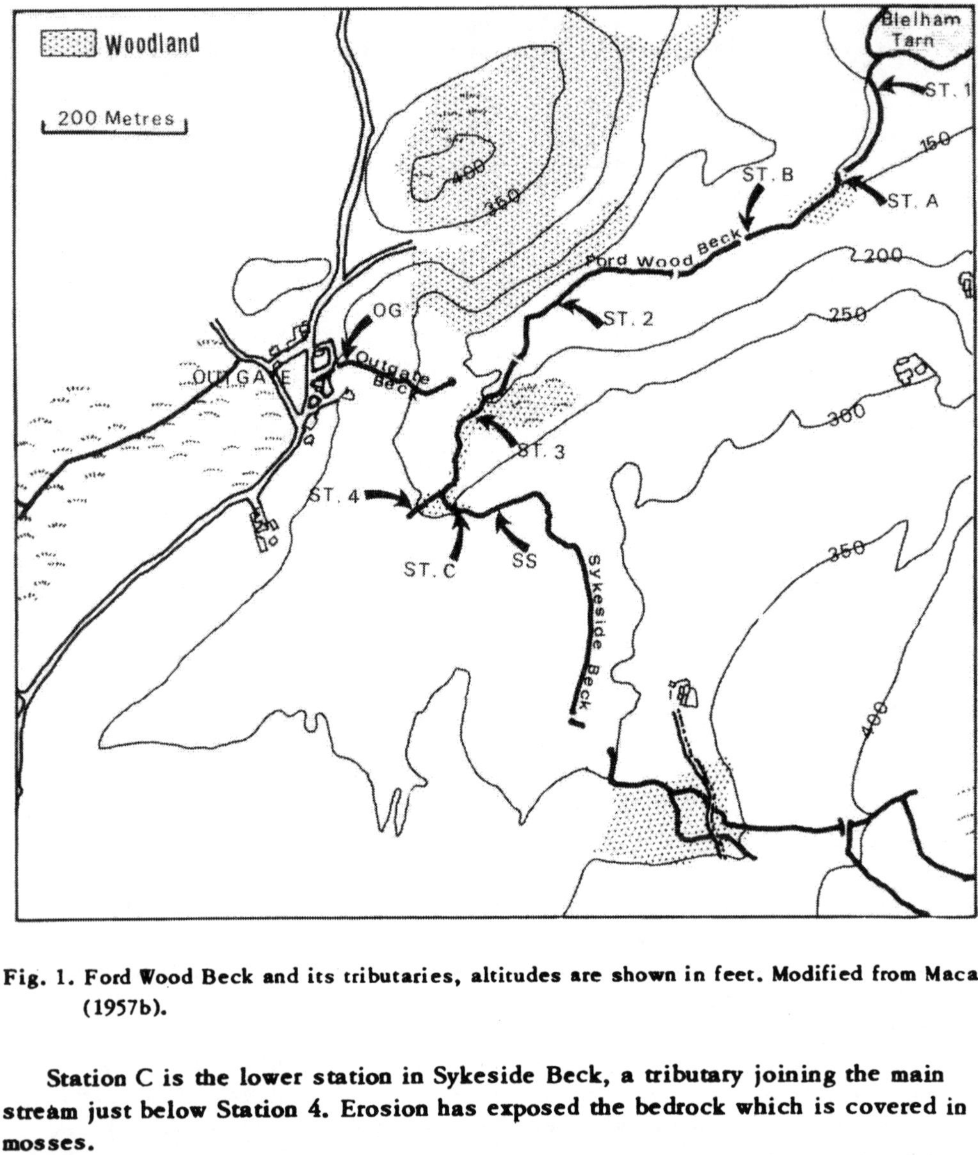

Fig. 1. Ford Wood Beck and its tributaries, altitudes are shown in feet. Modified from Macan (1957b).

Station C is the lower station in Sykeside Beck, a tributary joining the main stream just below Station 4. Erosion has exposed the bedrock which is covered in mosses.

Station SS in Sykeside Beck is situated just above Station C but unlike the latter, is not shaded by trees. The stream is narrow, c. 45cm, the current moderately swift and the bottom composed of stones of various sizes. The stream here may dry up during a fine summer.

Station OG in Outgate Beck, the other main tributary, is in an artificial walled channel. Flow is relatively slight and the stones small to medium sized.

Temperature

Water temperatures at the downstream end of an underground channel in Outgate Beck (just upstream of Station OG) were recorded for a period of 5 years (Macan, 1958). Winter temperatures were generally between 0°C and 6°C rising to 16-17°C in July and August (max. 19°C). Close to the mouth of Ford Wood Beck (Station 1) the temperature fluctuations were greater and the maximum temperature (maximum temperature recorded 21.5°C) always higher than at Outgate, whilst at Station 4 (close to the source of Ford Wood Beck) fluctuations were less than at Outgate, the temperature being higher in winter and lower in summer (Macan, op. cit.).

Methods

Small stones and gravel were scooped up with an enamel mug and passed through sieves of 1mm and 250 μm aperture. Some of the associated vegetation was removed and thoroughly washed over the sieves. Sampling at each station lasted until a suitable amount of material had accumulated in the fine aperture sieve. This material was then washed into a container and examined in the laboratory.

Species found

Twenty-eight species have been identified and these are listed in Table 1. One species, *Hydrovolzia placophora*, was recorded in 1958.

Numbers

Sampling was only possible at irregular intervals, during April, May, June, July, September and November in 1959 and during April, May, September and October in 1960, and it was not always possible to take samples from every Station during a sampling period. The results presented therefore refer only to those occasions on which samples were obtained and do not infer absence of species in the stream at other times of the year. Basic data on the numbers of each species found and the number of samples taken are presented in Table 2. More species and a greater number of individuals were found from the lower stations during both years and samples from these stations during 1960, although fewer than in 1959, yielded more species and more individuals than in the previous year. Excluding the Hydrovolziidae, 11 families were represented and the composition of each of these expressed as a percentage of the total number of mites found was:
Aturidae - 79.3 %; Sperchonidae - 7.1 %; Lebertiidae - 4.0 %; Thyasidae - 3.0 %; Hygrobatidae - 2.9 %; Axonopsidae - 2.4 %; Protziidae - 0.4 %; Torrenticolidae

Table 1. Families and species found in Ford Wood Beck and the tributaries Sykeside Beck and Outgate Beck

Hydrovolziidae	*Hydrovolzia (Hydrovolzia) placophora* (Monti 1905)
Protziidae	*Protzia eximia* (Protz 1896)
Thyasidae	*Panisus torrenticolus torrenticolus* Piersig 1898 *Panisellus thienemanni* (Viets 1920)
Sperchonidae	*Sperchon (Porosperchon) brevirostris brevirostris* (Koenike 1899) *Sperchon (Porosperchon) glandulosus glandulosus* (Koenike 1886) *Sperchon (Hispidosperchon) denticulatus* (Koenike 1895) *Sperchon (Hispidosperchon) clupeifer* (Piersig 1896) *Sperchon (Mixosperchon) setiger setiger* (Thor 1898) *Sperchonopsis verrucosa verrucosa* (Protz 1896)
Lebertiidae	*Lebertia (Lebertia) fimbriata fimbriata* Thor 1899 *Lebertia (Pilolebertia) insignis* (Neuman 1880) *Lebertia (Pseudolebertia) glabra* Thor 1897
Torrenticolidae	*Torrenticola (Torrenticola) anomala* (Koch 1837)
Mamersopsidae	*Bandakia concreta concreta* Thor 1913
Hygrobatidae	*Hygrobates (Hygrobates) fluviatilis* (Ström 1768) *Hygrobates (Hygrobates) nigromaculatus* Lebert 1879 *Atractides (Atractides) gibberipalpis* Piersig 1898 *Atractides (Atractides) nodipalpis nodipalpis* (Thor 1899) *Atractides (Atractides) tener* (Thor 1899) *Atractides (Octomegapus) octoporus* (Piersig 1904)
Feltriidae	*Feltria (Feltria) romijni* Besseling 1930
Pionidae	*Wettina podagrica* (Koch 1837)
Axonopsidae	*Vietsaxona lundbladi* (Motas and Tanasachi 1947) *Ljania bipapillata bipapillata* Thor 1898
Aturidae	*Aturus (Aturus) scaber scaber* Kramer 1875 *Aturus (Aturus) brachypus* Viets 1936 *Kongsbergia (Kongsbergia) materna* Thor 1899

0.3 %; Feltriidae 0.3 %; Pionidae - 0.1 % and Mamersopsidae 0.03 %. Of the 27 species found only 8 exceeded 1 % of the total catch for the years 1959, 1960. These species are shown in Table 3.

Notes on the species

Hydrovolzia placophora (Monti). One male specimen was found in March 1958 in a sample from Station OG in Outgate Beck taken by Dr T. T. Macan. This species was redescribed, figured and discussed by Gledhill (1960).

Protzia eximia (Protz). A species found at those stations which had aquatic mosses attached to stones and rocks although it was absent from Station 3. Ovi-

Table 2. Numbers of each species found at the different sampling stations during 1959 and 1960

SPECIES	1959 Stations 1	A	B	2	3	4	C	SS	OG	Total	1960 Stations 1	A	B	2	3	4	C	SS	OG	Total	Total 1959–60
Aturus scaber	287	774	109	26	84	–	28	6	11	1325	304	918	167	9	24	–	14	2	2	1440	2765
Sperchon clupeifer	21	33	8	–	1	–	–	–	–	63	6	49	9	1	–	–	–	–	–	65	128
Lebertia glabra	3	14	2	–	–	6	2	5	2	34	10	59	13	1	2	–	1	5	1	92	126
Panisus torrenticolus	–	16	–	–	–	–	1	–	–	17	6	76	1	–	1	–	5	1	–	90	107
Ljania bipapillata	3	13	1	2	4	1	6	–	9	39	1	13	3	1	1	–	–	1	1	21	60
Sperchonopsis verrucosa	3	12	1	–	13	–	–	–	1	30	11	11	4	–	2	–	–	–	–	28	58
Kongsbergia materna	–	13	–	–	4	–	–	–	–	17	6	20	–	–	–	–	1	–	–	27	44
Sperchon setiger	–	8	5	–	1	–	–	–	–	14	–	14	11	3	–	–	–	–	–	28	42
Atractides nodipalpis	1	–	2	–	2	1	–	–	1	7	3	17	6	–	–	–	–	–	–	26	33
Hygrobates fluviatilis	1	1	–	–	4	1	–	–	–	7	2	14	–	3	–	–	–	–	–	19	26
Vietsaxona lundbladi	8	–	2	2	–	–	–	–	–	12	8	–	–	4	–	–	–	–	–	12	24
Atractides tener	4	3	–	–	1	–	–	–	1	9	–	9	2	1	2	–	–	–	–	14	23
Sperchon glandulosus	–	4	–	–	–	–	–	–	–	4	4	9	6	–	–	–	–	–	–	19	23
Protzia eximia	1	5	–	–	–	–	–	–	–	6	–	3	4	1	–	–	1	–	–	9	15
Lebertia fimbriata	2	2	2	–	–	–	–	–	–	6	2	4	2	–	–	–	–	–	–	8	14
Aturus brachypus	5	1	1	–	–	–	–	–	–	7	3	–	2	–	–	–	–	–	–	5	12
Feltria romijni	2	2	–	–	–	–	–	–	–	4	3	5	–	–	–	–	–	–	–	8	12
Atract. gibberipalpis	–	3	–	–	–	–	–	–	–	3	1	8	–	–	–	–	–	–	–	9	12
Torrenticola anomala	–	6	–	–	–	–	–	–	–	6	2	1	2	–	–	–	–	–	–	5	11
Atractides octoporus	2	–	–	–	–	–	–	–	–	2	1	5	1	–	–	–	–	–	–	7	9
Wettina podagrica	–	–	–	–	–	3	–	–	1	4	–	–	–	–	–	–	–	–	–	–	4
Lebertia insignis	–	–	1	–	–	–	–	–	–	1	1	1	–	–	–	–	–	–	–	2	3
Sperchon brevirostris	–	–	–	–	–	–	–	–	–	–	2	–	–	–	–	–	–	–	–	2	2
Panisellus thienemanni	–	–	–	–	–	–	–	–	–	–	–	–	–	1	–	–	–	–	–	1	1
Hygrob. nigromaculatus	–	–	–	–	–	–	–	–	–	–	–	–	1	–	–	–	–	–	–	1	1
Sperchon denticulatus	–	–	–	–	–	–	–	–	–	–	–	1	–	–	–	–	–	–	–	1	1
Bandakia concreta	–	–	–	–	–	–	–	–	–	–	1	–	–	–	–	–	–	–	–	1	1
Hydrovolzia placophora	–	1 specimen coll. OG 1958																			
TOTAL	343	910	134	30	114	12	37	11	26	1617	377	1237	234	25	32	Nil	22	9	4	1940	3557
No. of species found	14	17	11	3	9	5	4	2	7	22	20	20	16	10	6	–	5	4	3	26	27
No. samples+specimens	5	4	3	2	3	2	2	2	3	26	3	3	3	2	1	–	1	1	1	15	
No. of samples taken	5	4	3	2	3	3	2	2	3	27	3	3	3	2	1	1	1	1	1	16	

Table 3. Species exceeding 1.0 % of the total catch for the years 1959 and 1960

	1959	1960	1959/1960
Aturus scaber	81.94	74.22	77.73
Sperchon clupeifer	3.90	3.35	3.59
Lebertia glabra	2.10	4.74	3.58
Panisus torrenticolus	1.05	4.63	3.00
Ljania bipapillata	2.41	1.08	1.68
Sperchonopsis verrucosa	1.86	1.44	1.63
Kongsbergia materna	1.05	1.39	1.23
Sperchon setiger	0.86	1.44	1.18

gerous females were found during April, May and October and nymphs during September. Lundblad (1968) records nymphs during May, July and October from some Swedish streams.

Panisus torrenticolus torrenticolus Piersig. Specimens occurred at those stations where aquatic mosses grew on the bedrock and were especially numerous where the stream formed cascades. Ovigerous females were found during April, September and October and some immature females occurred during April.

Penisellus thienemanni (Viets). A single male specimen was found at Station 2 in October 1960. It has been recorded from another stream, Holbeck, Westmorland and from a woodland helocrene in the English Lake District (Gledhill, 1960). Lundblad (1968) records specimens from limnocrenes, helocrenes and also from a spring-stream in Sweden.

Sperchon brevirostris brevirostris (Koenike). A species only taken twice: ♂, Station 1. 7.9. 1960; ♂, Station 1. 24.10. 1960. ♀

Sperchon glandulosus glandulosus (Koenike). The number of specimens caught (23) comprised only 9.1 % of all the sperchonids found. It was only present at the lower stations. Ovigerous females were observed in April.

Sperchon clupeifer (Piersig). The most common sperchonid comprising 50 % of all specimens found. Ovigerous females were found in April and nymphs were found in September and October. All males examined possessed a somewhat triangular chitinous plate on either side of the excretory pore as described for *Sperchon (Scutosperchon) thori* Koenike. This latter species is considered by Lundblad (1962) to be synonymous with *S. clupeifer.*

Sperchon denticulatus (Koenike). An immature female found at Station A in October 1960 was identified to this species.

Sperchon setiger setiger (Thor). Of the sperchonids found, 16.5 % belonged to this species. Ovigerous females were not observed but nymphs were present during October and November.

Sperchonopsis verrucosa verrucosa (Protz). One of the more common sperchonids, accounting for 22.8 % of those found. Immature females were found in April and October, ovigerous females in April and November and nymphs during April, May, June, September and October.

Lebertia fimbriata fimbriata Thor. A species found only at the lower stations. Nymphs were present during April, October and November.

Lebertia insignis (Neuman). Only 3 males of this species were found.

Lebertia glabra Thor. The most common species of *Lebertia* and recorded from every sampling station. Immature males were found in April and November; immature females in April, May and November; ovigerous females occurred in April and May and nymphs were present during April, September and October. The size range of 45 nymphs was 327 μm - 663 μm (mean 454 μm). The life-history of *L. glabra* has been studied by Efford (1966).

Torrenticola anomala (Koch). Although torrenticolid mites are very common in many streams and rivers of the Lake District they were scarce in Ford Wood Beck. Immature females and nymphs were recorded in September and October.

Bandakia concreta concreta Thor. A single female specimen was recorded from Station 1 in April 1960. Investigations on the distribution of interstitial (hyporheic, phreaticolous) water-mites (Gledhill, unpublished records) show this species to be a fairly regular member of this biotope. It is also recorded from limnocrenes and helocrenes (Gledhill, 1960; Hopkins, 1962).

Hygrobates fluviatilis (Ström). A species comprising 25 % of all the Hygrobatidae found. Nymphs were recorded during April, June, September, October and November. Lundblad (1968) recorded nymphs in May, June and October.

Hygrobates nigromaculatus Lebert. One nymph was recorded from Station B in October 1960.

Atractides gibberipalpis Piersig. Of the Hygrobatidae, 11.5 % belonged to this species as did 15.6 % of the *Atractides* spp. Nymphs were not recorded.

Atractides nodipalpis nodipalpis (Thor). The most numerous of the Hygrobatidae (31.7 %) and the most common *Atractides* species (42.9 %). Ovigerous females occurred in September and October. The number of eggs present in each of 5 females was counted and varied from 2 - 16 (mean 7.0). Nymphs were recorded during September. The life-history of this species has been studied by Efford (1966).

Atractides tener (Thor). Some 22.1 % of the Hygrobatidae belonged to this species and 29.9 % of the *Atractides* spp. Immature females were present in October and ovigerous females in April. Nymphs were recorded during April and October.

Atractides octoporus (Piersig). This species accounted for only 8.7 % of the Hygrobatidae and 11.6 % of the *Atractides* spp. Two ovigerous females were recorded, 1 in April 1960 and the other in September 1960. The September specimen contained 17 eggs. Nymphs were not recorded.

Feltria romijni Besseling. The taxonomy, distribution and habitat of this species were studied by Efford (1962) and data on its ecology have been presented by the same author (Efford, 1965). *F. romijni* was the only feltriid present in the stream. *Feltria minuta* is common in many of the fast flowing streams at higher altitudes and is also numerous in several cold *Sphagnum* springs where it is found with *Hydrovolzia placophora* (Gledhill, unpublished records). Two ovigerous females of *F. romijni* were found at Station A in April 1960.

Wettina podagrica (Koch). Station 4 and Station OG in Outgate Beck were the only sites where this species was found and it seems probable that its distribution in the stream is associated with lower water temperatures (Station 4 is close to the source and the stream at Station OG has passed through an underground channel). A male, female and nymph of *W. podagrica* were found at Station 4 in September 1959 and a nymph was found at Station OG in June 1959. *Hydrovolzia placophora*, another cold water stenotherm was also recorded from Station OG.

Vietsaxona lundbladi (Motaş and Tanasachi). A species recently added to the British fauna when it was redescribed and figured by Gledhill (1971). It occurred at those stations where the stones and gravels were more loosely packed and was absent from the cascade areas. A possible explanation is that this species may be associated with the hyporheic biotope as Gledhill (op. cit.) recorded specimens from this habitat. Two ovigerous females were recorded from Station 2 in April 1960.

Ljania bipapillata bipapillata Thor. Together with *Lebertia glabra* the only species to be recorded from all the sampling stations. Ovigerous females were present during April, June, September, October and November and the highest number of eggs recorded was 6. Nymphs occurred in April, May, June and September. Efford (1966) recorded nymphs and adults throughout the year in a small stream at Wytham Woods, Oxford. This species is common in small stony streams and has often been found in samples taken from the interstitial biotope (Gledhill, unpublished records).

Aturus scaber scaber Kramer. By far the most numerous species in the stream. Although some females examined resembled the subspecies *A. scaber rotundus* Romijn all males belonged to the former species. It was extremely common at those stations where water cascaded over moss-covered rocks. Ovigerous females were present during April and May and nymphs in April, May, June, September and October. The largest number found during 1959 occurred in September at Station A (80 ♂♂, 337 ♀♀, 17 nymphs) and in 1960 Station A again produced most but during April (14 ♂♂, 503 ♀♀, 37 nymphs). The overall ratio of males to females throughout the study period was 1 : 6.

Aturus brachypus Viets. A species not previously recorded from Britain although it has been recorded from Ireland (Halbert, 1944). *A. brachypus* has been found on several occasions in samples from the hyporheic biotope (Gledhill, unpublished records) and it is possible that in Britain this is the species' preferred habitat. Ovigerous females were present during April.

Kongsbergia materna Thor. Another species showing a preference for mossy cascade areas of the stream. Ovigerous females were found in April and May, all possessed a single egg.

Acknowlegements. I thank Dr T. T. Macan for his advice and encouragement when I began studying the water-mites. I also thank Dr C. Pinder for reading the manuscript and Mrs Peggy Davey for typing services.

References

Efford, I. E.: The Taxonomy, Distribution and Habitat of the Watermite, *Feltria romijni* Besseling, 1930. Hydrobiologia 19: 161-178, 1962.

Efford, I. E.: Ecology of the watermite *Feltria romijni* Besseling. J. Anim. Ecol. 34: 233-251, 1965.

Efford, I. E.: Observations on the life history of three streamdwelling watemites. Acarologia 8: 86-93, 1966.

Gledhill, T.: Some Water-mites (Hydrachnellae) from Seepage-water. J. Queckett Micr. Cl. 5: 293-307, 1960.

Gledhill, T.: The genera *Azugofeltria, Vietsaxona, Neoacarus* and *Hungarohydracarus* (Hydrachnellae, Acari) from the interstitial habitat in Britain. Freshwat. Biol. 1: 61--82, 1971.

Halbert, J. N.: List of Irish fresh-water mites (Hydracarina). Proc. Roy. Ir. Acad. 50: 39-104, 1944.

Hopkins, C.L.: The distribution of Hydracarina in the vicinity of Flatford Mill Field Centre, East Suffolk, Field Studies 4: 53-72, 1962.

Lundblad, O.: Die Hydracarinen Schwedens. II. Ark. Zool. 14: 1-635, 1962.

Lundblad, O.: Die Hydracarinen Schwedens. III. Ark. Zool. 21; 1-633, 1968.

Macan, T. T.: The Life Histories and migrations of the Ephemeroptera in a stony stream. Trans. Soc. Brit. Entomol. 12: 129-156, 1957 a.

Macan, T. T.: The Ephemeroptera of a stony stream. J. Anim. Ecol. 26: 317-342, 1957 b.

Macan, T. T.: The temperature of a small stony stream. Hydrobiologia 12: 89-106, 1958.

Mackereth, J. C.: Notes on the Plecoptera from a stony stream. J. Anim. Ecol. 26: 343-351, 1957.

Mackereth, J. C.: Notes on the Trichoptera of a stony stream. Proc. Roy. Entomol. Soc. London (A) 35: 17-23, 1960.

ON HALACARID MIGRATION INTO FRESHWATERS

A. PETROVA
Zoological Institute, Sofia, Bulgaria

Recent studies on subterranean littoral and continental waters in Bulgaria as well as certain facts already known enable us to explain some questions concerning the colonization of continental waters by Halacaridae,, the origin of freshwater genera and the systematic status of the Halacaridae and Limnohalacaridae. Some representatives of the genus *Halacarus* such as *H. subterraneus, H. capuzinus, H. basteri affinis, H. procerus, H. anomalus* and *H. bisulcus* serve as classical examples of subterranean migration towards continental waters. In our studies on the interstitial waters of the Black Sea littoral, we noted a gradual migration of different representatives of this genus towards the subterranean waters of the continent. Repeated sampling of the species *H. (Halacarellus) capuzinus* and *H. (Halacarellus) procerus* in different localities along the littoral zone seems to suggest that these species keep close to the coast, where they can be caught abundantly, while farther away from the coast they decrease in number. Although psammophilous, they maintain a close contact with the sea and are found less often in the "zone of mixing".. *H. (Halacarellus) subterraneus* and *H. (Halacarellus) phreaticus* behave quite differently (Petrova, 1971, 1972). These species show an even more pronounced tendency to colonize subterranean continental waters. They - abound in the interstitial waters of the littoral zone much closer to the coastline (up to 10m) and even penetrate the continental phreatic waters of wells and springs not far from the sea. Frequent sampling of these two species in several localities prompts us to assume that they are intermediate forms which have succeeded in overcoming the ecological zone of transition, i.e. the "zone of mixing" between the sea and the continental subterranean waters. An analogous case is that of *H. (Halacarellus) basteri affinis,* a species also found in a fresh-water spring near the sea, together with *H. (Halacarellus) phreaticus.* The genus *Halacarus* seems to be in "full bloom" as it contains a great variety of forms and shows extensive migratory tendencies.

A similar tendency for colonization of subterranean continental waters is observed in some representatives of *Copidognathus. Copidognathus tectiporus* is found in the coastal brackish lakes of the Black Sea, Lake Ohrid, and in the psammal of the sea. The discovery of another species of *Copidognathus,* most probably a new one, is also of considerable interest. Adults and nymphs were found some 30 km inland in a spring near Kačula, a locality in the Burgas district. This shows that the migration of this species towards continental waters is at an advanced stage. A similar species is *C. magnipalpus ponticus* (Viets, 1940), found in the coastal lake Diavolsko Blato, and also in a freshwater spring near the Black Sea coast at Kornia, near Sozopol). The representatives of this genus are localized in the Ponto-Caspian region and are most probably relicts of a fauna which populated the miocene basins, in present-day Panonia, the Black Sea, South Russian and Caspian regions. *Caspihalacarus hyrcanus* of the Caspian Sea, Lake Varna, the Danube and Dnieper Rivers and *Copidognathus oxianus* of the Aral and Caspian Seas also belong to this group of relict species.

Furthest removed from the sea are the freshwater genera *Lobohalacarus, Limnohalacarus, Soldanellonyx, Parasoldanellonyx, Porohalacarus* and *Porolohmannella*, especially *P. violacea*. Geologically these genera would seem to have lost their contact with the sea quite early. This is borne out by their wide geographic distribution, the wide ecological range and the presence of external genital acetabula which are considered to be chemoregulators. We are inclined to think that these organs are the results of a long evolution in freshwater conditions, representing an extreme form of adaptation to these conditions. The continental waters were colonized by these genera in epigeous and hypogeous ways. Sometimes an initial form marked the beginning of epigean and subterranean populations as, for example, representatives of the genera *Porolohmannella* and *Soldanellonyx*.

Coastal caves may also constitute a door for the migration of halacarids towards the subterranean waters of the continent, as is the case with *Lohmannella stammeri* (Viets, 1939).

Our knowledge of the Halacaridae and Limnohalacaridae is insufficient to form a comprehensive idea of the zoogeography, genesis and origin of these mites. The known facts suggest that despite their antiquity, Halacarids have preserved a vital potential expressed in their tendency to penetrate and occupy new biotopes. Species of *Halacarus* show recent migratory tendencies over a wide front. For instance, *H. (Halacarellus) capuzinus* and *H. (Halacarellus) subterraneus* are known as psammobionts from the North Sea, the Atlantic coast of North America and the Black Sea; however, only in the Black Sea region does *Halacarellus subterraneus* as well as the newly discovered species *Halacarellus phreaticus* penetrate the subterranean waters of the continent (wells and springs) while maintaining close contact with the sea (Petrova, 1971). Similar are those cases in which some species of the genus *Copidognathus* are found in the subterranean waters of the continent in springs near the Black Sea coast. The numerous changes that occurred in the formation of the ancient basins of the Ponto-Caspian region seem to have given a strong impetus to the processes of migration and species-formation. It is a question of intermediate forms which in their migration have barely started colonizing the continental waters. In their penetration of fresh waters, the marine species have undergone certain modifications but the wisdom of the separation of the freshwater forms from the marine forms into two independent families - Porohalacaridae and Halacaridae - is questionable. The reason for this division is the presence of external genital suckers in the Porohalacaridae which play the role of chemoregulators. These formations seem to be the result of a long evolution in freshwater conditions. However, we know that certain relict species such as *Caspihalacarus hyrcanus* and *Copidognathus tectiporus* possess intermediate characters. The former has external genital suckers, is classified as belonging in the family of freshwater Halacarids, but it also populates oligohaline waters and has passed into the freshwater of rivers, although only secondarily. Motas and Tanasachi drew our attention to the Danubian variety *Caspihalacarus hyrcanus danubialis*, which is distinguished from the nominate form by much larger posterior pairs of external genital suckers (Motas and Tanasachi, 1943). *Copidognathus tectiporus*, while actually belonging to the marine family, also has suckers on the thoracic epimera, which according to Viets perform a chemoregulating function. Other halacarid species belonging to the marine family of Halacaridae, such as *Halacarellus subterraneus*, *H. phreaticus*, *H. basteri affinis*, *Copidognathus magnipalpus ponticus*, *Copidognathus* n. sp., *Rombognathus uniscutatus*, have no external genital suckers, but live and develop normally in freshwater wells and springs in proximity to the sea. From the above it follows that one cannot draw a clearcut line between marine and freshwater Halacarids. Their common origin, the existence of intermediate forms and the absence of substantial differences between Halacaridae and Porohalacaridae show that their taxonomy is open to discussion and that consideration should be given to combining them into a common taxonomic category. Physiological and cytological research is of great importance for the proper solution of this question.

References

Motas, C., Tanasachi, J.: Un halacaride reliquat pontocaspien dans le Danube. Bull. Soc. Nat. Romania, 16: 1-7, 1943

Petrova, A.: La présence d' *Halacarellus subterraneus* Schulz et d' *Halacarellus phreaticus* n. sp. (Halacaridae, Acari) en Bulgarie. Acarologia, XIII: 2., 1971.

Petrova, A.: Sur la migration des Halacariens dans les eaux douces et la position systématiques des Halacariens et Limnohalacariens. (In press.)

Viets, K.: Halacariden (Acari) aus süditalienischen Höhlengewässern. Arch. Hydrobiol. 35, 4: 625-630, 1939.

Viets, K.: Hydrachnellae, Porohalacaridae und Halacaridae s. str. (Acari) aus Bulgarien. Zool. Anz. 130, 1/2: 36-41, 1940.

ON THE OCCURRENCE OF WATER MITES (HYDRACHNELLAE) IN THE SUBMERGED VEGETATION OF A CARP POND

P. PUNČOCHÁŘ
Hydrobiological Laboratory, Czechoslovak Academy of Sciences,
Prague, Czechoslovakia

Up to the present time little attention has been directed towards the quantitative evaluation of water mites in ponds and the seasonal dynamics of their occurrence. Most of the quantitative data on the fauna of water mites and their locomotive activity were obtained from lakes (Pieczynski, 1964, 1967, 1969; Mothes, 1967).

I therefore welcomed the possibility of working out the quantitative samples of water mites kindly supplied by dr. J. Kořínková (Department of Parasitology and Hydrobiology, Charles University, Prague) which had been collected during her study of the macrofauna from submerged plant stands in the Radov pond in south Bohemia.

The samples were taken by means of a littoral sampler consisting of a silk-covered framework (1 m x 1 m) with a folding bottom. The average mesh size was 250 μm. The height of the sampler exceeded the water depth at the sampling stations. After the sampler had been closed in the water, the plants were washed and picked out. Three graded sieves were used for the separation of the macrofauna from the vegetation and detritus. The animals were preserved and transferred to the laboratory for further treatment.

With the exception of *Brachypoda versicolor* (Müller, 1776), all species of water mites usually found in ponds in this region were present in the samples. *Brachypoda versicolor* passed through the sieves because of its small size. Smaller juvenile stages were also likely to escape.

Samples were collected from three kinds of aquatic plant stands in the Radov pond (Table 1.). In 1964, *Batrachium aquatile* (L.) Dum. was present in the pond. In the following year *Elodea canadensis* Rich. and *Potamogeton lucens L.* occurred simultaneously. The samples were always taken at several stations from the same kind of plant stand at intervals of 10 - 14 days. Further details about the methods used and the development of the macrovegetation are given by Kořínková (1971).

Table 1 also contains data on the numbers of the samples examined as well as the mean abundance of water mites in the stands. The smallest number of water mites calculated per 1 sq. m area of the overgrown bottom was found in the stand of *Elodea canadensis*. Since *Batrachium aquatile* occurred for a short

Table 1. Data on sampling stations, material, mean abundance of water mites and occurrence of juvenile stages in three kinds of plant stands in pond Radov (south Bohemia). The mean June-July values of abundance from both years are in parantheses

Kind of stand	*Batrachium aquatile*	*Potamogeton lucens*	*Elodea canadensis*
Season under study	1964 June – July	1965 May –	September
Number of sampling stations	4	2	4
Number of samples	18	20	44
Abundance of water mites per sq.m. area of the overgrown bottom	179	(160) 140	(129) 94
Abundance of water mites per g of dry plant weight	1.2	(2.4) 2.1	(0.8) 0.5
Percentage juvenile stages (nymphs + teliophane st.)	22.8	11.0	4.4

period only, the mean June-July values of abundance of water mites in stands of various plant species from both years of the study were compared. The differences between the values of this period were not statistically significant. If the entire season of growth of the plants of *Elodea* and *Potamogeton* was included, the difference was just above the 0.05 per cent level of statistical significance.

The great variability of the numbers of water mites between the sampling stations of the same plant stand also caused the statistical insignificance of the seasonal changes. Nevertheless, there was an apparent trend for a maximum abundance of water mites in June-July and a decline during August to a minimum in September in both *Elodea* and *Potamogeton*.

The abundance of water mites per gram of dry plant weight was lowest in *Elodea canadensis*, which also showed the lowest percentage of juvenile stages (i.e. nymphs and teleiophane stages).

In all 27 species and subspecies of water mites were found during this study. Twenty two different species and subspecies were found in each of the stands, 18 were common to all three plant stands. Only 3-5 species represented 80-90 per cent of all captured specimens, as is shown in Table 2.

Limnesia undulata was predominant in stands of *Elodea canadensis* and *Batrachium aquatile*. In *Potamogeton lucens*, there were four dominant species with almost the same 18 per cent relative abundance, namely *Limnesia maculata*, *Piona variabilis*, *Arrenurus globator* and *Arrenurus sinuator*. The share of *Limnesia undulata* in this stand was roughly a half (10 per cent). *Limnesia maculata* and the two above-mentioned species of the genus *Arrenurus* were almost equally numerous,

Table 2. The dominant species (relative abundance in per cent) in the studied stands and the diversity index of the fauna of water mites (D)

Kind of stand	*Batrachium aquatile*	*Potamogeton lucens*	*Elodea canadensis*
Limnesia undulata (Müller) 1776	37.6	10.0	31.4
Limnesia maculata (Müller) 1776	10.0	18.5	16.9
Piona variabilis (Koch) 1836	3.6	18.9	6.1
Arrenurus globator (s.str. + tub.) (Müller) 1776	11.9	19.8	19.4
Arrenurus sinuator (Müller) 1776	5.6	17.5	15.9
D (taxonomy diversity index)	1326	2531	1819

even if the density of the stand was different. *Piona variabilis* was evidently the most abundant in stands of *Potamogeton*.

The diversity of the fauna of water mites calculated as the taxonomy diversity index (modification of Goldman et al. 1968) was greatest in *Potamogeton lucens*, but smallest in *Batrachium aquatile*.

The results prove that plant density influenced the quantitative occurrence of some species of water mites; in particular *Limnesia undulata* preferred dense stands (*Elodea* and *Batrachium*), whereas *Piona variabilis* favoured sparse ones (*Potamogeton*). The mean abundance of water mites per gram of dry plant weight was 4 times higher in *Potamogeton lucens* than in *Elodea canadensis*.

This difference and also the seasonal occurrence of water mites and development of the plant stands indicate that sparse stands (e.g. *Potamogeton*) are more convenient for the development of a water mite fauna than the dense stands (e.g. *Elodea*).

References

Goldman, Ch. R. et al.: Primary productivity, bacteria, phyto- and zooplankton in Lake Maggiore: Correlations and relationships with ecological factors. Mem. Inst. Ital. Idrob. biol. 23: 49-127, 1968.

Kořínková, J. Sampling and distribution of animals in submerged vegetation. Věstník Čsl. spol. zool. 35, 3: 209-221, 1971.

Mothes, G.: Vergleich qualitativer Dredgenfänge mit quantitativen Bodengreiferproben.

Ergebnisse hydrachnologischer Untersuchungen. Limnologica (Berlin) 5, 3 : 435-440, 1967.
Pieczynski, E.: Analysis of numbers, activity, and distribution of water mites (Hydracarina), and of some other aquatic invertebrates in the lake littoral and sublittoral. Ekol. Pol. S. A. 12, 35 : 691-735, 1964.
Pieczynski, E.: The occurrence of water mites (Hydracarina) and some other invertebrates in the littoral and central part of Lake Sniardawy. Ekol. Pol. S.A. 15, 26: 544–551, 1967.
Pieczynski, E.: The trap method for ecological studies on water mites (Hydracarina) in lakes. Proc. the 2nd Int. Congr. Acar. 103–106 (Budapest), 1969.

PECULIARITIES OF THE FAUNA OF HYDRACARINA FROM LAKE BAIKAL

I. I. SOKOLOV and A. I. YANKOVSKAYA
Zoological Institute, Academy of Sciences, Leningrad, U. S. S. R.

Lake Baikal, one of the largest and deepest in the world, has a very diverse fauna which is uncommonly rich in endemic species. The study of Hydracarina, as well as other representatives of microbenthos in Lake Baikal, has received little attention.

In his work dealing with mites from Siberia, Sokolov (1930) listed 91 species of which only two were recorded from the shore of Lake Baikal: *Thermacarus thermobius* from the hot spring Khakusta on the north-eastern shore of the lake and *Pionacercus leuckarti* from the lake near Marytui. Small mites (about 1.5 mm) from the open part of Lake Baikal were first described by Sokolov (1944) as *Baicalacarus vermiformis* gen. n.sp.n. of the Fam. Protziidae (Hydrachnellae). Later, in 1948, Sokolov assigned them to the genus *Stygothrombium,* subg. *Cerberothrombium.* The group Thrombidiidae consists mainly of terrestrial forms, and only the genus *Stygothrombium* includes several freshwater species found in underground waters (wells) of Yugoslavia.

Due to the kindness of the late Prof. M. M. Kozhov we received for determination 35 samples (100 specimens) of water mites collected in 1965 and 1966 from Lake Baikal (near Solzan). In this collection were four species, one of them representing a new genus which we name *Pseudosoldanellonyx lobmanelloides,* since this form resembles *Soldanellonyx* because of the presence of numerous typical denticles on distally enlarged claws of the 1st pair of legs (Fig. 1 - 1) and *Porolohmanella* because of the presence of a long stylet-like rostrum, which together with considerably elongated palps forms a catching organ. However, the last character is convergent in its nature, and considering other characters, the new genus should be assigned to the Fam. Limnohalacaridae and placed in immediate proximity of the genus *Soldanellonyx* Walter.

Length of body of adult mites varies from 0.50 to 0.55 mm, width from 0.40 to 0.45 mm. Nymphs are smaller. Dorsal shield plates are poorly sclerotized, with delicate reticulation (Fig. 1 - 1 nymph). In adults the plates are similar, only somewhat larger. The genital plate with eight to nine suckers on each side (Fig. 1 - 2,3) resembles that of *Soldanellonyx.* In nymphs the genital plate is trapeziform as in nymph II of *Soldanellonyx chappuisi* and they have only three suckers on each side (Fig. 1 - 4). Ocular plates small, oval (Fig. 1 - 1). Ocular plates of male larger and without cornea and pigment. Behind ocular plates a small opening of the dermal gland on each side as in *Soldanellonyx* or *Parasoldanellonyx.*

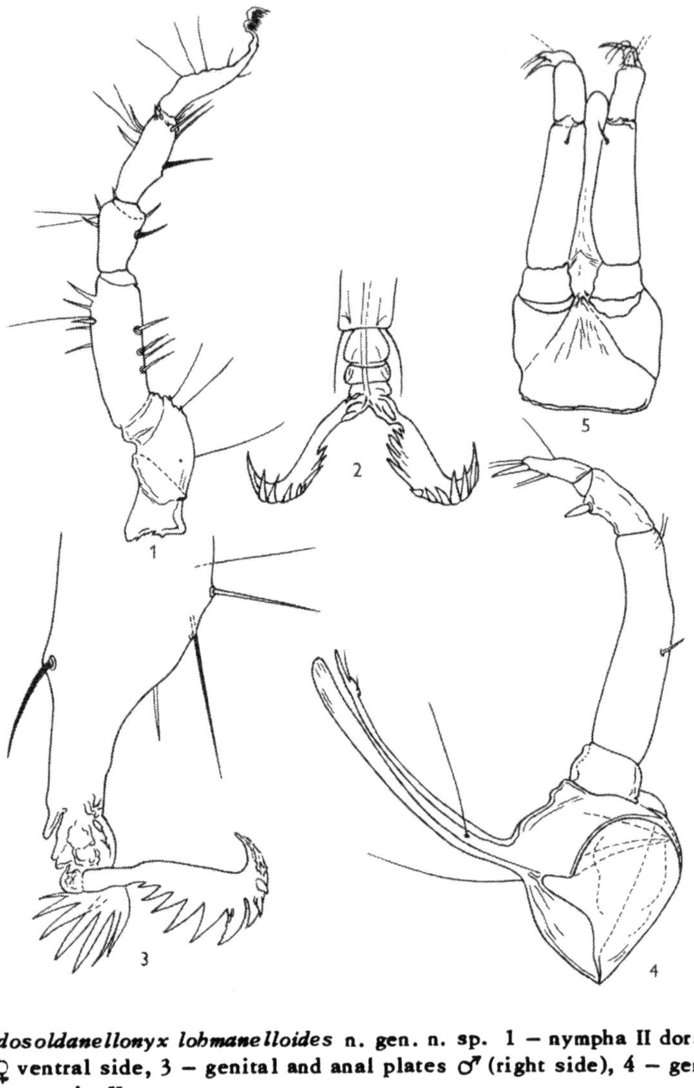

Fig. 1. *Pseudosoldanellonyx lohmanelloides* n. gen. n. sp. 1 – nympha II dorsal side, 2 – ♀ ventral side, 3 – genital and anal plates ♂ (right side), 4 – genital and anal plates nympha II.

Excretory opening on a small isolated genital plate separated from the posterior end of the latter by a narrow band of wrinkled cuticle. Maxillar organ (Fig. 2-4) comprising a relatively short and wide basal portion and a thin narrow rostrum which considerably protrudes forward (Fig. 1-4). Palps elongated, four jointed (Fig. 2-1). Together with the rostrum the palps form a kind of grasping apparatus like that of *Porolohmanella violacea* (Kramer). Cheliceras thin, slightly curved, with dilated basal portion inside the maxillary organ and a considerably narrowed stylet-like distal portion. The claw is poorly developed (Fig. 2-4). Legs simple, joints sub-

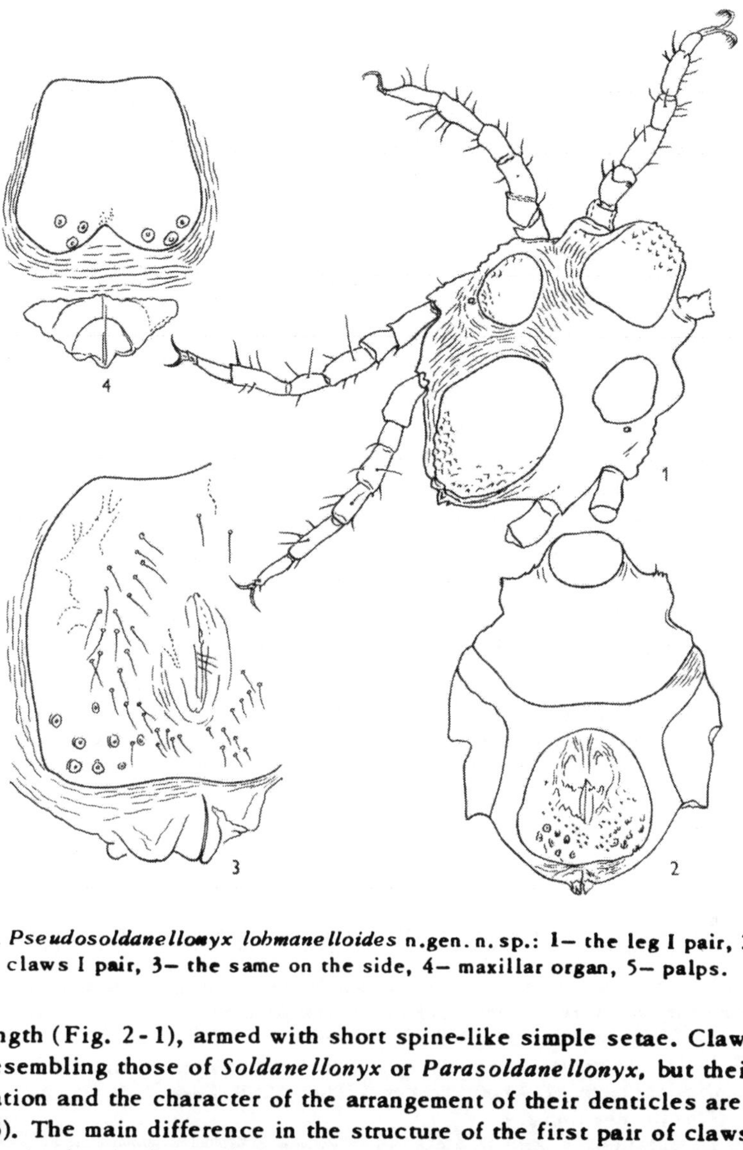

Fig. 2. *Pseudosoldanellonyx lobmanelloides* n.gen. n. sp.: 1– the leg I pair, 2–the claws I pair, 3– the same on the side, 4– maxillar organ, 5– palps.

equal in length (Fig. 2-1), armed with short spine-like simple setae. Claws of the first pair resembling those of *Soldanellonyx* or *Parasoldanellonyx*, but their general configuration and the character of the arrangement of their denticles are different (Fig. 2-2,3). The main difference in the structure of the first pair of claws of the Baikal forms is that in the Baikal species the stem of the claws enlarges downwards and forms a kind of irregular bowl with an edge bearing 10 to 12 denticles spaced equally (Fig. 2-3). Claws of the other legs simple, without additional denticles or pectination.

Members of the new genus and species *(Pseudosoldanellonyx lobmanelloides)* were found in 13 samples (26 specimens) taken at depths from 40 to 100 m on muddy grounds, sometimes with an admixture of sand, in May to September, 1964, 1965. Females were with eggs. In the same material members of the Fam. Limnohalacaridae were found, known already from lake Baikal (Sokolov, 1952) - *Parasoldanellonyx baicalensis* Sok., 1952.

In the material received by us from prof. M. M. Kozhov this species was most often encountered in samples from Solzan (74 specimens). In his 1952 monograph Sokolov mentioned for lake Baikal two varieties: *Parasoldanellonyx parviscutatus* (Walter) var. and *P. typhlops* Viets var., but they were not described by him. The typical species *Parasoldanellonyx parviscutatus* (Walter) and *Soldanellonyx chappuisi* Walter were also recorded from Baikal for the first time. All in all, for lake Baikal four species and two varieties of water mites (Hydracarina) from Fam. Limnohalacaridae are known: *Pseudosoldanellonyx lohmanelloides* gen. n. sp. n., *Parasoldanellonyx baicalensis* Sok., *P. parviscutatus* (Walter), *P. parviscutatus* (Walter) var., *P. typhlops* Viets var., *Soldanellonyx chappuisi* Walter and one species of Trombidiidae gen. *Stygothrombium*, subgen. *Cerberothrombium*, *vermiformis* (Sok.).

As far back as in 1948 Sokolov pointed to a phenomenon which is very typical of the Baikal fauna- the absence of representatives of the group Hydrachnellae (true freshwater mites).

In lake Baikal apparently exist some peculiar conditions which are on the one hand favourable for the development of mites of Limnohalacaridae closely related to the marine Halacaridae, and on the other hand some unknown factors limiting the distribution of fresh water mites of Hydrachnellae. Thus, at present five species and two subspecies of water mites are known from Lake Baikal, three of which are Baikal endemics.

Parasoldanellonyx, as well as *Stygothrombium*, apparently represent a relict of the tertiary fauna of Eurasia which is now preserved in only a few places.

References

Sokolov, I. I.: Beitrag zur Kenntnis der Hydracarinen Sibirien. Arch. für Hydrobiol., Bd. 22: 306-350, 1930.

Sokolov, I. I.: *Baicalacarus vermiformis* n. gen. n. sp. – the first representative of the Hydrachnellae from Lake Baikal. Uchioniye zapisky Leningradskogo Gosudarstvennogo Universiteta, vypusk 2: 46-53, 1944. (In Russian.)

Sokolov, I. I.: On the systematic position of *Baicalacarus vermiformis* Sokolov (Acari) and its relict character. Doklady Academii Nauk SSSR Tom 61, N° 1, Zoologiya : 46-53, 1948. (In Russian.)

Sokolov, I. I.: Water mites, part II, Halacarae. Fauna SSSR Paukoobrazniye, Tom 5, Moskva-Leningrad, 1952. (In Russian.)

Section III

PLANT MITES

Chairman
J. Boczek (Poland)

Assistant Chairman
J. Vaněk (Czechoslovakia)

Proceedings of the 3rd International Congress of Acarology, Prague, 1971

PHYSIOLOGICAL FUNCTIONS OF PLANTS AS AFFECTED BY DAMAGE CAUSED BY *TETRANYCHUS ATLANTICUS* MCGREGOR

N. ATANASOV
Plant Protection Institute, Kostinbrod, Sofia, Bulgaria

T. atlanticus is widely distributed in Bulgaria. Each year it multiplies to a great density and causes considerable reduction in the yields of cereal, leguminous and vegetable crops. Its great economic importance as an enemy of farm crops and the lack of full information concerning its distribution, harmfulness, bio-ecology and means of control prompted us to make it the object of our studies.

Sporadic information concerning the effect of damage caused by herbivorous mites on some plant physiological processes has been published by Nachev (1957), Boulanger (1958) and Balevski (1967). Apart from Tomašević's report (1965) on the transpiration changes in soybean plants resulting from damage caused by *T. atlanticus*, no other studies have been conducted on its influence on the physiological processes of plants. That is why we established, by means of experiments, the concrete effect of the damage caused by *T. atlanticus* on some major plant processes and, more precisely, the course of photosynthesis, transpiration and chlorophyll formation.

Materials and methods

Transpiration intensity of healthy plants and of those damaged by *T. atlanticus* was determined 20 days after infection, using Babushkin's method (1960). The experiment was carried out in four replications in an air-conditioned chamber at a constant air temperature of 25°C and a relative humidity of 65 per cent.

Photosynthesis of healthy and damaged bean plants was carried out 15 days after infection of the plants using Babushkin's (1960) micro-instruments (5 hrs of exposure) under natural conditions. The micro-instruments used made it possible to determine photosynthesis without disturbing the natural conditions of the plants.

The chlorophyll amount in healthy plants and in those damaged by *T. atlanticus* was determined by means of the chromatography-colorimetric method 28 days after infection of the plants.

Results and discussion

T. atlanticus causes damage by piercing the leaf epidermis with the stylets of its chelicera, introducing them into the cells of the palisade and spongy parenchyma and sucking up a portion of the cell sap with the chlorophyll grains. Tiny light patches appear on the pierced spot which are due to the diminished chlorophyll amount and coagulation of the cell sap as a result of the enzymes secreted by the mite. Air penetrates the damaged cells and this also affects the colour of the damaged leaves. The quantity of these patches rapidly increases in the consequence of which the leaf assimilation area decreases and its normal activity is disturbed.

The results from the studies show that up to 20 days after infection of the plants, increase in transpiration intensity is observed in the damaged plants compared with that of healthy plants. Thus healthy leaves, without mites, emit 0.011 g water per hour per sq cm, whereas infected leaves, 30 mites per leaf, emit 0.034 g water or 309 per cent (Fig. 1). This shows that heavily damaged leaves transpire 3 times as intensively as healthy plants. We believe, that the enhanced transpiration of damaged plants is due to the wide-open stomata of the leaf and to the openings resulting from the piercings of the mites.

Photosynthesis intensity is also seriously disturbed in plants damaged by *T. atlanticus*. Thus, 11.87 mg/hr carbon-dioxide was obtained from 1 sq. decimeter of healthy leaves and 10.21 mg from damaged ones, the difference being 1.66 mg. These results, converted into percentage, are graphically presented in Fig. 2.

In order to determine the chlorophyll content depending on *T. atlanticus* density, the plants were infected to varying extents 28 days prior to the experiment. The results, graphically presented in Fig. 3, show that leaf chlorophyll content sharply drops with increase in mite density. Thus, 25.8 mg chlorophyll were obtained from 1 gr fresh material from healthy leaves and 19.8 mg from heavily damaged leaves. The abrupt drop of plant chlorophyll, resulting from high *T. atlanticus* density, leads to disturbance in its assimilation activity. Mite damage is

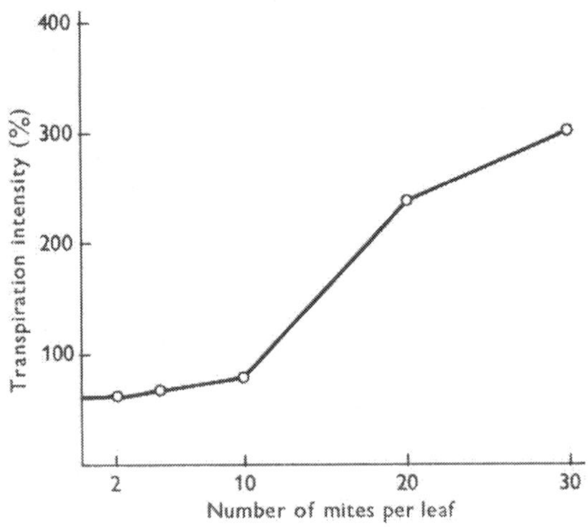

Fig. 1. Transpiration intensity at various levels of damage of bean leaves by *Tetranychus atlanticus*.

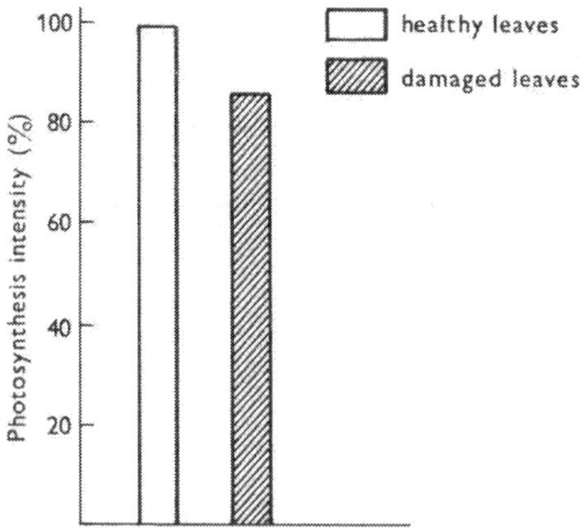

Fig. 2. Photosynthesis intensity in healthy leaves and those affected by *Tetranychus atlanticus*.

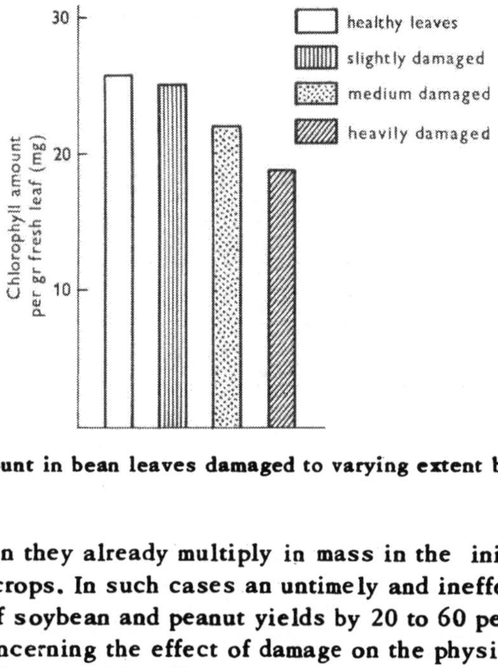

Fig. 3. Chlorophyll amount in bean leaves damaged to varying extent by *Tetranychus atlanticus*.

especially heavy when they already multiply in mass in the initial phases of the development of farm crops. In such cases an untimely and ineffective control results in a decrease of soybean and peanut yields by 20 to 60 per cent.

All these data concerning the effect of damage on the physiological processes and yields of attacked crops, depending on mite population density, served to determine the economic threshold of damage caused by *T. atlanticus*.

References

Balevski, A.D.: *Tetranychus* mites of fruit trees. Monograph, Academy of Agric. Sci., 1967. (In Bulgarian.)

B a b u s h k i n, A.N.,: Field micro-instruments for measuring photosynthesis and respiration. Proceedings of the Moldavian Institute of Irrigated Farming and Vegetable Growing, Vol. 2, 1960. (In Russian.)

Na ch e v, P.: *Brevipalpus levisi* McGregor, Proceedings of the Research Institute of Viticulture and Wine-producting, Vol. I, pp. 197–225, Pleven, 1957. (In Bulgarian.)

B o u l a n g e r, L.W.: The effect of European red mite feeding injury on certain metabolic activities of red Delicious apple leaves. Maine Agric. Exptl. Sta. Bull., 570, 1–34, 1958.

T o m a š e v i c, B.: Harmful mites of soybeans, Faculty of Agriculture, Novi Sad, 134, 1965. (In Serbian.)

Proceedings of the 3rd International Congress of Acarology, Prague, 1971

ON THE BIOLOGY OF *BREVIPALPUS CALIFORNICUS* BANKS

H. H. ATTIAH, A. A. SOLIMAN and M. L. WAHBA
Acarology Investigations, Crop Protection Department, Ministry of Agriculture, Dokki, Egypt, Faculty of Sciences, Ain Shams University, Egypt

Mites are among the most important pests affecting the citrus industry. Not only do they affect the quantity and the size of produced fruits, but they also reduce their quality due to the symptoms they cause. The citrus flat mite *Brevipalpus californicus* is one of the most injurious mites infesting all citrus species cultivated in both Lower and Upper Egypt. Studying the biology of that mite is no doubt an essential step towards achieving successful control.

Materials and methods

1. Duration of different developmental stages

Leaves and fruits of oranges highly infested with *Brevipalpus californicus* were brought from a citrus grove to the laboratory at Dokki. Green citrus fruits with rough rinds were chosen as hosts for rearing mites. Oranges were used for rearing from October till February, while lemons were used the rest of the year. The buttons of the fruits were first removed and the fruits were then washed with tap water, dried with a towel and examined with a hand lens to assure absence of any mites or other pests. An area around one third the size of the fruit surrounding the button was confined by tangle foot and a number of adult females of *Brevipalpus californicus* were transferred to this area by means of a camel hair brush. Each citrus fruit was placed on the top of a 250 ml. glass jar filled with water. Jars were put on a table inside the laboratory near a window kept constantly open to assure a seminatural fluctuation of temperature and relative humidity. The mites, each set, were left for a period of 24 hours to lay eggs and then transferred to a new fruit and kept for a second period and so on, so that a total number of around 100 eggs were laid during three or four days. The eggs were watched till hatching and the number of larvae were recorded each day. Statistical analysis was used to calculate the mean time and the standard deviation for the incubation period. The larvae hatching on the same date were all transferred to a new fruit and then watched daily till the first quiescent stage was reached by the first larva. The rest of the moving larvae were then transferred to a new fruit, so that all mites reaching the protochrysalis stage on a certain date remained on the same fruit. The mites were watched till they moulted and produced the protonymphs, the mean days required for the larval as well as the protochrysalis stages were calculated. The same procedure was then followed to calculate the time required for each stage until that of the imago. The previous steps were followed each month throughout the year to get the duration of each stage during the different months, while the temperature and percentage of relative humidity were recorded by means of a thermohygrograph.

2. Number of generations

A number of females (around 20) were transferred to a citrus fruit treated as previously mentioned. Females were kept for 24 hours to get eggs of the same age. The females were then discarded and the eggs were left to hatch and to develop. The moving stages were transferred to a new fruit whenever the old one became dry and unsuitable for the mites to feed on, while the quiescent stages were left and then transferred after moulting. When the first mite reached the adult stage the time required from egg to imago was recorded and the others were observed till the last one reached the adult stage. When 50 % of the eggs first laid reached the adult stage, a number of adult females (around 20) were then isolated on a new fruit and used to lay eggs in order to rear the following generation.

This procedure was followed for a whole year so that the number of generations as well as the duration of each generation could be estimated.

3. Longevity of females and number of eggs per female

Twenty females from each generation were transferred upon reaching the adult stage and were each confined to a citrus fruit. Each female was watched daily till death, so that the preoviposition period, the number of eggs per day, the total number of eggs laid and the life span could be calculated.

4. Parthenogenesis

As the absence of males was observed during rearing the mites, a number of deutonymphs were transferred separately, each to a separate fruit. After reaching the adult stage, the mites were left to lay eggs and to develop till the deutonymphal stage. Deutonymphs were then isolated separately and were watched for laying eggs.

Results and discussion

Fig. 1 represents the incubation period and the duration of different stages during different months of the year. It is clear that all stages required a shorter time to develop during late Spring and Summer (from May till September) than during early Spring and Autumn (March, April, October, and November), a condi-

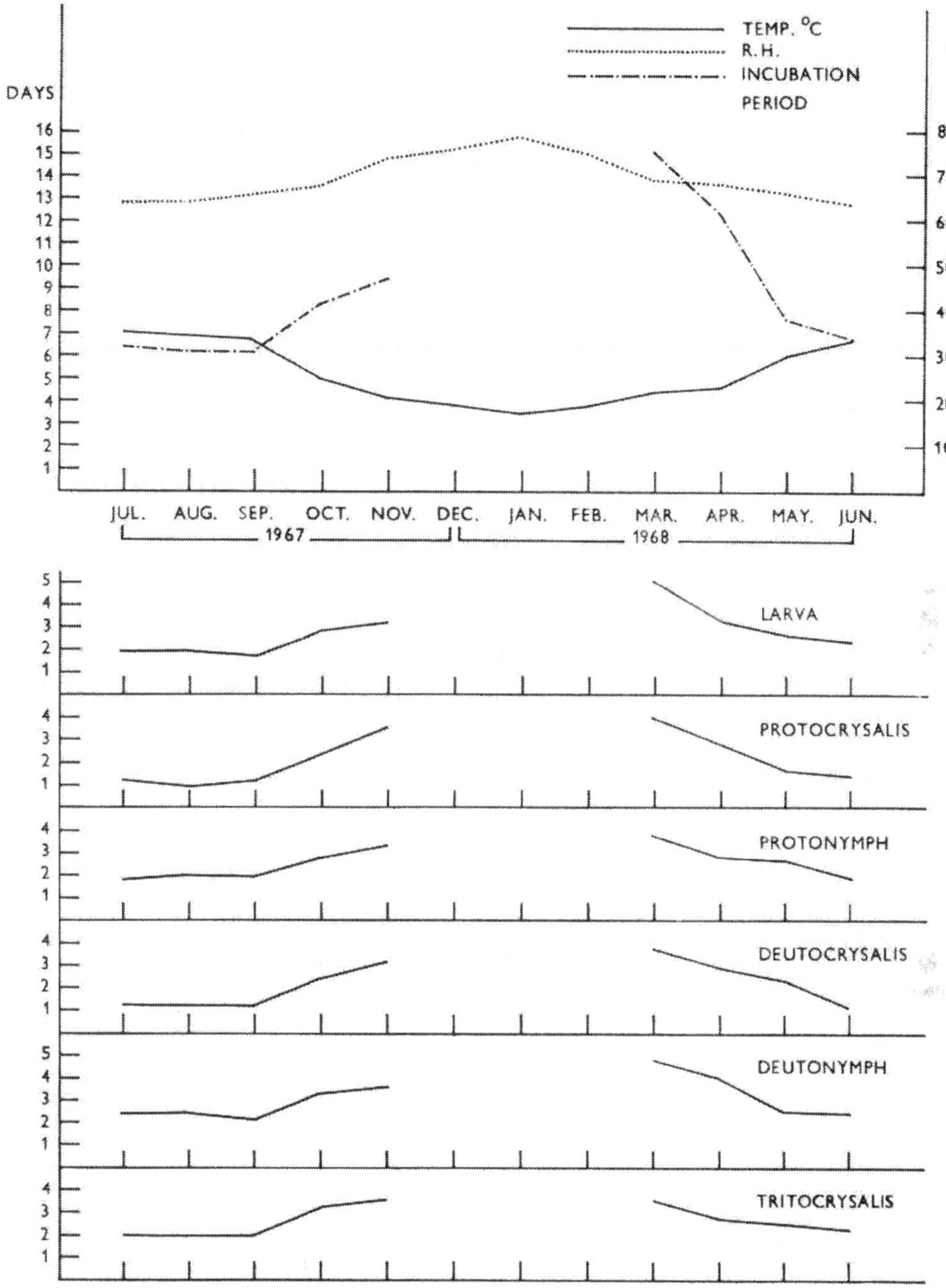

Fig. 1. Incubation period and duration of different stages of *Brevipalpus californicus.*

tion negatively correlated with temperature while positive with relative humidity. It was not possible to rear immature stages during winter time from December till February, a case supported by the study of duration and number of generations a year. As shown in Table 1, the number of generations amounted to 11. The last female reaching the imago stage in the tenth generation of that study acquired adulthood on 7.11. From around 80 eggs laid by some females of that generation on 12.11 to begin the eleventh generation only 11 eggs were able to develop to adults (last on 29.12.). The life span of these females was the longest of all generations averaging around three months, i.e., they were able to stand winter time and resume activity and laying eggs during March to begin the first generation of the second year. (N.B. study of the number of generations began on 1.7.67 and ended on 30.6.1968, but tabulated as shown in tables 1 and 2 for more convenience). The mean time required from egg to female was around 40 days for the first generation, decreased gradually and reached the minimum of around 17 days during summer and began to increase gradually to reach the longest time required (around 47 days) during November and December. The life span of the females (Table 1) and the oviposition period (Table 2) generally followed that same order. Number of eggs per female ranged around 15–33. Rearing deutonymphs separately assured parthenogenesis.

The following conclusions were reached from the previously described conditions:

1. Eleven generations are produced during one year, with a duration of about 17 days during summer till about 47 days during November and December. Winter is spent in the adult female stage and parthenogenesis is certain.

2. The preoviposition period is about 2 days during summer and 9 days during late autumn.

Table 1. No. and duration of generations and longevity of females of *Brevipalpus californicu*

Generation No.	No. of eggs reaching adult	Dates		Mean days to reach adult	Longevity of females (days)	Temperature range in °C	R.H. % range
		egg	latest adult				
1	60	1.3	9.4	39.95 ± 1.20	65.55 ± 7.72	18–24	55–85
2	98	16.4	15.5	29.97 ± 1.27	45.40 ± 7.49	21–33	45–80
3	102	19.5	8.6	21.96 ± 1.02	27.45 ± 6.62	27–35	45–80
4	70	11.6	30.6	18.77 ± 1.00	22.75 ± 5.35	29–37	40–80
5	80	1.7	19.7	17,06 ± 1.22	18.62 ± 6.20	32–36	40–80
6	52	21.7	8.8	17.71 ± 1.03	20.60 ± 4.77	31–36	45–80
7	72	10.8	28.8	17.75 ± 0.94	21.00 ± 6.19	31–36	40–80
8	74	30.8	17.9	17.62 ± 1.00	34.85 ± 11.42	29–36	45–80
9	78	20.9	10.10	19.82 ± 1.20	50.05 ± 5.59	21–35	45–80
10	48	14.10	7.11	23.97 ± 1.16	65.95 ± 12.27	20–27	50–85
11	11	12.11	29.12	47.09 ± 1.16	83.82 ± 7.11	16–23	65–95

Table 2. Preoviposition, oviposition and egg laying of females of *Brevipalpus californicus* in different generations

Generation No.	No. of females	Preoviposition period (days)	Oviposition period (days)	Eggs/Female	Eggs/Female day
1	20	7.20 ± 0.41	49.55 ± 4.58	20.65 ± 2.96	0.41 ± 0.05
2	20	4.80 ± 0.41	34.80 ± 4.95	22.50 ± 3.02	0.67 ± 0.10
3	20	2.75 ± 0.45	19.75 ± 4.69	18.35 ± 3.53	0.92 ± 0.14
4	20	2.00 ± 0.00	18.10 ± 3.96	21.75 ± 3.58	1.21 ± 0.13
5	40	2.17 ± 0.39	15.50 ± 6.03	20.55 ± 7.01	1.35 ± 0.14
6	20	2.00 ± 0.00	17.70 ± 4.51	22.75 ± 5.77	1.30 ± 0.20
7	20	2.20 ± 0.41	17.65 ± 5.51	19.15 ± 5.85	1.10 ± 0.14
8	20	3.00 ± 0.56	30.60 ± 10.66	31.40 ± 10.26	1.05 ± 0.14
9	20	4.35 ± 0.49	43.70 ± 5.35	33.35 ± 4.98	0.77 ± 0.10
10	20	5.25 ± 0.45	50.45 ± 9.43	23.90 ± 4.54	0.49 ± 0.08
11	11	9.36 ± 0.50	52.36 ± 3.32	15.63 ± 2.58	0.29 ± 0.04

3. The oviposition period is about 16 days during summer and 52 days during late autumn.

4. The longevity of females is about 19 days during summer and 84 days during late autumn and winter (hibernating females which resume activity during March).

5. Total number of eggs per female is about 15 eggs during late autumn and 33 eggs during late summer; number of eggs per day per female is about 0.29 eggs during late autumn and about 1 egg during summer.

6. In order to develop different immature stages require the minimum time during the summer, a longer time at the beginning of spring and the longest time, for a very few, in late autumn (second half of November) around 47 days from egg to imago. Required days for different stages during summer and at the beginning of spring are:

Stage life	Summer	Beginning of spring
Incubation period	6	15
Larva	2	5
Protochrysalis	1½	4
Protonymph	2	4
Deutochrysalis	1½	4
Deutonymph	2	5
Tritochrysalis	2	4
Total	17	41

PHOSPHORUS COMPOUNDS AS A CAUSE OF AN INCREASE IN NUMBERS OF THE FLAT MITE

H. H. ATTIAH and M. L. WAHBA
Acarology Investigations, Ministry of Agriculture,
Dokki, Egypt

Oil sprays as well as fumigation with hydrocyanic gas have long been used as the common means of control against scale insects on different species of citrus trees in Egypt. Applications are usually carried out late fall and winter. During the last few years some phosphorus compounds applied in summer were used for the same reason. Meanwhile it was noticed that infestation with citrus mites, especially the flat mite *Brevipalpus californicus*, has recently increased, a matter causing many shipments of citrus fruits to be rejected for export due to the infestation symptoms. Phosphorus compounds have been reported as a cause of an increase in tetranychid mites and of populations of aphid and white flies. The following experiment was undertaken to investigate the effort of some phosphorus compounds, including those recommended, against citrus scale insects, on the flat mite population in comparison with certain acaricides known to be effective against this mite.

Methods

Part of a lemon grove at Barrage (Lower Egypt) containing 64 trees in four rows of 16 trees each, was chosen. Seven pairs of trees from each row were sprayed by seven different compounds and the remaining two trees were left as control; each treatment was repeated three more times in a completely random block grouping in the remaining three rows. Spraying was undertaken on August 4th, 1969, by means of a motor sprayer. The compounds and dosages are as follows:
Kelthane 42 % e.c. at rate of 125 cc/100 litres of water;
Chlorobenzilate 50 % e.c. at a rate of 100 cc/100 litres of water;
Torak 90 % e.c. at rate of 250 and 300 cc/100 litres of water;
Diazinon 66 % e.c. at a rate of 100 cc/100 litres of water;
Dimethoate 40 % e.c. at a rate of 125 cc/100 litres of water;
Anthio 25 % e.c. at a rate of 100 cc/100 litres of water.

Sampling was done by picking 10 leaves at random from each tree on each sampling date, totalling 20 leaves from each repeated treatment and 80 leaves from each single treatment; samples were taken to the laboratory for counting, under a dissecting microscope, the moving stages only. The first count was done on

11.VIII. 1969, one week after spraying. Five other counts took place at different intervals, the last on 23.X.1969, about 80 days from spraying.

Results and discussion

As indicated in Table 1, seven days after treatment all compounds sprayed showed a reduction in mite infestation amounting to more than 90 % of the population as compared to control-trees treated with Kelthane and Chlorobenzilate; more than 80 % reduction in population occured in trees treated with Torak in both dosages, the higher concentration being somewhat more effective; moderate results were obtained with Anthio while both Diazinon and Dimethoate were the least effective.

Table 1. Populations of *Brevipalpus californicus* after spraying

Spray	Number of mites* and % of reduction or increase after days indicated, compared with control						Total
	7	14	24	34	58	80	
Kelthane 42 %	16	3	2	30	15	3	69[a]
125 cc	– 93.6	– 97.4	– 98.4	– 90.0	– 94.6	– 98.5	– 94.5
Chlorobenzilate	24	4	8	75	54	5	170[a]
50 %, 100 cc	– 90.3	– 96.6	– 93.7	– 74.9	– 80.6	– 97.5	– 86.6
Torak 90 %	37	13	22	416	373	292	1153[b]
250 cc	– 85.0	– 88.8	– 82.7	+ 39.6	+ 34.7	+ 49.0	– 8.5
	28	5	21	526	494	403	1477[b]
300 cc	– 88.6	– 95.7	– 83.5	+ 76.5	+ 78.3	+105.6	+ 17.2
Diazinon 60 %	137	47	69	718	658	535	2164[c]
100 cc	– 44.3	– 59.5	– 45.7	+140.9	+137.5	+173.0	+ 71.7
Dimethoate	122	28	32	783	714	594	2273[c]
40 %, 125 cc	– 50.4	– 75.9	– 74.8	+ 162.8	+ 157.8	+ 203.1	+ 80.4
Anthio 25 %	81	74	59	437	430	428	1509[b]
100 cc	– 67.1	– 36.2	– 53.6	+ 46.6	+ 55.2	+ 118.4	+ 19.8
Control	246	116	127	298	277	196	1260[b]

* No. existing on 80 leaves, 20 x 4 repeats on each sampling date.
a, b, c Significant statistical differences.

Counts of mites two weeks after spraying still revealed a high reduction of numbers in trees treated with Kelthane, Chlorobenzilate and Torak (both doses) while the effect of Anthio, Diazinon and Dimethoate was inferior, especially of the first two. These results remained the same for another ten days.

Thirty four days after sprying the trees, counts of mites revealed an increase in populations on all trees treated with phosphorus compounds compared with numbers existing in trees receiving no treatment. The population in Dimethoate plots was the highest with an increase amounting to 162.8 %, followed by Diazinon, Torak 300 cc, Anthio 250 cc and Torak 250 cc with increases of 140.9, 76.5, 46.6 and 39.6 per cent respectively. Trees treated with Kelthane or Chlorobenzilate still harboured a few mites, amounting to 10 % in the former and 25 % in the latter of the population existing in the control-trees.

The increase in flat mite population in plots treated with phosphorus compounds lasted till the end of the experiment, as shown by counts undertaken 58 days as well as 80 days after spraying. Dimethoate was still the least effective compound in that connection; followed by Diazinon, while Anthio and Torak, at a rate of 300 cc, replaced each other in order in these two counts. It is of interest to notice that the higher the dosage of Torak, the higher the increase in populations in all counts done after the populations began to show an increase in comparison with the untreated trees, i.e. after 34 days from spraying; moreover, the higher the dose, the higher the effect of the same compound in reducing the population as shown by counts done one week, two weeks and 24 days after spraying. This could be explained on the basis of a probable harmful effect on some predators, the higher the dosage used the higher the effect of the compound on both the prey and the predator. After the effect of the compound on the citrus mite had worn off, the effect on the predators revealed that the higher the dosage used, the higher the numbers of predators affected; ultimately, what was gained against the flat mite was offset by the adverse effect on the predator, resulting in a higher population of mites in trees treated with the higher dose during the duration of the experiment (1477 mites for Torak 300 cc versus 1153 mites for Torak 250 cc).

A comparison between the effect of Dimethoate and Diazinon might corroborate that result; while the former was more effective against the flat mite, causing a gradual reduction in its population 7,14, and 24 days after spraying, its possible greater effect on the predators caused a greater increase in numbers of the flat mite 34, 58 and 80 days from spraying.

Anthio might be considered less effective than both Dimethoate and Diazinon to both prey and predator, causing less reduction in flat mite populations during the first period and only a small increase of the mite during the second period.

Kelthane and Chlorobenzilate might be considered effective acaricides for *Brevipalpus californicus* while least harmful to predators, causing a high reduction in flat mite populations during the whole period of the experiment; however, the interpretations concerning the effect of different compounds on the predators of the flat mite and the relationships between both prey and predator under such chemical influences still have to be critically studied.

The fluctuation of the population of the flat mite harboured by the trees in the different treatment during the whole period of the experiment is shown in Figure 1. Statistical analysis for the total numbers of mites harboured by trees which have undergone different treatments during the whole time of experiment revealed no significant difference between mite populations existing on trees of the Kelthane and Chlorobenzilate plots, while plots treated with Torak in both doses and plots treated with Anthio did not differ from the untreated controls and those treated with Diazinon and Dimethoate were infested by significantly higher numbers than the controls without any significant difference between them.

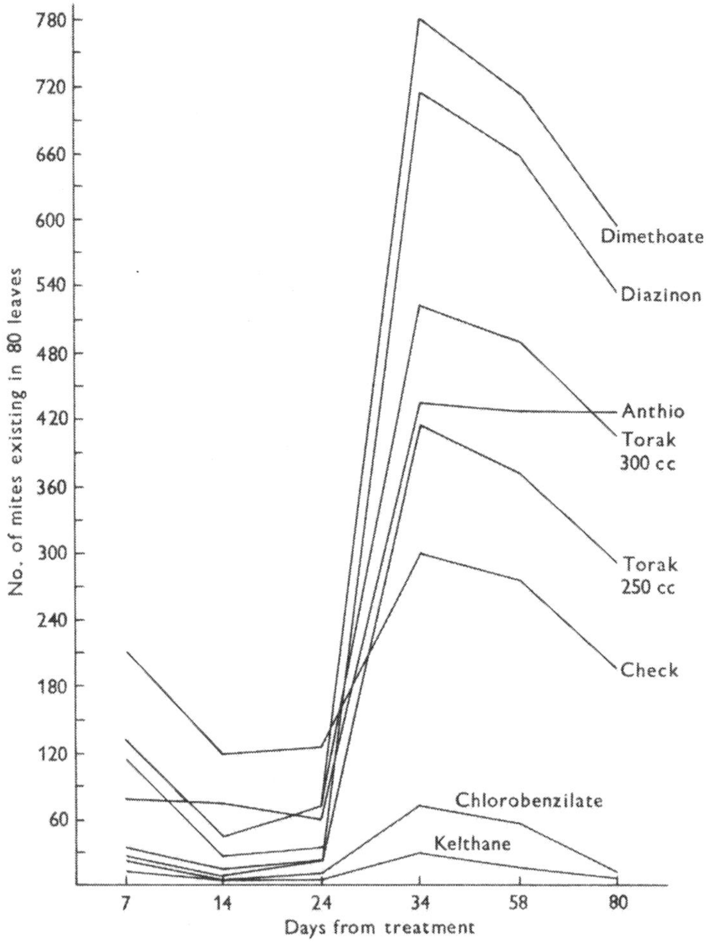

Fig. 1. Fluctuation of populations of *Brevipalpus californicus* on lemon trees following treatment.

Proceedings of the 3rd International Congress of Acarology, Prague, 1971

NOTES ON THE BIONOMICS OF *CALEPITRIMERUS VITIS* (NAL.) (ACARINA: ERIOPHYIDAE)

M. M. CARMONA
Estacao Agronomica Nacional, Oeiras, Portugal

Two distinct species of eriophyids damaging grapevines *(Vitis vinifera* L.) have thus far been found in Portugal: *Eriophyes vitis* (Pgst.) /(=*Colomerus vitis* (Pgst.) (Newkirk & Keifer, 1971)/, a gall mite causing erineum patches on the underside of the leaves, and *Calepitrimerus vitis* (Nal.), behaving basically as a rust mite; it is also known as *Epitrimerus vitis* Nal. or *Phyllocoptes vitis* Nal. (Carmona, 1966).

The latter species, *Calepitrimerus vitis* (Nal.) was described by Nalepa in 1905 as *Epitrimerus vitis* n. sp. and Keifer, in 1938, erected the genus *Calepitrimerus* which includes *E. vitis*. But Müller – Thurgau (1906) sent Nalepa specimens of eriophyid mites collected in Switzerland from grapevines; these were determined by Nalepa as *Phyllocoptes vitis* n. sp. Later Keifer (1952) suggested that *P. vitis* Nal. was possibly based merely on deutogynes of *Calepitrimerus vitis* (Nal.) and in 1964 (in litt.) gave us the confirmation of this interpretation.

The reason of Nalepa's misdetermination was, and still is, based on the mite's life history. This species has two kinds of females: one female resembles the males (summer female) and is the primary type on which the taxonomic concept of the species is based – protogyne; the other female type is structurally different from the primary type, has no male counterpart and is therefore the secondary female – deutogyne.

The *Calepitrimerus vitis* (Nal.) form comprises both sexes and is the perfect form of the mite to base the specific concept upon. The "Phyllocoptes" form consists of females only and is therefore an imperfect, taxonomic unit.

We studied the bionomics of *Calepitrimerus vitis* (Nal.) in open greenhouse conditions, on the vine-cultivation Borracal, in 1969. The weather conditions were similar to those encountered in the field, i.e., winter median temperatures between 16–20°C and summer median temperatures ranging from 24–30°C and relative humidities between 80–90 %. The observations were made every 10 days when the different vine organs were collected and the different mite forms and their frequency classified.

In October fully fed deutogynes could be found in the hibernation quarters. The selected places are: a) – the mass of bud hairs lining the inner surface of the outer scales of the buds; b) – the cottony protection between the primary growing point and the lateral growing ones; c) – the underside of the rhytidome, in the region where the annual shoots and the arms of the vines join.

In March – April, at the beginning of the growing season, the deutogynes start egg-laying. The new progeny lives around the basal area of the new shoots until these attain a length of 30–50 centimetres. Afterwards the mites migrate to the young leaves, the first migrations taking place in first days of May. In June most of the population is on the leaves. By the middle of June, however, a small population consisting of eggs, larvae, protogynes, deutogynes and males was found in the new buds. These mites fed especially on the basal parts of the scales. In August the great migration by deutogynes begins towards the winter buds and in October the entire population is in its hibernation quarters (Fig. 1).

The feeding habits of *Calepitrimerus vitis* (Nal.) correspond with a very definite symptomatology: a) – the winter buds show brown scarifications or necroses on the basal part of the outer scales, occasionally causing the death of the primary growing point or even of the whole eye; b) – in the beginning of the growing season some buds die before they open; sometimes the young shoot with 3–4 leaves also withers; occasionally these shoots continue to grow weakly and do not bear grapes; c) – a few new shoots develop into a witches-broom because of the death of the primary growing point and the development of the lateral growing ones which are usually not so strong; d) – when all buds in the spur are killed, either in winter or in spring, the plant sometimes reacts in the basal area of the spur and forms a dense growth of suckers and/or water sprouts. Both types of shoots originate from latent buds in the older wood; e) – in the developing leaves several types of malformations can be

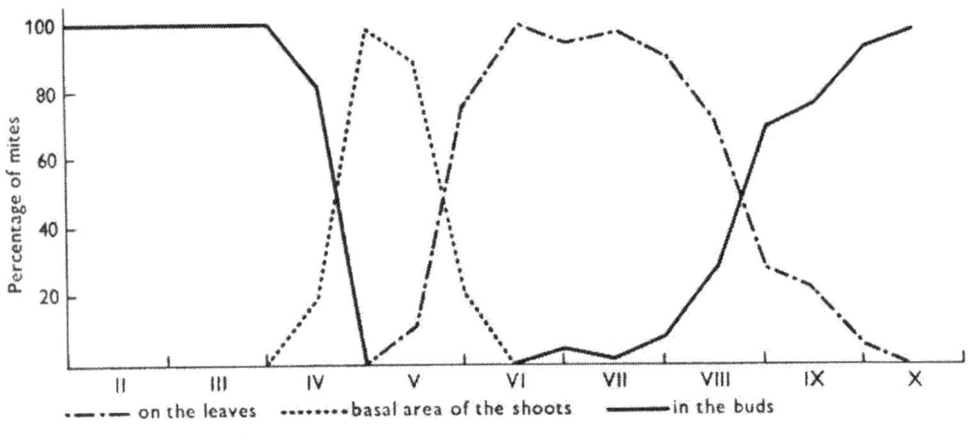

Fig. 1. Population distribution of *Calepitrimerus vitis* (Nal.) in 1969.

observed if they are attacked by relatively small populations of mites. These leaves turn either brown-reddish in red varieties or brown-yellowish in white ones and die within 1–2 months. Developing leaves or fully-grown leaves attacked by large populations of *Calepitrimerus vitis* (Nal.) at first retain the normal appearance without malformations but later on acquire the typical premature reddish or yellowish tones. These leaves die within 15–20 days after the attack; f) – in the beginning of the attack transparent areas can be found in the leaves resulting from the feeding of the mites. Necroses on the leaf petiole, on the green bark of the shoots and on the bunch of grapes, both in the stems and on the berryskin have also been observed; g) – the appearance of lateral shoots, normally never maturing; h) – as a general rule vines attacked by *Calepitrimerus vitis* (Nal.) are less vigorous and show an abnormal death rate. (Fig. 2.)

Our laboratory observations have been confirmed in the field in several instances.

References

Carmona, M.M.: Contribuição para o conhecimento dos ácaros das plantas cultivadas em Portugal. IV. Agronómia lusit. 26:175–203, 1966.

Keifer, H.H.: Eriophyid studies II. Calif. Dept. Agr. Bull. 27: 301–323. 1938.

Keifer, H.H.: The eriophyid mites of California. Calif. Insect Surv. Bull. 2:1 – 123, 1952.

Müller – Thurgau, H.: Die Milbenkrankheit der Reben. Zbl. Bakteriol., 2 Abt., 15:623–629, 1906.

Newkirk, R.A., Keifer, H.H.: Eriophyid studies C–5. ARS USDA, Spec. Pubs., 20 pp., 1971.

Winkler, A.J.: General viticulture. Univ. Calif. Press-Berkeley & Los Angeles, 1962.

Fig. 2. Some aspects of the attack of *Calepitrimerus vitis* (Nal.).

THE PRESENT STATUS OF OUR KNOWLEDGE OF INDIAN PLANT FEEDING MITES

G. P. CHANNABASAVANNA
University of Agricultural Sciences, Bangalore, India

Systematic studies on Indian plant feeding mites are still in their infancy and therefore our knowledge of this group is meagre. In the limited space available a brief review is attempted of what is known about this group of mites in India.

Taxonomy and faunistic studies

The first definite record of plant feeding mite in India was published by Wood-Mason (1884) who reported the occurrence of *Tetranychus bioculatus* on tea from Assam. By the end of the last century only a few species were known from this country. Contribution of Hirst (1924, 1926), Cherian (1933), Rahman and Sapra (1940 a), Puttarudraiah and ChannaBasavanna (1959) and a few others during the first three scores of this century gradually increased the knowledge of the plant feeding mite fauna and by 1964 about 35 species were known to occur in the country (Ghai, 1964). During the past seven years significant contributions have been published (ChannaBasavanna, 1966; 1968; ChannaBasavanna and Banu, 1971 a, 1971 b) which have increased the currently known species of Indian plant feeding mites to 111. Of these, Eriophyidae account for 70, Tetranychidae for 21, Tenuipalpidae for 18 and Tarsonemidae for 2 (Appendix 1). The number of different plant feeding mite groups does not indicate the relative representation of these groups in the country. Systematic surveys, if undertaken throughout the country, will reveal the real situation.

Biology and host records

Certain of the economic species have received some attention by workers in the past, and studies on the biology of these species, mostly confined to the laboratory, have been carried out. The biology of *Tetranychus cucurbitae* Rahman and Sapra, which is one of the most common species of spider mite infesting several vegetable plants and others, has been studied (Khot and Patel, 1956; Rahman and Sapra, 1946). The sorghum spider mite, *Oligonychus indicus* (Hirst), which frequently assumes localised importance on grain sorghum, has been studied for its bionomics (Cherian, 1933; Rahman and Sapra 1940 b). The tea crop often suffers from infestations of three species of mites, of which one is the spider mite *Oligonychus coffeae* (Nietner) and the other two are eriophyids, the pink mite *Calacarus carinatus* (Green) and the purple mite *Acaphylla theae* (Watt). Das and his associates (Das, 1959; Das and Sengupta, 1958, 1963; Das and Das, 1967) have investigated the biology of these species as occurring on tea in north-east India. *Tetranychus telarius* (L.) is another species which has received some attention in the past. Recently Banu and ChannaBasavanna (1971) studied the biology of another common spider mite, *Eutetranychus orientalis* (Klein).

Thus it can be seen that only some accounts on the bionomics of only a few Indian species of plant feeding mites are available, although there are several species of economic importance, such as the litchi mite, cotton woolly mite, jasmine erineum mite, fig spider mite and so on.

Control

Laboratory evaluations of several acaricides, including old and newer chemicals (the latter group with some organo-phosphorus insecticides), are being made especially at the Indian Agricultural Research Institute. Reports have been published on field tests of some of these acaricides for their efficiency in the control of some of the mite pests, like the sorghum spider mite, the common vegetable mite and the purple mite of tea. It is not possible to review the work carried out on this subject in a limited space. Those interested will get some idea of it by consulting the bibliography of the Indian plant feeding mites (ChannaBasavanna and Nageshchandra, 1971). The approach in the control of mite pests in this country appears to be mainly through application of chemicals rather than by building up basic information on pest species in a certain environment with the ultimate object of integrating different methods of control or attempting to contain the pest species. No specific resistance in any plant feeding species of mite against any chemical has been reported in the country, but critical observations and studies are required to obtain certainty. Some studies have been made on the resistance in sugarcane against the attack of *Oligonychus indicus*, while a few insect predators have been recorded as feeding upon certain spider mites. Relevant literature: Ghai (1964) and ChannaBasavanna and Nageshchandra (1971).

References

Banu K., ChannaBasavanna, G.P.: Plant feeding mites of India I. Mys. J. Agric. Sci., 1971. (In press.)

ChannaBasavanna, G.P.: A contribution to the knowledge of Indian eriophyid mites (Eriophyoidea, Trombidiformes, Acarina). Univ. Agric. Sci., Bangalore, Bull., 153 pp., 1966.

–, Banu, K.: Plant feeding mites of India II. Mys. J. Agric. Sci., 1971 a. (In press.)
– , Banu, K.: Plant feeding mites of India III. Mys. J. Agric. Sci., 1971 b. (In press.)
– , Nageshchandra, B. K.: Bibliography of Indian plant feeding mites. Univ. Agric. Sci., Bangalore, Bull. Res. Series, No. 8, 24 pp. 1971.
Cherian, M.C.: The cholam mites *(Paratetranychus indicus)* on sorghum. Madras Agric. J. 21: 1–6, 1933.
Das, G.M.: Bionomics of the tea red spider *Oligonychus coffeae* (Nietner). Bull. Entomol. Res. 50: 265–274, 1959.
– , Das, S.C.: Effect of temperature and humidity on the development of the tea spider mite *Oligonychus coffeae* (Nietner). Bull. Entomol. Res. 57: 433–436, 1967.
– , Sengupta, N.: Observations on the pink mite, *Acaphylla theae* (Watt) Keifer, of tea in North-East India. J. Zool. Soc., Bengal, 10: 39–48, 1958.
– , – : Biology and control of the purple mite, *Calacarus carinatus* (Green), a pest of tea in North-East India. J. Zool. Soc., India, 14: 64–72, 1963.
Ghai: Entomology in India. Entomol. Soc. India, 529 pp., 1964.
Hirst, S.: On some species of red spider. Ann. Mag. Nat. Hist. (Sec.9) 14: 522–527, 1924.
– Descriptions of new mites, including four new species of "red spider". Proc. Zool. Soc., London, 825–841, 1926.
Khot, N.S., Patel, G.A.: Life history of the gladiolus mite *Tetranychus equatorius* McGregor. Indian J. Entomol. 18: 149–160, 1956.
Rahman, K.A. and Sapra, A.N.: Mites of the family Tetranychidae from Lyallpur with descriptions of new species. Proc. Indian Acad. Sci. 11 (B): 177–196, 1940 a.
– , –: Biology of the mite *Paratetranychus indicus* Hirst a pest of sugarcane. Indian J. Entomol. 2: 201–212, 1940 b.
– , –: On the biology of the vegetable mite *(Tetranychus cucurbitae* Rahman and Sapra: Fam. Tetranychidae). Indian J. Agric. Sci. 15: 124–130, 1946.
Puttarudraiah, M. and ChannaBasavanna, G.P.: A preliminary account of phytophagous mites of Mysore. Proc. 1st All-India Congr. Zool., Calcutta 2: 532–540, 1959.
Wood–Mason, J.: Report on the tea bug of Assam, London, 20 pp. 1884.

Appendix

Eriophyidae

1. *Abacarus delhiensis* Basavanna
2. *Abacarus sacchari* Basavanna
3. *Acaphylla indiae* Keifer
4. *Acaphylla syzygii* Basavanna
5. *Acaphylla theae* (Watt)
6. *Aceria balanites* (Massee)
7. *Aceria bambusae* Basavanna
8. *Aceria barleriae* Basavanna
9. *Aceria cajani* Basavanna
10. *Aceria clerodendronis* Farkas
11. *Aceria cordiae* Basavanna
12. *Aceria crotalariae* Basavanna
13. *Aceria dalbergiae* Basavanna
14. *Aceria erythrinae* Basavanna
15. *Aceria ficus* (Cotte)
16. *Aceria gastrotrichus* (Nalepa)
17. *Aceria holopteleae* Basavanna
18. *Aceria granati* (Can & Mass)
19. *Aceria gossypii* (Banks)
20. *Aceria infectorriae* Basavanna
21. *Aceria jasmini* Basavanna
22. *Aceria justiciae* Basavanna
23. *Aceria litchii* (Keifer)
24. *Aceria lycopersici* (Wolff.)
25. *Aceria mangiferae* Sayed
26. *Aceria medicagenis* (Keifer)
27. *Aceria mori* (Keifer)
28. *Aceria nandiensis* Basavanna
29. *Aceria nerii* Basavanna
30. *Aceria pongamiae* Basavanna
31. *Aceria puttarudraiahi* Basavanna
32. *Aceria ruelliae* Basavanna
33. *Aceria sacchari* Basavanna
34. *Aceria sapindi* Basavanna

35. *Aceria sheldoni* (Ewing)
36. *Aceria sorghi* Basavanna
37. *Aceria tulipae* (Keifer)
38. *Aculus asperus* Basavanna
39. *Aculus colei* Basavanna
40. *Aculus indicus* Basavanna
41. *Aculus mckenziei* (Keifer)
42. *Aculus menoni* Basavanna
43. *Aculus moringae* Basavanna
44. *Anthocoptes adhatodae* Basavanna
45. *Calacarus carinatus* (Green)
46. *Calepitrimerus azadirachtae* Basavanna
47. *Calepitrimerus leucadis* Basavanna
48. *Calepitrimerus mysorensis* Basavanna
49. *Diptilomiopus assamica* Keifer
50. *Eriophyes canthii* Basavanna
51. *Eriophyes casuarinae* Basavanna
52. *Eriophyes cernuus* Massee
53. *Eriophyes cheriani* Massee
54. *Eriophyes cyperi* Basavanna
55. *Eriophyes ficivorus* Basavanna
56. *Eriophyes prosopidis* Saxena
57. *Eriophyes rubifolii* Keifer
58. *Eriophyes terminaliae* Basavanna
59. *Floracarus fleshneri* Keifer
60. *Keiferana neolitseae* Basavanna
61. *Metaculus mangiferae* (Attiah)
62. *Neocalacarus mangiferae* Basavanna
63. *Oxypleurites convolvuli* Basavanna
64. *Paraphytoptus chrysanthemi* Keifer
65. *Paratetra murrayae* Basavanna
66. *Phyllocoptruta oleivora* (Ashnead)
67. *Rhyncaphytoptus ficifoliae* Keifer
68. *Tegolopus bambusae* Basavanna
69. *Tetra bauhiniae* Basavanna
70. *Thamnacus euphorbiae* Basavanna

Tenuipalpidae

1. *Brevipalpus californicus* (Banks)
2. *Brevipalpus chilensis* Baker
3. *Brevipalpus deleoni* Pritchard & Baker
4. *Brevipalpus obovatus* Donn.
5. *Brevipalpus phoenicis* (Geijskes)
6. *Brevipalpus* sp.
7. *Cenopalpus pulcher* (Ganestrini & Fanzago)
8. *Colopalpus* sp.
9. *Dolichotetranychus floridanus* (Banks)
10. *Pentamerismus oregonensis* McGregor
11. *Phytoptipalpus albizziae* Pritchard & Baker
12. *Raoiella indica* Hirst
13. *Raoiella macfarlanei* Pritchard & Baker
14. *Raoiella* sp.
15. *Tenuipalpus crassus* Andre
16. *Tenuipalpus micheli* Lawrence
17. *Tenuipalpus* sp.
18. *Tenuipalpus granati* Sayed

Tetranychidae

1. *Bryobia eharai* Pritchard & Keifer
2. *Bryobia praetiosa* Koch
3. *Eotetranychus hirsti* Pritchard & Baker
4. *Eotetranychus suginamensis* (Tokoyama)
5. *Eutetranychus orientalis* (Klein)
6. *Oligonychus biharensis* (Hirst)
7. *Oligonychus iseilemae* (Hirst)
8. *Oligonychus indicus* (Hirst)
9. *Oligonychus coffeae* (Nietner)
10. *Oligonychus mangiferus* (Rahman & Sapra)
11. *Oligonychus oryzae* (Hirst)
12. *Oligonychus pratensis* (Banks)
13. *Oligonychus punicae* (Hirst)
14. *Petrobia harti* (Ewing)
15. *Petrobia latens* (Müller)
16. *Schizotetranychus andropogani* (Hirst)
17. *Schizotetranychus hindustanicus* (Hirst)
18. *Tetranychus cucurbitae* Rahman & Sapra
19. *Tetranychus fijiensis* Hirst
20. *Tetranychus ludeni* Zacher
21. *Tetranychus telarius* (Linnaeus)

Tarsonemidae

1. *Polyphagotarsonemus latus* (Banks)
2. *Tarsonemus kitchneri*

THE EFFECT OF TEMPERATURE ON HATCHING OF THE WINTER EGGS OF FRUIT TREE RED SPIDER MITE, *PANONYCHUS ULMI* (KOCH)

J. E. CRANHAM
East Malling Research Station, Maidstone, Kent, England

In south-east England, the winter (diapause) eggs of *Panonychus ulmi* hatch over several weeks, and the date of peak hatch varies considerably from year to year in the same orchard, and by as much as a month between different orchards in the same year. Light et al (1968) were not successful in forecasting peak hatch on the basis of heat sums.

Lees (1953) showed that diapause was broken by chilling at 1°, 5° or 9° for 150–200 days. The present paper briefly summarizes and discusses the results of a detailed study of the effects of chilling and incubation temperatures on hatching; a full account has been submitted for publication elsewhere.

Results

The effect of duration and temperature of chilling. As the chilling period was increased from 60 to over 200 days, the total % hatch on subsequent incubation at 21° increased, and the mean incubation time and its variance decreased. Total % hatch reached the recorded maxima after 200–230 days of chilling; temperatures of 0° and 5° were optimal and were more effective than –5° and 9°, however long the period of chilling. The limits for the reaction were close to –10° and 15° (Fig. 1, graph a).

The regression of the mean incubation time, t, is shown in Fig. 2, and for chilling temperatures of –5°, 0°, and 5°, t declined progressively to minima of 17, 9, and 6.5 days respectively. With chilling at 5°, the regression of log t could be fitted in linear form. By other direct experimentation the threshold temperature for complete development to eclosion was found to be 7°. After chilling at 0°, the minimal value of t was the same as for summer eggs at 21°; this indicates that diapause was fully completed but that no morphogenesis could occur at 0°. At 5°, the earliest stages could occur, to the extent of reducing t to 6.5 days. It appears that at –5° some final stages of diapause development could not take place, for the minimal value of t was 8 days more than the 9 days required for morphogenesis.

The effect of incubation temperatures. In the range of 9° to 27°, the total % hatch of winter eggs after chilling was higher at the lower temperatures of 9° and 15°; 27° was close to the upper limit (Fig. 1, graph b). The hatch at 21° was significantly lower than at 15° after 110 days chilling (in March) but the difference was non-significant after 230 days chilling (in July). Graph b shows the result after 170 days chilling (in May). The upper limit of 27° is notable because over 90 % of summer eggs hatched at this temperature, and the upper limit for them was between 32°–35° (Fig. 1, graph c). In further experiments, it was shown that 27° was lethal to most winter eggs even when, after prolonged chilling, the mean incubation time had declined to 6.5 days; but batches of eggs exposed first to 15° or 21° for a few days could finish their development and hatch at 27°. The results suggest that some final processes involved in terminating diapause occur concurrently with early morphogenesis.

With reference to Fig. 1, it will be seen that the ranges of (a) and (b) overlap around 9°, and it was found that some eggs stored continuously at 9° hatched after 200 days; but the total % hatch was smaller and occurred much more slowly than when eggs were chilled at 0° or 5° and then incubated at 9°. The latter was not optimal for chilling but it appears to be near optimal for the late stages of diapause development.

Early– and late-hatching strains. Samples were taken from six orchards, including some previously sampled by Light et al (1968), to provide a series which was found to differ over a range of three weeks in the dates of 50% hatch at "field" temperatures (in a north-facing shelter). Chilling and incubation

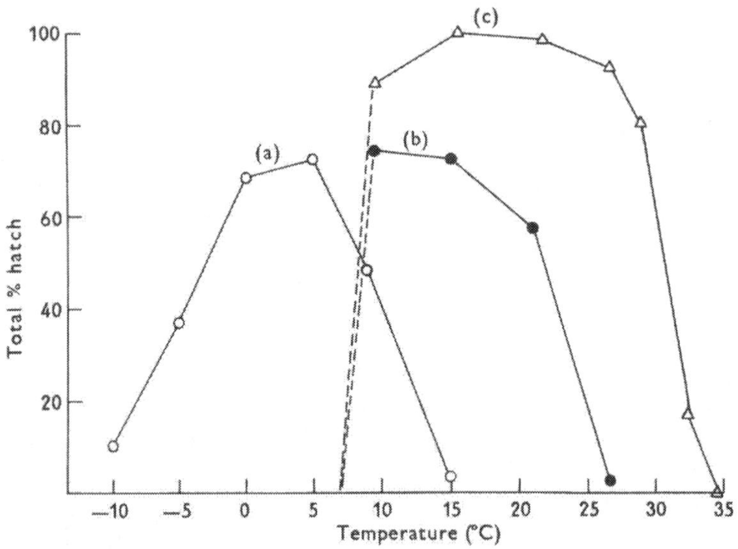

Fig. 1. The ranges of temperature effective (a) for the termination of diapause by chilling (% hatch on subsequent incubation at 21°), (b) for post-diapause development of winter eggs and (c) for development of summer eggs.

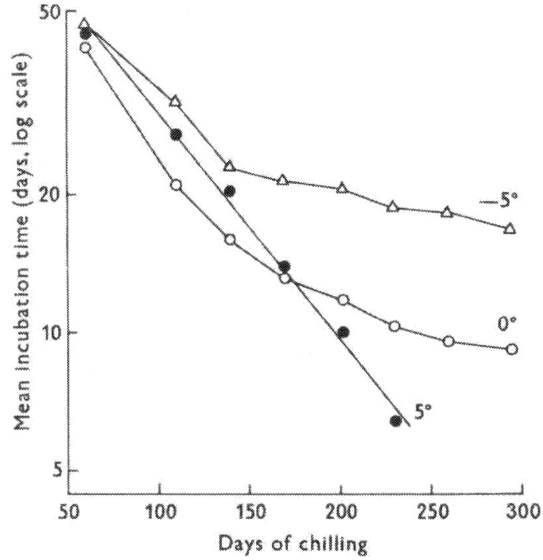

Fig. 2. The regression of log mean incubation time (t) at 21° on the duration of chilling at −5°, 0° and 5° C.

experiments showed essentially that late-hatching strains required longer periods of chilling to attain maximum % hatch and minimum incubation times, and the reduction of % hatch at higher incubation temperatures was even more pronounced than for early strains. Eventually, after extended chilling, mean incubation times were reduced to nine days after 0° chilling and seven days after 5° chilling, with all six strains.

Discussion

Diapause in *P. ulmi* is not a unitary process. The initial stage of diapause development occurs best at temperatures of 0° to 5°. The later stages can occur, at least in many eggs, at temperatures from 0° to 21°, i.e. it can be completed slowly at chilling temperatures or occur more rapidly at temperatures above that range.

Before morphogenesis commences, the regression of the mean incubation time reflects the diminishing heat sum required for completion of diapause development. Thus, if K_t is the heat sum required on incubation for development to eclosion, $K_t = K_d + K_m$ where K_d is that for completion of diapause and K_m that for complete morphogenesis. K_m appears to be virtually the same for summer and winter eggs (diapause occurs very early, at the blastoderm stage), and on the evidence obtained was the same for early- and late-hatching strains. Late-hatching strains were characterized by larger K_d values than early strains for the same duration of chilling.

We may now consider forecasting the dates of 50% hatch in the field from heat sums estimated by earlier incubation at constant temperature. Light et al (1968) looked for agreement with the accumulated heat sum from 1st January, and did not find it. In fact, from year to year, on any set date, the residual value of K_t depends both on the effectiveness of winter chilling and on the accumulated heat sum above the threshold. Also, differences in K_d due to genetic factors could change from year to year, if the genetic composition of populations changed under selection. However, for samples taken directly from the field, after 1st April, there was good agreement between (1) the estimated heat sums on incubation (not above 21°), and (2) the actual heat sums recorded in the field from the date of sampling to the date of 50 % hatch (Table 1). Before April there was poor agreement, due to the progressive reduction of K_t at chilling temperatures in the field.

Lees (1953), in discussing the phenology of the mite, noted that the intense diapause ensured survival by delaying hatch until there was adequate foliage on the trees. The extended time of hatch is an additional safety device, and this woul appear to be mainly due to genetically-based variation in K_d (i.e. in the intensity of diapause) within a population. It is reasonable to suppose that K_d has evolved as a device to control hatching time, and may be influenced both by natural selection (e.g. climatic factors) and artificial selection (e.g. choice and timing of acaricides). Thus, if the chilling requirement provides the "coarse adjustment" of the mechanism controlling time of hatch, the "fine adjustment" is provided by the gradually declining heat sum required to complete diapause development. The effec of this is to apply a brake to development which is gradually released by rising field temperatures, and only more slowly at low temperatures. This acts to prevent premature hatch if the weather is unseasonably warm in March or April; alternatively, if cold weather is unusually prolonged, the hatch occurs faster when it does turn warm (Table 2). This exemplifies once again how subtly the mite can adapt to the environment.

Table 1. Estimated heat sums for 50% hatch of *Panonychus ulmi* eggs sampled from the field and incubated at 21°, compared with actual heat sums recorded in the field (both above a threshold of 5.6°)

Heat sums (days °C) for 50% hatch					
1968			1969		
Sampling date	Estimated	Actual	Sampling date	Estimated	Actual
30.I.	638	312	15.I.	793	357
29.II.	451	301	3.III.	451	312
1.IV.	233	232	1.IV.	311	284
3.V.	109	103	1.V.	172	178

Table 2. The effect of the declining heat sum required to terminate diapause, as a "brake" on the rate of development of winter eggs

Temperature of chilling	Temperature of incubation	Days to 50% hatch, t		
°C	°C	3.III.	1.IV.	1.V.
5	9.5	140	–	55
5	15	53	–	21
5	21	28	20	14
Field temperatures	21	29	20	11

References

Lees, A.D.: Environmental factors controlling the evocation and termination of diapause in the fruit tree red spider mite, *Metatetranychus ulmi* Koch. Ann. Appl. Biol. 40, 449–86, 1953.

Light, W.I. St. G., John, Margaret E., Gould, H. J. Coghill, K. I.: Hatching of the winter eggs of the fruit tree red spider mite, *Panonychus ulmi* (Koch). Ann Appl. Biol. 62, 227–39, 1968.

FEEDING RESPONSE OF THE TWO-SPOTTED SPIDER MITE TO SOME PLANT COMPONENTS

Z. T. DĄBROWSKI
Department of Entomology, Warsaw Agriculture University,
Warsaw, Poland

Current knowledge of the plant food acceptance by the spider mites and the mechanism underlying the tetranychid mites-plant interaction is inadequate in many important aspects. There is a great need for fundamental information on all of the phases of mite behaviour as well as nutritional requirements of these mites.

The research reported herein is a continuation of study of the host plant resistance problem started by the author in Dr Rodriquez' Acarology Laboratory (Rodriquez et al. 1971; Dąbrowski and Rodriquez, 1972). The main objective of the author's present research is to determine the existence and concentration of the chemical plant stimuli influencing different feeding behavioral responses of the females of *Tetranychus urticae* Koch.

The feeding rate of *T. urticae* females on various solutions was estimated by allowing the mites to feed on a water solution of plant extracts or chemical compounds made radioactive with 32 P. The experimental system consisted of open-end glass tubes with two Parafilm sachets attached at the top (for detailes see papers mentioned above).

The preliminary experiment showed that the water solution of residues of lyophilized bean (var. Saxa), strawberry (var. Dixieland), tobacco (*Nicotiana tabacum* L.) and ginkgo (*Ginkgo biloba* L.)eaves elicited various feeding response of *T. urticae* females (Fig. 1). The bean and strawberry foliage extracts elicited marked positive gustatory response of *T. urticae* females. The tobacco and ginkgo leaves showed to possess some negative evoking chemical stimulus which significantly reduced the *T. urticae* feeding ratio in comparison to destiled water as a control.

The high ratio of *T. urticae* feeding on the bean and strawberry foliage extracts suggests that only one or a group of chemicals can act as a feeding stimulant or deterrent. According to Dr Beck's definition stimuli tending to promote continuous feeding are termed "feeding stimulants", and those preventing continuous feeding are designated as "feeding deterrents" (Beck 1965).

In the following investigations nine sugars, seventeen amino acids and nine vitamins, previously described by biochemists in plant tissues, were used to determine their effect on the mite feeding behaviour. Concentrations of chemicals studied were based on the chemical analyses of plant material as well the concentration of the chemicals in diets used to rear *T. urticae* (Rodriquez 1969).

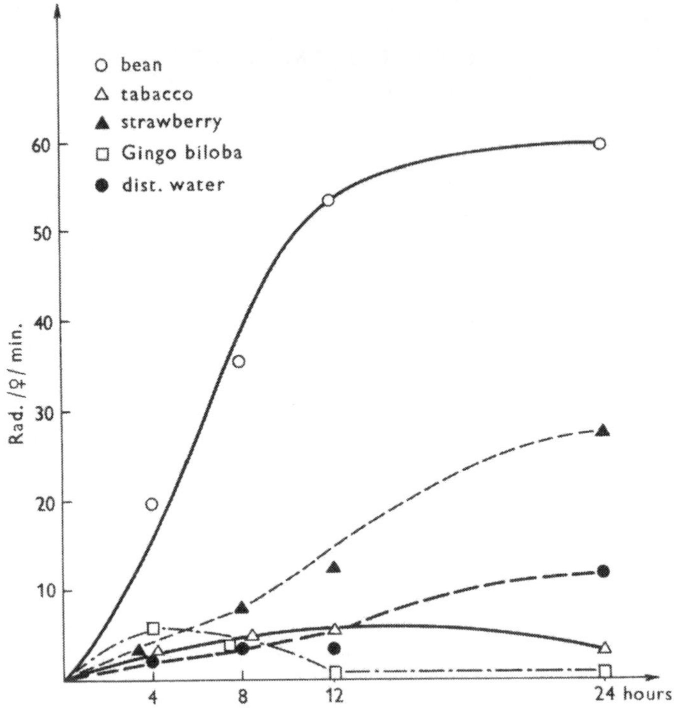

Fig. 1. Gustatory responses of *Tetranychus urticae* feeding on ^{32}P–labeled water extracts from foliage of four plant species.

Most of the sugars showed a stimulative effect on the mite feeding ratio. The highest ingestion was observed on sucrose and fructose solutions, used in 2 and 1 per cent concentrations. Various concentrations of one sugar elicited significant differences in the response of *T. urticae*, as is the case, for example, for sucrose solution (Fig. 2). It is of interest that sucrose in 1 and 2 per cent concentrations acted as a better feeding stimulant than the water solution of lyophilized bean leaves. Classification of other sugars in regard to their effect on *T. urticae* food ingestion is given in Tab. 1.

Most of the water-soluble vitamins studied in lower concentrations showed a stimulative or neutral effect on the food ingestion by the mites; for example: folic acid, inositol, riboflavin and amino benzoic acid. Some other vitamins, as pyridoxine hydrochloride, tiamine, hydrochloride and ascorbic acid did not significantly effect the *T. urticae* feeding ratio when they were tested in the lower concentration (Tab. 1.).

A solution of 100 mg of most tested amino acids in 100 ml destiled water acted as strong feeding deterrents. Some other compounds used in a lower concentration even stimulated the mites. For example: L–tyrosine, L–lysine or L–phenylalanine.

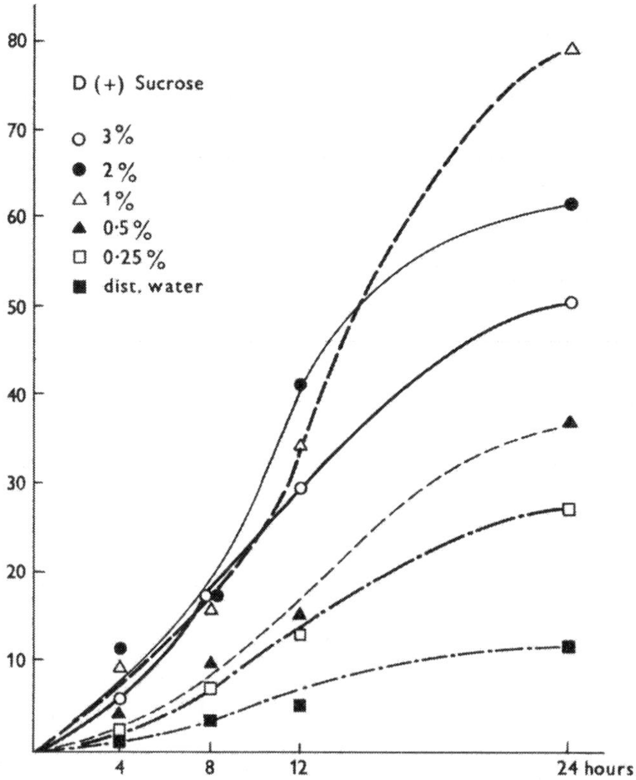

Fig. 2. Effect of water sucrose solutions on gustatory response of *Tetranychus urticae* females.

Table 1. Classification of chemicals influencing different feeding behaviour responses of *Tetranychus urticae* females

Sugars

Concentration:	3%	2%	1%	0,5%	0,25%
d /+/ Sucrose	SS	SS	SS	SS	SS
d /–/ Levulose	S	SS	SS	SS	SS
Rafinose	S	S	S	S	S
Mannose	S	S	S	S	S
d /+/ Maltose	SS	S	S	N	N
d /–/ Sorbitol	S–D	S	S	S	N
d /+/ Xylose	S–D	S–D	S	S	S
d /+/ Arabinose	S–D	S–D	S–D	S–D	S
Cellobiose	D	D	N	N	N

Vitamins:

Concentration	1 mg/100 ml	0,1 mg/100 ml	0,01 mg/100 ml
i – Inositol	N	N	N–S
Riboflavin	N–DD	N–S	N–S
Thiamine hydrochloride	N–D	N–D	N
Folic acid	D–N	D–S	S
p–Amino benzoic acid	D	S	S
Nicotinic acid	DD	S	N
Pyridoxine hydrochloride	D	N	N
Ascorbic acid	DD	N–D	N
Biotin	DD	DD	N

Amino acids

Concentration:	1000 mg/100 ml	100 mg/100 ml	10 mg/100 ml
L–tyrosine	S–D	SS	S
L–lysine	N–D	S	SS
L–phenylalanine	N–D	S	S
L–arginine	N–D	S	N
L–serine	N–D	N–D	SS
L–alanine	DD	SS	N
L–isoleucine	DD	S	S
L–leucine	DD	S	S
L–histidine	DD	N	S
L–methionine	DD	N	N
L–aminoacetic acid	D	D	S
L–glutamic acid	DD	N–D	S
L–aspartic acid	N–DD	N–DD	N–D
L–cystine	DD	N–DD	N
L–proline	DD	D	N
L–threonine	DD	DD	S
L–valine	DD	N	N

SS – strong stimulant, S – stimulant, N – neutral,
D – deterrent, DD – strong deterrent,
S–D means that after 8–12 hours the compound showed stimulant effect and after 24 hrs the deterrent effect, etc..

L–glutamic acid, L–aspartic acid and L–cystine acted during the first 12 hrs of mite feeding as the gustatory neutral chemicals and as strong deterrents after 24 hours. The phenomenon could also be effected by some toxic action on the mites of the single compound used. Accumulation of these tasteless chemicals could restrain the mites from food ingestion. Some other aspects of the amino acids effects on the *T. urticae* feeding behaviour during 24 hours test will be commented upon in the author's future papers.

Conclusions

1. The water extracts of bean and strawberry plants contained a substance (s) which acted as a powerful feeding stimulant to the *T. urticae* females.
2. *T. urticae* exhibited differences in gustatory response for the extracts from different plant species. The order of preference was the same as the mite fecundity observed in another experiment on fresh leaves from these plants.
3. The most obvious stimulative differences exhibited were between the sugars. Some of them possess, however, a significant deterrent effect.
4. The gustatory response of the *T. urticae* females to amino acids and water-soluble vitamins depended upon the concentration used.
5. The stimulus-response mechanisms underlying the so-called botanical instinct of *T. urticae* females is not completely clarified. The data presented here suggest that the feeding stimulants at the tissue level play an important role in the mite's host plant utilization. Some of the feeding stimulants are nutritionally important substances (Rodriquez 1969). On the other hand some of the so-called secondary plant chemicals /phenolic compounds/ provide gustatory stimuli which acted as deterrents to *T. urticae* females (Dąbrowski and Rodriquez 1972).

References

Beck, S.D.: Resistance of plants to insects. Ann. Rev. Entomol. 10: 207–232, 1965.

Dąbrowski, Z.T., Rodriquez, J.G.: Gustatory responses of *Tetranychus urticae* Koch to phenolic compounds of strawberry foliage. Zesz. Prob. Post. Nauk Roln., Proc. VIIth European Mite Symposium, 129, 1972.

Rodriquez, J.G., Dietetics and nutrition of *Tetranychus urticae* Koch. Proc. 2nd Int. Congress Acarol., 469–475, 1969.

– , Dąbrowski Z.T., Stiltz L.P., Chaplin C. E., Smith W.O. Jr.: Studies on resistance of strawberries to mites. 2. Preference and nonpreference responses of *Tetranychus urticae* and *T. turkestani* to water-soluble extracts of foliage. J..Econ. Entomol. 64: 383–387, 1971.

SOME PROBLEMS IN THE CONTROL OF COTTON MITES IN TURKEY

Z. DÜZGÜNEŞ
University of Ankara, Turkey

Cotton has been the leading export item of Turkey in the last five years. About 510 thousand tons of lint is produced from of 750 thousand hectares. The yield per hectare has increased 2.5 times in the last 20 years as a result of better varieties and techniques.

The two main cotton growing regions of Turkey are the Mediterranean and the Aegean regions. A small area in the east, near the Russian and Iranian borders, is also suitable for cotton production. It is planned to move the cotton area to the south-east when the new Keban dam is completed. As the cotton area extends and the interest of the growers in pest control increases, some new problems appear. Those related to spider mites will be reported upon in this paper.

The first observation in Turkey on a mite occurring on cotton was made in 1934 by Iyriboz (1937) in the Aegean region. The same mite *(Tetranychus urticae)* was also found by Gassner on cotton in the east, damaging up to 30 % of the crop in 1937 (Bodenheimer, 1941).

However the history of mite-damage in the important cotton region in the south and the west started in about 1960. In these regions only local damage of no importance had been reported before. For the southern region even a special cotton dust without acaricide effect had been manufactured.

Following the extensive use of this and other insecticides, especially against some lepidopterous larvae, the populations of *Tetranychus* species increased to a destructive level; the most prevalent species now being *T. urticae* Koch, *T. atlanticus* McGregor, *T. cinnabarinus* Boisduval. For the western region *T. desertorum* should also be added to the list (Düzgüneş, 1962).

The first problem arises from the fact that in some fields two or three species are found together. Because of the differential effects of acaricides used, the total population of mites cannot be reduced to the economic threshold. The effective time and population density for applying acaricides also differ from species to species. The diversity of ecological conditions between as well as within these large regions creates another problem. An acaricide or insecticide with acaricide effect gives significantly different results in different fields.

Yet another problem is related to the appearance of a resistant strain of *T. cinnabarinus* in the south. This was first observed in 1969 when surprisingly bad results were obtained in several fields from the pesticides used earlier with an effectiveness of 99 %. A test with the slide-deep method showed a five-fold resistance. Observations in the same region in the succeeding years indicate that the resistance is becomming stronger and the area affected by resistant strain wider.

In the western cotton region, however, no such case has been reported so far despite the application of the same pesticides for an even longer time. This is another indication of the fact that the ecological conditions also play an important role in the development of resistance (Düzgüneş, 1965).

It may also be interesting to note here that the resistant and normal strains of *T. cinnabarinus* which have been reared on bean leaves and plants in my laboratory under identical conditions, have shown some disparities in their biological characteristics as also observed by Dittrich (1961, 1962, 1963), Lehr and Smith (1957), Taylor and Smith (1956), Kenster and Streu (1967). The most important changes found in the resistant strain as compared with the normal one are: a) The period from egg to adult is longer; b) The number of eggs per female is less; c) The percentage of hatching eggs is smaller; d) The life of adult females is shorter; e) The natural mortality in all stages is higher. From these results it is possible to postulate that the resistant strain is a mutant which can only increase under favourable conditions.

In order to deal with these problems we started a comprehensive project which aims at an effective system of integrated control. Meanwhile we have been studying the biology of each species under different conditions in the laboratory and the field. This study provides us with information on the changes in population density as related to generations, seasons, years, hosts, climatic factors, predators, cultivation technique (such as irrigation, soil management, sowing time) and application of pesticides.

In some sections of western region we already started with an experimental integrated control against mites, aphids and thrips. Preliminary results of the continuous survey organized to determine the main pests, the activity of predators and the results of the application of pesticides seem to be highly promising.

References

Bodenheimer, F.S.: Türkiyede ziraate ve ağaçlara zararlı olan böcekler ve bunlarla savaş hakkenda bir etüd. 1941, 347 pp., 1958.

Dittrich, V.: Populationsgenetische Untersuchungen an normalen und phosphorsäureester – resistenten Stämmen von *Tetranychus urticae* Koch. Z. angew. Entomol. 48: 34–57, 1961.

– A comparative study of toxicological test methods on a population of the two-spotted spider mite (*Tetranychus telarius* (L.). J. Econ. Entomol. 55: 644–648, 1962.

– Genetics and dynamics of organophosphate – resistance in strains of *Tetranychus urticae* Koch. Adv. Acarology I: 238–247, 1963.

Düzgüneş, Z.: Pamuk akarları (Cotton mites). Türkiye ziraatına zararlı olan böcekler ve mücadelesi. Fas. 6:70–77 (Bölüm III), 1962.

– Ziraatta akarlar ve bunlarla mücadele (The mites in agriculture and their control). XI. Türk Mikrobiyoloji Kongresi 283–304, 1965.
Iyriboz, N.: Pamuk hastalıklari (Cotton pests). Ziraat Vekâleti 139 pp., 1937.
Kenster, Jr. D.L., Streu H.T.: A biological and toxicological study of strains of two-spotted spider mites. J. Econ. Entomol. 60: 1073–1078, 1967.
Lehr, R., Smith F.F.: The reproductive capacity of the two-spotted spider mite complex. J. Econ. Entomol. 50: 634–636, 1957.
Taylor, E.A., Smith F.F.: Transmission of resistance between strains of two-spotted spider mite. J. Econ. Entomol. 49: 858–859, 1956.

A CONTRIBUTION TO THE STUDY OF *ACERIA OLEAE* (NALEPA, 1900). (ACARINA: ERIOPHYIDAE)

E. N. HATZINIKOLIS
Institute of Vegetable Crops, Halandri, Athens, Greece

The knowledge concerning *A. oleae* is very limited. In 1900 Nalepa described the new species *Eriophyes oleae*. It is possible that the Greek research workers Kavadas (1927), Anagnostopoulos (1937), and Pelecassis (1962) referred to the same mite which is found on leaves of olive trees in Greece. In 1959 and 1954 Georghiou reports that this mite is the cause of damage to olive fruits in Cyprus. Recently Hatzinikolis (1969 a, b, c, 1970) gave some information on the pathogenesis and distribution of *A. oleae*. The investigations reported herein were conducted to increase the limited knowledge of *A. oleae* which is very important for the olive tree plantations in Greece.

Methods

The technique of breeding *A. oleae* in the laboratory on olive tree leaf discs is based on the methods of Davis (1964) and Rice and Strong (1962). The environmental conditions in the breeding chamber were: temperature 21–25°C, relative humidity 80–90 %, light intensity 3800 lux and day-length 14 hrs.

Identity

Nalepa (1900) described the new species *Eriophyes oleae*. Georgiou later (1958) refers to the mite as *Aceria oleae* after personal communication with H. Keifer. In our studies the mite is referred to as *A. oleae*. No morphological differences were observed among individuals in summer and winter, except for a difference in length and a slight change in colour ranging from white in summer to pink in winter.

Distribution

The mite was found everywhere in Greece. It was found in 38 out of 40 olive growing areas of Greece.

Economic importance

A. oleae has a great economic importance for the olive tree plantations in Greece. The mite attacks mainly the young trees or the new growth which follows a serious pruning for the renovation of old trees. The result of this attack is an inhibition of growth because of the partial or total damage of the foliage of trees. Under favourable conditions it is possible to cause a total fruitlessness after destruction of flowers and young fruits.

Pathogenesis

A. oleae attacks the leaves, the flowers and fruits of olive trees. The leaf attack is causing characteristic changes of leaf shape and leaf colour and eventually defoliation. The flower attack is causing flower drop or the formation of small fruit which do not grow any further. The olive fruits are attacked severely in the young stage and this is causing the premature dropping of these fruits.

Ecology

A. oleae overwinters as female adults under the star-shaped hairs of the lower surface of the leaves. Before flowering the majority of the population is still in these places. During the flowering period 90 % of the population is found in the flower clusters mainly on the ovaries of the flowers. Later on the same proportion is found also on the small fruits. After that stage the mites leave the fruits and return to the young leaves where they stay until the following spring. The percentage

Table 1. Percentage of the population of *Aceria oleae* on olive tree

Months	Leaves			Flowers			Fruit		
	Larvae & Nymphs %	Pupae %	Adults %	Larvae & Nymphs %	Pupae %	Adults %	Larvae & Nymphs %	Pupae %	Adults %
January	2	28	70						
February		4	96						
March	2	I	97						
April	49	20	25	2		4			
May	5	12	1	24	11	47			
June			2	13	14	2	25	14	30
July	6	2	34				21	25	12
August	10	52	38						
September	22	27	51						
October	18	20	62						
November	10	76	14						
December	3	28	69						

proportion of the population of *A. oleae* on different organs of the olive tree is given in Table 1. Fig. 1. shows the variation in population density of *A. oleae* on growing olive tree leaves.

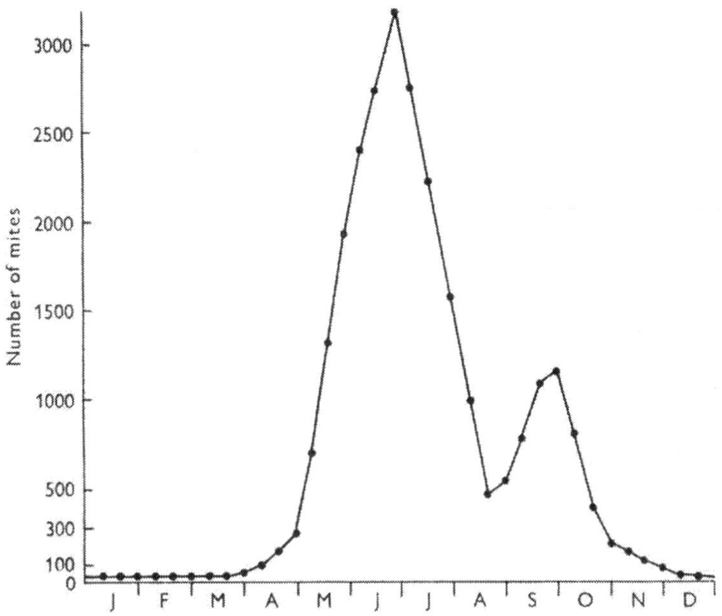

Fig. 1. The variation in population density of *Aceria oleae* on six young olive tree leaves.

Biology

The results of the biological studies of *A. oleae* are given in Table 2.

The eggs hatched after incubation during a period ranging from 56 to 84 hrs with an average of 60 ± 9 hrs. The preovipositional period ranged from 43 to 77 hrs with an average of 58 ± 12 hrs. The ovipositional period ranged from 8.3 to 16.0 days with an average of 12.1 days. The duration of the female feeding stage and the quiescent larval stages were 1.5 and 2.6 days respectively and 3.6 ± 0.6 days in all The female feeding stage and the quiescent nymphal stages lasted 2.1 and 2.8 days respectively and 5.4 ± 0.5 days in all. The total developmental time ranged from 10.8 to 15.2 days with an average of 13.9 ± 0.8 days.

The number of eggs laid by one female varied from 7 to 42 and the maximum of eggs laid in one day was 6. The number of successive generations in the year 1969 was 12 to 15.

Table 2. Developmental times of immature stages, duration of pre-ovipositional and ovipositional periods of *Aceria oleae* at 21-25°C

Stage	No. obs.	Number of days									Standard deviation
		Feeding			Quiescent			Total			
		Max.	Min.	Avg.	Max.	Min.	Avg.	Max.	Min.	Avg.	
Egg	30							3.5	2.3	2.5	0.4
Larva	20	2.3	1.3	1.5	3.8	2.3	2.6	6.1	3.3	3.6	0.6
Nymph	22	3.3	2.0	2.1	3.6	2.6	2.8	6.6	4.6	5.4	0.5
All stages (egg to adult)	20							15.2	10.8	13.9	0.8
Pre-ovipositional period	28							3.2	1.8	2.2	0.5
Ovipositional period	40							16.0	8.3	12.1	-

References

Anagnostopoulos, P.: The enemies of fruit trees. 1– 642, 1939.

Davis, E.N.: Autecological studies of *Rhynacus breitlowi*, (Acarina: Eriophyidae). Florida Entomol. 47:113–121, 1964.

Georghiou, C.P.: Reported of the Plant Protection for the year 1954. 95, 1954.

– Plant feeding mites of Cyprus. Bull. F.A.O., VII: 153–160, 1959.

Hatzinikolis, E.N.: Preliminary notes on Tetranychoidea and Eriophyidae mites infesting cultivated plants in Greece. Proc. 2nd Int. Congress Acarology: 161–167, 1969 a.

– Acariens phytophages signalés en Grèce sur l'olivier *(Olea europaea L.)* 8th F.A.O. Meeting on the Control of the Olive Pests, held Athens, Greece, 1969b.

– Acariens Eriophyoidea signalés sur des plantes cultivées en Grèce. Ann. Inst. Phytopathol. Benaki 2:54–56, 1969c.

– Lutte chimique contre un acarien méconu, mais très dangereux, sur l'olivier *(Tegonotus bassani* Keifer, 1959). VIIe Congrès International de la protection des Plantes, Paris, France, 1970.

Kavadas, D.: *Phytoptus oleae* un acarien de l'olivier. Agric. Bull., N. 66:36–38, 1927. (En Greque.)

Nalepa, A.: Neue Gallmilben. Anz. Akad. Wiss., Wien, 137 (15):154–156, 1900.

– Systematik der Eriophyiden. Akad. Wiss., Wien, 2:139–140, 1905.

Pelekassis, C.D.E.: A catalogue of the more important insects and other animals harmful to the agricultural crops of Greece during the last thirty-year period. Ann. Inst. Phytopathol. Benaki 5:1–104, 1962.

Rice, R. and Strong, F.E.: Bionomics of the tomate russet mite *Vasates lycopersici* (Masses). Ann. Entomol. Soc. Am. 55:431–435, 1962.

STUDIES ON THE FEEDING OF FOUR SPECIES OF PHYTOSEIID MITES (ACARINA: PHYTOSEIIDAE)

D. KROPCZYNSKA–LINKIEWICZ
Warsaw Agricultural University, Department of Applied Entomology,
Warsaw, Poland

Data dealingwith food requirements of phytoseiid mites are very scarce. The food preference of some species was studied by Collyer (1956), Dosse (1956), Herbert (1956, 1959) and Chant (1959). A more detailed study has been made by McMurtry and Scriven (1964, 1965). Their results concerning feeding, reproduction and development of phytoseiid mites on different foods were restricted to species which commonly occur in California.

The studies reported in this paper were conducted to acquire information on feeding responses of four of the most common predatory mites occurring in Polish orchards. Experiments were carried out in the laboratory. We have been trying to determine food preferences for three species of phytoseiids: *Phytoseius macropilis* (Banks), *Typhlodromus potentillae* Garman and *Typhlodromus pyri* Sch. The following species of mites were offered as prey: *Panonychus ulmi* (Koch), *Bryobia rubrioculus* (Sch.), *Tetranychus urticae* Koch and *Aculus schlechtendali* (Nal.). The data were compared with results obtained earlier for *Typhlodromus finlandicus* (Oud.) (Kropczyńska 1970).

Results

Nor one of the examined species of phytoseiids appears to be adapted to feeding upon only certain species of mites. Differences were found between species only in the number of prey attacked and consumed. Females of *T. finlandicus* killed daily the lowest number of mites: on average 15 eriophyids and 6 spider mites per day. The remaining species consumed much larger numbers daily: 40 eriophyids and 14 immature stages of tetranychids on average. All phytoseiids showed a preference for immature stages of prey; this was especially true in the case of *B. rubrioculus* where predators were able to consume only larvae and nymphs of this mite. (Table 1.)

Females of *T. finlandicus* were found to lay more eggs when they had gall mites as food. The fecundity of the three other species was higher when they fed upon tetranychids. *T. potentillae* and *Ph. macropilis* had higher rates of oviposition than the remaining predators when fed upon the tested animal foods. The ovi-

Table 1. Biological data for females of phytoseiids reared on different foods

Mite	Food	Time of development /days/	No. eggs laid per female	Daily prey consumption
Typhlodromus finlandicus	*Aculus schlechtendali*	10.8	15.6	15.3
	Panonychus ulmi	7.6	10.8	6.1
	Tetranychus urticae	8.9	7.2	6.0
	Bryobia rubrioculus	7.7	9.9	1.6
	Apple mildew	10.7	9.5	
	Honeydew		1.1	
Typhlodromus potentillae	*Aculus schlechtendali*	8.8	12.1	40.3
	Panonychus ulmi	9.5	17.7	14.6
	Tetranychus urticae	10.8	14.0	13.9
	Bryobia rubrioculus	11.7	4.3	4.5
Typhlodromus pyri	*Aculus schlechtendali*	10.4	9.9	39.6
	Panonychus ulmi	10.9	10.1	13.2
	Tetranychus urticae	9.5	7.1	13.2
	Bryobia rubrioculus	13.7	8.9	4.1
Phytoseius macropilis	*Aculus schlechtendali*	8.9	16.2	35.7
	Panonychus ulmi	9.6	16.9	13.8
	Tetranychus urticae	10.3	16.3	12.5
	Bryobia rubrioculus	11.4	5.8	3.2

position rates for *T. finlandicus* and *T. pyri* were similar. Phytoseiids fed on *B. rubrioculus* did not reach as high a level of reproduction as on other food.

We also tried to determine the number of prey necessary for egg production. The data obtained showed differences. *Ph. macropilis* females who consumed 5 spider mites and 10 eriophyids per day ceased oviposition. *T. finlandicus* had the lowest daily food requirements necessary for egg laying – 1 tetranychid and 4 eriophyids per day.

All species studied fed and reproduced readily on pollen from apple trees. Only *T. finlandicus* fed on the spores and hyphae of apple mildew (*Podosphaera leucotricha* Elliset Everh); on this food the average number of eggs laid per one female during the lifetime was 9.5. This same species can feed on honeydew, but the quality of this food is very low. We did not observe that other species could feed upon fungus.

From these data the following conclusions can be drawn:

Of the four investigated mites, *T. finlandicus* seems to be the most polyphagous species. It can feed and reproduce on all different kinds of animal food tested and is also able to utilize pollen, honeydew, fungal spores and hyphae to prolong survival and in some cases even for egg production. The food preferences of *T. potentillae*, *Ph. macropilis* and *T. pyri* indicated that these species are better adapted to feeding upon phytophagous mites. From the food preferences and higher reproductivity of *T. potentillae* and *Ph. macropilis* it would appear that these are more effective predators of phytophagous mites than certain other species.

References

Chant, D.A.: Phytoseiid mites (Acarina: Phytoseiidae). Part I. Bionomics of seven species in south-eastern England. Part II. A taxonomic review of the family Phytoseiidae, with descriptions of 38 new species. Can. Entomol. 91, Supplement 12, 166 pp. 1959.

Collyer, E.: Notes on the biology of some predacious mites on fruit trees in south-eastern England. Bull. Entomol. Res. 47 /2/: 205–214. 1956.

Dosse, G.: Über die Entwicklung einiger Raubmilben bei verschiedenen Nahrungstieren (Acarina, Phytoseiidae). Pflanzenschutzber. 16: 122–136, 1956.

Herbert, H.J.: Laboratory studies on some factors in the life history of predacious mite *Typhlodromus tiliae* Oudms. (Acarina: Phytoseiidae). Can. Entomol. 88/12/:701–704, 1956.

– Note on feeding ranges of six species of predacious mites (Acarina: Phytoseiidae) in the laboratory. Can. Entomol. 91/12/, 1959.

Kropczyńska, D.: Biologia i ekologia drapieżnego roztocza *Typhlodromus finlandicus* (Oud.) (Acarina: Phytoseiidae). Zesz. Probl. Post. Nauk. Roln. 109:11–42, 1970.

McMurtry, J.A., Scriven, G.T.: Studies on the feeding, reproduction, and development of *Amblyseius hibisci* (Acarina: Phytoseiidae) on various food substances. Ann. Entomol. Soc. Am. 57:649–655, 1964.

– Life-history of *Amblyseius limonicus*, with comparative observations on *Amblyseius hibisci* (Acarina: Phytoseiidae). Ann. Entomol. Soc. Am., 58:106–11, 1965.

A CONTRIBUTION TO THE FAUNA AND BIOLOGY OF TETRANYCHID MITES OF THE CRIMEA (ACARIFORMES, TETRANYCHOIDEA)

I. Z. LIVSHITS AND V. I. MITROFANOV
The State Nikita Botanical Gardens, Yalta, Crimea, USSR

Fauna

On the 160 species of Tetranychidae occurring in the USSR, 81 have been found in the Crimea; they belong to 15 genera of the families Bryobiidae, Tetranychidae and Tenuipalpidae.

Family Bryobiidae. Genus *Bryobia: B. graminum* Schr., *B. watersi* Manson, *B. lagodechiana* Reck, *B. vasiljevi* Reck, *B. macrotibialis* Mathys, *B. kissophila* Eyndh., *B. praetiosa* Koch, *B. rugosa* Liv. et Mitr., *B. ulmophila* Reck, *B. confusa* Liv. et Mitr., *B. centaureae* Liv. et Mitr., *B. redikorzevi* Reck, *B. longisetis* Reck, *B. nikitensis* Liv. et Mitr., *B. gusbariensis* Liv. et Mitr.; genus *Tetranycopsis: T. hostilis* Reck, *T. matikashviliae* Reck; genus *Petrobia: P. latens* (Müll.), *P. zachvatkini* (Reck et Bagd.), *P. sp.;* genus *Neopetrobia: N.* sp.; genus *Tauriobia: T. insueta* Liv. et Mitr.; genus *Reckia: R. samgoriensis* (Reck).

Family Tetranychidae. Genus *Schizotetranychus: Sch. tiliarium* Herm., *Sch. pruni* Oudms., *Sch. fraxini* Reck, *Sch. jachontovi* Reck, *Sch. schizopus* (Zacher), *Sch. prunicola* Liv., *Sch. brachypodii* Liv. et Mitr., *Sch. euphorbiae* Liv. et Mitr., *Sch. luteolus* Liv. et Mitr., *Sch. carpini* Oudms., *Sch.* sp., *Sch. tbilisensis* Reck; genus *Panonychus: P. ulmi* (Koch), *P. citri* (McGregor); genus *Oligonychus: O. longiclavatus* (Reck), *O. piceae* (Reck), *O. ununguis* (Jacobi), *O. lagodechii* Liv. et Mitr., *O. pini* (Hirst), *O. subnudus* (McGregor), *O. sp., O. bischi* Reck, *O. brevipilosus* (Zacher); genus *Tetranychus: T. atlanticus* McGregor, *T. similis* Wainst., *T. savenkoe* Reck, *T. frater* Wainst., *T. viennensis* Zacher, *T. cinnabarinus* Boisd., *T. sp.;* genus *Eurytetranychus: E. buxi* Garman.

Family Tenuipalpidae. Genus *Brevipalpus: B. recki* Liv. et Mitr., *B. lineola* (Can. et Fanz.), *B. wainsteini* Liv. et Mitr., *B. longirostris* Liv. et Mitr., *B. pulcher (Can. et Fanz.), B. carpini* Liv. et Mitr., *B. mespili* Liv. et Mitr., *B. pseudospinosus* Liv. et Mitr., *B. cuneatus* (Can. et Fanz.), *B. olearius* Sayed, *B. russulus* (Boisd.), *B. obovatus* Donn., *B. thelicraniae* Liv. et Mitr.; genus *Pentamerismus: P. taxi* (Haller), *P. oregonensis* McGregor, *P. juniperi* Reck, *P. foliisetis* Liv. et Mitr., *P. tauricus* Liv. et Mitr.; genus *Tenuipalpus: T. rosae* Kadzhaja, *T. kobachidzei* Reck, *T. cheladzeae* Gomelauri, *T. granati* Sayed, *T. ephedreae* Liv. et Mitr.; genus *Aegyptobia – A. exarata* Liv. et Mitr., *A. beglarovi* Liv. et Mitr., *A. wainsteini* Bagd.

Biological characters

From a study of the biology of 13 species of Tetranychidae, the following conclusions can be drawn:

1. All Tetranychid mites are oviparous. Normal reproduction involves fertilization, but parthenogenesis can also take place; in spider mites it is arrhenotokous, in false spider mites and Bryobiidae thelytokous. In the latter, the type of parthenogenesis evidently does not enable the formation of races which are resistant to acaricides. Fertility of females is dependent upon specific features, nutrition and length of life. A total of 40 to 400 eggs is produced, the daily range varying from 1–3 to 6–8 eggs per female. *Tetranychus atlanticus* is remarkably fertile: a female can lay up to 15 eggs in 24 hours at a temperature of 27°C.

2. In the species of Tetranychidae studied, the life history entails the following stages: egg-larva-protonymph-deutonymph-adult mite (♂ or ♀). The assumption that the life history of the male has one stage less than that of the female is erroneous. The zone of temperature inducing normal development lies within the range of 7–35°C. In mites, as in insects, the inverse proportion between temperature and length of time of development is observed, and geographical changes in phenology and number of generations are to a considerable extent subordinated to the so-called rule of the sum of effective temperatures. Under equal temperature conditions, spider mites complete the developmental cycle more rapidly than the Bryobiidae. Development of false spider mites takes place at a considerably slower rate. The constancy or correlation between the duration or development of separate stages of mites has been stated to be independent of ambient temperature while 40–50% of the time necessary for the development from egg to imago is spent in embryogenesis.

3. Tetranychid mites overwinter in the egg-stage or as fecundated female. Diapause is induced by a gradual shortening of day-light towards autumn, by a lowering of temperature as well as by a deteriorating of food supply. Bryobiidae are especially sensitive to the food factor but spider mites less so. Commencement of diapause in false spider mites under unfavourable food conditions was not observed in nature. (Table 1.)

Table 1. Indices of biological characters of some species of Tetranychidae

Name of species	Developmental cold threshold, t°C	Sum of effective temperatures needed for development of one generation	Number of generations	Fertility of one female, no. of eggs	Overwintering stage
Eurytetranychus buxi	9.0	203	9	49	egg
Bryobia redikorzevi	7.2	340	5–6	40	egg
Tauriobia insueta	10.7	–	1	?	egg
Tetranychus viennensis	10.0	185	7–9	390	female
Tetranychus atlanticus	7.8	182	10	200	female
Schizotetranychus pruni	12.0	170	6–7	60	female
Schizotetranychus prunicola	10.3	200	7–8	50	female
Oligonychus ununguis	5.8	309	7–8	27	all stages
Oligonychus brevipilosus	6.8	220	8	44	all stages
Panonychus ulmi	8.0	210	7–9	60	egg
Brevipalpus pulcher	10.0	500–600	2	34	female
Brevipalpus obovatus	12.0	390	3–4	27	egg, female
Tenuipalpus cheladzeae	11.2	544–638	2	36	all stages
Pentamerismus taxi	10.5	567–682	1–2	32	nymph, female

STUDIES ON ERIOPHYID MITES IN BULGARIA

P. D. NATCHEFF
Higher Agricultural Institute "G. Dimitrov", Department of Entomology,
Sofia, Bulgaria

This paper deals with some of the species of eriophyid mites found on wild and cultivated plants.

Of the 1250 annual and perennial species of plants examined, 71.4 % was populated with eriophyid mites, which shows that these species are widely distributed. Most affected were the perennial species which therefore probably create conditions which favour the breeding of mites. With some cultivated plants, such as plum-trees, cornel-trees and others, the rate of infection reaches 95 % and the harm caused by eriophyid mites is considerable and clearly visible.

As for the economic importance of eriophyid mites, this is considerably less compared with that of other species. It can clearly be seen in young saplings in seed-plots, where the limited number of leaves on each plant is of great importance for its growth. For example, the young pear and apple saplings and vine-twigs in the seed-plot, infested with *Eriophyes piri* (Pgst), *Eriophyes mali* (Nal.) Liro and *Eriophyes vitis* (Pgst), lag behind in their growth, have weak outgrowths and are almost unfit to be used for planting material (Natcheff, 1963). The same applies to plum trees infested with *Aculus fockeui* (Nal. and Trt.) and this species causes greater damage. Its mass propagation causes an early leaf-drop, and in fruit-trees the quality of fruit becomes poor. During the following year the damaged trees become sterile. Of a great importance is also *Aceria phloeocoptes* (Nal) which develops small round galls upon the plumtree's branches, thus hampering its normal growths and causing deformation of branches, which in some varieties (Renglotte) are greatly shortened and die off at a high rate (Natcheff, 1967). The hazel-bush is strongly damaged by *Phytoptus avellanae* (Nal.), which develops in the buds, causing deformations which make them resemble a tiny cabbage, untill death. Thus, annual growths become leafless.

As regards annual and herbaceous plants, *Epitrimerus boczeki* Natcheff grows freely upon pepper leaves. It often multiplies to a dense population up 25–30 individuals per square cm, and causes yellowing and an untimely death of young pepper plants in nurseries. The effects of this mite upon large permanently planted pepper plants are less severe.

Special attention should be paid to *Aceria tulipae* (Nal.) which, together with *Abacarus hystrix* (Nal.), is very often found in all regions of the country on a large number of wild cereals. The former is of particular importance for the seed of *Bromus inermis* L. where damages lead to weed growing and sterility. In this case losses reach 90 %.

Many of the plants damaged by eriophyid mites show structural and morphological modifications, such as galls, deformations and others. The growth of these formations is at the expense of the modifications occurring most often in the parenchymal cells. As a result from the feeding by mites the latter transform into meristem (Natcheff, 1968), start fissioning and in the long run form galls on the tissue from which the mites obtain their food. Some of the plants later on show red pigmental matter. On analysis the damaged tissues prove to contain a larger quantity of free amino acids, which indicates a disturbed protein synthesis. With the ageing of the galls the pigmental matter increases and gradually penetrates deeply into the tissues. Each of the plant varieties reacts in a different way, for example in the Renglottes 99% of the side twigs die when affected with *Aceria phloeocoptes* (Nal.).

Similar modifications are observed with galls caused by *Eriophyes vitis* (Pgst.) on the leaves of the grapevine. Here the modification affect mainly the cells of the epidermal tissues, which become filiform and coil irregularly, making a white erineum. With damages of this kind the lack of some pigmental substances can always be observed.

Analogical modifications occur in the weeds of *Bromus inermis* L. when damaged by *Aceria tenuis* (Nal.) (Ivanova, Natcheff, 1967) as well as in the bud leaves of hazel bushes (*Corylus avellana*).

The following species have so far been identified:

1. ***Phytoptus avellanae*** Nal.
2. ***Aceria seventini*** Natcheff
3. ***Aceria mali*** (Nal.)
4. ***Aceria tristriata tipica*** (Nal.)
5. ***Aceria tristriata erinea*** (Nal.)
6. ***Aceria phloeocoptes*** (Nal.)
7. ***Eriophyes vitis*** (Pgst.)
8. ***Eriophyes similis*** (Nal.)
9. ***Eriophyes mali-marginemtorquens*** Liro.
10. ***Eriophyes piri*** (Pgst.)
11. ***Coptophylla lamimani*** (K.)
12. ***Phyllocoptes gracilis*** (Nal.)
13. ***Phyllocoptes coryli*** Liro.
14. ***Phyllocoptes abenus*** K.
15. ***Phyllocoptes malifoliae*** (Parrot) n. comb.
16. ***Epitrimerus piri*** (Nal.)
17. ***Calepitrimerus vitis*** (Nal.)
18. ***Oxypleurites bulgariensis*** Natcheff
19. ***Anthocoptes ribis*** Massee
20. ***Anthocoptes loricatus*** (Nal.)
21. ***Tegonotus oleae*** Natcheff
22. ***Aculus juglandis*** Natcheff
23. ***Aculus comatus*** (Nal.)
24. ***Aculus cornutus*** (Banks)
25. ***Aculus schlechtendali*** (Nal.)
26. ***Aculus fockeui*** (Nal. et Trt.)
27. ***Aculus balevskii*** Natcheff
28. ***Diptacus giganthorhynchus*** (Nal.)
29. ***Rhyncaphytoptus ficifoliae*** K.
30. ***Rhyncaphytoptus popovi*** Natcheff
31. ***Aceria tenuis*** (Nal.)

References

Bembasat – Ivanova, Natcheff, P.: Recherches sur les modifications des Glumelles de *Bromus inermis* L. (Gramineae provoquées par l' *Aceria tenuis* (Nal.) (Acarina – Eriophyidea). Marcelia, t. 34, fasc. 3–4, 1967.

Natcheff, P.: Documentation sur la biologie de l'*Eriophyes vitis* Pgst. et le combattre. Bull. de l'Institut de Vitic. et Vinic., t. 4, 1963.
– *Epitrimerus boczeki* new Eriophyid mite species of pepper in Bulgaria. C.R. Acad. Sci., t. 18, No. 9, 1965.
– Studies on Eriophyid mites of Bulgaria II. Acarologia, t. III, fasc. 3, 1966a.
– Studies on Eriophyid mites of Bulgaria V. C.R. Acad. Sci., t. 9, No. 12, 1966b.

phyoidea). Landwirtschaftliche Hochschule "G. Dimitrov" Sofia, Serie Pflanzenbau, Bd. XVIII (XLVI), 1967a.
– Studies on Eriophyid mites of Bulgaria VI. Wissenschaftliche Arbeiten, Bd. XVIII(XLVI), Serie Pflanzenbau, 1967b.
– Studies on Eriophyid mites of Bulgaria VII. Changes in plum plants under the influence of *Aceria phloeocoptes* (Nal.) (Acarina – Eriophyoidea). Wiss. Arbeiten, Bd. XIX (XLVII), Serie Pflanzenbau, 1968.

POPULATION DYNAMICS OF SOME TETRANYCHIDS IN SUBTROPICAL FLORIDA

SABA, F.
Vero Beach Laboratories, Vero Beach, Florida, USA

There are a number of Tetranychid species in Florida that have a wide range of host plants. Of these *Tetranychus tumidus* Banks, *Tetranychus urticae* Koch, and *Tetranychus yusti* McGregor (Saba, 1971) are polyphagous and can survive throughout the year on a number of host plants, especially on those belonging to the following families: Compositae, Leguminosae, Malvaceae and Graminaea. Since the winters of the humid sub-tropics are mild and with temperatures usually above the freezing point, several suitable host plants may be encountered at any time on which those species may feed. All three species were found almost exclusively on low growing shrubs, bushes and grasses. Nutrition is therefore not a limiting factor for any of these three species in the sub-tropics during any season of the year. It was observed over several years for all three species that under field conditions in Vero Beach where the sub-tropical climate begins, the spider mite populations increase in spring and autumn, but decline and virtually disappear in summer and winter.

The study of field population densities of *Tetranychus tumidus* in relationship to temperature, rainfall, predators and parasites resulted in understanding the factors influencing the population dynamics of this spider mite species. As to temperature, the population starts increasing in spring when the morning low temperature remains in the upper teens (>15°C) and the daily high in the twenties. Those temperatures, and higher ones, prevail throughout summer and autumn until October, after which they drop below those limits again. If other limiting factors were not envolved, the population growth would have been continuous throughout summer. In winter the indirect affect on the population is by nutritional influence: the change in temperature and day-light hours terminates summer host plant growth, and those are gradually replaced by winter plant species; thus spider mites will then migrate, seeking new host plants. The direct affect of those low temperatures is the prolonging of the embryonic and post-embryonic development as well as a reduction in oviposition so that a pronounced decline in the number of spider mites is noted (Saba, 1972). *Tetranychus tumidus* and *T. yusti* have no form of diapause as *T. urticae*, yet if on several consecutive days in winter temperatures rise to levels mentioned above, spider mites of all three species, usually females, can be seen feeding and already ovipositing on succulent plant leaves.

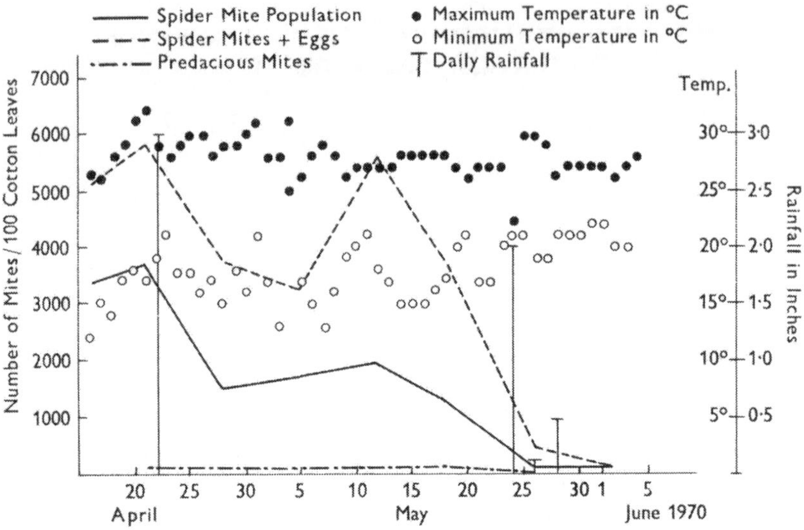

Fig. 1. Heavy rainfall in April and plant wilting due to drought in May-decreased spider mite density,

Rainfall may occur in any month of the year, but in southern Florida precipitation is heavy and often in the summer months, usually from May until September. Heavy rains of 70mm or more within 24 hours cause a mechanical reduction of the population: quiescent, larval and nymphal stages falling of the leaves, are either injured or unable to reach the same host (Fig. 1). Late summer rains, followed by low temperatures may occur in some years; the spider mite population declines and remains at low densities (Fig. 2).

However, the indirect effect of rainfall is of greater significance. When, during the warmer periods of the year precipitation occurs, it raises the humidity to form optimal conditions for *Entomophthora* species, a parasite which can practically decimate a thriving spider mite population within days. In one example a field infestation of 220 tumid spider mites per one cotton leaf (not counting eggs) was reduced down to less than one spider mite per leaf within two weeks (Fig. 3).

Several predators were found in field populations of *T. tumidus*. *Phytoseiulus macropilis* (Banks) was the most numerous, reaching up to 100 predaceous mites per 200 cotton leaves collected at random, followed by 12 per 200 leaves of *Scolothrips sexmaculatus* (Perg.). Both these species fed also on *T. urticae* and *T. yusti* in the laboratory. Also 2 Coccinellids were found in the field at the rate of 4 of each species per 200 leaves; they were *Hippodamia convergens* (Guer.) and *Cyclonedia immaculata* (Fab.). However, laboratory trials were not conclusive as to their predation on spider mites, both Coccinellids have been recorded as aphid predators (Muma, 1953).

Conclusion: Spider mites in the humid sub-tropics can thrive on a number of host plants throughout the year. Limiting factors have been the fungus *Ento-*

mophthora, which exerts the strongest influence on spider mite populations during the rainy warm season, causing a practically complete decimation of the population. In winter time the limiting factor is low temperatures that reduce oviposition and prolong development of the spider mite so that a heavy population build-up does not occur.

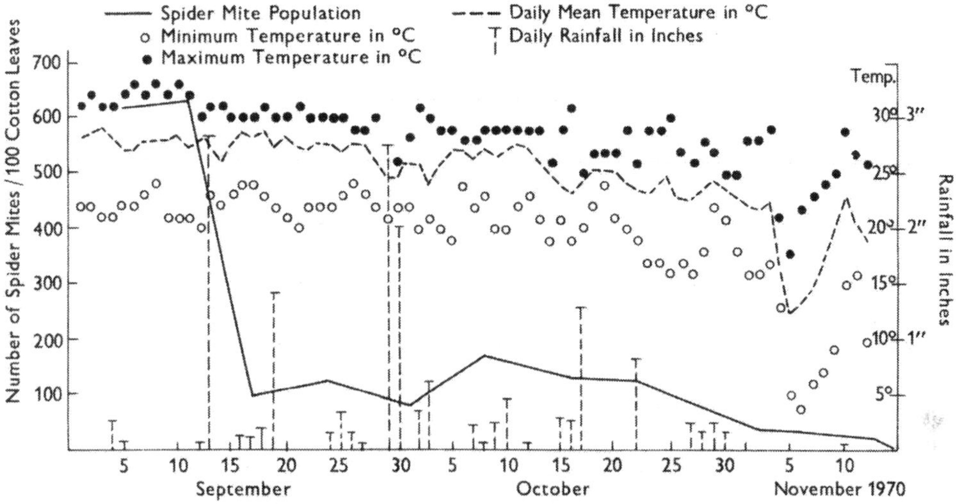

Fig. 2. Frequent precipitation in September followed by morning low temperatures as of late October, inhibited high population densities.

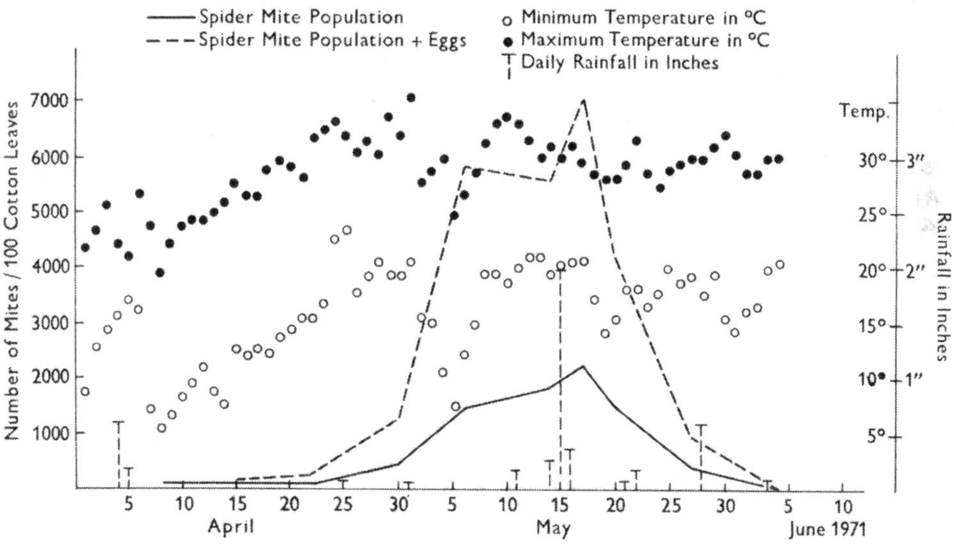

Fig. 3. Precipitation together with warm temperature in mid May established optimal conditions for *Entomophthora* infestation and dessimation of *Tetranychus tumidus*.

References

Muma, M.H.: Ladybeetle predators of citrus aphids. Cit. Mag. 15 (8): 32–33, 1953.
Saba, F.: *Tetranychus yusti* McGregor, a spider mite of potential economic importance. J. Econ. Entomol. 64 (1): 141–4, 1971.
Life history and population dynamics of *Tetranychus tumidus* in Florida (Acarina: Tetranychidae). Ann. Entomol. Soc. Am. 1972. (In print.)

MITES AND THEIR ECONOMIC IMPORTANCE IN IRAN

H. SEPASGOZARIAN
Plant Medicine Department, Agricultural College of Tehran University,
Karadj, Tehran, Iran

Iran is a large country of 1.645.200 km^2 with 12.000.000 hectares of forests and 14.000.000 hectares under cultivation, half of which is worked annually. The most important agricultural products are: cereals, rice, cotton, sugar beet, sugar cane, melons, tobacco, oil seed crops and pulse crops. A large area is also occupied by orchards of stone fruit, apples, pears, dates, citrus, olive and pistachio trees.

Climatically Iran is usually warm and dry. The maximum precipitation is 2000mm in the western part of the Caspian Sea. The amount of precipitation in agriculturally important areas fluctuates between 200 to 600 mm per year and therefore about half of the total area of cultivated lands is irrigated.

Until twenty years ago there was no problem of mites affecting our agricultural crops. Since then, the biological equilibrium has been disturbed and the population of useful predators decimated because of agricultural mechanization and the use of manufactured pesticides such as organochlorides. The population of mites increased gradually and this caused problems in agricultural areas.

During the last ten years the problem of mites affecting agricultural plants has been carefully studied in Iran. At the same time, the Institute of Public Health started work on ticks transmitting human diseases. Along with this, the College of Veterinary Science and Razi Institute have carried out research on ticks associated with live-stock.

Thus Iran has found her way towards the organization of a scientific approach to mite problems. We are eagerly ready to cooperate with scientists and research workers throughout the world in order to elaborate more intensive research on mites which are of economic importance to Iran.

Mites of economic importance which have been collected in Iran

1. *Tetranychus urticae* Koch; this mite was collected from about 60 different host plants such as cotton, sugar beet, ornamental plants, melons, vegetables and many weeds.

2. *Tetranychus cinnabarinus* Boisduval; this mite is mostly associated with *Tetranychus urticae* on many plants such as cotton and ornamental plants in green houses and in fields.

3. *Bryobia rubrioculus* Scheuten; this mite has been found on apple, pear, plums and prunes. Another species of *Bryobia* has been collected from *Buxus*. A description of this species will be published in near future.

4. *Panonychus citri* McGregor; collected from citrus trees in northern and southern parts of the country. Different experiments have been carried out annually to control these mites and the most satisfactory chemical had been recommended to the citrus growers.

5. *Oligonychus afriasiaticus* McGregor; this lives on date plants and is controlled every year by the Plant Protection Organization.

6. *Eotetranychus pomi* Sepas; attacks apple trees. Its biology and control measures have been worked out in Iran as well as in Germany. (Sepasgozarian, 1956).

7. *Oligonychus bicolor* Banks; collected on sour cherry, quince and mulberry.

Some species of other genera, such as *Schizotetranychus* and *Neotetranychus*, were also found but these are of no economic importance. We have also collected *Panonychus mori* Yokoyama on mulberry trees, but *Panonychus ulmi* Koch has not yet been found in Iran. Prof. Dosse claims that he collected *Panonychus ulmi* on mulberry trees, but I believe that these specimens are *Panonychus mori*.

Tenuipalpidae

1. Different species of *Cenopalpus* such as *Cenopalpus pulcher* Can. and Fanz., *C. pritchardi* Baker, *C. lanceolatisetae* Attiah and *C. pennatisetis* Wainstein have been collected on apple trees. Of these, *Cenopalpus pulcher* is of greater economic importance than the others. Sepasgozarian (1970) studied the biology and control measures of this species. He recommended Supracid and Moracid to control this pest.

2. In 1966 *Brevipalpus obovatus* Donn. was found on tea plants in northern Iran. There was an outbreak of this pest during this year and it was also found on many other host plants such as citrus and *Ligustrum vulgare*. No chemical control was applied against this pest. Fortunately, natural enemies and agricultural measures have brought this pest under control. Khalilmanesh (1969) has already studied the biology and ecology of this mite.

3. On pomegranate trees two species of mites have been collected: *Tenuipalpus granati* Sayed and *Tenuipalpus punicae* Pritch. and Baker. In some years the populations of these two species increase, but the use of pesticides such as Dimethoates, Supracid, Kelthane and Morocid will bring them under control.

4. Along with *Tenuipalpus granati* a new species of *Tenuipalpus* is found on pistachio trees. The description of this species will be published in due time.

5. *Tenuipalpus eriophyoides* Baker has been found on date plants. Due to the control measures against *Panonychus afriasiaticus* and other insect pests, the population of this mite does not increase very much. Other species of Tenuipalpid mites which have been collected in Iran and are of no economic importance are:

Cenopalpus bakeri McGregor – on apple and quince,
Brevipalpus olearius Sayed – on olive,
Brevipaltus orchidarum Parfit and *Brevipalpus phoenicis* Geijskes – on cyclamen.

Eriophyidae

1. The most important species in this family in Iran is *Phyllocoptruta oleivora* Ashm. which attacks citrus trees in the northern part of the country. Until 1956 the symptoms of damage due to mite were mistaken for a cooper deficiency, but since then it became clear that the symptoms are related to the damage caused by this mite. For control measures wettable sulphur was recommended (Sepasgozarian 1958). Now a 5 percent solution of Dithane is used against this pest.

2. Two years ago *Vasates lycopersici* Massee was found in Iran on tomato plants, which were seriously damaged. This mite has been successfully controlled with Chlorbenzilates and Dimethoates, but its biology and ecology in Iran has yet to be studied.

3. Different species of mites of this family such as *Eriophyes piri* Pagst. and *Eriophyes piri* var. *mali* Nal. have been collected on apple, pear, plum and prune trees. It is very important to study the biology and ecology of these mites. However, a new taxonomic revision based on the research of Keifer seems to be necessary for these species. At present, these mites are controlled by wettable sulphurs and systemic pesticides such as Dimethoates and Metasystox.

4. More than seventy varieties of grapes are grown in Iran. Some of them are attacked by *Eriophyes vitis* Nal. Due to the spraying of the plants with wettable sulphur against *Uncinula necator*, the populations of these mites does not increase much. Yet during some years, they cause damage to the vines.

Mites attacking stored products

Unfortunately insufficient research has been carried out on this problem in our country. Studies by Sepasgozarian in 1966 have shown that four different species are associated with stored products in Iran. These are:

Acarus siro L.
Glycyphagus cadaverus Schrank
Glycyphagus domesticus De Geer
Tyrophagus demidiatus Herm.

Natural enemies

Ecological factors in Iran are very favourable for the development of mites as well as for the development of natural enemies (predators). Unfortunately, due to the use of organochlorides, the population of these natural enemies is continuously adversely affected. Until few years ago not enough information was available on the importance and effect of these organochlorides on predators. But today the effect of organochlorides on mites and their predators is known and we are trying to use methods of application which spare the predators. The predators of mites which have been collected in Iran are discussed below. We believe that more comprehensive research should be conducted in this field.

Predatory mites

1. *Zetzellia mali* Bwing; this mite shelters and overwinters with *Tetranychus urticae* and *Eotetranychus pomi* under apple bark. It does not produce webs but hibernates in the webs produced by the species mentioned. Its colour is dark orange at this time. It leaves the shelter when the above mentioned mites appear in spring-end, feeds at first on their eggs, later on attacking the other developmental stages of its prey. On an average this mite destroys about 30 eggs and other developmental stage daily.

2. *Typhlodromus vulgaris, Typhlodromus rhenanus* Oudm. and about 10 other species of Phytoseiidae have been collected in Iran up till now. These mites are known to be predators, but unfortunately their population is very low in orchards which are continuously treated with pesticides.

Predatory insects

Hemiptera: *Anthocoris nemorum* L. and *Orius minutus L.* are known as the most important predators of mites. They feed on all stages ot them. Quite a few species belonging to other families of Hemiptera, e.g., Anthocoridae, have been collected. Although they may play a very important role as predators, they have not yet been identified.

Coleoptera

The fauna of coccinellid beetles in Iran has been very well studied and described by Vodjdani (1965). The biology of these beetles is known but their effect on mite populations is not yet understood. The most important predacious coccinellid beetles which have been collected in Iran are:

Stethorus punctillum Weisse, *Coccinella 7–punctata* L., *C. 11–punctata* L., *C. bipunctata* L., *C. 14–punctata, C. 10–punctata* L., *C. conglobata* Horv., *Chilocorus bipustulatus* L. and some other species belonging to the genera *Exochomus, Brumus* and *Rodalia.*

Some of the predatory species of thrips, and Diptera larvae belonging to the families Syrphidae and Cecidomyidae, are important in decreasing the mite populations. Their biology and behaviour should be studied in the future.

Mites resistant to pesticides

Tetranychus urticae along with *Tetranychus cinnabarinus* are very active in cotton in northern Iran. Our observation shows that these mites are strongly resistant to organophosphates such as Metasystox, Dimecron, Ekatin etc.

Up to eighteen years ago the farmers could control the spider mites on cotton with 250 cc of Metasystox per hectare. But now they do not get a satisfactory result even with 2 liters of this pesticide per hectare. Based on the experiments conducted by different experts, the selective acaricides such as Morocid and Kelthane are recommended for such situations.

Organochlorid compounds such as DDT have been used extensively on our cotton fields to control bollworm. At present, due to their side effects on predators, parasites and mite populations, there is a tendency to decrease the usage of these pesticides and to replace them with other compounds.

References

Khalilmanesh, B.: Proceedings of the Second Congress of plant medicine. Tehran, 1969.

Sepasgozarian, H.: Morphologie und Biologie der gelben Apfelspinmilbe *Eotetranychus pomi* n. sp. Z. angew. Zool. 4:435–491, 1956.

– Spinnmilben und Akarizide. Elmi and Fanni 2, 3, 55–72, 1958.

– Storage Pests of Iran and their control. Publication Nr. 1025 of Tehran University, 1966.

– The red false Spider Mite of Apple. *Cenopalpus pulcher* Can. Fanz. Bull., Nr. 11, College of Agriculture of Tehran University, 1970.

Vodjdani, S.: Les Coccinelles utiles et nuisibles de l'Iran. Bulletin Nr. 65 Faculté d'Agronomie de l'Université de Téhéran, 1965.

ORIGIN AND EVOLUTION OF GALL MITES (ERIOPHYOIDEA) IN THE LIGHT OF THEIR INTERNAL ANATOMY

A. P. SILVERE
Institute of Experimental Biology, Estonian Academy of Sciences,
Tallin, USSR

While studying the problems of phylogeny and evolution of organisms, data on their comparative morphology and anatomy are usually applied. Such an approach to the studies on evolution of gall mites, however, is hampered on one hand by specific characteristics of their body structure (which distinguish this homogeneous group of tiny terrestrial arthropods from the general complex of mites) and on the other, by the absence of data on the internal anatomy of gall mites. Due to both facts not a single extant group of mites can be regarded to be closely related to the gall mites. The exceptional position of gall mites within Acari and even arthropods becomes especially evident after their anatomy, histology and cytology have been studied by electron microscopy.

The results of electron-microscopic studies of the anatomy of the gall mite may help us to understand correctly the origin of the primary primitive features in the structure of gall mites as well as their association with the phylogeny of arthropods studied and may elucidate some common trends in the evolution of gall mites. As for the latter problem it has been presumed that the actual ecological relationship in which the evolutional result, namely the divergence and specialization of organisms, becomes manifest, is determined in stenotopic organism by correspondence of its specific characteristics with certain (limiting) factors of environment.

Some primary-primitive features in gall mites have been detected which are overstepping the limits of characteristics of arthropods and which make the gall mites related to lower invertebrates (Fig. 1); however the structure of the extremities, mouth apparatus and other features of external morphology leave no doubt about the affinity of gall mites with other arthropods. This has been confirmed by the muscle structure of gall mites which, despite the absence of cross-striations, in our experiments proved to be sensitive to adrenalin, the mediator typical for the musculature of arthropods, but not of worms.

While interpreting the mentioned discrepancy it may be useful to consider the primitive organs and tissues in connection with the ontogeny of gall mites, and the specialization of external morphology - in connection with phylogeny. The correctness of this approach is evident as some primitive structural features of gall mites are to be found in developmental or embryonic stages of typical arthropods. Hence

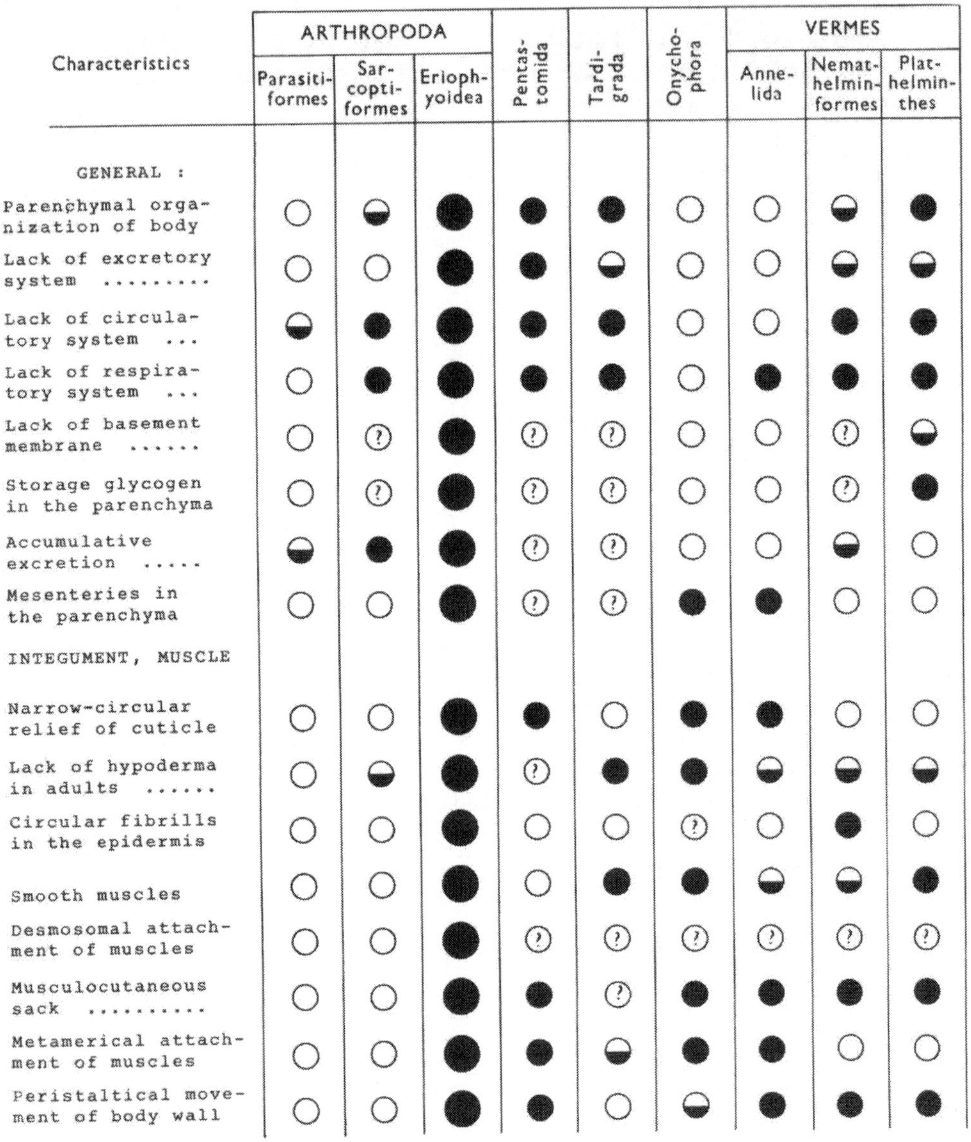

Fig. 1. Distribution of characteristics of the gall-mites in some groups of invertebrates.

it follows that the discrepancy in the structure of gall mites may be regarded as the preservation of developmental or embryonic organs and tissues in adult forms of arthropods. This characteristic is typical for neotenic organisms in which gall mites may undoubtedly be included. This fact allows us to assume that the origin of gall mites is connected with paedomorphosis. The characteristics of the internal structure of gall mites, however, should be primarily connected with developmental or embryonic stages of ancestral forms of gall mites, possibly some other mites, but not with primitive or lower invertebrates. Thus the primitive features in the structure of gall mites acquire a special significance as relics of different stages in the morphogenesis of organs and tissues of the present arthropods and elucidate the evolution of corresponding structures.

The thin, slightly differentiated integument of primitive gall mites gives little protection against dessication during insufficient humidity of the environment and determines further evolution of gall mites and their limited dependence on factors of environment. The dependence of gall mites on the environmental humidity as a leading factor in their evolution is further enhanced by other primitive characteristics: extremely small size of body, complicated relief of cuticula and parenchymatous body structure. The latter conditions the absence of regulation of fluid distribution in the organism. On this basis two main trends in the divergence of gall mites may be distinguished. The first trend is more specific and characteristic of gall mites: in most primitive slightly differentiated forms it is the capability for gall-formation. This physiological trend of divergence forms the basis of development of primitive mite groups, phytoptids, and eriophyids, which create galls, the artificial biotopes on the organs of the host plant, with favourable conditions for the development of numerous mite populations.

The second trend of evolution of gall mites is based on the morphological changeability which causes the development of the so-called free-living gall mites, phyllocoptids. This trend of evolution of gall mites affected primarily their integument, smoothing away the relief of cuticula, diminishing the surface of evaporation, as observed in some eriophyids, the inhabitants of semiclosed biotopes. In free-living gall mites the differentiation of integument becomes also manifest in the solidity of relief elements of cuticula, the thickening of the cuticula and in the development of special protective structures, decreasing effectively the dessication through the integument. This morphological differentiation, without involving the general body structure of mites, considerably promotes the possibility of distribution of free-living four-legged mites on the surface of plants, irrespective of the air humidity.

Among slightly differentiated primitive gall-forming mites monophagous forms occur whose nutrition is specialized and often associated with gall formation; in free-living gall mites the food specialization is less distinct and the development of a more powerful buccal apparatus considerably promotes the possibilities of feeding on various organs and tissues of plants.

PRELIMINARY BIOLOGICAL STUDIES IN *TETRANYCHUS NEOCALIDONICUS* MARC ANDRÉ

A. A. SOLIMAN, H. H. ATTIAH AND M. L. WAHBA
Faculty of Science, Ain Shams University, Egypt, Acarology Investigations, Crop Protection Dept., Min. Agriculture, Dokki, Egypt

Attiah (1969) recorded the presence of *Tetranychus neocalidonicus* on citrus trees in a grove at Abo Homnos (Beheirah Governerate) in Egypt. The mites were found more often on ripe fruits than on green ones causing yellowish blotches on the surface of orange and mandarin fruits.

The present work represents a preliminary study on the biology of the mite, which according to Rahman and Sapra (1946) is omnivorous species infesting vegetables and many crops in the Punjab and previously known as *Tetranychus cucurbitae*.

Materials and methods

1. Duration of different stages

During November 1968, citrus leaves infested with *Tetranychus neocalidonicus* were collected from the infested grove. The leaves were taken to the laboratory at Dokki and a considerable number of females were transferred to sweet potato leaves, placed on wet cotton in Petri dishes, and kept at room temperature. After laying eggs the females were removed and the eggs were kept on the host plant leaves in order to hatch and to develop to the adult stage. After copulation a number of fertilized females were transferred by means of a camel hair brush to a sweet potato leaf placed on wet cotton in a Petri dish, left for a period of 24 hours to lay eggs and then transferred to a new leaf and kept for a second period and so on. The eggs were watched till hatching and the number of larvae were recorded each day. Statistical analysis was followed to calculate the mean time and the standard deviation for the incubation period. All the larvae which hatched on the same days were transferred to a new sweet potato leaf and were watched daily till the first quiescent stage was reached by the first larva (larvae.) The rest of the moving larvae were then transferred to a new leaf, so that all mites reaching the protochrysalis stage on a certain date remained on the same leaf. The mites were watched till they moulted and produced the protonymphs, the mean days required for the larval as well as the protochrysalis stages were calculated. The same procedure was then followed to calculate the time required for each stage to reach the imago. The same work was repeated during December, but around the end of that month the room temperature as well as other climatic conditions were not suitable for the mites causing the death of all different stages and so the rearing of mites was then undertaken in a laboratory where the constant temperature was 25–27°C, the relative humidity fluctuated between 50–70 per cent. Duration of the different stages was again calculated under these conditions using the same technique.

2. Longevity of females and number of eggs per female

As soon as twenty (presumably fertilized) females had reached the adult stage, they were each transferred to a separate sweet potato leaf and placed on wet cotton in a Petri dish. Each female was watched daily until death, so that, the preoviposition period, the number of eggs per day, the total number of eggs laid and the life span could be calculated. Results were obtained at room temperature as well as at the constant temperature and relative humidity previously mentioned.

3. Longevity of males

When twenty males reached the adult stage they were transferred each to a separate sweet potato leaf. Each male was watched daily till death.

4. Sex ratio

Ten newly mated females were transferred separately, each on a separate potato leaf in a Petri dish in the laboratory under the constant conditions of temperature and humidity previously mentioned. The eggs laid by each female during its whole life were allowed to develop to the adult stage using as host sweet potato leaves which were placed on wet cotton in Petri dishes. Upon reaching the imago stage, the sex of each was recorded and therefore the sex ratio of the progeny of each female was obtained.

Results and discussion

Table 1 represents the duration of the preoviposition period and of the different stages under three different standards of temperature and relative humidity. It is clear that the higher the temperature (accompanied by a low relative humidity) the more rapidly the mite developed. Females as well as males passed through two nymphal stages, though males emerged constantly before females by 0.7–0.9 days.

The oviposition period, egg laying and longevity of females and that of males were studied at a temperature of 25–27°C and relative humidity of 50–70 %. Results were:

Oviposition period	17.00 ± 3.57 days	(20 females)
Eggs (female)	105.30 ±19.87	(20 females)

Eggs (female) day	6.27 ± 0.81	(20 females)
Longevity of female	19.50 ± 3.53	(20 females)
Longevity of male	14.50 ± 3.10	(20 males)

Ten females were used separately to determine the sex of their progeny during their whole life. The total was 1209 of which 310 were males, i.e. 25.6 % males and 74.4% females.

Table 1. Duration of preoviposition period and different stages of *Tetranychus neocalidonicus*

Stages	15 – 19°C 65 – 95 % R.H.	16 – 21°C 65 – 90 % R.H.	25 – 27°C 50 – 70 % R.H.
Preoviposition period	Days not studied	Days not studied	Days 2.00 ± 00 (20)
Incubation period	9.63 ± 0.70 (100)	5.94 ± 0.19 (100)	5.31 ± 0.40 (601)
Larva	2.10 ± 0.39 (100)	1.65 ± 0.37 (100)	1.11 ± 0.32 (434)
Protochrysalis	2.12 ± 0.33 (100)	1.08 ± 0.26 (100)	1.10 ± 0.32 (406)
Protonymph	1.87 ± 0.34 (100)	1.08 ± 0.26 (100)	1.05 ± 0.22 (262)
Deutochrysalis	1.93 ± 0.26 (100)	1.13 ± 0.32 (100)	1.08 ± 0.28 (252)
Deutonymph	1.83 ± 0.38 (100)	1.46 ± 0.41 (100)	1.13 ± 0.33 (222)
Tritochrysalis ♂	2.07 ± 0.26 (38)	1.27 ± 0.45 (33)	1.14 ± 0.36 (75)
♀	2.90 ± 0.30 (62)	1.97 ± 0.17 (67)	1.93 ± 0.26 (152)

Number between brackets = number of observations

The following conclusions were reached:

1. Under a controlled temperature, ranging from 25–27°C and relative humidity, between 50–70 %, the preoviposition period, incubation period and the larval, protochrysalis, protonymphal, deutochrysalis and deutonymphal stages lasts about 2,

5, 1, 1, 1, 1, and 1 day respectively. During tritochrysalis all males emerge before females. The mean days of this stage is about 1 day for males and 2 days for females.

2. At a temperature of 16–21°C and relative humidity between 65–90 %, the incubation period, larval stage, protochrysalis stage, protonymphal stage, deutochrysalis stage and deutonymphal stage require about 6, 2, 1, 1, 1 and 1 days respectively. The mean days of the tritochrysalis stage is 1 day for males and 2 days for females.

3. At a temperature of 15–19°C and relative humidity between 65–95 % about 10, 2, 2, 2, 2 and 2 days are required for hatching, and for the successive stages from larva to deutonymph respectively, whereas the mean days of the tritochrysalis stage is about 2 days for males and about 3 days for females.

4. At a controlled temperature ranging from 25–27°C and relative humidity from 50–70 %, the oviposition period is about 17 days, the total number of eggs per female is about 105 eggs, the average number of eggs per day per female about 6 eggs, and the longevity of females is about 19 days and that of males about 14 days. Under the same conditions the sex ratio is about 75 % females and 25 % males.

References

Attiah, H.H.: The tetranychini of the U.A.R. – the genus *Tetranychus* Dufour (Acarina, Tetranychidae). Acarologia, t. XI, fasc. 4, 1969.

Rahman, K.A. and Sapra, A.N.: On the biology of the vegetable mite (*Tetranychus cucurbitae* Rahman and Sapra: Fam. Tetranychidae). Indian J. Agric. Sci. 15, pt. 3, pp. 124–130, Delhi, 1946.

Section IV

TAXONOMY OF HIGHER CATEGORIES, PHYLOGENY AND ZOOGEOGRAPHY

Chairman
G. O. Evans (U.K.)

Assistant Chairman
V. Halašková (Czechoslovakia)

IDIOSOMAL SIGILLOTAXY OF GAMASIDS. PRELIMINARY OBSERVATIONS TOWARDS THE KNOWLEDGE OF THE MORPHOLOGY OF THESE ARACHNIDA

C. ATHIAS-HENRIOT
Laboratoire de Faune du Sol de l'I.N.R.A., Dijon, France

Acarization is a set of features, each of which is devoid of primacy but is, probably, a necessary condition. These are, for instance:
a – more or less complete disappearance of intersegmental joints and loss of intersegmental apodemes, b – complete coalescence between cephalothorax and abdomen, c – gnathosomization (a local complex process), d – rearrangement of the cephalothoracic tergosternal musculature, etc.

Further, with the preceding properties are statistically linked the comparatively small size and the gradual disappearance of leg IV at the time when chitin biosynthesis is recovered in the ontogeny.

A peculiar feature of gamasids is the sclerotization of the idiosomal cuticle. Except a few forms (such as the males of some parasitid genera), gamasids have undergone desclerotization processes in response to functional necessities. Desclerotized areas serve as substitute for the lost intersegmental joints.

It is likely that acarization includes such compelling conditions that different arachnid phyla, which have undergone it, exhibit a lot of convergent characters.

Valuable contributions towards gamasid anatomomorphology have been published (see references in Athias, 1971). But the actual significance of much of the data has often been overlooked or misinterpreted.

The present paper reports part of a lengthy investigation concerning gamasid morphology. Many features have not yet been interpreted satisfactorily.

General Remarks

In Arachnida, there are archaeotaxically 19 pairs of tergosternal muscles, among which 11 are in the abdomen (owing to the reduction of segment VII and modification of segment VIII). The series of 11 abdominals is retained, for instance, in *Opilioacarus*, the last segment of which is the XIXth.

The arachnidian kind of cephalization has led to an odd bulkiness in the cephalothorax. The use of the space is facilitated by the development of a tendinous relay, the endostylus, which divides the cephalothorax into two stores and exhibits a front notch embracing the brain. Some of the tergosternal muscles of the appendage bearing segments are used as extrinsic appendage muscles. Except tergocoxal muscles which have kept a straight course, all other tergosternal cephalothoracic muscles are suspensors.

The cephalothoracic dorsum carries the sigilla of the tergosternal muscles of the 6 appendage bearing segments. In many arachnids, the abdominal musculature is numerically reduced. There are zero (some spiders) to 8 tergosternal pairs (Uropygi); the last segment provided with such muscles (pillars) bears number XVI, ... XII, according to the orders or suborders.

Chaudonneret (1964) is of the opinion that a sigillum never leaves its segment and that the shifting of a sigillum would be achieved by distortion of the segment or by fusion of muscles of successive segments accompanied by the disappearance of intermediate sigilla.

Owing to the loss of intersegmental joints and to uniform sclerotization, the body of ancestral (?) gamasid became a rigid box. Apart from the fact that sclerotization is kept by sigillar cuticle, the places where secondary desclerotization occurs are largely a matter of contengency. This is obvious from the literature concerned with body shields of gamasids; the distribution of these shields is the result of various processes, although its primacy seems to be admitted as a postulate by many authors.

Material

The tergosternal sigillotaxy has been worked out from the survey of 50 genera belonging to 5 cohorts, excluding uropods. The aim was - supposing it to be feasible - to define an "archaeotypical" pattern and to point out systematic and functional variations. It is still only partly possible to disclose the correlations between these variations and other (morphological, functional, biological) characters of the animal, because our knowledge of the "integrated" gamasid is still scarce.

The taxonomical order followed is given in the Table 1.

Idiosomal Sigillotaxy

1 – Cephalothorax

1.1 – Dorsal shield – A special feature of gamasids is the shifting backwards of the sigilla of the straight cheliceral muscles which are in parasagittal situation. The sigilla of other straight muscles or suspensors exhibit a typical arachnidian pattern. From the axis to the antiaxial edge, there are 5 series of sigilla corresponding to : a – the straight muscles of various gnathosomal parts, b – the parasagittal suspensor, c – the straight tergocoxal leg muscles, d – the lateral suspensors which, as the preceding ones, are metamerically arranged, e – the antiaxial suspensors, the

Table 1.

Insemination	Tocospermy			Podospermy	
Dorsal adenotaxy of cephalothorax	neoadeny		orthoadeny or deficiency		
Movable digit	Presence of flagella (♀♂)		sexual dimorphism (♂)		
	+	– (1)	no	spermatotrema	spermatodactylus
	Antennophorina	Liroaspina	Zerconina	Parasitina	Dermanyssina

(1) In Discozerconidae, chelicerae are highly modified (♀) and exhibit a strong sexual dimorphism; the male is provided with an elaborate injection apparatus.

attachments of which comprise an anterior group of 3 sigilla and a posterior group of 2 sigilla.

It seems that the sigillum of the posterior suspensor is situated on the front part of the opisthosoma.

1.2 – Sternum – The sternum exhibits 2 sigilla. The anterior corresponds to a sternosubcoxal (leg IV) muscle; the posterior to a ventral suspensor.

2 – Ventral genital region

This area bears a series of 6 sternogenital sigilla (genital muscles which are probably not tergosternal in origin),

3 – Opisthosoma

3.1 – Dorsum – The sigilla are distributed into 3 groups. The front group probably includes the sigillum of the 1 st tergosternal (VIII ?), that of a tergosubcoxal muscle (leg IV) and that of the posterior dorsal suspensor.

The next group is a series of 4 sigilla which are the attachments of 4 abdominal tergosternals (NO. IX – XII ?). The caudal group is formed by the 2 sigilla of the 2 anal pillars (NO. XIII – XIV ?).

3.2 – Venter – A series of 4 inguinal sigilla fits the tergosternal muscles IX – XII (?) a 5th one (at least in some females) is the origin of a sternosubcoxal muscle (leg IV). The anal pillars are inserted on 2 perianal sigilla.

4 – Other sigilla

The abdominal venter carries mostly a preanal sigillum. This could be the caudal end of an oblique preanal muscle originating dorsally near the front edge of the opisthosoma. Dorsally, some species carry a paraxial abdominal sigillum which could correspond to a heart muscle (?).

An undescribed rhodacarid genus exhibits, when compared with all other observed gamasids, several extra sigilla.

5 – Variations

The variations of the tergosternal sigillotaxy include variations of locations and superficies; there is a high degree of functional adjustment. Some sigilla may be deficient (indicating probably the disappearance of the corresponding muscle); as, for instance, one of the 2 perianal ones. Others may break into 2 ...n pieces.

On the dorsum, left and right cheliceral sigilla frequently join on the sagittal line; the sigilla of the cheliceral retractors shift into the abdomen in species provided with elongated chelicerae; the abdominal tergosternal sigilla X – XII (?) can move towards the front opisthosomal sigillar group., etc.

On the venter, the sigilla of the inguinal series undergo fusions; the preanal sigillum can be lacking; the perianal sigillar set can be reduced; reduction can also affect the sternogenital sigillar series.

Each of the 5 major groups (see Table) possesses special features (Athias, 1971) and within each of these groups, the sigillar patterns exhibit familial, generic or specific variations.

Discussion

Chief questions are not yet given accurate answers (number of tergosternal abdominal muscles, composition of the front dorsal opisthosomal sigillar group, anatomical nature of the sternogenital muscles, that of muscles attached on the preanal sigillum or on the various extra sigilla which have been seen, etc.). When one takes in account other anatomomorphological uncertainties, the balancesheet of our knowledge about gamasid architecture and about their systematic status does not seem to be a heavy one.

Chaudonneret's hypothesis about sigillar shifting probably cannot account for the moving into the abdomen of the attachments of the cheliceral retractor of gamasids.

The definition of the suborder Gamasida is to be made by comparison with other Arachnida and other Parasitiformes.

References

Athias-Henriot, C.: Un progrès dans la connaissance de la composition métamérique des gamasides: leur sigillotaxie idiosomale (arachnides) – Bull. Soc. Zool. France 96. (In press), 1971.

Chaudonneret, J.: Une unité structurale, le métamère. Cahiers Etudes Biol. n°(13/15): 77 – 101, 1964.

Proceedings of the 3rd International Congress of Acarology, Prague, 1971

SOME CONSIDERATIONS ON THE SYSTEM AND PHYLOGENY OF GAMASID MITES

N. G. BREGETOVA
Zoological Institute, Academy of Sciences of the USSR, Leningrad, USSR

Reuter (1909) was the first acarologist who regarded the Gamasid mites as a superfamily Gamasoidea or Parasitoidea. In my former papers I used this rank for Gamasid mites. But later it became evident that some of the families belonging to this complex may be united into several natural groups of a higher rank, namely superfamilies (Evans, 1955; Lange, 1957; Bernhard, 1963; Karg, 1965; Bregetova, 1967, 1969). This means that the former superfamily Gamasoidea should be given a higher rank. Some acarologists (Vitzthum, 1943; Camin and Gorirossi, 1955; Radovsky, 1967; Krantz, 1970) regarded the Gamasid mites as a cohort Gamasides or Gamasina Leach, 1815. At the present state of our knowledge the rank cohort for the Gamasid mites may be accepted as quite reasonable.

In my previous papers (1966, 1967, 1969) the comparative study of the ontogeny of Gamasid mites was proposed as a basis for the elaboration of their natural system. Developing this principal idea, I should like to give a short account of my considerations on a structure of the system of Gamasid mites and on some supposed phylogenetic relationships within this group.

Now I recognize the supra-familial units within Gamasides mentioned below. I agree with Athias-Henriot that the family Parasitidae Oudms, 1901 (syn. Poecilochiridae Willm., 1940 and Saprogamasidae Willm., 1948) occupies an isolated position among other families of the Gamasid mites. In this family the females have a single compact ovary containing nutrient materials; the genital opening leads to a large vagina which serves as a receptaculum of a sperm packet and also as on oviductus; the genital shield in the female is triangular with a pointed anterior margin. In the male the spermatodactyl is fused distally with a movable digit of the chelicerae. This family merits the rank of a separate superfamily Parasitoidea Reuter.

In contrast with this, the majority of Gamasid mites belonging to other families, in particular to Laelaptidae, Haemogamasidae, Macrochelidae, etc., have a different type of genital organs: "the lyreate organ" (Michael, 1892) provides the ovary with nutrient materials; the genital orifice leads to the tubular vagina serving as an oviduct; the male sexual products are introduced into the female through paired copulatory orifices and deposited into the sacculus femineus through the tubulus annulatus. In the male the movable digit is never fused distally with the spermatodactyl; the latter is free or absent.

A number of families having such a type of genital organs can be united in several superfamilies. The first of them, Rhodacaroidea superfam. nov., includes free-living and mainly predacious mites. Their larvae and protonymphs resemble one another: on the dorsal surface they have a carapace and a pygidial shield with twin sclerites between them. Deutonymphs and adults have one or two dorsal shields. Chelicerae in all developmental stages usually have well developed chelae with denticles and are adapted for capturing a prey. Palpal apotele is three- or two-tined. Families: Rhodacaridae Oudms, 1902 (incl. Ascaidae Oudms, 1905; Allolaelaptidae Oudms, 1939, partim; Digamasellidae Evans, 1957; Halolaelaptidae Karg, 1965), Aceosejidae Baker et Wharton, 1952 (sensu Evans, 1957), Phytoseiidae Berl., 1913 (syn. Typhlodromidae Bernhard, 1955 in Karg, 1962) and ? Ameroseiidae Berl., 1918 (sensu Evans, 1961).

The second group consists of the superfamily Macrocheloidea. Free-living mites. Females are often phoretic (Phoresy on insects and other animals). Larvae and protonymphs are quite dissimilar. Larvae are not able to feed; their chelae are poorly developed, the idiosoma has no sclerotized shields. Protonymphs have well developed chelae with denticles and are adapted for capturing a prey, (in deutonymphs as well as in adult mites), have two large dorsal shields, viz. the carapace and notogaster. Deutonymphs have only one dorsal shield with lateral incisions. In adult mites the dorsal shield is entire. Palpal apotele threetined. Males with apophyses on legs II and/or on legs IV. Families: Macrochelidae Vitzhum, 1930(incl. Parholaspidae Evans, 1956; Neoparasitidae Oudms, 1939, partim), Pachylaelaptidae Berl., 1913 (Vitzthum, 1931).

The third superfamily is Laelaptoidea Berl., 1892. A large group where very intensive adaptive radiation took place and in which parasitic forms are very common. Larvae and sometimes protonymphs are embryonized. Larvae without shields, protonymphs as a rule with shields as in Rhodacaroidea. Deutonymphs always very active, chelicerae as in females, one dorsal shield with lateral incisions. Adults with one dorsal shield. Palpal apotele two-tined (seldom three-tined). Males without apophyses on their legs, but sometimes with thick spines. Families: Laelaptidae Berl., 1913 (incl. Hypoaspidinae Vitzthum, 1941, Raillietinae Vitzthum, 1941, Myonyssinae Breg., 1956,Omentolaelaptidae Fain, 1961, Iphiopsidae Kramer, 1886 (Greenidae Berl., 1910), Pseudoparasitidae Vitzthum, 1941, partim), Eviphididae Berl., 1913 (= Hyletastinae Vitzthum, 1941), Entonyssidae Ewing, 1923, Haemogamasidae Oudms, 1926, Hirstionyssidae Evans et Till, 1966, Dermanyssidae Kolenati, 1859, Hystrichonyssidae Keegan, Yunker et Baker, 1960, Ixodorhynchidae Ewing, 1923, Otopheidomenidae Treat, 1955, Dasyponyssidae Fonseca, 1940, probably Spinturnicidae Oudms, 1901 and Spelaeorhynchidae Oudms, 1902.

The fourth superfamily is Macronyssoidea Breg., 1967. Parasitic mites. The type of their ontogenetic development is unusual for other Gamasid mites: larva and deutonymph have no shields, or the latter has two hardly visible shields; their chelicerae are very feeble, and both these stages do not feed. But protonymph and adults are very active in bloodsucking. Dorsal shields of protonymph as in Laelaptoidea mites. Families: Macronyssidae Oudms, 1936 (sensu Radovsky, 1967) and Ptilonyssidae Castro, 1948.

The fifth superfamily is Rhinonyssoidea superfam. nov. Parasitic mites. They have a peculiar type of their ontogenetic development: their protonymphs and deutonymphs are very similar. Families: Rhinonyssidae Trouessart, 1895 and Halarachnidae Oudms, 1906.

It is still rather difficult to find a certain position in the system for the following families: Podocinidae Berl., 1916 (sensu Evans et Hyatt, 1958), Ameroseiidae Berl., 1918 (sensu Evans, 1961), Antennoseiidae Karg, 1965, (= Vitzthumiidae Thor, 1930), Veigaiaidae Oudms, 1939 (syn. Metaparasitidae Oudms, 1904, Gamasolaelaptidae Oudms, 1939, Cyrtolaelaptidae Cooreman, 1943), Arctacaridae Evans, 1955, Zerconidae Berl., 1892, Ologamasidae Ryke, 1962, Megalolaelaptidae Fonseca, 1946 and Laelaptonyssidae Womersley, 1956 (the true rank of the last two is unclear for me).

In the tentative phylogenetic scheme (Fig. I) 27 families are shown that are acacepted by me (some of them include several families of other authors).

In the absence of fossil Gamasid mites nothing can be said about the time of their origin. Only a study of the comparative morphology and ontogeny can help us

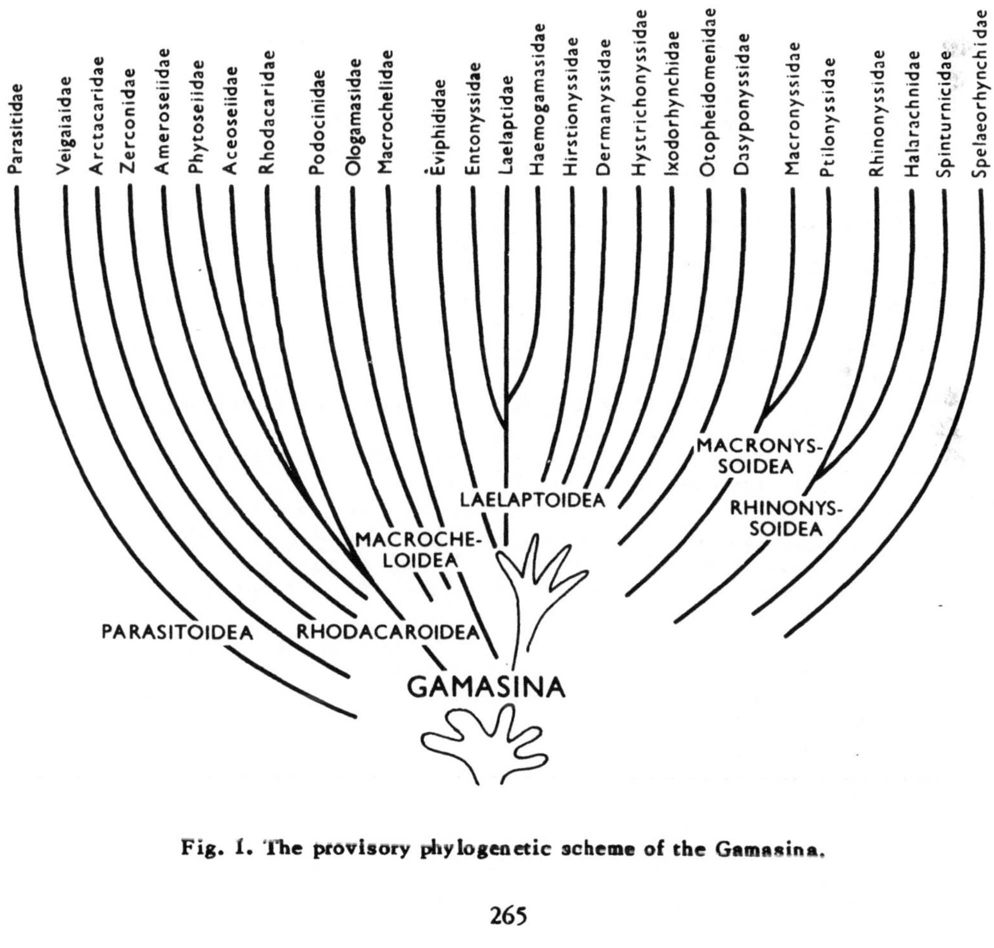

Fig. I. The provisory phylogenetic scheme of the Gamasina.

in the construction of a more or less plausible phylogenetic tree. In the case of parasitic mites some helpful data can be obtained from the study of the evolution of the host-parasite relationships.

I consider the superfamily Rhodacaroidea, which has close soil connections, as an ancient group. At the same time the more recent origin of parasitic mites is evident. For instance, the formation of Rhinonyssidae – (nasal parasites of birds) took place before the formation of the recent orders of birds, namely before the Upper Cretaceous.

Macronyssidae are ectoparasites of small mammals, birds and reptiles. As is known, they are especially closely associated with bats, and more primitive mites parasitize more primitive bats (Radovsky, 1967). This enabled Radovsky to make conclusions about the parallel evolution of bats and mites and determine approximately the beginning of Cenozoic as the time of origin of the Macronyssidae.

References

Bernhard, F.: Die Überfamilie Laelaptoidea Bernhard nov. superfam. In: Beiträge zur Syst. u. Ökol. mitteleurop. Acarina. Bd. 2: Mesostigmata I: 17 – 31, 1963.

Bregetova, N.G.: Principles of the classification of the Gamasid mites. I-st Acarological Conterence. Leningrad, "Nauka": 41–42, 1966 (In Russian).

Bregetova, N.G.: Ontogenesis in the Gamasid mites as a criterion for the erection of their natural system. Parazitologia, vol. I, No 6: 465–479, 1967. (In Russian, English summary).

Bregetova, N.G.: Ontogeny of Gamasid mites as a basis of their natural system. Proc. 2nd Internat. Congress of Acarology, 1967: 289–295, Budapest, 1969.

Camin, J.H. and Gorirossi, F.E.: A revision of the suborder Mesostigmata (Acarina), based on new interpretations of comparative morphological data. Chicago Acad. of Sci. Special Publication No 11. 70 p., 1955.

Evans, G. Owen: A collection of Mesostigmatid mites from Alaska. Bull. Brit. Mus. (Nat. Hist.). vol. 2, No 9: 17–307, 1955.

Evans, G. Owen: A revision of the British Aceosejinae (Acarina: Mesostigmata). Proc. Zool. Soc. London, vol. 131, pt. 2: 177–229, 1958.

Evans, G. Owen. The Genus Neocypholaelaps Vitzthum (Acari: Mesostigmata). Ann. and Mag. Nat. Hist., 13th Ser., vol. 6(64): 209–230, 1963.

Evans, G.Owen and K.H.Hyatt: The genera Podocinum Berl. and Podocinella gen. nov. (Acarina: Mesostigmata). Ann. and Mag. Nat. Hist., 12th Ser., vol. 10:913–932, 1958.

Karg, W.: Larvalsystematische und phylogenetische Untersuchung sowie Revision des Systems der Gamasina Leach, 1815 (Acarina, Parasitiformes). Mitt. Zool. Mus. Berlin, Bd. 41 (2): 193–340, 1965.

Krantz, G.W.: A manual of acarology. Oregon State University, Corvallis. 335 p., 1970.

Lange, A.B.: On a system of mites of the order Parasitiformes. Theses of the reports of the IX All-Union Conference on parasitological problems. Acad. of Sci., Moscow-Leningrad; pp. 135–136, 1957 (In Russian).

Michael, A.D.: On the variations in the internal anatomy of the Gamasinae, especially in that of the genital organs, and on their mode of coition. Trans. Linn. Soc. London, 1888–1894, vol. 5: 281–324, 1892.

Radovsky, F.J.: The Macronyssidae and Laelaptidae (Acarina: Mesostigmata) parasitic on bats. Univ. of California Publ. in Entomology, vol. 46: 1–288, 1967.

Reuter, E.: Zur Morphologie und Ontogenie der Acariden. Acta Soc. Sci. Fennicae, Bd. 36, No 4: 1–288, 1909.

Vitzhum, H.: Acarina. Bronn's Klassen und Ordnungen des Tierreiches. 5. Bd., 4. Abt., 5. Buch, 1–7 Lief., S. 1–1011. Leipzig, 1940–1943.

ONTOPHYLOGENETIC CONSIDERATIONS ON THE CHAETOTAXY OF CAECULIDAE (ACARI – PROSTIGMATA)

Y. COINEAU
Laboratoire Arago, Banyuls sur Mer, France

On the basis of a comparative study of the post-larval development of the chaetotaxy of many species of Caeculidae we may distinguish two basic chaetotactic stages:
I. The initial ontogenetic stage of the larval chaetotaxy. This is identical for all species whose larvae are known, the difference consisting only in the relative structure of setae. The implantations of the larval setae are marked with simple black dots in Fig. 1.
II. The initial phylogenetic stage. This corresponds to the accumulation of dots of the chaetotactic pattern in different species. It consists mainly of the setae which are considered as primitive and such chaetotaxy is qualified as palaeotrichy. In order to define it, the primitive setae must be recognized and this is a difficult task. Some of them may have disappeared during evolution, while secondary setae may have evolved. The latter are qualified as neotrichy. The primitive setae have kept their original position and their homologues can be recognized without hesitation. Sometimes they are the secondary setae, but these are usually unstable. The primitive setae, the alveoli of which are marked by black dots in the figures, are very easy to distiguish if they exist. They are noted both for their form and size. In general they are larger than the secondary setae; *Allocaeculus pilosus* Lawrence can serve as an example. Sometimes they are hardly discernible among the secondary setae, as is the case in *Allocaeculus sclerodermatus*, but it has always been possible to detect them, because they are usually larger than the others and their basal bulb is larger in diameter. However, this rule is not generally valid, because the primitive setae of *A. kalabariensis* Coineau, 1971, on the contrary, are shorter than the others, but have a special form. This is a case of idiophylaxy.

All the primitive setae which we have identified do not exist in the larvae possessing the segments of body which carry them. Some of these primitive setae which appear during the postlarval development are also absent in the nymphs and adults of some species. This reduction of chaetotaxy is the result of the regressive evolution the course of which is shown in Fig. 1.

The ovoid forms correspond to the outline of the dorsal side of the body, the anterior part being turned up, the lines representing the wrinkles which delimit the sclerites. The black dots mark the implantations of setae, simple dots correspond to the larval setae, those which are surrounded by a crown appear during the deve-

lopment. Each of these three columns corresponds to the nymphal stage, i.e. proto-, deuto- and tritonymph of three species belonging to the genus *Microcaeculus: M. hispanicus, M. sabulicola,* and *M. franzi.*

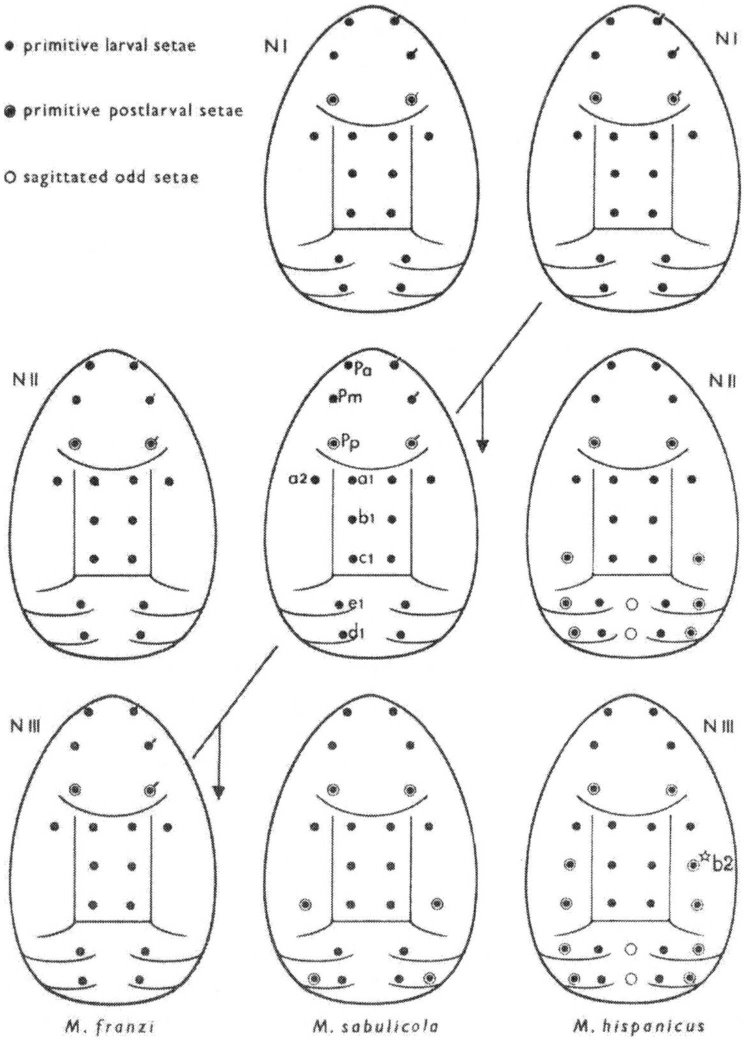

Fig. 1. The example of the evolution by retardation in the genus *Microcaeculus* (Caeculidae). The characters correspond with the scheme of dorsal chaetotaxy in the group of nymphs of three species (according to Coineau, 1971).

The lateral row is completed during the development of *M. hispanicus:* seta b_2, which appears to be the last, is often absent in the adults or is often subjected to malformation.

On the Morroccan coast exists a similar species, *Microcaeculus insularis* Franz, in which this absence is connected with the seta of a stable character, the seta b_2 usually being absent.

In *M. sabulicola* this reduction of chaetotaxy is still more marked and it was observed that the setae which in the previous species are present in the deutonymph, develop in this species as late as in the tritonymph. These setae are not present in the adults of *M. franzi.*

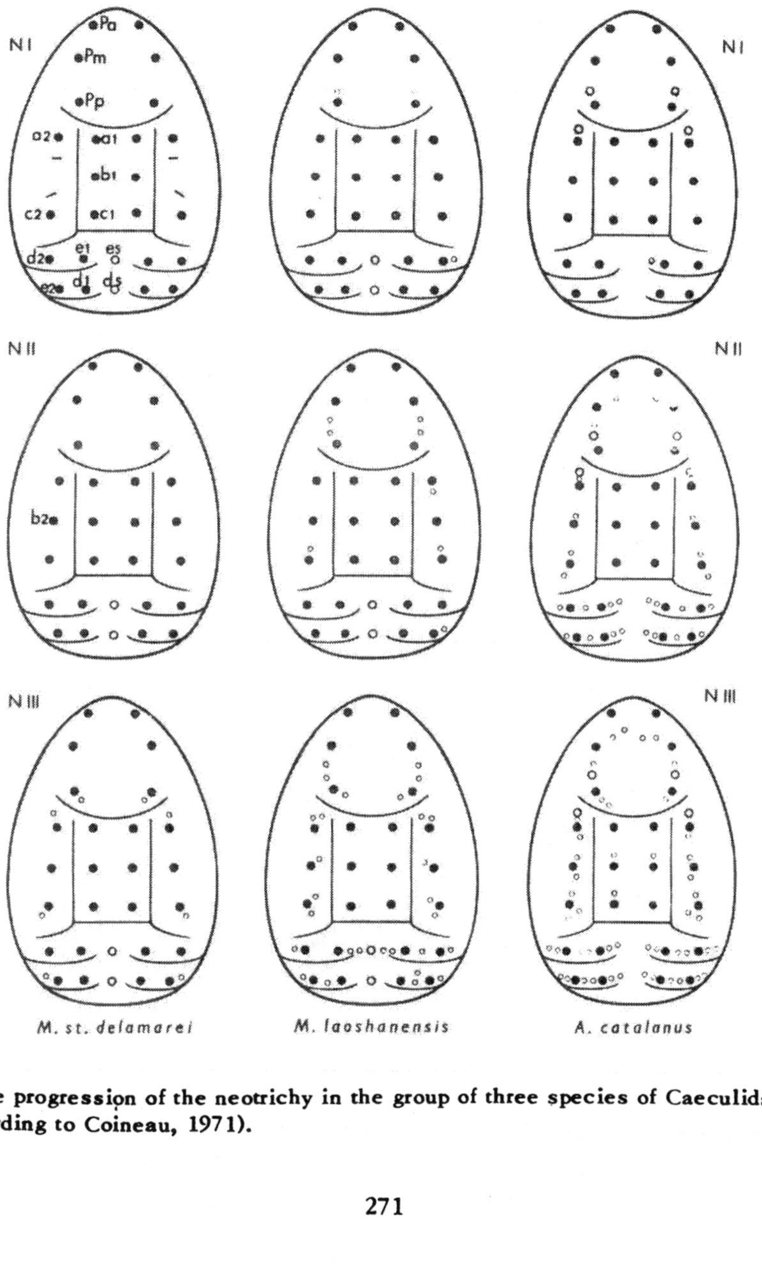

Fig. 2. The progression of the neotrichy in the group of three species of Caeculidae (according to Coineau, 1971).

When going from one species to the other starting from the right, we observe the same chaetotactic pattern appearing gradually in a later phase of ontogenesis. The occurrence of setae found in one stage become more and more random, till they appear in the next stage only and are finally completely eliminated when these deviations ("vertitions") occur in the adult. This is an example of regressive evolution by retardation.

Fig. 2 shows the phenomenon of evolution which produces a contrary result, i. e. the progressive evolution. A secondary multiplication of setae occurs – neotri-

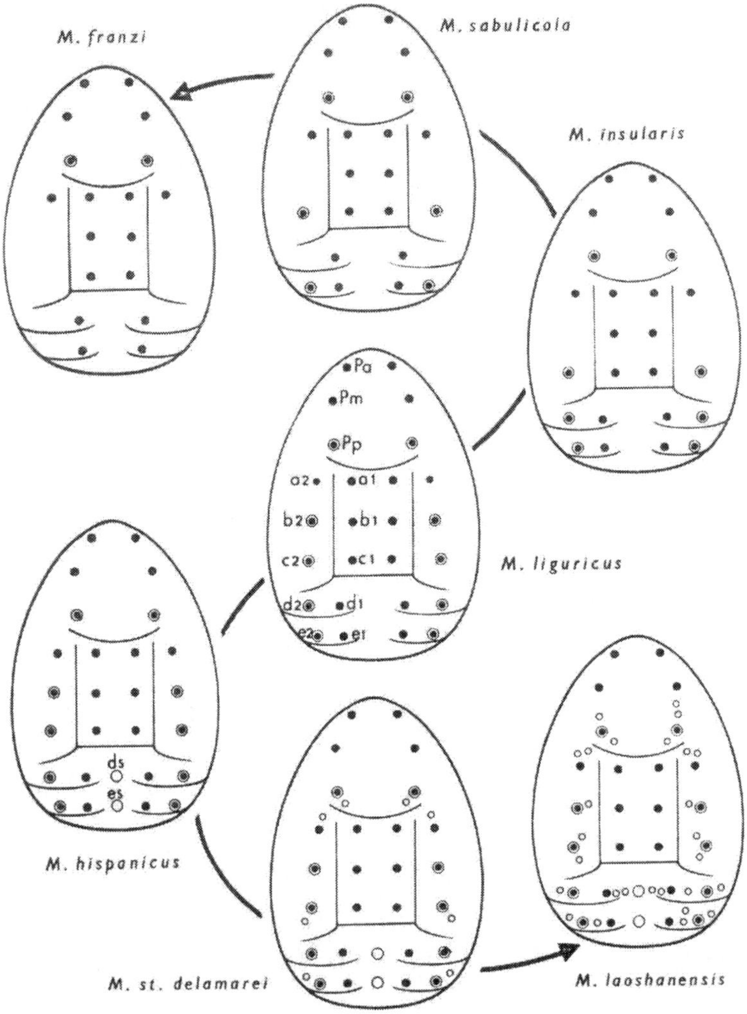

Fig. 3. Two evolutional tendencies, regression by retardation and neotrichy in the genus *Microcaeculus* Franz 1952 (according to Coineau, 1971).

chy. This figure illustrates, from the left to the right, nymphs of three species with their neotrichy gradually more and more apparent: *M. steineri delamarei*, *M. laosbanensis* and *Allocaeculus catalanus*.

M. steineri delamarei has already in the tritonymph stage the identical chaetotaxy as the adult of *M. hispanicus*. We can observe even in this stage and in particular in the adult stage the supplementary setae of a neotrichic nature, placed near certain primitive setae. The presence of the setae is very unstable and they correspond actually with the deviations ("ecarts") by their excess. That is in fact the first manifestation of the neotrichy. This early stage of neotrichy appears already in the protonymph in *M. laosbanensis* and then neotrichy increases during ontogenesis. In *A. catalanus*, the intensity of this phenomenon is even more accentuated.

The adult neotrichy is the more intense, the more precocious it is during the ontogenesis. In this case the ontogenesis parallels the phylogeny in this progressive evolution.

We are just witnessing the first steps of neotrichy in *M. steineri delamarei*, and its explosion in *A. catalanus*. On the one hand, we are following the full progressive evolution, on the other hand, we can observe the loss of the primitive post-larval setae, caused by retardation, which is the last stage of this depilation represented by *M. franzi*. In this case the subject under discussion is on the contrary a regressive evolution. The genus *Microcaeculus* shows two divergent evolutional tendences, the simplifying and the multiplying one, whose starting point is a medium hypothetical limit. The double arrow-head of this figure symbolizes these two types of evolution. Of course we do not demonstrate an affinity, although a close relationship does exist, but a juxtaposition of the state which derives from the identical primary motive and indicates the trend of evolution.

The simplicity of the diagram (Fig. 3) cannot convey the complexity of the evolutional phenomenon. The same species may manifest, in series of homeotypic organs, the result of the action of two evolutional tendencies as antagonistic as neotrichy and the regressive evolution by retardation. We can note this fact in *M. graecus* Franz, whose chaetotaxy is analogous to the chaetotaxy of *M. steineri delamarei*, i.e. with a small neotrichy, but without seta b_2.

Neotrichy appears to be primarily the manifestation of a local character, whose position of the primitive seta creates an universal basis. The relative independence of neotrichy has its source in the diversity of expression of the chaetotaxy.

CLASSIFICATION AND PHYLOGENY OF MITES

L. VAN DER HAMMEN
Rijksmuseum van Natuurlijke Historie, Leiden, Netherlands

The problem of classification and phylogeny of mites starts with the problem of their monophyletic or polyphyletic origin. In order to solve this first problem we must begin with a reconsideration of the characters by which the mites as a group, a subclass, are distinguished. This concerns only two characters: the presence of a gnathosoma, and the occurrence of a larva with only three pairs of legs. The latter character is not present in all mites, because the number of legs can (in rare cases) be further reduced, whilst in Holothyrida the larva appears to be still unknown. Apart from mites, a hexapod larva is also known from Ricinuleida, in which group this character has certainly developed by parallel evolution. The absence of a fourth pair of legs in the larva must be regarded as a regression of the posterior part of the prosoma, and is comparable with the regression of the number of posterior opisthosomatic segments in Actinotrichida. It is possibly connected with the small size of the larva, and could in the two superorders of mites, recognized here, have developed by parallel evolution.

In order to arrive at a more definitive conclusion on mono- or polyphyletic origin, it will evidently be necessary to consider the gnathosoma. In a recent study on the phylogeny of the Opilioacarida and their affinities with other groups of mites (Van der Hammen, 1971), I pointed to a very important difference between Anactinotrichida and Actinotrichida (the two superorders into which mites are subdivided here). In Actinotrichida, the gnathosomatisation (i.e. the development, in the course of evolution, of a gnathosoma) is connected with the development of a podocephalic canal (originally external, but in some groups secondarily internal) which runs from the podosoma to the dorsal surface of the infracapitulum, ventral of the bases of the chelicerae. This canal carries the products of several glands, among which some can apparently be regarded as coxal (cf. Wainstein, 1965; Prasse, 1968). In Opilioacarida, i.e. in primitive Anactinotrichida, the coxal glands (cf. With, 1904) secrete into a pair of sternal taenidia (in the sternal region of the first pair of legs) which unite between the sternapophyses, and continue into the sternal taenidium of the infracapitulum (a groove which generally has been called deutosternum or hypostome).

When we compare the phylogenies of podocephalic canal and sternal taenidia, we must arrive at the conclusion that in a theoretical common ancestor of Anactinotrichida and Actinotrichida, both the sternal taenidia and the podocephalic canal must have been absent; this, however, means at the same time that the gnathosoma of this (theoretical) ancestor was still at the beginning of its evolution, and this ancestor itself still at the very beginning of its evolution as a mite. This leads to the conclusion that, in reality, the gnathosoma of mites is the result of parallel evolution, which means that the two groups of mites have, independently, developed a gnathosoma out of homologous parts. In a recent study of the phylogeny of the gnathosoma (Van der Hammen, 1970a), I have indeed demonstrated that this phylogeny must have been different in the two superorders, especially as to the phylogeny of the tegulum which in Anactinotrichida extends more or less horizontally in a forward direction. Evidently, the subclass of mites is not of monophyletic origin, and a further study of the distinguishing characters of the two superorders will underline its biphyletic nature.

Of the characters that serve to distinguish Anactinotrichida from Actinotrichida, I mention the following.

(1) The stigmata belong to a homologous opisthosomatic series: four pairs in Opilioacarida, two or three pairs in Holothyrida (two pairs is the general number for this group but, according to unpublished investigations by Mrs. Athias, a third pair is present in a New Zealand species of *Allothyrus*), and one pair in Gamasida and Ixodida.

(2) The setae present no birefringent axis of actinopilin.

(3) The genital opening consists of a transverse fissure that secondarily can be covered by a number of shields.

(4) A sternum and free coxae are generally present.

(5) There is no sejugal furrow or interval separating legs II and III.

(6) There are often sternapophyses (or their homologues) situated in the sternal region of the first pair of legs.

(7) Primitively the coxal glands secrete into sternal taenidia; there is no podocephalic canal.

(8) The chelicera generally presents a dorsal lyrifissure beside an antiaxial.

(9) The palp often presents apotelic claws or homologues of apotelic claws.

(10) There are apparently no trichobothria.

(11) The femur of the legs is generally subdivided into two parts (basi- and telofemur) by a secondary articulation associated with lyrifissures.

(12) The tarsus of the legs often presents secondary articulations. Generally, there is a basi- and a telotarsus, although this subdivision is often lacking in the case of leg I. (It may be remarked, however, that a number of Prostigmata present also a secondary subdivision of the tarsus).

(13) The legs present several lyrifissures beside those associated with secondary articulations.

(14) The pretarsus sometimes presents setae or rudiments of setae.

The Actinotrichida, when compared with the Anactinotrichida, differ by the following characters.

(1) The tracheae (when present) belong to two groups: (a) there are genital tracheae in some representatives of three of the four orders in which the superorder is

subdivided here; and (b) there are often tracheae, in several regions of the body, which have probably developod from porose areas of the cuticle.

(2) The setae present an actinopilinous axis.

(3) Originally, there are six pairs of prodorsal setae.

(4) The primary genital opening is trifid; it often protrudes into a progenital chamber which extends over three segments; in that case, the secondary genital opening is longitudinal.

(5) There is a podocephalic canal into which primitively three pairs of glands secrete.

(6) There are no free coxae and no sternum, but generally there are epimera or coxisternal regions.

(7) There is a sejugal furrow or interval.

(8) The chelicera presents no dorsal lyrifissure.

(9) The palp presents no apotelic claw (and no apotele).

(10) There are often trichobothria on idiosoma and appendages.

(11) If a femur consists of two parts, these are primary segments connected by a primary articulation which is not associated with lyrifissures.

(12) The legs present a characteristic basic chaetotaxic pattern which, in principle, facilitates the recognition of a number of homologues in all four orders.

(13) There is a maximum of one lyrifissure on each of the appendages.

(14) Regressive stases (calyptostases as well as elattostases) are rather common.

When we compare these two lists now, we can easily see that they are nearly exclusively composed of characters which cannot be arranged in pairs of homologous origin. Evidently, this means that the fundamental differences in the two groups of mites concern differences between groups of a high category, differences referring to strongly diverging evolutions of the general plan of construction of the Arachnidea.

The Opilioacarida, here classified with the Anactinotrichida, are still regarded as a separate (third) group by several authors. It is important to deal here briefly with their relationships. Opilioacarida have a number of characters in common with Actinotrichida, which are rarely or never found in other groups of Anactinotrichida. This concerns: (1) The soft cuticle. (2) The presence of pigment grains under the cuticle (although these also occur in species of *Holothyrus*). (3) The presence of a rostral lobe (although different from a naso). (4) The presence of a disjugal furrow. (5) The shape of one of the paralabial setae, which functionally is a rutellum, although it is homologous with the corniculus of Holothyrida and Gamasida. (6) The position of the paired orifice of the infracapitular glands in the cervix, just as in some Actinotrichida. (7) The birefringency of some tarsal setae, although their true nature has not been studied. (8) The rapid backward gait, when trying to escape danger. (9) The occurrence of autotomy.

As mentioned before, some of these characters indeed occur in some other Anactinotrichida (or are homologous with other Anactinotrichid characters); others (such as autotomy and rapid backward gait) are functional characters which can be the result of convergence, or are of uncertain value (such as the birefringency of some setae). The remaining characters are ancestral ones which have no decisive value in a classification, the primary importance being that of shared possession of secondary characters.

Special value is attributed here to the important series of characters which Opilioacarida share with Holothyrida; of these characters I mention the following: (1) The infracapitular setae are numerous; although they are slightly variable in number, the basic pattern is in all cases the same, so that the same notation can be applied to both groups, a notation different from that of other Anactinotrichida; of special value is the occurrence of laterodorsal infracapitular setae, unknown in other groups of mites. (2) The tegulum is not prolonged into a suprachelіceral limb i.e. into a structure generally named tectum or epistome (although this structure occurs aldo in some *Uropodella*, a primitive genus of Gamasida). (3) The large lateral lips surround the preoral cavity ventrally as well as laterally and partly dorsally; they are hardly subdivided into specialized parts (only a dorsal labellum can be recognized). (4) The apotelic claw of the palp is still easily recognizable as such. (5) The palpal tarsus is characterized by the presence of a dorsal area with hollow setae. (6) The pretarsus still bears one or two pairs of true setae.

In my opinion the distinct relationship between Opilioacarida and Holothyrida justifies the classification of the former with Anactinotrichida. Opilioacarida constitute in fact the most primitive group of this superorder, a group of which the morphology is of considerable value for the comprehension of Holothyrida, Gamasida and Ixodida.

A better comprehension of the biphyletic origin of mites could be obtained by comparison with other groups of Arachnidea. Data in literature, however, are not sufficient for this purpose, because many details need reinvestigation. I point, for instance, to the problem of the identity of the first segment of the appendages; it is very important for us to know whether this is a coxa or a trochanter. In one case recently reinvestigated by me (Van der Hammen, 1969a), that of chelicera and palp of the Palpigradida, the first segment appeared to be a trochanter, although it had been named coxa in literature. In some groups of Arachnidea the occurrence of free coxae and a sternum has become complicated because of the secondary enlargement of the coxae in medial direction, at the cost of the sternum. This process has, by many authors, been misunderstood as the only sclerotized process of the ventral surface. In reality it has not occurred in mites. Other important problems in this connection are: (1) the possible homology of the sternapophyses with certain apophyses in Scorpionida, Amblypygi, and Opilionida; (2) the formation, in some higher developed groups, of a chitinous tube at the base of the acetabula, which does not represent the original coxa; (3) the recognition of laterocoxal setae, which are often reduced and can often only be understood in a comparative study. It seems that such groups as Ricinuleida and Opilionida are more related to Anactinotrichida, whilst Palpigradida could be related to Actinotrichida. It seems that Actinotrichida are more strongly different from most other groups of Arachnidea than is the case with Anactinotrichida. These, however, are merely opinions. What we need, in reality, is a new comparative morphological investigation of all groups of Arachnidea. In fact, comparative Arachnid morphology is our fundamental problem, the study of which is the condition for further conclusions on Acarid morphology.

After these observations on the two superorders of mites, we shall now proceed with the subdivision of each of these groups. As mentioned before, the Anactinotrichida are subdivided here into four orders: Opilioacarida, Holothyrida, Gamasida, and Ixodida. It appears that any attempt to arrange these four orders in two groups is artificial. Opilioacarida constitute the group with the greatest number of ancestral characters; as mentioned before, they are at the same time distinctly related to Holothyrida. *Allothyrus*, a genus of Holothyrida, shows even closer relationships with Opilioacarida (it is characterized by the presence of a pair of sternapophyses, more than one pair of setae on the principal cheliceral segment, several infracapitular setae which are lacking in *Holothyrus*, and in the New Zealand species three pairs of stigmata). On the other hand, *Allothyrus* shows distinct relationships with Gamasida (because of the development of the lateral lips). According to recent, unpublished investigations by Mrs. Athias, some striking similarities exist between Holothyrida or Opilioacarida, and Ichthyostomastogasteridae, a family of Gamasida, mainly South American in distribution; of these I mention the absence of a suprachelliceral limb *(Uropodella)*, and the presence of a rutellum-shaped corniculus together with the ingestion of solid food *(Asternolaelaps)*. Evidently the four orders of Anactinotrichida are all distinctly related, although separated from each other by distinct differences.

In the course of several morphological investigations of groups of Actinotrichida, I recognized four orders which I named Actinedida, Oribatida, Acaridida, and Tarsonemida. Although authors nowadays usually classify Tarsonemida with Trombidiformes or Prostigmata, I have recently separated them from this group (Van der Hammen, 1970). (The remainder of Trombidiformes or Prostigmata corresponds with my Actinedida). The presence of ambulacra with primitively two claws and a pulvillus, the small number of opisthosomatic segments (C-H), and the presence of a single pair of stethosomatic stigmata, constitute characters which separate Tarsonemida from Actinedida and Oribatida. The Tarsonemid ambulacrum is reminiscent of that of many Acaridida because of its pulvillus, whilst the number of opisthosomatic segments is apparently the larval number in Acaridida (as far as is known). The shape of some Tarsonemid solenidia is reminiscent of that of several Acaridida. Unpublished stereoscan micrographs of the palp of a Tarsonemid recently described by me show a striking similarity to the palp of the Acaridida.

Actinedida and Oribatida constitute two groups which are for the greater part very distinctly and very widely separated; some of their primitive representatives, on the contrary, are very closely related. Sometimes it seems even difficult to delimit the two groups at their bases. Within each group, however, the species show distinct relationships, so that there can be no doubt as to their classification. In a recent paper (Van der Hammen, 1969) I enumerated a number of characters which serve to distinguish primitive Actinedida from Oribatida. Some of these characters apply to all or nearly all Actinedida, others have only a restricted validity; all of them, however, can serve to distinguish Actinedida from Oribatida. It concerns the following characters: (1) a rostral tectum is absent, and the naso extends unprotected between the chelicerae; (2) the prodorsum often has two pairs of bothridial setae; (3) although there can be opisthosomatic microsclerites, a notogaster or notogastral sclerite is not present; (4) the femora can present solenidia; (5) the claws can present tenent hairs; (6) the median claw can, in some or in all stages, be replaced by an empodium; (7) the number of nymphs tends to be reduced.

The four orders of Actinotrichida could be arranged in two cohorts: one comprising Actinedida and Oribatida, the other comprising Acaridida and Tarsonemida. I have not named these groups here because, in my opinion, it is still too early for it.

A few words must be added on the nomenclature used here for the higher groups. These names are not submitted to the Code of Zoological Nomenclature. Nevertheless, the names chosen here partly represent the oldest available names. This concerns the names Holothyrida Thon (1905), Gamasida Leach (1815), Ixodida Leach (1815), Oribatida Dugès (1833), Acaridida Latreille (1802), Tarsonemida Canestrini & Fanzago (1877); of these names, the original endings only have partly been corrected. The name Opilioacarida Zakhvatkin (1952) is preferred here instead of Notostigmata With (1904) for the sake of uniformity. The name Actinedida was recently introduced by me (Van der Hammen, 1968) for two reasons: (1) for the sake of uniformity, and (2) in order to avoid misunderstanding (the group having received a new interpretation, without Tarsonemida). Because of the great variety within the group, I have looked for a name without any sense. I have chosen the generic name *Actineda* as root, which just as *Oribata* (the root of Oribatida) is no more in use, and of which the translation has not much meaning. As far as ordinal names are concerned, I have rejected the names ending with stigmata. When these names were reintroduced, they were for the greater part no more in general use; furthermore they are far from being the oldest available names, and they include several incorrect meanings (some Tetrastigmata have six stigmata; several Prostigmata have no stigmata or they have genital tracheae; many Cryptostigmata have no stigmata; and some Astigmata have genital tracheae). In fact this nomenclature has numerous disadvantages and not a single advantage.

As for the names Anactinochaeta and Actinochaeta, sometimes used instead of Anactinotrichida and Actinotrichida, these must be regarded as incorrect, because Zakhvatkin (1952) introduced the name Actinochaeta for a group of Arachnid orders, among which Actinotrichida.

The phylogeny of the mites could now be illustrated by the dendrogram of Fig. 1. It represents the mites as biphyletic. The dotted vertical line in the centre of the drawing represents the archetype (i.e. a synthetic combination of ancestral characters), used here as standard of reference. Because comparative Arachnid morphology is still insufficiently known, the archetypical values and the average divergence of the orders are mere estimations. In a further stage of the investigation, this will pre-eminently be a numerical problem. The Opilioacarida are supposed (in the dendrogram) to have not very much diverged from the ancestor of the Anactinotrichida.

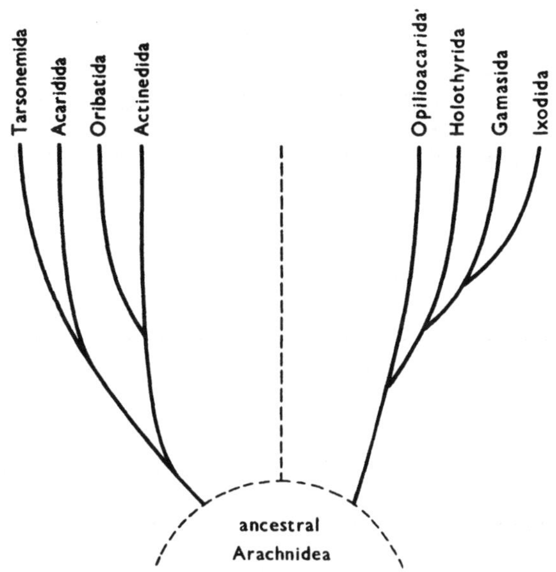

Fig. 1. Dendrogram representing the phylogeny of mites.

I have supposed that Gamasida and Ixodida derived from the ancestral line of the Holothyrida. The distribution of Opilioacarida and Holothyrida is more or less Gondwanian. The closest representatives among the Gamasida, according to unpublished investigations by Mrs. Athias (as mentioned before), are species belonging to Ichthyostomastogasteridae, a family which is centralized in South America. Consequently, the origin of the Anactinotrichida could be Gondwanian. As to the phylogeny of the Actinotrichida, the recognition of two groups is also represented in the dendrogram. It is supposed that primitive Actinedida are closest to the common ancestor of the Actinotrichida.

The dendrogram illustrates at the same time schematically the present state of my investigations of evolution and comparative morphology. It represents only a stage in the development of acarology. As mentioned before, one of the most important subjects is now the study of comparative Arachnid morphology. Besides, a further elaboration of Acarid morphology, especially a study of Ixodida and Acaridida in comparison with other groups of mites, is also a subject with a high priority. Our glossary of acarological terms, which is now in an advanced state, will certainly be an important contribution to this comparative study. The results of all these subjects together, will close a first round in the study of classification and phylogeny.

References

Hammen, L. van der: Introduction générale à la classification, la terminologie morphologique, l'ontogénese et l'évolution des Acariens. Acarologia 10 : 401–412, 1968.

– Notes on the morphology of *Alycus roseus* C.L. Koch. Zool. Med. 43 : 177–202, 1969.

– Notes on the mouthparts of *Eukoenemia mirabilis* (Grassi) (Arachnidea: Palpigradida). Zool. Med. 44: 41–45, 1969a.

– *Tarsonemoides limbatus* nov. spec., and the systematic position of the Tarsonemida (Acarida). Zool. Verh. 108: 1–35, 1970.

– Remarques générales sur la structure fondamentale du gnathosoma. Acarologia 12: 16–22, 1970a.

– La phylogénèse des Opilioacarides, et leurs affinités avec les autres Acariens. Acarologia 12: 465–473, 1971.

Prasse, J.: Zur Anatomie und Histologie der Acaridae mit besonderer Berücksichtigung von *Caloglyphus berlesei* (Michael 1903) und *C. michaeli* (Oudemans 1924) III. Die Drüsen und drüsenähnlichen Gebilde, der Podocephalkanal. Wiss. Z. Mart. Luther Univ. Halle, Math. Naturw. R. 17: 629–646, 1968.

Wainstein, B.A.: Die Coxaldrüsen bei *Hydrachna* Müll. (Hydrachnellae). Zool. Anz. 175: 437–440, 1965.

With, C.J.: The Notostigmata, a new suborder of Acari. Vidensk. Medd. Naturh. Foren. Kjøbenhavn, 137–192, 1904.

Zakhvatkin, A.A.: /Subdivision of the mites (Acarina) into orders and the position of these in the system of the Chelicerata/. Parazitol. Sbornik 14: 5–46, 1952. (In Russian.)

AN ANALYSIS OF THE CLASSICAL TAXONOMIC METHOD WITH PARTICULAR REFERENCE TO THE FAMILY PHYTOSEIIDAE (ACARINA: MESOSTIGMATA)

R.I.C. HANSELL, D.A. CHANT
Dept. Zoology, University of Toronto, Ontario, Canada

There is a great need in taxonomy today to develop an understanding of the methods by which classical taxonomists construct classifications. Of particular interest are the methods by which characters are weighted, that is, placing different emphasis on different characters in making taxonomic decisions. Numerical taxanomists have suggested that all characters should be given equal weight, while most classical taxonomists feel that differential weighting is desirable; however, classical taxonomists often do not agree about which characters should be given more emphasis. And indeed, most serious disagreements about the classification of mites can be traced to this problem.

We will give a simplified example of how we might represent the effect that differential weighting can have on taxonomic decisions. If we have three animals A, B, and C which differ by two characters – say leg length and arm length; if A has extremely short arms and legs, B has long arms and short legs, and C has short arms and long legs, then considering measured leg length only, A and B are most similar and may be classified together. Considering arm length only, A and C are most similar. But, considering both characters equally together, the short lengths of A makes B and C most similar. Hence, all three possible *taxonomic* decisions can be made merely by changing the emphasis from equal consideration of arms and legs to marked weighting of either arms or legs.

The amount of weight necessary to give the same decision as that which the classical taxonomist made, can be determined by comparison with a numerical model – such as Euclidean distance in a multidimensional character space, in which the scale of the character axes is changed by multipliers which are equivalent to character weights. These methods are defined fully in Hansell and Chant (1972) and Hansell and Ewing (1972).

We may consider that differences in character weighting have originated in some of the following ways: (i) perceptual and methodological differences between taxonomists, (ii) disagreements as to the phylogenetic significance of a character, (iii) disagreement as to the stability of a character within a specified group of taxa, and finally, (iv) historical considerations, involving different interpretations of previous classifications and the character weightings implied in them.

If we consider a complex or qualitative character such as the shape of the ventrianal shield in the family Phytoseiidae, we find that its shape can be sub-quadrate, subtriangular, elongate, figure eight shaped, or, it may consist of two separate shields. A numerical taxonomist might express this variation minimally by three lateral measurements across the long axis plus the length; he might then give all characters equal weight. However, a combination of the character values to give an index of mediolateral indentation, and separately to give an index of elongation, each mimics a complex character which has been weighted in different ways by different taxonomists. For instance, in the formation of the generic concepts for *Amblyseialus, Amblyseiella* and *Amblyseiulus* Muma (1961) gave positive weights to these characters – which we have represented by combined character indices; but, when the four measurements are analyzed independently, we find that Muma did not give them positive weights. In the case of *Amblyseialus,* in order to place the member species together, the individual widths and lengths must have been given slightly positive emphasis and the degree of indentation must have been given relatively heavy positive weight.

The implication here is that, given the same specimens and using the same unit characters, different taxonomists can form entirely different classifications depending upon the way in which the unit characters are combined to form compound characters and by the level of character complexity at which the taxonomists work. Differences in perception and method of working may produce large differences in the resultant classifications.

Another instance of the complexity of the factors involved in character definition is the references to the "dorsal hexagonal area" in mesostigmatic mites and particularly in the family Phytoseiidae. This is an area delineated by arbitrary lines drawn between specific dorsal setae, and it forms a pattern which is easily recognized by a visual inspection of a mounted specimen. Chant's (1959) use and positive weighting of characters of length and position of setae in the "dorsal hexagonal area" is based on the character's taxonomic utility, but the possibility remains that the ease of perception of this pattern directed his attention to them and that other dorsal characters were not given equal attention and analysis. It is perhaps significant that Pritchard (Schuster and Pritchard 1963) recognized an alternative but still easily perceived pattern of dorsal setae in a paper which provided yet another possible arrangement of genera. That individuals reveal differences in perception which may influence the characters they select for particular study has been recognized in the work of Thurstone (1944), Vernon (1947) or Woltmann (1952) and it would be of particular interest to know what extent the variations in classifications are due to this factor.

A character may be emphasized or de-emphasized depending upon its "stability" within some group of species. The sample of species over which it is considered depends upon the previous grouping decisions of the taxonomist, and, in effect, the previous decisions will bias the result. The weight given to a character in a taxonomic decision may differ from that given by another taxonomist solely because of this self imposed sampling problem. For instance, Chant (1959) gave heavy emphasis to the presence of dorso-lateral sclerites in defining his concept of the genus *Iphiseius;* and, indeed, the character was stable within the small number of species recognized as belonging to the genus. Muma (1961) defined concepts of the genera *Amblyseius* and *Amblyseiulus* which were unique in that they had not previously existed with those sets of members. In discussing the genus *Iphiseius,* he commented that the degree of sclerotization varies within the genera *Amblyseius* and *Amblyseiulus,* and hence that lateral sclerotization is not a good generic character. Accordingly, he de-emphasized this character in taxonomic decisions on the placement of the *Iphiseius* species.

Another situation in which characters are considered only over a limited number of species arises due to the regional location of collections and the infrequency with which types are exchanged between countries. The independent, but overlapping concepts of the genera *Amblyseius sensu* Berlese (1914), and of *Typhlodromus sensu* Oudemans (1930) may be attributed to this type of regional isolation. The resulting taxonomic disagreements were not solved for another thirty years.

An example of the development of a narrow generic concept due to a limited faunal sample, is Muma's (1955) erection of the genus *Amblyseiella* to distinguish a new species of Phytoseiid from the very few other known species of this family occurring in Florida citrus groves.

A fundamental influence on the development of group concepts, and on the relative weightings given to taxonomic characters is the order in time in which species are discovered and described. In a sense, a classification attempts to be predictive of new species which will be discovered, so that they can rapidly be placed in a suitable genus. Effectively, this is predicting from a small sample onto a large sample. To do this effectively, a concept of character weights which can be applied is necessary. However, predictive errors revealed by the discovery of intermediate forms between genera may not be recognized by the taxonomist if he unduly retains character weightings in the case of highly split classifications or of group concepts in less divided classifications.

The classical taxonomic method is ideally described as a process, self-correcting through time as new data from described species and newly perceived characters are introduced into the system. However, where the conservation of previous group concepts, or previous character weightings, arbitrarily over-rules the weight of new information then the taxonomic method ceases to be self-correcting, and a distorted classification results.

References

Berlese, A.: Acari nuovi. Redia. 10: 113–150, 1914.

Chant, D.A.: Phytoseiid mites. Parts I and II. Canad. Ent. Suppl. 12, 166 pp., 1959.

Hansell, R.I.C., Chant, D.A.: A method for estimating relative weights applied to characters by classical taxonomists in the genus *Iphiseius* Berlese (Acarina: Phytoseiidae). Systematic Zoology. (In press), 1972.

Hansell, R.I.C., Ewing, B.: The detection and estimation of character weighting in

classification J. Theor. Biol. (in press), 1972.
Muma, M.M.,: Phytoseiidae associated with citrus in Florida. Ann. Ent. Soc. America. 48: 262–272, 1955.
– Subfamilies, genera, and species of Phytoseiidae. Bull. Florida State Museum. 5: 251–302, 1961.
Oudemans, A.C.,: Acarologische Aanteekeningen 56. Ent. Bericht. 4: 180–188, 1930.
Schuster, R., Pritchard, E.: Phytoseiid mites of California. Hilgardia. 34: 191–285, 1963.
Thurstone, L.L.: A Factorial Study of Perception. Univ. Chig. Press: Illinois. 635 pp., 1944.
Vernon, M.D., A further study of visual perception. Brit. J. Psychol. 38:79, 1947.
Woltmann, A.G.: The Bender visual motor Gestalt test, in Projective Psychology. Ed. Lawrence E. Abt and Leopold Belcak. New York. Alfred A. Knopf. 847 pp., 1952.

"GANGSYSTEMATIK" OF THE PARASITIFORMES AND THE FAMILY UROPODIDAE BERLESE

W. HIRSCHMANN
Fürth i.B., German Federal Republic

In 1964 Dr. Irene Zirngiebl-Nicol and I published the "Gangsystem" of the family Uropodidae Berlese, 1892. In this work 78 genera or subgenera were considered to be synonyms. The remaining 28 genera were placed in one family which was divided into 3 subfamilies: Uropodinae, Oplitinae and Uroactiniinae. The Uropodinae was further divided into 3 tribes: Uropodini, Dinychini, Trichouropodini; the Oplitinae into 2 tribes: Polyaspidini, Trachyuropodini; the Uroactiniinae into 2 tribes: Diarthrophallini, Uroactiniini.

Postembryonic "Gangsystematik" of the Uropodidae (Fig. 1)

The postembryonic development of the Uropodidae comprises 4 stages: larva

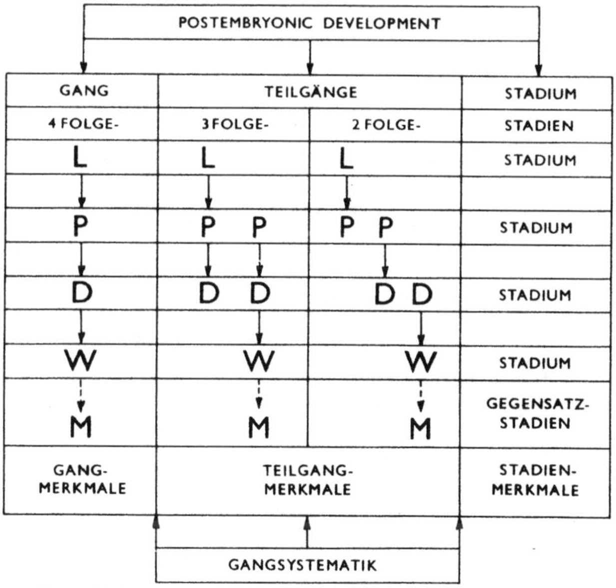

Fig. 1. L = larva; P = protonymph; D = deutonymph; W = female; M = male.

(L) – protonymph (P) – deutonymph (D) – adult (female (W) or male (M)).

A"Gang" consists of 4 stages: L–P–D–W or M and a "Teilgang" of 3 or 2 stages: 3: L–P–D, P–D–W or M; 2: L–P, P–D, D–W or M. "Gegensatzstadien" can be female or male; phoretic- or normal deutonymph. In the "Gangsystematik" we distinguish 3 types of characters:

(1) "Gangmerkmale"; all 4 stages show the same character.
(2) "Teilgangmerkmal": 3 or 2 stages show the same character.
(3) "Stadiummerkmal": the character is present only in one stage.

"Gangmerkmale", "Teilgangmerkmale", "Stadiummerkmale" (Fig. 2)

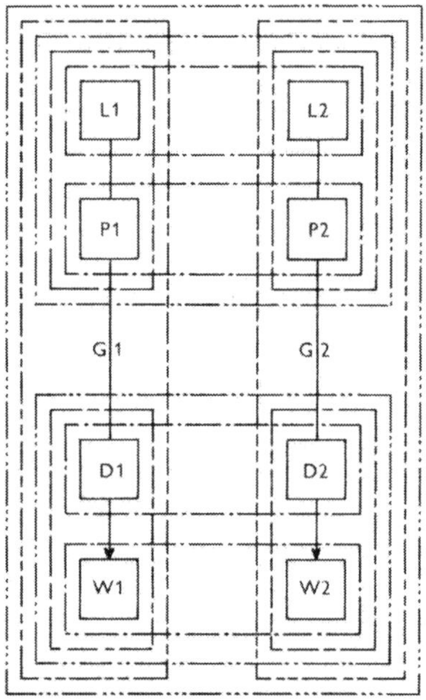

Fig. 2. L = larva; P = protonymph; D = deutonymph; W = female; G = Gang.

= "gangspezifisch,"
= "teilgangspezifisch,"
= "stadiumspezifisch,"
= "gängesammelnd,"
= "teilgängesammelnd,"
= "stadiensammelnd,"

The character can be specific or collective. In the "Gangsystematik" we distinguish:

(a) "gangspezifische, teilgangspezifische, stadiumspezifische" characters
(b) "gängesammelnde, teilgängesammelnde, stadiensammelnde" characters.

"Gangmerkmale"

(a) A character is "gangspezifisch" when it occurs in all stages of a species.
(b) A character is "gängesammelnd" when it occurs in all stages of two or more species.

"Teilgangmerkmale"

(a) A character is "teilgangspezifisch" when it is present in one species only.
(b) A character is "teilgängesammelnd" when it is present in two or more species.

"Stadiummerkmale"

(a) A character is "stadiumspezifisch" when it is present in one stage of one species only.
(b) A character is "stadiensammelnd" when it is present in equivalent stages of two or more species.
(c) A character is "gegensatzstadiumspezifisch" when it is present in the male or female, or in the phoretic or normal deutonymph.
(d) A character is "gegensatzstadiensammelnd" when it is identical in the partners of the "Gegensatzstadien".

"Gangart" entails a comparison of characters between all the stages of one species, "Ganggattung" entails the comparison of the "Gangen" of different species. The comparison of different "Ganggattungen" leads to "Gangfamilie".

Thus, the "Gangsystematik" is based on two principles: (a) the systematic characters must be recognisable in every stage of each "Gang" and (b) the system is based on the "Gangmerkmale".

"Gangsystematik" of the Uropodidae

We consider the most important "Gangmerkmal" characters in the Uropodidae to be associated with the chelicerae:
"Sie sind fast körperlang, schmal, bandförmig, mit im Verhältnis kleinen Laden. Ein ventraler Sinneszapfen ist meist vorhanden. Das distale Sinneshaar ist als Pinselpilus ausgebildet, dorsales laterales Spaltorgan und ventraler Sinneshaarsaum fehlen".

Diagnostic characters of the subfamilies and tribes of the Uropodidae

Subfamily Uropodinae:	fixed digit with knobbed seta (Kolbenpilus)
Tribe Uropodini:	chelicerae without "Rollplatte"
Tribe Dinychini:	movable digit unidentate, "Rollplatte" present
Tribe Trichouropodini:	movable digit bi-or multi-dentate, "Rollplatte" present
Subfamily Oplitinae:	fixed digit with double seta ("Doppelpilus")
Tribe Polyaspidini:	chelicerae without "Rollplatte"
Tribe Trachyuropodini:	chelicerae with "Rollplatte"

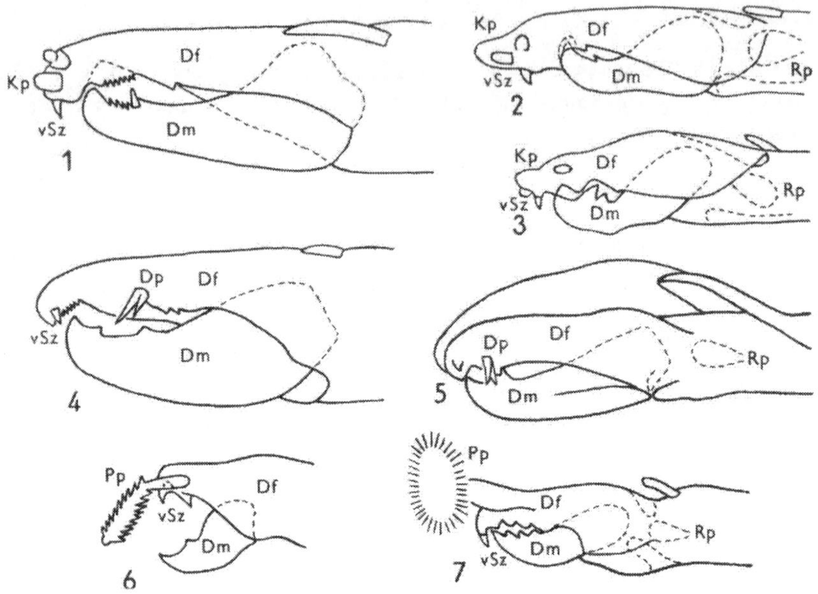

Fig. 3. Df = Digitus fixus; Dm = Digitus mobilis; Dp = Doppelpilus," Kp = "Kolbenpilus;" Pp = "Pinselpilus;" Rp = "Rollplatte;" vSz = "Ventraler Sinneszapfen."
1, 2, 3: Uropodinae; 4, 5: Oplitinae; 6, 7: Uroactiniinae.
1 = Uropodini, 2 = Dinychini, 3 = Trichouropodini, 4 = Polyaspidini, 5 = Trachyuropodini, 6 = Diarthrophallini, 7 = Uroactiniini.

Subfamily Uroactiniinae: fixed digit with "Pinselpilus"
Tribe Diarthrophallini: chelicerae without "Rollplatte"
Tribe Uroactiniini: chelicerae with "Rollplatte".

In addition to these "gängesammelnden" characters, the chelicerae also show species specific and "Gang" specific characters, such as the ratio of the length of the movable digit to the length of the extension of the fixed digit beyond the tip of the movable digit. This character has been used for the identification of species in the following genera: *Uroobovella, Uropoda, Discourella, Uroseius, Metagynella, Dinychus, Deraiophorus, Cyllibula* and *Nenteria*.

Diagnostic characters of the venter of the gnathosoma

An important "familienspezifisches, gängesammelndes" character is the longitudinal arrangement of the coxal setae C1, C2, C3, C4. The coxal setae are thicker than in the Gamasina and the majority are serrated. The different shapes of the corniculi are partly "gattungsspezifisch, artenspezifisch, gängspezifisch and partly gangesammelnd".

Three structures are visible on the hypostome:

(a) L a c i n i a e: acute, smooth, serrated or fimbriated paired hyaline mouth lobes.

(b) Anterior hypostomal groove: a hyaline connection between the mouth opening and the first row of transverse denticles.
(c) Posterior hypostomal region: formed by the fusion of the palpcoxae posterior to C3 and including 2–5 transverse rows of denticles or arches (male) or denticle-fields.

The characters of the hypostome are partly "gattungspezifisch artenspezifisch, gangspezifisch, teilgangspezifisch" and partly "gängesammelnd".

Dorsal and ventral "Gangmerkmale" and "Teilgangmerkmale"

The most important "familienspezifische, gängesammelnde Merkmal" is the wide distance between setae J 3. This is because the pygidial shields of the larva and protonymph lack setae; the setae being arranged around the shield. Further setae J 2 and J 3 form more or less a transverse line. In the deutonymph and adults the pygidial region remains free of setae and may be occupied by a discrete post-dorsal shield in the adult. A "Teilgang L–P–D" is seen in the strip of weakly sclerotized integument between the pygidial and the posterior margin of the body and the "Teilgang D–W or M" by the hypertrichy between the inner and marginal setae. In the protonymphs, the lateral setae are not situated on the podosomatal shield but are on 32 platelets; the latter fuse in the deutonymph.

A characteristic feature of the venter of the idiosoma is the absence of setae V1 ("gängesammelnd, familienspezifisch").

References

The following list of publications includes the majority of the works hitherto published on the "Gangsystematik" of the Parasitiformes by the authors.

Hirschmann, W.: Gangsystematik der Parasitiformes Teil 1: Rumpfbehaarung und Rückenflächen. - Acarologie Folge 1, 3–20, I–V., Taf. 1–26, 1957.

– Gangsystematik der Parasitiformes Teil 2: Mundwerkzeuge, Hypostombestimmungstafeln, Synonymie der Haarbenennung. Acarologie Folge 2, 2–23, I, II, Taf. 1–7, 1959.

– Gangsystematik der Parasitiformes. Entomol. Z. 73, 1/2, 4, 1963.

– Neue Gedanken zur Systematik der Milben. Ber. Nat. Ges., Bayreuth, 1961/63, XI, 221–225, 1963.

– Gangsystematik der Parasitiformes Teil 90: Katalog der Gattungen (Untergattungen) und übergeordneter Taxa. Acarologie Folge 16, 28–31, 1971.

Hirschmann, W., Zirngiebl–Nicol, I.: Gangsystematik der Parasitiformes Teil 4: Die Gattung *Trichouropoda* Berlese 1916, Cheliceren u. System der Uropodiden, Entwicklungsstadien von *Microgynium, Uroseius, Polyaspis, Trachytes, Uropoda, Urosternella, Dinychus, Oplitis, Trachyuropoda, Celaenopsis, Liroaspis*. Acarologie Folge 4, 1–41, Taf. 1–14, Taf. Anh. Nr. 408–549, 1961.

– Gangsystematik der Parasitiformes Teil 6: Uropodiden Die Gattung *Uroobovella* Berlese 1903 nov. comb. Teilgänge von *Nenteria* nov. comb. Erstversuch der Aufstellung eines Gangsystems der Uropodiden aufgrund der Gnathosoma-Unterseite u. Chelicere. Acarologie Folge 5, 58–80, Taf. 22–32, 1962.

– Gangsystematik der Parasitiformes Teil 7: Uropodiden Das Gangsystem der Familie Uropodidae (Berlese 1892) Bestimmungstabellen, Kurzdiagnosen, Operculum-Bestimmungstabellen. Acarologie Folge 6, 1–22, Taf. 1–5, 1964.
– Gangsystematik der Parasitiformes Teil 37: Die Geschichte der Uropodidensysteme. Acarologie Folge 12, 3–6, 1969.
– Gangsystematik der Parasitiformes Teil 73: Neuzeichnung bekannter Uropodiden-Arten, Gänge, Teilgänge, Stadien, Chaetotaxie, Literatur, Synonyma, Fundorte, Grösse. Acarologie Folge 12, 125–132, Taf. Nr. 72–188, 1969.

Other references to the genera Uropodidae in Acarologie Folge 8, 10, 11, 12, 15 and 16.

OBSERVATIONS ON THE GENERIC CLASSIFICATION OF TARSONEMID MITES (PROSTIGMATA)

E. E. LINDQUIST
Entomology Research Institute, Ottawa, Canada

This paper is part of a larger study that reviews the classification of the Tarsonemidae and demonstrates that more characters can be used in defining supraspecific groups in this family. Other parts of this study deal with structures overemphasized as generic characters, and with potentially useful characters of the leg setation. This part considers some gnathosomal features.

Gnathosomal characters are particularly useful because they are usually diagnostic for both sexes and often for the larva. Thus, the form and orientation of the palpi, correlated with the shape of the gnathosomal capsule, distinguishes *Rhynchotarsonemus*, with an elongate beak-like gnathosoma, and *Steneotarsonemus*, with a broad subcircular gnathosoma, from other tarsonemid genera.

The number of dorsal gnathosomal setae, one pair, is apparently constant and distinguishes the tarsonemids as a family from the scutacarids and nearly all pyemotids. Lindquist (1969a) noted a second pair of setae, appressed to the dorsolateral surfaces of the gnathosoma. They were thought to be the homologues of the external dorsal setae of scutacarids and pyemotids. But alternatively they may be the "postpalpal" setae described by Cross (1965) for the Pyemotidae, and termed the "laterocoxal" setae by van der Hammen (1970) who did not discern them for the Tarsonemidae. The presence or absence of these setae as a supraspecific character is of limited diagnostic value because of the difficulty in discerning them. I have reported them present on adults, but not on larvae, of species of *Iponemus*, *Tarsonemus*, and *Heterotarsonemus* (Lindquist 1969a, 1969b). I here report them present on adults and larvae of *Steneotarsonemus*, on females of *Pseudotarsonemoides*, and conspicuous (as long as the dorsal gnathosomal setae) on females of a species of *Tarsonemus* in the *pauperoseatus*-group of Suski (1968), and on males and females of two species of *Ununguitarsonemus*. They are absent on adults of *Tarsonemus indoapis* Lindquist, *Xenotarsonemus belemnitoides* (Weis-Fogh), two species of *Polyphagotarsonemus*, and one of *Coreitarsonemus*.

The form of the pharynx and of the associated pair of gland-like structures (probably salivary glands) in the gnathosoma offer a complex of potentially diagnostic characters. Pharyngeal structures were used as generic-level characters in the Pyemotidae by Cross (1965). They were used to a limited extent in the Tarsonemidae by Lindquist (1969a, 1969b). Occasionally, because of sexual dimorphism, pharyngeal structures are diagnostic for one sex only, as in *Iponemus*;

this reflects different feeding habits between the sexes. More commonly, there is little or no dimorphism, and the pharyngeal characters are useful to recognize, and to confirm the conspecificity of, both sexes and often the larva of a species.

The diagnostic value of pharyngeal structures varies among different groups in the Tarsonemidae. The greatly enlarged, muscular pharynx in *Iponemus* (Fig. 4) and the strongly sclerotized, horseshoe-shaped pharynx in *Pseudotarsonemoides* (Fig. 1) are highly diagnostic at the generic level, but for the female only. A slender but strongly sclerotized pharynx, with a relatively large pair of glands, is characteristic for both sexes and the larva of all known species of *Heterotarsonemus*, but is not very diagnostic because certain species-groups of *Tarsonemus* have identical pharyngeal and glandular structures (Fig. 5). Similarly, the *subcorticalis* species-group in *Tarsonemus* is characterized by a moderately muscular pharynx with a pair of small glands (Fig. 6), but identical structures are found in some other species-groups in *Tarsonemus*. Other distinctive types of pharyngeal structure occur among species of *Tarsonemus*, but their diagnostic usefulness at the species-group level cannot be evaluated yet (Figs. 2, 3, 7, 9).

In some cases the salivary glands are as diagnostic as the pharynx. For example these glands are distinctively node-like spheroids in *Tarsonemus apis* Lindquist (Fig. 2).

The structure of the pharynx is correlated with its function. Those tarsonemids with little or no musculature around the pharynx usually have instead a strongly sclerotized pharynx, the walls of which appear to be inflexible and of little or no function as a pump. Strong-walled pharynges are usually slender, as in species of *Heterotarsonemus* and many species of *Tarsonemus* (Fig. 5), but some are shorter and thicker, tending toward the massive horseshoe-shaped form in *Pseudotarsonemoides*. The salivary glands are conspicuously well developed and distinct from such pharynges (Figs. 1, 2, 7, 8).

In contrast, a variety of forms of pharynges with a well developed muscular sheath occur in different groups of Tarsonemidae, and all have thin, apparently elastic internal walls to facilitate a pumping action. Their salivary glands are often smaller, and usually indistinct because of being attached or appressed to the pharyngeal musculature (Figs. 3, 4, 6, 9).

The form of the cheliceral stylets and cheliceral bases ("Stechborstenhebel" of Sachs, 1951) distinguish a few tarsonemid genera. Adults and larvae of *Steneotarsonemus* are characterized by unusually long, stout stylets that are strongly out-curved proximally and attach to a pair of enlarged, conspicuous cheliceral bases. The bases are oriented roughly transversely to the longitudinal axis of the gnathosoma, at lateral right angles to the axis of the stylets when retracted. The cheliceral stylets of adults and larvae of *Acarapis* also are distinctively long and strong, but they are upcurved proximally and the cheliceral bases are oriented along the longitudinal axis of the gnathosoma, roughly at dorsoventral right angles to the axis of the stylets (Sachs, 1951). The cheliceral structures are usually small and inconspicuous in the other known genera, except in the coreitarsonemine genera of Fain (1971).

To conclude, I would urge those who study tarsonemid mites to give more careful attention to the characters of the gnathosoma.

References

Cross, E.A.: The generic relationships of the family Pyemotidae (Acarina: Trombidiformes). Univ. Kansas Sci. Bull. 45: 29–275, 1965.

Fain, A.: Note sur la repartition geographique des Coreitarsoneminae parasites des hémiptères coreides avec description de taxa nouveaux (Acarina: Trombidiformes). Bull Ann. Soc. Roy. Entomol. Belg. 107: 81–88, 1971.

Hammen, L. van der: *Tarsonemoides limbatus* nov. spec., and the systematic position of the Tarsonemida (Acarida). Zool. Verh. 108: 1–35, 1970.

Lindquist, E. E.: Review of holarctic tarsonemid mites (Acarina: Prostigmata) parasitizing eggs of ipine bark beetles. Mem. Entomol. Soc. Can. 60: 1–111, 1969a.

— : New species of *Tarsonemus* (Acarina: Tarsonemidae) associated with bark beetles. Can. Entomol. 101: 1291–1314, 1969b.

Sachs, H.: Zur Morphologie von *Acarapis*. I. Bau und Funktion der Mundwerkzeuge der Tracheenmilbe *Acarapis woodi* Rennie, 1921. Z. Bienenforsch. 1: 103–112, 1951.

Suski, Z.W.: Polish mites of the family Tarsonemidae (Acarina, Heterostigmata). *Tarsonemus idaeus* n. sp., Bull. Acad. Polon. Sci., Ser. Sci. Biol. 16: 637–642, 1968

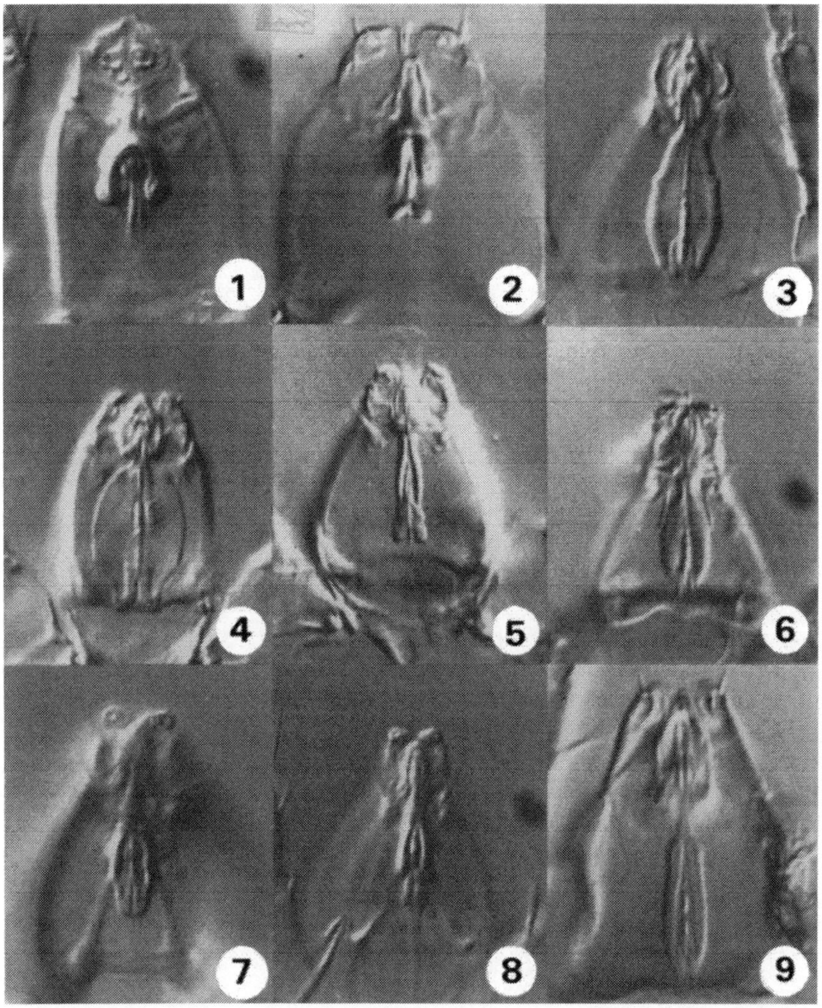

Figs. 1—9. Ventral view of pharyngeal structures in gnathosoma: 1. *Pseudotarsonemoides* sp.; 2. *Tarsonemus apis* Lindquist; 3. *Tarsonemus* sp.; 4. *Iponemus plastographus* (Lindquist and Bedard); 5. *Tarsonemus ips* Lindquist; 6. *Tarsonemus* sp. in *subcorticalis*-group; 7. *Tarsonemus* sp. in *pauperoseatus*-group; 8. *Tarsonemus fusarii* Cooreman; 9. *Tarsonemus pallidus* Banks.

ON THE TAXONOMY OF THE BRYOBIIDAE (ACARIFORMES, TETRANYCHOIDEA)

V. I. MITROFANOV

State Nikita Botanical Gardens, Yalta, USSR

The taxonomic status of Bryobiidae is characterized by the unstability and absence of a common approach to the conception of the amount of suprageneric categories. It appears to us that it would be impossible to refuse completely involving chaetoma and chaetotaxy features for the suprageneric taxonomy which reflects the structural plan of the nervous system. Following Reck (1952, 1959), we single out Bryobiidae into an independent family consisting of three subfamilies and nine tribes taking as a basis the morphological features used in the classification key below.

Key for the subfamilies and tribes of Bryobiidae.

1 (10) Claws or empodium uncinate.

2 (5) Claws II –IV uncinate. Empodium I–IV pad like. Bryobiinae Berlese

3 (4) Hypostome ventrally with two pairs of setae, opisthosoma with two pairs of pregenital setae... Ştrunkobiini tribus nov.

4 (3) Hypostome ventrally with one pair of setae, opisthosoma with one pair of pregenital setae... Bryobiini Berlese

5 (2) Claws II–IV pad like. Empodium I–IV uncinate... Petrobiinae Reck

6 (7) Hysterosoma dorsally with 9–10 pairs of humeral, 7–8 pairs of prelumbal and 5–6 pairs of lumbal setae in curved rows... Dasyobiini Strunkova (in litt.)

7 (6) Hysterosoma dorsally with three pairs of humeral, two pairs of prelumbal, and two pairs of lumbal setae in transverse rows.

8 (9) Ventrum of body with many setae plumose... Neotrichobiini Tuttle et Baker

9 (8) Ventrum of body with only three pairs of intercoxal setae setiforme... Petrobiini Reck

10 (1) Claws and empodium pad like... Hystrichonychinae Prit. et Baker, stat. nov.

11 (16) Empodium with two rows of tenent hairs.

12 (15) Hysterosoma with three pairs of dorsomedial setae.

13 (14) Humeral, prelumbal and lumbal setae drawn close in pairs. Middle humeral setae in prolongal row with outer humeral and outer prelumbal setae ... Beerellini Wainst.

14 (13) Humeral, prelumbal and lumbal setae in even distance. Middle humeral setae in transverse row with outer and inner humeral setae ... Monoceronychini Wainst.

15 (12) Hysterosoma with four pairs of dorsomedial setae ... Aplonobiini tribus nov.

16 (11) Empodium with one pair of tenent hairs... Hystrichonychini Prit. et Baker

Being a type of Bryobiidae, the genus *Bryobia* has the largest number of species. Of the 89 species and forms described during more than 200 years study, 41 of them are valid, 35 of which were found in the USSR. All these belong to five natural subgenera (Mitrofanov, 1969) which are determined as follows:

Key to subgenera of *Bryobia*

1 (8) Middle humeral setae arranged sublaterally.

2 (5) Propodosomal lobes well developed, their width at the base is 1,5 – 2 times their length. Outer lobes great, separated from inner ones by a deep depression.

3 (4) Anterior propodosomal angulations strongly developed; tarsus IV dorsally with duplex setae ... *Bryobia* s. str.

4 (3) Anterior propodosomal angulations absent, tarsus IV dorsally without duplex setae ...*Lyobia* Liv. et Mitr.

5 (2) Propodosomal lobes absent or developed feebly. Outer and inner propodosomal lobes, if present, not separated by deep depression.

6 (7) Distance between bases of inner sacral setae less than that between bases of outer sacral setae; inner sacral setae contiguous with outer sacral setae and arranged sublaterally. ...*Periplanobia* Liv. et Mitr.

7 (6) Distance between bases of inner sacral setae bigger than that between bases of outer sacral setae or equal; inner sacral setae bases are not close to outer setae and arranged laterally... *Allobia* Liv. et Mitr.

8 (1) Middle humeral setae arranged laterally... *Ebarobia* Liv. et Mitr.

Fourteen species of this genus, mainly of subgenera *Lyobia* and *Bryobia*, are agricultural pests.

References

Mitrofanov, V. I.: Mites of genus *Bryobia* C. L. Koch. 1836 (Acariformes, Bryobiidae). Thesis synopsis for the degree of Ph.D., Leningrad, 1969. (In Russian.)

Reck, G. F.: On some principles for classification of Tetranychidae. Rep. Acad. Sci., Georgian SSR, 13 (7):419–425, 1952. (In Russian.)

– Key of Tetranychidae. Tbilisi, Acad. Sci. Georgian SSR, 1959. (In Russian.)

SOLENIDIAL TRANSPOSITION IN *PIMELIAPHILUS* (ACARI, PTERYGOSOMATIDAE) AND ITS POTENTIAL SYSTEMATIC SIGNIFICANCE[1]

I. M. NEWELL
Department of Biology and Department of Entomology, University of California, Riverside, California, USA

In the development of Acari from larva to adult, there is a progressive increase in the total number of setae on the body and the appendages. There is also a definite pattern of increase: certain setae are added at each molt. In oligotrichous forms, these setae can be named or numbered. Grandjean (1964) summarized his extensive observations on the ontogeny of chaetotactic patterns in the solenidia of the Oribatei, and showed that these patterns have systematic significance in many cases, at least at supraspecific levels.

Evans (1969) identified chaetotactic patterns in the Mesostigmata, and showed that these could be divided into two groups, namely the Sejina + Antennophorina, and the Gamasina + Uropodina + Diarthrophallina + Microgyniina. The former are characterized by the addition, in the protonymphal stage, of an extra pair of ventral setae in the region of the circumsegmental fissure of tarsus IV.

The diversity in leg chaetotaxy in the Trombidiformes is greater than that encountered in the other major orders of Acari, and perhaps partly for this reason, has not been an active field of study for determining relationships within this group. Adult chaetotaxy has, of course, been used extensively to differentiate species and genera. In the Parasitengona, larval chaetotaxy is also important. But I have not encountered any cases in which an ontogenic chaetotactic character has been employed to aid in differentiating supraspecific taxa, other than the development of extreme hypertrichy in the transition from the larval to the deutonymphal stage in the terrestrial Parasitengona. The observations described here show that such characters are potentially useful in solving systematic problems in the Trombidiformes.

In typical acarine development, chaetotactic changes from one instar to the next involve the addition of setae, in a definite order. Setae appear never to be lost, and their positions usually do not undergo significant changes from one instar to the next. A major exception to this is the companion setae of the hypertrichous Parasitengona. Here the companion setae are recognizable in the larval stage only, where they accompany solenidia or eupathidia on the tibia or tarsus (e.g., see

[1] This study was supported by Grant No. GB 15840 from the National Science Foundation (USA) to the University of California, Riverside.

Newell, 1957). In the deutonymph (the protonymphs lack setae in the Parasitengona), the companion setae are either lost, or more likely become transformed and are unrecognizable. In oligotrichous groups (Tetranychidae, certain Pterygosomatidae et al.), the companion setae retain their identity throughout ontogeny.

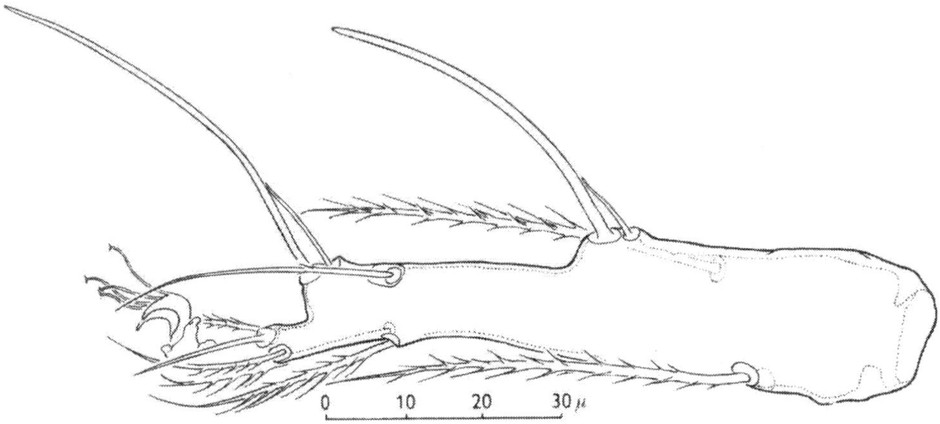

Fig. 1. Tarsus I, *Pimeliaphilus gloriosus*, showing dorsal solenidion and eupathid, each with a companion seta.

In *Pimeliaphilus*, there is a unique chaetotactic pattern in the ontogeny of the basal solenidion of tarsus I of the adult. This consists of the transposition of this solenidion from the tibia to the tarsus of leg I in the change from larva to deutonymph. As in the Parasitengona, the protonymph in *Pimeliaphilus* is a calyptostase in which solenidia are unrecognizable. In the larva of *P. gloriosus* Newell and Ryckman 1966, tarsus I has a single solenidion + companion seta at .41d (Fig. 1). There is also a dorsal eupathid + companion seta at .82d. The deutonymph has two solenidia, the proximal one without, the distal one with a companion seta (Fig. 3, in which the dorsal eupathid is omitted). Tibia I of the larva also has a solenidion (at .83d), a bothridion at .25d, plus four normal setae (Fig. 2). Tibia I of the deutonymph has the bothridion and four normal setae, but the solenidion is gone.

The solenidion of the larval tibia I and the proximal solenidion of tarsus I of the deutonymph are identical in size and structure. Evidently the solenidion of tibia I of the larva has become transposed into the tarsus during the transformation from larva to deutonymph. The alternative to this interpretation is that, simultaneously, the tibia has lost its solenidion while the tarsus has gained one of identical size and structure. It is more logical to assume that the formative cells for this solenidion have left their original position in the tibia and moved into the basal whorl of setae in the tarsus, along with their accompanying neural components. Associated with this, there is a noticeable distad displacement of the distal solenidion + companion seta of the tarsus. This is not a simple anomaly, but a normal phenome-

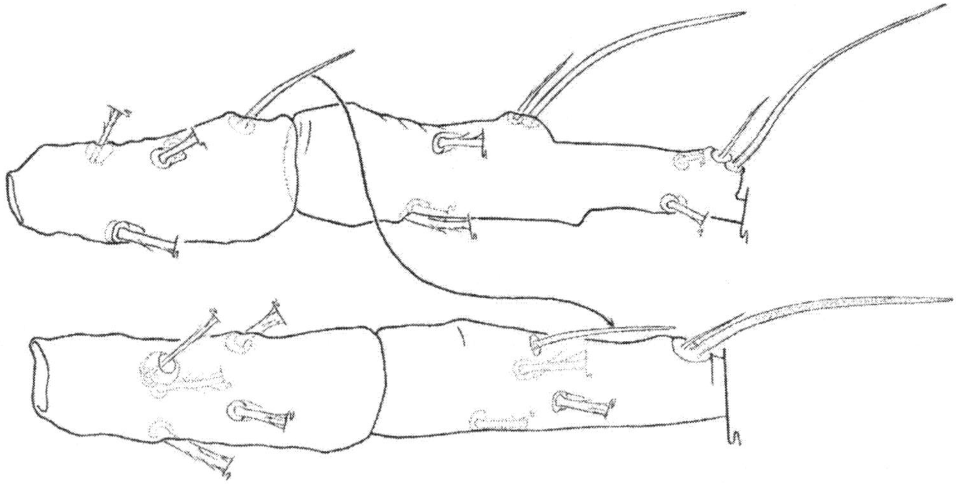

Fig. 2. Tibia and tarsus I, larva.

Fig. 3. Tibia and tarsus I, deutonymph, showing two solenidia in tarsus II.

non, seen in many individuals of *Pimeliaphilus gloriosus*, and in several other species of Pterygosomatidae.

Further evidence that transposition is the logical explanation for this is found in species which have only the distal dorsal solenidion on tarsus I of the deutonymph and adult. In several such species which the writer has studied, the larva lacks a solenidion on tibia I. In line with the foregoing interpretation, no tibial solenidion is available to move into the tarsus, and none appears there.

Jack (1964) made an extensive study of the chaetotaxy of the legs of adult Pterygosomatidae. He showed the following taxa to have a proximal solenidion (Jack's W_2) on tarsus I: *Hirstiella trombidiformis, insignis, stamii,* and *Geckobiella texana* (to which can be added *H. pyriformis* Newell and Ryckman 1964). Taxa lacking the proximal solenidion include *Hirstiella tenuipes, Geckobiella harrisi,* and all species of *Ixodiderma, Scaphothrix, Zonurobia, Gerrhosaurobia, Pterygosoma* and *Geckobia.* Examination of the larvae of the species in the first group will probably show that tibia I of the larva has a distidorsal solenidion which undergoes transposition, while the larvae in the second group do not. Further, the ambiguous position of *Geckobiella harrisi* and *Hirstiella tenuipes* would suggest that the systematic positions of these should be reviewed.

The solenidial transposition described here is very likely a unique phenomenon, under a complex genetic control which is unlikely to have evolved more than once in the Eleutherengona. Therefore it is potentially useful in determining relationships within that group. The limits of the Pterygosomatidae are not yet clear, since some of the species presently placed there appear to intergrade imperceptibly with certain Raphignathidae on the one hand and with certain Cheyletidae on the other, at least insofar as these families are presently constituted. Further study of the distribution of the *Pimeliaphilus* type of solenidial transposition will probably be useful in answering some of the questions remaining there.

References

Evans, G.O.: Observations on the ontogenic development of the chaetotaxy of the tarsi of legs II – IV in the Mesostigmata (Acari). Proc. 2nd Int. Congr. Acarology, pp. 195–200, 1969.

Grandjean, F.: La solenidiotaxie des Oribates. Acarologia 6(3):529–556, 1964.

Jack, K.M.: Leg chaetotaxy with special reference to the Pterygosomidae (Acarina). Ann. Natal Mus. 16:1–20, 1964.

Newell, I.M.: Studies on the Johnstonianidae (Acari, Parasitengona). Pacific Science 11: 396–466, 1957.

Newell, I.M. and Ryckman, R. E.: *Hirstiella pyriformis* sp. n. (Acari, Pterygosomidae), a new parasite of lizards from Baja California. J. Parasitol. 50: 163–171, 1964.

Newell, I.M. and Ryckman, R. E.: Species of *Pimeliaphilus* attacking insects, with particular reference to those parasitizing Triatominae (Hemiptera, Reduviidae). Hilgardia 37(12): 403–436, 1966.

Proceedings of the 3rd International Congress of Acarology, Prague, 1971

THE SPECIFIC STABILITY AND THE CHAETOTACTIC VARIATIONS IN *MUCRONOTHRUS NASALIS* WILLM. (ORIBATEI)

J. TRAVÉ
Laboratoire Arago, Banyuls sur Mer, France

Mucronothrus nasalis (Willm. 1929) is an Oribatid mite related to the genus *Trhypochthoniellus* Willm. The systematic position of these two genera is not quite clear and should be defined. It is necessary to place them either in the family Trhypochthoniidae or Malacanothridae, or in a separate family Trhypochthoniellidae.

M. nasalis lives in very humid or aquatic biotopes, which are cold or very cold. I have collected it in the Pyrenées, at an altitude of 2,300 metres. It is known from Scandinavia and from the north of Europe, from the Alps, Greenland, the Cordilleras and New Zealand (Hammer, 1965).

The probable world-wide distribution of this species and its biotope, which seems to exclude every possibility of a long distance dissemination, prove that a very ancient species is involved. It already existed before the dislocation from primitive Pangea, that is, according to Dietz and Holden (1970), in the Permian era, 200 million years ago.

The specimens from various populations do not differ in their morphology and M. Hammer attributes this stability to the cold climate. It is probable that parthenogenesis, which seems to be the rule in this species (I did not find a single male in 100 specimens examined, originating from different regions), has supported this stability as well.

It was interesting to study in detail the chaetotaxy of these specimens, collected in very distant research centres and to observe if any constant variations appear on the level of the phaners. Such variations would justify the differentiation of races or sub-species. By the courtesy of Mrs. M. Hammer, who provided me with a large part of her important material, which I compared with my Pyrenean collection, I was enabled to carry out this study.

The chaetotactic variations of the legs

The chaetotaxy of legs in *M. nasalis*, such as in many oribatids, has undergone a regressive evolution. This denudation is progressive. The setae are not affected at random, but follow an order of priority. Moreover, the disappearance of a seta occurs slowly. At first this happens in a small number of specimens, then increases till eventually the seta completely disappears. The changes that we can observe during these evolutional phenomena were named in French "écarts",

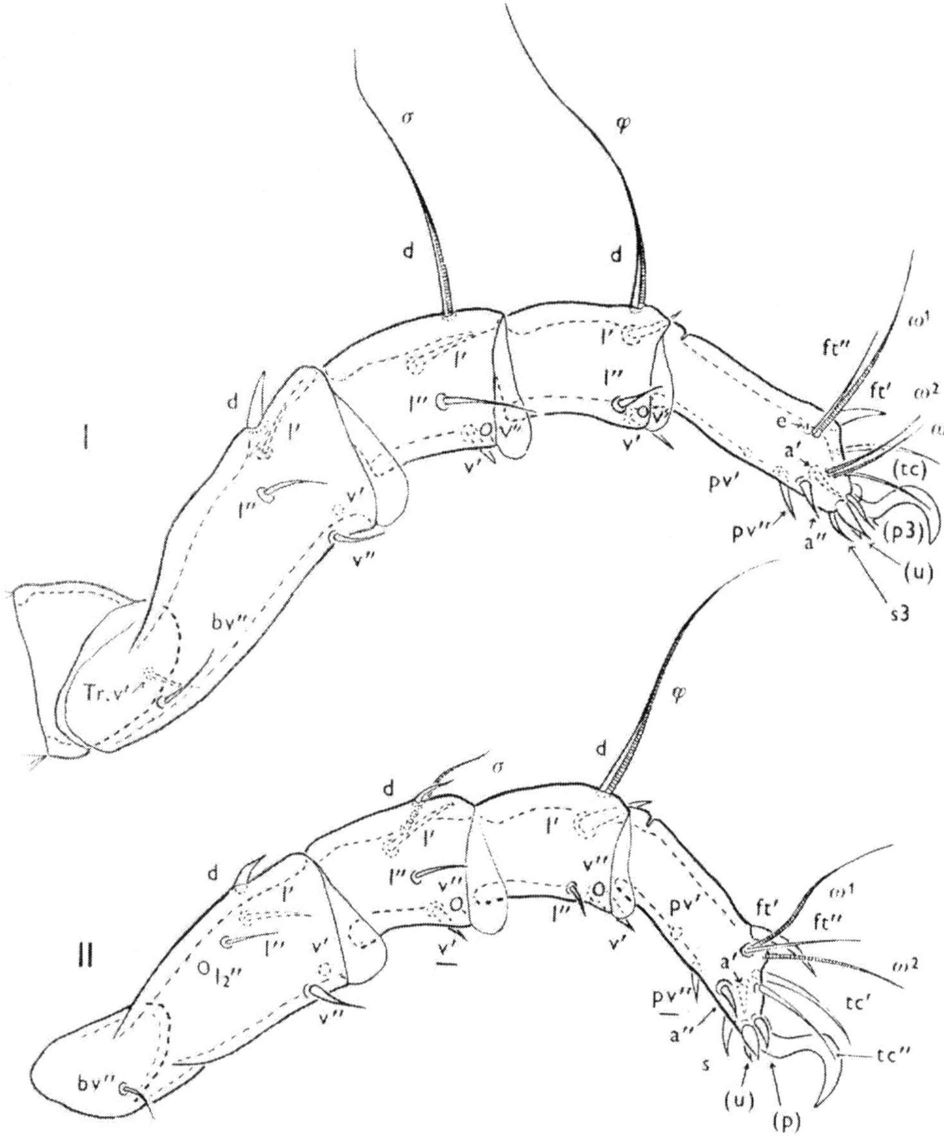

Fig. 1. Legs I and II of adults *Mucronothrus nasalis*, lateral view. Specimen from the Pyrenees. The setae with deviations are underlined. Setae very rare or present only in other populations are indicated by unbroken circles, if they are paraxial, by broken circles if they are antiaxial.

or "vertitions", i.e. deviations. (Grandjean, 1938). These deviations, which are very valuable for the study of evolution, differ from mutations in a certain number of characters, e.g. asymmetry (Grandjean, 1939; 1949; 1952).

Nothroidea offer a good example of a very dense chaetotaxy of legs (Camisiidae, Nothridae), which allowed an extensive study of accessory tarsal setae in *Platynothrus peltifer* (C. L. Koch) by Grandjean (1958), or of a very sparse chaetotaxy (Trhypochthoniidae, Malacanothridae) in the case of *M. nasalis*, which is

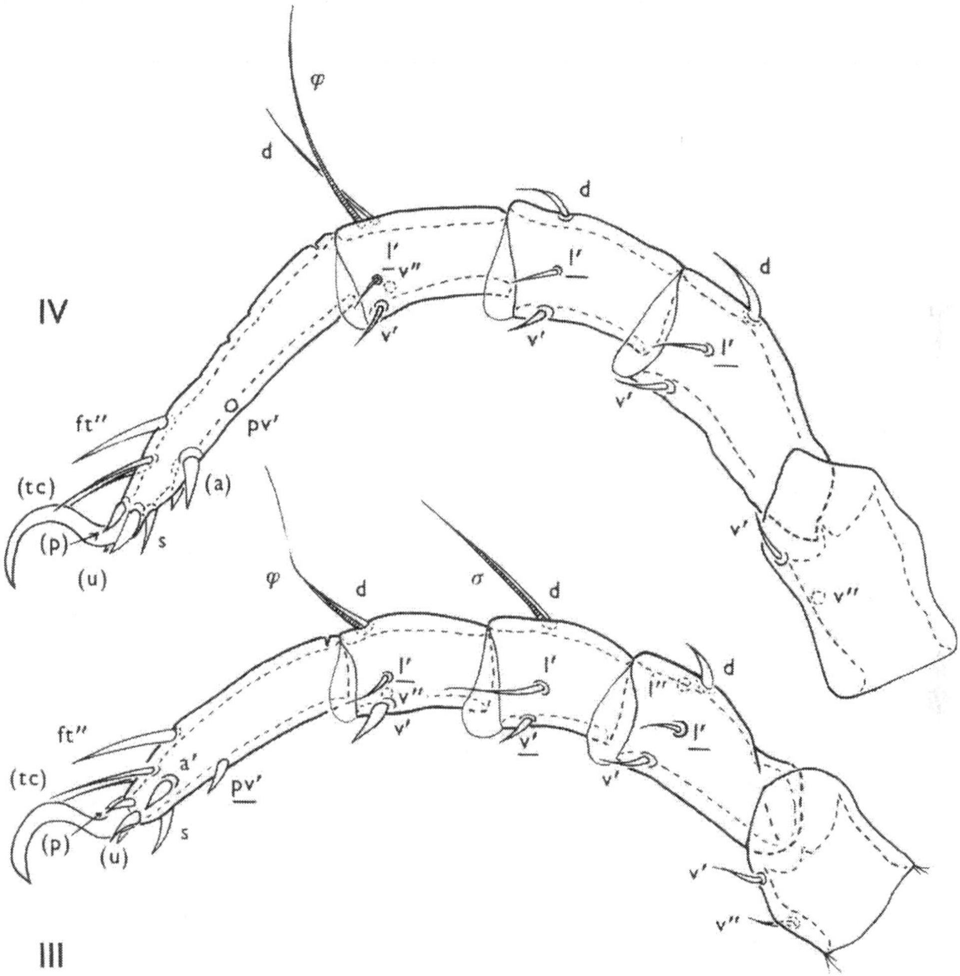

Fig. 2. Right legs III and IV of adult *Mucronothrus nasalis* in lateral view. Specimens from the Pyrenees. The setae showing the deviations are underlined. Setae very rare or present only in other populations are indicated by unbroken circles if they are paraxial, by broken circles if they are antiaxial.

one of the most sparse known in oribatids. Here the complementary tarsal setae have disappeared and some rudimentary setae are regressing, or are even nonexistent. That is the case of certain setae pv or v (Figs. 1 and 2).

The chaetotactic analysis concerns 82 specimens, all adult females, which were available for examination: 25 from the Pyrenées, 25 from New Zealand, 13 from Argentina, 13 from Denmark and 6 from Greenland, the latter unfortunately damaged. All these specimens came from just one collection from a centre. As the observations were made from two sides, the right and the left, every seta was examined 164 times, except the case where some segments were lacking (i.e. some specimens from Greenland). To simplify the comparison the examinations of the specimens from Argentina and from Denmark were performed with 12 complete specimens from both sides and merely from one side with 13th specimen, chosen at random. 162 observations were used for statistical evaluation: 50 from both the Pyrenées and New Zealand, 25 from both Argentina and Denmark, 12 from Greenland.

Results

Although the number of specimens seems to be small, more than 15,000 setae were examined. Considering some absent setae, the number of observations approaches 20,000.

Anomalies

There were few anomalies; 6 setae were doubled and one specimen, not used in the account, showed a tibio-tarsal coalescence accompanied with an adherent phanerotaxy.

Deviations

The results have to be considered in two ways: first globally, taking into account the data from 5 research centres, then from each research centre separately. This is very different, because the usual patterns of each of the centres could be very divergent.

a) The global analysis. The maximum number of setae on the four legs of the adults is 114. This chaetotaxy, that may be qualified as elementary, we can find of course in no specimen. Every research centre has a "normal" pattern, i.e. the most frequent. It is identical for the Pyrenées and Denmark and different for the three other centres. (Table 1.) Of these 114 setae 69 (60 %) are very frequent, always present, 15 are frequent, absent only once, 7 are very sporadic (they are present only once) and 23 (20 %) demonstrate some deviations, whose number varies from 2 to 80. These 20 % of setae are responsible for 470 deviations from 492, that is more than 95 %. These are above all the lateral or ventral setae of femur or genua, as well as the setae of tarsus (pv). The leg III is the most affected.

	I	II	III	IV
Pyrenées	1–5–4–4–13–1	1–5–4–4–12–1	2–3–3–3–10–1	1–3–3–3–10–1 2
Denmark	1–5–4–4–13–1	1–5–4–4–12–1	2–3–3–3–10–1	1–3–3–3–10–1
Greenland	1–5–4–4–14–1 13	1–5–4–4–12–1 6 13	2–3–3–3–10–1	1–3–2–3–10–1
Argentina	1–5–4–4–13–1	1–6–4–4–12–1	2–4–3–3–9–1	1–3–3–3–10–1
N.Zealand	1–5–5–4–13–1	1–6–4–4–12–1	2–4–3–3–9–1	1–3–3–3–10–1
Maximum	1–7–5–5–14–1	1–7–5–5–13–1	2–4–4–4–12–1	2–4–4–4–12–1

Table 1. Chaetotactic patterns of legs of adults in 5 populations examined and the maximum theoretical pattern. If two patterns are identical, two numerical figures are placed one above the other.

b) Analysis according to research centres. The deviations studied on the level of every centre are less numerous as if considered globally on the level of species, the chaetotactic pattern being different for each of the centres. The total number of deviations is 259 for 15,255 setae, that is 1.69 %. The largest number of deviations appear in 6 specimens from Greenland (2.95 %), if compared with those from Pyrenées (1.97 %), New Zealand (1.50 %), Denmark (1.48 %) and Argentina (1.12 %). The deviations from different centres do not apply to the same setae, to the same segments or legs. Table 2 shows the most interesting results.

It is interesting to follow the presence of certain setae: the setae Ge.v "I and Ge v" II have practically disappeared, with the exception of the New Zealand population (98 and 24 %). The setae Ta. pv. 'I and Ta.pv.' II are almost non-existent, with the exception of 6 specimens from Greenland (54.5 % and 41.6 % respectively). The specimens from the Pyrenées diverge by a progressive disappearance of setae l' of the femora and of genua of the leg IV. Finally we can differentiate the southern and the northern populations by the setae Fe.l "2 II, Fe.l" III on the one hand and on the other hand by the seta Ta. pv' III. While the southern populations retained the two femoral setae and lost the tarsal seta, the opposite is the rule for the northern populations. The data obtained are very significant.

Legs and setae		Pyrénées			Denmark			Greenland			Argentina			N. Zealand			Total	
		+	−	+ %	+	−	+ %	+	−	+ %	+	−	+ %	+	−	+ %	+	−
I	Fe. v'	7	43	14	4	21	16	0	12	0	2	23	8	3	47	6	16	146
	Ge. v"	1	49	2	0	25	0	0	12	0	1	24	4	49	1	98	51	111
	Ti. v"	0	50	0	0	25	0	0	12	0	1	24	4	7	43	14	8	154
	Ta pv'	7	43	14	3	22	12	6	5	54,5	2	23	8	1	49	2	19	142
II	Fe. l" 2	0	50	0	0	25	0	4	8	33,3	22	3	88	45	5	90	71	91
	Fe. v'	2	48	4	0	25	0	0	12	0	2	23	8	6	44	12	10	152
	Ge. v"	0	50	0	0	25	0	0	12	0	0	25	0	12	38	24	12	150
	Ge v'	48	2	96	25	0	100	12	0	100	23	2	92	50	0	100	158	4
	Ti. v"	3	47	6	1	24	4	0	12	0	0	25	0	0	50	0	4	158
	Ta pv"	48	2	96	25	0	100	12	0	100	25	0	100	42	8	84	152	10
	Ta pv'	1	49	2	1	24	4	5	7	41,6	0	25	0	1	49	2	8	154
III	Fe l'	49	1	98	24	1	96	10	0	100	25	0	100	50	0	100	158	2
	Fe l"	0	50	0	1	24	4	0	10	0	25	0	100	39	11	78	65	95
	Ge. v'	44	6	88	24	1	96	7	3	70	25	0	100	50	0	100	150	10
	Ti. l'	46	4	92	22	3	88	8	2	80	25	0	100	50	0	100	151	9
	Ta. pv'	46	4	92	23	2	92	10	0	100	1	24	4	0	50	0	90	80
IV	Tr v"	3	47	6	5	20	20	0	11	0	3	22	12	3	47	6	14	147
	Fe. l'	26	24	52	25	0	100	8	1	88	20	5	80	44	5	88	123	36
	Ge. l'	32	18	64	23	2	92	3	6	33,3	24	1	96	49	1	98	131	28
	Ti. v"	0	50	0	0	25	0	0	9	0	2	23	8	3	47	6	5	154
	Ta. pv'	3	47	6	3	22	12	0	9	0	0	25	0	0	50	0	6	153

Table 2. Primary setae, representing the deviations. + indicates the presence, + indicates the absence, + % indicates the presence-percentage. The most significant values are underlined, encircled or marked by a rectangle, according to the frequency.

Conclusions

This analysis is very brief and will be developed elsewhere. However, it demonstrates the interest in the study of chaetotactic variations in different populations. The populations studied belong incontestably to a single species, *M. nasalis*, which was revealed to be very stable in spite of its very ancient origin. The differentiation took place probably very slowly, because the chaetotaxy is basally identical.

The chaetotactic variations are important and refer to the specific setae. The populations have developed differently and distinctly. Tarsus III with setae e.g. is specific for two southern populations, that one with 10 setae is typical for the northern populations. This evolution is not yet complete and some exceptions exist. In fact, while the regression in one population is hardly starting, it is nearly finished in the other. Anyway, it seems that there is a balance between the disappearance of a seta on a segment and the preservation of another seta on the other segment or the other leg at the same time.

If these findings are confirmed by an other study dealing with more numerous and more various populations, or with other species, they will enable us to solve more easily the difficult problems concerning the formation of species..

References

Dietz, R.S., Holden, J.C.: The Breakup of Pangea. Sci. Am. 223 (4):30–41, 1970.

Grandjean, F.: La suppression d'organes dans l'évolution homéotype. C.R. Acad. Sci. Paris, 206: 1853–1856, 1938.

– La répartition asymétrique des organes aléatoires. C.R. Acad. Sci. Paris, 208:861–864, 1939.

– Sur les rapports theoriques entre écarts et mutations. C.R. Acad. Sci. Paris, 228:1675–1678, 1949.

– Sur les variations individuelles. Vertitions (écarts) et anomalies. C.R. Acad. Sci. Paris, 235: 640–643, 1952.

– Sur le comportement et la notation des poils accessoires postérieurs aux tarses des Nothroides et d'autres Acariens. Arch. Zool. Exptl. Gén. 96 (4): 277–308, 1958.

Hammer, M.: Are low temperatures a species preserving factor ? Illustrated by the Oribatid Mite *Mucronothrus nasalis* (Willm.) Acta Univ. Lund. II (2): 1–10, 1965.

PRELIMINARY NOTE TO A TENTATIVE NEPOPHYLOGENY OF TROMBICULIDS[1]

P. H. VERCAMMEN–GRANDJEAN
G. W. Hooper Foundation, University of California, San Francisco, California, U.S.A.

The Trombiculidae and Leeuwenhoekidae are two strong and thriving families, stubbornly defensive and brilliantly aggressive. They follow a straight orthogenesis, favored by advantageous faculties of adaption, a good capacity of survival and an undeniable pereniality. No marked deviations and few cases of regression. Their generous nature, their aptitude to change hosts at opportune times, and the large range of their hosts – spreading from molluscs to any kind of vertebrates, passing through arthropods – open the doors of the world to their unlimited dispersion. This being so, it is common sense to admit that their classification would be only slightly influenced by contingencies barriers, and other limits applied to other animal groups. In other words, once the morphological and ecological identities are duly established, differences in location will constitute only a minor segregative reason. However, one must face the exception of bat parasites, that the peculiar biology of their hosts destines them to a diversifying confinement (see the *Trisetica* line).

My first "Essai de classification des Trombiculidae larvaires (1960)" was essentially based on the aspect of the scutum, joined to a combination of eleven characters, or "Synthetic Identification Formula – SIF" (V.–G., 1965). The first and most edifying of these characters is the palpotarsal pilous formula (fT), of which the barbed setae constitute the principal variable, correlated directly with an evident progressive evolution. These variations are observed repeatedly in various groups as shown below, ranging from 3 to 7 barbed setae (B) to which can occasionally be added a nude subterminala (S); the basal solenidion being neglected as an omnipresent constant.

Trombiculinae : fT = 3B, 4B, 4B.S, 5B, 5B.S, 6B, 6B.S, 7B, or 7B.S.
Gabrliepinae : fT = . . 4B, 4B.S, 5B . . or 6B.
Apoloniinae : fT = ... 4B.S, 5B, 5B.S, . . 6B.S, . . or 7B.S.
Leeuwenhoekinae: fT = . . 4B, . . . 5B, . . . 6B, . or 7B; it is remarkable that this group never possesses a subterminala (S). The second of the eleven SIF characters regards the nature, nude or barbed, of the galeal seta or galeala; whilst the

[1] This work was supported by Research Grant AI–03793 from the National Institute of Allergy and Infectious Diseases, National Institutes of Health, U.S. Public Health Service.

third concerns the number of prongs on the palpotibial claw. As for the eight subsequent digits, they relate to the numbers of special setae on the legs.

Until two years ago, curiosity had never been particularly stimulated by the amazing regularity of the numbers of ordinary or barbed setae on the legs. Many authors reported these numbers, but without any other significant application. When studying the *plecoti* group in the subgenus *Chiroptella (Oudemansidium)*, I noticed that the posterior femur possessed only one nude femoral seta and six ordinary setae (p.o.), while the members of the other group *(muscae)* show two femoralae and 5 p.o. Was it a mutation p.o.→fp, or the reverse, fp→p.o.? I am inclined to the second interpretation, fp→p.o., because the group *plecoti* assembles species reasonably more advanced than the members of the *muscae* group. *Miyatrombicula scottae* lacks a nude genuala on the posterior leg. On the other hand, the number of p.o. is four instead of three. This feature is without doubt a mutation, gp→p.o. Similar cases are not rare, indicating the gain of a nude seta is intermediate in the final acquisition of an ordinary barbed seta. A succession of such mutations would explain the succession of palpotarsal formulae from 3B to 7B.S. However, the passage through a nude form does not seem obligatory. In fact, the leeuwenhoekins never show a subterminala on the palpotarsus, although one process may not necessarily exclude the other.

Assembling in a compact frame the numbers of barbed setae on tarsus (T), tibia (t), genu (G), and femur (F) (this latter being divided or not) of the anterior (a), median (m), and posterior (p) legs, one enters in possession of a "global character" easy to gauge at first glance. We will call it "Leg Setae Tabulation" or "L.S.T." (see tables 1–4). Comparative study of it can be conducted globally, or in series on the same rank, or upon a single numeral. Trombiculids plus apoloniins show a genual formula, G = 4.3.3, while leeuwenhoekins, in general, have 4.4.4; for the rest the general numerals are in common. Species of the genus *Chatia* show a median femoral number, Fm = 7, whereas all the members of the two families trombiculids and leeuwenhoekids show only 6 or 5. At the intra-generic level the L.S.T. shows already a real comparative value, but what to say about inter-generic variation and inter-familial ones! Markedly high variations in the tarsal, and often tibial ranks, respond invariably to gigantism (*Chatia, Brennanella, Multigniella, Whartonia*, etc. ..). Variations of one or two units in the genual rank may reflect subgeneric or species-group status (*Sasatrombicula, Neoschoengastia, Odontacarus*, etc.). In the same way, inter-generic variations are also manifested by differences of one or two units in the genual and femoral ranks.

To an increase of the numbers corresponds frequently a parallel raise of the palpotarsal formula, and also a tangible progressive evolution. At the inter-familial level, the L.S.T. is more than a simple taxonomical character. For instance, primitive morphologies of homeoparasites,, like cheyletids and pterygosomids, correspond to the lowest L.S.T. The L.S.T. of semiparasites, like trombidiids, is a little higher and even higher are those of trombiculids sens. lat.; then the johnstonianids, closely followed by the hydrachnids. Finally come the very evolved erythraeids. Heteroparasites of the ixodid type show an L.S.T. of a different orderliness, but anyway witnessing an advanced evolution, and leaning on greater sizes. To persuade oneself of the chronographic properties of the L.S.T., it suffices to compare that of modern erythraeids with those of the archaic Erythraeidae n°8 Ewing, 1937, and of a johnstonianid *(Lassenia scutellata)*, a classical trombiculid, a *Microtrombidium*, a pterygosomid, and a cheyletid *(Cheletogenes* sp.).

	Modern erythr.			Erythr. n°8			*Lassenia scutell.*			Trombic. sp.			*Microtr.* sp.		
	a	m	p	a	m	p	a	m	p	a	m	p	a	m	p
T	28	28	28	34	30	30	36	29	24	21	16	15	18	14	12
t	18	18	18	10	8	8	8	8	8	8	6	6	6	5	5
G	12	12	12	5	4	4	5	4	4	4	3	3	4	2	2
F	9	9	7	7	6	5	7	6	6	6	6	5	5	5	4

	Geckobia sp.			*Cheletog.* sp.			*Rhipicephalus evertsi* (Ixodid)		
	a	m	p	a	m	p	a	m	p
T	10	10	10	7	7	5	21	18	18
t	5	5	5	4	4	3	7	6	6
G	2	2	1	1	1	1	8	8	8
F	4	4	3	2	2	1	10	10	9

One sees immediately that Erythraeidae n°8 "marks the same hour" as *Lassenia scutellata*, not of the modern erythraeids. Difference: a little more than 60 million years. It becomes clear that the number of ordinary barbed setae on the appendages reflects the degree of evolution, and this bestowed to the larval stage. In addition, homeoparasites, such as cheyletids and pterygosomids, present post-larval or parental L.S.T. close to the larval ones; capacity not allocated to nymphs and imagoes of semiparasites, such as the predatory and hyperpilose trombiculids, trombidiids, etc. This allows to enunciate that: "If no noticeable change disrupts the way of living and the ecology of a biont, the larval morphology remains the directing pattern during its entire ontogeny, in other words, that the initial post-embryonic structures remain common throughout all of the stages."

What precedes reinforces the thesis developed in my publication of 1969 about "The larval stage, repository of phylogeny and taxonomy in acarines , stipulating that the ambulatory stage, directly post-embryonic, shows the taxonomical and phylogenical characters in their simplest, purest, and more constant display. From which derives the name of "nepophylogeny" (nepo = infant, early stage). This phylogeny and ten more years of studies enable me to present a new tentative classification of Trombiculidae and Leeuwenhoekidae, appreciably different from that of 1960, and in the form of a genealogic tree. This taxonomic edifice necessitated the erection of five new genera and subgenera:

1) *Electronella* n.sg.: Type species (T.S.): *Euschoengastia (Electronella) electron* (Brennan & Yunker, 1966). SIF = 5B.S–B-3–3111.oooo.Telestigmal scutum, pleated, wider than long with anterolateral shoulders. Eyes 2 x 2. PL > AL > AM. fCx = 1.1.3 & fSt = 2.4. Anterior and median leg tibiala gathered at the apex. Genualae stumpy. Leg claws with onychotriches. Host : bird.

2) *Preutrombicula* n.g. : T.S. : *Preutrombicula ignota* (Brennan, 1971). Linked to *Novotrombicula*, with a large scutum and a bifurcate palpotibial claw.
SIF = 6B–N–2–2111.1ooo. Eyes 2 x 2.

3) *Matacarus* Vercammen-Grandjean, 1956: T. S.: *Matacarus (M.) buretti* (V.–G. 1956) admitted as full genus out of *Acomatacarus* (syn. of *Odontacarus*) for two new subgenera. Principally African, with femoral pilosity F = 6.6.5 (rarely 6.6.4) and not 6.5.4 as in *Odontacarus* sens. lat. : a) *Matacarus* sg. : SIF = 7B–B–$^{2}_{8}$–$^{2}_{1}$111.$^{00}_{12}$oo, b) *Afracarus* n.sg. : SIF = 7B–B–3–2001.1ooo, T.S. : *Matacarus (Afracarus) maroccanus* (Taufflieb, 1958), c) *Arabacarus* n.sg. : SIF = 6B–N–$^{3}_{4}$–2111.$^{1}_{0}$ooo, T.S. : *Matacarus (Arabacarus) arabicus* (Radford, 1954).

4) *Anouracarus* nom. nov.: for *Batracarus* V.–G., 1968, preoccupied by *Batracarus* Fain.

5) *Fascutonia* n.sg. : for a group of *Whartonia* without solenidion on posterior leg tarsus, and with a femoral pilous number F = 6.6.5, proper to the Ethiopian Region, the others customarily showing 6.5.4. SIF = 7B–N–3–2111.$^{0}_{1}$ooo. T.S. : *Whartonia (Fascutonia) lepidopteriscuta* (Vercammen-Grandjean, 1965).

N.B.: On the genealogic tree an abbreviation followed by an asterisk represents an unpublished taxon. This will be described in the revision "The Chigger Mites of the World".

(1) TABULATION of ORDINARY LEG SETAE – L.S.T.

	Pterygos. "p"			*P.* "t"			*Geckob.* "g"			*Hoplothr. quinquesc.*			*Atomus rhopal.*			*Microtr.* "d"			*Parathr.* sp. (4)			*Podothromb. sanremense*		
	a	m	p	a	m	p	a	m	p	a	m	p	a	m	p	a	m	p	a	m	p	a	m	p
T	10	10	10	10	10	10	10	10	10	15	14	9	15	14	9	18	14	12	17	14	13	32	22	20
t	5	5	5	5	5	5	5	5	5	6	5	5	6	5	5	6	5	5	5	5	5	5	5	5
G	1	1	0	1	0	0	2	2	1	3	2	2	4	4	3	4	2	2	4	3	3	4	3	3
F	3	2	2	4	4	2	4	4	3	7	5	4	5	5	4	5	5	4	5	4	4	5	4	4

	Podothr. "m"			*P.* "o"			*P.* "s"			*P.* "d"			*P. svalbard.*			*Durenia* sp. (2)			*Womersl. minuta*			*Audyana thompsoni*		
T	32	22	20	21	15	13	36	24	21	46	28	22	31	18	16	28	21	19	28	24	22	22	16	15
t	5	5	5	5	5	5	5	5	5	5	5	5	5	5	5	7	6	6	7	6	6	8	6	6
G	4	3	3	4	3	3	4	3	3	4	3	3	4	3	3	4	4	4	4	4	4	4	4	4
F	5	4	4	5	4	4	5	4	4	5	4	4	5	4	4	6	6	6	6	6	6	6	6	4

	Centrotr. distans			*Johnston. latiscut.*			*Diplothr. monoense*			*D. cascad.*			*Nothotr. deiniacr.*			*Grossia onychia*			*Pteridop. pseudhan.*			*Lassenia scutell.*		
T	26	22	15	37	28	23	34	27	16	32	23	16	36	32	24	34	31	25	57	51	63	36	29	24
t	4	4	4	6	6	6	6	6	6	6	6	6	6	6	6	6	6	6	8	8	8	8	8	8
G	5	4	4	4	4	4	4	4	4	4	4	4	4	4	4	4	4	4	4	4	4	5	4	4
F	6	6	6	6	6	6	6	6	6	6	6	6	7	6	5	6	6	5	7	6	6	7	6	6

	Lassenia lasseni			*Monunguis streblida*			*Neotromb.* "a"			*N. bengalen.*			*N.* "t" + *samsinaki*			*N. tenuipes*			*N. tricusp.*			*N. barring.*		
T	50	43	40	20	15	15	20	10	10	21	12	12	21	15	15	21	15	15	20	10	10	24	19	19
t	16	17	15	8	7	7	8	6	6	8	7	7	8	7	7	8	6	6	7	6	6	8	6	6
G	8	8	8	4	4	4	4	4	4	4	4	4	4	4	4	4	4	3	4	4	4	4	4	4
F	10	10	10	7	6	6	7	6	6	7	6	6	7	6	6	7	6	6	7	6	6	7	6	6

	Hydrovol. "j"			*Limnoch.* sp. (3)			*Piersigia limophila*			*Euthyas truncata*			*Panisops.* sp. (4)			*Thyas* sp. (2)			*Lundblad. musicola*			*Panisus* "218"		
T	18	18	17	17	16	16	18	17	17	20	20	20	20	20	20	20	19	19	20	19	19	19	18	18
t	9	9	7	9	9	9	9	9	9	10	10	10	10	10	10	10	10	10	10	10	10	10	10	10
G	4	4	4	4	4	4	4	4	4	4	4	4	4	4	4	4	4	4	4	4	4	4	4	4
F	7	7	6	7	7	6	7	7	6	7	7	6	7	7	6	7	7	6	7	7	6	7	7	6

	Hydryph. sp. (2)			*Hydrodrom.* sp. (2)			*Thermac. nevadens.*			*Polydiscia squamata*			*Proterythraeus southcotti*		
T	20	19	19	20	20	20	20	20	20	21	21	20	34	30	30
t	10	10	10	10	10	10	10	10	10	9	9	9	10	8	8
G	4	4	4	4	4	4	4	4	4	4	4	4	5	4	4
F	7	6	6	7	7	6	7	7	7	6	7	6	7	6	5

	Charletonia sp. (25)			*C. domrowi*			*Bochartia adrassus*			*Caeculis. darwinien.*			*Sphaerotarsus leptopilus*		
T	28	28	28	28	28	28	28	28	28	28	28	30	21	21	21
t	18	18	18	12	13	14	16	16	14	17	18	18	14	14	14
G	12	12	12	10	14	12	8	8	8	12	12	12	12	8	8
F	9	9	7	9	9	7	7	9	7	9	9	7	7	7	7

N.B.: (Between parentheses, the # of species seen).

(2) TABULATION of ORDINARY LEG SETAE – L.S.T.

	a	m	p	Pp	a	m	p	Pp	a	m	p	Pp	a	m	p	Pp	a	m	p	Pp
CHEYLETIDAE																				
					Cheletogenes sp.				*Ornithocheyletia* sp.											
T					7	7	5	3	7	7	5	4								
t					4	4	3		4	4	4									
G					1	1	1		1	1	1									
F					2	2	1		2	2	1									
PTERYGOSOMIDAE																				
					Pterygos. " p "				*P.* " t "				*Geckobia* " g "							
T					10	10	10		10	10	10		10	10	10					
t					5	5	5		5	5	5		5	5	5					
G					1	1	0		1	0	0		2	2	1					
F					3	2	2		4	4	2		4	4	3					
IXODOIDEA																				
	Argas cooleyi				*A. arboreus*												*Otobius megnini*			
T	17	16	16	7	17	16	16	8									16	14	13	7
t	8	7	7	4	8	8	8	4									6	6	6	2
G	8	8	8	4	9	8	8	4									6	6	6	4
F	7	8	7	5	9	9	8	5									8	9	7	3
	Ornithodoros capensis				*O. amblus*				*O. coriaceus*				*O. hermsi*				*O. parkeri*			
T	17	15	15	7	17	15	15	7	16	15	15	7	16	15	15	7	17	15	15	7
t	6	6	6	2	6	6	6	2	6	6	6	2	6	6	6	2	6	6	6	2
G	5	5	5	5	5	5	5	5	6	6	6	4	6	6	6	5	6	6	6	5
F	7	8	6	4	7	8	7	4	8	8	8	4	8	9	7	4	8	8	8	4
	Dermacentor occidentalis				*D. andersoni*				*Rhipiceph. evertsi*								*Haemaphysalis leporispalustris*			
T	22	18	18	7	22	18	18	7	21	18	18	8					24	18	18	9
t	7	6	6	3	7	6	6	3	7	6	6	4					7	7	7	4
G	8	8	8	5	8	8	8	5	8	8	8	5					8	8	8	10
F	10	10	9	6	10	10	9	6	10	10	9	7					10	9	9	3

N.B.: Pp, or palpal pilosity, is exceptionally added here to show the expectable increase relationship between leg and palpal setae.

(3) L.S.T. of TROMBICULIDAE & LEEUWENHOEKIDAE

a m p a m p a m p a m p a m p a m p a m p a m p

	TROMBICULINAE general.:	sometimes:	*Neotromb. autumn.*	*N. (A.) sciuri*	*N. (P.) microti*	*N. (N.) tamiyai*	*N. (Nds.) clavigl.*	*Trombicula minor*
T	22 16 15	21 15 14	21 16 15	22 16 15	21 16 15	22 16 16	22 16 15	22 16 15
t	8 6 6	8 6 6	8 6 6	8 6 6	9 6 6	8 6 6	8 6 6	8 6 6
G	4 3 3	4 3 3	4 3 3	4 3 3	4 3 3	4 3 3	4 3 3	4 3 3
F	6 6 5	6 6 5	6 6 5	6 6 5	6 6 5	6 6 5	6 6 5	6 6 5

	Chiroptella (O.) muscae	*C. (Oudm.) plec. gr.*	*Miyatrom. scottae*	*Sasatrom. (M.) from.*	*S. (Marc.) boedonia*	*S. (Traub.) multister.*	*S. (Sas.) hexast.*	*S. (Sas.) sp.*
T	22 16 15	22 16 15	22 16 15	22 16 15	22 15 14	28 19 18	25 20 18	22 16 15
t	8 6 6	8 6 6	8 6 6	8 6 6	8 6 6	8 6 6	8 6 6	8 6 6
G	4 3 3	4 3 3	4 3 4	4 3 2	4 3 2	4 4 3	4 3 2	4 3 2
F	6 6 5	6 6 6	6 6 5	6 6 5	6 6 5	6 6 5	6 6 5	6 6 5

	Hexidionis (H.) sp.	*H. (Zost.) sp.*	*H. (Pent.) sp.*	*Multign. cosmetor.*	*M. "t"*	*Neosch. (N.) sp.*	*N. (Hypog.) sp.*	*Trisetica (T.) rhin.*
T	23 16 15	21 16 15	20 16 15	25 22 19	26 22 17	22 16 15	22 16 15	22 15 14
t	8 6 6	8 6 6	8 6 6	8 7 7	8 6 6	8 6 6	8 6 6	8 6 6
G	4 3 3	4 3 3	4 3 3	4 4 3	4 4 3	4 4 3	4 3 3	4 3 3
F	6 6 5	6 6 5	6 6 5	6 6 5	6 6 5	6 6 5	6 6 5	6 6 5

	T. (Rudn.) sp.	*Riedlinia sp.*	*R. (Trg.) wilsoni*	*Myxacarus oscillat.*	*Eutromb. sp.*	*E. "e"*	*Bab. (V.) ipoides*	*B. (Vatac.) kuntzi*
T	22 16 15	22 16 15	21 15 14	20 14 13	22 16 15	21 16 15	21 16 15	21 17 17
t	8 6 6	8 6 6	8 6 6	7 6 6	8 6 6	8 6 6	8 6 6	8 6 6
G	4 3 3	4 3 3	4 3 3	4 3 3	4 3 3	4 4 4	4 3 3	4 4 4
F	6 6 5	6 6 5	6 6 5	6 6 5	6 6 5	6 6 5	6 6 5	6 6 5

	GAHRLIEPINAE general.:	sometimes:		*Apolonia tigipioens.*	*Sauracar. sp.*	*Straelen. sp.*	*Womersia sp.*	*Vargatula hispida*
T	22 16 15	21 15 14		21 16 15	22 16 15	20 16 13	18 16 13	20 14 14
t	8 6 6	8 6 6		8 6 6	8 6 6	8 6 6	8 6 6	8 6 6
G	4 3 3	4 3 3		4 3 3	4 3 3	4 3 3	4 3 3	4 3 3
F	6 6 5	6 6 5		6 6 5	6 6 5	6 6 5	6 6 5	6 6 5

	LEEUWENHOEKINI several:	others:	*Odontacar. (O.) sp.*	*O.(Leog.) sp.*	*O.(Acom.) sp.*	*O.(Orocb.) micheneri*	*Cordiset. sp.*	*Albechia senase*
T	23 16 15	25 17 17	23 18 16	23 17 15	23 16 15	23 16 15	23 16 15	23 16 15
t	8 6 6	8 6 6	9 6 6	8 6 6	8 6 6	8 6 6	8 6 6	8 6 6
G	4 4 4	4 4 4	5 5 5	4 4 4	4 4 4	5 4 4	4 4 4	4 4 4
F	6 6 5	6 5 4	6 5 4	6 5 4	6 5 4	6 5 4	6 5 4	6 5 4

	Matacarus (Arab.) sp.	*M.(Mat.) sp.*	*M.(Afr.) sp.*	*M.(Afr.) cooki*	*Sasacarus (S.) whar.*	*S.(Mult.) sp.*	*S.(S.) vercam.*	*Wagenaaria similis*
T	23 17 15	25 17 15	25 17 15	23 16 15	26 18 16	24 18 16	26 18 16	28 18 17
t	8 6 6	8 6 6	8 6 6	8 6 6	8 6 6	8 6 6	8 6 6	8 6 6
G	4 4 4	4 4 4	4 4 4	4 4 4	4 4 4	4 4 4	4 4 4	4 4 4
F	6 6 5	6 6 5	6 6 5	6 6 4	6 6 6	6 6 5	6 7 5	6 7 5

N.B.: Underlined figures represent numbers unusual for the group.

(4) L.S.T. of LEEUWENHOEKIDAE

	Hyracarus (H.) sp.			*H. (Hex.) natalen.*			*Comatacar. (C.) amer.*			*Com. lemnis.*			*Com. stewarti*			*Com. dolosa*			*Chatia setosa*			*C. (C.) ochotona*		
	a	m	p	a	m	p	a	m	p	a	m	p	a	m	p	a	m	p	a	m	p	a	m	p
T	26	17	17	25	17	17	26	17	15	26	17	17	56	32	30	33	17	15	49	31	31	58	44	33
t	9	6	6	9	6	6	8	6	6	9	6	6	9	7	7	9	6	6	12	9	9	8	6	6
G	4	4	4	4	4	4	4	4	4	4	4	4	5	4	4	5	4	4	5	4	4	5	4	4
F	6	6	5	6	6	5	6	6	5	6	6	5	6	6	5	6	6	5	6	7	5	6	7	5

	C. (Pars.) barunaen.			*C.(Pars.) "r"*			*C.(Ptil.) wissemani*			*C. (Ptil.) "b"*			*C. (Ptil.) murahmadi*			*Brennan. glenni*			*Whartonia (W.)pachyw.*			*W. (Asol.) trinidad.*		
T	36	24	20	66	43	43	34	21	21	47	30	30	32	19	19	25	17	16	34	24	22	22	16	15
t	8	6	6	9	7	7	9	7	7	11	8	8	8	6	6	8	6	6	8	6	6	8	6	6
G	4	4	4	5	4	4	4	4	4	4	4	4	4	4	4	4	4	4	4	4	4	4	4	4
F	6	7	5	6	7	5	6	7	5	6	7	5	6	7	5	6	5	4	6	5	4	6	5	4

	W. (A.) womersl.			*W. (F.) lepidopt.*			*W. (F.) caobang.*			*W. (F.) maai*			*W. (F.) prima*			*Austrac. procav.*			*A. (Aus.) dendrohyr*			*A. (Atr.) theileri*		
T	21	16	15	23	16	15	24	17	16	24	18	17	54	44	42	25	17	16	25	17	15	25	16	16
t	8	6	6	8	6	6	8	6	6	8	6	6	8	6	6	8	6	6	8	6	6	8	6	6
G	4	4	4	4	4	4	4	4	4	4	4	4	4	4	4	4	4	4	4	4	4	4	4	4
F	6	5	4	6	6	5	6	6	5	6	6	5	6	6	5	6	6	5	6	6	5	6	6	5

	A. (Ech.) polydisc.			*A. (E.) wittebol.*			*Anourac. rouxi*			*A. (Dum.) lygosom.*			*Xenacarus plumosus*			*Hanneman. hylodeus*			*H. edwardsi*			*H. hegen. + newst.*		
T	24	17	16	25	16	16	24	16	15	24	16	15	25	17	15	24	17	16	26	17	16	24	18	16
t	6	6	6	7	6	6	8	6	6	8	6	6	8	6	6	8	6	6	8	6	6	8	6	6
G	4	4	4	4	4	4	4	4	4	4	4	4	4	4	4	4	4	4	4	4	4	4	4	4
F	6	6	5	6	6	5	6	6	5	6	6	5	6	5	4	6	5	4	6	5	4	6	5	4

	H. bobdayi			*H. pattoni*			*H. penetr.*			*H. "c"*			*H. "d"*			*Brennan. longisp.*			*B. scarcella*			*B. carpent.*		
T	23	17	15	24	18	16	26	17	16	31	20	18	34	22	20	64	43	43	51	33	28	21	17	15
t	8	6	6	8	6	6	8	6	6	8	6	6	8	6	6	16	12	12	11	8	8	8	6	6
G	4	4	4	4	4	4	4	4	4	4	4	4	4	4	4	5	4	4	5	4	4	4	4	4
F	6	5	4	6	5	4	6	5	4	6	5	4	6	5	4	6	5	4	6	5	4	6	5	4

"Last minute additions"

	Ps'blanka "s"			*Microtr. "e"*			*Chat. (Sh.) tarsalis*		
T	18	14	13	18	14	13	62	39	32
t	6	5	5	7	5	5	8	6	6
G	4	2	2	4	2	2	5	4	4
F	5	4	4	6	4	4	6	7	5

N.B.: Underlined figures represent numbers unusual for the group.

References

Vercammen-Grandjean, P.H.: Introduction à un essai de classification rationnelle des larves des Trombiculinae Ewing 1944 (Acarina-Trombiculidae). Acarologia 2: 469–471, 1960.

– The chigger mites of the Far East (Acarina: Trombiculidae & Leeuwenhoekiidae). An illustrated key and a synopsis; some new tribes, genera and subgenera. U.S. Army Med. Res. & Dev. Command, Washington, D.C., 135 p., 1968.

– Repository of phylogeny and taxonomy in acarines. In: Evans, G.O., ed., Proceedings of the 2nd International Congress of Acarology, Sutton Bonington (England) 19th–25th July, 1967. Akadémiai Kiadó, Budapest, Hungary, pp. 107–112, 1969.

Additional material from Proceedings of the 3rd International Congress of Acarology,
ISBN 978-94-010-2711-3, is available at http://extras.springer.com

OF TECHNIQUES AND ORTHO-ICONOGRAPHY[1]

P.H. VERCAMMEN-GRANDJEAN
G.W. Hooper Foundation, University of California, San Francisco, California, U.S.A.

New methods, even if they must not be accepted or adopted, are to know, just in case . . .

I – Techniques

A – Collection: One is not always allowed to sacrifice a valuable host, difficult to approach, in order to collect its parasities. In Africa, I utilized frequently an enema of sodium pentothal in order to anesthetize precious mammals (monkeys, procavids . . .). Today an easy intramuscular injection with a 10 % solution of phencyclidine hydrochloride (1 mg./kg.) insures a one to two hour peaceful sleep, and still keeps the animals alive.

When a mite is found solitarily attached on the skin of its host, I cut off a little piece of the skin bearing it, thanks to a fine pair of ophtalmical scissors commonly used for retina operations (pl. E).

B – Conservation: Recently, in the course of inventory of a collection made more than four years ago by Dr. D. M. Minter (London School of Hygiene & Tropical Medicine), I found some slides bearing a drop of dried Hoyer. These drops contained a certain number of chigger-mites. This is an easy and expeditive method, no liquid, no vials, no possible leak or loss; a simple slide easy to mail, less manipulation, indelible marking directly on the glass with a diamond pencil. One drop of cedar oil on the lacquered Hoyer, and one is able to study a specimen at random under the immersion objective. For permanent preparations the slide is deposited in a Petri dish and covered with hot water. After a few minutes one is able to pipette selectively the desired specimen, with a little of the surrounding medium, brought to the consistency of jam. Then it is immersed in the dilating solution "A" of Marc André, for one or two hours, in order to mount it finally in the permanent medium "B" of Marc André, or in Faure, Hoyer . . .

C – Mounting: Searching for a stray mite within the preparation constitutes a tense and time-consuming activity. It can be avoided by purposely centering the

[1] This work was supported by Research Grant AI–03793 from the National Institute of Allergy and Infectious Diseases, National Institutes of Health, U.S. Public Health Service.

mite, thanks to a glass template, sort of manipulating the plate, guiding and facilitating the mounting (pl. A).

MOUNTING TEMPLATE A

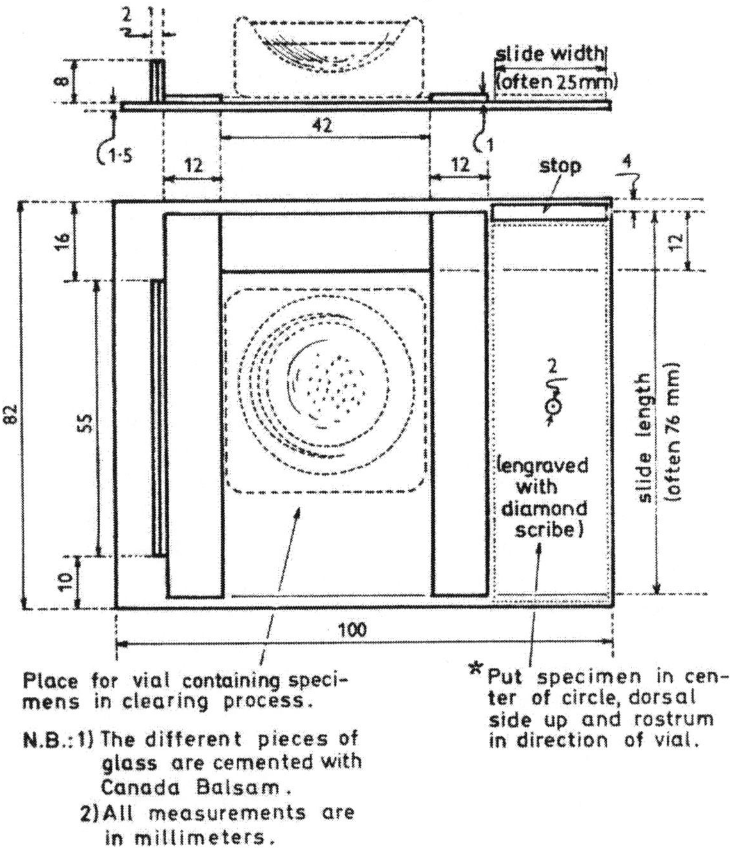

D – L u t i n g: It is not easy to find a satisfactory lute, resistant and easy to remove for eventual remounting. Cellulose-based varnishes deceived me as they are not tight. The white glue available in most markets proves to be a suitable and economical compound (of polyvinyl acetate). Easy to apply with a thin brush, it seals tightly and durably, and is not attacked by the mounting media. In drying, it becomes translucent and acquires a rubber-like consistency which protects the mounting against possible shocks. Finally, it is soluable in hot water (water bath at 60–70 °C).

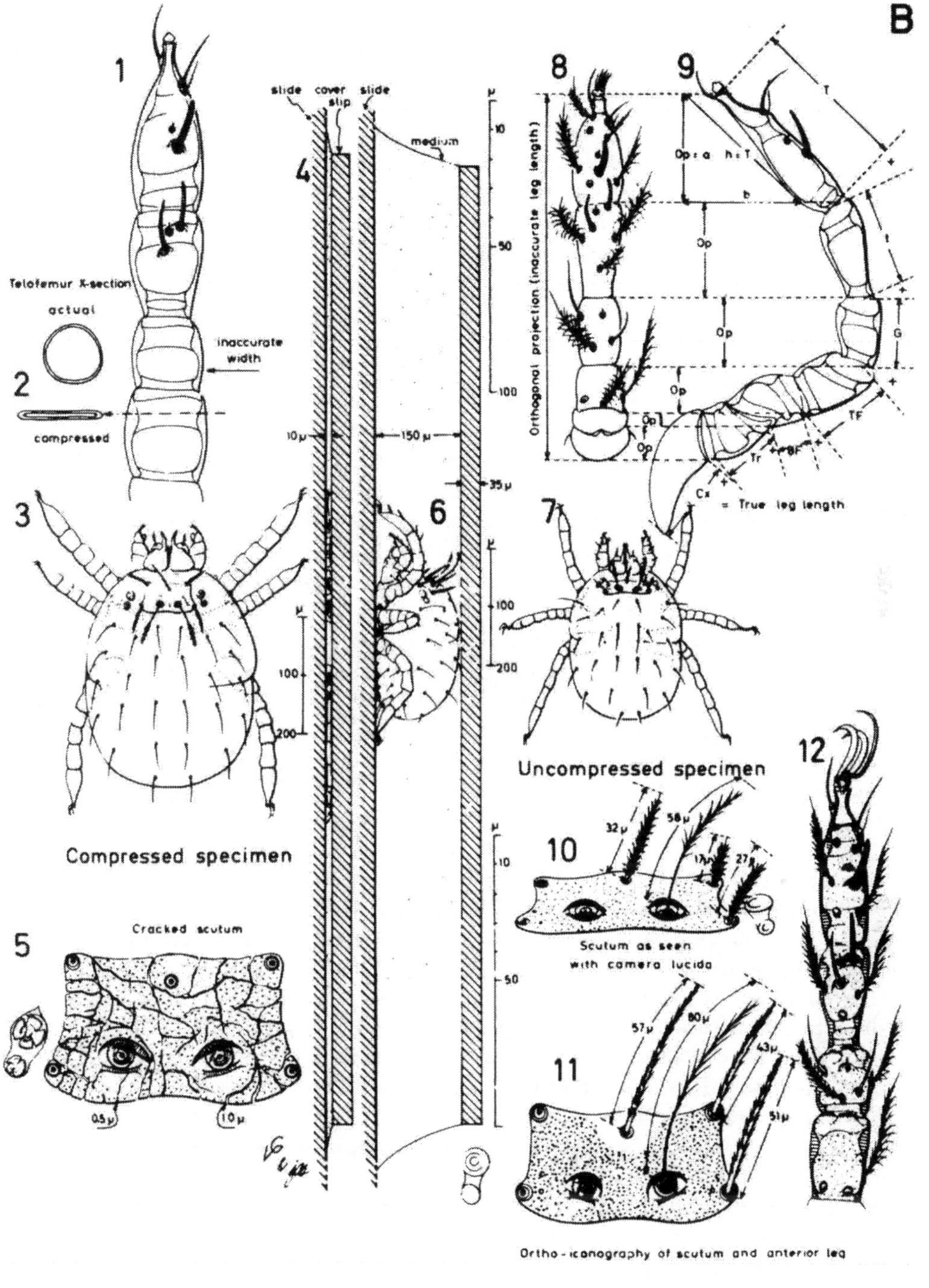
B
1
2
3
4
5
6
7
8
9
10
11
12
Telofemur X-section
actual
inaccurate width
compressed
slide cover slip slide
medium
10μ
150μ
35μ
Compressed specimen
Cracked scutum
0.5μ
1.0μ
Uncompressed specimen
Orthogonal projection (inaccurate leg length)
Op
Op = a h = T
T
t
G
TF
BF
Tr
Cx
= True leg length
32μ
58μ
17μ
27μ
Scutum as seen with camera lucida
57μ
80μ
43μ
51μ
Ortho-iconography of scutum and anterior leg

II – Ortho–iconography

Consists in illustrating a text with realism, precision and conformity. The archaeologist deprived of basis and cornerstones, can not do better than draw the fossil as is in situ. On the contrary, the acarologist is expected to be aware of the anatomy of a species by comparison with others, related and known –– above all when these are numerous ––, and in consequence is supposed to rectify the defects of mountings by appropriate correctives. After all, no taxonomist is validly excusable when he produces unrealistic or repellent illustrations (pl. B).

A – What is to be drawn? To show essential parts for identification of a species: scutum, rostrum, and legs, I use the same scale, 1 mm = 1μ, which proves ideal and useful for comparative morphology. The body setae are advantageously depicted at a scale of 2 mm = 1μ. Except for taxonomical necessity, the idiosoma is rarely represented, as it takes too much room and sometimes requires drastic typographical reductions going so far as to reduce the legibility of more fundamental characters. The scutum is shown in the same plane with the eyes. The setae and sensillary organs are lowered to the same plane and oriented in a manner not to hide other structures, or to cross each other (pl. B–11). The gnathosome is also shown in the same plane. The palps are chelostyles (cheliceral blades) often orient themselves naturally in the course of mounting, in a profile position which ideally shows basic structures and for that reason has been adopted classically by illustrators (pl. B–3 & 7). Anterior and median legs of the left side are depicted in a dorsal position, while the posterior is shown in profile.

B – Rectification: Camera lucida and other instruments of projection considerably ease the work of the illustrator. However, they deliver only orthogonal projections (Op), and therefore objects oriented in a steep position will foreshorten the image (pl. B–8 & 10). Let us contemplate an uncompressed individual (pl. B–6), and another of the same species but squeezed to 10μ (pl. B–4). The result: two different images (pl. B–7 & 3) –– this explains how certain colleagues published up to four times the same species under different names.

For objects mounted in a steep position, the remedy is triangulation. If the orthogonal projection is represented by "a", and if "b" is the second leg of a right-angle triangle and represents the distance between the upper and lower limits of the object to draw (read on the graduations of the fine-adjustment knob), we establish $a^2 + b^2 = h^2$, and thus $h = \sqrt{a^2 + b^2}$. This hypotenuse "h" is close to the actual dimension (to be noted that it is easier to obtain by geometrical means). A curved object may justify the use of multiple triangulation (pl. B–9).

In cases of excessive distention, crevices may appear in the armor of the coxae, gnathobase, and scutum (pl. B–5), necessitating the linear deduction of each crack in view of correct measurement and illustration (pl. B–11).

When, as in pl. B–2, the width of the leg segments is exaggerated by excessive compression, the real width can be obtained by the formula: $M = (Mc - C)\frac{\pi}{5} + C$, in which "M" is the measure to find, "Mc" that one read under compression, and "C" the coefficient of compression. Values up to 40μ are tabulated on pl. C.

MEASUREMENT RECTIFICATIONS OF OBJECTS UNDER COMPRESSION

M — C: Coefficient of Compression (in micra)

Formula: $M = (Mc - C)\frac{\pi}{5} + C$

Mc: Measurements (μ)	2	4	6	8	10	12	14	16	18	20	22	24	26	28	30	32	34	36	38	40
6	5	5	6																	
8	6	7	7	8																
10	7	8	9	9	10															
12	8	9	10	11	11	12														
14	10	10	11	12	13	13	14													
16	11	12	12	13	14	15	15	16												
18	12	13	14	14	15	16	17	17	18											
20	13	14	15	16	16	17	18	18	19	20										
22	15	15	16	17	18	18	19	20	21	21	22									
24	16	17	17	18	19	20	20	21	22	23	23	24								
26	17	18	19	19	20	21	22	22	23	24	25	25	26							
28	18	19	20	21	21	22	23	24	24	25	26	27	27	28						
30	20	20	21	22	23	23	24	25	26	26	27	28	29	29	30					
32	21	22	22	23	24	25	25	26	27	28	28	29	30	31	31	32				
34	22	23	24	24	25	26	27	27	28	29	30	30	31	32	33	33	34			
36	23	24	25	26	26	27	28	29	29	30	31	32	32	33	34	35	35	36		
38	25	25	26	27	28	28	29	30	31	31	32	33	34	34	35	36	37	37	38	
40	26	27	27	28	29	30	30	31	32	33	33	34	35	36	36	37	38	39	39	40

Explanation: Mc = 32 μ, is the width of a compressed leg segment read on the micrometer,
C = 8 μ, or distance between upper and lower walls of the segment (read on the graduations of the fine adjustment knob),
and $(32 - 8)\frac{\pi}{5} + 8 = 23$ μ, which can be found at the crossing point of Mc/32 and C/8 lines, on the table.

C – Body measures and size: It is customary to express the size of a being by a single numerical data, forcefully relative. The dimensions of an object (body for instance) have always to be expressed by three co-ordinates, conventionally called: length (L), width (W), and depth (D).

1) Several authors report the length and width of a species but not its depth, as if organized beings were flat, immaterial ectoplasm. They simply ignore the third dimension. Well, the two others are a function of this depth, that the state of compression varies so considerably. There is, thus, no doubt left concerning the poor value of these data, which, on the other hand, vary even more considerably with the progress of engorgement.

If the larva is unfed and not compressed: The length (L) and the

D

Body Reconstitution

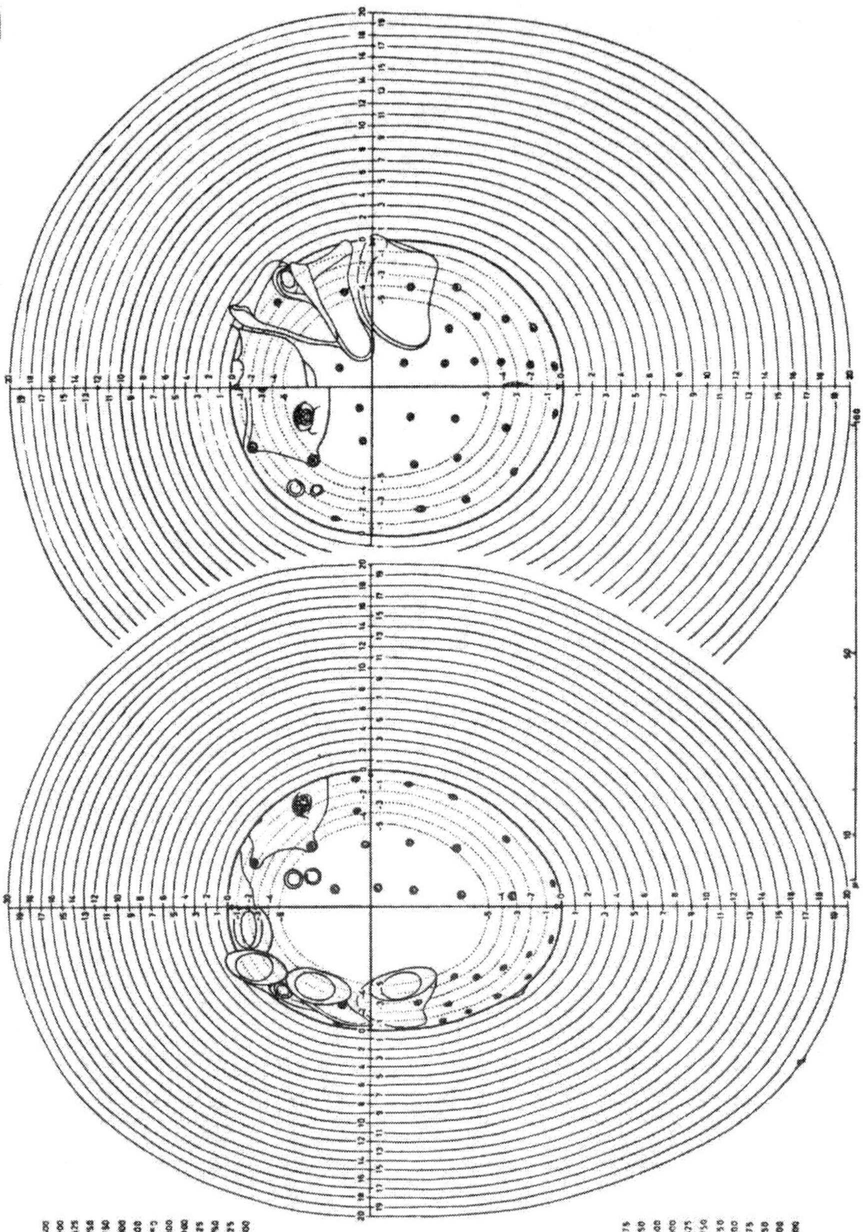

	L'	W'	D'	P
-5	46.25	40	35.25	65,217,500
-4	52.00	45	39.50	92,664,000
-3	57.75	50	43.95	126,905,625
-2	63.50	55	48.30	168,687,750
-1	69.25	60	52.65	218,760,750
0	75.00	65	57.00	277,875,000
1	80.75	70	61.35	346,837,420
2	86.50	75	65.70	426,278,[illegible]
3	92.25	80	70.05	516,968,000
4	98.00	85	74.40	619,752,000
5	103.75	90	78.75	735,328,125
6	109.50	95	83.10	864,447,750
7	115.25	100	87.45	1,007,861,125
8	121.00	105	91.80	1,166,298,000
9	126.75	110	96.15	1,340,571,375
10	132.50	115	100.50	1,531,388,750
11	138.25	120	104.85	1,739,461,500
12	144.00	125	109.20	1,965,600,000
13	149.75	130	113.55	2,210,514,625
14	155.50	135	117.90	2,475,015,750
15	161.25	140	122.25	2,759,793,750
16	167.00	145	126.60	3,065,619,000
17	172.75	150	130.95	3,393,243,875
18	178.50	155	135.30	3,743,412,750
19	184.25	160	139.65	4,116,892,000
20	190.00	165	144.00	4,514,400,000

W I L D Microscope M/20 n° 41712

Oc. 10x $f = \frac{\mu}{s}$ $s = \mu$

Obj.:			
6x	$\frac{2160}{100}$	21.60	
20x	$\frac{580}{100}$	5.80	
40x	$\frac{290}{100}$	2.90	
100x	$\frac{119}{100}$	1.19	
10x	$\frac{1180}{99}$	11.90	

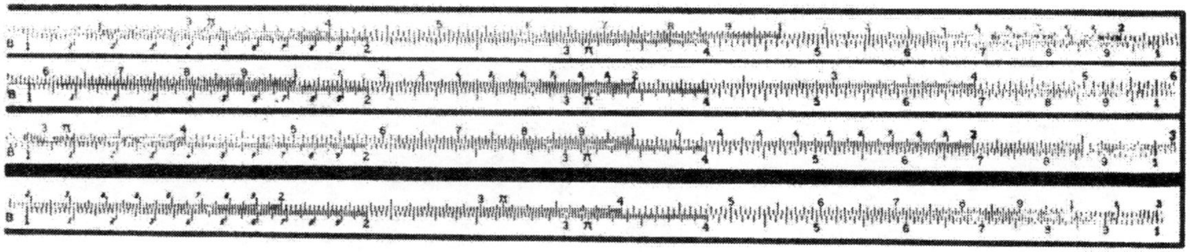

μ = micra

s = micrometric stripes

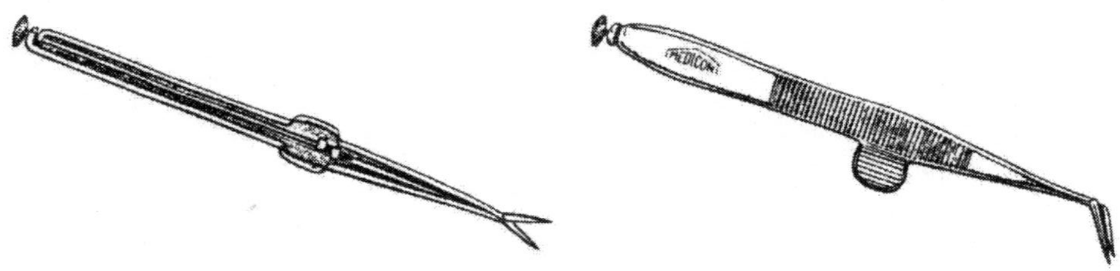

F

E

width (W) are easily measured. As for the depth (D), it is obtained by the fine adjustment knob graduations.

If the larva is unfed but compressed: Its form is that of an elliptic cylindroid of which the volume is $Vc = \pi \left(\frac{L}{2} \times \frac{W}{2}\right) \times D$, and that volume is naturally equal to that of the initial ovoid or ellipsoid idiosoma, before compression, i.e., $Vo = \frac{4\pi}{3}\left(\frac{L'}{2} \times \frac{W'}{2} \times \frac{D'}{2}\right)$, from which $\frac{3}{2}(LWD) = L' W' D' = P$. On the abacus (pl. D), the three dimensions to obtain are figured in front of the value of P, itself obtained by the dimensions L, W, and D of the compressed specimen. This abacus shows in addition the models of corresponding idiosomes.

If the larva is engorged: The results obtained would be of slight interest, and it is better to get the probable dimensions of the unfed larva of that species, dimensions which would be sensibly the product of multiplication of the length of the posterior coxa (LCxp) by appropriate coefficients as follows: L' = LCxp x 2.3., W' = LCxp x 2.0 and D' = LCxp x 1.75.

2) What precedes concerns the body in its three dimensions, although the flat graphic representation of it requires the use of two other coefficients multiplying also the length of the posterior coxa: L = LCxp x 3.45 and W = LCxp x 3.0.

3) As a consequence of the undependability of the body dimensions, but recognizing the necessity of a notion of relative size, I have utilized for a long time the leg length (coxa included, pretarsus and claws excluded). These lengths are constant, and, after all, legs are directly proportional to the idiosoma they have to carry. This relative size is figured in a single number easy to remember and called "Index pedibus", which consists of the sum of the three leg lengths, pa + pm + pp = Ip.

D – Conversion of ocular-micrometer lines into microns: Several colleagues, and I the first, committed errors in the course of conversion of oculomicrometic lines into micra . . . it is classical. However, it is possible to eliminate certain causes. Some calculate; others edify tables or utilize slide-rules. The simplest, most convenient, and surest method is to have photostats of the slide-rule setting co-ordinated to the magnification used.

STUDY OF THE "ERYTHRAEIDAE, R.O.M. NO. 8" OF EWING, 1937*

P.H. VERCAMMEN–GRANDJEAN
G.W. Hooper Foundation, University of California, San Francisco, California, U.S.A.

In 1937, Dr. H.E. Ewing described briefly a larval mite embedded in Cretaceous amber from Cedar Lake, Manitoba (Canada). He stated that "Eventually it may be possible to place this specimen to genus, but at present our knowledge of the larval forms of the Erythraeidae is not sufficient."

Today, greater knowledge, improved techniques, and increasing quality in microscopy enable us the undertaking of delicate tasks deemed impossible in the past.

Contrary to the opinion of my predecessors, I consider that: 1. The semiprofile position of the mite's body and the doubled-up position of the legs were advantageous.

2. The larva was unfed according to the proportion of the idiosoma compared with that of the capitulum, the position of the posterior coxae, and the state of the more or less pointed pygosoma.

3. The psychodoid nematocerous Diptera cohabiting the piece of amber is not that much larger than the erythraeid to sustain the assumption of strict parasitism; it appears much more to be an aspect of predation, providing that the midge is really the victim of the mite, and not merely a simple neighbor, imbedded in the same flood or resin.

In recognition of his remarkable studies on the family Erythraeidae, the species is dedicated to Dr. R.V. Southcott, of Adelaide (South Australia). I thank Dr. David Barr, Assistant Curator of the Department of Entomology and Invertebrate Zoology, Royal Ontario Museum, Toronto (Canada) for the loan of the precious amber piece containing the Erythraeidae no. 8.

Proterythraeus southcotti n.g., n.sp.
(Figs. 1, 2, & 3)

A – Ecological data and type material: One specimen, embedded in a fragment of amber found on the west shore of Cedar Lake near the mouth of the

* This work was supported by Research Grant AI–03793 from the National Institute of Allergy and Infectious Diseases, National Institutes of Health, U.S. Public Health Service.

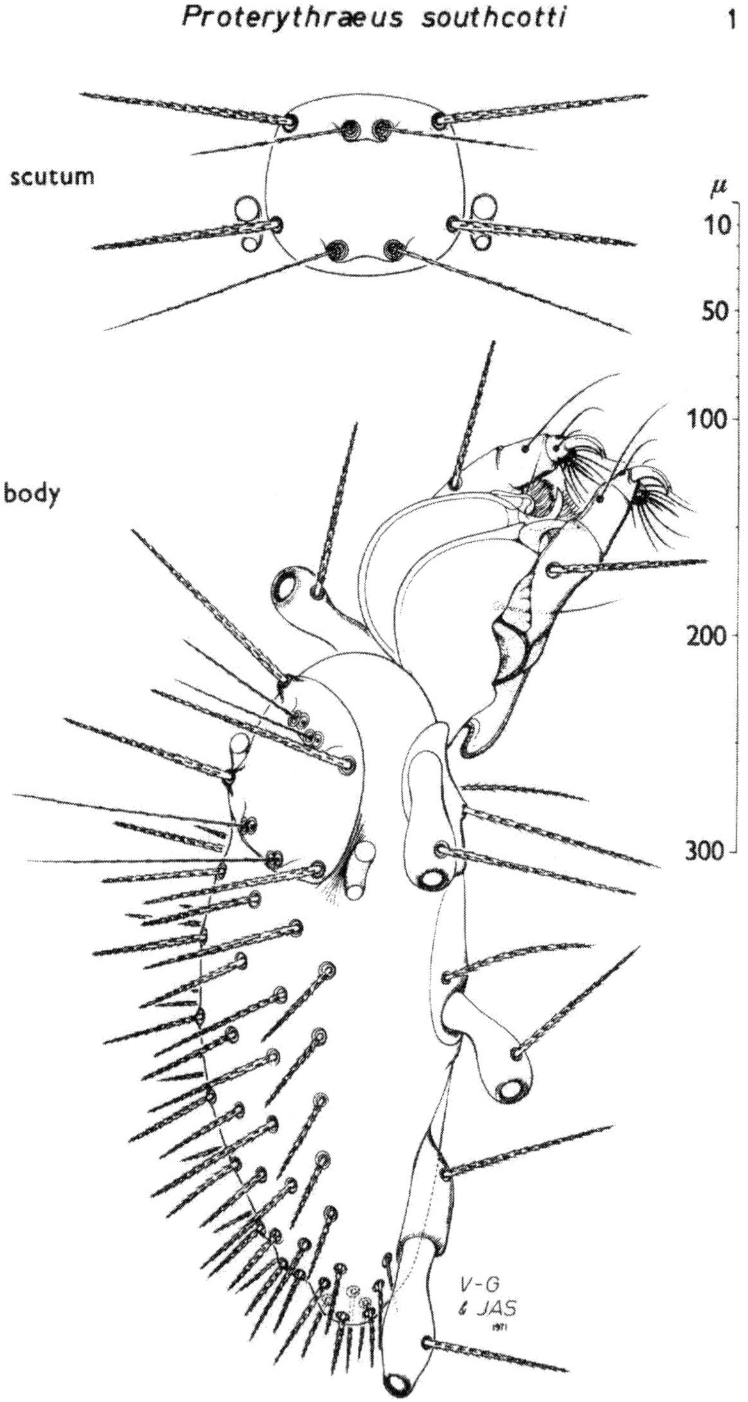

Proterythraeus southcotti
1
scutum
body
μ
10
50
100
200
300
V-G
& JAS
1971

Proterythraeus southcotti 2

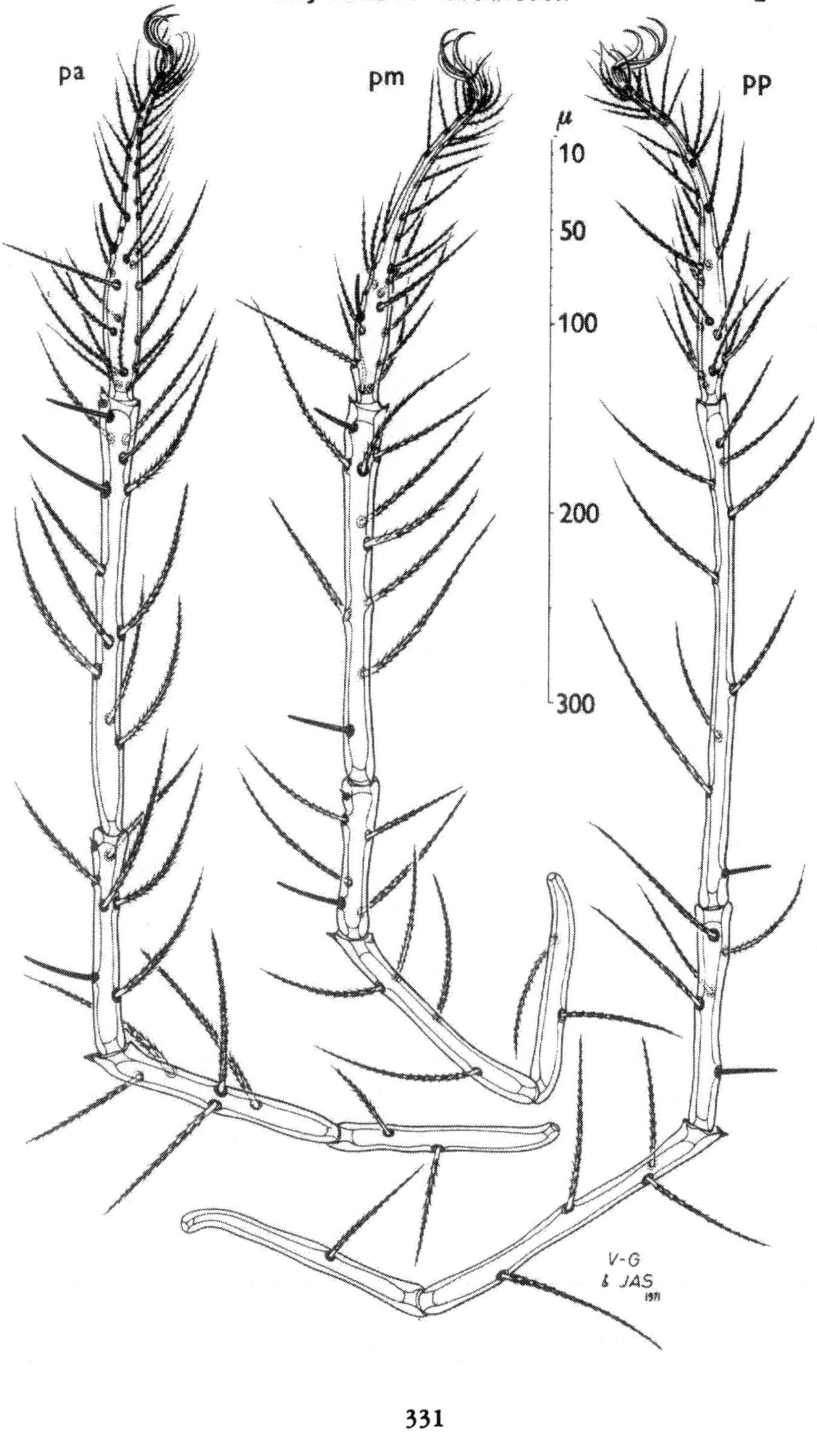

Saskatchewan River, Manitoba (Canada). This amber was proved by Carpenter, 1937 to be from Cretaceous strata. That specimen is deposited in the Royal Ontario Museum (Canada).

B – Description: 1. Scutum: bears two pairs of submedial sensillae and two pairs of lateral barbed setae. The eyes are contiguous to the scutum and composed of two corneas (11 and 6μ); the left pair being clearly silhouetted in the prepa-

Proterythraeus southcotti 3

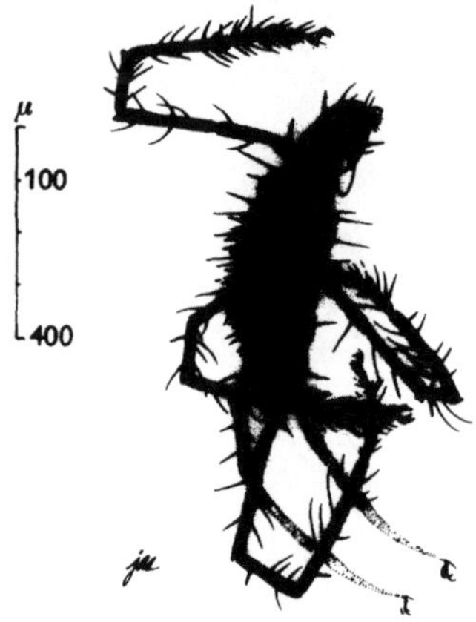

ration, the right visibly shadowed under refracted light. Left lateral scutal setae arising from two silhouetted insertion buds, the right pair being faintly depicted under strong illumination. Refracted light combined with strong illumination enabled the depiction of the trichobothrial pits, which joined to the direction taken by the branched setae and the sensillae enable the elaboration of a morphological graphism. The work of reconstitution enabled the drawing of the scutum as shown on pl. 1. Due to the opaque background its texture cannot be shown, nor can that of the body, or of the legs' integument. The only place where cuticular pleats can be discerned is between the scutal edge and right eye.

2. Body and body setae: Body not quite twice as long (310μ) as gnathosoma (160μ). Its pygosome slightly pointed, exceeding only slightly the apex of the posterior coxa. Body setae arduous to detect; the number given in the "numerical taxonomy" being tentative. These setae are unbent and covered with slight spicules, like those of the scutum.

3. Gnathosoma: Only the base of the right palpus, a folding of the veil, and the palpotarsi are translucent; the rest of the capitulum is opaque. Bold chelobases bearing typically recurved chelostyles. Galeala, or protorostral, nude. Deutorostral apparently nude. Palpus with thick and barbed femorala (74μ), one long nude genuala (58μ), and three shorter nude tibialae (24, 16 & 28μ), bifurcate tibial claws, strongly curved. Tarsus with one slender and tapering solenidion, one short but

pointed terminala (10μ), and six ordinary setae (18 to 28μ).

4. Legs: All legs seven-segmented, long, thin, and rodlike; tibiae being the longest articles and the three tarsi similar to shape: visibly swollen on their basal two-fifths and their apex tapering in an elegant curvature. Except for the more or less halteriform trochanterae, the genua are the shortest mobile segments of the legs. The Ip = 2759μ, corresponding to that of a very large species on the scale of chigger-mites, although it is proved that an Ip of 3400 is not uncommon for erythraeids (Southcott, 1966). The length of the leg segments is reported under the "numerical data" and also are reported the numbers of the ordinary barbed setae, the length of the special nude setae, and of the tarsal claws. Each of the three tarsi is provided with a nude, subapical pretarsala (PT = 28 μ)* and a short pretarsus (pedicle = 14 μ) bearing two subequal claws, framing a longer, but thinner empodium; all three are nude. The exact shapes of the coxae are impossible to discern. It can be stated definitely that there is no intercoxal structure suggesting the existence of urstigma, or Claparéde organ. Only one thin anterosternal seta is visible.

C – Numerical data: SIF = 6N.T–N-2-1111.oooo (Vercammen-Grandjean, 1968)
fPp = (B)–(N)–(N).N.N

No ST or pST   3 PT = 28μ   fsp = 7.7.7.   fCx = 98/74/85μ
fSt = 74μ (?)   Gr = 24μ   Ch = 30μ   So = 18μ   Protorostral = 34μ
Deutorostral = 62μ; Tritorostral, not seen; Idiosoma: L = 310 μ, W = 150μ, D = 120μ
fD = 6.6.6.6.6.6.4.2. = 42 (?)   fV = . . . 4.2 (?)

AW	PW	SBa	SBp	ASB	SS	PSB	SD	AP	AL	PL
70	84	15	26	16	55	12	83	52	102	188

Sa	Sp		Da		P	V	pa	pm	pp	Ip
76	122	58	76	48	28	40	894	832	1033	2759

N.B.: "SS" stands for distance between the anterior and posterior sensillary bases.
(?) = tentative approach.

	Length of leg segments							Barbed leg setae							Special nude leg setae								Tarsal claws		
															Tarsus			tibia			Genu				
	T	t	G	TF	BF	Tr	Cx	T	t	G	TF	BF	Tr	Cx	Pt	S	μs	μs	t'	t"	μs	G	In.	Emp.	Out
a	178	230	116	130	120	58	62	34	10	5	5	2	1	1	28	25	4	7	37	56	7	42	17	24	20
m	178	200	82	130	120	54	68	30	8	4	4	2	1	I.	28	24	6	–	24	35	9	38	18	26	22
p	186	267	117	180	140	68	75	30	8	4	4	1	1	1	28	–	–	–	24	–	–	32	20	28	24

N.B.: μs = microspur (vestigiala). S = solenidia S_1 or S_2. t', t" = distal, proximal tibialae.

---

* Pretarsalae are normally existing on the anterior and median tarsi only, never on the posterior tarsus, except on the present species and on *Multigniella cosmetornis*, an unusual Trombiculidae.

D – **F a m i l y s t a t u s:** Considering the leg setation, the nature of the leg claws, and several other characters, it is obvious that the genus *Proterythraeus* is obsolete among the modern erythraeids and deserves a distinguishable position (see also L.S.T. hereafter). Therefore, we suggest the family name of Proterythraeidae n.f.

Leg setae as an indicator of evolution

In another study (Vercammen–Grandjean, in press), I reported about numerical tabulation of ordinary leg setae, and discussed the efficiency of that arrangment as a global taxonomic character, allowing fruitful comparisons at first glance. This frame of twelve numbers involved the tarsus (T), tibia (t), genu (G), and femur (F) of the anterior (a), median (m), and posterior (p) legs, and is called Leg Setae Tabulation (L.S.T.) Comparing the L.S.T. of *Proterythraeus southcotti* with that of a modern erythraeid (*Charletonia* sp.), of a johnstonianid (*Lassenia scutellata*), a classical trombiculid (*Neotrombicula autumnalis*), a trombidiid (*Microtrombidium* sp.), a pterygosomid (*Geckobia* sp.), a cheyletid (*Cheletogenes* sp.), and, to conclude, that of an ixodid (*Rhipicephalus evertsi*) showing a different arrangement, but as simple and with the same remarkable constancy.

	Charlet. sp.			*Proteryt. southcot.*			*Lassenia scutell.*			*Neotromb. autumnal.*			*Microtr.* sp.			*Geckobia* sp.			*Cheletog.* sp.			*Rhipiceph. evertsi*		
	a	m	p	a	m	p	a	m	p	a	m	p	a	m	p	a	m	p	a	m	p	a	m	p
T	28	28	28	34	30	30	36	29	24	21	16	15	18	14	12	10	10	10	7	7	5	21	18	18
t	18	18	18	10	8	8	8	8	8	8	6	6	6	5	5	5	5	5	4	4	3	7	6	6
G	12	12	12	5	4	4	5	4	4	4	3	3	4	2	2	2	2	1	1	1	1	8	8	8
F	9	9	7	7	6	5	7	6	6	6	6	5	5	5	4	4	4	3	2	2	1	10	10	9

One sees immediately that *Proterythraeus southcotti* "marks the same hour" as *Lassenia scutellata* and not that of modern erythraeids. the difference: a little more than 60 million years . . . It is obvious that the numbers of ordinary setae adorning the appendages reflect the degree of evolution of the acarines, preordained in the larvae. Considering the family group showing the lowest L.S.T., the cheyletids (*Cheletogenes, Ornithocheyletia*), their morphology corresponds to a very primitive status which is confirmed by a low palpotarsal formula (fT = 3). After them come the pterygosomids *(Pterygosoma, Geckobia, Zonurobia, Ixodiderma . . .)* showing more evoluted morphological characters and a numerically higher L.S.T. Next follow trombidiids *(Hoplothrombium, Microtrombidium, Parathrombium, Podothrombium . . .).* The trombiculids come after with 68 genera and more than 2000 species, some of them presenting various deviations. They are followed almost immediately by trombellids *(Durenia, Womersleyia, Audyana . . .)* and a little later by johnstonianids, which present a rather large range of variations from *Centrotrombidium* to *Lassenia;* the neotrombidines *(Neotrombidium, Monunguis)* being not far behind. *Proterythraeus southcotti* is in the middle of these groups as concerns comparative L.S.T. and degree of evolution. Later come water mites, which show a narrow range of variations. Much later come the erythraeids, which occupy a summit among the Trombidiformes and sustain the comparison with the L.S.T. of ticks.

References

Ewing, H.E.: Insects and arachnids from Canadian amber: Arachnida, Order Acarina. Univ. Toronto Stud. Geol. 40:56–62, 1937.

Carpenter, F.M.: Insects and arachnids from Canadian amber: Introduction. Univ. Toronto Stud. Geol. 40:7–13, 1937.

Southcott, R.V.: Revision of the genus *Charletonia* Oudemans (Acarina: Erythraeidae). Australian J. Zool. 14:687–819, 1966.

Vercammen–Grandjean, P.H.: The chigger mites of the Far East (Acarina: Trombiculidae & Leeuwenhoekiidae). An illustrated key and a synopsis; some new tribes, genera and subgenera. U.S. Army Med. Res. & Dev. Command, Washington, D.C., 135 p., 1968.

– Preliminary note to a tentative nepophylogeny of trombiculids. (In press.)

TAXONOMY OF ORIBATID MITES – RETROSPECT AND PROSPECT

T. A. WOOLLEY
Department of Zoology, Colorado State University, Fort Collins, Colorado, U.S.A.

Ross (1971) considers systematics as the oldest of the sciences because it began with man's lingual classification of things in his experience. Even though classification is arbitrary, biosystematics is an attempt to determine what to classify in nature, comprising both speciation – a natural process, and phylogeny – an artificial fabrication of the human mind subject to varying opinions and personal interpretations. In his view, taxonomy, biosystematics and phylogeny must be modified as new data and interpretations are disclosed.

Yochelson (1971) assumes all categories of hierarchal classification are "philosophical constructs without physical features." He views all levels of classification as dealing with products of the imagination, not with specimens, yet essential to taxonomic arrangement.

The objectives of taxonomic research are the facilitation of identifications and the hierarchal ordering of organisms. Whenever animals are studied, they should be identifiable in as simple a system as possible. Complex systems with restrictive definitions and critically finite characteristics understood only by specialists are not helpful to the uninitiated student nor to the general researcher.

Most oribatid families are only vaguely understood; they need desperately to be better defined. With the establishment of definite, inclusive familial characteristics, we can place proper genera within these limits and establish more correct superfamilies or other higher categories. It may be that in oribatid taxonomy we are really working at the generic level when we describe families. One conclusion is sure: we cannot continue to erect ill-defined families without suffering the natural consequence of complexity and confusion.

Compared to other suborders of acarines, oribatid taxonomy has a very complex system. Because of excessive splitting at the familial level, the taxonomy has moved from a single family, e.g. Oribatidae, listed by Banks in 1908, through revisions by Ewing, Jacot, Sellnick, Willmann, to more complex systems (cf. Balogh, 1965, Grandjean, 1954, 1969) with 39–48 superfamilies and over 100 families. In contrast, Mesostigmata has 18 superfamilies, 65 families; Prostigmata 16 and 61; Hydrachnellae 11 and 50 (?); Astigmata 9 and 36–50 (?). Even though the oribatids may be considered the "Coleoptera" of the mite world, this number of superfamilies and families does not seem warranted or necessary.

Until fairly recently, oribatid taxonomy was based entirely on the adult stages. Grandjean (1953) incorporated some immatures in his system and changed the system in 1969. While it is important to correlate characters of immatures in any taxonomic system, so many oribatid immatures are unknown that the present system is naturally incomplete. In the interim, the adults should be thoroughly characterized and classified. Oribatid research should desert the worship of complex, split systems and strive toward a better delineation of taxonomic categories (particularly families and genera) in relatively simple form. Grandjean (1965) expresses a similar concern.

Balogh's (1965) provisory system is an excellent summary of the genera of oribatids, exclusive of immatures, and forms a basis for generic determination that otherwise would remain unknown except for extensive research of the literature. Even so, familial definition is still incomplete or overlapping in many instances.

It is the intent of this paper to 1) stimulate and solicit cooperation of oribatidologists toward a greater correlation of taxonomic work by encouraging intensive investigation of the genera and definition of the families with a view toward sim-

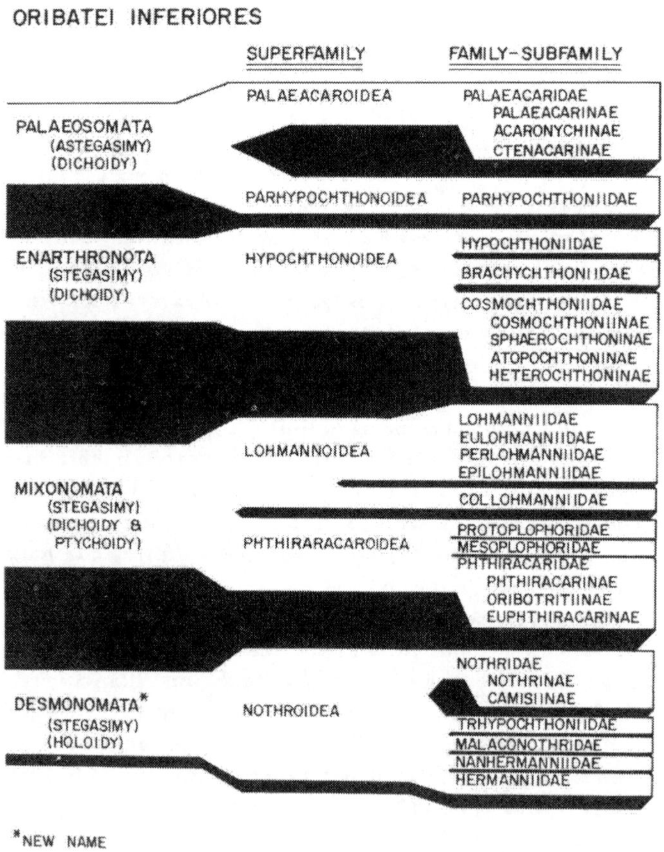

Fig. 1. Proposed scheme of familial arrangement for Oribatei Inferiores. (After Grandjean, 1959, 1969.)

plification of the system at the familial and higher category levels, and 2) to suggest consideration of an exemplary, provisory system as a possible means of reaching the first objective.

The system that follows is admittedly artificial (as Balogh (1965) stated for his), but it represents a possible plan for general recognition and ordering of oribatids. It involves reduction of the numbers of superfamilies, expansion of some familial categories by subordination to subfamilies of some current "families". It is an attempt to make an arrangement somewhere between the earlier simplistic situation (Woolley 1958) and current conditions of excessive splitting into monotypic and monogeneric families. The accompanying figures (Figs. 1, 2) are illustrative of possible arrangements and employ Grandjean's (1969) basic groups with some modifications. Some familial designations have been omitted.

Space permits only limited examples within the proposed scheme. More details and examples will be developed subsequently. For instance, it appears to me that such families as Lohmanniidae, Epilohmanniidae, Perlohmanniidae and Eulohmanniidae in Myxonomata possibly represent "experimentation" in the evolutionary sense with arrangements of the ventral plates along with other features, hence the difficulties attendant to their delineation. The same may also be said for the families in the Desmonomata*–Nothroidea. The variety of structural relations makes them difficult to order within a simplified hierarchy. Collohmanniidae, which is somewhat between the dichoid Enarthronota and the ptychoid Myxonomata in Grandjean's scheme, seems to be possibly transitional between the dichoid and the ptychoid forms. Relationships between dichoid-ptychoid forms and the holoid forms still remain obscure.

Ptychoidy as a phenomenon is unlikely to have occurred more than once, even though the three expressions of the phenomenon are rather different. The suggested arrangement (Fig. 1) assumes that the Protoplophoridae, Mesoplophoridae and Phthiracaridae are more closely allied because of ptychoidy, but other aspects need to be considered, as well. Splitting of the families in Phthiracaridae is considered unwarranted although structures show some similar divergences for each of the proposed subfamilies.

Carabodoid, Eremaeoid, Oppioid and Liacaroid families (Fig. 2) (Woolley, 1969) have each been consolidated because the splitting of these into ill-defined families has contributed to some of the existing confusion. Other examples of consolidation are noticeable in the Pycnonoticae and Poronoticae as well.

By restricting more general characters to broader categories, the superfamilial list is reduced. By consolidating overly split families and employing subfamilial designations, the numbers of families is limited. Subfamilial designations, though provisory, will be studied further to enlarge and clarify the scheme, to help define descriptive terms and taxonomic categories, to delineate families by comparative morphology, chaetotaxy and other means.

Until some strenuous effort is made by intensive study to define families, the

* nom. novum, implying joined.

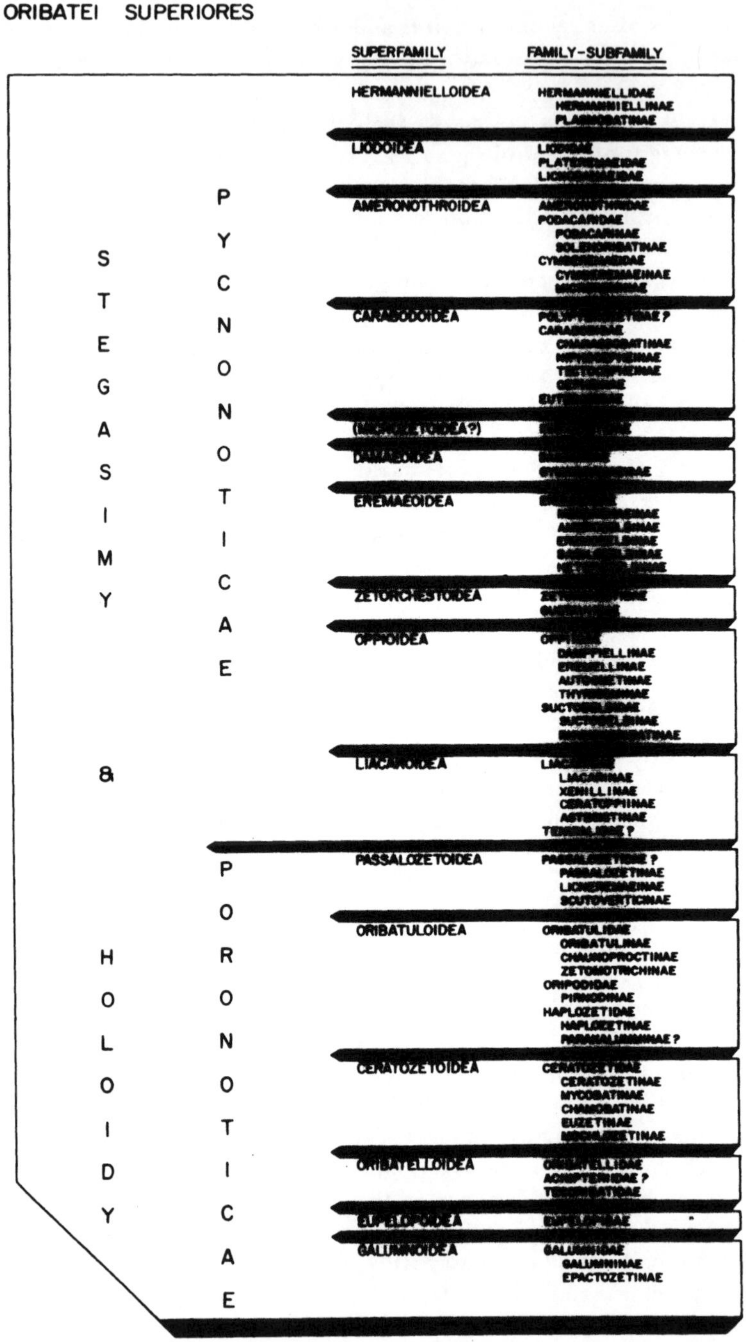

Fig. 2. Proposed scheme of familial arrangement for Oribatei Superiores. (After Grandjean, 1953, 1969; Woolley, 1958, 1969.)

taxonomy of oribatid mites will become more unwieldy than it presently is. We desperately need to delimit the present families and not to continually erect new families just because the new genera and species we discover do not fit the ill-defined categories in the present complex system. The prospects for oribatid taxonomy are rather grim unless we make such an effort.

References

Balogh, J.: A synopsis of the world oribatid genera. Acta Zool. XI (1/2): 1–99, 1965.

Grandjean, F.: Considérations sur le classement des oribates-leur divisions in 6 groupes majeurs. Acarologia 11(1): 125–153, 1969.

– Essai de classification des oribates. Bull. Soc. Zool. France 78 (6/5): 421–446,1953.

– Nouvelles observations sur les oribates (4e Serie), Acarologia 7 (1): 91–112, 1965.

Ross, H. H.: Systematics and scientific understanding. Bull. Entomol. Soc. Am. 17 (1): 15–17, 1971.

Woolley, T.A.: A preliminary account of the phylogeny of the oribatei. In Proc. Xth Int'l Congr. Entomol. 1: 867–874, 1958.

– The Liacaroidea in the taxonomy of oribatids. Proc. North Central Branch, Entomol. Soc. Am. 24 (2): 146–149, 1969.

Yochelson, E. L.: Phylum and class nomenclature in systematics. Syst. Zool. 20 (2): 245–249, 1971.

Proceedings of the 3rd International Congress of Acarology, Prague, 1971

SOME ASPECTS OF THE MORPHOLOGY OF THE GNATHOSOMA OF HETEROZERCON (ACARI: MESOSTIGMATA)

G.O. EVANS
Agricultural Zoology Department, University College, Dublin, Eire

Abstract

An account is given of the morphology of the gnathosoma of *Heterozercon audax* Berl. with particular reference to the form and probable function of the highly specialised malae internae (hypostomal processes), the structure of the labrum-epipharynx complex and the nature of the cheliceral sheaths. The structure of the gnathosoma in *Heterozercon* is compared with that of other Mesostigmata. The systematic position of the Heterozerconidae is discussed.

IMPRIMATUR FOR TAXONOMICAL PUBLICATIONS

E. PIFFL
Institute of Zoology, University of Vienna, Austria

Abstract

In 1896 Berlese selected erroneously his *Oribates dentatus*, 1883 as a typus of a monodactylous group, which in 1908 was named *Protoribates*. Although Michael as early as in 1898 doubted about the monodactylous phenomenon, Berlese corrected his error only in 1916, but this time he committed two mistakes. He changed the typus and selected once more a tridactylous species: *O. monodactyla* Haller 1884. In 1929 Jacot discovered this error, nevertheless *O. monodactyla* existed until 1968 as typus. Regardless of the number of claws, Jacot placed the monodactylous *O. lophotrichus* Berl. 1904 in the genus *Xylobates*, which was distinguished only by the described lamellae. In 1936 Grandjean pointed out the correct differentiating characters in his universal morphological method for *Haplozetes*, *Protoribates* and *Xylobates*, while erecting the family Haplozetidae. In 1934 these characters were partly unknown to Jacot and therefore his *Styloribates pectinatus* was distinguished according to a general character: the styles correspond to the doubled horn. The Jacot's areae porosae are sacculi and *Styloribates* is a *Haplozetes*, probably even dentatus. If the observation of Coetzer in 1968 is correct and the typus *pectinatus* is congeneric with his new *Muliercula*, then *Muliercula*, *Haplozetes* and *Protoribates* are synonyms, but in no case *Muliercula* is a member of Oribatulidae. Neither is it a *Xylobates* sensu Woolley 1968; besides his *X. prionotus* seems to be *lophotrichus* of Berlese.

In order to avoid similar mistakes which arise for taxonomy and systematic primarily from unsatisfactory information, an imprimatur is proposed by which only publication achieves nomenclatory status. Therefore in future only publications with imprimatur should be taken into account. The imprimatur ought to be given by a body founded by the International Congress of Acarology and supplied with sufficient documentation. As date for the priority the drawing day of the imprimatur should count.

Section V

MITES OF STORED PRODUCTS

Chairman
V. I. Volgin (USSR)

Assistant Chairman
E. Žďárková (Czechoslovakia)

CHELACHECAROPSIS BAKERI, A NEW GENUS AND SPECIES ASSOCIATED WITH STORED FOOD MITES IN EGYPT (ACARINA, CHEYLETIDAE)

H.H. ATTIAH
Acarology Investigations, Crop Protection Department, Ministry of Agriculture, Dokki, Egypt

Mites of the family Cheyletidae are often found in association with stored food mites. Their potential ability to eradicate a population of tyroglyphid mites is debatable in spite of the remarkably fierce-looking palpi they possess.

Summers and Price, 1970, reviewed the work previously done on the family and erected some new genera. The present work adds one new genus, represented by one new species, collected from rice mills in Egypt.

Chelachecaropsis g.n.

Palp tarsus with inner comb modified into a simple plain seta, hence tarsus provided with 2 sickle-like, one comb-like, one simple and one short clublike setae. Body pyriform, constricted between propodosoma and hysterosoma, completely covered dorsally and ventrally by fine striae, without differentiated shield or plate; with 12 pairs of dorsal fusiform pilose setae and one pair of humeral smooth ones; a pair of eyes present. Legs I–II and III–IV widely separated but by less than body width at constriction; tarsal claws and a padlike empodium with tenent hairs present on all legs. Only females are known.
Type species: *Chelachecaropsis bakeri* sp. n.
Remarks: The genus *Chelachecaropsis* is related to both *Chelacheles* and *Chelacaropsis*.

Chelacheles as described and illustrated by Baker, 1958, has two comb-like setae, fourteen dorsal pairs of setae (including the humerals) and with the distance between coxae II and III more than body width; *Chelacaropsis*, on the other hand, has 14 pairs of dorsal broadly clavate setae (Baker 1949).

Chelachecaropsis bakeri sp. n. (Figs. 1,2)

Female: Length including rostrum 353 μ, breadth 195 μ at widest part of propodosoma. Rostrum somewhat elongate, with longitudinal striae; peritremes simple, inverted U – shaped, with six pairs of segments; palpal femur longer than wide; palpal thumb provided dorsally with one comb-like and one simple seta and ventrally with two sickle-like and one club-like setae, the comb-like setae with eight strong fingerlike teeth; the palpal claw with four basal teeth. Body dorsally completely covered with longitudinal fine striae; area between propodosoma and hysterosoma with transverse striations, dotted in some places; dorsum with six

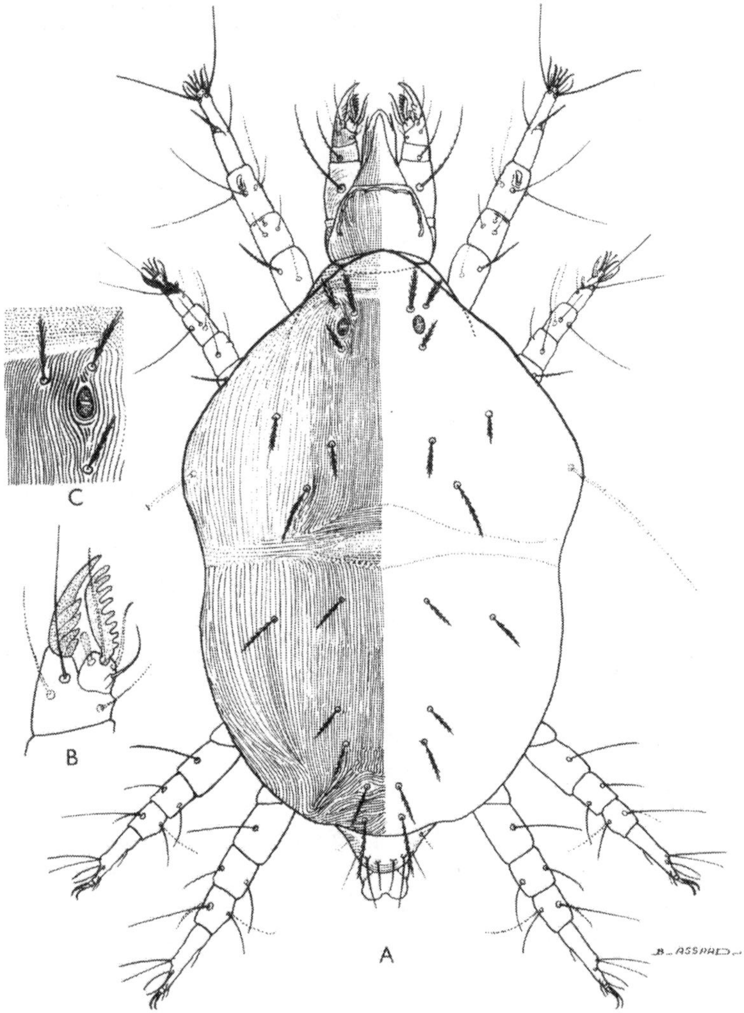

Fig. 1. *Chelachecaropsis bakeri* n.sp. A – female, dorsal; B – tarsus and tibia of palp; C – eyes and surrounding setae

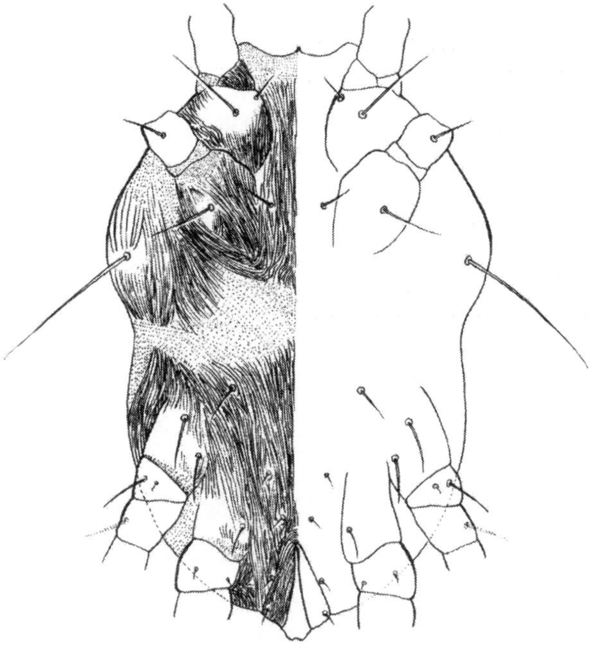

Fig. 2. *Chelachecaropsis bakeri* n.sp., female, ventral

pairs of propodosomal and six pairs of hysterosomal short fusiform pilose setae, all of subequal length except for the posterior propodosomal and posterior hysterosomal setae which are a little longer than the others; humeral seta long, smooth, about 3.5 times as long as the posterior propodosomal seta; a single pair of lens-like eyes present. Coxae II and III widely separated but by less than body width at constriction; venter of body including coxae with longitudinal striations except for the area between propodosoma and hysterosoma which has almost transverse dotted striations; 3 pairs of ventral setae, one pair of paragenital and two pairs of genital setae; coxae I, II, III, IV with 2, 1, 2, 1, setae. Three dorsal pairs of anal setae. Tarsus I with a set of 4 paraterminal and infraterminal setae, 2 addorsals arising from a terminal process of tarsus (pedicel of Summers and Price, 1970), one tapered long solenidion (quard seta missing) and one ventral seta; tarsus II with same chaetotaxy as tarsus I but solenidion II differs in situation, size and shape while the addorsal setae arise from same process with which is only 0.6 the length of solenidion 1 and has a rounded apex; tarsi III and IV with a normal set of 6 terminal, addorsal and one ventral setae. Tibia I with 4 setae and one peg, tibia II, III, IV each with 4 setae. Genu I with 2 setae and one peg, genua II, III, IV each with 2

setae. Femora with 2, 2, 1, 1; trochanters with O, O, 2, 1 setae for legs I, II, III,IV respectively.
Holotype: One female collected from rice mills at Cairo, deposited in the collectior of Acarology Investigations, Ministry of Agriculture, Dokki, Egypt.

References

Baker, E.W.: A review of the mites of the family Cheyletidae in the United States National Museum. Proc. U.S. Nat. Mus., No. 99 (3238) : 267–320, 1949.
Baker, E.W.: *Chelacheles strabismus*, a new genus and species of mite from Portugal (Acarina, Cheyletidae). Proc. Entomol. Soc. Wash., Vol. 60, No. 5, 1958.
Summers, F.M., Price, D.W.: Review of the mite family Cheyletidae. Univ. California Publ. Entomol. Vol. 61, 153 pp., 1970.

DEMOGRAPHIC PARAMETERS OF *TYROPHAGUS PUTRESCENTIAE* (SCHR.) REARED ON VARIOUS FOODS*

J. BOCZEK, T. LEGAT
Warsaw Agricultural University, SGGW, Ursynów, Warsaw, Poland

Food, like temperature and humidity, affects fecundity, duration of development, mortality, longevity and age structure of a population. According to Gołębiowska (1963) the egg-production of *T. putrescentiae* (Schr.) was increased by a fifth and its longevity by half when reared on flaked oats as against powdered milk. The development of one generation of *Acarus farris* on wheat germs lasted 16 days with a mortality of 57 %; on yellow cheese the development took 19.5 days with 2% mortality (Jakubowska, 1967).

Materials and methods

In our studies three generations of *Tyrophagus putrescentiae* were raised on fruits of *Coriandrum sativum*, two generations on dried apricots and one generation on plums. Mites for laboratory experiments originated from infested food stuff and were kept in mass cultures with rye germs as food. All experiments were conducted with mites of one population under a constant temperature of 25°C and relative humidity of 85 %. Rearing cages were kept in dessicators over saturated solutions of KCl.

Time of development and mortality were calculated from data relating to the development of 200 eggs (40 rearing cages with 5 eggs each). Fecundity and longevity were compared for 20 pairs. The data obtained were tabulized and demographic parameters were computed. Using r_m and the survival rate, age structure was also determined.

Results

Data concerning the duration of consecutive pre-imaginal stages on various foods are presented in table 1. The longest developmental period for pre-imaginal stages, 27.8 days, was observed when reared on raisins while the shortest, 14.3

* The study is supported by a grant from the U.S.D.A. under PL 480 (FG–Po–195).

Table 1. Duration of consecutive developmental stages of *Tyrophagus putrescentiae* on various foods

Food	Generation	% of time							Whole development days
		E	La	Lr	Pa	Pr	Ta	Tr	
Coriandrum	F 1	18.1	22.5	5.6	17.5	7.6	21.6	7.0	14.3
	F 2	14.8	17.9	5.6	20.9	6.2	29.7	4.7	19.4
	F 3	14.4	27.9	5.5	16.5	5.5	25.2	5.0	18.1
Almonds	F 1	21.4	25.0	3.6	19.6	5.4	18.4	6.5	20.8
	F 2	16.5	21.8	4.8	22.6	3.3	26.3	5.2	25.2
Apricots	F 1	21.6	37.9	3.5	18.9	5.9	7.3	4.9	20.4
	F 2	18.1	30.3	4.2	17.2	6.0	14.7	9.5	14.3
Plums	F 1	17.1	46.1	4.2	17.1	3.9	6.5	5.2	25.7
Raisins	F 1	14.4	34.6	4.6	22.6	1.9	18.6	3.1	27.8
Rye germs	F 1	18.1	30.2	7.3	14.1	8.0	15.4	6.9	14.9

days. was observed with the F 1 on *Coriandrum* and with the F 2 on apricots and this is shorter than when reared on rye germ, when 14.9 days are required. The most distinct differences in the percentage of time spent by consecutive generations on various foods seems to be in the active tritonymph and in the time of development of the whole generation.

The survival rate of pre-imaginal stages (table 2) was very low on plums (6.5 %) and on raisins (11.5 %). In five cases it was, however, higher than on rye germs (52.5 %). The highest rate of survival was in the F 2 on apricots (74.6 %) although on the same food the F 1 reached only 19.7 % survival.

It was observed that the longer the duration of development of the pre-imaginal stages, the higher the mortality became during development.

A comparison of survival rate and fecundity of females showed that on *Coriandrum* females lived longer and laid more eggs in the F 2 while on almonds and apricots females lived longer and laid more eggs in F 1. The data for survival during the F 1 on apricots and on plums are very similar.

Demographic parameters for different foods and generations are presented in table 3. It can be seen from the table that the lowest instrinsic rate of population increase (r_m) is on dried plums. The development of *T. putrescentiae* lasted longest on this food. All diets provided poorer food than did rye germs (r_m = 1.09). Data for the F 1

Table 2. Survival rate of pre-imaginal stages (%)

Food	Generation	E–La	La–Lr	Pa–Pr	Ta–Tr
Coriandrum	F 1	85.6	70.7	67.9	66.5
	F 2	90.0	76.0	73.6	72.5
	F 3	92.0	59.0	56.6	56.0
Almonds	F 1	95.0	71.0	65.5	63.0
	F 2	94.6	62.1	56.1	52.6
Apricots	F 1	91.5	27.5	20.3	19.7
	F 2	94.7	80.0	76.0	74.6
Plums	F 1	92.2	9.8	6.5	6.5
Raisins	F 1	95.7	17.4	12.4	11.5
Rye germs	F 1	85.9	58.2	53.3	52.5

and F 3 on *Coriandrum* are almost identical. The results for the F 1 and F 2 on almonds are also similar. The value for the F 2 on apricots is a few times higher than for F 1. It appears that mites of the F 2 have become accomodated to this new food; hence their development lasted much shorter than did that of the F 1. Net reproductive rates (Ro) indicate that the mite population during one generation increased

Table 3. Demographic parameters of *Tyrophagus putrescentiae* raised on various foods

Food	Generation	Demographic parameters			
		r_m	R_0	T	
Coriandrum	F 1	0.60	11.76	4.14	1.82
	F 3	0.60	6.51	3.36	1.82
Almonds	F 1	0.51	27.72	4.50	1.67
	F 2	0.74	31.57	5.63	2.08
Apricots	F 1	0.28	6.99	6.97	1.32
	F 2	0.98	29.82	4.06	2.67
Plums	F 1	0.09	2.10	8.61	1.09
Rye germs	F 1	1.09	99.03	5.55	2.97

Table 4. Age structure of *Tyrophagus putrescentiae* raised on various foods

Food	Generation	Age structure as % of the whole	
		pre-imaginal stages	imaginal stages
Coriandrum	F 1	86.9	13.1
	F 2	94.7	5.3
Almonds	F 1	84.5	15.5
	F 2	94.7	5.3
Apricots	F 1	92.7	7.3
	F 2	95.5	4.5
Plums	F 1	85.3	14.7
Rye germs	F 1	97.3	2.7

only 2.1 times on dried plums (F 1), almost 32 times on almonds (F 2) but as much as 99 times on rye germs.

The data dealing with the age structure of populations of *T. putrescentiae* kept on these foods are presented in table 4. The percentage of adults fluctuates distinctly from 2.7 % on rye germs to 15.5 % in the F 1 on almonds. It was usually higher on the less suitable foods like plums but it was also high on *Coriandrum* and on almonds.

Conclusions

(1) Dried fruits seem to be much less suitable food for *T. putrescentiae* than rye germs. Unsuitable food (e.g. raisins, plums) causes a distinct decrease of population, delays development of a whole generation, increases mortality of preimaginal stages, and sometimes also increases the percentage of imaginal stages in the population.

(2) Data for consecutive generations can differ very markedly. On almonds and on apricots females lived longer and laid more eggs in the F 1 than in the F 2; on *Coriandrum* females of the F 3 lived longer and laid more eggs than did females of F 1.

References

Gołębiowska, Z.: *Tyrophagus putrescentiae* (Schr.), morphology, biology and ecology. Prace Nauk. IOR 5:29–88, 1963.

Jakubowska, J.: Occurrence and ecology of mites of genus *Acarus* (Acarina). Ph. D. thesis, SGGW, Warszawa, 108 pp., 1967.

A STUDY ON THE INFLUENCE OF SOME ECOLOGICAL FACTORS ON THE HYPOPUS FORMATION OF STORED PRODUCT MITES

W. CHMIELEWSKI
Institute of Plant Protection, Laboratory of Acarology, Poznan, Poland

The phenomenon of the presence of a hypopial stage in mites has long attracted the interest of acarologists. The literature contains many papers (Schulze, 1924 a,b; Oboussier, 1939; Ushatinskaya, 1945, Sorokin 1951, etc.) in which the authors, on the basis of their own observations and experiments, have tried to define the conditions under which this stage is produced. Data concerning this problem may be also found in most papers dealing with acarid mites (Zakhvatkin, 1941; Türk & Türk, 1957; Hughes, 1961; Boczek, 1966). Opinions and suggestions of different authors concerning the causes of the hypopus formation often seem to be controversial, but predominant is the view that the presence of a hypopial stage is caused by adverse environmental conditions. Comprehensive studies on various factors causing the hypopus formation are obviously required. Recently such studies concerning nutrition as factor influencing hypopus formation in *Acarus siro* L. have been carried out by Griffith (1966). Preliminary experiments and observations have been conducted by the author of the present paper (Chmielewski, 1967), and his subsequent studies have been aimed at assessing the influence of temperature and relative humidity on hypopus formation.

Methods

Four values of temperature (15,20,25,30°C) and of the relative humidity (55, 75, 85, 95 %) were taken into account respectively. In each of 16 combinations of the experiment a vessel was placed containing the following cultured species of mites each: *Carpoglyphus lactis* (L.), *Glycyphagus domesticus* (DeGeer), *Lardoglyphus konoi* (Sasa et Asanuma) and *Acarus farris* (Oudemans), i.e. 4 vessels in each variant. *C. lactis* was cultured on baker's yeast, *L. konoi* on fish flour, *G. domesticus* and *A. farris* on wheat sprouts. A series of detailed investigations was carried out with *C. lactis* only, in 16 combinations of temperature and relative humidity, with 4 repeats in each combination and with 2, 4, 8 and 16 pairs of adult mites in each vessel. The experiments lasted 90 days and the hypopi produced were counted and removed daily or at 2–3 day intervals. The results obtained were processed by computer. Further studies are being continued.

Results

Observations of the mites *C. lactis*, *G. domesticus*, *L. konoi* and *A. farris* in mass cultures under different conditions showed great deviations in numbers of hypopi produced depending on temperature and relative humidity of the air (Table 1). Detailed investigations carried out first with *C. lactis* under identical environmental conditions confirmed the results of the initial experiments. It became evident that at some temperature (20, 25°C) the mean numbers of hypopi obtained in one repeat were considerably higher than at other temperatures (15, 30°C). The same applies to relative air humidity. It was established that there is an interaction between temperature and relative humidity during hypopus formation, i.e. some combinations of these factors proved especially favourable for the presence of this stage. The highest numbers of hypopi were produced at a temperature of 20–25°C and a relative air humidity of 75–85 % (Tables 2, 3, 4; Figs. 1, 2, 3). In order to achieve a better comparison (in a graph) of the numbers of hypopi obtained under these conditions, the Freeman–Tukey's transformation was used ($\sqrt{X} + \sqrt{X+1}$). The formation of the mean number of *C. lactis* hypopi (x) in one repeat, at different temperatures (T°C) and relative air humidities (RH) is reminiscent of Gaus' distribution. The comparison of results obtained after 30, 60 and 90 days of experiment showed that the difference in the numbers of hypopi produced was increasing in the course of time. The results obtained in the investigations on the influence of nutrition upon hypopus formation indicate that nutrition conditions also play an important role in

Table 1. Hypopus formation in 4 mite species (*C. lactis*, *G. domesticus*, *L. konoi*, *A. farris*) depending on temperature (T°C) and relative air humidity (RH %)

Number of hypopi						
T°C		RH %				
		65	75	85	95	Σ
15	Number %	598 0.78	1.769 2.31	1.414 1.85	2.598 3.39	6.380 8.33
20	Number %	2.165 2.83	8.392 10.96	10.153 13.25	6.541 8.54	27.251 35.58
25	Number %	4.736 6.18	3.993 5.21	13.736 17.93	3.437 4.49	25.902 33.82
30	Number %	2.403 3.14	3.366 4.39	7.723 10.08	3.570 4.66	17.062 22.27
Σ	Number %	9.902 12.93	17.520 22.87	33.027 43.12	16.146 21.08	76.595 100.00

Table 2. Formation of the mean number of *C. lactis* hypopi (x) depending on temperature (T°C) and relative air humidity (RH %) after 30 days of experiment

T°C		Mean number of hypopi RH %			
		65	75	85	95
15	X	0.00	1.25	1.50	0.50
	$\sqrt{X}+\sqrt{X+1}$	1.00	1.92	2.02	1.54
20	X	0.00	107.25	481.50	197.25
	$\sqrt{X}+\sqrt{X+1}$	1.00	14.61	37.68	24.40
25	X	0.00	172.25	1,052.50	120.25
	$\sqrt{X}+\sqrt{X+1}$	1.00	17.81	61.10	16.88
30	X	0.00	64.75	443.50	72.00
	$\sqrt{X}+\sqrt{X+1}$	1.00	12.39	37.89	13.98

$\sqrt{X}+\sqrt{X+1}$ – Freeman–Tukey's transformation

Table 3. Formation of the mean number of *C. lactis* hypopi (x) depending on temperature (T°C) and relative air humidity (RH %) after 60 days of experiment

T°C		Mean number of hypopi RH %			
		65	75	85	95
15	X	0.75	107.75	49.25	155.00
	$\sqrt{X}+\sqrt{X+1}$	1.68	17.12	8.51	21.41
20	X	4.00	1,100.25	1,252.50	445.75
	$\sqrt{X}+\sqrt{X+1}$	2.78	60.28	64.24	39.12
25	X	0.50	533.25	1,586.25	253.25
	$\sqrt{X}+\sqrt{X+1}$	1.53	33.40	76.55	24.42
30	X	0.00	274.00	761.00	134.25
	$\sqrt{X}+\sqrt{X+1}$	1.00	27.29	53.01	21.81

$\sqrt{X}+\sqrt{X+1}$ – Freeman-Tukey's transformation

Table 4. Formation of the mean number of *C. lactis* hypopi (x) depending on temperature (T°C) and relative air humidity (RH %) after 90 days of experiment

T°C		Mean number of hypopi RH %			
		65	75	85	95
15	X	48.25	259.00	225.00	258.00
	$\sqrt{X}+\sqrt{X+1}$	11.89	29.73	23.58	27.49
20	X	70.00	1,253.00	1,368.50	497.75
	$\sqrt{X}+\sqrt{X+1}$	9.63	66.68	70.00	42.02
25	X	8.25	577.00	1,775.75	284.00
	$\sqrt{X}+\sqrt{X+1}$	3.64	35.06	80.99	25.94
30	X	0.00	332.00	886.50	157.50
	$\sqrt{X}+\sqrt{X+1}$	1.00	30.23	57.30	23.86

$\sqrt{X}+\sqrt{X+1}$ – Freeman-Tukey's transformation

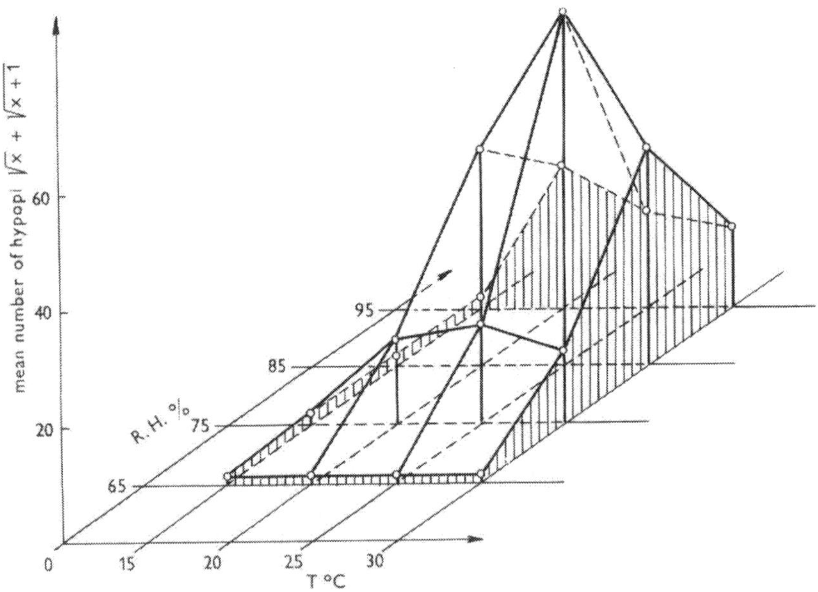

Fig. 1. Formation of the mean number of hypopi (x) of *C. lactis* depending on temperature (T°C) and relative air humidity (RAH %) during 30 days of experiment $\sqrt{x}+\sqrt{x+1}$ – transformation of Freeman-Tukey.

the formation of this stage. The number of *C. lactis* hypopi obtained on baker's yeast (1532 specimens) was about 66 times higher than that obtained on figs (23 specimens) under identical environmental conditions and during the same period (3 months).

Further detailed studies on hypopus formation are being continued, but even now it may be stated that environmental conditions play a great role in this phenomenon.

Conclusions and discussion

The present results obtained in the investigations on the hypopus formation indicate that major ecological factors (temperature, relative humidity, nutrition) exert great influence upon the hypopus formation of acarid mites. The studies have shown that an interaction between temperature and relative humidity influences this phenomenon, in other words, that some temperatures and humidities

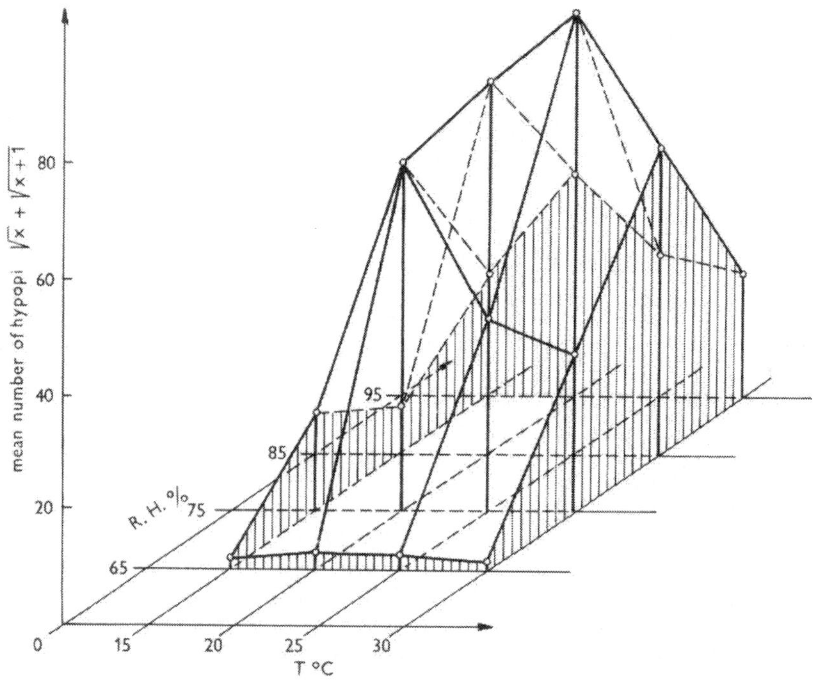

Fig. 2. Formation of the mean number of hypopi (x) of *C. lactis* depending on temperature (T°C) and relative air humidity (RAH %) during 60 days of experiment $\sqrt{x}+\sqrt{x+1}$ – transformation of Freemau-Tukey.

are specially favourable for the hypopus formation. With *C. lactis* the most favourable conditions proved to be a temperature of 20–30°C and a relative air humidity of 75–85 %. While comparing these results with those obtained in biological-ecological investigations of *C. lactis*, from which tables of the life-cycle of this species could be arranged (Chmielewski, 1970a, b), it should be noted that the most intensive hypopus formation is induced by conditions under which the factors influencing the population development (speed of nett reproduction,

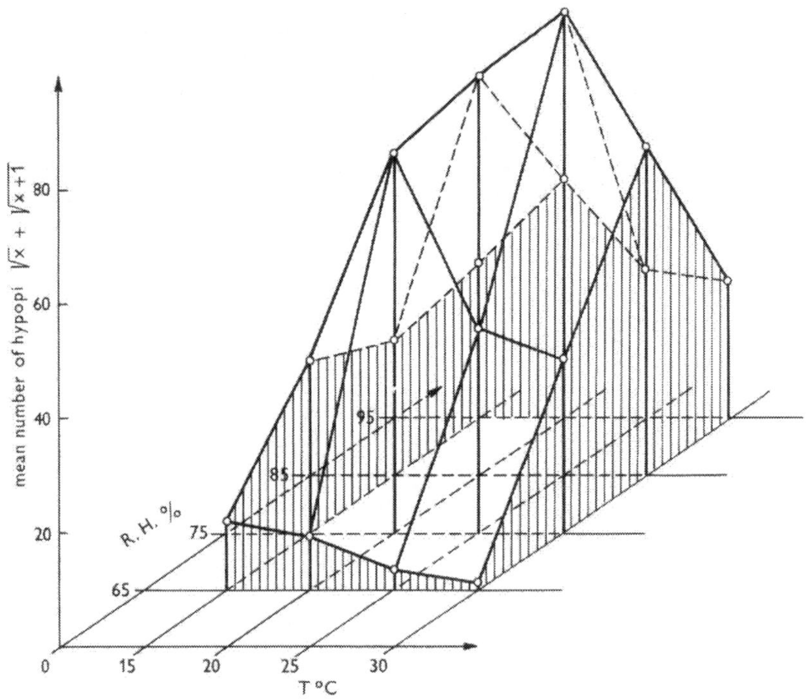

Fig. 3. Formation of the mean number of hypopi (x) of *C. lactis* depending on temperature (T°C) and relative air humidity (RAH %) during 90 days of experiment $\sqrt{x} + \sqrt{x+1}$ – transformation of Freeman-Tukey.

congenital speed of population growth, final speed of population growth and mean time of development of one generation) have shown the highest values. This means that due to the influence of favourable conditions on the mite population their numbers increase alarmingly, causing subsequent gradual deterioration of the environmental conditions. As a result of an extremely dense population and adverse environmental conditions the mites show a tendency to migrate and then often produce hypopi. In this stage the mites neither feed nor reproduce. These facts, as well as the migrational character of the hypopial stage of some mite species, cause a retardation in the increase of the population and a reduction in intraspecies competition. This mechanism of self-regulation of population density protects the population against its annihilation under environmental conditions.

References

Boczek, J.: Roztocze szkodniki roślin i produktów przechowywanych, Warszawa, 246 pp., 1966.

Chmielewski, W.: O powstawaniu i roli stadium hypopus u roztoczy. Ekol. Pol., B 13/4: 325–338, 1967.

– Rozprzestrzenianie się roztoczy przechowalnianych i wzrost ich populacji w określonych warunkach ekologicznych. Biuletyn Inst. Ochr. Roślin 47:293–305, 1970a.

– Life cycle tables of *Carpoglyphus lactis* L. (Paper presented at the 7th European Symposium of Acarology, 15–17 September 1970, Yablonna–Warsaw); 1970b. (In Russian.)

Griffiths, D.A.: Nutrition as a factor influencing hypopus formation in the *Acarus siro* species complex (Acarina, Acaridae). J. Stored Prod. Res. 1: 325–340, 1966.

Hughes, A.M.: The mites of stored food. London, 287 pp., 1961.

Oboussier, H.: Beiträge zur Biologie und Anatomie der Wohnungsmilben. Z. angew. Entomol. 26; 253–296, 1939.

Schulze, H.: Zur Kenntnis der Dauerformen (Hypopi) der Mehlmilbe *Tyroglyphus farinae* L. Zbl. Bact. Infektionkrkh. II, 9 (22–24): 536–549, 1924a.

Schulze, H.: Die Hypopi der Mehlmilbe *Tyroglyphus farinae* L. Entomol. Ber. 6 (138): 286–292, 1924b.

Sorokin, S.V.: Causes for the hypopus formation in grain mites. Zool. Z. 30:523–539, 1951. (In Russian.)

Türk, E., Türk, F.: Systematik and Ökologie der Tyroglyphiden Mitteleuropas. Zool. Inst. Friedrich-Alex. Univ. Erlangen, Akad. Verlag Leipzig, Bd. 1, Teil 1, Ab. 1, 231 pp., 1957.

Ushatinskaya, R.S.: Influence of humidity and temperature on the hypopus formation *Glycyphagus destructor* Schr. Zool. Z. 24:165–174, 1945. (In Russian.)

Zakhvatkin, A. A.: Acarid mites (Tyroglyphoidea). Fauna SSSR, Paukoobraznye, T. 6, Vyp. 1, Moskva-Leningrad, pp. 476, 1941. (In Russian.)

THE INFLUENCE OF SOME ACTIVE SUBSTANCES OF MEDICINAL HERBS ON STORED PRODUCT MITES (ACAROIDEA)

B. CZAJKOWSKA
Department of Applied Entomology, Warsaw, Agricultural University, Ursynów, Warsaw, Poland

Medicinal herbs, which are being used more and more in many countries of the world, are very often infested by stored product mites. These pests contaminate and feed on these products, thereby decreasing their value. Sasa (1964) and Sinha (1968) mention mites as pests of medicinal herbs in Japan. Boczek et al. (1961) studied the fauna and ecology of these mites in medicinal herbs.

Material and methods

29 volatile oils, 18 alcaloids and 11 glycosids, each in three doses, were added to powdered rye germs. *Acarus siro* L., *A. farris* (Oud.). *Tyrophagus putrescentiae* (Schr.), *Rhizoglyphus echinopus* (F. et R.) and *Carpoglyphus lactis* (L.) were then observed upon these mediums. Length of development and mortality and, in the case of *T. putrescentiae*, fecundity and longevity were compared for each experiment. For the last mentioned species tables were prepared and demographic parameters computed.

All rearing was done in sterilized rearing cages which were stored in dessicators with a constant humidity of 85–89 %. These were in turn kept in constant temperature chambers at 25°C.

Results

It was found that the best indicators of diet suitability are the number of eggs laid by a female during its life time, the generation developmental duration, and especially the mortality rate during development.

The development of *A. siro* proceeded succesfully on all diets with oils with the exception of that containing the highest dose of parsley oil, but dinstinct fluctuations in mortality were observed during development (Table 1.).

The survival rate of this mite was low when kept on diets with alkaloids. Addition of caffeine and quinine and the highest doses of ergotamine caused 100% mortality.

The development was completely inhibited only on the diet containing the highest dose of arbutine. The time of development of this species was usually not shortened and the mortality increased on diets with glycosids.

The mortality of *A. farris*, when reared on diets containing vegetable oil, was always higher than on the control diet: the lowest mortality relates to spruce *(Abies alba)* oil.

Table 1. Influence of some active substances of medicinal herbs on *Acarus siro* L.

Active substances	Plant species or its compounds	% of active substances	Development days	Mortality %
Oils	*Petroselinum sativum*	4.2 16.8	12 0	46 100
	Thymnus vulgaris	0.4 1.6	10 11	45 33
	Mentha piperita	0.9 3.6	10 10	68 64
Alkaloids	caffeine	2.0 8.0	0 0	100 100
	opium	1.0 4.0	12 15	56 80
	quinine	10.0 40.0	0 0	100 100
	ergotamine	0.5 2.0	11 0	32 100
Glycosids	arbutine	10.0 40.0	12 0	68 100
	rutyne	1.5 6.0	10 10	64 85
	lanatozyde	0.2 1.0	11 10	70 73
	ursulen acid	0.3 1.2	10 10	32 32
	control: rye germs		10	28

Table 2. Influence of some active substances of medicinal herbs on *Tyrophagus putrescentiae* (Schr.)

Active substances	Plant species or its compounds	% of active substances	Development days	Mortality %	Longevity days	Fecundity	R_o	r_m
Oils	*Petroselinum sativum*	8.4 16.8	0 0	100 100	28	24	–	–
	Matricaria chamomilla	0.7 2.8	11 10	12 1	35	274	122.45	1.332
	Salvia officinalis	1.4	12	21	27	115	40.01	1.036
Alkaloids	colchicine	0.6 2.4	0 0	100 100	–	–	–	–
	caffeine	2.0 8.0	16 0	82 100	–	–	–	–
	quinine	10.0 40.0	17 0	97 100	–	–	–	–
	chelidonine	0.3 1.2	11 12	11 36	39	280	97.60	1.255
Glycosids	digitoxine	0.1 0.4	16 0	38 100	28	26	3.58	0.219
	arbutyne	10.0 40.0	13 15	43 54	45	163	33.63	0.875
	ursulen acid	0.3 1.2	9 8	19 23	52	236	90.20	1.249
	rutyne	1.5 6.0	9 9	23 26	48	297	109.83	1.331
	control: rye germs		10	12	75	204	99.03	1.090

Several vegetable oils were shown to have a stimulating effect on *T. putrescentiae* (Table 2). This species did not develop on a diet of parsley oil. Great differences in mortality during development were found. Only 1 % mortality was observed on a diet of *Matricaria chamomilla* oil. Three alkaloids (colchicine, caffeine, quinine) made development of this species impossible. Mortality and length of development fluctuated very greatly: longevity was usually short but females often laid many eggs. Only digitoxine *(Digitalis purpurea)* stopped the development of this species. Other glycosids increased mortality but females laid relatively large numbers of eggs. On a diet with rutine *(Ruta graveolens)* the population of this species increased 110 times during a single generation.

All diets with oils enabled the development of *R. echinopus* (Table 3). Many of them caused a mortality which was lower than that of the control diet.

Table 3. Influence of some active substances of medicinal herbs on *Rhizoglyphus echinopus* (F. et R.)

Active substances	Plant species or its compounds	% of active substances	Development days	mortality %
Oils	*Petroselinum sativum*	4.2	12	31
		16.8	14	28
	Thymus vulgaris	0.4	12	16
		1.6	12	17
	Matricaria chamomilla	0.7	12	31
		2.8	13	24
	Mentha piperita	0.9	11	7
		3.6	11	25
Alkaloids	quinine	10.0	0	100
		40.0	0	100
	berberine	10.0	16	64
		40.0	18	67
	colchicine	0.6	12	43
		2.4	11	62
	sparteine	1.0	14	51
		4.0	19	95
	caffeine	2.0	15	77
		8.0	23	80
Glycosids	arbutyne	10.0	11	22
		40.0	14	26
	ursulen acid	0.3	12	9
		1.2	12	15
	rutyne	1.5	13	28
		6.0	12	30
	control: rye germs		11	27

The alkaloids studied also affected the development of this species positively; only quinine made its development impossible. Several alkaloids shortened the time of development of one generation and lowered the mortality rate but some of them, like berberine *(Berberis vulgaris)*, colchicine *(Colchicum autumnale)*, higher doses of sparteine *(Sarothamnus scoparius)* and caffeine, decreased the survival rate of this species. All glycosids showed a strong inhibitory effect on the development of this root mite.

The development of *C. lactis* was inhibited both on diets containing oils and alkaloids. This effect was greatest in the case of oils of *Archangelica officinalis* and *Citrus lemon*.

Conclusions

1. Alkaloids showed the strongest inhibitory effect upon *C. lactis*, *A. siro* and *T. putrescentiae* respectively. Only the higher doses of quinine made the development of all studied species impossible. Some alkaloids (e.g. scopolamine) sometimes produced a stimulatory effect.

2. Glycosids also strongly inhibited the development of *C. lactis*. Higher doses of some glycosids also inhibited the reproduction of *A. siro* and of *T. putrescentiae*.

3. Some compounds known to be very toxic to man and higher animals (alkaloids: scopolamine, opium; glycosids: digitoxine) appeared to have no harmful effect on these mites.

4. Oils contained in the diet often inhibited the development of *A. farris* and sometimes also of *C. lactis* and *A. siro*. The remaining two species studied usually developed better on these foods with oils than upon the control diet. Many vegetable oils stimulate the development of these mites.

5. These results suggest that conditions for the development of mites, especially of *A. siro* and *A. farris*, in stored medicinal herbs, are not optimal. However, in practice these mites occur commonly in these products. This could be due to the fact that these pests, feeding and developing in medicinal herbs for many generations, have adapted themselves to these foods, thereby increasing fecundity and decreasing mortality.

References

Boczek, J., Filipek, P., Cołebiowska, Z.: Characteristic of mite fauna of stored medicinal herbs of Poland. Prace Nauk. IOR 3:1–64, 1961.

Sasa, M.: Special problems of mites in stored food and drugs in Japan. Proc. I. Int. Congr. Acarol., Acarologia 6:380–391, 1964.

Sinha, R.N.: Climate and potential range of distribution of stored product mites in Japan. J. Econ. Entomol. 61:70–75, 1968.

THE FAMILY ACARIDAE OF THE CAUCASIAN FAUNA

G. SH. KADZHAYA
Research Institute of Zoology, Georgian Academy of Sciences, USSR

The Caucasus is a part of the USSR situated between the Black and Caspian seas. This territory includes the Republics of Azerbaidzhan, Armenia, Georgia as well as a section of RSFSR in its northern region. Due to an extraordinary variety of natural conditions and diversity of climatic zones, the Caucasus is often compared to a gigantic polyhygrothermostat providing extraordinary conditions for the formation of a rich flora and fauna.

Collections for our studies on the family Acaridae were made in almost all landscape zones of the Caucasus – from humid zones to sub-Alpine subtropics regions. More than 40 species were found, of which 5 genera and about 15 species are new for science.

From faunistic and systematic aspect not only the new species deserve attention, but also those which were previously known from this territory, occurring in several or many populations. Major species are: *Acarus siro*, *A. farris*, *Aleuroglyphus ovatus*, *Tyrophagus putrescentiae*, *T. perniciosus*, *T. silvester*, *T. mixtus*, *T. molitor*, *T. longior*, *Mycetoglyphus fungivorus*, *Forcellinia diamessa*, *F. wasmanni*, *Rhizoglyphus echinopus*, *Caloglyphus rodionovi*, *C. sphaerogaster* and *Mesorhizoglyphus colchicus*.

At present there are very controversial views among scientists concerning the polytypic nature of species as regards the family Acaridae, but the geographical variability of its species has not yet been completely studied. It is our opinion that the polytypic nature and geographical variability of a species can be studied only on the basis of a precise knowledge of the population level. In order to obtain a complete as well as an overall view of all populations mentioned above, we used a statistical method which was also applied when the populations of the same species were being compared with one another. In order to characterize a population, about 15–25 morphological features of ♂♂ and ♀♀ were measured and subsequently mean arithmetic, mean quadratic deviations, and coefficients of a variation for each feature were deduced. Some features of a population were figured. In comparing the individual populations it became clear which of the differences between them were reliable or not.

In order to obtain a picture of the chorological and biotopic variability of a species, various generations of populations and various micropopulations were also statistically assessed.

The results obtained in comparing the species populations are as follows: Most populations of the same species overlap one another and it is often impossible to draw a boundary line between them. Not only the mean values of size, but also the indices of other features are constantly changing. It becomes evident that differences between generations are often as great as those between different populations.

There are rather great differences between extreme populations; however, they seem to be connected by a number of intermediate populations. Such a variability, when the populations gradually overlap one another, thus forming a constant chain of variability, is called a normal individual variability.

There are also various differences between populations. If a population differs statistically in most of its features from less than a third of other populations, such a difference is designated to be of degree I; if a population differs from more than one third (up to 2/3), the difference is of degree II; if a population differs from more than two thirds of other populations, the difference is regarded to be of degree III.

According to the normal variability and to the degree of difference between populations, all 16 species mentioned above are divided into 3 groups.

The first group includes species whose normal variability is negligible, and whose populations basically differ in degree I (*Acarus siro, Aleuroglyphus ovatus, Tyrophagus silvester, Mycetoglyphus fungivorus, Mesorhizoglyphus colchicus).*

The second group includes species whose normal variability fluctuates within a large range and whose populations differ basically from others in degree II *(Acarus farris, Rhizoglyphus echinopus, Tyrophagus molitor, Forcellinia).*

The third group includes species whose normal variability, like that of the species of the first group, is rather low; the difference between populations on the whole does not exceed degree I. However, single populations exceed the limits of a normal variability and they differ from other populations not in degree I, but in III. Such cases are regarded as resulting from a different reaction to the environment as compared with other populations, and the populations themselves are subdivided into subspecies *(Tyrophagus putrescentiae, T. perniciosus* etc.).

In a geographical respect the most constant features are: relative lengths of chelicerae, tarsi and preceding segments, anus, distance between setae, form of penis, etc. More variable are: relative lengths of setae p_1 and p_2 (♂♂), ad_2 (♀♀), d_1 d_2, etc.

In different parts of the Caucasus are several regions which are noted for the discoveries of large numbers of new genera, species and subspecies. Future studies may show that such regions are important centres for the formation of new forms.

THE ENERGY BUDGET OF *TYROPHAGUS PUTRESCENTIAE* (SCHR.) (ACARIDAE)*

Z. A. STĘPIEN, W. GOSZCZYNSKI, J. BOCZEK
Warsaw Agricultural University, SGGW, Ursynów, Warsaw, Poland

There are very sparse data dealing with the consumption rate and efficiency of conversion of consumed energy in acarid mites. Energy consumption by mites of the genus *Acarus* was measured by Boczek (1957), and Jakubowska (1967). Similar data were also presented by Pimentel et al. (1960) and Solomon (1946). The energy budget of *Rhizoglyphus echinopus* was given by Stępień (1970). All these data indicate that acarid mites are very voracious animals. The present study deals with *Tyrophagus putrescentiae*, a very important species in food storage.

An energy budget is an equation expressed in identical energy units (e.g. calories) per unit of time in which the amount of energy flowing in as food to the organism is expressed as a sum of the elements representing further uses of this energy.

Symbols and equations of an energy budget are:

$$C = P+R+U+F = D+F \quad (1),$$
$$D = P+R+U \quad (2),$$
$$A = P+R \quad (3),$$ where:

C= consumption, food intake; D = digestion, part of the food that is digested and absorbed; P = production (body, exuvia, reproductive products, etc.); R = respiration, cost of maintenance; U = urinary and other "waste" part of digestion; F = unabsorbed part of food consumed, faeces; A = assimilation, i.e., the sum of production and respiration.

The efficiency of an organism as an energy convertor is expressed by the following indices:

$$U^{-1} = \frac{A}{C} \cdot 100 \qquad (4)$$ – the index of feeding efficiency;

the rate of food assimilation, as a percentage of the energy consumed;

$$K_1 = \frac{P}{C} \cdot 100 \qquad (5)$$ – the gross production efficiency

* The study is supported by a grant from the U.S.D.A. under PL 480 (FG–Po–195).

index: the energy used for growth, as a percentage of the energy consumed;

$$K_2 = \frac{P}{P+R} = \frac{P}{A} \cdot 100 \qquad (6)$$ – the net production efficiency

index: the energy used for growth as a percentage of the energy assimilated.

The energy budget and the efficiency indices may be calculated in two different ways: (I) all parameters of the budget and the efficiency are calculated for a short period of an animal's life (e.g. day, week) – such a budget is said to be instantaneous; (II) – all the elements of the budget and the efficiency are calculated for the whole period from the beginning of life of the specimen (or the average specimen) to every moment of its life – such a budget is said to be a cumulative energy budget (first proposed by Klekowski, Prus and Żyromska–Rudzka, 1967).

An instantaneous energy budget and efficiences illustrate well the physiological properties of a given species at an individual level, and also the changes in these properties in relation to developmental stage, physiological state and environmental conditions.

A cumulative energy budget and efficiences allow to assess for each moment of the life of a specimen (or an average specimen) the amount of energy that will be released through food consumption, dissipated as heat (respired) and degraded as FU, in order to enable the animal to attain a certain size of developmental stage.

Materials and methods

Mass cultures of mites were maintained in large rearing cages made of plaster of Paris, or in the device designed by Woodring (1968), with rye germs (of a known calorific value) as food. The latter device allows the gathering of large amounts of clean mites. Eggs and resting stages were collected by means of finely meshed nets.

At each successive stage of development all developmental stages of mites were weighed on a quartz microbalance. Dry weight was obtained after mites were dried at 105°C for 3 hours.

The average daily consumption rates were calculated for consecutive developmental stages and for the whole life-span of the mites by means of a gravimetric method.

Oxygen consumption of mites was measured with a Cartesian diver microrespirometer (Klekowski, 1967). The calorific values of the food, the mites and their eggs were established in Phillipson's bomb calorimeter. The calorific value of exuvia was accepted as 4.98 Kcal/g.

Results

The calorific value of active stages of *T. putrescentiae* is 4.98 cal/mg, and of the rye germs – 5.05 cal/mg. The consumption rate of all stages is given in Table 1. As can be seen in this table, one specimen consumes almost 600 cal. 10^{-3} which means ± 118 μg of the rye germs during its life. One adult "average specimen" consumes on average 8.42 cal. 10^{-3} per day.

The instantaneous energy budget (per day) for the "average specimen" of *T. putrescentiae* is presented in Table 2. This table reveals that the index of food

Table 1. Consumption of various stages of *Tyrophagus putrescentiae*

Stage	Duration of development days	Cal. 10^{-3}	μg
La – Pa	4.5	6.1	1.2
Pa – Ta	2.1	9.2	1.8
Ta – A	2.3	28.7	5.7
A	65.5	551.6	109.2
Total	74.4	595.7	117.9

Table 2. Instantaneous (daily) energy budget of *Tyrophagus putrescentiae*

Stage	Cal. 10^{-3}/day				%		
	Consumption C	Production P	Respiration R	Assimilation A	U^{-1}	K_1	K_2
La	1.4	0.8	0.52	1.31	96.7	58.2	60.5
Pa	4.4	1.6	0.78	2.38	54.5	36.6	67.1
Ta	12.5	9.6	1.03	10.64	85.2	76.9	90.3
A	8.4	4.7	1.40	6.08	72.2	55.6	77.0

assimilation U^{-1} is very high in the larva and in the active tritonymph stage; assimilation by adult mites is also quite high (72.25 %). The index of net production efficiency K_2 fluctuates from 90.3 % in the active tritonymph to 60.49 % in the larva.

The elements of cumulative energy budget are presented in Table 3 and in Fig. 1. This table shows that the index of cumulative food assimilation (U_c^{-1}) during development drops from 96.16 % in the larva to 67.01 % in the protonymph, then increases again in the tritonymph (77.12 %) and in the adult (71.96 %). The indices of cumulative net and gross production efficiency (K_{1c}, K_{2c}) are also extremely high.

Table 3. Elements of cumulative energy budget of *Tyrophagus putrescentiae* (Schr.)

Stage	Time days	Cal. 10^{-3}				%		
		Consumption C_c	Production P_c	Respiration R_c	Assimilation A_c	U_c^{-1}	K_{1c}	K_{2c}
E	2.7	–	–	–	–	–	–	–
La	7.2	6.14	3.57	2.33	5.90	96.16	58.17	60.49
Lr	8.2	6.14	3.27	2.63	5.90	96.16	53.25	55.37
Pa	10.3	15.32	5.98	4.28	10.26	67.01	39.06	58.29
Pr	11.4	15.32	5.49	4.77	10.26	67.01	35.85	53.51
Ta	13.7	44.06	26.83	7.14	33.97	77.12	60.90	78.97
Tr	14.7	44.06	25.98	7.99	33.97	77.12	58.97	76.47
A (♀)	80.2	595.67	328.78	99.87	428.65	71.96	55.20	76.70

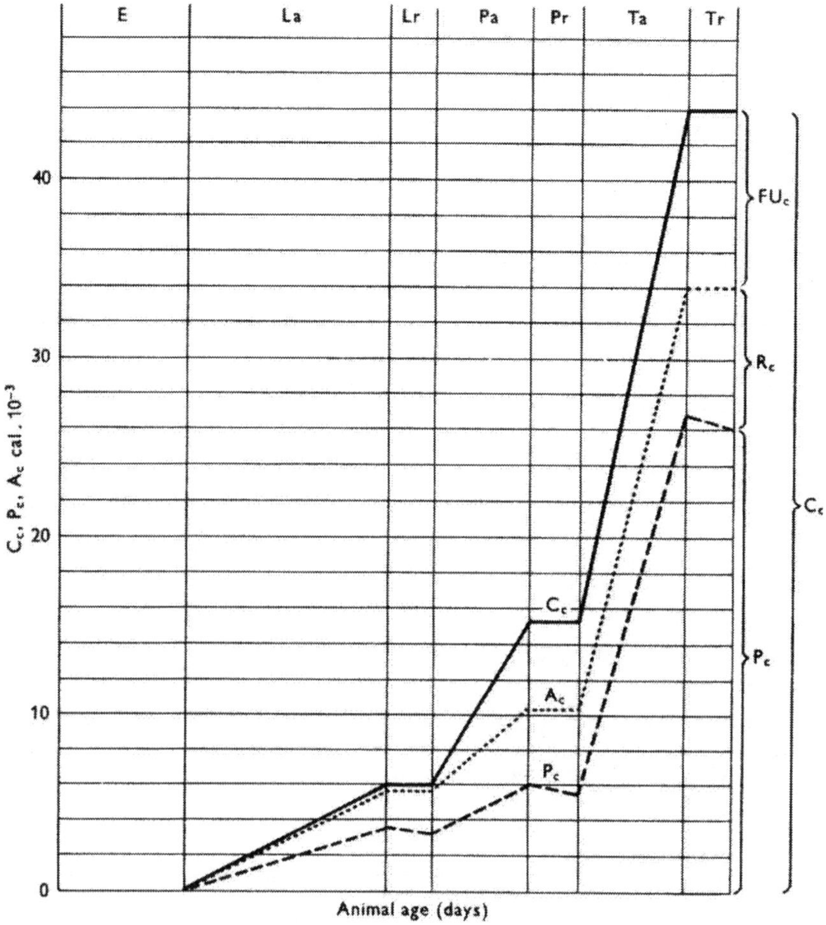

Fig. 1. Elements of cumulative energy budget for pre-imaginal stages of *Tyrophagus putrescentiae* (Schr.).

Conclusions

(1) The consumption rate of all pre-imaginal stages of *Tyrophagus putrescentiae* is 1.5 times lower and of the adult 2.5 times lower than the rates for *Rhizoglyphus echinopus*. *T. putrescentiae* is, however, a smaller animal than *R. echinopus*. The highest consumption rate takes place in the tritonymph stage.

(2) The indices of instantaneous and of cumulative food assimilation (U^{-1}, U_c^{-1}) in *T. putrescentiae* are very high, higher than in *R. echinopus*. Also the indices of instantaneous and cumulative gross and net production efficiency (K_1, K_{1C} and K_2, K_{2C}) are very high.

(3) In *T. putrescentiae* from about 40 % to about 61 % of food consumed is converted into body and egg production, and from about 58 % to about 79 % of assimilated energy is incorporated into body and eggs. Both these acarid mite species utilize food consumed (and especially assimilated) very efficiently for body and egg production.

(4) In comparison with other mites (e.g. *Oribatei* – Engelmann, 1967; *Steganacarus magnus* – Berthet, 1964) *T. putrescentiae* and *R. echinopus* act very efficiently as "energy carriers" to the consumers of the next trophic level (predators).

References

Berthet, P.: L'activité des *Oribatides (Acari, Oribatei)* d'une chenaie. Mém. Inst. Roy. Soc. Nat. 152:152, 1964.

Boczek, J.: The flour mite, *(Tyroglyphus farinae)*. Morphology, biology and ecology. Roczn. Nauk Roln. A, 75:559–644, 1957.

Engelmann, M.D.: The role of soil arthropods in the energetics of an old field community. Ecol. Monog. 31:221–238, 1967.

Jakubowska, J.: The occurrence and ecology of mites of genus *Acarus (Acarina)*. Ph.D thesis, 108 pp., 1967.

Klekowski, R.Z.: Cartesian diver technique for microrespirometry. Int. Meeting on "Methods of assessment of secondary production in freshwaters", Praha, 26 pp., 1967.

Klekowski, R.Z., Prus, T., Żyromska–Rudzka, H.: Elements of energy budget of *Tribolium castaneum* (Hbst) in its developmental cycle. In: K. Petrusewicz (ed). Secondary productivity of terrestrial ecosystem, 859–879, PWN, Warszawa, Kraków, 1967.

Pimentel, D., Ramsay, M.W., Streams, F.A.: Rearing tyroglyphid mites on *Neurospora*. Ann. Entomol. Soc. Am. 53:549, 1960.

Solomon, M.E.: Tyroglyphid mites in stored products. Nature and amount of damage to wheat. Ann. Appl. Biol. 33:280–289, 1946.

Stępień, Z.A.: Energy budget of *Rhizoglyphus echinopus (Acarina, Acaridae)* during its development. Ph. D. thesis, 103 pp., 1970.

Woodring, J.P.: An automatic collecting device for tyroglyphid *(Acaridae)* mites. Ann. Entomol. Soc. Am. 61:1030–1031, 1968.

Proceedings of the 3rd International Congress of Acarology, Prague, 1971

FAUNA OF ACARID MITES OF THE FAR EAST

V. N. TAREEV
Institute of Agriculture, Ussuriysk, USSR

Although there has been an increased awareness of the importance of acarid mites (Acaroidea), their fauna has been studied rather incompletely. Vast territories have not been investigated yet, e.g. the Far East which is characterized by peculiar natural conditions. The main subjects and tasks of the present paper therefore have been: to determine the species composition of acarid mites; to study the type of their distribution in different localities; to discover economically harmful species and routes of their penetration into storehouses; to define some natural laws as reflected in the geographical distribution of acarid mites under conditions of the Far East.

In order to collect the mite material under synanthropic conditions the grain and seeds of different crops and the products of their processing, the root-crops, the dried fruits, cheeses, yeast, egg powder and fish dried in the sun etc. as well as hay, straw, litter in living quarters and store houses, spilled grain and the soil under store houses, bee-hives etc. were investigated.

Under natural conditions collections were made from nests and burrows of birds and rodents; from tree hollows and decaying wood; from outflowing tree juices and bark saturated by them; from mosses and lichens; from a large number of insects (primarily Hymenoptera and their nests, Coleoptera etc.) and other biotopes.

Collections and laboratory treatment of material were carried out by the method commonly used in the studies of the given group of mites, with additional detailed investigations according to specific requirements.

After analysing the collected material an abundant and diverse fauna of acarid mites, including 62 species, was found. They all belonged to 4 families: Acaridae (34 species); Saproglyphidae (5 species); Pyroglyphidae (1 species); Glycyphagidae (22 species). Of the 18 species which are new for fauna of the USSR, 11 were new for science.

The various habitats in which the acarid mites occur may be grouped into two categories: synanthropic (economic) and field (natural) habitats. The habitats of the first category may be again subdivided into stored materials of plant and animal origin, and materials accumulating due to economic activities of man, but not subject to storage. The habitats of the second category may be divided into 4 groups: materials produced due to the influence of various abiotic factors; produced due to animal activities, produced in the field by man; different animal and plant organisms. This classification of habitats is of course of a relative character.

About half the acarid mites found occurred under synanthropic conditions, damaging stored forage crops and food products. 20 species infested grain and seeds; 10-the products of their processing; 13-the root-crops, 5 dried fruits; 17 – litter in habitable premises and storehouses, spilled grain and soil under storehouses; 14 – bee-hives.

As for natural habitats, 31 species of acarid mites were found in nests of rodents (27 in nests of flying squirrels, 14 in nests of chipmunks, 9 in nests of mice and voles), 18 in nests of birds; 13 in tree hollows; 8 in decaying wood; 16 on insects and in their nests; 15 in soil.

An analysis of the distribution of stored product mites, taking into account the data on the occurrence and abundance of particular species, makes it possible to outline a certain geographical character in the synanthropic distribution of acarids in the USSR. In this connection it is expedient to divide them into 7 groups: 1. very common species characterized by cosmopolitan distribution; 2. species of wide, but sporadic distribution, but not of mass occurrence; 3. species of moderate climate preferring northern latitudes; 4. species of moderate climate showing propensity for southern latitudes; 5. species of distinctly southern orientation; 6. the Far-East species; 7. species known from single finds. The fauna of the Far East is characterized by representatives of six of the above mentioned ecological-geographical groups (group 5 excepted). This fact emphasizes the richness and peculiarities in the formation of the Far-East faunistic complexes. There are also peculiarities in the distribution of representatives of genera such as *Gohieria, Tyrolichus, Acotyledon, Caloglyphus, Glycyphagus, Diamesoglyphus* etc.

The distribution of different species of stored product mites in nature differs in scale and character according to different parts of their distribution area. Some species are to be found both in store houses and in nature; others are widespread under synanthropic conditions and relatively rarely exceeding the limits of the synanthropic area; the last but not least occur now and then in store houses.

On the whole, the species composition of acarid mites of the Far East fauna is characterized by a high level of endemism, greatly differing from that of Europe, North America and other zones of their general area of distribution.

Unfortunately, the incomplete knowledge of the fauna of acarid mites occurring in regions adjoining to the territory of the Soviet Far East does not allow a full assessment of the peculiarities of the acarid fauna in this region which is of great interest from a zoogeographic point of view.

Proceedings of the 3rd International Congress of Acarology, Prague, 1971

THE HYPOPUS AND ITS MAIN TYPES

V. I. VOLGIN
Zoological Institute, USSR Academy of Sciences, Leningrad, USSR

The life cycle of primitive Acariformes is believed to have consisted of 7 stages: egg, two larval stages and three nymphal stages, adult. Evolution leads to reduction of the cycle, mainly due to the embryonization of the first larva, named "praelarva", and to the modification or exclusion of the deutonymph, the function of which in primitive forms is growth only. The deutonymph of the free-living Acaridiae has been most affected by changes. In this group this stage differs greatly in morphology and function from other nymphal stages, lacking mouth parts and usually being provided with different kinds of fixators on the ventral side of the opistosome, on leg segments etc. Such a deutonymph, named "hypopus", needs no nutrition, does not grow, is highly stable and serves the conservation and dissemination of the species. Unfortunately, no forms transitional from the primitive deutonymph to the specialized hypopus have survived.

A hypopus is known in the majority of species of Acaridae (Tyroglyphidae), Saproglyphidae, Glycyphagidae, Anoetidae and Canestriniidae. However, there is no hypopus in the cycle of a number of widely spread and common species of Acaridae pertaining to the genera *Tyrophagus*, *Tyrolichus* and *Aleuroglyphus*. In these species the development proceeds according to the simple binymphal scheme, omitting the deutonymph. Hypopi are not found or identified as yet in a number of species of the rather heterogenous family Glycyphagidae, especially in nidicole species, but their discovery will in many cases be highly probable.

Hypopi are very diversified, but on the basis of their structure and biological peculiarities they can be placed in natural groups. The first classification, at the time sufficiently clear and complete, was proposed by Zakhvatkin (1941). He distinguished two main types of hypopi: the resting and the disseminating one. To the first type were assigned the immobile hypopi of the genera *Glycyphagus*, *Lepidoglyphus* and *Chaetodactylus* as well as the hypopi of limited mobility encountered in the genus *Acarus*. The second type of hypopus forms a large and varied category and was divided into two subtypes: a) entomochoral which included hypopi of Acaridae, Saproglyphidae, Anoetidae, Canestriniidae and some Glycyphagidae and b) teriochoral with hypopi of the subfamily Labidophorinae (Glvcyphagidae). In recent years, due to the work of A. Fain, our knowledge of the hypopal stage has been greatly increased. Carrying out intensive studies of parasitic mites, including Acaridiae, he described a number of interesting and original hypopi from hair follicles of Mammalia and proposed his own scheme of classification. He distinguished 6 different types of hypopi, naming them acaridal, anoetidal, labidophoridal, rodentopidal, echimyopidal and hypodectidal. Most of these types appear to be new, compared with the classification of Zakhvatkin, the resting hypopus alone is here absent. Among the new types, the rodentopidal and echimyopidal ones are worth special attention. Hypopi of these types live mostly in hair follicles and are provided with a variety of specific peculiarities. We do not regard the anoetidal type as a separate one – it is a mere variant of the entomochoral type according to the scheme of A.A. Zakhvatkin. As for the deutonymph of *Hypodectes propus* (Nitzsch) of the family Hypoderidae, distinguished by A. Fain as a separate type, this cannot be considered a hypopus at all, regardless of its superficial resemblance. This deutonymph is the only and strictly obligatory prae-imaginal stage in the cycle of *Hypodectes;* living under the skin of young pigeons it feeds directly through its integument and grows intensively, contradicting the hypopal concept.

A critical analysis of the schemes of classification proposed by A.A. Zakhvatkin and A. Fain and the study of a large factual material lead me to the conclusion that all known varieties of hypopi can be divided into 4 basic and equivalent morphobiological groups or types: acaroidal (entomochoral), labidophoridal (teriochoral), follicular (rodentopidal and echimyopidal) and resting.

The acaroidal type is most widely spread. It is common within the families Acaridae, Saproglyphidae, Anoetidae and Canestriniidae and characterized by the presence of more or less developed sucker plates on the ventral side of the opistosome which are used for the attachment to small Arthropoda, mainly insects. The degree of development of the sucker plate and suckers is not the same in all above mentioned families. Suckers are very weakly developed in *Acotyledon paradoxa* Oudms. and in some forms of the genus *Acarus;* large plates and well developed suckers are found in hypopi of the genera *Caloglyphus, Rodionovia, Carabidobius, Horstia* and some others. Species of the genera *Sennertia* and *Chaetodactylus* possess, in addition to rather weakly developed sucker plates, large, spirally curved claws on tarsi I–III. Hypopi appear as soon as conditions become unfavourable and usually have no direct relationship with the actual cycle of the insect or other arthropod used for dissemination. A closer connection is observed between hypopi of the genus *Caloglyphus* and beetles of the family Scarabaeidae.

The labidophoridal hypopus is characterized by the presence on the opistosome of a very peculiar apparatus for fixation to mammalian hair. It consists of a pair of movable valves with a more or less ribbed surface, and of two pairs of transversely ribbed clutches situated under them. This apparatus forms a kind of vice in which the hair is grasped firmly. The degree of development of the fixating apparatus is not the same in the different groups of this type. It is most powerful in species of *Labidophorus, Dermacarus* and some others. The valves and clutches are most weakly developed in *Metalabidophorus* and *Microlabidophorus*. The latter has thickened and widened serrated chaetae on the tibiae and genual joints of the legs. The labidophoridal type of hypopus is characteristic for the subfamily Labidophorinae (Glycyphagidae). The entire life cycle of the mites of this subfamily is closely connected with mammals. In the majority of species a definite host-specifity is observed. The appearance of hypopi in populations is probably seasonally determined.

The follicular hypopus is also common in the family Glycyphagidae. Its body lacks an apparatus for fixation, but the tibiae and tarsi of the hind legs are provided with powerful thorns and widened chaetae. A primitive form of this hypopus (*Balogbella melis* Mah. = *Melesodectes auricularius* Fain et Luk.) was found in secretions of a badger's ear, but usually the follicular hypopus lives in hair follicles, leaving them before moulting become to a tritonymph. This follicular type is most widely spread in the subfamily Ctenoglyphinae (= Rodentopinae). Mature mites and other mobile stages inhabit burrows and nests of mammals. Their biology has not been studied. The appearance of hypopi will doubtless be connected with the host's cycle of development.

The resting hypopus is notable for a considerable reduction of legs, fixing and sensory organs. This type is immobile and remains enveloped in the densified exuvia of the protonymph. It is found in species of the genera *Glycyphagus*, *Lepidoglyphus* and *Chaetodactylus* and can probably be found in some related groups of mites. The resting hypopus is much less susceptible to unfavourable conditions than hypopi of other groups.

References

Zakhvatkin, A. A.: Acarid mites (Tyroglyphoidea). Fauna SSSR, Paukoobraznye, T. 6, vyp. 1, Moskva – Leningrad, pp. 476, 1941. (In Russian).

ORIENTATION OF TYROPHAGUS PUTRESCENTIAE (SCHRANK) TOWARDS OLFACTORY STIMULI

E. ŽĎÁRKOVÁ
Research Institute of Food Industry, Department of Entomology, Prague, Czechoslovakia

Hitherto the orientation of stored food mites has been studied from three different aspects: a) their ability to choose an optimal substrate for development, b) their preference for certain nutrients, c) their preference for favorable relative humidities of the air.

Żyromska-Rudzka (1959) studied the colonization of three different kinds of fruit jams by an overcrowded population of *Carpoglyphus lactis* (L.). In experiments lasting four and half months the mites left the overcrowded substrate, choosing the only one which was suitable for their development.

During studies of the food preference of *Tyrophagus putrescentiae* (Schrank) the present author (Žďárková, 1969) found that the mites can select the more suitable of two offered substrates within 48 hours. Knülle (1961) studied the orientation of *Acarus siro* L. towards different air humidities. The mites preferred humidities ranging between 75–85 % to higher or lower ones. Within this range of preference the sensitivity of mites to different humidities is low, whereas that in humidities outside that range is extremely high, when the mites can distinguish as little as 0.1 % difference in relative humidity.

In all described tests the mites had an opportunity to recognize the offered possibilities and than choose one of them. The present paper gives data on the immediate responses of mites towards different humidities and odour stimuli. The results indicate that the mites can distinguish immediately between different olfactory stimuli which may play an active role in selecting favourable food for development.

Material and methods

Tyrophagus putrescentiae (Schrank) was used for all experiments. The mites were reared under constant climatic conditions (26°C, 85 % RH) on hard cheese. (The culture originated from a sample of mixed cereals from the university silo in Urbana, Ill., U.S.A. Our specimens differed in ratio d_2d_1 from the specimens of *Tyrophagus putrescentiae* given by Johnston (1965). The ratio was 2/1 compared with the other one of 2. 5–3/1).

A modified olfactometer according to Lener (1970) was used. The olfactometer consisted of a plastic testing chamber 2.5 cm in diameter and 20 cm long with a single hole (1 cm in diameter) in the middle of the chamber and two narrower tubes fixed by a tape on each side of the testing chamber. Their inner ends were covered with black lens paper. Each of the outer ends was plugged with a rubber stopper holding the sensor for measuring the humidity and pierced by a hypodermic needle for air supply. The point of the needle was directed to the walls so that the incoming air could not strike directly against the sensor. The sensors were connected with the humidity indicator.* The amount of air was checked by letting it bubble through water. The humidity was controled by calcium chloride or moist cellulose cotton.

One to two hundred mites, irrespective of age and developmental stage, were brushed out of rearing cells on to a small piece of weighing paper, which was then placed into the central hole of the testing chamber. The hole was than closed with a rubber stopper with an openning covered by nylon cloth. The conditioned air was running through the apparatus approximately one hour before the mites were introduced in order to equilibrate the humidities on both sides of the testing chamber. After the mites had been inserted, the testing chamber was covered with a non-transparent black cloth. The behaviour of mites was checked at five minutes intervals. The air could leave the test chamber only through the central hole where the conditioned air from both arms was mixed. Therefore the chamber was divided into four equal units and only those mites which entered the outer units on either side were considered to have responded affirmatively. Each test lasted 30 minutes and was repeated twice if both replications coincided, otherwise three or four times.

Two series of experiments were performed. In the first series the orientation responses to different relative humidities of the air were tested. In the second series olfactory responses to various natural and artificial materials were examined. The results were statistically evaluated.

Results

I. Orientation of mites towards different relative humidities (Table 1). The mites were enabled to choose humidities ranging from 14.3 % to 89 %. Fifty different combinations of relative humidities were tested. In extremely dry (under 22 %) or humid (above 78 %) air the mites proved to be unable to make a choice, no matter how far the two compared humidities were apart. If one of the humidities was higher than 22 % or lower than 78 %, the mites always preferred the higher one. Within this range they were able to distinguish as little difference as 1 % relative humidity. The behaviour of mites in very dry (under 27 %) and very humid air (89 %) was similar. They

---

* Manufactured by Hygrodynamics, Inc. Silver Springs, Md., U.S.A.

Table 1. Orientation of mites towards different relative humidities

Test n.	RH in %		Behaviour at start	Time in minutes 5		10		15		20		25		30		Signification
	A	B		A	B	A	B	A	B	A	B	A	B	A	B	
1a	21.5	14.3	A/o/B	1	0	3	3	8	7	10	7	10	7	10	7	–
1b	21.1	17.1	A/o/B	0	0	4	4	11	10	12	16	16	20	16	20	–
2a	20.4	18.3	A/o/B	did not leave the paper												
3a	26.8	22.5	A/o/B	0	0	7	0	20	0	25	10	25	10	25	10	+
3b	27	22.8	A/o/B	0	0	10	2	15	6	28	7	30	9	30	9	+
4a	27.3	24.6	A/o/B	3	0	4	0	10	0	10	0	20	0	20	0	+
4b	27.5	24.5	A/o/B	0	0	7	0	15	0	15	3	25	3	25	3	+
5a	30.5	28	A/o/B	25	4	75	20	80	25	80	25	80	25	80	25	+
5b	31	28	A/o/B	15	4	28	5	60	15	75	20	75	20	75	20	+
6a	32	28.4	A/o/B	8	7	35	20	55	20	90	25	90	25	90	25	+
6b	32	28.5	A/o/B	10	4	25	15	45	15	80	20	80	25	80	25	+
7a	31.5	30.5	A/o/B	25	5	45	7	45	10	45	10	45	10	45	10	+
7b	31.5	30	A/o/B	15	0	45	20	70	20	70	20	70	20	70	20	+
8a	33	31	A/o/B	6	4	35	15	70	35	80	35	80	35	80	35	+
8b	33.5	31	A/o/B	10	3	25	10	65	30	75	30	80	30	80	35	+
9a	43	41	A/o/B	30	7	50	10	60	20	60	20	60	20	60	20	+
9b	43	41	A/o/B	15	4	40	10	50	10	70	10	70	10	70	10	+
10a	51	50	A/o/B	25	15	55	25	100	60	110	65	120	60	120	60	+
10b	51	50	A/o/B	45	40	60	45	120	70	120	70	120	70	120	70	+
11a	52	50.5	A/o/B	13	0	25	13	40	25	45	25	60	40	60	40	+
11b	52	50.2	A/o/B	27	13	25	13	45	20	45	20	45	20	45	20	+
12a	61.5	58	A–. B	25	11	50	10	75	10	80	10	80	10	80	10	+
12b	60	58	A–.–B	50	25	80	30	80	30	80	30	80	30	80	30	+
13a	78.	61	A–. B	90	10	95	3	90	4	95	5	95	5	100	5	+
13b	78	60	A–. B	90	2	90	2	100	2	100	2	100	2	100	2	+
14a	79.5	66	A–.–B	60	40	60	40	80	20	90	15	90	15	100	10	+
14b	79.5	67	A–.–B	70	30	70	30	80	20	80	20	80	20	80	20	+
15a	80	69	A–.–B	100	60	100	60	100	60	100	60	100	60	100	60	+
15b	80	69	A–.–B	60	40	60	40	60	40	60	40	65	35	65	35	+
16a	79	72	A–.–B	70	40	75	45	75	45	75	45	75	45	75	45	+
16b	79.5	72	A–.–B	65	45	70	55	70	55	85	50	85	50	85	50	+
17a	80	76	A–.–B	80	20	75	25	70	30	70	30	70	30	70	30	++
17b	80	75	A–.–B	60	30	60	40	70	30	70	30	70	30	70	30	+
18a	81	78	A–.–B	70	35	80	35	80	35	80	35	80	35	80	35	+
18b	80	78	A–.–B	50	40	60	40	60	50	70	40	80	40	80	40	+
18c	80	78	A–.–B	50	40	80	40	80	40	85	35	80	35	80	35	+
19a	79.5	77	A–.–B	100	20	100	25	100	30	100	40	100	40	100	40	+
19b	79.5	77.5	A–.–B	80	30	80	30	90	35	100	45	100	45	100	45	+
20a	80.	79	A–.–B	70	65	70	65	70	75	70	75	70	75	70	75	–
20b	80	79	A–.–B	70	60	80	75	80	75	85	80	85	80	85	80	–
21a	81	80	A–.–B	80	75	65	85	75	80	75	80	75	80	75	80	–
21b	81	80	A–.–B	75	65	80	75	80	75	85	95	85	95	85	95	–
22a	82	80	A–.–B	40	20	70	55	70	55	70	60	70	65	70	65	–
22b	82.5	80.5	A–.–B	50	45	50	40	50	40	60	50	75	65	75	65	–
23a	85	80.5	A–.–B	20	25	40	35	40	35	40	35	40	35	40	35	–
23b	85	81.5	A–.–B	30	35	40	45	45	50	45	50	45	50	45	50	–
24a	87.5	81	A–.–B	30	30	60	60	60	60	60	65	60	65	60	65	–
24b	89	81	A–.–B	4	4	8	9	10	9	10	9	17	20	17	20	–
25a	85	83	A–.–B	40	35	50	35	65	60	65	60	65	60	65	60	–
25b	85	83	A–.–B	50	40	50	40	50	40	50	40	50	40	50	40	–

Explanation

A B – two fields of testing chamber.

A /o/B – no response of mites.

A–. B – immediate response of mites towards A.

A–.– B – immediate simultaneous response of mites towards A and B.

The figures relate to the numbers of mites counted on field A or B when checked.

Table 2. Orientation of mites towards different odours

Test n.	RH+tested material	RH	Behaviour at start	Time in minutes 5		10		15		20		25		30		Signification
	A	B		A	B	A	B	A	B	A	B	A	B	A	B	
1a	80 check	80.5	A–.–B	30	30	45	50	80	75	80	75	80	75	80	75	–
1b	80 check	80.5	A–.–B	40	50	45	55	60	60	80	85	80	85	80	85	–
2a	81+brick cheese	81	A–.–B	35	20	45	20	60	25	80	30	80	30	100	35	+
2b	79.5+ brick ch.	80	A–.–B	45	25	55	25	60	20	65	25	65	25	70	30	+
3a	79.5+blue ch.	79	A–.–B	70	40	100	40	120	50	120	50	120	50	120	50	+
4a	80+blue ch.	80+ date	A–.–B	70	60	100	80	100	80	100	80	120	70	120	70	+
4b	80+blue ch.	80+ date	A–.–B	75	65	80	70	95	70	95	70	95	70	95	70	+
5a	82+1% lactic acid	82	A–.–B	80	60	75	55	75	55	70	50	70	50	70	50	+
5b	81+1% lactic a.	80	A–.–B	80	60	80	60	90	65	90	65	100	70	100	70	+
5c	78+1% lactic a.	78	A–.–B	50	40	100	80	110	80	120	75	130	70	100	70	+
6a	79+5% lactic a.	80	A–.–B	60	20	70	20	80	20	80	20	80	20	80	20	+
6b	78.5+5% lactic a.	78	A–.–B	20	10	30	20	25	20	42	15	50	15	50	15	+
6c	78.5+5% lactic a.	79	A–.–B	70	50	80	60	80	60	80	60	80	60	80	60	+
6d	80+5% lactic a.	80	A–.–B	10	1	35	8	40	10	50	10	70	20	70	20	+
7a	81.5+10% lactic a.	80.5	A–.–B	40	80	50	70	50	70	50	70	50	70	50	70	–
7b	81+10% lactic a.	81.5	A–.–B	60	65	80	65	80	65	100	85	100	85	100	85	–
8a	80+15% lactic a.	81	A–.–B	15	20	30	40	40	80	40	80	40	80	40	80	+
9a	78+40% lactic a.	78.5	A–.–B	20	80	15	90	15	90	25	80	25	80	25	80	+
10a	78.5+one drop cinnamaldehyd	78	A–.–B	40	20	60	25	80	40	100	40	130	40	150	50	+
10b	79.5+5% cinnal.	79.5	A–.–B	30	50	40	60	40	60	40	70	40	70	40	70	+
11a	79+ one drop anisaldehyde	78.5	A–.–B	40	20	70	20	100	30	100	30	100	30	100	30	+

remained in the middle of the chamber with only a few of them making any orientation movements. A close observation of the behaviour of the mites revealed that in cases where the differences between the tested humidities were rather great, the mites found their way immediately (in the first 5 minutes) to the higher one, otherwise they spread first all over the chamber and eventually they concentrated at the more favourable humidity.

II. Orientation of mites towards different odours (Table 2). When testing olfactory responses of mites, the same testing chamber was employed as used in the humidity tests. Different natural and synthetic materials were used. The relative humidity on each side was fixed at 78–82 %. The odour of the tested materials was introduced into the air stream on one side only while the air stream on the other side had the same humidity but was odourless. The tested materials were: brick cheese, blue cheese, dates, 1, 5, 10, 15, 45 % solution of lactic acid, cinnamaldehyde and anisaldehyde. All materials had a distinct odour. In all cases the mites spread in both directions at first, later they concentrated at one end of the chamber. They preferred both cheeses to pure air or to dates. Lactic acid was attractive in concentrations 1 and 5 %. Attractiveness of 10 % lactic acid was not significant, whereas 15 and 45% concentrations were strongly repelling. Similarly, both aldehydes acted as strong repellents in high concentrations, but were attractive as a mild odour.

Conclusions

The mite *Tyrophagus putrescentiae* (Schrank) can recognize a difference in relative humidity of the air as low as 1 %, within a range of 22 % to 78 %. The higher the difference, the faster the response. In extremely high or low humidities there was no response to any choice test.

Preliminary experiments with odorous substances indicate that the mites discriminate between different olfactory stimuli by being attracted or repelled by different odours.

References

Johnston, D.E.: *Tyrophagus neiswanderi*, a New Acarid Mite of Agricultural Importance. Agric. Res. Develop. Center, Wooster, Ohio 977, 17 pp., 1965.

Lener, W.: Economical and Disposable Olfactometer. J. Econ. Entomol. 63 (6), 1618, 1970.

Knülle, W.: Die Luftfeuchte-Unterschiedsempfindlichkeit der Mehlmilbe *(Acarus siro* L.) Z. vergl. Physiol. 44(5):463–477, 1961.

Żyromska-Rudzka, H.: Wstepne obserwacje nad zajmowaniem nowego podloza przes *Carpoglyphus lactis* L. (Acarina, Glycyphagidae). Ekol. polska A7 (13):339–356, 1959.

Žďárková, E.: Mites of stored food in Czechoslovakia. Proc. of the 2nd Inter. Congress of Acarol. 1967. Publ. by Hungarian Academy of Sciences 261–264, 1969.

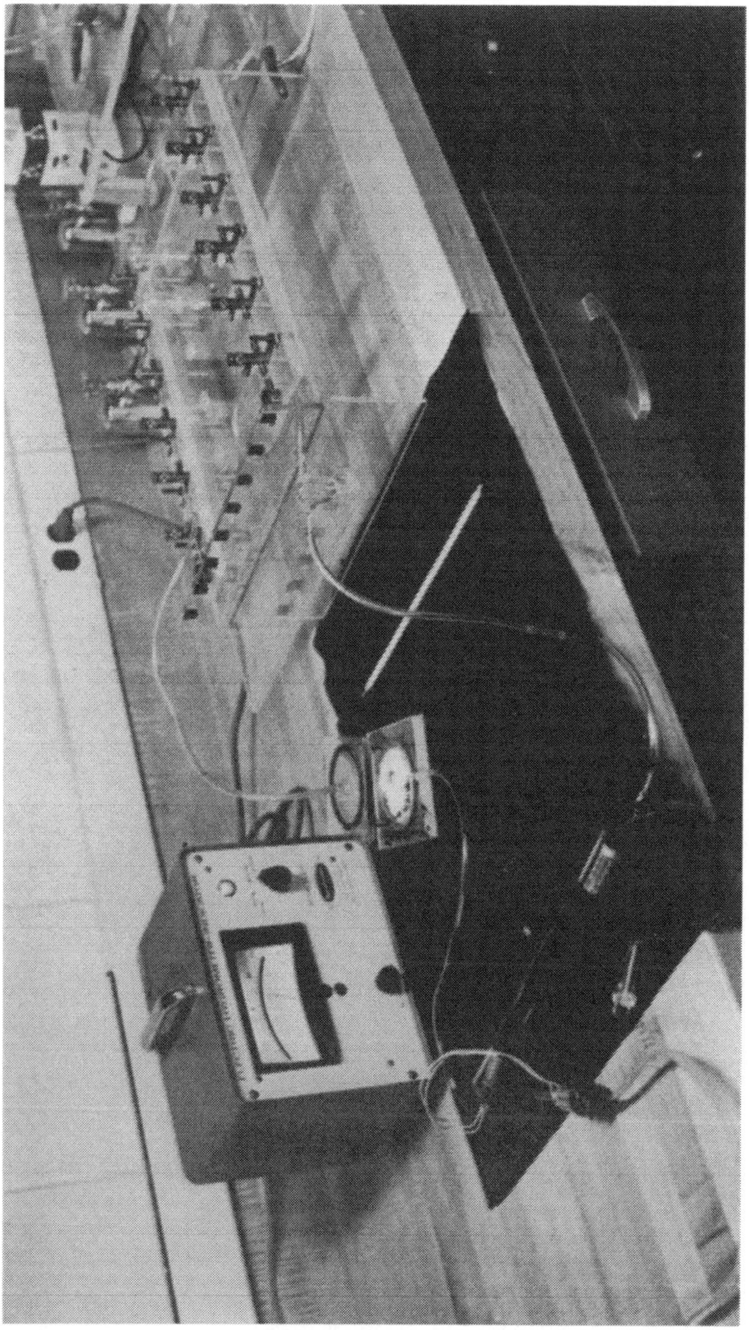

Fig. 1. The olfactometer used in the experiments.

Section VI

PARASITIC MITES, FAUNISTICS AND TAXONOMY

Chairman

N. G. Bregetova (USSR)

Assistant Chairman

M. Mrciak (Czechoslovakia)

Proceedings of the 3rd International Congress of Acarology, Prague, 1971

CHIGGERS IN THE HIGH ALTITUDE REGION OF THE EAST HINDU-KUSH

M. DANIEL
Institute of Parasitology, Czechoslovak Academy of Sciences,
Prague, Czechoslovakia

The Hindu-Kush mountain range, namely its highest eastern part, is not only an important orographic, but also zoogeographic junction. Its main mountain ridge can be regarded in this part of Asia as the dividing line between the Oriental and Palearctic regions, and as a crossroad of many other lower zoogeographic categories. Despite of this the information on the fauna of this extensive region is scanty or practically missing at all.

The material, on which the present paper is based, was collected during the First Czechoslovak Expedition to Hindu-Kush (1965), who visited the Afghan part (Vakhan region) north of the main mountain ridge of the East Hindu-Kush. The Vakhan region is filled out by the northern part of this high mountain massive with dominant altitudes of peaks about 5 – 6,000 m, some of them even exceeding 7,000 m. The main ridge which is the borderline between Afghanistan and Pakistan, is running approximately in the direction of the parallel line, while the whole northern section of the massive is divided by short steep valleys running in the meridian direction and covered with ice in their upper parts. One of them, the Ishmurkh Darrah Valley, served as the main working area of the expedition. It intersects all vertical zones occurring in the East Hindu-Kush for a distance of 15 km and is therefore suitable for the studies on the altitudinal distribution of small mammals and their ectoparasites. During July and August small mammals were trapped in the said valley as well as in the side valley of Chap Darrah at altitudes within the range of 2,750 and 4,550 m, where traces of the animals' existence were found or where their occurrence at least could be anticipated.

Climatically and geobotanically the Vakhan region belongs to the province of Central-Asiatic deserts (Linchevsky and Prozorovsky, 1949). It is characterized by predominant desert and steppe vegetation which reaches high altitudes without any interruption. The Ishmurkh Valley may be divided into three ecologically different zones and this division served as a basis for the evaluation of zoological findings:

Zone I (up to 3,200 m) is characterized by a wide, flat and moderately ascending bottom. The slopes are formed by strongly eroded walls of conglomerate. The major part of this section is devoid of vegetation, which is concentrated in places where the flow of glacial streams is slower and aggrading fine sand. The vegetation is of steppe character and so is the fauna, namely entomofauna.

Zone II (about 3,200 – 3,800 m). The valley starts to rise abruptly and its higher and steeper slopes are formed by crystalline rocks. Under the rock walls there are extensive scree fields which are in frequent motion. The vegetation is limited to small patches in places where the scree motion is stopped. The flora and fauna in this zone are of transitory character and gradually change into montane type. The collection site of the side valley of Chap Darrah belongs to the same type.

Zone III (3,800 m and higher) starts from the front of glacier and continues to higher elevated places of the valley; this part is of true high altitude character. Favourably exposed places abound in flora and also fauna (namely entomofauna) which are both quantitatively and qualitatively richer than at lower zones, apparently due to the height of precipitation (see Römer, 1964). The collection site was on a rock islet at the confluence of two arms of glacier; the rock terraces, however, had the traces of previous glacial activity. It was the highest place where the occurrence of small mammals *(Alticola roylei)* has been ascertained.

A total of 128 small mammals belonging to the species *Marmota caudata, Ochotona roylei, Alticola roylei, Apodemus sylvaticus, Cricetulus migratorius* and *Crocidura russula* were trapped and parasitologically investigated, and 996 chiggers belonging to 11 species of the genera *Neotrombicula, Leptotrombidium, Microtrombicula, Chatia* and *Euschoengastia* were collected from them. Only 4 species were found to be more abundant and from their occurrence conclusions on their vertical distribution may be drawn. 35% of larvae found belonged to the species *Chatia (Shunsenia)* sp., which is closely related to *Ch. (S.) wissemani.* Other species were represented as follows: *Leptotrombidium (Leptotrombidium) dihumerale (18 %), Mi-*

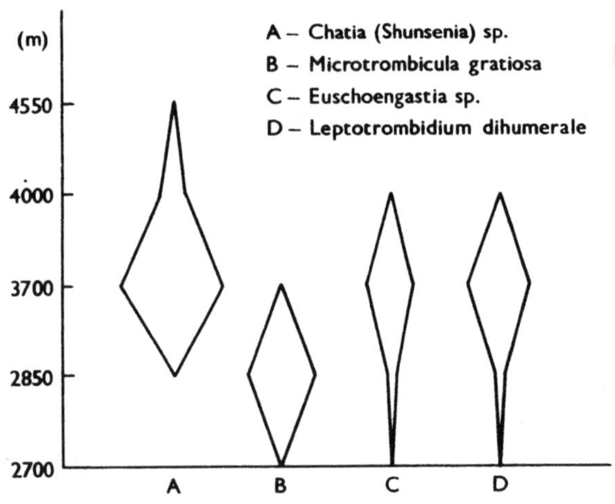

Fig. 1. Altitudinal stratification of particular species of chiggers in the Ishmurkh Darrah Valley (East Hindu-Kush).

crotrombicula gratiosa (18 %) and *Euschoengastia* sp. (15 %). The vertical distribution is evident from Fig. 1, in which the width of column is directly proportional to total number of larvae found. The above mentioned 4 species were most abundant at the dividing line between the upper and middle zone of the valley, i.e. in places which had high altitude characteristics. The occurrence of the species above and below that line was attenuated except for the representative of the genus *Cheatia (Shunsenia)*, which was the only one found as high as 4,550 m.

In order to estimate the vertical changes in the extensity (expressed by percentage of infested hosts) and intensity (expressed by average number of larvae on infested host) of chigger infestation, *Alticola roylei* (which was the only one host species found in all collection sites) was selected from the whole host group and

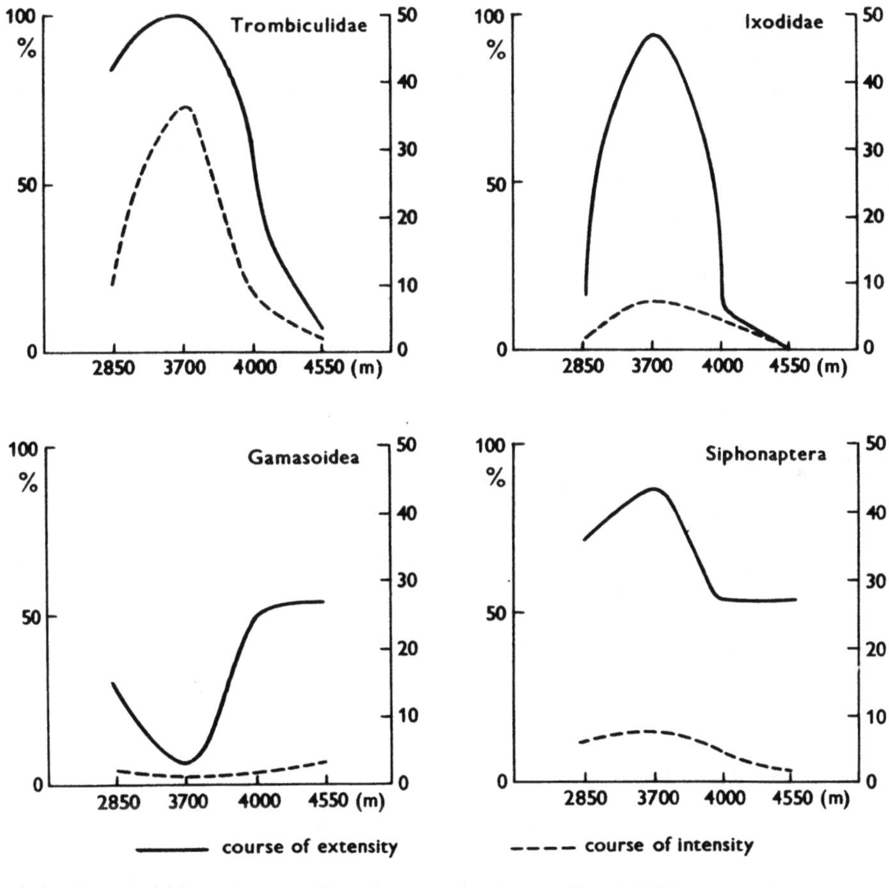

Fig. 2. Changes in extensity and intensity of infestation of *Alticola roylei* by particular groups of ectoparasites depending on altitude.

its infestation by chigger larvae was considered apart. Fig. 2. shows the highest values of the extensity and intensity curve at the dividing line between the third and second zones of the valley. For comparison analogical graphs are added showing further groups of ectoparasites. Ticks have the extensity curve similar to that of chiggers, but the intensity is universally low in all zones of the valley. Gamasoid mites, on the other hand, have kept high extensity and intensity values at higher altitudes and a similar tendency may be observed in fleas.

The outlined results may be well compared with the conclusions of American authors (Traub et al., 1967) based on data obtained in the Kaghan Valley, situated in the northern part of the West Pakistan south of the East Hindu-Kush ridge. At lower altitudes the species *L.(L.) dibumerale* and *Ch. (S.) wissemani* were pronounced components of the chigger fauna. These species had been also found in the West Pamir (Shluger and Kudryashova, 1969). From the West Pamir (2,400 m) also *Microtrombicula gratiosa* was described. The species discussed may be therefore regarded as representatives of high altitude fauna inhabiting the above mentioned mountain systems, characterized by a considerable ecological amplitude. Other sporadically found species whose placement in taxonomy has not been fully elucidated yet and whose relationship to the faunas of the compared areas has not been revealed, should be probably regarded as local endemites. This problem might be solved by the forthcoming studies of chiggers collected by the Second Czechoslovak Expedition in the Tirich Mir mountains massif (the Pakistani part of the East Hindu-Kush).

References

Linchevsky, A., Prozorovsky, A.V.: The basic principles of the distribution of the vegetation of Afghanistan. Kew Bull. 214, 1949.

Römer, H.: Vegetation im Wakhan. Deutsche Wakhan-Exped. 1964, pp. 41–43, München, 1964.

Traub, R., Wisseman, C.L.Jr.: The occurrence of scrub typhus infection in unusual habitats in West Pakistan. Trans. Roy. Soc. Trop. Med. Hyg. 61:23–57, 1967.

Shluger, E.G., Kudryashova, N.I.: New and for the first time found in the USSR species of Chiggers (Acariformes, Trombiculidae). Parazitologia, Leningrad, 3:115–122, 1969. (In Russian.)

PECULIARITIES OF THE DEVELOPMENT OF MORPHOLOGICAL STRUCTURE IN GENERA OF THE SUBFAMILY LISTROPHORINAE

H. V. DUBININA
Zoological Institute, Academy of Sciences of the USSR,
Leningrad, USSR

The fur mites of the subfamily Listrophorinae collected from mammals of the Soviet Union have been studied. Hitherto the subfamily contained three genera, *Listrophorus* Pagenst., *Eurychiroides* Trt. and *Lynxacarus* Radf. About thirty species were assigned to the genus *Listrophorus* which comprised ectoparasites of Rodentia, Insectivora, Lagomorpha and small Carnivora. The monotypic genus *Eurychiroides*, described by Trouessart from the rodent *Sigmodon*, was erected on the basis of the presence of strongly enlarged and fused coxae of legs III; the latter form almost a round shield. In Trouessart's opinion this organ functions as a sucker. Apparently, evolutionally it was an anterior ventral shield which is strongly sclerotized in these mites in contrast to the sclerotized, poorly visible and usually not described ventral shields of *Listrophorus*. Since Trouessart's description (1912) this mite has not been found again. The genus *Lynxacarus* was also monotypic. A detailed study of the paratype of *Lynxacarus morlani* Rad. from the lynx enabled me to specify the characteristics of the genus, to assign to it a new species, *Lynxacarus dubinini* H. Dub., and thus to augment the generic diagnosis. At the time the collective nature of the genus *Listrophorus* became clear. In morphological details and degree of development of some structures the species of this genus at my disposal represent two natural groups which can be raised to generic status, i.e. *Listrophorus* proper and a new genus, *Listracarus*. Fain (1970) gave a brief diagnosis of nine genera (of which seven were new) and 33 new species of the subfamily Listrophorinae. This work became known to me after abstracts of this paper had been already sent to the congress. Since the separation of the genus *Leporacarus* was based mainly on the species *Listrophorus gibbus* Pagenst., a typical species of my new genus *Listracarus*, the latter should possibly be considered as a synonym of the genus *Leporacarus* Fain. However, *Listracarus myospalacis* H. Dub. sp. n. assigned by me to the same genus should be referred, according to Fain, to one of his other new genera, probably *Hemigalichus*. This problem can be solved only when descriptions of new genera and species will be published by Fain. For the time being I retain my genus *Listracarus* without denying the possibility that it may be a synonym of *Leporacarus* Fain.

In this paper an attempt was undertaken to make a comparative analysis of some morphological structures of species found in the USSR. They belong to the three genera in question, *Listrophorus* s. str., *Listracarus* and *Lynxacarus*. I agree with Prof. Fain that such characters as the nature of sclerotization of the dorsal integument of the idiosoma, the structure of the genital apparatus, and some others are most important for the classification of these mites. The majority of free-living mites of the suborder Sarcoptiformes have extensible, smooth or plicate integument. With the development of parasitism changes proceed, as a rule, in the direction of adaptive strengthening of the integument, i.e. sclerotization. This tendency can be traced on integument of the dorsal side; the ventral one, due to the coxo-sternal skeleton, is normally well sclerotized. In the subfamily Listrophorinae the integument evolution is clearly seen in males, whereas in females, due to peculiarities of reproduction, the hysterosomal integument remains, as a rule, elastic.

By the degree of sclerotization of the male dorsal integument *Listracarus* proved to be most primitive in the subfamily (Fig. 1) (this conclusion is confirmed by the analysis of other structures). Species of this genus have plicate dorsal epicuticle under which there are dark procuticular plates of propodosomal and hysterosomal "shields". Such is the case in the male of *L. myospalacis*, whose dorsum is covered as if by an armour of "shields". In *L. gibbus*, the propodosoma is covered by shields: Their epicuticle is smooth and detached from the plicate epicuticle of the surrounding integument. On the hysterosoma the plicate epicuticle undergoes considerable changes and differs much from that surrounding the "shield". Since in *L. gibbus* changes involve the epicuticle we may suppose that a greater degree of sclerotization is characteristic of this species as compared to *L. myospalacis*. Despite the fact that only two species of this genus were studied, their heterogeneity in contrast with two genera, representing very homogeneous groups also indicates the primitiveness of the genus *Listracarus*. In both species of the genus *Lynxacarus* the propodosomal shields are fully formed; there are two or three procuticular plates in the area of the hysterosomal "shields" and their epicuticle shows a tendency for smoothening though a distinct shield is not yet formed. There are analogous structures in some analgesid mites. The process of shield formation here seems to be at a more advanced stage than in *Listracarus*. The greatest sclerotization rate is observed in the genus *Listrophorus* where all propodosomal and hysterosomal shields are already formed. Thus, *Listracarus* and *Lynxacarus* represent separate stages of the sclerotization process which in the subfamily Listrophorinae finds its completion in the genus *Listrophorus*. It should be noted that the sclerotization of integument parallels a process of differentiation of the "shield" complex, the armour of *L. myospalacis*, into separate shields which do not cover the whole dorsum. Mites of *Listrophorus* have three propodosomal and two hysterosomal shields. The morphology of the ventral side is constant, even in the larger taxonomic units. However, the main sternal shield, which is of great importance, as one of the organs of attachment to the host's hair, has much evolved in the subfamily though retaining its characters in the genera. In a series of *Listracarus* – *Listrophorus* – *Lynxacarus* a gradual reinforcement of its development takes place and this is probably associated with an increasing importance of this organ (Fig. 2). This

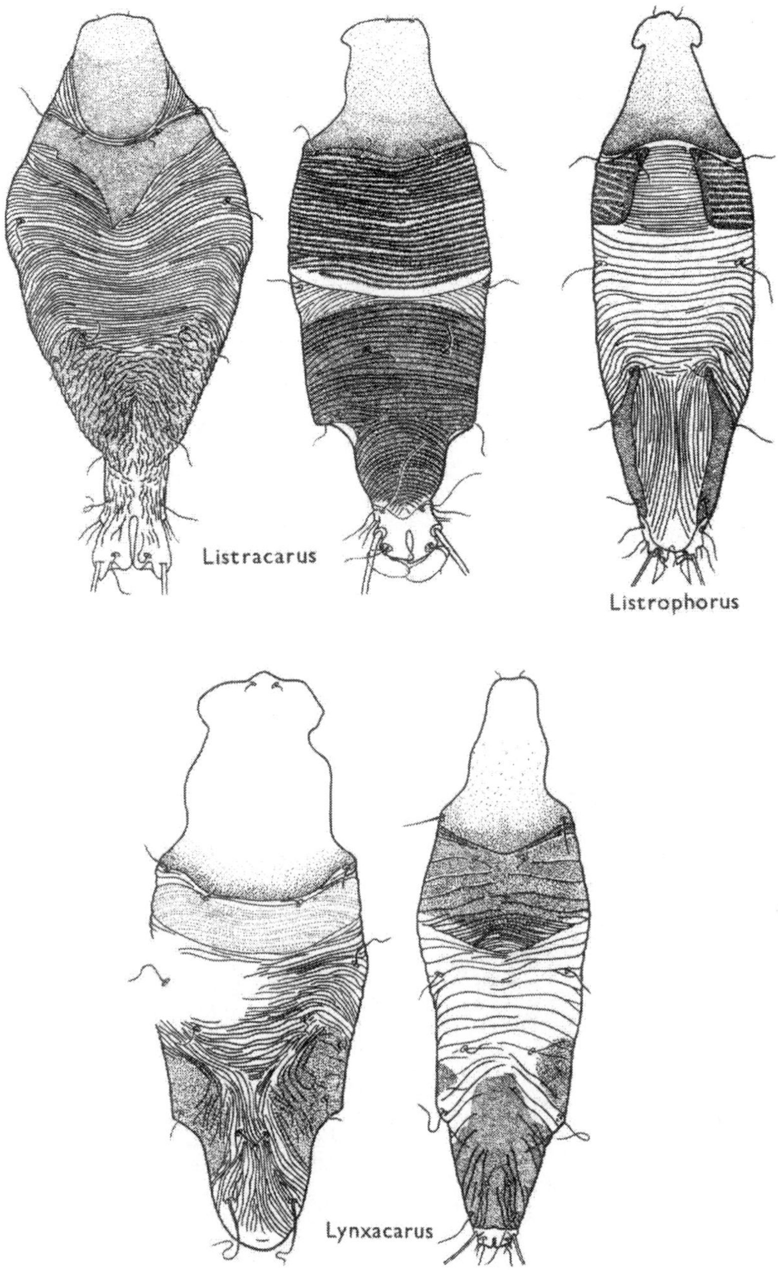

Fig. 1. Dorsal view of the male idiosoma

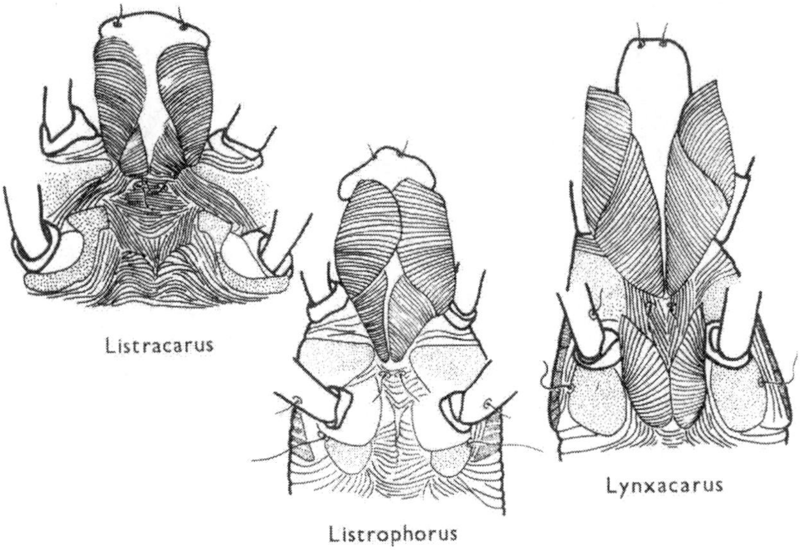

Fig. 2. Ventral view of the male proterosoma

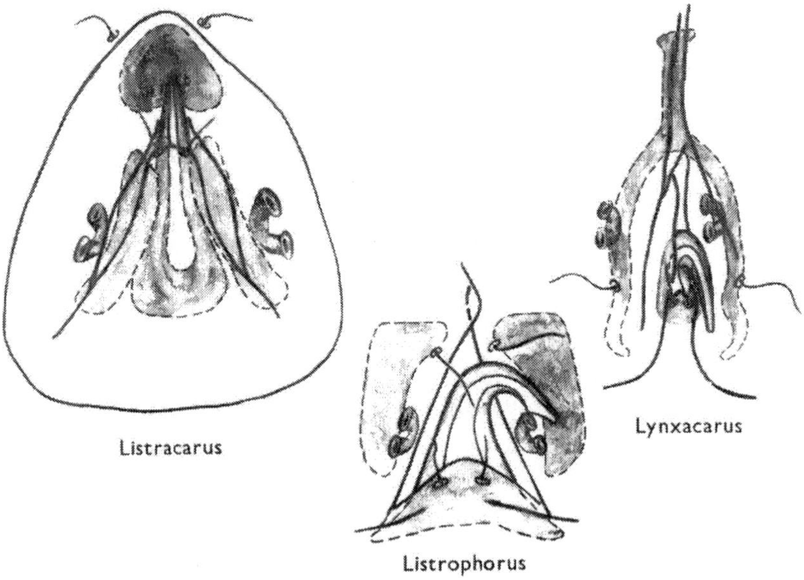

Fig. 3. Male genital apparatus

organ is most poorly developed in *Listracarus*, possessing a single anterior pair of very short attachment clamps. *Listrophorus* has more developed, longer clamps and a plicate integument between the bases of legs II form something like an additional pair of clamps. This second pair of fully developed clamps is present only in *Lynxacarus* and this indicates a great adaptation of these mites to parasitism.

Great specialization of *Lynxacarus* is also expressed in the structure of III and IV pairs of legs. In two other genera these legs structurally closely resemble a typical walking limb. In *Lynxacarus* an adaptive modification results in their enlargement and curvature and in males in a reduction of the chaetom of legs IV.

One of the most important criteria for the separation of the genera is a structure of the male genital apparatus (Fig. 3). In this case *Listracarus* also shows all signs of a primitive structure; as well as in most free-living mites the male genital apparatus is situated on a separate genital plate on the ventral shield. Such a plate is absent in males of two genera and their genital apparatus is covered by medial folds of this shield. As regards sclerites surrounding the copulatory organ, *Listracarus* has one gonobasal and two lateral sclerites. In males of *Listrophorus* the gonobase is either completely reduced or is partially fused with the lateral sclerites. *Lynxacarus* has only horseshoe-shaped sclerite surrounding the copulatory organ from all sides. It follows from the above that the characters of *Listracarus* seem to be initial ones for two directions in the evolution traced, on the one hand in the genus *Listrophorus*, highly specialized, and on the other hand, in the genus *Lynxacarus*, whose morphology testifies its profound adaptation to parasitism. The scheme of the structure of the subfamily Listrophorinae presented here is not definitive.

References

Dubinina, H.V.: The fur mite of the genus Lynxacarus (Sarcoptiformes, Listrophoridae). Parazitologia, 3(5): 440–445, 1969.

Fain, A.: Diagnoses de nouveaux Lobalgides et Listrophorides (Acarina, Sarcoptiformes). Rev. Zool. Bot. Afr., 81 (3–4): 271–300, 1970.

Radford, Ch.D.: Two new genera of parasitic mites (Acarina: Laelaptidae and Listrophoridae). Parasitology, 41 (1–2): 102–104, 1951.

Trouessart, E.: Troisième note sur les Sarcoptides pilicoles et déscription des genres nouveaux. Bull. Soc. Zool. France, 42: 151–158, 1917.

ECTOPARASITIC MITES (GAMASOIDEA) ON SMALL MAMMALS IN NORTHERNMOST SWEDEN

A. EDLER
Department of Animal Ecology, University of Lund, Sweden

In September 1968 and April and July 1969 small mammals were collected in northernmost Sweden at about 68° N, 18° E and altitudes 400–600 m. This preliminary communication summarizes certain results relating to populations of ectoparasitic mites on the small mammals.

Results

622 small mammals were collected (Tab. 1). 335 of them were infested by gamasid mites. Laelaptidae was the most common family, comprising 72.1 % of all mites (Fig. 1). Only 8.6 % were Haemogamasidae but 18.4 % Liponyssidae. The remaining 1 % belonged to various other families. In other parts of Sweden Haemogamasidae comprised a greater part of the gamasid mites than Liponyssidae (Edler, in litt.). In all Swedish samples Laelaptidae are predominant. In the southernmost parts of the country, less than 50 m a.s. 1., Haemogamasidae comprised 15.4 % and

Table 1. Small mammals collected and infested by gamasid mites

Species	Collected specimens	Infested specimens	No. of mites	Infestation %	Mean value per infested host animal
Sorex minutus L.	17	2	2	*	*
Sorex araneus L.	112	19	31	17.0	1.6
Clethrionomys rutilus Pallas	84	24	147	28.6	6.1
Clethrionomys glareolus Schreb.	1	1	1	*	*
Clethrionomys rufocanus Sundevall	188	118	545	62.8	4.6
Microtus agrestis L.	220	171	2,272	77.7	13.3
All host species	622	335	2,998	53.9	8.9

(* = Too small material).

Liponyssidae less than 1 %. In central Sweden, altitude 25–500 m, the two families are about equally common parts of the material investigated, viz 11.6 % and 10.3 % respectively. In a material from northern Sweden at 66° N and 340–820 m the relation was 17.5 % to 15.8 %.

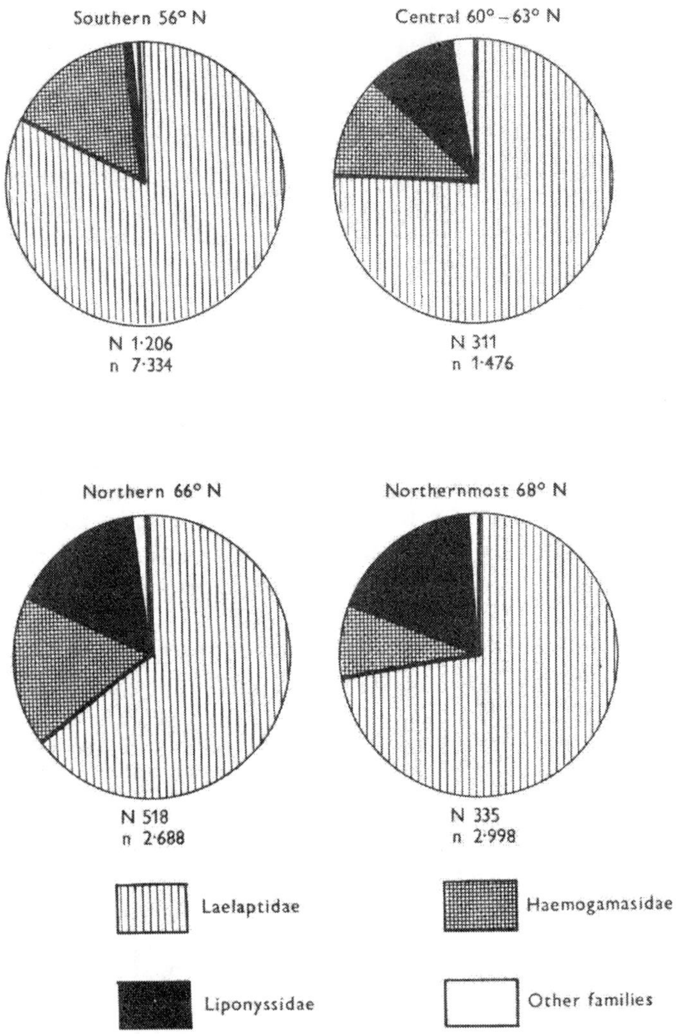

Fig. 1. Relative frequencies of families of gamasid mites on small mammals collected at various localities in Sweden. (N = no. of host animals; n = no. of mites.)

The infestation frequency and the mean value differed in some cases between the seasons (Tab. 2). When considering all hosts together, significantly more hosts were infested in April and July than in September. The number of mites per infested host animal was also significantly higher in April and July than in September.

Table 2. Infestation frequency and mean number of mites per infested individual for all host species in different seasons

Collecting time	Infestation %	Mean value
September 1968 (autumn)	32.5 ± 3.7	3.6 ± 0.6
April 1969 (winter/spring)	64.0 ± 6.8	16.6 ± 3.9
July 1969 (summer)	60.9 ± 2.4	10.4 ± 1.7

As regards infestation the host animals may be classified in three groups (Tab. 1). Weakly infested hosts (group 1) are *Sorex araneus* and *Clethrionomys rutilus*, moderately infested (group 2) *C. rufocanus*, and highly infested (group 3) *Microtus agrestis*. Regarding the mean number of mites per infested host animal, there is no significant difference between *C. rutilus* and *C. rufocanus*. Nor is the difference between *C. rutilus* and *S. araneus* significant.

Table 3. Infestation of two groups of host animals with the most common mite species – expressed as I/M ± m

Host / Parasite	*Clethrionomys* spp.	*Microtus agrestis*
Laelaps clethrionomydis	31.6/ 4.5 ± 0.6	0.9/ *
Laelaps hilaris	4.4/ *	67.3/ 11.3 ± 1.1
Hyperlaelaps arvalis	0.4/ *	27.3/ 3.7 ± 0.5
Haemogamasus nidiformis	10.7/ 5.2 ± 2.2	11.4/ 1.6 ± 0.3
Hirstionyssus isabellinus	12.1/ 2.1 ± 0.4	22.7/ 11.3 ± 7.7

I = % infested animals, M = number of parasites per infested host, m = the deviation (* = too small material).

Clethrionomys spp. and *Microtus agrestis* were compared regarding the infestation and mean value of the five most common gamasid mite species (Tab. 3). Only *Haemogamasus nidiformis* Bregetova infested the hosts with the same frequency. The others differed significantly between the two groups of hosts. *Laelaps clethrionomydis* Lange was found mostly on species of *Clethrionomys*. More commonly found on *M. agrestis* than on *Clethrionomys* were *Laelaps hilaris* Koch, *Hyperlaelaps arvalis* Zachvatkin, and *Hirstionyssus isabellinus* Oudemans. *L. hilaris* and *H. arvalis* were especially specific to *M. agrestis*. This agrees with previous investigations (Mrciak, 1959 and Edler, 1972). Two mite species were found on both groups of hosts in numbers sufficient to calculate the mean numbers per infested host. No significant differences were found between the numbers of *H. nidiformis* and *H. isabellinus*, respectively.

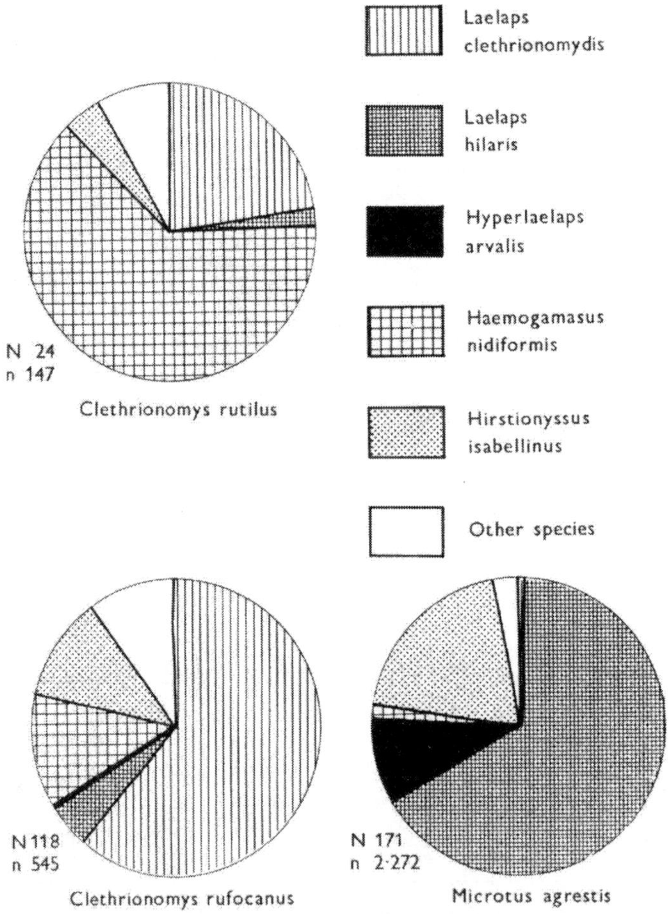

Fig. 2. Dominance of well-represented mite species on various host animals. (N = no. of host animals; n = no. of mites.)

As regards the dominance of the most common gamasid mites (Fig. 2) on *C. rutilus* 63.3 % were *H. nidiformis*, 22.4 % *L. clethrionomydis* and 4.1 % *H. isabellinus*. On *C. rufocanus* 61.7 % were *L. clethrionomydis*, 12.3 % *H. nidiformis* and 11.5 % *H. isabellinus*. On both species of *Clethrionomys* less than 1 % were *H. arvalis*. On *M. agrestis* 66.3 % were *L. hilaris*, 9.4 % *H. arvalis*, 20.1 % *H. isabellinus*, 1.7 % *H. nidiformis*, and 1 % *L. clethrionomydis*.

Discussion

Normally the families Laelaptidae, Haemogamasidae and Liponyssidae dominate the ectoparasitic mite material from small mammals (Mrciak, 1963, Edler,, in litt.). Laelaptidae is the most common because species of *Laelaps* are permanent parasites. The differences between the families in samples from various parts of Sweden (Fig. 1) are to some extent due to the host material. *Apodemus* spp. are found in southern and central Sweden, but are absent further north. Some voles, e.g. *C. rutilus* and *C. rufocanus*, are recorded only in the north. The material of Liponyssidae mainly consists of two species, viz. *H. isabellinus*, mostly found on species of *Microtus*, and *Hirstionyssus soricis* Turk, specific to Soricidae. These hosts are found all over the country, so the proximate reason for the differences seems to be the climate. In southern Sweden the winter is short, some years without snow. In northern Sweden and particularly inland, where the present material was collected, the winter is long and cold with much snow, often extending from September to May or June. The high infestation in winter/spring, i.e. in April, may be related to the decreasing host population. Animals still alive in April had numerous contacts with nests, nest-parasites and dead mammals. The high infestation frequency in July is correlated with a great number of juveniles and subadults, which still had numerous contacts with their birth nests. Juveniles and subadults made up about 60 % of the infested hosts in July.

The small mammals appeared to be divisible into three groups as regards infestation. The reason for this grouping is not clear. Insulating properties of the fur of small mammals may be of some importance (cf. Gebczyński and Olszewski, 1963). The high infestation of *M. agrestis* agrees with conditions in southern Sweden (Edler, 1972).

In agreement with the above findings *L. clethrionomydis* is stated to be specific to *Clethrionomys* spp. (Mrciak, 1960), and *L. hilaris* and *H. arvalis* to species of *Microtus* (and *Pitymys*) (Mrciak, 1959).

H. nidiformis is a mountain species, so far known from a few countries in Europe and Asia. It usually makes up only a small part of the gamasid mites in collections from small mammals. The similar species *H. nidi* Michael is usually much more

Table 4. Frequency of *Haemogamasus nidi* and *Haemogamasus nidiformis* in different areas

	Haemogamasus nidi	*Haemogamasus nidiformis*	Number of gamasid mites
Bulgaria (Mrciak 1959)	11.3	0.1	920
High Tatras (Mrciak 1958)	13.0	0.2	2,069
Central Sweden (Edler 1969)	7.8	0.1	1,476
North Sweden (Edler 1968)	6.6	4.8	2,688
Northernmost Sweden (Present paper)	0.3	6.7	2,998

100 % = all gamasid mites in the investigations cited.

abundant, though *H. nidiformis* tends to make a greater part of the gamasid mites in mountain districts towards the north (Tab. 4). Most records are from species of *Clethrionomys*, *Microtus* and *Pitymys*, but in Sweden it has a wide host spectrum (Bregetova, 1956, Mrciak, 1958, Edler, 1968 and 1969). In the present material two new hosts are recorded, viz. *S. minutus* and *C. rutilus*.

References

Bregetova, N.G.: Gamasid mites (Gamasoidea). Opred. po faune SSSR, Izd. Akad. Nauk SSSR, Moscow, 247 pp., 1956. (In Russian.)

Edler, A.: Ectoparasitic mites (Acarina: Parasitiformes) from small mammals from the Tärnasjö and Ammarnäs areas in Swedish Lapland. Entomol. Tidskr., Stockholm, 89: 269–277, 1968.

– Ectoparasitic mites (Acarina) from small mammals in Central Sweden. Entomol. Tidskr., Stockholm, 90: 272–284, 1969.

– Ectoparasitic mites (Acarina) from small mammals in southern Sweden. Entomol. Tidskr., Stockholm, 93. (In press.)

Gebczyński, M., Olszewski, J.: Katathermometric measurements of insulating properties of the fur in small mammals. Acta Theriologica, Bialowieza, 7: 369–371, 1963.

Mrciak, M.: Parasitische Milben (Parasitiformes, Acari) von kleinen Säugetieren der Hohen Tatra. Zool. listy, Brno, 7: 65–86, 1958.
– Ein Beitrag zur Kenntnis der Milben (Parasitiformes) von Kleinsäugern aus dem Gebiet Bulgariens. Práce Brněnské základny ČSAV, Brno, 4: 365–376, 1959.
– Contribution à la connaissance des acariens de l'ordre Parasitiformes trouvés sur les petits mammifères en Roumanie. Biológia, Bratislava, 15: 498–507, 1960.
– Über Beziehungen mancher Milbenarten (Gamasoidea) zu Kleinsäugetieren und Biotopen des Ondaver Hügellandes. Acta F.R.N. Univ. Comen., Bratislava, 7: 437–532, 1963.

CLASSIFICATION AND GEOGRAPHICAL DISTRIBUTION OF LISTROPHOROIDEA

A. FAIN
Institute of Tropical Medicine, Antwerp, Belgium

Classification of the Listrophoroidea

The name Listrophorinae was utilized for the first time by Mégnin and Trouessart in 1884.

Canestrini (1892) ranged in the Listrophoridae all the fur mites (= the "gliricoles" of Mégnin). They belonged to three genera: *Listrophorus* Pagenstecher, *Myocoptes* Claparede and *Criniscansor* Poppe Canestrini did not include in this family the genus *Chirodiscus* Trouessart and Neumann, probably because *Chirodiscus amplexans*, the type species, had been described from a bird, and therefore could hardly been considered as a fur mite.

Trouessart (1892) erected the subfamily Chirodiscinae for the genus *Chirodiscus*. In 1895, he described the new genus *Labidocarpus*, which he also placed in this subfamily. In 1896, he added a third genus *Schizocarpus* Trouessart (= *Histiophorus* Friedrich) to the Chirodiscinae.

Gunther (1942) divided the Listrophoridae into 4 subfamilies:

1. Listrophorinae: which attach themselves to the host by means of their modified gnathosoma; the legs being normal.
2. Myocoptinae: which have the posterior legs modified as attaching organs; the anterior legs and the gnathosoma being normal.
3. Atopomelinae: which are attached to the host by means of their anterior legs only slightly modified and still bearing ambulacra. The gnathosoma being normal.
4. Labidocarpinae: with the same characters as the Atopomelinae but with strongly modified anterior legs that have lost their ambulacra.

In 1951, Dubinin erected the superfamily Listrophoroidea for the single family Listrophoridae and in 1968, McDaniel raised to family rank Gunther's 4 subfamilies.

Recently (1971) we proposed a new definition of the Listrophoroidea and of the families forming this group. We have retained in this superfamily only the true pilicolous mites and we have therefore excluded the hypopi of the Glycyphagidae and the Rhyncoptidae.

The following families have been retained in the Listrophoroidea (see Fain, 1971):

1. Listrophoridae Mégnin and Trouessart, 1884:

Presence of a strongly sclerotized and ventrally concave tegmen covering completely the gnathosoma. There are two large striated sternal membranes which pro-

trude to form a tunnel in which the hair of the host is tightened. All the legs normal, with 5 free segments and ending in a sucker. Body generally subcylindrical.
Type genus: *Listrophorus* Pagenstecher, 1861.
2. Myocoptidae Gunther, 1942:
Gnathosoma normal, without tegmen. Sternum without membranes. Anterior legs normal, bearing suckers. Legs III and IV in the female and III in the male strongly modified into powerful claspers. Body flattened or subcylindrical.
Type genus: *Myocoptes* Claparède, 1869.
3. Atopomelidae Gunther, 1942:
Tegmen absent. Gnathosoma normal without ventral membranes. Anterior legs with 5 free segments, the distal segments being slightly thickened and modified, the tarsi bear a small sucker and are devoid of striated attaching membranes. Posterior legs without attaching organs but with tarsi and tibiae always completely fused. Body either flattened or subcylindrical, rarely compressed laterally.
Type genus: *Atopomelus* Trouessart, 1917.
4. Chirodiscidae Trouessart, 1892.
Tegmen absent. Gnathosoma without ventral striated membranes. Anterior legs variably modified, without suckers but bearing a large striated chitinous membrane, simple or twofold. Posterior legs either normal or with the genu and the femur fused. Body variable: flattened, subcylindrical or laterally compressed.
Type genus: *Chirodiscus* Trouessart and Neumann, 1889.
We have divided the Chirodiscidae into 4 subfamilies:
a) Chirodiscinae Trouessart, 1892:
Body strongly flattened and without dorsal shields. Vulva longitudinal. Epimeres and epimerites I and II fused. Anterior legs with most of the segments fused and ending into a long striated membrane.
Type genus: *Chirodiscus* Trouessart and Neumann, 1889.
b) Labidocarpinae Gunther, 1942:
Body compressed laterally. Anterior legs very short with most of the segments fused and ending into a large striated membrane divided into two equal or subequal leaves.
Type genus: *Labidocarpus* Trouessart, 1895.
c) Schizocoptinae Fain, 1970:
Body flattened. Anterior legs normal with all the segments free and bearing apically a large striated membrane divided into two very unequal leaves. Vulva in the shape of an inverted Y.
Type genus: *Schizocoptes* Lawrence, 1946.
d) Lemuroeciinae Fain, 1968:
Body flattened. Anterior legs modified but with all the segments free, and ending into a large striated single membrane. Anterior legs and gnathosoma with strong retrorse processes. Vulva transverse.
Type genus: *Lemuroecius* Fain, 1968.

Geographical distribution of the Listrophoroidea

The different families of the listrophorids are not equally distributed through the world.

1. The family Atopomelidae contains 37 genera and 183 species. All these species except one, are confined to tropical or subtropical regions. This family is specially well represented in Africa (south of the Sahara), Madagascar, Australia and Neotropical America. In the Palaearctic region it is represented by a single cosmopolitan species, *Chirodiscoides caviae*, living on the guinea-pig.

2. The family Listrophoridae, containing only 20 genera and 61 species, has a more uniform distribution. It is represented in all geographical regions, except in Madagascar and Australia where it is completely absent.

3. The family Chirodiscidae with 30 genera and 75 species is almost completely confined to the bats. Its distribution corresponds roughly to that of these hosts.

4. The family Myocoptidae groups 16 genera and 49 species. Half of these species have been described from Africa (south of the Sahara), but this family is also represented in the Palaearctic, the Nearctic and the Oriental regions. So far, this family has not been recorded from Madagascar and Australia.

Most of the listrophorid species are endemic. A few species are represented either in two continents or in several parts of the world. Only the species parasitizing the domestic animals are really cosmopolitan. Among them the most important are:

Chirodiscoides caviae Hirst, on the guinea-pig
Leporacarus gibbus (Pagenstecher), on the rabbit
Myocoptes musculinus (Koch), on mice
Trichoecius romboutsi (Van Eyndhoven), on mice
Listrophoroides cucullatus (Trouessart), on *Rattus rattus* and *Rattus norvegicus*, mainly in tropical countries.

Geographical distribution of the Listrophoroidea

(N.B.: the cosmopolitan species are between brackets)

	Total number	Palaearctic Region	Nearctic Region	Africa South of Sahara	Madagascar	Oriental Region	Australian Region	Neotropical Region
Listrophoridae								
genera	20	4	3	3	–	5	–	5
species	61	12	10	17	–	11	–	11
		(1)	(1)	(1)				
Myocoptidae								
Myocoptinae								
genera	15	5	3	4	–	3	–	–
species	48	12	6	24	–	6	(1)	–
		(2)	(2)	(1)				
Dromiciocoptinae								
genera	1	–	–	–	–	–	–	1
species	1	–	–	–	–	–	–	1
Atopomelidae								
genera	37	1	–	5	3	1	20	8
species	183	(1)	–	61	30	4	59	29
				(2)		(1)	(1)	(2)
Chirodiscidae								
Chirodiscinae								
genera	1	–	–	–	–	–	1	–
species	1	–	–	–	–	–	1	–
Labidocarpinae								
genera	27	4	2	11	–	2	1	7
species	71	16	2	31	–	3	1	18
Schizocoptinae								
genera	1	–	–	1	–	–	–	–
species	2	–	–	2	–	–	–	–
Lemuroeciinae								
genera	1	–	–	–	1	–	–	–
species	1	–	–	–	1	–	–	–

References

Canestrini, G.: Prosp. Acarof. 5 : 567, 1892.

Fain, A.: Les Listrophoridés en Afrique au Sud du Sahara (Acarina: Sarcoptiformes). II. Familles Listrophoridae et Chirodiscidae. Acta Zool. Pathol. Antverp. 54: 1 – 231, 1971.

Gunther, C.E.M.: Notes on the Listrophoridae (Acarina: Sarcoptoidea). Proc. Linn. Soc. N.S.W. 67 : 109–110, 1942.

Trouessart, E.: Considérations générales sur la classification des acariens suivies d'un essai de classification nouvelle (suite). Rev. Sci. Nat. Ouest 2 : 49, 1892.

A PARTICULAR WAY OF PIERCING THE NASAL MUCOUS MEMBRANE USED BY MITES OF THE FAMILY RHINONYSSIDAE TROUESSART, 1895

Z. FEIDER, I. MIRONESCU
University "Alex I. Cuza", Museum of Natural History, Iaşi, Roumania

Abstract

The parasitic mites generally use the cheliceral digits for piercing the tegument or the mucous membrane in order to get the blood of the host.

We observed on the histological sections the nasal mucous membrane of the species *Ardea purpurea*. To its epithelium a species of *Mesonyssus* Fain, 1960, was attached by the claws of the three last pairs of legs. The claws of the first pait of legs, transformed into a piercing organ, penetrate as far as the level of corium, which was thickened by inflammation and its vessels were dilated, sclerotic and full of erythrocytes.

The species of *Mesonyssus* have the cheliceral digits transformed into leaf-like formations, unable to pierce the nasal mucous membrane. Hence it follows that in this species as well as in the other species of *Mesonyssus*, the claws of the first pair of legs have been adapted to a new function.

While studying the characteristic form of the claws of the first pair of legs in the species of the genus *Rhinonyssus*, whose cheliceral digits are also leaf-like, we may suppose that these claws too are adapted for an identical function of a piercing organ.

It is possible that the mentioned way of piercing the mucous membrane may be common in other genera of this family.

GAMASID MITES (GAMASOIDEA, PARASITIFORMES) OF THE CAUCASUS

A. T. GADZHIEV
Institute of Zoology, Azerbaijan Academy of Sciences, Baku, USSR

The natural conditions of the Caucasus, where large areas of semi-deserts and mountain steppes alternate with sub-alpine and alpine meadows and snow-capped peaks, are extremely varied. Nearly all climatic and vegetation zones are represented here and due to this fact the fauna of gamasid mites of all the Caucasus includes 200 species.

The most numerous and widespread are the representatives of the families Macrochelidae Vitzth., Phytoseiidae Berl., Laelaptidae Berl., Liponyssidae Ewing. Apart from them, there are also the following families: Rhodacaridae Oudmns., Parasitidae Oudmns., Veigaiidae Oudmns., Aceosejidae Baker et Wharton, Pachylaelaptidae Vitzth., Eviphididae Berl., Haemogamasidae Oudmns., Dermanyssidae Kol. and Spinturnicidae Oudmns.

The structure of the gamasid fauna in different ecological conditions of the Caucasus is not homogeneous. A wide ecological plasticity is characteristic of a large group of species which populate vast areas (*Macrocheles matrius* Hull., *M. glaber* Mull., *Amblyseius finlandicus* Oudmns., *A. aberrans* Oudmns., *Dermanyssus gallinae* Red., *D. hirundinis* Herm., *Hirstionyssus musculi* Johnst.) stretching from semi-desert lowland to high mountains.

The second group of species tends to be associated with certain zones and landscapes. They are the species: *Haemolaelaps glasgowi*, *Hl. androgynus*, *Hl. angustiscutis* Breg., *Laelaps hilaris* Koch, *L. pitymydis* Lange, *Myonyssus rossicus* Breg., *Haemogamasus nidi* Mich.

The distribution of specific parasites of animals depends on the range of habitat of their hosts; such species are: *Laelaps muris*, *L. algericus*, *L. jettmari*, *L. echidninus*, *Sauronyssus saurarum*, *Steatonyssus musculi*, *St. murinus* Lucas, *Spinturnix psi* Kol., *Eyndhovenia euryalis* Can. etc.

Of the total number of species recorded from the Caucasus, only 62 belong to the parasitic species, of which 2 are peculiar to reptiles, 14 to bats, 6 to birds and 40 to rodents. The parasitic gamasids show a different degree of specificity to their hosts. The most distinct specificity is found in bats and reptiles.

An intensive exchange of parasitic gamasids exists among different animals and birds. The intensity of exchange depends on the living conditions of host, on the degree of the parasite's adaptation to the host and on the relationships among hosts. In animals which live under similar ecological conditions this exchange of parasites may be observed more frequently and to a greater extent than in animals living under different ecological conditions. Parasites with a slight adaptation to their host take part in the exchange more often than those with a distinct degree of adaptation. In closely related species of animals, under other favourable conditions (if they live in similar ecological zones) the exchange of parasites takes place more frequently than in unrelated animal species.

About one third of all gamasids in the Caucasus constitute members of the family Phytoseiidae, living on plants. Among them more than 10 common species play an important role in the regulation of the number of plant-feeding mites.

Mites of the families Parasitidae, Veigaiidae, Macrochelidae, Pachylaelaptidae and Aceosejidae, which inhabit the soil surface and are very numerous, play an important role in the process of soil formation.

The altitudinal and landscape zones are distinctly reflected in the distribution of gamasid mites as well as of other elements of biocenosis. The landscape-ecological zones are characterized not only by the species composition of gamasid mites, but by the qualitative proportion of leading faunal components as well. Thus, the smallest number of species were found in the semi-desert lowland (69), a considerably larger number occurred in the foothills (105) and forests (92), while in the zone of high-altitude meadows only 78 species were recorded.

An analysis of gamasids' relationships with their habitat indicates that the principle of their adherence to habitat is not absolute (as in other organisms) and holds true usually in a small space and time. The zonal exchange of habitats is well evident in the example of *Hl. glasgowi* and *Hl. ellobii.* The first species is widespread in meadows with dense vegetation cover and considerable humidity. However, in the zone of semi-deserts and arid steppes *Hl. glasgowi* selects humid biotopes only, such as riverbanks and lakeshores. *Hl. ellobii* is a typical inhabitant of semi-deserts and open steppes. In its migration to other zones characterized by dense vegetation cover and high humidity, this species selects dry, well warmed and open habitats with thinned-out vegetation cover. The vertical and seasonal exchange of habitats takes place along with the zonal one.

The wealth and the exceptional variety of natural conditions of the Caucasus have provided the presence of nearly all zoogeographical complexes. However, the basis of the fauna is composed of widespread Palaearctic species. Due to their great species variety and high numbers the gamasids constitute a serious menace to man.

In the Caucasian settlements of urban and rural type 41 species of gamasid mites have been found. The basic core constitute parasites of synanthropic rodents: *L. algericus, L. echidninus, L. jettmari, O. bacoti;* parasites of bats: *St. murinus, Spinturnix kolenatii* and of birds: *D. gallinae, D. hirundinis, D. passerinus, Steatonyssus viator*. Also species parasitizing field rodents (*Hl. glasgowi, Hl. casalis, Hl. angustiscutis, Hl. longipes, Haemogamasus nidi,* take part in the formation of gamasid fauna in the settlements.

In the Caucasus the gamasids often become infected with causative agents of various diseases. For example, from gamasids collected from *Microtus arvalis* and *Meriones persicus* Blanford and from their nests more than 20 strains of the agent of erysipeloid were isolated. From the mites collected from *Microtus arvalis* alone 12 strains of the agents of tularemia were isolated. The gamasids collected from *Meriones erythrourus* proved to be infected with the agent of plague in more than 20 cases.

Frequent cases were also recorded of *D. gallinae, St. viator* and *O. sylviarum* attacking man. In all these instances the source of infection were the nests of synanthropic birds (the rock pigeon and the sparrow).

PRESENT SITUATION IN THE STUDIES OF CHIGGER MITES IN THE USSR

G. I. HUSHCHA
Institute of Zoology, Ukrainian Academy of Sciences, Kiev, USSR

The present paper summarizes the studies on chiggers carried out in various regions of the Soviet Union. Up till now there have been no general papers concerning the trombiculid fauna in the USSR. The author outlines the general trends of studies, but cannot analyse all papers published on the subject in this short communication. The studies on trombiculids in the Soviet Union were started not long before the second World War. Although previous sporadic findings of chiggers had been recorded, the information concerning this problem was very scant. In 1898 Ju. Wagner was the first to report mite larvae from Kiev, which Oudemans (1902) described as *Trombidium russicum*. In 1938 Y.D. Kirshenblat, while studying the parasite fauna of rodents, observed larvae of *T. autumnalis* in the Transcaucasus, Leningrad region and Uzbekistan.

The systematic studies of chiggers were connected with those carried out by E.N. Pavlovsky on the natural foci of transmissive diseases in the USSR. They were started by E.G. Shluger, who published a brief paper in 1947 on the material collected by the Far East expedition. It contained a list of 17 species of trombiculids. In the following papers (1948–1950) Shluger extended the range of her studies and in her dissertation described 35 species of mites of the USSR fauna. This was the first compendium on the trombiculid fauna of the USSR. In the above papers only *T. zachvatkini*, and *T. autumnalis* were mentioned with regard to the European part of the Soviet Union.

The first data on ecology of trombiculids were elucidated in the Soviet literature. In 1953 S.O. Vysotskaya and E.G. Shluger complemented the data on ecology of *T. zachvatkini* based on materials from the Leningrad region.

The starting point for further studies of trombiculids in the Soviet Union was the volume "Mites of the rodent fauna of the USSR" published in 1955. In the section entitled "Superfamilies Trombae", written by Shluger, the first keys to genera and species of larvae of the mites of the family Trombiculidae were given. The section included diagnoses of 32 species as well as brief data on morphology and development of mites. Due to further intensive studies of trombiculids the system presented in Shluger's key is of historical interest at present. It reflects the situation of systematics of 1943–1947, but it constituted the only key of chiggers in the Soviet literature and contributed considerably to the development of trombiculid studies in the USSR.

G.I. Huscha

The studies on the natural foci of transmissive diseases and the elucidation of the role of ectoparasites in them have lately considerably expanded the field of parasitological investigations in different regions of the USSR. Many new data have been obtained concerning the species composition and distribution of trombiculids. The studies of this nature, as a rule, are coordinated by the Gamaleya Institute of Epidemiology and Microbiology. All determination work, especially determination of the species composition of trombiculids in the Soviet Union has been done by Shluger.

The intensive studies on chiggers in the Far East were continued by the Irkutsk State Research Anti-Plague Institute of the Siberia and the Far East carrying out parasitological investigations in the territory from Tuva and Altai to the island of Sakhalin. Data were obtained on the species composition, distribution, range of hosts and seasonal changes in the parasitation of trombiculid larvae (Zhovtyi, Yemel'yanova, Fedorova et al., 1956–1960; Slonov, 1959; Zemskaya, Grokhovskaya, 1959; Kon'kova, 1962; Khudyakov, 1964).

These studies became specially important after the natural foci of tsutsugamushi fever had been discovered (Tarasevich et al., 1964; Kudryashova, Tarasevich, 1964; Kudryashova et al., 1968; Gopachenko, 1970).

The first results of the studies on the trombiculid fauna in the Central Asiatic republics were published in 1950. E.F. Sosnina reported the distribution of 23 species of chiggers in Tadzhikistan (1950–1963). Due to the work of the Central-Asiatic Anti-Plague Institute many new data from this region were obtained in a short period. In the following years the research was continued and complemented (Bibikova et al., 1956–1959; Korotkova, 1959; Rachinina et al., 1959; Dzhanokmen, 1967; etc.).

The first brief information on the trombiculids in the Caucasus and Transcaucasus was given by Kirshenblat (1938). Later Shluger (1955) and Razumova (1954–1957) recorded the distribution of trombiculids in various regions of the Northern Osetia and Georgia. The chiggers of Daghestan, Northern Osetia and Transcaucasus were reported by Kudryashova (1965–1967). The most successful and systematic research on trombiculids in the Transcaucasus (the Azerbaidzhan and Nakhichevan region) has been carried out since 1965 by L.V. Mulyarskaya who recorded in this region about 100 chigger species of the families Trombiculidae and Leeuwenhoekiidae and at present studies the chiggers in different landscape zones of this region (Mulyarskaya et al., 1965–1970).

In comparison with other regions of the USSR the territory of the Russian Soviet Federative Republic has been studied less. There has been scanty information on trombiculids from this Republic by Shluger and some other authors. The distribution of these mites was reported from the Voronezh region, the Krasnoyarsk region, the Omsk Reserve, the Tartar Republic, the Buryat region, the Yakut region, the Magadan region and the Kaliningrad region.

Studies on chiggers parasitizing rodents have been carried out also in Latvia (Grinbergs 1959; Lapin' 1956, 1970), in Byelorussia (Arzamasov et al., 1962, 1969). In Moldavia more than 30 species of chiggers parasitizing mammals and birds have been found in the short period from 1965 onwards (Andreyko, 1965–1968; Kudryashova, 1965; Pinchuk, 1968, etc.).

The territory of the Ukraine has been studied most completely. This is to the credit of many authors (Simonovich, Vshivkov, Hushcha, Sosnina, Vysotskaya etc.) as well as the Institute of Zoology of the Ukrainian Academy of Sciences which started a systematic research of chigger mites in 1956. An analysis of the results obtained and a list of recorded species were published by the author in 1971, and a repetition in this communication seems therefore to be unnecessary. A review of the data concerning the trombiculids is being prepared for publication in the volume "Fauna of trombiculids in the Ukraine".

The epidemiological role of trombiculids is being studied with regard to their role in the transmission of pathogen in natural foci of tsutsugamushi fever. No common cases of trombiculid infection in man have been reported in the Soviet Union. Some cases of trombidiosis in birds were reported in the south of the Ukraine.

The research on morphology, laboratory colonization and systematic studies are primarily conducted in Moscow, Kiev and Baku.

Finally the following conclusions may be drawn:

1. The majority of papers on the studies of trombiculids in the Soviet Union is of an ecological-faunistic character. These studies have made it possible to accumulate a large faunistic material which may serve as a basis for further, more complicated work on taxonomy.

2. At present more than 200 mite species of the families Trombiculidae and Leeuwenhoekiidae are known in the fauna of the Soviet Union. This number is not final and will undoubtedly increase after more advances have been made in the studies on the host range.

3. The territory of the USSR has not been studied evenly. Most completely studied have been the Ukraine, Azerbaijan and some regions of the Far East and Central Asia.

4. In the fauna of the Soviet Union the most numerous mites are those of the genera *Neotrombicula, Leptotrombidium, Neoschoengastia* and *Guntherana*. The last two mentioned genera are distributed in the southern regions of the USSR.

References

Andreyko O.F., In: Parazity zhivotnykh i rasteniy, Kishinev, Izd. "Kartya Moldavenyaske" 1: 53–60, 1965. – Andreyko O.F., In: ibid, Kishinev 4: 90–101, 1968. – Andreyko O.F., Pinchuk L.M., In: ibid 1: 43–52, 1965. – Andreyko O.F., Skvortsov V.G., Pinchuk L.M., In: ibid, Kishinev 3: 8–23, 1968. – Arzamasov I.T., In: Fauna i ekologia parazitov gryzunov, Minsk, Izd. "Nauka i tekhnika" 138–235, 1963. – Arzamasov I.T., In: Fauna i ekologia zhivotnykh Byelorussii, Minsk, Izd. "Nauka i tekhnika" 212–221, 1969. – Arzamasov I.T., Kurskov A.N., Doklady AN BSSR 4: 202–203, 1962. – Arzamasov I.T., Merkusheva I.V., Mikholap O.N., Chikilevskaya I.V., Minsk, Izd. "Nauka i tekhnika" 73–75, 1969. – Bibikova V.A., Gorbunova A.N., Koptsev L.A., Kudryavtseva K.F.,

Trofimova R.K., Trudy Sredne-Aziatskogo nauch.-issled. protivochumnogo instituta 2: 11–17, 1956. – Bibikova V.A., Kaluzhenova Z.I., Ibid 6: 291–294, 1959. – Dzhanokmen K.A., Avtoreferat dissertatsii, Alma-Ata 23 pp., 1967. – Gopachenko I.M., Avtoreferat dissertatsii, Vladivostok, Dal'nevost. nauch. centr AN SSSR, 22 pp., 1970. – Grinbergs A., Izv. AN Latv. SSR 12: 123–132, 1959. – Hushcha G.I. In: Paraziti, parazitozy ta shlyakhi ikh likvidatsii, Kiev "Naukova dumka" 196–212, 1972. – Khudyakov I.S., Dokl. na 15–m ezhegodn.chtenii pam. N.A. Kholodkovskogo 1962, M.–L. "Nauka" 12–21, 1964. – Kirshenblat Ya.D., Izd. LGU 92 pp., 1938. –Kon'kova K.V., Doklady Irkutskogo protivochum. instituta 4: 189–192, 1962. – Korotkova V.S., Ibid 6: 295–300, 1959. – Kudryashova N.I., Avtoreferat dissertatsii, Moskva, the Gamaleya inst., 14 pp., 1965. – Kudryashova N.I., Zool. zhurnal 44: 133–135, 1965a. – Kudryashova N.I., Med. Parazit. i parazitar. bol. 6: 726–730, 1967. – Kudryashova N.I., Mirolyubova L.V., Tarasevich I.V., Plotnikova L.F., Egorova A.L., Ibid 37: 302–305, 1968. – Kudryashova N.I., Tarasevich I.B., Ibid 33: 718–721, 1964. – Lapin' J., Izv. AN Latv. SSR 9: 111–122, 1956. – Lapin' I.M., Ibid 11: 25–32, 1970. – Mulyarskaya L.V., Trudy Inst. Zool. AN AzSSR 24: 162–182, 1965. – Mulyarskaya L.V., Parazitologia 2: 137–141, 1968. – Mulyarskaya L.V., Zool. zh. 48: 1097–1099, 1969. – Mulyarskaya L.V., In: Vopr. parazitologii, Baku "Elm" 225–235, 1969a. – Mulyarskaya L.V., Vestn. zool. 5: 42–47, 1970. – Mulyarskaya L.V., Verdieva Z.F., Dokl. AN Azerb. SSR 23: 92–95, 1967. – Mulyarskaya L.V., Gafarova F.G., Izv. AN AzSSR, seria biol. 2: 53–60, 1969. Oudemans A.C., Entomol. Berichten, I, 7, Amsterdam, 43–45, 1902. – Pinchuk L.M., In: Parazity zhivotnykh i rasteniy, Kishinev, Izd. "Kartya Moldovenyaske" 4: 184–190, 1968. – Razumova I.N., Avtoreferat dissertatsii, Leningrad, ZIN AN SSSR, 13 pp., 1954. – Razumova I.N., Parazitol. sbor. 17: 229–236, 1957. – Rachinina N.A., Malygina Z.G., Trudy Sredne-Aziatskogo nauch.-issled. protivochumnogo inst. 6: 301–306, 1959. – Slonov M.N., Med. parazitol. i parazitar. bolezni 28: 175–178, 1959. – Sosnina E.F., Soobsch. Tadzh. fil. AN SSSR 28: 7–9, 1950.–Sosnina E.F., In: Ushchelye Kondara, Izd. AN SSSR M.–L. 228–239, 1951. – Sosnina E.F., Trudy Inst. zool. i parazitol. im. E.N. Pavlovskogo AN Tadzh. SSR 64: 156, 1957. – Sosnina E.F., Ibid 115: 111–140, 1959. – Sosnina E.F., Shluger E.G., Ibid 24: 184–206, 1963. – Tarasevich I.V., Kulagin S.M., Kudryashova N.I., et al. ZhMEI 5: 19–23, 1964. – Shluger E.G., Novosti meditsiny 5: 18, 1947. – Shluger E.G., Entomol. obozr. 30: 157–164, 1948. – Shluger E.G., Rukopis dissertatsii, 120 pp., 1949. – Shluger E.G., Avtoreferat diss., ZIN AN SSSR, L. 7 p., 1950. – Shluger E.G., In: Kleshchi gryzunov fauny SSSR, AN SSSR 188–217, 1955. – Vysotskaya S.O., Shluger E.G., Parazitol. sbornik 15: 345–352, 1958. – Zemskaya A.A., Grokhovskaya I.M., Med. parazitologia i parazit. bolezni 28: 152–157, 1959. – Zhovtyi I.F., Emelyanova N.D., Kopylova O.A., Pokopyev V.N., Izv. Irkutskogo Gos. nauch. issled. protivochumnogo instituta Sibiri i Dal'nego Vostoka 17: 219–226, 1958. – Zhovtyi I.F., Emelyanova N.D., Fedorova L.V., In: Problemy parazitologii, Trudy 2. nauchnoy konferentsii parazitologov USSR, Kiev, Izd. AN USSR 240–241, 1956. – Zhovtyi I.F., Emelyanova N.D., Fedorova L.V. et al., Izv. Irkutskogo Gos. nauch.–issled. protivochumnogo inst. Sibiri i Dal'nego Vostoka 16: 156–172, 1957. –

Proceedings of the 3rd International Congress of Acarology, Prague, 1971

ERADICATION OF MYOBIID MITES INFESTING LABORATORY MICE

T. KOBULEJ, S. CZER
Veterinary College, Institute of Veterinary Zoology and Parasitology,
Budapest, Hungary

Laboratory mice are very often infested with mites of the genus *Myobia* to such an extent that they cannot be used for experimental work. This fact consequently hampers the accomplishment of scientific tasks set up at individual research centres. No wonder that considerable efforts have been made to work out an effective but simple method of control. Among numerous methods recommended, however, only few have reliable killing effect on myobiids. The common disadvantage of these methods is the fact that they are based on individual dipping of infested animals. Such a measure cannot be widely used in practice as it is clumsy and not too safe for the mice.

As the mice in our laboratory are often infested with myobiid mites we have started to work out a reliable method of eradicating the mites. Finally we have elaborated two methods, consisting in spraying the animals with 1 percent solution of trichlorphon (= Ditriphon, Neguvon) or allowing the mice to drink 0.1 percent solution of the same chemical.

We carried out two series of experiments. In the first series we studied the action on myobiids of sprays of different concentration (1%, 0.1%, 0.18%) of trichlorphon solved in water, in the second series we traced the action of the same compound applied per os. The experiments were carried out with common laboratory white mice naturally infected with mites of the species *Myobia musculi*. Heavily infested adult mice of both sexes as well as young individuals were used in the experiments. The effects were checked under stereomicroscope in mice treated with the acaricide and anaesthesized by Nembutal. When no myobiids were found on the mice after one month, the application was considered effective.

Trichlorphon proved to be an effective acaricide. Its 1 percent solution can be successfully applied as spray which should be repeated after a 12–day interval. Its 0.1 percent solution applied as drinking water is even more effective; such water should be consumed every alternate day, during four days and the treatment repeated after an interval of one week. The intervals are necessary because the acaricide has no action on eggs and larvae.

As a result, 1600 animals treated with 0.1 percent solution of trichlorphon per os, were rid of myobiid mites. No deaths were involved. Even after 30–40 days this treatment caused neither clinical symptoms nor histologically evidenced changes in liver and kidneys. The mice under treatment were able to produce young which subsequently developed into healthy adults.

NEST MITES OF SMALL MAMMALS IN THE WESTERN CARPATHIANS

L. KRAMÁROVÁ, M. MRCIAK
Natural History Faculty, Comenius University, Bratislava, Czechoslovakia

In recent years we have studied the nest fauna of small mammals in some mountain regions of Czechoslovakia (Vsetínské Beskydy Mountains, High Tatras, Spišská Magura Mountains). The aim of these studies is to ascertain the faunal composition of nests as a whole and to describe the representation of individual parasitic groups in the total nest fauna. Some results from the Vsetínské Beskydy Mts. and High Tatras appeared in press (Mrciak, Daniel, Rosický, 1969; Daniel, Mrciak, Rosický, 1970; Mrciak, Rosický 1970), but the results obtained from the mountain range Spišská Magura have so far remained unpublished. This mountain range is a part of the Carpathian system stretching to the north-east of the High Tatras. The highest peak Repisko reaches an altitude of 1259 meters (Fig. 1.).

During our studies we used the same methods which had been employed in our research conducted in the Vsetínské Beskydy Mts. and High Tatras. We focussed our attention to nest collection in all seasons of the year and to various types of biotope (forest, field, meadow, banks of streams). While analyzing the nest fauna two main factors were taken into account: nest localization (on the ground, in the ground) and nest building material (grass, moss or grass-moss).

A total of 303 nests of small mammals was collected (*Sorex araneus* 9, *Apodemus flavicollis* 29, *A. sylvaticus* 8, *Microtus arvalis* 213, *Pitymys subterraneus* 11, *Clethrionomys glareolus* 20 and *Arvicola terrestris* 13) in the region studied. A high population of *M. arvalis* which penetrated all biotopes, was recorded during our studies and consequently the highest number of nests collected was of this species.

From these 303 nests of small mammals a total of 185,100 animals were collected belonging to the following groups: Vermes 1,295, Mollusca 30, Aranea 574, Pseudoscorpionidea 71, Gamasoidea 110,575, Tyroglyphoidea 20,556, Bdellidae 614, Trombiculidae 4, Trombidiidae 61, Ixodidae 232, Oribatei 13,364, Chilopoda 90, Collembola 22,313, Coleoptera 2,656, Diptera 271, Siphonaptera 2,136, larvae of different groups of insects 9,858. The analysis shows that the dominant group was the gamasoid mites (59 %). If we also consider other mite groups, then the mites as such constitute as much as 78% of the total nest fauna of small mammals. The next abundant group represented in the nests was Collembola (12 %) (Fig. 2).

These results roughly correspond with those obtained in the High Tatras and Vsetínské Beskydy Mts, i.e. in all mountain regions of the Western Carpathians studied. In regions of lower altitudes the representation of individual animal groups in the nests of small mammals does not undergo any marked changes. Our results

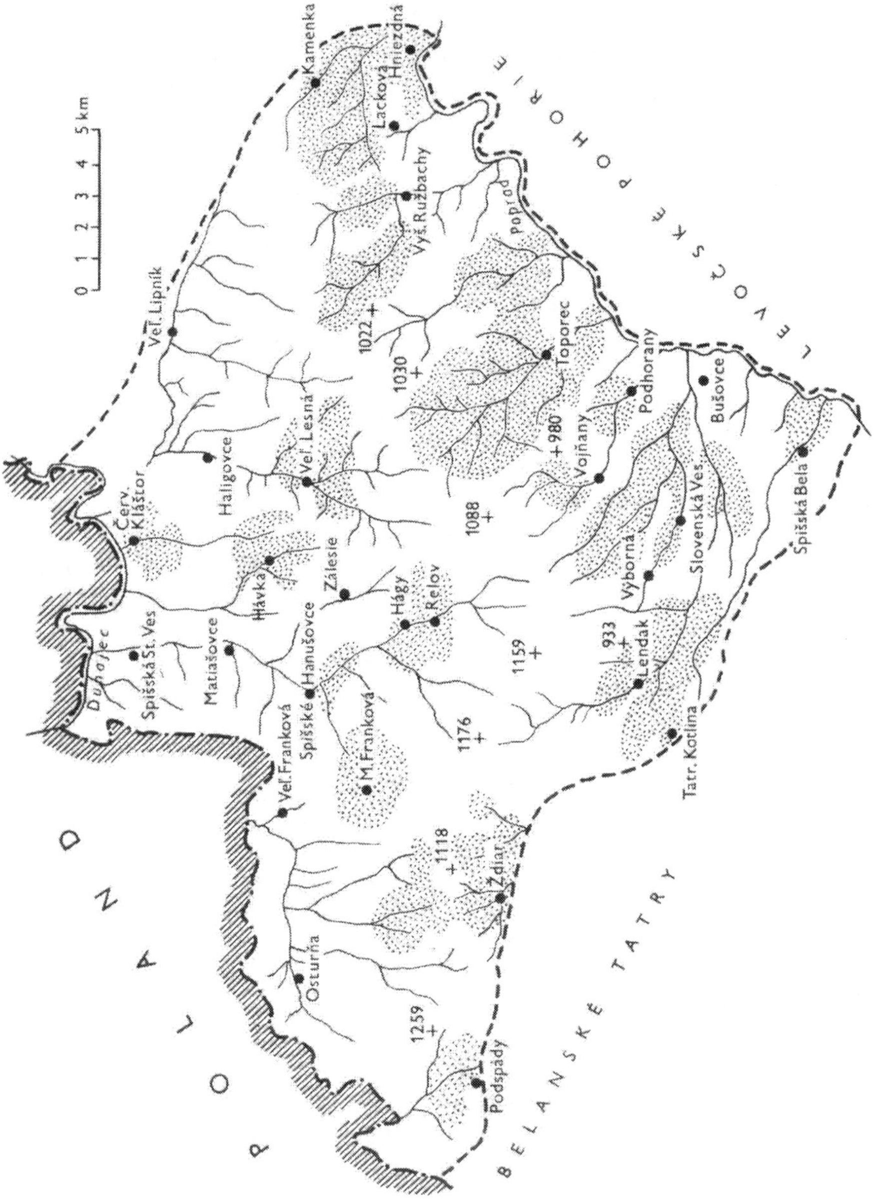

Fig. 1. A map of the Spišská Magura mountain range with marked localities where nests of small mammals were collected (horizontal shading).

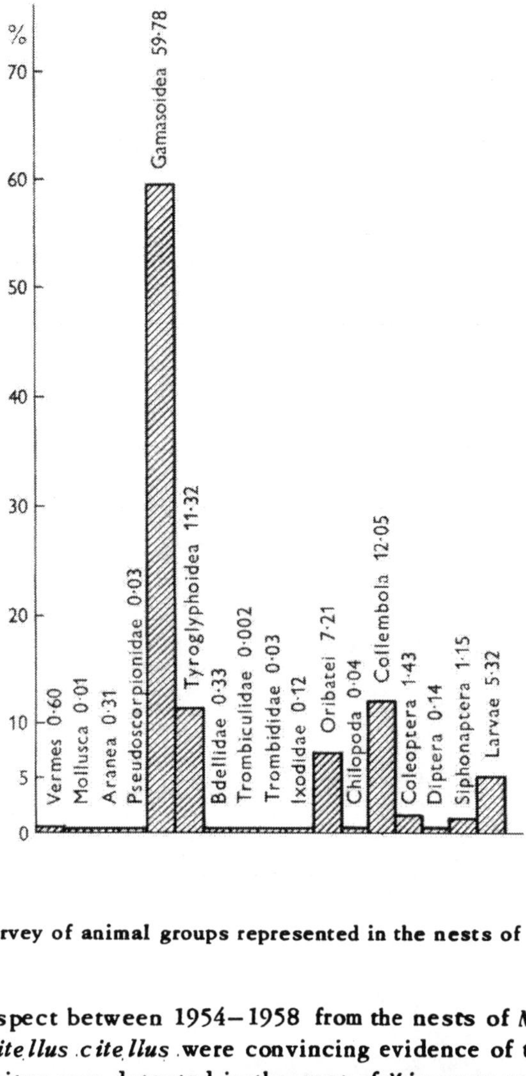

Fig. 2. A survey of animal groups represented in the nests of small mammals.

obtained in this respect between 1954—1958 from the nests of *Microtus arvalis*, *Talpa europaea* and *Citellus citellus* were convincing evidence of this fact. The highest representation of mites was detected in the nest of *Micromys minutus* collected in the nature reserve Jurský Šúr near Bratislava in 1964, reaching almost 90 % of the total nest fauna.

While analyzing the nest fauna from the Spišská Magura Mts. we ascertained that gamasoid mites were most abundant in the nests of all species of small mammals. For example, in the nests of *Microtus arvalis* they were represented by 70 %, of *Clethrionomys glareolus* by 68 %, of *Sorex araneus* by 65 %, of *Arvicola terrestris* by 37 %. The lowest percentage was found in the nests of *Pitymys subterraneus* (28 %), differing from the higher percentage of mite representation in the *P. subterraneus* nests collected in Vsetínské Beskydy Mts. and the High Tatras.

Our studies showed that the mite fauna in the nests of small mammals is greatly affected by the nest building components. Nearly every nest of the mentioned small mammals consists of two or three components. They are: the external part of the nest consisting of a rough material and the internal part, consisting mostly of a fine vegetation material. This internal part is of greatest importance for the fauna of ectoparasites, especially mites of the order Parasitiformes which constitute 80 – 90 % here. Oribatei, which occur in the external part of the nest characterized by a higher humidity, are rarely found here. While assessing the nest fauna of small mammals from the Spišská Magura Mts. in separate seasons of the year (spring, summer, autumn, winter) as well as in all types of biotopes (forest, meadow, field, banks of streams) the most dominant among all animal groups were Gamasoidea, Tyroglyphoidea, Oribatei and Collembola. The remaining groups were represented by a low percentage, or were lacking altogether.

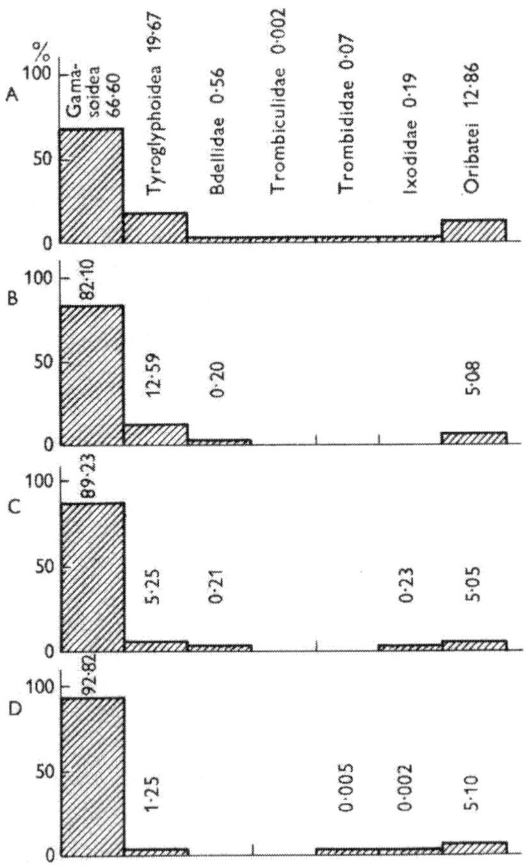

Fig. 3. A survey of mite groups represented in the nests of small mammals during a year: A = spring; B = summer; C = autumn; D = winter.

Due to the high percentage of mites represented in the nest fauna of all species of small mammals we would like to point out in this paper the occurrence of mite groups only (Gamasoidea, Tyroglyphoidea, Bdellidae, Trombiculidae, Ixodidae, Oribatei), irrespective of other animal groups. In separate seasons of the year certain changes in the representation of mite groups could be observed. The occurrence of gamasoid mites was lowest in spring (66 %), but in the succeeding seasons it gradually increased (in summer 82 %, in autumn 89 % and in winter 92 %). In contrast, Tyroglyphoidea were most numerous in nests collected in the spring; the same applies to Oribatei. Ixodidae were most abundant in autumn and in spring (Fig. 3).

As regards biotopes, the highest representation of gamasoid mites was found in the biotopes of cultivated fields (87 %) and in the biotope of forest (79 %). In the biotopes of stream and marshy places they reached only 59 %. The highest repre-

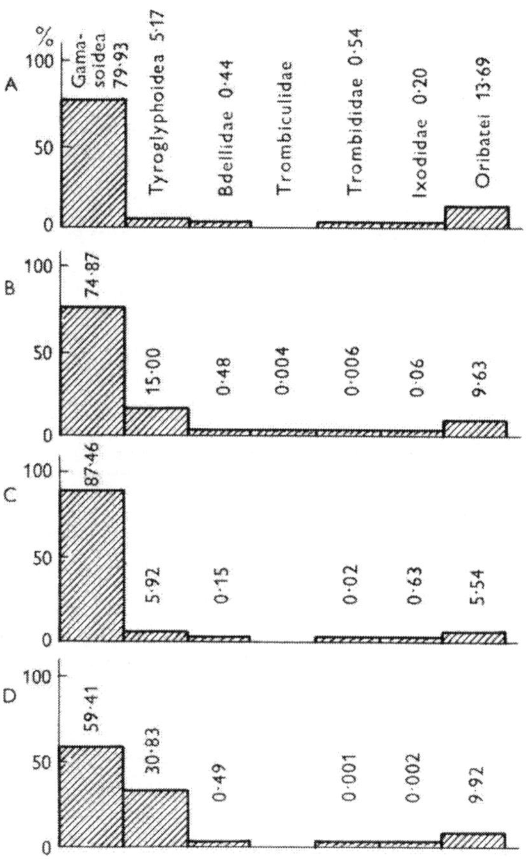

Fig. 4. A survey of mite groups represented in the nests of small mammals with regard to biotopes: A = forest; B = meadow; C = field; D = stream banks.

sentation in all biotopes was held by tyroglyphoid mites. Oribatid mites were most abundant in the forest biotope (Fig. 4).

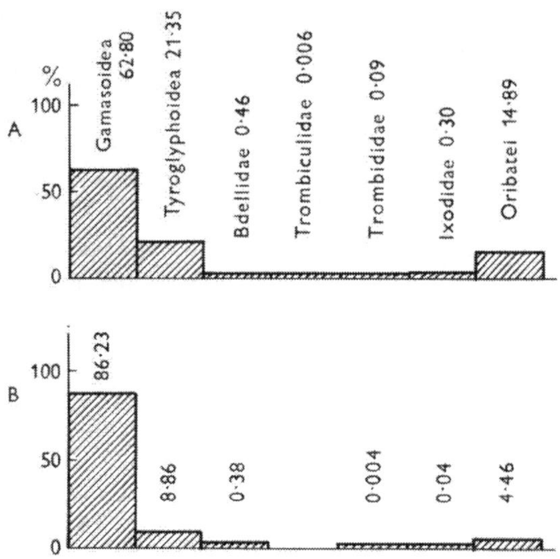

Fig. 5. A survey of mite groups represented in the nests of small mammals with regard to the localization of nest: A = nests on the ground; B = underground nests.

Some differences in the mite fauna were revealed in nests located on the ground and in underground nests. Gamasoid mites were represented by 86 % in underground nests, by 62 % in nests on the ground and the representation of tyroglyphoid mites was reversed. The representation of oribatid mites was considerably higher in nests collected on the ground than in underground nests. Trombiculids were also found in nests situated on the ground. (Fig. 5)

The nest-building material (grass, moss or grass-moss mixture) proved to be of no importance in the numerical changes of gamasoid mites, because the representation values in all cases were approximately equal (75 %). The highest number of tyroglyphoid mites was found in the grass nests (65 %), the lowest in the moss nests (15 %). The oribatid mites, on the other hand, were most represented in the moss nests (15 %), least in the grass nests (8 %). The ticks were most abundant in the grass nests as well as in the grass-moss nests. No ticks were found in the moss nests (Fig. 6).

The mite fauna of nests of small mammals requires further and more detailed studies of an ecological nature and these will be continued by us in the future.

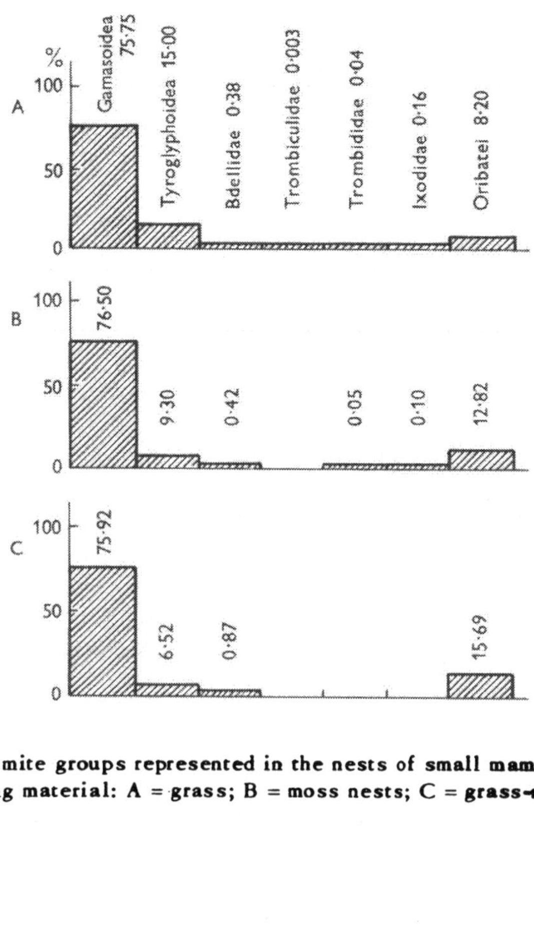

Fig. 6. A survey of mite groups represented in the nests of small mammals with regard to the nest-building material: A = grass; B = moss nests; C = grass-moss nests.

References

Daniel, M., Mrciak, M., Rosický, B.: Location and composition of nests built by some central European insectivores and rodents in forest biotopes. Acta F.R.N. Univ. Comen. – Zoologia XVI, 1 – 36, 1970.

Mrciak, M., Daniel, M., Rosický, B.: Parasites and nest inhabitants of small mammals in the western Carpathians. Acta F.R.N.Univ. Comen. 12 – Zoologia 13: 81–116, 1969.

Mrciak, M., Rosický, B.: Ekologicke študie o nidikolnej faune drobných cicavcov vo Vysokých Tatrách. Národné parky – bohatstvo civilizácie, pp. 360–383, 1970.

CHIGGERS (FAMILY TROMBICULIDAE) OF THE NORTH AMERICAN DESERTS*

R. B. LOOMIS
Department of Biology, California State College, Long Beach,
California, U.S.A.

Of the more than 285 species belonging to 44 genera of trombiculid mites in North America (Brennan & Jones, 1959, etc.), 135 species (47 %) of 31 genera (70 %) have been found in the deserts. Some species occur only at marginal or isolated localities, but most records are based on several to many samples.

Chiggers regularly parasitize amphibians (2 genera and 4 species, especially *Hannemania)*; reptiles (13 genera of 26 species, largely *Acomatacarus*, *Eutrombicula*, *Hexidionis*, *Hyponeocula*, *Kayella* & *Odontacarus)*; birds (9 genera of 16 species, especially *Neoschoengastia* and *Toritrombicula)* and mammals (29 genera and 108 species). Most mammalian hosts were rodents and the common genera included *Euschoengastia*, *Euschoengastoides*, *Hexidionis*, *Hyponeocula*; *Otorhinophila*, *Pseudoschoengastia* and *Sasacarus*. Bats possessed very few chiggers (7 genera and 13 species, including *Microtrombicula*, *Perissopalla*, *Speleocola*, *Tecomatlana* and *Whartonia)*.

The deserts

The three major hot subtropical to temperate deserts of North America are the Mojave, Sonoran and Chihuahuan deserts (Figure 1). They all have hot summers, but they differ in rainfall patterns: Mojave with winter precipitation, Chihuahuan with summer rainfall, the eastern Sonoran with both summer and winter rains, the western part with mostly winter rainfall. The Sonoran is the lowest, hottest, and driest, has tropical influence and the greatest diversity of plants, hosts and chiggers. To the north is the higher Mojave Desert, largely covered by Yuccas, Creosote and Burro Bush. It has fewer kinds of chiggers and many of these are shared with the Sonoran Desert.

The high Chihuahuan Desert is separated from the Sonoran and Mojave deserts by mountains and plateaus and many of its chiggers also occur throughout the adjacent grasslands.

---

* Studies upon which this paper is based were supported by the U.S. Public Health Service Research Grant AI03407.

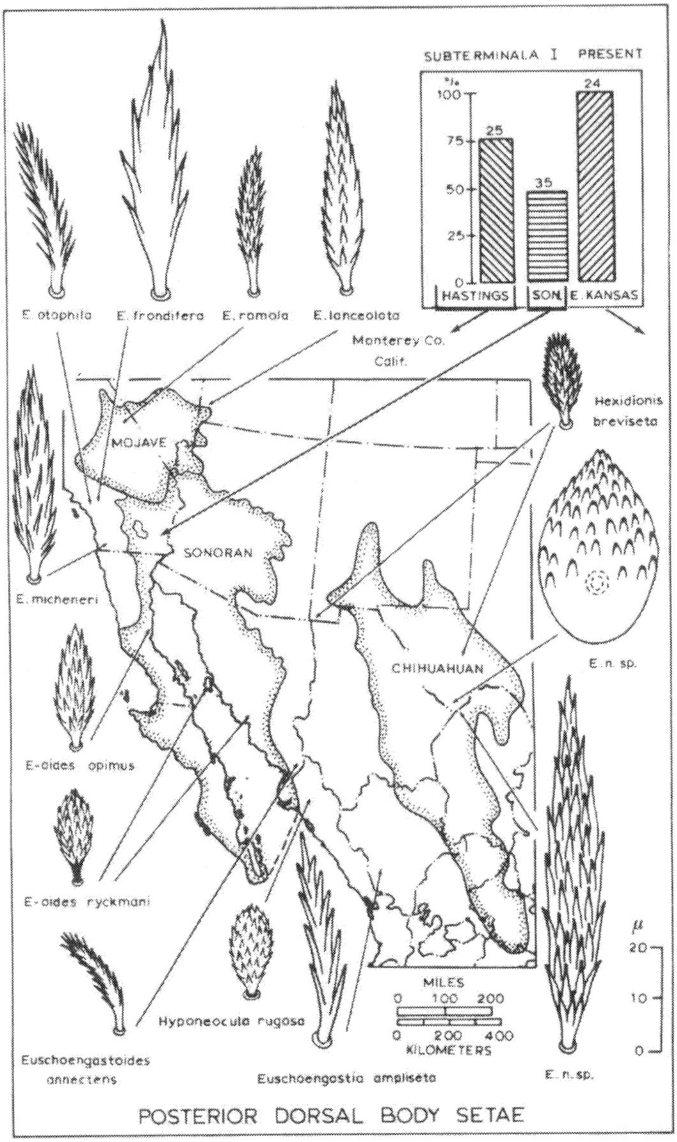

Fig. 1. The North American deserts and certain characteristics of larval trombiculids. The Mojave, Sonoran and Chihuahuan deserts of North America are outlined and certain setae on larvae are shown. The percentages of subterminala I on larvae are shown (upper right) for the Hastings Natural History Reservation, Monterey Co., in coastal California (Brennan & Jones, 1954); the Sonoran Desert of southeastern California, and eastern Kansas (Loomis, 1956). Selected expanded posterior dorsal body setae are depicted for eight species of *Euschoengastia*, three species of *Euschoengastoides* and one each of *Hexidionis* and *Hyponeocula*. Each arrow indicates a known locality.

All three deserts contain isolated mountain ranges supporting chiggers of non-desert groups.

The unfed larva

The hot, dry desert represents one of the most challanging habitats to the unfed active larva. Even though these chiggers do not actively search for hosts on the desert surface, the habitats in crevices, subsurface burrows and nest cavities have a low humidity. Therefore these larvae must have the ability to avoid desiccation,

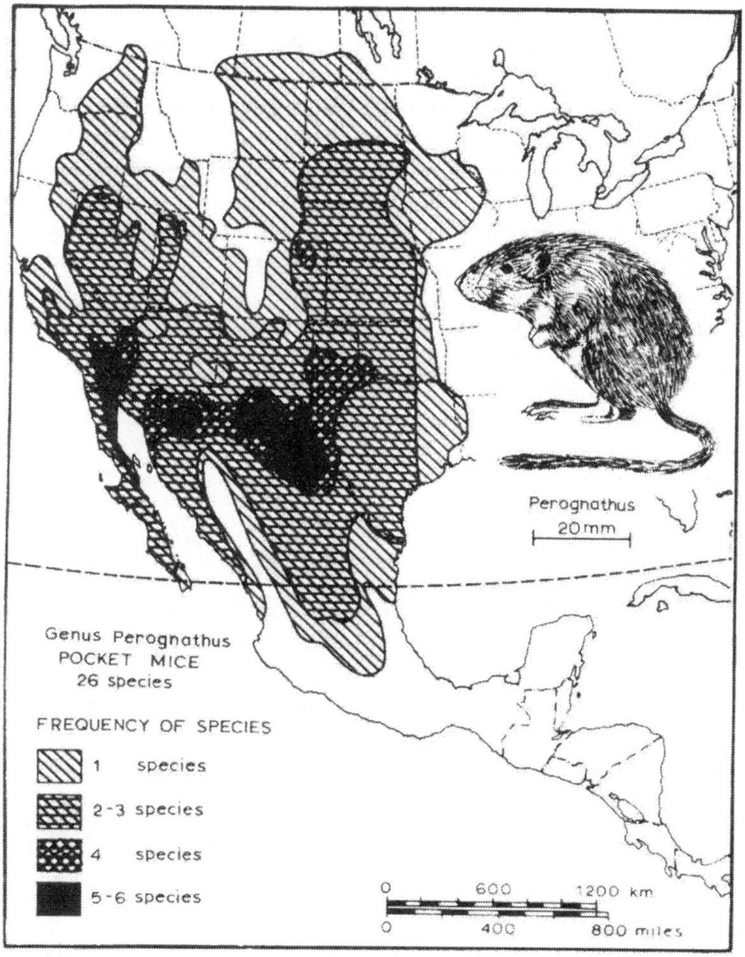

Fig. 2. The distribution and frequency of 26 species of heteromyid pocket mice of the genus *Perognathus*. Note the centers of abundance in the three deserts.

both in behavior and in morphology, such as having expanded body setae, smaller body and legs, less permeable cuticle, and reduction in size and numbers of porous setae.

A survey of the larvae with expanded body setae have revealed that: 1) Most species (especially in the genus *Euschoengastia)* are found in peripheral areas, rather than in the desert (Figure 1). (Those two new species of *Euschoengastia* which seem to be from the Chihuahuan Desert actually are from isolated mountain ranges.) 2) *Euschoengastoides* (Loomis, 1971), *Hexidionis*, and *Hyponeocula* (Tanigoshi, 1968), all common desert genera, have some species with inflated body setae, and several species are shown in Figure 1.

These setae are believed to form a protective cover or fluid reserve for the active unfed larva to prevent desiccation in a dry habitat.

Among the numerous leg setae is a hollow seta with an open tip on tarsus I called a dorsal eupathid or subterminala I. Fowler (1968) found that it is a hygroreceptor and will lose and gain water through this open tip. The percentages of occurrence of this seta are shown in Figure 1. Where the mean annual rainfall is high (over 30 inches) such as in eastern Kansas, all 24 species possess this seta (Loomis, 1956). Where rainfall is moderate (about 20 inches), at the Hastings Natural History Reservation in coastal Monterey County, California (Brennan & Jones, 1954), the positive percentage is 76 % and all 6 species lacking this seta were taken on *Perognathus*, and also from several desert localities. In the Sonoran Desert of southeastern California, which has less than 5 inches of rainfall, only 49 percent of the chiggers had this seta. In addition, many of these desert chiggers have fewer and shorter leg setae. Therefore it appears that the reduction of nude leg setae, especially the lack of subterminala I, is correlated with arid habitats and water conservation.

Perognathus as a chigger host

The genus *Perognathus* consists of 26 species of small to medium sized heteromyid rodents or pocket mice of the North American grasslands and deserts, as shown in Figure 2. These pocket mice are nocturnal, frequently are abundant and prefer dense stands of grass and shrubs. A single mouse may possess hundreds of attached larvae of one to several species on the ears, nose, cheek pouches and genitalia. They are known hosts for 81 species of 18 genera of chiggers. Some percentages of parasitism on *Perognathus* are: *Euschoengastoides* (Loomis, 1971) all 11 species, 40 % of all hosts; *Hyponeocula* (Tanigoshi, 1968) 10 species, 7 on *Perognathus*, 33 % of all hosts; *Otorhinophila* (Wrenn & Loomis, 1967) all 7 species, 59 % of all hosts. All these genera (Figure 3) average 44 % (573 *Perognathus* of 1306 hosts).

These chiggers and the host pocket mice have the same general pattern of species abundance (Figures 2 & 3). Their centers of abundance are in the three major deserts and the greatest frequency reflects overlapping faunas originating in two or more different desert regions.

Perognathus live in many different habitats. Therefore not only do they pick up their regular chiggers, they also acquire many of those regularly on *Dipodomys*, a common heteromyid of the dunes and open deserts, as well as some of those trombiculids usually on cricetid rodents of the genera *Neotoma* and *Peromyscus* from rocky and forested habitats.

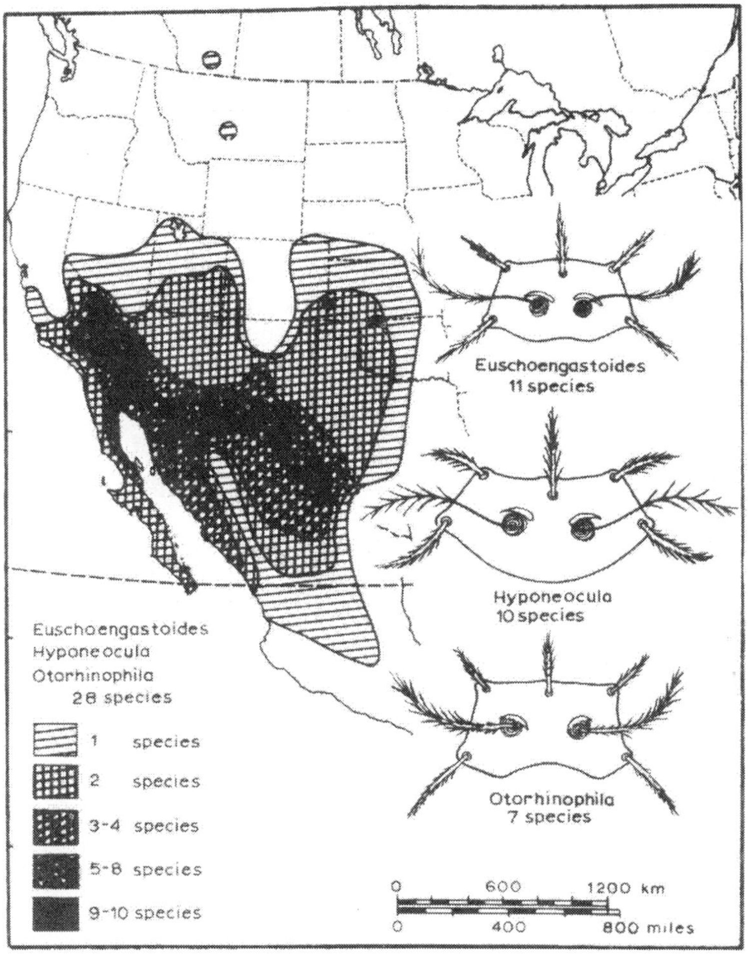

Fig. 3. The distribution and frequency of 28 species of chiggers belonging to the three genera *Euschoengastoides* (Loomis, 1971), *Hyponeocula* (Tanigoshi, 1968) and *Otorhinophila* (Wrenn & Loomis, 1967). The scutal plates are shown on the right. Note the greatest number of species in the Sonoran Desert of southeastern California.

Summary

The unfed larva in the desert is able to conserve water by the absence and/or reduction in size of the porous leg setae, inflated body setae, a smaller body and shorter legs, nocturnal and underground activity in close proximity to a suitable host (such as *Perognathus*), usually stimulated by any precipitation in the appropriate season. Once attached externally or in the skin or nasal passage, the engorging larva derives a new source of water which must maintain it until it feeds again, usually as the nymph.

References

Brennan, J.M., Jones, E.K.: A report of the chiggers (Acarina: Trombiculidae) of the Frances Simes Hastings Natural History Reservation, Monterey County, California. Wasmann J. Biol. 12 (2): 155–194, 1954.

Brennan, J. M., Jones, E. K.: Keys to the chiggers of North America with synonymic notes and descriptions of two new genera (Acarina: Trombiculidae). Ann. Entomol. Soc. Am. 52 (1): 7–16, 1959.

Fowler, J. L.: Observations on the morphology of selected setae on larval trombiculid mites. MA Thesis, California State College, Long Beach, 78 p., 1968.

Loomis, R. B.: The chigger mites of Kansas (Acarina, Trombiculidae). Univ. Kansas. Sci. Bull. 37: 1195–1443, 1956.

Loomis, R. B.: The genus *Euschoengastoides* (Acarina, Trombiculidae) from North America. J. Parasitol. 57 (4), 1971. (In press.)

Tanigoshi, L.K.: Revision of the genus *Hyponeocula* (Acarina, Trombiculidae) from western North America. MA Thesis, California State College, Long Beach, 122 p., 1968.

Wrenn, W.J., Loomis, R.B.: *Otorhinophila*, a new genus of chiggers (Acarina, Trombiculidae) from western North America. Acarologia 9 (1): 152–178, 1967.

TO THE ZOOGEOGRAPHY OF SOME MITE SPECIES (GAMASOIDEA) PARASITIZING SMALL MAMMALS IN EUROPE

M. MRCIAK
Natural History Faculty, Comenius University, Bratislava, Czechoslovakia

Ecological-faunistical data on gamasoid mites have lately attracted attention as regards the distribution of the mites. As a rule, they are the results obtained only from very small or larger localities. The data concerning large territorial sections over which the gamasoid mites are distributed and elucidating the routes and causes of this distribution, are scarce. Of interest is the vertical but primarily the horizontal spread which often occurs over long distances from their original habitats to different geographical regions. Such a distribution of gamasoid mites depends on the exchange of the original host or phylogenetically distant hosts. This fact indicates that the distribution of gamasoid mites associated with small mammals as their hosts depends on great adaptive changes. Primarily they are the climatical or microclimatical changes and changes of host species.

In this brief contribution I would like to point out some new results concerning the distribution of some mite species of montane type in Europe. One such species is *Laelaps clethrionomydis*, first found in the Soviet Union, occurring in different regions of the taiga zone where it has adapted itself to various host species. Goncharova (1956) reports examples of findings in the East-Siberian region where this mite is parasitizing *Microtus gregalis*. In places where this host is missing, the mite has adapted itself to *Microtus ungurensis*. In the Ala Tau region beyond the river Ili it parasitizes mainly *Clethrionomys frater* (Bibikova, 1956). In the taiga zone between the Kola Peninsula across Siberia to southern Primorye region its main host is *Clethrionomys rufocanus*. In the taiga region of the European part of the Soviet Union where *C. rufocanus* is missing, the mite has adapted itself to *C. glareolus*. In the forest zones of Central Europe it is distributed along with its host *C. glareolus*. Its more abundant occurrence has been almost always recorded in the mountain or high altitude regions (the High Tatras, Belanske Tatras, Western Tatras, the Polish section of the High Tatras, Low Tatras, Spišská Magura, Babia Gora, Beskydy, Jeseníky, Krkonoše, Šumava Mountains). This fact indicates that *L. clethrionomydis* has preserved its affinity to the fauna of boreo-montane type. The results of our studies in the Kamniške Alps (Yugoslavia) indicate that the southern parts of the Alps where a single specimen (1 female) was found mark the distribution boundary of this species in the south. However, our further studies conducted in the Julian Alps and in many mountain regions of the Dinaric Alps as well as in the Pirin mountain range and the Rhodope Mountains (Bulgaria) where rich acarological material was obtained, have confirmed that the distribution of *L. clethrionomydis* extends towards the Balkan Peninsula over the Carpathian mountain system in the east and over the Alpine mountain systems, and Dinar mountain range to the Prokletije mountain range in the west (Mrciak, Brelih in lit.). In northern Europe its occurrence is known on *C. glareolus* and *C. rufocanus* and to a lesser degree also on other species of small mammals (Edler, 1968, 1969; Mrciak, Viramo, in litt.). Its occurrence is more abundant in the northern parts of Finland, Sweden and Norway. The studies conducted in many localities of south-western Finland yielded negative results. On numerous specimens of *C. glareolus* trapped in this part of Finland only the mite *Laelaps hilaris* was found (Mrciak, Brander, 1965). The occurrence of *L. clethrionomydis* in the northern part of Scandinavia is relatively abundant irrespective of mountain ranges of higher altitudes. In this fact the results differ from those obtained in the Balkans and Central Europe as well, where it is strictly associated with mountain localities only. It is regrettable, however, that data from the Polish and German lowlands are lacking.

With respect to the European mite fauna, apart from the above mentioned species, some other gamasoid mites may be considered as species of montane type previously referred to (Rosický, Mrciak, 1970). They are: *Laelaps pitymydis, Haemogamasus bregetovae, Haemogamasus nidiformes* and *Hirstionyssus tatricus*. The last mentioned species, however, reveals characteristics of the boreo-alpine fauna. Among the mentioned species the rarest in the acarine fauna of Europe is *L. pitymydis*, known from the Soviet Union (Caucasus, Georgia, southern Osetia). In European countries only one female specimen was reported from the High Tatras (Mrciak, 1958); this may be assumed to be a case of relict occurrence. The remaining species, of which *H. tatricus* and *H. nidiformes* in the southern and central Europe are of distinct montane character, are relatively more abundant. On the other hand, in the northern parts of Scandinavia its occurrence was often confirmed in the forest zones of low altitude. In such regions of Finland, Sweden and Norway also sporadic findings of *Haemogamasus liponyssoides*, a species of boreal character, were recorded from different species of small mammals (*Sorex araneus, Clethrionomys glareolus, C. rufocanus, Microtus agrestis*) (Mrciak, Viramo, in litt.; Edler, Mrciak, in litt.). In the southern regions of Scandinavia it was not found and its occurrence in other European countries seems to be improbable. Although Serbova (1961, 1962) reported numerous findings of this species in the environs of Burgas, Petrič and Goce Delčev, these findings in Bulgaria should be verified in view of the very different geographical conditions. The distribution of this species has, until now, been reported from Canada, USA, Mongolia, China, Northern Korea and several regions of the USSR (Tomsk, Primorye, Karelia, Khabarovsk region etc.).

The acarine fauna of Europe is further enriched with a new mite species of boreal type, *Laelaps dolomysi* (Mrciak, in litt.), described from the host species, the hitherto preserved endemite *Dolomys bogdanovi* (V. et E. Martino, 1922), an endemic vole which has survived only in several mountainous limestone localities of Yugoslavia. The distribution of this mite depends on the distribution of its host *Dolomys bogdanovi*, on which the mite occurs in large numbers. By the courtesy of Dr. Hanzák, Prof. Kratochvíl and Dr. Brelih I was able to obtain and determine some rich acarological material collected from *D. bogdanovi* in all hitherto recorded localities (Malá Kapela, Čakor, Durmitor, Bjelasica, Čapljina-Karaotok, Vranjak, Karadžiča, Šar-Planina, Popova Šapka, Orjen) as well as the material obtained during our own expeditions. In a single case 1 female specimen of *Laelaps dolomysi* was collected from *Microtus nivalis* trapped in Šar-Planina, near the habitat of *Dolomys bogdanovi*. The occurrence of this mite on *M. nivalis* may be considered as accidental.

The opposite of montane mite species are those species which have adapted themselves to lowland and warmer regions. Well known are also a few species with great ecological valency, whose distribution underwent a long and complicated process, depending on the gradual adaptation of these species to new ecological conditions.

The studies on the distribution of mites are of manifold importance, be it in the economy or in the problems of epidemiology and epizootology.

References

Bibikova, V. A.: Gamasoid mites of the south-eastern Kazakhstan. Trudy Zool. Inst. AN Kazakhskoy SSR 5: 125–160, 1956. (In Russian.)

Goncharova, A. A.: To the fauna of gamasoid mites of Eastern Siberia. Parasitol. Sb. Zool. Inst. AN SSSR 10: 198–205, 1956. (In Russian.)

Edler, A.: Ectoparasitic mites (Acarina: Parasitiformes) from small mammals from the Tärnasjö and Ammarnäs in Swedish Lapland. Entomol. Ts. Arg. 89.H. 3–4, 1968.

Edler, A.: Ectoparasitic mites (Acarina) from small mammals in Central Sweden. Entomol. Ts. Arg. 90.H. 3–4, 1969.

Edler, A., Mrciak, M.: K faune roztočov (Gamasoidea) Škandinávie. Acta Fac. Rer. Nat. Univ. Com. Zool. (In press.)

Mrciak, M.: Roztoče z radu Parasitiformes (Acari) z drobných cicavcov Vysokých Tatier. Zool. listy 1, 65–86, 1958.

Mrciak, M.: The new species of the mite *Laelaps dolomysi* (Gamasoidea Parasitiformes). Anotaciones Slov. Nar. Mus. (In press.)

Mrciak, M.: Brander, T.: Beitrag zur Kenntnis der Milben (Acarina) einiger Säugetiere in Finnland. Lounais-Hämeen Luonto 15: 23–25, Forssa, 1965.

Mrciak, M., Brelih, S.: K poznaniu fauny roztočov (Acarina) drobných cicavcov v Juhoslávii. Acta Fac. Rer. Nat. Univ. Com. Zool. (In press.)

Mrciak, M., Viramo, J.: K faune foztočov (Acarina) drobných cicavcov Fínska. Acta Fac. Rer. Nat. Univ. Com. Zool. (In press.)

Rosický, B., Mrciak, M.: Zoogeographia parazitov Vysokých Tatier. Národné parky – bohatstvo civilizácie, 445–459, 1970.

Serbova, S.: A contribution to the studies of mite fauna (Parasitiformes) from small mammals in the district of Burgas. Complex research on natural foci of diseases in Strandja planina, BAN, Sofia, 81–88, 1961. (In Bulgarian.)

Serbova, S.: A contribution to the studies of mite fauna (Parasitiformes) from small mammals in the district of Petrič and Goce Dolčev. Natural foci of Diseases in Petričko and Gocedolčevsko, BAN, Sofia, 75–80, 1962. (In Bulgarian.)

THE DISTRIBUTION OF GAMASOID MITES PARASITIZING THE HOUSE MOUSE *(MUS MUSCULUS)* IN THE UKRAINE

G. I. SHCHERBAK
The Institute of Zoology, the Ukrainian Academy of Sciences, Kiev, USSR

While studying the gamasoid mites of the Ukraine a special attention was directed to the study of the gamasoid fauna of the house mouse. Our interest in the gamasoid mites of this rodent has been aroused by the fact that the house mouse lives in the immediate vicinity of man and is the reservoir animal of pathogens inducing a number of diseases (vesiculous rickettsiosis, Q-fever etc.), in whose dissemination gamasoid mites also take part. The gamasoid fauna of *M. musculus* was studied in its different habitats of all landscape-climatic zones, in the Ukrainian woodlands, forest-steppe, steppe, the Carpathians and the Crimean highlands. Besides our own material we analysed data from literature on gamasoid mites parasitizing the house mouse (Antonenko, 1970, Belokon, 1965, Zhuravleva, 1964, Naglov, Naglova, 1963 etc.). More than 40 species of gamasoid mites were found on the house mouse throughout the territory of the Ukraine (Table 1). In different zones sporadic deutonymphs of the family Parasitidae, and females of the families Phytoseiidae, and Veigaiidae were found on individual rodents; the species *Laelaps pavlovskyi, L. hilaris, L. jettmari, Haemolaelaps semidesertus, Hirstionyssus criceti, H. isabellinus, Haemogamasus pontiger, Macrocheles decoloratus, M. matrius, Hyperlaelaps arvalis, Myonyssus decumani* which usually occur on many other small mammals and in their nests, were also found.

More frequent, although not numerous were the mites *Haemolaelaps glasgowi* throughout the territory of the Republic, the species *Hirstionyssus latiscutatus, H. apodemi, Haemogamasus nidi, Eulaelaps stabularis* in all zones except the woodlands and *Laelaps agilis* except the Crimean zone. The absence of the mites *H. latiscutatus, H. apodemi, E. stabularis* on the house mouse in the woodlands may be explained by their low numbers in this zone on other rodents, too. *L. algericus* was the most frequent species which occurred in higher numbers in all zones, both in natural habitats and in houses of rural and summer type.

The most complete studies were carried out on the gamasoid mites of the house mouse under conditions of the forest-steppe, where the mice were trapped in multistoreyed houses of a large town, in one-storeyed rural houses, in bungalows and finally, in natural habitats. A total of 25 gamasoid mites was found, but the species composition found on the house mice trapped under different ecological conditions was varied (Table 2). The table shows that the lowest number of species was found on mice trapped in multistoreyed houses, where only 3 species occurred; they were: *Parasitus consanguineus* (found once), *Ornithonyssus bacoti* (extensity of infesta-

Table 1. Gamasoid mites found on the house mouse in the Ukraine

Species of mites	Wood-lands	Forest steppe	Steppe	The Carpa-thians	Crimea
Cyrtolaelaps minor Willm.		⊕	⊕		
Euryparasitus emarginatus (C.L. Koch)		⊕			
Poecilochirus necrophori Vitzth.			⊕		
Pergamasus crassipes L.		⊕			
Parasitus remberti Oudms.		⊕			
P. consanguineus Oudms.		⊕			
P. distinctus (Berlese)		⊕			
P. magnus (Kramer)				⊕	
Parasitidae gen. sp. NII		⊕	⊕		⊕
Gamasolaelaps excisus C.L.Koch		⊕			
Macrocheles matrius Hull		⊕	⊕		
M. decoloratus C.L.Koch			⊕		
Ameroseius plumosus Oudms.		⊕		⊕	
Ameroseius sp.		⊕			
Amblyseius sp.		⊕			⊕
Blattisocius sp.		⊕			
Garmania hypudaei Oudms.		⊕	⊕		⊕
Alliphis siculus Oudms.		⊕			
Eulaelaps stabularis C.L.Koch		⊕	⊕	⊕	⊕
E. kolpakovae Breg.		⊕			
Myonyssus decumani Tirab.			⊕		
Hypoaspis sp.			⊕		
H. lubrica Voigts et Oudms.			⊕		
H. miles Berl.		⊕	⊕		
Haemolaelaps semidesertus Breg.			⊕		
H. casalis Berl.		⊕	⊕		
H. glasgowi Ew.	⊕	⊕	⊕	⊕	⊕
Laelaps jettmari Vitzth.		⊕			
L. pavlovskyi Zachv.		⊕			
L. hilaris C.L.Koch	⊕	⊕		⊕	
L. agilis C.L.Koch	⊕	⊕	⊕	⊕	
L. algericus Hirst	⊕	⊕	⊕	⊕	⊕
Hyperlaelaps microti Ew.		⊕	⊕		
Haemogamasus pontiger Berl.		⊕			
H. hirsutus Berl.				⊕	
H. nidi Mich.		⊕	⊕	⊕	⊕

The distribution of gamasoid mites

Species of mites	Woodlands	Forest steppe	Steppe	The Carpathians	Crimea
Hirstionyssus isabellinus (Oudms.)	⊕	⊕			
H. latiscutatus Meil. et Lav.		⊕	⊕	⊕	⊕
H. apodemi Zuew.		⊕	⊕	⊖	⊖
H. criceti Sulz.			⊕		
Ornithonyssus bacoti Hirst		⊕	⊕	⊕	⊕
Allodermanyssus sanguineus Hirst		⊕	⊕	⊕	⊕
Dermanyssus gallinae Redi			⊕		

Table 2. Gamasoid mites found on the house mouse under different ecological conditions

Species of mites	Multistoreyed houses	Rural houses	Bungalows	Natural habitats
Euryparasitus emarginatus				⊖
Pergamasus crassipes				⊖
Parasitus remberti				⊖
P. consanguineus	⊖			⊖
P. distinctus			⊖	
Ameroseius plumosus			⊖	⊖
***Ameroseius* sp.**		⊖	⊖	
***Blattisocius* sp.**			⊖	
***Amblyseius* sp.**				⊖
Garmania hypudaei			$\frac{2,6}{0,06}$	$\frac{10,3}{1,2}$
Macrocheles matrius				⊖
Eulaelaps stabularis			⊖	⊖
E. kolpakovae		⊖		
Haemolaelaps casalis		⊖	⊖	
H. glasgowi			⊖	⊖
Laelaps pavlovskyi			⊖	⊖
L. hilaris			⊖	⊖
L. agilis			⊖	⊖
L. algericus		$\frac{9,6}{1,4}$	$\frac{28,2}{0,6}$	$\frac{28,6}{1,0}$
Haemogamasus nidi			⊖	⊖
Hirstionyssus latiscutatus		$\frac{8,4}{1,0}$	⊖	$\frac{5,1}{0,1}$
H. apodemi				
Ornithonyssus bacoti	$\frac{7,7}{0,7}$	$\frac{31,3}{2,1}$	$\frac{4,1}{0,1}$	⊖
Allodermanyssus sanguineus	$\frac{21,6}{1,5}$	$\frac{3,6}{0,07}$	⊖	

(The fraction means: $\frac{\text{extensity of infestation}}{\text{intensity of infestation}}$; ⊖ – sporadic findings).

tion (e.i.) 7.7 %, intensity of infestation (i.i.) 0.07) and *Allodermanyssus sanguineus* (e.i. 21.6, i.i. 1.5). The species composition of gamasids found in rural settlements includes 7 species, the dominant one being *O. bacoti* (e.i. 31.3 %, i.i. 2.1), second in number was *L. algericus* (e.i. 9.6 %, i.i. 1.4) and in the third place *H. latiscutatus* and *H. apodemi* (total extensity of infestation 8.4 %, i.i. 1.0). The most abundant in species was the fauna of gamasoid mites parasitizing house mice trapped in bungalows where permanent contact with many other species of small mammals is usually maintained. Eighteen species of mites were found, of which 60 % of the specimens were represented by the species *L. algericus* (extensity of infestation 28.2 %, i.i. 0.6), the second most numerous species was *O. bacoti* (e.i. 4.1 %, i.i. 0.1). There was almost no difference between the fauna of gamasoid mites from house mice trapped in bungalows and that from house mice trapped in natural habitats. Under natural conditions 16 species of gamasids were found, out of which 10 were common species; the dominant species was *L. algericus* (e.i. 28.6 %, i.i. 1.0), and next was *Garmania hypudaei* (e.i. 10.3 %, i.i. 1.2). The absence of the mite *A. sanguineus* from house mice trapped in natural habitats under conditions of the forest-steppe was remarkable.

The seasonal fluctuations in the gamasid infestation of house mice trapped in bungalows and in natural habitats have many features in common. The extensity of infestation during the year varied slightly, representing in autumn 45.9 %, in houses, 57.6 % in natural habitats, in winter 32.9 % and 45.6 %. and in summer 43.4 % and 48.8 % respectively.

As for the habitat association of individual species of gamasoid mites it should be noted that under the conditions of the Ukraine the species *O. bacoti* and *A. sanguineus* are, as a rule, synanthropic. In more southern republics (e.g. Armenia) these mites have been found on a wide range of hosts, in human dwellings as well as natural habitats, throughout the year. To summerize it may be said that in the Ukraine the house mice are associated with a wide range of gamasids, and are characterized by a relatively high infestation with them. The specificity of the gamasoid fauna associated with the house mice is not an absolute value and may vary within wide limits, depending on the actual conditions (primarily on the living conditions) of the house mouse.

References

Antonenko, V.V.: Gamasoid mites (Gamasoidea, Parasitiformes), the dominant parasites amongst the hair of rodents in the middle Dniepr region and their nests. Vestnik zool. 4: 21–24, 1970. (In Russian.)

Belokon' E. M.: Gamasoid mites and fleas of small mammals in the western regions of the Ukraine. Avtoreferat dissertatsii, 5–10, 1965. (In Russian.)

Naglov, V. A., Naglova, G. I.: Ectoparasites of mouse-like rodents inhabiting the Don forests. Tr. Ukrainskogo resp. nauch. obshch. par. 2: 146–153, 1963. (In Russian.)

Zhuravleva, L. A.; Gamasoid mites of rodents in the steppes of southern Ukraine. Avtoreferat dissertatsii, 6–9, 1964. (In Russian.)

Section VII

TICKS, FAUNISTICS AND TAXONOMY

Chairman

Yu. S. Balashov (USSR)

Assistant Chairman

E. Honzáková (Czechoslovakia)

ON THE ECOLOGY OF *IXODES RICINUS* L. NYMPHS (IXODIDAE)

L. V. BABENKO
The Martsinovsky Institute of Medical Parasitology and Tropical Medicine,
Moscow, USSR

Ixodes ricinus covers a number of climatic zones and is characterized by a geographic variability of many of its life cycle phenomena. Studies of this variability may elucidate the conditions required by the species for its existence within its area. Comparative studies will only be possible when data are available concerning the variability of certain tick life phenomena within the same geographical areas during many years. This paper offers the results of many years of observations on the ecological features of the nymph of *I. ricinus* in Moscow region, USSR, in comparison with some literature data.

The seasonal activity of *I. ricinus* nymphs can be divided into 2 basic periods – a) spring-summer activity of specimens moulting from larvae in the preceding season and hibernating in unfed condition (April-July) and b) autumn activity of non-hibernating specimens moulting from larvae in the current season (August – October). Each of these periods is not uniform. As regards the first period, the density of active specimens may according to the weather in spring and the type of forest litter either increase gradually, reaching its maximum in June or the beginning of July, or have two peaks – in April-May, on the account of specimens hibernating in the very top layers of the loose litter with the activation occurring immediately at the onset of sunny days, and in June-July on the account of specimens hibernating in deeper layers of the litter. (Fig. 1.)

The spring-summer period of activity is characteristic of almost the whole range of the species. In the autumn either one peak in August or two peaks in August and September may also occur in the vicinity of Moscow. The activity in August is apparently characteristic of specimens which have undergone the larval morphogenetic diapause, whereas in September it is seen in those ticks which moult without diapause from larvae of the current feeding year. It would be logical to assume that in areas characterized by only a September-October peak and the absence of peaks in August (Zasukhin et al., 1936; Rosický, 1953; Loew et al., 1963; Uspenskaya, 1966 et al.) nymphs which have undergone the larval diapause constitute a smaller portion in the population as against those in the vicinity of Moscow. In the north of the range of *I. ricinus* (Edwards and Arthur, 1947; Kheisin et al., 1955) the autumn peak may be not pronounced at all. In those conditions there is evidently not enough time for the activation of the newly moulted nymphs (Serdyukova, 1952; Lees and Milne, 1951). High autumn peaks are not infrequently followed by an increased abundance of active nymphs in spring as was also reported by Černý (1961). (Fig. 2.)

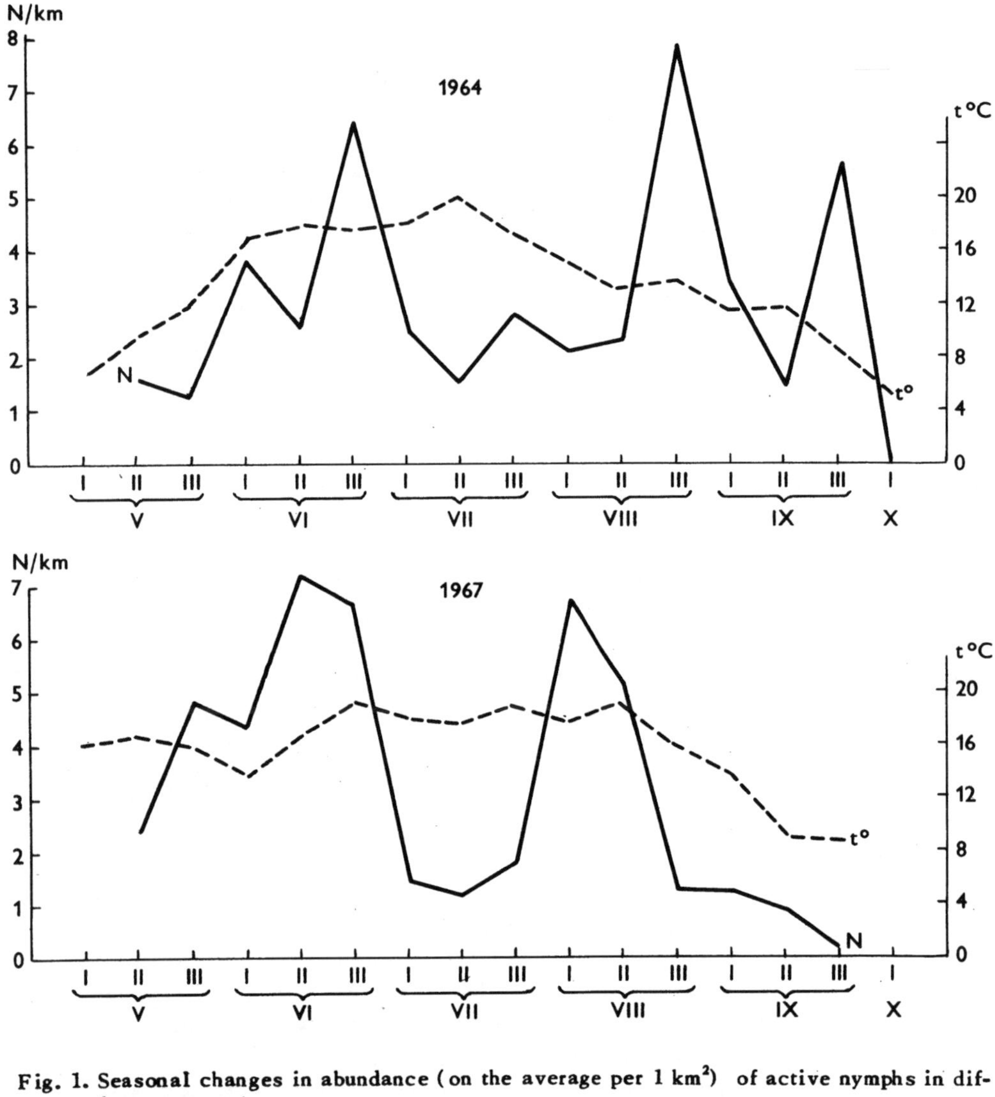

Fig. 1. Seasonal changes in abundance (on the average per 1 km^2) of active nymphs in different years.

The comparison of the abundance indices of active nymphs with those of small mammals as well as of the indices for feeding nymphs has shown that host abundance is the leading factor for the estimation of the density of fed nymphs; however, there may be sharp deviations in the number of active nymphs either increasing or

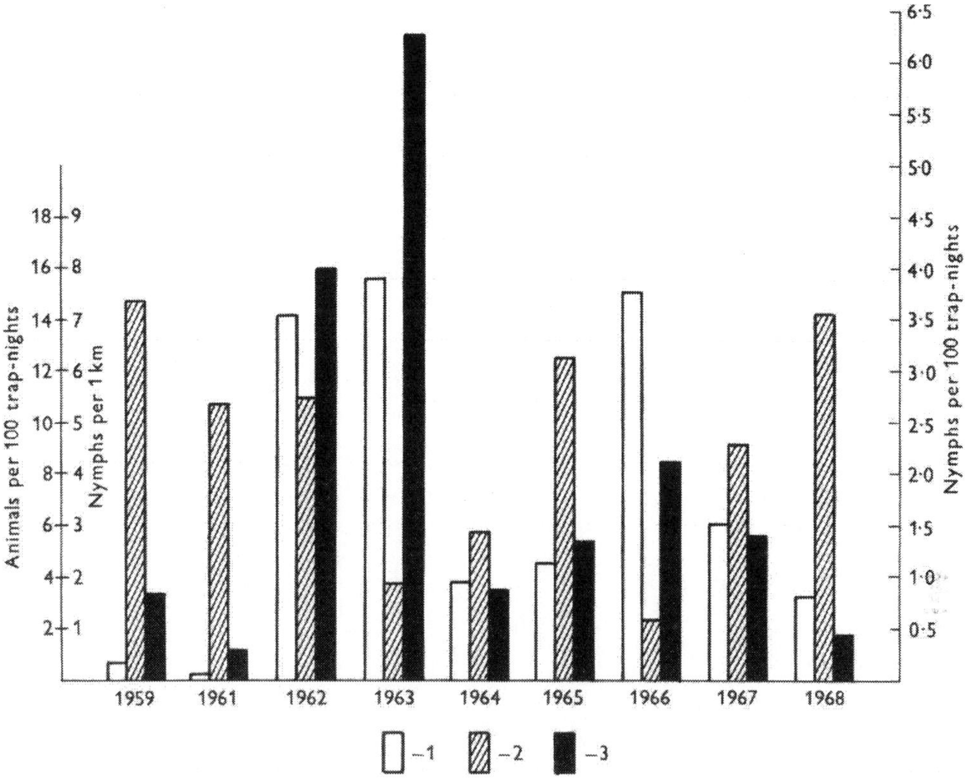

Fig. 2. Comparison of average abundance indices of animals and nymphs (in vegetation and on animals) for May-July: 1 – Animals per 100 trap-nights, 2 – Nymphs per 1 km^2, 3 – Nymphs on animals per 100 – trap-nights.

decreasing. In the years of a low density of murine rodents, i.e. the primary nymph hosts, the abundance indices of *I. ricinus* nymphs parasitic on animals and those of active specimens in the vegetation layer are increasing. The first phenomenon was reported also for *I. persulcatus* (Naumov, 1955; Shilova, 1960) and attributed to a higher activity of animals in years of low population density. The reason for the second point undoubtedly lies in host shortage. (Fig. 3.)

Observations of the behaviour of hungry nymphs in cages have shown that the activity of specimens is highest from 4 to 10 p.m. (peaks of activity at 8–10 p.m.) and lowest from 8 to 12 a.m.. Sometimes a slight increase of activity occurs from 4 to 8 a.m. (Babenko, 1970). This activity rhythm has a biological basis, since it agrees with the behaviour of the primary hosts, i.e. small forest mammals. It is influenced by abiotic factors including especially photoperiodic conditions as well as weather. (Fig. 4.)

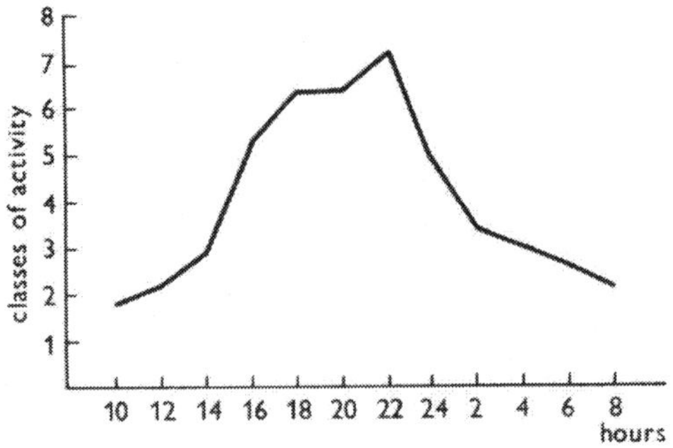

Fig. 3. Diurnal activity rhythm of nymphs of *Ixodes ricinus* (in cages).

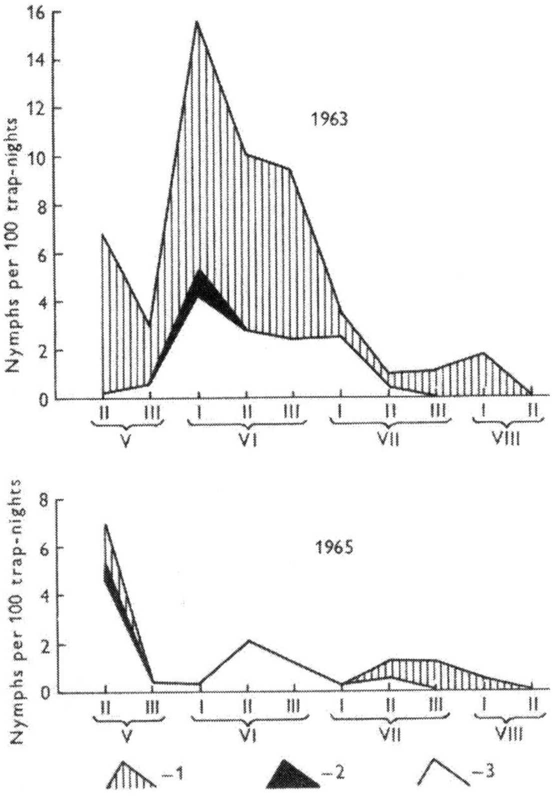

Fig. 4. Specific features of nymphal development in different feeding periods: 1 – Nymphs in diapause, 2 – Nymphs having insufficient time for complete moulting, 3 – Nymphs which completed their development during the feeding season.

Table 1. Duration of development (without diapause) of nymphs of *Ixodes ricinus* in some parts of their range in the USSR (days)

Observation site	Summarized average monthly temperatures for May–September	Time of nymph feeding					Authors
		April	May	June	July	August	
Karelia	60.5	–	106–129	76–94	general diapause	general diapause	Kheisin et al., 1955
Moscow region	70.8	138–140 and over	62–118 (92.0)	48–104 (70.7)	50–81.2 (71.4)	general diapause	The author's data
Byelorussia	75.8	99–175	75–156	66–104	59–64	general diapause	Arzamasov, 1967
Moldavia	93.3	115	60–102	53	47–75	75–105	Uspenskaya, 1967
Azerbaijan Lenkoran district	113.6	–	55–60	46–59	32–48	32–65	Babenko, Gusein-zade, 1971

Note: Average data are in brackets.

A comparison of data on the duration of the nymphal development of *I. ricinus* in Karelia (Kheisin, 1955), Byelorussia (Arzamasov, 1961), Moldavia (Uspenskaya, 1967), Azerbaijan (Babenko, Gusein-zade, 1971) with data obtained during many years in the vicinity of Moscow (Babenko, 1970) and their comparison with temperature and photoperiodic conditions in each of these areas have shown that the basic distinctions lie in the time of emergence of diapausing specimens: in the south, where the days are shorter, they appear much later, than in the north. (Tab. 1.)

The difference in the duration of nymphal development in different ranges of *I. ricinus* proved to be not so essential as might be expected from a comparison of the temperature conditions in various areas (Table 1). The specimens of southern populations seem to require more heat than those of the northern ones.

From the data obtained it may be assumed that the conditions in the Moscow vicinity are not quite favourable for nymphs of *I. ricinus*. The reasons are as follows: a weak and irregular increase in their activity in the autumn, the significance of a larval diapause for autumn populations, frequent deviations from the regular seasonal course of the diapause and the presence of a large number of engorged nymphs which are unable to finish their development safely as has already been mentioned. Apparently, these are the factors which limit the distribution of *I. ricinus* to northern areas of the central part of the USSR (Korenberg et al., 1971).

References

Arzamasov, I. T.: Ixodid ticks, Publ. BSSR Acad. Sci., Minsk, 1961.

Babenko, L. V.: Some features of *Ixodes ricinus* nymph ecology. Second Acarol. Meeting, Theses of papers, P.I: 43–46, Publ. Naukova dumka, Kiev, 1970. (In Russian.)

– Gusein-zade, A.: Comparative ecological study of different geographical populations of *Ixodes ricinus L.* (Parasitiformes, Ixodidae). Transactions of the 13th Int. Entomol. Congress, 1971.

Černý, V.: Über die Möglichkeit einer Prognose der Anzahlen von *Ixodes ricinus* im Freiland. Angew. Parasitol. 2(4): 106–109, 1961,

Edwards, E., Arthur, D.: The seasonal activity of the tick *Ixodes ricinus* in Wales. Parasitol. 38 (1–2): 72–85, 1942.

Kheisin, E.M.: Duration of development of *Ixodes ricinus* and *I. persulcatus* P. Sch. larvae and nymphs in different seasons of the year (on the problem of diapause). Transactions of the Karelo-Finnish St. University, Petrozavodsk, 6: 28–44, 1955. (In Russian.)

Kheisin, E. M., Bochkareva, K., Lavrenenko, L., Mikhailova, T.: Oviposition and development of *Ixodes ricinus* L. in nature in Karelo-Finnish SSR. Transactions of the Karelo-Finnish St. University, Petrozavodsk, 6: 28–44, 1955. (In Russian.)

Korenberg, E. I., Dzyuba, M. I., Zhukov, B. I.: Area of *Ixodes ricinus* L. ticks in the USSR. Zool. Zh. 50: 41–50, 1971. (In Russian.)

Lees, A.D. and Milne, A.: The seasonal and diurnal activities of individual sheep ticks (*Ixodes ricinus*). Parasitol. 41 (3–4): 189–208, 1951.

Loew, J., Radda, A., Pretzmann, G., Groll, E.: Untersuchungen in einem Naturherd der Frühsommer-Meningoencephalitis (FSME) in Niederösterreich. I. Mitt. Ökologie und Saisondynamik von *Ixodes ricinus*. Zbl. f. Bakt., I, orig. 190. (2): 183–206, 1962.

Naumov, N. P.: A study of mobility and density of small mammals by means of trap ditches. In: Problems of reg. gen. experim. parasitology and med. zoology, M. v. 9, 1970. (In Russian.)

Rosický, B.: Bionomicko-faunistický nástin klíšťat (Ixodidae) z území ČSR. Zool. entomol. listy 2 (2): 120–130, 1953.

Serdyukova, G.V.: New data on the development of *Ixodes ricinus* larvae and nymphs in nature. Doklady Akademii Nauk 83 (5): 769–772, 1952. (In Russian.)

Shilova, S. A.: Biological substantiation of epidemiological prognoses of tick-borne encephalitis. Bull. Med. Soc., Inst. Nat. Biol. 65(1): 37, 1960. (In Russian.)

Uspenskaya, I. G.: Biology of *Ixodes ricinus* L. in Moldavia. First acarological meeting, Theses of papers, Moskva, Leningrad, p. 213–214, 1966. (In Russian.)

Uspenskaya, I. G.: Seasonal changes of periods of embryogenesis and metamorphosis of *Ixodes ricinus* L. larvae and nymphs in Moldavia. Probl. Parasitol., Kiev, p. 72–73, 1967. (In Russian.)

Zasukhin, D. N., Lonzinger, G. K., Okrokvertsova, L. A.: Materials for studies of *I. ricinus* in south-east of RSFSR. Transactions of Saratov Res. Inst., V. 3, p. 148–154, 1936. (In Russian.)

EXPERIMENTAL INTERSPECIFIC HYBRIDIZATION OF TICKS OF GENUS *ORNITHODOROS*

YU. S. BALASHOV
Zoological Institute, Academy of Sciences of the USSR, Leningrad, USSR

We have carried out hybridizations between 5 species of ticks of the genus *Ornithodoros* Koch out of which 3 species of the subgenus *Pavlovskyella* Posp. – Shtr. (*O. papillipes* Bir., 1895, *O. tartakovskyi* Ol., 1931 and *O. verrucosus* Ol., Sas. et Fen., 1934) had identical sets of 16 chromosomes. *O. papillipes* and *O. tartakovskyi* are widely spread in Middle Asia; sometimes they settle together on the same microbiotopes. *O. verrucosus* is known in Zakavkazye, in Northern Caucasus and in South Ukraina but its area may coincide with areas of two other species on the territory of Iran and Turkey. *O. nereensis* Pavl. 1941 and *O. alactagalis* Issak., 1936 belong to subgenus *Theriodoros* Posp. – Shtr. and morphologically are very similar but differ greatly in sets of chromosomes. *O. nereensis* which has been found on a limited territory of south-west Turkmenia only has 24 or 26 chromosomes and *O. alactagalis* from Zakavkazye has 32 or 34 chromosomes.

In each series of experiments 10–30 male and female ticks were taken. After feeding they were set in pairs. Then we noted the percentage of females of the parental generation which had laid eggs, the average number of eggs per female, the number of batches in which larvae had hatched and the percentage of larvae hatched in relation to the total number of eggs laid. The same figures were used to estimate fertility F_1 of interspecific hybrids.

O. papillipes and *O. verrucosus* seem to be more similar in their morphological and ecological characteristics and as they are allopatric species one can obtain an impression that they may be vicariating forms. At the same time *O. tartakovskyi* seem to be more isolated from *O. papillipes* due to morphological and ecological characteristics as they are sympatric species. Results of the hybridizations (Table 1) affirmed the self-dependency of each of these three species, they also revealed mechanisms of reproductive isolation between them, but still they did not confirm similarity of *O. papillipes* and *O. verrucosus*.

O. verrucosus turned out to be genetically absolutely isolated from *O. papillipes* and *O. tartakovskyi*. During the hybridization of *O. tartakovskyi* with two other species it was noted that 40–70 % or more of hybrid eggs perished. Larvae of the F_1 obtained from the batch developed into males and females but the latter were absolutely sterile in cases of intrahybrid and reciprocal hybridizations with parental species. During these hybridizations the majority of females did not lay eggs at all and of the ± 2000 eggs laid by all groups only some non viable larvae hatched.

Table 1. Fertility of hybrids of the first filial generation

Parents	Females		The number of eggs in the batch	The number of batches in which larvae appeared	The percentage of hatched larvae
	The total number	The number which laid eggs			
Ornithodoros verrucosus x *Ornithodoros papillipes*	10	1	47	0	0
Ornithodoros papillipes x *Ornithodoros verrucosus*	8	2	73	0	0
Ornithodoros verrucosus x *Ornithodoros tartakovskyi*	10	2	40	0	0
Ornithodoros papillipes x *Ornithodoros tartakovskyi*	10	10	192	6	29
Ornithodoros nerrensis x *Ornithodoros alactagalis*	10	8	54	6	27
Ornithodoros alactagalis x *Ornithodoros nereensis*	10	6	41	5	16

On the contrary only partly genetical incompatibility is peculiar to *O. papillipes* and *O. tartakovskyi*. During the hybridization of *O. papillipes* females with *O. tartakovskyi* males the majority of females laid normal batches of eggs 30 – 60 per cent of which had viable larvae. When crossing hybrids of the first filial generation between themselves or with parental species a large number of eggs was obtained, out of 30 per cent of which viable larvae of fertile second filial generation also hatched. Apparently formation of interspecific hybrids of these ticks under natural conditions is hindered by some other isolation mechanisms, otherwise cases of introgression should be observed in the areas of their common inhabitancy.

It is of interest that the hybridization of *O. tartakovskyi* with *O. papillipes* is possible in only one combination, namely when males of the first species copulated with females of the second one. When females *O. tartakovskyi* are hybridized with males of *O. papillipes* the females mostly perish as a result of mechanical injury during coupling. A similar situation can be observed during hybridization of *O. tartakovskyi* with *O. verrucosus*. In this case we observe isolation caused by discrepancy in size of the copulatory apparatuses.

The degree of interspecific genetical incompatibility is not always correlated with the differences in morphology and in the number of chromosomes. Hybridizations of *O. nereensis* with *O. alactagalis* were successful despite great differences in the sets of chromosomes. From eggs laid by females of parental generations 37 – 49 per cent of viable larvae hatched. The sex ratio in the hybrid F_1 was close to 1. Hybrid males and females turned out to be fertile in intrahybrid and recurrent hybridizations with parents. In the batches of hybrid females 16 – 27 per cent of the total number of eggs produced hatched which later on developed into fertile individuals of the second filial generation.

Summing up the experiments of interspecific hybridizations of ticks we consider it necessary to point out that the possibility of obtaining fertile hybrids of some species in the experiment does not contradict the obligatory existence of reproductive isolation between them. The latter can be achieved in different ways and may not be connected with interspecific genetical incompatibility. Numerous mechanisms of reproductive isolation and the inevitable elimination by means of natural selection of any hybrid off-spring excludes an increase of introgression even in the case of joint inhabitation of species of incomplete genetical isolation. At the same time the very possibility of interspecific hybridizations shows the preserved genetical relationship of the species being hybridized. Therefore the ease of obtaining fertile interspecific hybrids, making it possible to obtain them experimentally, is of value in estimating relationships within the limits of genera or subgenera.

SOME DISTRIBUTIONAL PECULIARITIES OF SPECIES BELONGING TO THE *IXODES PERSULCATUS* GROUP

N. A. FILIPPOVA
Zoological Institute, Academy of Sciences of the USSR, Leningrad, USSR

As a result of a revision of the species of the genus *Ixodes* Latr. of the USSR fauna it has been found that to the name *Ixodes persulcatus* several distinct species were referred in literature before 1967. These are: *I. persulcatus* Schulze, 1930, *I. pavlovskyi* Pom., 1948, *I. kazakstani* Ol. et Sor., 1934, *I. nipponensis* Kitaoka et Saito, 1967, *I. kashmiricus* Pom., 1948, *I. sachalinensis* Fil., 1971 (Filippova, 1969, 1971).

At present the main ecological and distributional characters of the above mentioned species have been elucidated (Filippova et Ushakova, 1967; Ushakova et Filippova, 1968, Filippova et Belaev, 1971).

I. persulcatus is one of the most flourishing species of its genus. Its distribution is very extensive and the numerosity is high. It spreads from Karelia and the Baltic Sea in the west to the Sea of Okhotsk and Sea of Japan in the east. Its distribution stretches over a considerable part of the subzone of the middle taiga and over the whole subzone of the south taiga. In the zone of European mixed forests this species is very rare in deciduous components and it has not been found in Asian ones. Its distribution spreads throughout the tertiary taiga formations of the Far East and the post glacial period taiga in the European part of the USSR (Filippova, 1971). *I. persulcatus* is a mesohygrophilous species. The annual precipitation fluctuates mainly between 600 – 900 mm in its area and only in some regions exceeds this level. The predominant species are those which are typical of the taiga zone. Both the larvae and nymphs of *I. persulcatus* feed on mammals and birds. As a result of studying a vast collection of material we can affirm that the imago of *I. persulcatus* is mainly a parasite of mammals. A special research on the trophic relations of *I. persulcatus* in the Altai showed that the imago of this species parasitizes mammals almost exclusively, including cattle.

The distribution of *I. pavlovskyi* is disjunctive. In the west it is found in the mountainous Altai and the high regions of southern Siberia. In the east it is present in the mountains of Sikhote-Alin and at other elevations i.e. the river basins of the Amur and Ussuri rivers. The following recent and palaeogenetic common features are typical of both parts of the area: these are mountain-taiga territories. Dark coniferous forests in the range of *I. pavlovskyi*, 500 – 800 m above sea level, contain a considerable number of deciduous trees and have microclimatic conditions favourable for this species. The relict tertiary character of mountain-taiga forests of the Far East is well known. But in the composition of the mountain taiga forests of the southern region of west Siberia there is also a relict element - these are fir-tree and linden forests. Recent mountain- taiga (fir-tree) forests covering a large territory of the Altai are considered by botanists as reduced tertiary broad-leafed and coniferous forests which spread continuously all over the south of mountainous Siberia in the Late Tertiary including the Far East. This can be proved by the fact that in the contents of the floristic formations of southern Siberia and the Altai there are many relict species of plants which are not typical of eastern Siberia but which can be found again, in the Far East. Pollen and lynological analyses also prove this fact (Filippova, 1971).

I. pavlovskyi is more hygrophilous, the quantity of annual precipitation being about 1500 mm in the zone where it is most numerous in the Altai. Larvae and nymphs of *I. pavlovskyi* feed on birds and small mammals and imago feeds mainly on birds. Many species of birds from the ecological group which gather food on the ground are registered as hosts of the imago of *I..pavlovskyi*. All the hosts known (now dozens of species) of *I. pavlovskyi* feed *I. persulcatus* during different phases of its life cycle.

Thus we have determined not only the sympatric character of *I. pavlovskyi* in relation to *I. persulcatus*, but also their joint use of the same biotopes and species of birds and mammals and often the same individuals. The zones of joint inhabitation and high numerosity of these two species coincide with the natural foci of tick-borne encephalitis.

The distribution of *I. kazakstani* as a whole lies within the limits of the Tien-Shan mountains. This species lives in forests in the valleys of mountain rivers and on the banks of mountain lakes. The conditions in such forests are characterized by a high humidity as a result of the high level of ground-water and the well developed meadow vegetation. In spite of the close proximity of mountain taiga and forests in Tien-Shan *I. kazakstani* and *I. persulcatus* are separated biotopically. Larvae and nymphs of both species feed partly on the same species of rodents which have in comparison with each species of ticks a wider range of vertical distribution.

I. nipponensis is widely spread on the Japanese Islands. It also inhabits the southern part of the Primorsky Krai, USSR in the sea shore zone. There are no taiga forests here and open meadows and steppes dominate, alternating with swamps in the lowland and oak forests on hills and in some places with broadleafed forests in valleys. The second research on a large amount of material showed that this species was not found in the taiga forests in the south of Primorsky Krai. *I. persulcatus* can be found in biotopes of *I. nipponensis* but very rarely.

Areas of *I. pavlovskyi* and *I. kazakstani* have a relict character. Their successful development can be connected with that of the pliocene warm-temperate foliage forest. Formation of the area of *I. ricinus* (L.) should be refferred to the same period. The origin of the area of *I. persulcatus* is connected with the origin of the taiga and with the replacement of the warm-loving pliocene broad-leafed forests by the coniferous forest during the Pleistocene age as a result of a drop in temperature and humidity of the territory considered.

References

Filippova, N. A.: Taxonomical aspects of studies of ticks of genus *Ixodes* Latr. (Ixodoidea, Ixodidae), vectors of viruses of tick-borne encephalitis. Rev. Entomol. USSR 48 (3): 675–688, 1969. (In Russian.)

Filippova, N.A.: On species of *Ixodes persulcatus* group (Parasitiformes, Ixodidae). VI. Parazitologia 5 (5): 385–391, 1971. (In Russian.)

Filippova, N. A.: A new species of ixodid ticks, *Ixodes sachalinensis*, sp. n. (Ixodoidea, Ixodidae) from Sakhalin. Rev. Entomol. USSR 50 (1): 236–239, 1971. (In Russian.)

Filippova, N. A., Ushakova, G.V.: On species of *Ixodes persulcatus* group (Parasitiformes, Ixodidae). I. Parazitologia 1 (4): 269–278, 1967.(In Russian.)

Filippova, N. A., Belaev, V. G.: On species of *Ixodes persulcatus* group (Parasitiformes, Ixodidae). V. Parazitologia 4 (6): 515–523, 1970. (In Russian.)

Ushakova, G. V., Filippova, N. A.: On the species of the *Ixodes persulcatus* group (Parasitiformes, Ixodidae). Parazitologia 2 (4): 334–338, 1968. (In Russian.)

THE INTRODUCTION INTO DENMARK OF THE KENNEL TICK [*RHIPICEPHALUS SANGUINEUS* (LATR. 1806)] WITH REMARKS ON ITS REACTIONS TO DIFFERENT HUMIDITIES

N. HAARLØV
Zoological Institute, Royal Veterinary and Agricultural University,
Copenhagen, Denmark

With the recent rapid development in intercontinental traffic the risks of introducing parasites have been aggravated and especially such which infect man and domestic animals. In Denmark *Rhipicephalus sanguineus* is a fairly good example of this phenomenon; it was observed for the first time in 1966 in a house near Copenhagen (Winding et al. 1970, Willeberg 1970). Since then the species has been found each year but with a pronounced increase in numbers in 1969 which from June to September was characterized by an exceptionally high number of sunshine hours.

In the following an attempt will be made to analyze the reason for the occurrence of *R. sanguineus* in Denmark and its reaction to different levels of humidity (R. H.) motivated by its transport from the dry, continental climate of its native biospheres to the humid and Atlantic one in Denmark. As a standard of reference *Ixodes ricinus* (Linnaeus 1758) will be included as this represents a common species typically adapted to Danish habitats and occupying almost the same niche as that of *R. sanguineus* in Africa (Tambs-Lyche 1943, Milne 1950, Hoogstraal 1956).

Methods

Relative humidities were measured with Cobalt Thiocyanate Papers (Solomon 1957) and standardized with water-saturated filter paper (100 % R. H.), KCl (85 % R. H.), $CaNO_3$ (50.5 % R. H.), NaOH (7 % R. H.) (Winston and Bates 1960, Wharton 1963). Temperature was 23° – 25°C, and the light was as slight and diffuse as possible. Tolerance experiments were conducted in plastic boxes (diam. 10 cm) with no possibility of direct contact between ticks and salt/water. Preferences were studied in a circular apparatus (diam. 20 cm) in which the following percentages of relative humidities per cm were measured radially from centre to periphery: 0–10, 0–10, 10–20, 20–40, 40–60 (link between NaOH and KCl), 60–70, 70–80 (link between KCl and H_2O), 80–90, 90–100, 90–100. The ticks were placed on a fine-meshed nylon net. Their positions were registered every 15 seconds in one of the ten above-mentioned concentrically developed zones of humidity. Experiments lasted from 15–60 minutes.

Material

Ixodes ricinus were taken from Danish mammals and *R. sanguineus* from dogs imported into Denmark. Males, females and larvae included individuals more or less engorged.

Results

The tolerances of both species are summarized in Table 1.

Table 1.

Spp. / R. H. (23–25 °C)	*Rhipicephalus sanguineus*			*Ixodes ricinus*		
	Hatching % (3 ♀♀)	Larval life (days)	Adult life (2 ♀♀) (days)	Hatching % (1 ♀)	Larval life (days)	Adult life (2 ♀♀) (days)
100 %	78 (1306 eggs)	20–60–?	>30	95 (146 eggs)	c. 360–?	>20
85 %	84 (1344 eggs)	10–60	20–25	25 (129 eggs)	10–15	5–10
50.5 %	13 (1308 eggs)	1–20	5–10	0 (119 eggs)	0	2– 3
7 %	2 (1359 eggs)	7–10	3–5	0 (178 eggs)	0	2– 3

Mutual differences are esp. evident in relation to percentages of hatching with *R. sanguineus* being able to hatch under very dry conditions, and with an optimum lower than that of *I. ricinus*. In the plastic boxes it was easy to see how a relatively large percentage of the larvae of *R. sanguineus* stuck to the water-filled and gelatineous surfaces of the eggs at 100 % R.H. and later succumbed, and also that after some days those moving freely in the boxes showed abnormal mobility as if their pulvilli stuck to the substratum. At 85 % R.H. about one tenth of the unhatched eggs of *I. ricinus* had fully developed embryos, and it was evident that the individuals hatched had difficulties in getting through the egg shell. Larvae of both species aggregated and were rather inactive and thigmotactic before they had emptied their gut for guanin. Maximal length of larval life at 100 % R. H. could not be stated, as the larvae of both species were evidently killed by fungi. Incubation took approx. 25 days at all humidities.

When moving in the preference apparatus, the ticks were apparently orientating themselves with Haller's organ on tarsus 1 which was moved forward and backwards almost continually (Lees 1948). In a control experiment with the same humidity in the apparatus, the individual moved in straight lines contrary to the more circular movements when the apparatus was divided into 8 concentric zones of different humidities (Figures 1–3). During the period of the experiment a tick did not seem to change in its reactions. Larvae, males and females of *I. ricinus* which had been exposed to 85 % R. H. for approx. 24 hours before the experiment started appeared to prefer 90 – 100 % R. H. When exposed to 100 % R. H. beforehand, larvae of both species preferred 40 – 70 % R. H., females 60 – 80 % R. H. while males of *R. sanguineus* preferred 10–40 % R. H. and those of *I. ricinus* did not indicate any preference.

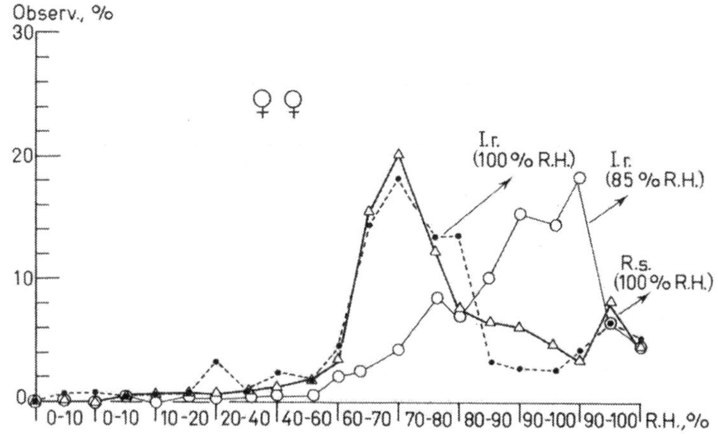

Fig. 1. Preferences of *Ixodes ricinus* (I.r.) with specimens (♀♀) before experiment and exposed to 100 % R. H. (n = 3, obs. = 673), to 85 % R. H. (n = 7, obs. = 702) and for *Rhipicephalus sanguineus* (R.s.) to 100 % R. H. (n = 4, obs. = 842). The figures on the abscissa refer to the humidity gradient in the preference apparatus.

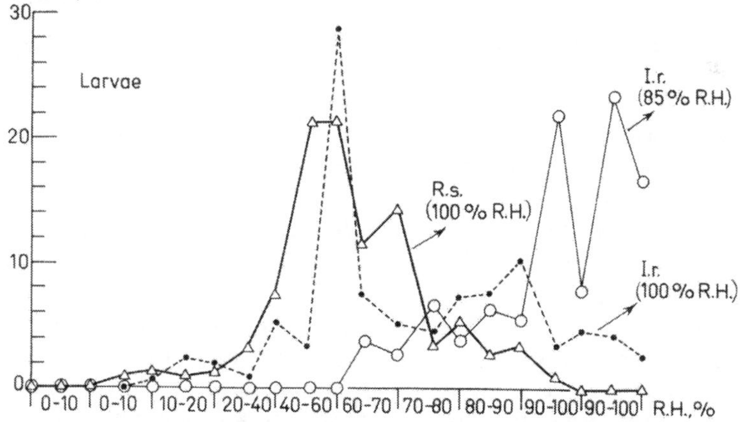

Fig. 2. Preferences of *Ixodes ricinus* (I.r.) with specimens (larvae) before experiment and exposed to 100 % R. H. (n = 4, obs. = 452), to 85 % R. H. (n = 6, obs. = 177) and for *Rhipicephalus sanguineus* (R.s.) to 100 % R. H. (n = 1, obs. = 145). The figures on the abscissa refer to the humidity gradient in the preference apparatus.

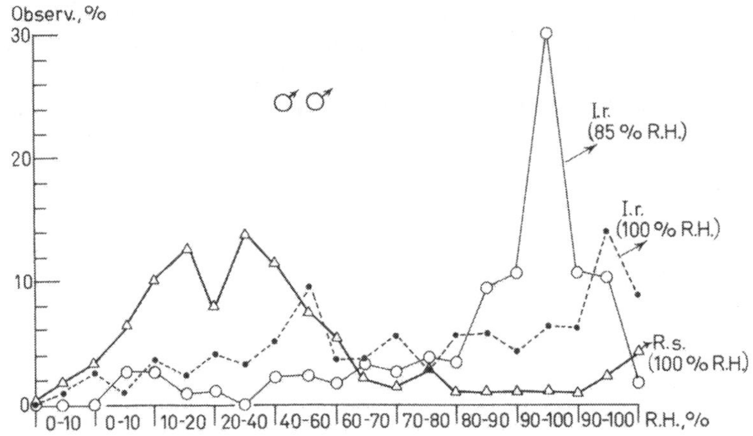

Fig. 3. Preferences of *Ixodes ricinus* (I.r.) with specimens (♂♂) before experiment and exposed to 100 % R. H. (n = 2, obs. = 527), to 85 % R. H. (n = 2, obs. = 238) and for *Rhipicephalus sanguineus* (R.s.) to 100 % R. H. (n = 1, obs. = 251). The figures on the abscissa refer to the humidity gradient in the preference apparatus.

Discussion

It must be stressed that the available material is small and heterogeneous, and the prospects of drawing conclusions are therefore correspondingly limited. Ecological factors other than humidity such as e.g. temperature should of course also be included in the discussion in order to indicate more definitely why *R. sanguineus* ticks are not found as commonly in this country as in the more southern ones, the more so because Sweatman (1967) has convincingly shown its dependency on relatively high temperatures. The congruent occurrence of exceptionally high numbers of sunshine hours and registered cases of *R. sanguineus* in 1969 point in the same direction. These statements, however, seem to show that in Denmark temperatures in microclimate at ground level will only be optimal at intervals of several years and thus prevent establishment of stable populations outside houses. In houses, on the other hand, where temperatures are favourable, humidity may be the limiting factor. In centrally heated houses there may be about 40 % R. H. or less during most of the year (Haarløv and Alani, 1970). Following Sweatman (1967) that the egg stage of ticks in general is highly susceptible to desiccation and thereby possibly being the stage limiting the distribution of the species, it is evident from the tolerance experiments that *R. sanguineus* will be able to hatch indoors. This is the more important as the females may lay eggs in places "with a completely dry atmosphere" (Sweatman, l.c.). Furthermore, larvae will tolerate similarly dry conditions for some days and thus get a chance to move to more suitable microhabitats. From the tolerance experiments and from results by other research workers the non-parasitic stages of *I. ricinus* will neither be able to hatch nor to live in houses (Lees, 1946, 1948, 1969, Milne, 1950, Arthur, 1963, Knülle, 1965).

In view of the results of the experiments on preferences (Figures 1–3) it is apparent that dissimilarities in tolerances may well be followed by similarities in preferences as exemplified by the larvae and females of both species. For the females of *I. ricinus* (Fig. 1) their preferences correspond well with the results of Lees (1948) obtained in an alternative chamber where the specimens definitely avoided the most humid part. Further it does not seem unnatural with preferences 60 – 80% R. H. for the females of both species when it is realized that an ectoparasite like the louse has its preference at 75 % R. H. (Buxton, 1947). The relatively low preference of humidity for the larvae (Fig. 2) may be an important factor in the capacity for dispersal and as mentioned above esp. for *R. sanguineus*. Krijgsman (1937) found for *R. sanguineus* too that desiccated larvae were strongly positively hydrotactic contrary to the reactions of the non-desiccated ones, but only after the end of the period of aggregation, just after hatching. To draw conclusions from the reactions of an only male of *R. sanguineus* (Figure 3) is of course somewhat hazardous. Yet it had a very pronounced preference for remarkably low humidities and this fact in connection with the pronounced non-preference for the males of *I. ricinus* may be correlated to the greater risks for these animals to be exposed to varying and often low humidities in their search after females. At Nordic latitudes human houses will thus be the only habitat for *R. sanguineus* in which to make stable populations and also the place where they have the greatest chance of contact with dog, which is their favourite host i.e. when veterinary precautions for their eradications have not been adopted.

References

Arthur, Don R.: British Ticks, London, Butterworths, 1963.

Buxton, P. A.: The Louse. An account of the lice which infest man, their medical importance and control. Sec. Ed., London, Edward Arnold & Co., 1947.

Hoogstral, H.: African Ixodoidea. Vol. 1. Ticks of the Sudan. Cairo, U.S. Naval Med. Res. Unit., 1956.

Haarløv, N., Alani, M.: House-dust Mites (*Dermatophagoides pteronyssinus* (Trt.), *D. farinae* Hughes, *Euroglyphus maynei* (Cooreman) Fain. in Denmark (Acarina). Entomol. Scand. 1 (4): 301–306, 1970.

Knülle, W.: Equilibrium Humidities and Survival of Some Tick Larvae. J. Med. Entomol. 2 (4): 335–338, 1965.

Krijgsman, B.J.: Reizphysiologische Untersuchungen an blutsaugenden Arthropoden im Zusammenhang mit ihrer Nahrungswahl. Arch. Neerl. Zool. 2: 401–413, 1937.

Lees, A. D.: The Water Balance in *Ixodes ricinus* L. and certain other Species of Ticks. Parasitology 37 (1–2) : 1–20, 1946.

Lees, A. D.: The Sensory Physiology of the Sheep Tick, *Ixodes ricinus* L. J. Exptl. Biol. 25(2): 145–207, 1948.

Lees, A. D.: The Behaviour and Physiology of Ticks. Acarologia 11(3): 397–410, 1969.

Milne, A.: The Ecology of the Sheep Tick, *Ixodes ricinus* L. Parasitology 40 (1–2): 14–34, 1950.

Solomon, M.E.: Estimation of Humidity with Cobalt Thiocyanate Papers and Permanent

Colour Standards. Bull. Entomol. Res. 48 (3): 489–506, 1957.

Sweatman, K.: Physical and Biological Factors Affecting the Longevity and Oviposition of Engorged *Rhipicephalus sanguineus* Female Ticks. J. Parasitol. 53 (2): 432–445, 1967.

Tambs–Lyche, H.: *Ixodes ricinus* og piroplasmosen i Norge. Norsk Veterinaer-Tidsskrift, 9, 10, 11, og 12: 337–542, 1943.

Wharton, G.W.: Equilibrium Humidity. Adv. Acarol. 1:201–208, 1963.

Willeberg, P.: Forekomst af husflåt, *Rhipicephalus sanguineus*, i Danmark. Ugeskr. Læg 132, 27: 1270–1271, 1970.

Winding et al.: Husflaten *(Rhipicephalus sanguineus)*, en aktuel snylter hos hunden. Nord. Vet.–Med., 22:48–58, 1970.

Winston, P.W., Bates, D. H.: Saturated Solutions for the Control of Humidity in Biological Research. Ecology 41 (1): 232–237, 1960.

TICKS (IXODOIDEA) ON MIGRATING BIRDS IN CYPRUS, FALL 1967 AND SPRING 1968, AND EPIDEMIOLOGICAL CONSIDERATIONS*

M. N. KAISER, H. HOOGSTRAAL, & G.E. WATSON
Medical Zoology Department, United States Naval Medical Research Unit No 3, Cairo, Egypt,
Department of Vertebrate Zoology, Smithsonian Institution, Washington, D. C., U.S.A.

Studies in Egypt on migrating birds and their viruses and tick parasites, initiated by the NAMRU–3 Medical Zoology Department in 1956, were extended to Cyprus in fall 1967 and spring 1968 by the Palearctic Migratory Bird Survey.

In the fall, 2,580 birds of 92 species, mostly purchased from native collectors on Akrotiri Peninsula of southcentral Cyprus, were examined. In the spring, 22,015 birds of 62 species, mostly purchased at Paralimni village, were examined. Paralimni is at about 250 ft. altitude and 1 mile inland. Various biotopes at Paralimni, cultivated fields and citrus orchards and low Mediterranean scrub and juniper woodland, are ideal for bird trapping and netting.

Warbler and related birds of the family Muscicapidae comprised the largest number of bird species taken (34 in fall, 33 in spring), and the largest number of bird specimens taken (41.9 % in fall, 93.9 % in spring).

Since examination for ticks was sometimes done in the field with the naked eye, quantitative host-parasite data are given here with reservation.

In the fall, 115 southward migrating birds (29 species of 9 families) yielded 167 ticks representing 5 ixodid species common to Eurasia: *Hyalomma m. marginatum*, *Haemaphysalis punctata*, *H. concinna*, *Ixodes frontalis* (= *pari*), and *I. ricinus* (Tables 1 & 3). In the spring, 177 northward migrating birds (37 species of 11 families), yielded 797 ticks representing 7 ixodid and 1 argasid species: *Hyalomma marginatum rufipes*, *Amblyomma variegatum*, *A. nuttalli*, *A. lepidum*, *Ixodes eldaricus*, *I. frontalis*, *I. redikorzevi*, and *Argas (Persicargas) streptopelia* (Tables 2 & 3). Each of these is Ethiopian except the three *Ixodes* species, which attached to the birds in Cyprus or in eastern Mediterranean countries while en route to Cyprus.

* From Research Project MR005.09–0037, Bureau of Medicine and Surgery, Department of the Navy, Washington, D. C., U.S.A. The opinions and assertions contained herein are the private ones of the authors and are not to be construed as official or as reflecting the views of the Department of the Navy or of the naval service at large. The work reported here was accomplished under the auspices of the Palearctic Migratory Bird Survey, a collaborative program between the Smithsonian Institution, NAMRU–3, and the Yale Arbovirus Research Unit (WHO Virus Reference Center).

Table 1. Ticks from Fall Migratory Birds in Cyprus, 1967

Bird hosts			Tick species					
Family	No. spp.	No. birds infested	*Hyalomma marginatum marginatum*			Other species		
			N	L	Ticks/Bird	♀	N	L
Columbidae (Dove)	1	1	0	2	2.0	0	0	0
Strigidae (Owl)	1	4	5	2	1.8	0	0	0
Picidae (Wryneck)	1	1	0	2	2.0	0	0	0
Alaudidae (Lark)	1	2	1	2	1.5	0	0	0
Motacillidae (Pipets & Wagtails)	4	15	11	8	1.4	0	1	0 *Ixodes frontalis*
Muscicapidae (Warbles, etc.)	16	77	32	49	1.4	3	3	12 *Ixodes frontalis*
						0	0	4 *H. punctata*
						0	0	1 *H. concinna*
						0	2	4 *Ixodes ricinus*
Emberizidae (Buntings)	2	2	6	4	5.0	0	0	0
Fringillidae (Finches)	2	12	7	3	1.0	2	0	0 *Ixodes frontalis*
Ploceidae (Weaver finch)	1	1	0	1	1.0	0	0	0
Total	29	115	62	73		5	6	21

Hyalomma marginatum subspecies were the most prevalent in both seasons. They accounted for 80.8 % and 95.0 % of the total ticks during fall and spring, respectively (Tables 1 & 2). In the fall, 62 nymphs and 73 larvae of *H. m. marginatum* were taken and all tick-infested bird species carried *marginatum*. In the spring, 540 nymphs and 217 larvae of *H. marginatum rufipes* were taken and all but two tick-infested bird species carried *rufipes*.

Hyalomma marginatum subspecies are two-host ticks. Immature stages feed chiefly on birds or hares, seldom on burrowing animals. Larvae moult to nymphs without detaching from the host. Adults infest wild and domestic artiodactyl mammals. *H. m. marginatum* occurs in the southwestern Palearctic Faunal Region, from southwest USSR and northern Iran through much of the Mediterranean Basin to the Atlantic. *H. marginatum rufipes* occurs in all of Africa (Ethiopian Faunal Region) except in extremely dry and wet areas.

In the fall, immature *Haemaphysalis punctata* were collected from a willow warbler *(Phylloscopus trochilus)* and a flycatcher *(Muscicapa striata)*, and a larval *H. concinna* was taken from a willow warbler.

Ixodes frontalis females, nymphs, and larvae were collected in the fall from a wagtail *(Motacilla flava)*, 6 species of warblers *(Acrocephalus arundinaceus, Phylloscopus trochilus, P. collybita, Saxicola torquata, Erithacus rubecula,* and *Turdus philomelos)*, and a green finch *(Carduelis chloris)*. In the spring, 2 nymphs of this tick species, most probably acquired locally, were found on 2 species of warblers: *Erithacus rubecula* and *Turdus merula*. *Ixodes frontalis*, exclusively a parasite of birds, is reported sporadically from western Europe, Greece, Ukraine, and Caucasia.

Ixodes eldaricus nymphs (2) were collected in the fall from a tree pipit *(Anthus trivialis)* and a thrush *(Turdus merula)*. *I. eldaricus* was originally described from a female from a partridge, *Perdix perdix*, in Georgian SSR and later from females and nymphs from the rock partridge, *Alectoris graeca*, in Armenian SSR and Azerbaijan SSR. In Iraq, it has been reported as *Ixodes* sp. and as *I. tatei*, from the rock partridge and fox, *Vulpes vulpes* (female ticks), and from the field mouse, *Apodemus* sp., and jungle cat, *Felis chaus furax* (nymphs). The biology of *L. eldaricus* has not been studied.

Ixodes ricinus nymphs and larvae were carried in the fall by 3 species of the warbler family: *Erithacus rubecula, Luscinia luscinia* and *Turdus philomelos*. *I. ricinus* larvae and nymphs feed mostly on small mammals and birds. Adults and fewer numbers of nymphs feed mostly on large domestic and wild animals. *I. ricinus* occurs in forests and humid grasslands in Europe and western USSR, Northwest Africa, Turkey, Lebanon, and Iran.

Ixodes redikorzevi adults (2 females and 1 in copula male) were collected from 2 species of thrushes, *Turdus merula* and *T. philomelos*. All stages of this three-host tick infest small mammals and birds; there are few records from domestic animals. This tick ranges from Southeast Europe and the Near East through Caucasia and Tadzhikistan to Afghanistan and West Pakistan.

Three African *Amblyomma* species were taken from northward migrating birds.

An *A. variegatum* nymph was collected from a tree pipit *(Anthus trivialis)*. Im-

Table 2. Ticks from spring Migratory Birds in Cyprus, 1968

Bird Hosts			Tick species								
Family	No. spp.	No. birds infested	*Hyalomma marginatum rufipes*			Other species					
			N	L	Ticks/Bird	♂	♀	N	L		
Falconidae (Falcon)	1	7	28	8	5.1	0	0	0	0	
Columbidae (Dove)	1	4	0	0		0	0	0	30	*Argas (P.) streptopelia*
Cuculidae (Cuckoo)	1	1	1	0	1.0	0	0	0	0	
Strigidae (Owl)	1	7	16	2	2.6	0	0	0	0	
Upupidae (Hoopoe)	1	1	1	0	1.0	0	0	0	0	
Alaudidae (Lark)	1	3	3	4	2.3	0	0	0	0	
Motacillidae (Pipet & Wagtail)	2	11	9	3	1.3	0	0	1	0	*Amblyomma (P.) variegatum*
						0	0	1	0	*Ixodes eldaricus* [a]
Laniidae (Shrike)	1	6	16	2	3.0	0	0	0	0	
Muscicapidae (Warbles, etc.)	23	132	463	194	5.3	0	0	0	0	
						0	0	2	0	*Ixodes frontalis* [a]
						0	0	1	0	*Amblyomma nuttalli*
						0	0	1	0	*Amblyomma lepidum*
						0	0	1	0	*Ixodes eldaricus* [a]
						1	2	0	0	*Ixodes redikorzevi* [a]
Emberizidae (Buntings)	2	2	0	4	2.0	0	0	0	0	
Fringillidae (Finches)	3	3	3	0	1.0	0	0	0	0	
Total	37	177	540	217		1	2	7	30	

[a] *Ixodes* infestations were probably acquired in Cyprus or nearby Eastern Mediterranean countries.

Hlouskova 7. sekce Tab. 3 str. 36

Table 3. Tick–infested Migratory Birds in Cyprus

Bird Hosts		Fall 1967			Spring 1968		
Family	No. spp.	Number birds					
		examined	infested	%	examined	infested	%
Falconidae (Falcon)	1	0	0		33	7	21.2
Columbidae (Dove)	1	14	1	(7.1)*	4	4	(100.0)*
Cuculidae (Cuckoo)	1	0	0		58	1	1.7
Strigidae (Owl)	1	4	4	(100.0)*	109	7	6.4
Upupidae (Hoopoe)	1	0	0		335	1	0.3
Picidae (Wryneck)	1	3	1	(33.0)*	0	0	
Alaudidae (Lark)	1	22	2	9.1	13	3	(23.1)*
Motacillidae (Pipets & Wagtails)	5	408	15	3.7	58	11	18.9
Lannidae (Shrike)	1	0	0		467	6	1.3
Muscicapidae (Warbles, etc.)	29	975	77	7.9	18,805	132	0.7
Emberizidae (Buntings)	3	4	2	(50.0)*	42	2	4.8
Fringillidae (Finches)	3	212	12	5.7	59	3	5.1
Ploceidae (Weaver finch)	1	35	1		0	0	
Total	49	1,677	115		19,983	177	

*Less than 20 birds examined.

mature stages of *A. variegatum* infest small mammals and birds. Adults and nymphs are common on cattle and other artiodactyl mammals. *A. variegatum* occurs in much of Africa.

An *A. nuttalli* nymph was found on a thrush nightingale (*Luscinia luscinia*). All stages of *A. nuttalli* infest tortoises and reptiles; immature stages feed also on birds and small mammals. Adults also infest cattle and goats. Although *A. nuttalli* is widely distributed in tropical, subtropical and savanna regions of Africa it is seldom common.

An *A. lepidum* nymph was taken from a thrush (*Turdus merula*). This three-host tick is common on cattle and sheep in the semidesert and savanna of eastern Africa. Adults, sometimes associated with smaller numbers of nymphs and larvae, infest domestic animals and large wild animals. Immature stages infest small mammals and birds.

Argas (Persicargas) streptopelia larvae (30) were collected from 4 northward migrating turtle doves (*Streptopelia turtur*). This recently described species parasitizes doves in Cyprus, Egypt, Sudan, Senegal and South-West Africa. In a special study in Cyprus in 1969, 218 migrating *Streptopelia turtur* were examined; the infestation rate with *A. (P.) streptopelia* larvae was 15.6 % and mean infestation load was 17.2 larvae per dove. Free-living populations of this tick have not yet been discovered in Africa or Eurasia.

Epidemiological considerations

The presence of immature stage *Hyalomma m. marginatum* on 30 species of fall migratory birds in Cyprus, together with *Ixodes frontalis* on eight, *I. ricinus* on three, *Haemaphysalis punctata* on two, and *H. concinna* on one of these species, points to the potential for viral and rickettsial transmission and spread. Equal challenging to the epidemiologist is the spring migration, in which *H. marginatum rufipes* is recorded from 36 host species together with *Amblyomma variegatum*, *A. lepidum*, and *A. nuttalli* each on one, and *Ixodes eldaricus*, *I. frontalis*, and *I. redikorzevi* each on two of these species. One other bird, the turtle dove, was infested on the spring migration only by *Argas (P.) streptopelia*. Of the 37 fall bird species, 17 were the same species as those found infested in spring.

Proceedings of the 3rd International Congress of Acarology, Prague 1971

SOME CHARACTERISTIC FEATURES OF THE LIFE HISTORY, ECOLOGY AND BEHAVIOUR OF THE TICKS *HAEMAPHYSALIS INERMIS, H. CONCINNA* AND *H. PUNCTATA*

J. NOSEK
Institute of Virology, Slovak Academy of Sciences, Bratislava, Czechoslovakia

The species mentioned in the title are structurally quite different and represent different degrees of phylogenetic trends in haemaphysalids. *Haemaphysalis inermis* represents structurally a very primitive species with a laterally projecting basis capituli and elongate palpi without a palpal spur in all stages. The basis capituli are rectangular in *Haemaphysalis punctata*, posterior cornua present in male only, palpi basolaterally salient, palpal spur developed. *Haemaphysalis concinna* has a rectangular basis capituli with a posterior cornua in both sexes, more basolaterally salient palpi and a well developed palpal spur.

The type and degree of palpal and spur development is closely correlated with the type of host preferred by each species (Hoogstraal, 1965). Immature stages of *H. inermis* are frequently found on steppe animals such as lizzards and the common vole, while those of *H. punctata* and *H. concinna* are found on birds and small mammals occurring in their habitats. It seems that *H. concinna* prefers birds as hosts.

The feeding period of *H. inermis* on its natural hosts is relatively short; in the female 6 – 7 days, very short in nymphs and larvae: 4 hours and 1 – 2 hours respectively. This period of females of *H. punctata* and *H. concinna* lasts 10 – 11 days and 8 – 10 days respectively. The feeding period of nymphs and larvae is approximately the same, 3 – 4 days as can be seen in Fig. 1.

For adults of all three species free living ungulates are the important hosts.

The structurally primitive species *H. inermis* is of South Asian tropical origin and requires a very high degree of humidity. Its distribution in the temperate zone is confined to places with mild winters e.g. sea coast and river basins.

H. punctata is ecologically very adaptable, tolerating the mild humid climate of pastures of southern England and persisting into the karst region in Slovakia and Hungary and semi-desert zones of Central Asia.

H. concinna is a widely distributed species in the forest of temperature Eurasia. Its distribution in the temperate zone is confined to the deciduous and mixed forests and to relatively humid places like shores of lakes and river basins.

H. inermis occurs in Slovakia in a region approximately delimited by the 8°C annual isotherm and 600 – 700 mm isohyets. The substrate in nearly all localities is andesite and andesite pyroclasts. *H. inermis* here inhabits the forest-steppe communities (*Quercetum pubescentis, Corneto-Quercetum, Festucetum pseudodalma-*

ticae), thermophilic oak wood *Quercetum petraeae* and hornbeam-oak forest *Carpino-Quercetum*. The distribution of *H. concinna* in Czechoslovakia is limited to the same annual isotherm and isohyets as that of *H. inermis* but has different habitats. It inhabits the light humid leafy forests and mixed hornbeam-oak forests with bush-undergrowth, forest clearings and the edge of forest in lowland and hilly regions.

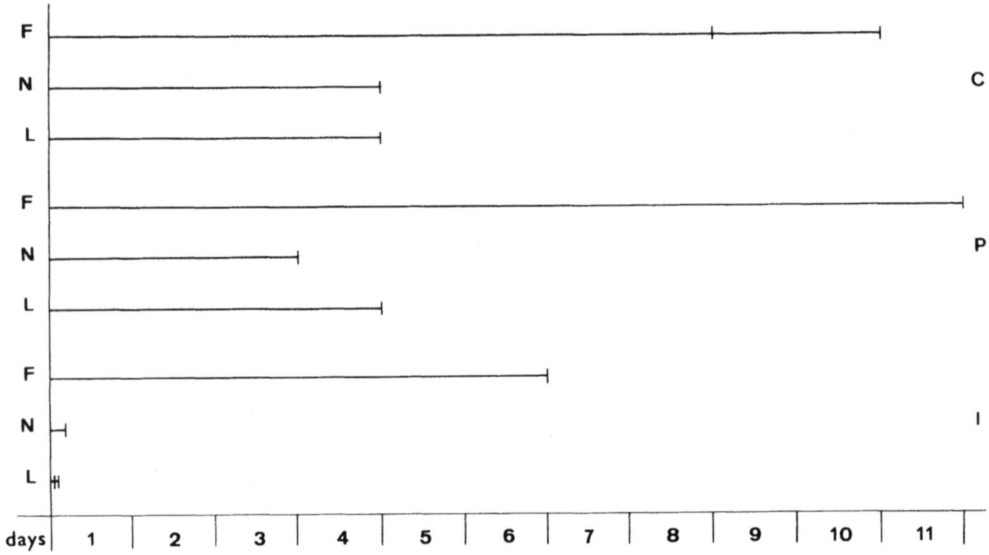

Fig. 1. Feeding periods of females (F), nymphs (N) and larvae (L) of the ticks *Haemaphysalis concinna* (C), *H. punctata* (P), *H. inermis* (I).

H. punctata ticks occur in Slovakia in a region approximately delimited by 7°C–9°C annual isotherm and 650 – 1000 mm isohyets. *H. punctata* inhabits here the dry habitats of pastures, the edge of oak woods and the wood-steppe *(Corneto-Quercetum)*. Geological substrates in these localities are andesite and andesite pyroclasts or Triassic limestone (grikes).

The seasonal incidence of the species mentioned is shown in Fig. 2. Daily incidence: the influence of environmental factors - temperature and humidity - is expressed in the daily activity of individual stages of these species. The adults of *H. inermis* have a diurnal activity; however, the number of ticks on vegetation changes during the day depending on the season, weather and host condition. During favourable conditions on sunny days after a thaw the ticks wait for their hosts on the apical part of twigs or plants, e.g. *Seseli osseum*. During the winter they are found nearer the ground between the stalks and although they change their position, they stay on the vegetation during fog, rain and snow. Larvae and nymphs are most active in the morning after the disappearance of dew and before the evening. They occur on grass, on bushes, treestumps and boulders.

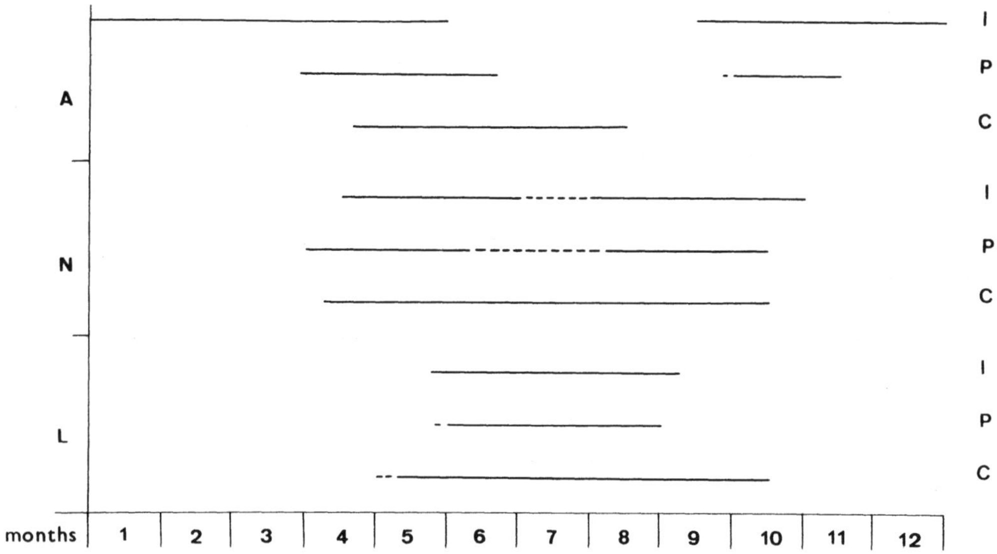

Fig. 2. Seasonal incidence of adults (A), nymphs (N) and larvae (L) of *Haemaphysalis inermis* (I), *H. punctata* (P), *H. concinna* (C).

Adults of *H. punctata* wait for their hosts in approximately the upper third of grass-stalks, with palpae directed downwards, and they are active during day-time. Larvae and nymphs survive the dry period (at the beginning of summer) under stones in bushes, under tussocks of *Festuca sulcata* and *pseudodalmatica*, etc. The nymphs are less sensitive to dry conditions. They occur in the shade of bushes and on grass, frequently in open areas.

Adults of *H. concinna* have a diurnal activity and the larvae and nymphs are frequently active during the entire day. The nymphs await their hosts in groups, on the underside of leaves of shore vegetation. The development of *H. inermis* in nature is often completed within two years, that of *H. punctata* and *H. concinna* cover three vegetation periods. Three different hosts, either of the same or different species, are required to complete the life history of all three species. The number of eggs in fully fed females reaches 3000 – 5000 in *H. punctata* and *H. concinna* while in those of *H. inermis* it can vary from 100 to 2600.

Females outnumber males in all three species. Copulation takes place on the host and experimental evidence suggests that, as females outnumber males, the latter may copulate with several females. The ratio of ♂♂ : ♀♀ changes during the seasonal activity and during successive years. In *H. punctata* and *H. concinna* the ratio 1 : 1.4 was observed, but in *H. inermis* this can vary from 1 : 1.4 to 1 : 10. In mixed populations in nature (bisexual and parthenogenetic), the ratio ♂♂ : ♀♀ = 1 : 40 was observed.

References

Hoogstraal, H.: Phylogeny of *Haemaphysalis* ticks. Proc. XIIth Int. Congress Entomol., London, 8 – 16 July, 1964, pp. 760 – 761, 1965.

Nosek, J.: The ecology, bionomics and behaviour of *Haemaphysalis (Haemaphysalis) concinna* tick. Z. Parasitenk. 36, 233 – 241, 1971.

– The ecology, bionomics and behaviour of *Haemaphysalis (Aboimisalis) punctata* tick in Central Europe. Z. Parasitenk. 37, 198 – 210, 1971.

– The ecology, bionomics and behaviour of *Haemaphysalis (Alloceraea) inermis* tick. (In press.)

BIOLOGY OF *HAEMAPHYSALIS (ALLOCERAEA) INERMIS* IN MOLDAVIA

I. G. USPENSKAYA
Institute of Zoology, the Moldavian Academy of Sciences, Kishinev, USSR

Haemaphysalis inermis occurs in deciduous forests, mainly in southern Europe (Pomerantsev, 1950; Mačička, 1958; Feider, 1965). In Moldavia this species inhabits the forests of the central and south-western parts of the Republic, but its numbers are not very high. Usually in spring and autumn single specimens of adult ticks are to be found on cattle. On the other hand, there are sites with increased numbers when up to 6 adult ticks, in the period of intensified spring activities, occur on vegetation per 200 m of collecting route. The frequency index on cattle in April is 25 %, while in other forest sections it does not exceed 5 % and is usually 0 to 1–2 %.

The activity of adult ticks in southern areas of its range of occurrence is limited to the autumn-winter-spring period (Pomerantsev, 1950). In Moldavia the curve of adult activity is of a two-peak character with maxima in April and October. In summer no adult ticks have been found either on cattle or on vegetation. In winter (December - February) single specimens of ticks are to be found on animals. It is evident that the type of activity has been preserved, but the proximity of the northern boundary of the range of occurrence changes and conditions the shifting of maximum activity to the spring and autumn months.

Larvae and nymphs are active in the summer months, but are seldom detected. We have obtained data on the activity of immature stages during our experiments of rearing ticks under natural conditions. In our studies on the biology of *H. inermis* the same methods have been used as those in the studies on *Ixodes ricinus* (Uspenskaya, 1969).

In the period of spring activity of adult ticks (March-May) males and females moulted in the previous autumn have been observed to be active. The females which found a host and fed on it, lay eggs after 2 – 4 weeks. Oviposition lasts 3 – 4 weeks Embryogenesis lasts 25 – 35 days, depending on the feeding period of the female (in March, April or May). Larvae hatch by the end of July or the beginning of August and after 7 – 10 days some of them become active; they feed from 2.5 – 3.5 hours up to 2 days.

The larval-nymphal metamorphosis lasts from 40 to 290 days, depending on the period of feeding of unengorged larvae. The metamorphosis of longest duration is typical for larvae which have fed in September – October and overwintered in an engorged state, the one of shortest duration is characteristic of larvae which have fed in August (Table 1).

Table 1.

Period of feeding of unengorged LL and NN	Periods of larval nymphal metamorphosis in days	Period of nymphal imaginal metamorphosis in days
May	102 – 115	110 – 120
June	80 – 90	80 – 90
July	63 – 78	55 – 65
August	40 – 50	30 – 40
September	270 – 290	12 – 14
October	270 – 290	310 – 315

The feeding of nymphs lasts somewhat longer than that of larvae, namely from 4 hours up to 2 days. The nymphs of a new generation usually do not feed in the autumn of the year of moulting; they become engorged during the warm period of the succeeding year, from April to September inclusively.

The nymphal-imaginal metamorphosis lasts for a considerable time, depending on the periods of feeding of unengorged nymphs (Table 1). The shortest metamorphosis has been observed in nymphs of the generation of the previous year which survived till September and engorged in the second decade of that month. Imagoes emerged from them in 12 – 14 days on 3 – 6 October. The longest nymphal metamorphosis has been observed in nymphs which engorged in October of the year of moulting, namely 310 – 315 days.

The developmental cycle of *H. inermis*, under conditions prevailing in Moldavia, lasts from 15 months to 3 years. The shortest cycle begins from the moment of the female's engorgement during April, May. Larvae emerge from the end of July to the beginning of August. After 7 – 10 days about 20 % of them become active. Later, during August and the first half of September, 6 – 10 % of larvae become active. Those which find a host and feed until mid-September moult into nymphs in October, but these, as a rule, do not become active (only 2 – 3 % of young nymphs are able to feed). The majority of nymphs feed during the warm period of the following year and in August – September, i.e. 16 – 17 months after the engorged females drop off their host, they moult into imagoes.

The two-year cycle is the most common one in Moldavia. It begins from the engorgement of females during the warm period and the hatching of larvae in August-September. As most larvae of the new generation (about 80 %) overwinter in the unengorged state and feed in April-September of the following year, the nymphs emerge from them in August-September, i.e. one year after the hatching of the larvae. Some of these nymphs engorge, but most of the others overwinter in the unengorged state; they feed during the following year and all of them complete their metamorphosis, producing a new generation of imagoes, in September – October, 2 years after the hatching of larvae or 2 years and 3 – 4 months after the dropping-off of females. Such a type of development is possible even if the females feed in autumn, assuming that larvae during the coming warm season manage to engorge and moult into nymphs.

The majority of ticks, originating from females which have fed in autumn, take 3 years for their development.

Hence, the development of *H. inermis* in Moldavia is characterized by the emergence of a new generation in August – September, the considerable activation of larvae and the slight activation of nymphs in the autumn of the year of hatching, the speeding-up of development of immature stages during the warm season and by the diapause of most ticks of all developmental stages in winter in the unengorged, less frequently in the engorged state. Not a single stage overwinters twice in the unengorged state. The age grouping of the population changes once a year, namely in the autumn.

References

Pomerantsev, B. I.: Ixodid ticks (Ixodidae). Fauna SSSR. Paukoobraznye, vol. 4. 1950. (In Russian.)

Mačička, O.: K bionomii *Haemaphysalis inermis* u nás. Čs. parasitologie, Vol. 2, 121–124, 1958.

Feider, Z.: Acaromorpha, Superfamily Ixodoidea. Fauna Republicii Populare Române, Arachnida, V. 5, Fasc. 2, 1965.

Uspenskaya, I. G.: Biology of *Ixodes ricinus* L. under conditions of Moldavia. Parasites of vertebrates. Kishinev, AN USSR, 1969. (In Russian.)

Section VIII

MITES AND TICKS OF MEDICAL AND VETERINARY IMPORTANCE

Chairman
H. Hoogstraal (USA)

Assistant Chairman
V. Černý (Czechoslovakia)

Proceedings of the 3rd International Congress of Acarology, Prague, 1971

DIAPAUSE PHENOMENA IN TICKS (IXODOIDEA) AND THEIR REGULATION

V. N. BELOZEROV
Biological Research Institute of Leningrad University, Leningrad, USSR

Seasonal cycles in Arthropods are based on an alternation of periods of active development and dormancy (e.g. diapause). The problem of diapause has been studied for a long time but mainly by entomologists. The problem has now attracted the attention of acarologists dealing with ixodid ticks, although an instance of tick diapause (the seasonal delay in moulting of engorged larvae of *Ixodes ricinus* L.) was described by S.K. Beinarowitch in Russia as early as 1907. The first special consideration of tick diapause was undertaken by Alfeev (1948, 1954) who distinguished four categories of diapause characteristic for these bloodsuckers:
1. a non-activity of unfed ticks, 2. a delay of engorgement, 3. a delay of metamorphosis in engorged larvae and nymphs, 4. a delay of oogenesis in engorged adult females. An egg diapause has also been described (Serdjukova, 1951) causing a delay in embryogenesis. Seasonal adaptations of the same type were ascertained recently in a number of ticks from moderate climatic zones.

About all cases of diapause known in ticks can be classified into two basic types, behavioural diapause and morphogenetic diapause, which are characteristic for different stages (the former for unfed ticks and the latter for engorged ticks).

The behavioural diapause is characterized by an absence of aggressiveness of unfed ticks. It consists in blocking some specific links within a chain of reactions which lead to host-finding. A distinctive seasonal character of activation and tick-incidence is connected with behavioural diapause and its investigation is of primary importance in medical acarology. Diapause of this type occurs in larvae and nymphs (North-American ticks *Dermacentor variabilis* and *D. albipictus*, the European tick *Ixodes ricinus)* and in adult ticks (Euroasian ticks *D. marginatus* and *D. pictus).* Undoubtedly, in ixodids this type of diapause is quite usual and widespread in connection with their ability for prolonged survival. Recently we have demonstrated the presence of such a dormancy in every stage of the life cycle in *Haemaphysalis concinna* and in nymphs of *H. longicornis*.

A characteristic feature of the morphogenetic diapause is a delay of morphogenetic processes in engorged ticks (a delay of metamorphosis in pre-imaginal stages and a delay of oogenesis in adult females) or a delay of embryogenesis in the egg stage also. The diapause in this case is caused by the blocking of some essential links within a chain of events or morphogenesis. This diapause made the strict synchronization possible of an appearance of the following stages with a seasonal climatic rhythm (egg hatching in *I. ricinus*, moulting of larvae and nymphs in some species of *Ixodes*, e.g. *I. ricinus*, *I. persulcatus*, *I. apronophorus*, *I. trianguliceps*, and in some species of *Haemaphysalis*, e.g. *H. concinna*, *H. japonica* and *H. punctata*) and at least egg-laying in *D. marginatus*, *D. pictus*, *I. trianguliceps* and *Hyalomma asiaticum*.

The reactions of ticks to day-length are of primary importance in the regulatory mechanisms of both types of diapause and, therefore, of seasonal cycles of their activity and development. A dependance of the activity of ticks upon day-length was discovered by Smith and Cole (1941) for larvae and nymphs of *D. variabilis*. Subsequently a participation of photoperiodic reaction in the control of metamorphosis of *I. ricinus* larvae (Loew, 1962) and in that of oogenesis in adult females of *D. marginatus* was discovered (Belozerov, 1963). Recently it has been shown that practically every case of diapause in ixodids is regulated photoperiodically.

In ticks with the long-day type of photoperiodic reaction (larval stages of *I. ricinus* and *I. persulcatus*, larval and nymphal stages of *H. concinna*) the long-day conditions enable the non-diapause development, but short-day conditions determine the induction of the morphogenetic diapause. Vice versa, in ticks with the short-day reaction (adults of *D. marginatus* and *D. pictus*) the diapause is induced by long-day influences, while a non-diapause development is induced by short-day treatment. In nymphs of *I. ricinus* (Belozerov, 1966) and *I. persulcatus* (Babenko, 1967) an inversion of photoperiodic reaction occurs, in other words the short-day long-day reaction. Here a non-diapause development is possible only after the change of a short-day regime for the long-day regime. In this case the delay of metamorphosis of engorged nymphs into the adult stage can be caused by two forms of morphogenetic diapause, a short-day one and a long-day one, which are different in their inductive mechanisms but similar in their final result.

A remarkable feature of the reactions to day-length in ticks is extremely long duration of the photoperiodic sensitivity which is maintained in the course of starvation, during bloodfeeding and even after engorgement. The photoperiodic reaction in ticks therefore participates not only in diapause induction, but also in the maintenance of the stage of diapause. The last event is of special importance for the regulation of the behavioural diapause.

References

Alfeev, N.I.: On diapause in ixodid ticks. Tr. Voenno-med. Akad. 44: 50–60, 1948. (In Russian.)

Alfeev, N.I.: On duration and forms of diapause in ticks in relation to exogenous conditions. Ibid. 58: 121–138, 1954. (In Russian.)

Babenko, L.V.: Diapause in nymphs of *Ixodes ricinus*. Wiad. Parasitol. 13: 517–524, 1967.

Beinarowitch, S.K.: Ticks of the North - West Russia as mediators of enzootic hemoglobinuria in the live-stock. Arch. vet. nauk I: 7–43, 1907. (In Russian.)

Belozerov, V.N.: Day-length as a factor governing the delay of egg-laying in females of *Dermacentor marginatus* Sulz. Med. Parasitol. (Moskva) 32: 521–526, 1963. (In Russian.)

Belozerov, V.N.: Nymphal diapause in tick *Ixodes ricinus*. I. Behaviour of unfed nymphs and development of engorged nymphs in dependence on photoperiod. Ibid. 35: 723–729, 1966. (In Russian.)

Loew, J.: Untersuchungen ueber den Einfluss exogener Faktoren auf die Metamorphose von *Ixodes ricinus*. XI. Int. Congress Entomol., I: 774, 1962.

Serdjukova, G.V.: Hibernation of eggs of *Ixodes ricinus* in conditions of Karelian peninsula. Dokl. Akad. Nauk SSSR 81: 1171–1173, 1951. (In Russian.)

Smith, C.N., Cole, M.: Effect of length of day on the activity and hibernation of the american dog tick, *Dermacentor variabilis*. Ann. Entomol. Soc. Am. 34: 426–431, 1941.

HIBERNATION OF THE TICK *IXODES RICINUS* IN THE LIGHT OF CONTINUOUSLY MEASURED MICROCLIMATE ELEMENTS

M. DANIEL, E. HONZÁKOVÁ, V. ČERNÝ, F. DUSBÁBEK
Institute of Parasitology, Czechoslovak Academy of Sciences,
Prague, Czechoslovakia

The developmental cycle of ticks is influenced and modified by biotic conditions on one hand and abiotic ones on the other. To the biotic conditions may be primarily listed the presence and numbers of host in a given biotope. To the abiotic conditions belong above all temperature and humidity of environment, although other physical factors such as light regimen etc. play a role here, too. In one of our previous papers (Daniel and Černý, 1967) we pointed out the necessity of exact studies on microclimate for the research of bionomy of ticks and proposed to use the methods elaborated by Daniel (1965) in his studies on nest microclimate of mammals and birds. These methods meet several main requirements:

1. to conduct all measurements under field conditions continuously and for a long period
2. to arrange the probes so that they be capable to register the most subtle changes in actual environment
3. to take the measurements at a distance and make recordings suitable for computer calculations.

Such measurements can be made by electrothermometers, primarily thermistors. Humidity is measured by unventilated thermistorpsychrometer which has been developed by us to this purpose and in which the principle of August's psychrometer has been applied. Our measurements and recordings were carried out by means of multirecorders which marked the data on 6 dotted lines, with intervals between bandswitching of 20 sec., i.e. the separate lines were dotted every two minutes.

On the basis of these methods we decided to analyze the effects of microclimate on the developmental cycle of the tick *I. ricinus* L. First we devoted our attention to the analysis of overwintering conditions of ticks, aiming to apply the results and experience obtained in further investigations of the whole cycle.

The present paper is based on our studies conducted between November 1966 and March 1967 and attempting to elucidate the overwintering of ticks under conditions of South-Moravian thermophilic oak forest (Valtice locality, district of Břeclav).

A total of 500 unfed and engorged imagoes, engorged larvae and an unspecified number of unfed larvae of *Ixodes ricinus* were used in the experiment. All ticks came from laboratory colonies, reared under standard conditions (+22°C/90%RH). The ticks were placed in cylinder-like cages 50 x 25 mm large, each made from a wire skeleton covered with silon fabric. The cages were deposited 0, 10, 20, 30, 40 and 50 cm deep in undisturbed soil layers. Pedologically the soil was classified from the surface level to the depth of 40 cm as loamy sand (pH 7.50 – 7.85), at 50 cm as loam with pH 7.90 (classification after Kopecký's scale). In order to trace the influence of different cages, additional cages of similar shape and size were also used; they were made from plastic, metal, glass and wood and contained 380 specimens of *I. ricinus* at the depths of 10 and 20 cm.

Of the microclimatic elements primarily temperature of particular soil layers was taken, as well as the temperature inside the cages at given depth. The soil humidity was measured by means of plaster blocks. The data obtained were calculated on a computer. Further on observations were made of mesoclimate in the experimental area using a standard meteorological box. The data on macroclimate of the area were obtained from the State Meteorological Station 7 km away from the experimental area.

After closing the experiment all cages were taken out and the number of ticks counted. The engorged specimens which survived, were observed as regards their ability for subsequent metamorphosis. The unfed specimens were allowed to feed and their ability for further development was determined. The data obtained served as input data for calculations on the MINSK computer.

The methods used in mathematical-statistic analyses may be briefly characterized as follows::

While studying the survival of ticks it was necessary to determine its dependence on two factors – the depth of soil and the developmental stage (including different physiological state) of the tick. The most convenient method was the analysis of variance with two-way classification. The data given in Table 1 were taken as a basis. Each

Table 1. Results of hibernation
Survival of the tick *Ixodes ricinus* in particular soil layers

Developmental stage of the tick \ Depth at which the cage was located	0	10	20	30	40	50
Engorged larvae	12/24	16/25	9/25	19/25	14/25	23/25
Unfed nymphs	23/30	15/15	27/30	28/30	20/30	26/30
Engorged nymphs	–	10/10	9/10	10/10	10/10	9/10
Unfed females	5/10	5/10	6/10	8/10	8/10	7/10
Unfed males	1/10	3/10	8/10	5/10	8/10	7/10

(Number of surviving specimens are given in the numerator, number of released specimens in the denominator).

square contains data in the form of a fraction. Its denominator indicates the number of ticks released, the numerator – the number of surviving specimens. The same method was used in further evaluation of the development of experimental ticks (the initial data are given in Table 2 which is arranged in a similar way as Table 1). It was revealed that the probability of survival of *I. ricinus* under conditions of our experiment depended significantly on the developmental stage and its physiological state (engorged – unfed). The dependence on the depth at which the ticks were located has not been demonstrated. Only a hint of such a dependence was detected in unfed males. For this category also the method χ^2 was used which demonstrated on the level of 5 % significance the dependence of survival of unfed males on the depth: at the depth of 20 cm and deeper more males significantly survived than at the depth less than 20 cm.

The studies on the further development of larvae and nymphs from the experiment yielded better results. The same procedure was used for the assessment. The results show that further development towards successful metamorphosis greatly depends on the developmental stage and its physiological state; the dependence on the depth has not been demonstrated. However, in this case a descending series may be arranged according to the share of specimens with developmental disturbances:

Unfed nymphs – engorged larvae – engorged nymphs, there being a significant difference on the 5 % significance level between unfed nymphs on one hand and engorged nymphs and larvae on the other. The ticks overwintering in unfed state showed more disturbances in their development.

The results of the experiment show that *Ixodes ricinus* has a wide ecological amplitude which has apparently not been exceeded in the experiment conducted. The successful overwintering did not mostly depend on particular soil layer, only

Table 2. Results of hibernation
Negative developmental effects on the tick *Ixodes ricinus* overwintering in particular soil layers

Developmental stage of the tick \ Depth at which the cage was located	0	10	20	30	40	50
Engorged larvae	8/24	10/25	9/25	16/25	6/25	10/25
Unfed nymphs	16/30	13/15	27/30	24/30	16/30	16/30
Engorged nymphs	–	6/10	4/10	3/10	2/10	4/10

(The numerator indicates the number of ticks with disturbed development, the denominator the number of released specimens).

in unfed males the situation was found to be reversed. For this reason the microclimatic conditions of the experiment are presented in the form of a Table only, in order to document the conditions of the experiment. Further development of nymphs and larvae was influenced by the fact whether they passed the winter in unfed or engorged state (Tables 3, 4).

In conclusion, the main result of the experiment i.e. the finding that a successful overwintering (in the sense of survival and the subsequent development) does not depend on the soil layer, should be compared with the conditions in open nature. As our experiment with the release of ticks of *I. ricinus* (engorged larvae and nymphs) in two consecutive autumn seasons in the same experimental site has shown, most ticks pass the winter in the ground layers (0 – 5 cm), while in lower layers they occur to a lesser degree or occasionally only. Many specimens pass the winter even on the ground among dry leaves and other remains of vegetation and do not penetrate the soil at all.

Table 3. Mean monthly temperatures in particular soil layers (in °C)

Month \ Layer	Surface	10 cm	20 cm	30 cm	40 cm	50 cm
XII	1,57	1,65	2,11	2,67	3,31	3,59
I	–0,40	–0,29	0,14	0,71	1,42	1,50
II	–0,18	–0,42	–0,15	0, 30	0, 91	0, 92
III	3,28	2,48	2,62	2,32	2,57	2,55

Table 4. Mean humidity in particular soil layers (in %)

Soil layer	Mean humidity (in %)
surface	49,9
10 cm	24, 8
20 cm	36, 4
30 cm	30, 4
40 cm	34, 6
50 cm	35, 8

Our results obtained in methods used are also important. The suitability of methods used has been confirmed for the field experiment as well as for subsequent mathematical–statistic evaluation. These methods were applied in a long-term experiment attempting to elucidate the effects of microclimate in particular microbiotopes within the framework of the biotope of thermophilic oak forest upon the development of ticks. Our preliminary results indicate that the microclimate is an important and even limiting factor in the distribution of ticks in the terrain. We assume that computer calculations will facilitate to determine in detail the mechanism of these effects.

References

Daniel, M.: The method of exact continuous measurements of the microclimate in nests of reservoir animals for research of natural disease foci. Proc. Sympos. Theor. Quest. Nat. Foci Diseases, pp. 497–506, Publ. House Czechoslov. Acad. Sci., Praha, 1965.

Daniel, M., Černý, V.: To the methods of studying the environmental temperature of the tick *Ixodes ricinus* L. Folia parasitol. (Praha) 14: 177–183, 1967.

DEMODICIDS (TROMBIDIFORMES: DEMODICIDAE) OF MEDICAL AND VETERINARY IMPORTANCE*

C.E. DESCH, W.B. NUTTING
Department of Zoology,
University of Massachusetts, Amherst, Massachusetts, USA

Incomplete descriptions have led to considerable confusion in the literature concerning species identification within the genus *Demodex*. Concomitant with a paucity of morphological information is an incomplete knowledge of the life cycle of most of these species. Phase microscopy has revealed useful characteristics not reported or incorrectly interpreted by earlier workers. It is the intention of this paper to present pertinent information and a key which will aid in identification of species and of life stages of demodicids from human and domestic mammalian hosts.

Non-meristic characters: Because of small size and general similarity in shape among members of the genus *Demodex*, species identification requires careful observation employing oil immersion and phase optics. Gnathosomal structures of importance are: number, position and configuration of palpal claws (setae), shape of supracoxal spines and pharyngeal bulb, and position of subgnathosomal setae relative to this bulb. These structures are generally identical within the species.

In the podosoma, one should note the shape of the aedeagus and the shape and position of the genital orifice (operculate vs. non-operculate) and the two pairs of dorsal podosomal setae. These highly modified setae also have been observed in females of *D. folliculorum* var. *equi* (Bennison, 1943), *D. canis* (Desch, unpublished) and *D. folliculorum* (Desch and Nutting, MS submitted). The structure and position, relative to epimeral plates IV, of the female genitalia is also a specific character. Cognizance of the arrangement of the epimeral plates and of the leg spacing along the podosoma may prove useful.

The proctodeum, when present, is a key character located in the opisthosoma (Desch, et al., 1970). Its shape and the position of the proctodeal pore are intraspecifically stable. For a given species the proctodeum may be present in females only, in both sexes, or absent in both sexes. Consideration also should be given to the general body contours with emphasis on the opisthosoma.

Previous workers, in general, paid scant attention to the morphology of deve-

* This investigation was supported in part by NIH Grant 5–TOIAI–226–08.

lopmental stages among demodicids. However, immature stadia exhibit distinct taxonomic features which will prove especially valuable as more species are named to the genus. Ova are of consistent shape, and may or may not possess an operculum. Immatures display gnathosomal structures homologous to adults, but lack subgnathosomal setae. Besides general body contour, one should note number, position and configuration of legs, leg claws and solenidia, and epimeral scutes. Taking into account these criteria, a protonymphal stage has been recognized in most species examined by us.

A number of mammals are known to host two distinct and ecologically separated species of *Demodex*. In this respect niche may be diagnostic especially when mites are examined in histological sections. Skin sections of human alar region have shown *D. folliculorum* and *D. brevis* to occur less than 1 mm apart. However, *D. brevis* has been observed always at the follicle base or deep within the sebaceous glands of the large vellus hairs. On the other hand, *D. folliculorum* inhabits the simple hair follicles at or above the level of the sebaceous gland duct.

Meristic data: Due to considerable intraspecific variation, especially in opisthosomal length, and interspecific overlap, especially in total body length and width, meristic data alone are inadequate criteria for species discrimination (see also Lombardini, 1942). However, when employed in conjunction with the above mentioned non-meristic characters, meristics are of value for this purpose. An aspect of meristics overlooked by a number of previous workers is adequate sample size. A sample size of 20 specimens of each life stage and sex is most desirable and practical when determining means and standard deviations of body dimensions (Desch and Nutting, 1969). Important measurements for species identification and description are length and width of major body regions, total body length, and length of vulva and aedeagus. Ova and immatures should be measured for total body length and maximum width.

Redescriptions: Employing the above taxonomic aids we have reported elsewhere on a number of demodicids: *Demodex caprae* (Desch and Nutting, 1969), *D. canis* (Nutting, 1971), and *D. folliculorum* and *D. brevis* (Desch and Nutting, MS submitted). Below we present recent observations on two more species: *D. bovis* and *D. phylloides*.

Demodex bovis Stiles, 1892

(Meristic data in Table 1)

Male (Fig. 1.A): Gnathosoma trapezoidal, widest at base. One single and two double tined palpal claws. Rounded pharyngeal bulb open posteriorly; subgnathosomal setae anterior to bulb (Fig. 2.A). Supracoxal spines directed medially; bifurcated distally (Fig. 2.B – 2.C).

Dorsal genital orifice between two prominent protuberances at level of posterior margin of legs I. Both pairs of dorsal podosomal setae equidistant from midline; first pair at level of genital orifice. Prominent, nearly semicircular fold an-

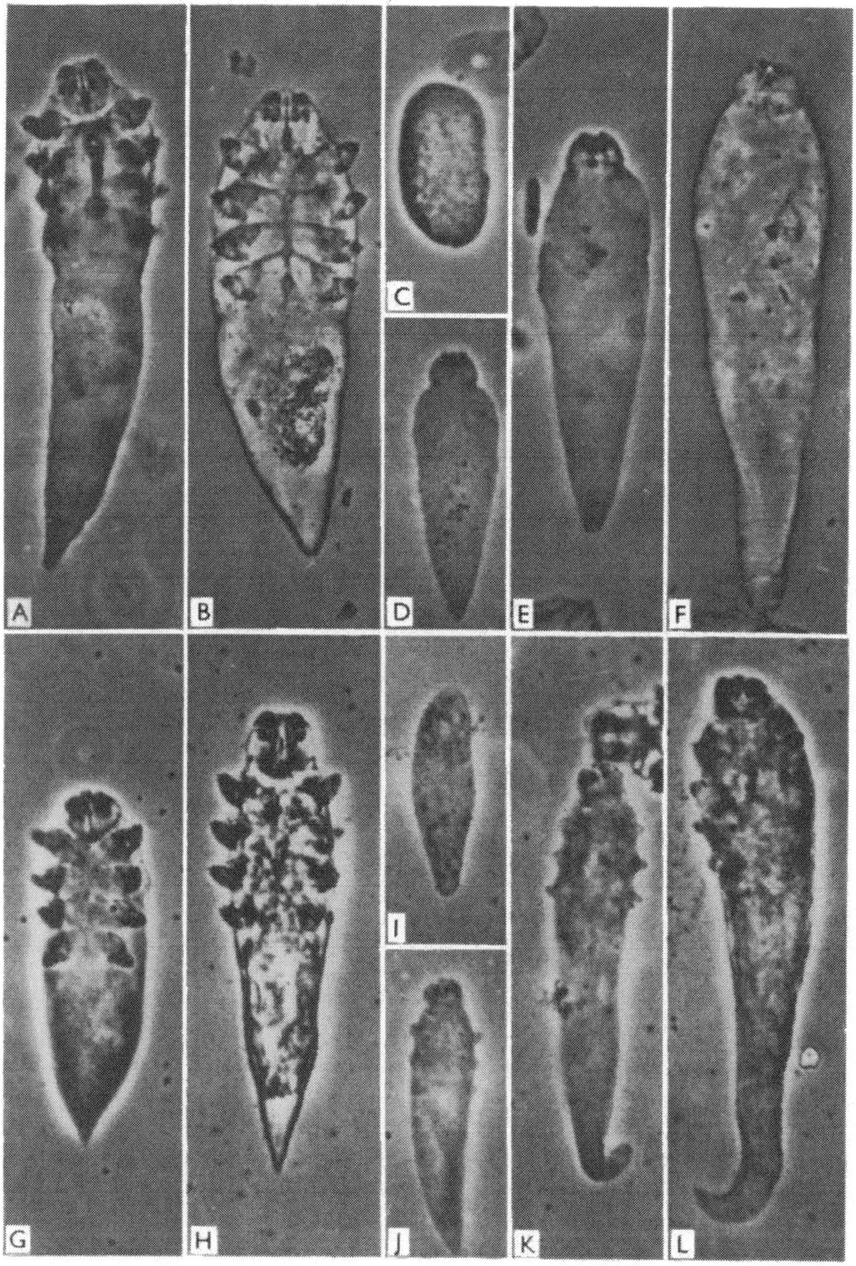

Fig. 1. A)—F): *Demodex bovis* 269×. A) male; B) female (slightly compressed); C) egg; D) larva; E) protonymph; F) nymph.

G)—L): *Demodex phylloides* 269×. G) male; H) female; I) egg; J) larva; K) protonymph; L) nymph.

Table 1. Meristic Data: Means and standard deviations in measurements of 20 specimens each (except where noted) of *Demodex bovis*, *D. phylloides*, *D. caprae* and *D. canis* (All measurements in micrometers)

Species		*D. bovis*	*D. phylloides*	*D. caprae*[+]	*D. canis*[++]
Female					
Gnathosoma	length	31.7 ± 0.4	35.1 ± 0.6	32.6 ± 0.7	24.0 ± 2.0
	width	35.0 ± 1.0	32.8 ± 1.3	32.7 ± 0.9	22.6 ± 2.9
Podosoma	length	67.6 ± 1.8	70.3 ± 3.4	70.6 ± 1.7	63.9 ± 3.7
	width	56.3 ± 3.4	54.2 ± 3.1	56.7 ± 3.5	36.9 ± 1.9
Opisthosoma	length	118.7 ± 10.7	116.5 ± 6.0	128.5 ± 9.7	135.9 ± 17.7
	width	46.7 ± 5.4	47.2 ± 4.6	52.5 ± 4.9	---
Total length		218.0 ± 10.7	221.8 ± 8.3	230.8 ± 10.6	224.3 ± 18.3
Vulva		10.2 ± 0.4	4.7 ± 0.2	9.9 ± 0.4	4.5 ± 0.5**
Male					
Gnathosoma	length	26.8 ± 0.9	27.7 ± 0.8	21.0 ± 0.8	20.4 ± 1.9
	width	31.0 ± 1.3	28.0 ± 0.8*	31.2 ± 0.8	22.9 ± 1.9
Podosoma	length	63.2 ± 3.6	55.6 ± 4.6	69.5 ± 1.6	55.8 ± 5.2
	width	57.5 ± 2.8	49.7 ± 3.8*	58.5 ± 3.2	32.9 ± 3.3
Opisthosoma	length	170.9 ± 11.1	88.1 ± 5.5	125.0 ± 8.1	91.6 ± 3.9
	width	50.0 ± 4.1	44.7 ± 4.0*	52.2 ± 6.7	---
Total length		259.2 ± 14.4	171.5 ± 7.6	217.8 ± 8.6	167.8 ± 5.3
Aedeagus		38.9 ± 1.1	24.6 ± 1.5	24.3 ± 0.6	20.0 ± 1.6**
Ovum	length	78.5 ± 3.8	99.2 ± 3.3	83.9 ± 4.4	81.5 ± 3.5
	width	40.3 ± 2.1	27.6 ± 2.7	46.5 ± 4.7	26.6 ± 2.4
Larva	length	116.7 ± 9.6	135.1 ± 11.3	154.9 ± 12.3	91.0 ± 5.9
	width	37.4 ± 3.8	32.8 ± 3.6	49.5 ± 4.7	27.2 ± 3.5
Protonymph	length	184.2 ± 14.7	207.2 ± 14.8	234.0 ± 18.5	130.7 ± 10.6
	width	47.0 ± 5.0	42.5 ± 3.5	58.2 ± 6.8	29.2 ± 2.0
Nymph	length	288.6 ± 29.9	322.0 ± 15.6	297.1 ± 30.7	201.2 ± 21.9
	width	60.3 ± 8.6	52.2 ± 3.2	69.0 ± 5.8	33.3 ± 3.1

[+] Desch and Nutting, 1969,
[++] Nutting, 1971,

* 9 specimens,
** 10 specimens.

terior to genital orifice. Aedeagus with large sheath and bulbous base (Fig. 2.D). Legs evenly spaced along podosoma.

Opisthosoma terminates in a blunt point; annulation very faint. Proctodeum absent.

Female (Fig. 1.B): Gnathosomal shape and structures similar to male.

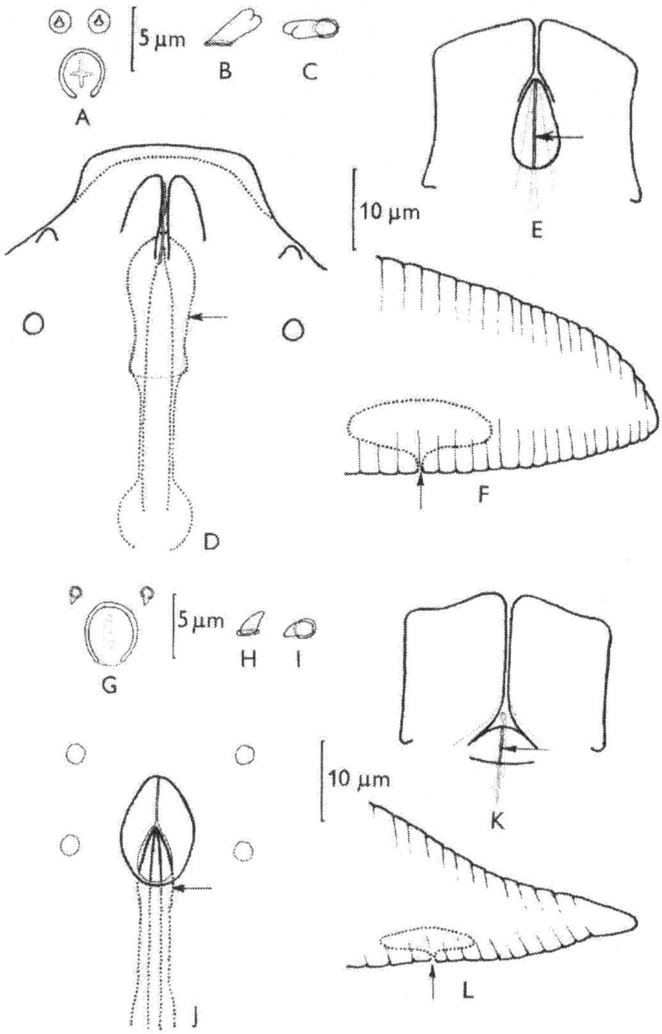

Fig. 2.A–F: ***Demodex bovis:*** **A) pharyngeal bulb and subgnathosomal setae. Cross-shape configuration within bulb is an apodeme for origin of associated muscles; B) frontal view of right supracoxal spine; C) dorsal view of right supracoxal spine; D) male genitalia; arrow indicates aedeagal sheath; E) female genital aperture and epimeral plates IV; arrow indicates vulva. F) lateral view of proctodeum; arrow indicates proctodeal pore.**

Fig. 2.G–L: ***Demodex phylloides:*** **G) pharyngeal bulb and subgnathosomal setae; H) frontal view of right supracoxal spine; I) dorsal view of right supracoxal spine; J) male genitalia; arrow indicates aedeagal sheath;;K) female genital aperture and epimeral plates IV; arrow indicates vulva; L) lateral view of proctodeum; arrow indicates proctodeal pore.**

Legs evenly spaced along podosoma. Slit-like vulva in a shallow, oval-shaped depression between epimeral plates IV (Fig. 2.E).

Opisthosomal shape similar to male but, shorter. Opisthosomal annulation prominent. Proctodeum as an antero-posterior directed, blunted fusiform sac, 17.0 ± 1.3 μm long (20 specimens), connected at its midpoint to the proctodeal pore which lies midventrally 28.1 ± 4.1 μm (20 specimens) from posterior tip of opisthosoma (Fig. 2.F).

Ovum (Fig. 1.C): Oval-shaped; non-operculate.

Larva (Fig. 1.D): Fusiform. Four single tined palpal claws. Supracoxal spines peglike; 2 μm long. Horseshoe-shaped pharyngeal bulb open posteriorly; subgnathosomal setae absent. Three pairs of trifid claws lateral along body. Epimeral scutes and legs absent.

Protonymph (Fig. 1.E): General shape and gnathosomal structures similar to larva. Claws on body similar to larva, with greater antero-posterior spacing. Leg plates absent. Three pairs of epimeral scutes; scutes I smaller than II and III.

Nymph (Fig. 1.F): General shape and gnathosomal structures similar to larva. Four pairs of lateral plates (legs) each with two trifid claws. Four pairs of epimeral scutes; scutes I smaller than others.

Demodex phylloides Csokor, 1879

(Meristic data in Table 1)

Male (Fig. 1.G): Gnathosoma trapezoidal, widest at base. On double and three singletined palpal claws. Rounded pharyngeal bulb open posteriorly; subgnathosomal setae anterior to bulb (Fig. 2.G). Small, peg-like supracoxal spines turned medially; 2 μm long (Fig. 2.H – 2.I).

Dorsal genital orifice oval-shaped. Aedeagus without prominent bulbous base. Dorsal podosomal tubercles very indistinct; apparently equidistant from midline. Posterior pair at level of genital orifice (Fig. 2.J). Legs evenly spaced along podosoma.

Opisthosoma distinctly pointed; annuli very faint. Proctodeum absent.

Female (Fig. 1.H): Gnathosoma longer than wide; associated structures similar to male.

Leg spacing as in male. Vulva as a longitudinal slit at posterior margin of epimeral plates IV (Fig. 2.K).

Opisthosoma attenuate; terminates in acute point. Pronounced transverse annulation. Proctodeum fusiform, 12.2 ± 0.8 μm long (20 specimens); connected to proctodeal pore by a short tube 7 – 8 μm from its anterior end. Pore located midventrally 20.3 ± 2.6 μm from posterior tip of opisthosoma (Fig. 2.L).

Ovum (Fig. 1.I): Fusiform; blunted anteriorly. Non-operculate.

Larva (Fig. 1.J): Fusiform. One single and one triple tined palpal claw. Horseshoe-shaped pharyngeal bulb open posteriorly; subgnathosomal setae absent. Three

Table 2. Key to known demodicids of common domestic animals and man

1 a	Average total length of female greater than that of male	3
b	Average total length of male greater than that of female	2
2 a	Supracoxal spines with subterminal bifurcation	*Demodex bovis*
b	Supracoxal spines without subterminal bifurcation	*Demodex folliculorum* var. *equi*
3 a	Epimeral plates of male meet at midline	4
b	Epimeral plates of male do not meet at midline	*Demodex brevis*
4 a	Legs of immatures extend lateral from body; claws distal	5
b	Legs of immatures as flat plates, not extending from body; claws attached to these plates	*Demodex caprae*
5 a	Protonymphs and nymphs without posterior end of body recurved	6
b	Protonymphs and nymphs with posterior end of body recurved	*Demodex phylloides*
6 a	Average opisthosomal length comprising more than ½ total length	7
b	Average opisthosomal length comprising less than ½ total length	*Demodex equi*
7 a	Subgnathosomal setae lateral to pharyngeal bulb	8
b	Subgnathosomal setae anterior to pharyngeal bulb	*Demodex ovis*
8 a	Proctodeum absent in males	9
b	Proctodeum present in males as anteriorly directed tube	*Demodex cati*
9 a	Dorsal podosomal setae of male dumbbell (or figure eight) - shaped; spindle-shapedl ovum	*Demodex canis*
b	Dorsal podosomal setae of male rounded; arrowhead-shaped ovum	*Demodex folliculorum*

pairs of lateral legs each with a single, quatrifid claw. Two pairs of very minute epimeral scutes at level of legs II and III.

Protonymph (Fig. 1.K): Gnathosomal structures similar to larva. Body attenuate with posterior prominently recurved. Two pairs of epimeral scutes at level of legs II and III. Three pairs of legs each with two quatrifid claws.

Nymph (Fig. 1.L): Gnathosomal structures similar to larva. General body shape similar to protonymph. Four pairs of legs each with two quatrifid claws. Four pairs of epimeral scutes; scutes I and IV smaller than II and III.

Key: – Even though demodicids are highly host specific (see Nutting, 1965), a number of mammals are parasitized by two or more species of *Demodex* (e.g., man, horse, golden hamster, red-backed vole). Thus, one may not assume, with certainty, species identity simply by knowing the host. This is especially true, as noted above, in tissue sections. Therefore, a key is presented in Table II to the known demodicids of common domestic mammals and man. The number of steps in the key require multiple specimens, however, this should present little difficulty because these species often occur in moderate to very large numbers on a single host.

We should keep alert to the fact that, as with man and horse, other domestic animals may be harboring two or even more species of this unusual genus of mites.

References

Bennison, J.C.: Demodicosis of horses with particular reference to equine members of the genus *Demodex*. J. Roy. Army Vet. Corps 14: 34–49, 66–73, 1943.

Desch, C.E., O'Dea, J., Nutting, W. B.: The proctodeum – a new key character for demodicids (Demodicidae). Acarologia 12: 522–526, 1970.

Desch, C.E., Nutting, W.B.: *Demodex caprae* – structural features of use in demodicid systematics. 44th Annual Meeting of Am. Soc. Parasitologists, Washington, D.C., 1969.

Desch, C.E., Nutting, W.B.: *Demodex folliculorum* (Simon) and *D. brevis* (Akbulatova) of man: redescription and re-evaluation. (MS submitted).

Lombardini, G.: Contributo alla conoscenza della morfologia dei Demodicidae. Redia 28: 89–102, 1942.

Nutting, W.B.: Host-parasite relations: Demodicidae. Acatologia 7: 301–317, 1965.

Nutting, W.B.: *Demodex canis* – a biologist's viewpoint. Annual Meeting of Am. Anim. Hosp. Assoc., Miami Beach, Florida 1971.

Proceedings of the 3rd International Congress of Acarology, Prague, 1971

EMERGING PATTERNS OF TICK-VIRUS RELATIONSHIPS*

H. HOOGSTRAAL
Medical Zoology Department, United States Naval Medical Research
Unit Number Three (NAMRU-3), Cairo, Egypt

The foundation of knowledge regarding tick-virus relationships is the comprehensive investigation into physiochemical properties and serological classification of arboviruses by Jordi Casals and his colleagues at the Yale Arbovirus Research Unit. Basic epidemiological knowledge derives from a few long-term or intensive field and laboratory research programs devoted to Russian spring-summer encephalitis (RSSE), European tickborne encephalitis (TBE), and Indian Kyasanur Forest disease (KFD). Work with these pathogens also led to recovery of numerous different viruses. Biological and epidemiological knowledge, still in the pioneer stage for most of these viruses, reflects the specific concern, location and range of movement, imagination, and competence of scientists in a few widely scattered virological and biomedical research institutes.

Even in this pioneer stage, some 70 different viruses in at least 12 serological groups have already been recovered from more than 20 argasid and 50 ixodid species infesting practically every major vertebrate animal group. These viruses and ticks are distributed on every continent and many islands, in arctic marine bird colonies, through temperate conifer and mixed forests, plains, steppes, and mountain peaks, to equatorial primary and secondary forests and savannahs. The number and variety of poorly understood interactions between these viruses, reservoirs, vectors, and ecological situations indicates the generally primitive state of basic epidemiological concepts concerning arthropod-associated organisms infectious for man and other vertebrate animals.

Viruses satisfying every criterion of the term tickborne are being discovered in an increasing variety of ecological and epidemiological situations. A notable example, African swine fever (ASF) virus, contains DNA and is unrelated to any other known virus. Long associated with native African warthogs, ASF infection is asymptomatic in these animals but may be catastrophic among domestic swine in Africa or when the virus spreads to European piggeries. Owing to negative results of very li-

* From Research Project MR005.09-0037, Bureau of Medicine and Surgery, Department of the Navy, Washington, D. C., U.S.A. The opinions and assertions contained herein are the private ones of the author and are not to be construed as official or as reflecting the views of the Department of the Navy or of the naval service at large.

mited earlier research, ticks were considered to have no role in ASF epidemiology. Recently Plowright and others have found *Ornithodoros moubata porcinus* to be abundant in East African warthog burrows and often naturally infected with ASF virus. Though the level of viremia in infected warthogs may be inadequate to produce persistent infection in the tick, high rates of transovarial passage of the virus from naturally infected female ticks to F_1 nymphs, and the ability of the new tick generation to transmit the virus when biting susceptible pigs, assures a maintenance mechanism for ASF virus in nature.

Group B viruses of the RSSE complex represent a pattern of intimate tick-virus relationship in which the virus multiplies in one or several tick species and is transmitted to a susceptible vertebrate host in the tick's oral secretions. Transstadial transmission of these viruses appears to be the rule but transovarial transmission remains a poorly understood and inadequately investigated phenomenon.

Relatively much is known regarding RSSE, TBE, and KFD epidemiology. Louping ill of Britain, Ireland, (? and Europe), Langat encephalitis (LANE) of Malaya, Omsk hemorrhagic fever (OHF) of Siberia, and Powassan encephalitis (POWE) of Canada and northern USA require much more epidemiological investigation. Nothing is known about the vectors and natural hosts of the Japanese Negishi virus.

RSSE complex viruses were long considered to be temperate Eurasian (Palearctic Faunal Region) in distribution. Discovery in tropical Asian forests (Oriental Faunal Region) of LANE and KFD viruses, and in temperate America (Nearctic Faunal Region) of POWE virus has greatly increased the range of geographical, ecological, and biological factors requiring epidemiological consideration.

Earlier concepts of few tick reservoir and vector species for RSSE viruses are also changing. In Primorye (Soviet Far East) and in southern Republics of USSR *Haemaphysalis* and *Dermacentor* ticks have been shown to serve as RSSE virus vectors in addition to the well known *Ixodes persulcatus*. KFD virus naturally infects at least 11 tick species in the genera *Ornithodoros, Ixodes, Haemaphysalis, Dermacentor,* and *Rhipicephalus*. POWE virus occurs in at least five tick species in the genera *Ixodes, Haemaphysalis,* and *Dermacentor.*

Group B viruses not members of the RSSE complex have recently been discovered in remarkably diverse situations. West Nile virus, generally considered to be mosquitoborne, is easily isolated in summer and winter from *Argas hermanni* in Egyptian pigeon houses and is also reported from *Hyalomma m. marginatum* infesting birds in Southwest USSR. Within the past few months, a new Group B virus has been obtained from pigeon-infesting *Argas hermanni* in Afghanistan, Kadam virus has been reported from cattle-infesting *Rhipicephalus pravus* in Uganda, and Tyuleniy virus was reported simultaneously by Soviet and American researchers from *Ixodes uriae* parasitizing marine birds in northeastern USSR and northwestern USA. The biological and epidemiological significance of these discoveries remain to be investigated.

Interest in other virus groups equals that in Group B as the Casals classification system reveals group properties, or intrageneric relationships, among many earlier and recently recovered viruses.

The Hughes group presently consists of five viruses from *Ornithodoros* ticks infesting marine bird colonies. Three of these viruses (Hughes, Farallon, and Soldado) are from *O. denmarki* in USA, Mexico, and Trinidad. The fourth and fifth are from *O. amblus* at Puenta Salinas, Peru, and from *O. muesebecki* at Zirqa Island, Arabian Gulf. Despite the great geographical distance separating them, *O. amblus* and *O. muesebecki* are closely related and their viruses are equally closely related. When *O. amblus* and *O. muesebecki* bite people, they produce fever and other disagreeable symptoms. The viruses recently isolated from these ticks are likely candidates as the cause of these symptoms.

One of the most scientifically challenging developments of the past three years has been the considerable extension of knowledge regarding Crimean hemorrhagic fever. As soon as the virus was characterized, after many years of effort, it was shown to be identical from Central Asia through Southwest USSR to eastern Europe and also identical to Congo virus of Nigeria, Congo, Uganda, and Kenya. The newly designated Crimean-Congo hemorrhagic fever (CCHF) virus has been isolated from eight kinds of ticks in the genera *Hyalomma*, *Amblyomma*, and *Boophilus*. Almost simultaneously, Hazara virus, from *Ixodes redikorzevi* infesting a rodent in an alpine meadow in northern Pakistan, was found to be closely related. Thus a CCHF Group of two viruses is now recognized. Medical and virological surveillance for hemorrhaging clinical cases and for CCHF virus is critically needed in large areas of Eurasia and Africa. The natural history of the virus requires elucidation throughout its extensive geographical range and variety of ecological and zoogeographic regions.

Certain human diseases appearing to be caused by viruses transmitted during tickbite also require more serious investigation. In northern Europe, erythema chronicum migrans, a distinctive and not uncommon erythema, moves centrifugally over the skin from the site of an *Ixodes ricinus* bite. In Germany, *I. ricinus* bites are associated with various dermatological symptoms which may also develop as serious neurological complications. These were described in 1967 by Dr. G. Schaltenbrand of the Wurzburg Neurological Clinic and Polyclinic. The frequently fatal Himalayan hemorrhagic disease is said by Indian military physicians to follow tickbite and to be caused by a long, lowgrade infection that supresses the bone marrow, thus allowing development of platelet antibodies and uncontrolled development of the agent in the capillaries. The agent, vectors, and transmission and epidemiological factors of this remarkable disease have not been studied.

In retrospect, certain patterns of tick-virus interrelationships are beginning to become apparent. However, unsolved problems, a very few of which we have mentioned here, are so numerous that most current concepts in this field can be only tentative or speculative.

BIOLOGICAL PATTERNS IN THE AFGHANISTAN TICK FAUNA*

H. HOOGSTRAAL
Medical Zoology Department, United States Naval Medical Research
Unit Number Three (NAMRU–3), Cairo, Egypt

Thirty-nine tick taxons (37 species, 2 *Hyalomma* species each with 2 subspecies) are recorded from Afghanistan (Table 1). Almost all are representatives of southern areas of the Palearctic Faunal Region. A few representing the western area of the Oriental Faunal Region are from Indian plains or western Himalayan foothills or highlands. Ecologically, Afghan ticks are restricted to steppes, semideserts, or mountains. None is characteristic of a forest biotope. Additional tick parasites of wild hosts undoubtedly remain to be discovered.

The most widely distributed are 5 bat-parasitizing species. *Argas vespertilionis, A. boueti, Ixodes vespertilionis*, and *I. simplex* occur in the Ethiopian, Palearctic, and Oriental Faunal Regions; *A. confusus* occurs in the Ethiopian and Palearctic Regions. Host migration and the sheltered cave environment account for the extraordinary range of bat parasites.

Two undescribed *Argas* species, parasites of lizards and hamsters, are unknown elsewhere. The exotic *Boophilus microplus* has been introduced with cattle or camels into a few eastern areas. *B. annulatus* (? = *B. calcaratus*) infests cattle in the north but appears to be rare. Notably, the African *Rhipicephalus sanguineus* has not been taken in Afghanistan.

Reptiles have specialized ticks, 2 of which are southern Palearctic in distribution. The steppe tortoise is host to immature and adult *Hyalomma aegyptium* and various reptiles support immature *Haemaphysalis sulcata*. Two other tick species represent the Oriental Region. The Indian desert monitor lizard is parasitized by immature and adult *Aponomma gervaisi* of the Indian plains and *Agama* lizards are infested by immature *Haemaphysalis kashmirensis* of the western Himalayan highlands.

Domestic pigeons and chickens are common in urban and rural Afghanistan. Pigeons are heavily infested by *Argas hermanni*, which occurs in a continuous belt from central Nepal to Northwest Africa. Quaranfil virus and a new group B virus have recently been isolated from *A. hermanni* from Kabul and the nearby Royal Farm; other viruses infect this tick elsewhere. Pigeon houses in Kabul Province

* From Research Project MR005.09–0037, Bureau of Medicine and Surgery, Department of the Navy, Washington, D. C., U.S.A. The opinions and assertions contained herein are the private ones of the author and are not to be construed as official or as reflecting the views of the Department of the Navy or of the naval service at large.

Table 1. Ticks of Afghanistan

Family Argasidae

1. *Argas (Argas) hermanni* Audouin, 1827
2. *Argas (Persicargas) persicus* (Oken, 1818)
3. *Argas (Carios) vespertilionis* (Latreille, 1802)
4. *Argas (Chiropterargas) boueti* Roubaud and Colas-Belcour, 1933 [= *A. (C.) afghanistaniensis* Santos Dias, 1961]
5. *Argas (Chiropterargas) confusus* Hoogstraal, 1955
6. *Argas* (? Subgenus) sp. n. (1)
7. *Argas* sp. n. (2)
8. *Ornithodoros (Alveonasus) lahorensis* Neumann, 1901
9. *Ornithodoros (Alectorobius) coniceps* (Canestrini, 1890)
10. *Ornithodoros (Pavlovskyella) tartakovskyi* Olenev, 1931
11. *Ornithodoros (Pavlovskyella) tholozani* (Laboulbène and Mégnin, 1882)

Family Ixodidae

12. *Ixodes (Pholeoixodes) arboricola* Schulze and Schlottke, 1929
13. *Ixodes (Pholeoixodes) crenulatus* Koch, 1844
14. *Ixodes (Ixodes) redikorzevi* Olenev, 1927
15. *Ixodes (Ixodes) himalayensis* Dhanda and Kulkarni, 1969
16. *Ixodes* (? Subgenus) *simplex* Neumann, 1906
17. *Ixodes (Eschatocephalus) vespertilionis* Koch, 1844
18. *Haemaphysalis (Aboimisalis) cornupunctata* Hoogstraal and Varma, 1962
19. *Haemaphysalis (Herpetobia) kashmirensis* Hoogstraal and Varma, 1962
20. *Haemaphysalis (Herpetobia) sulcata* Canestrini and Fanzago, 1878
21. *Haemaphysalis (Rhipistoma) erinacei turanica* Pospelova-Shtrom, 1946
22. *Aponomma gervaisi* (Lucas, 1847)
23. *Dermacentor (Dermacentor) raskemensis* Serdyukova in Pomerantsev, 1948
24. *Dermacentor (Dermacentor) marginatus* (Sulzer, 1776)
25. *Dermacentor (Dermacentor) niveus* Neumann, 1897
26. *Hyalomma (Hyalommasta) aegyptium* (Linnaeus, 1758)
27. *Hyalomma (Hyalommina) kumari* Sharif, 1928
28. *Hyalomma (Hyalomma) anatolicum anatolicum* Koch, 1844
29. *Hyalomma (Hyalomma) anatolicum excavatum* Koch, 1844
30. *Hyalomma (Hyalomma) asiaticum asiaticum* Schulze and Schlottke, 1929
31. *Hyalomma (Hyalomma) detritum* Schulze, 1919
32. *Hyalomma (Hyalomma) dromedarii* Koch, 1844
33. *Hyalomma (Hyalomma) schulzei* Olenev, 1931
34. *Hyalomma (Hyalomma) marginatum turanicum* Pomerantsev, 1946
35. *Hyalomma (Hyalomma) marginatum isaaci* Sharif, 1928
36. *Rhipicephalus (Rhipicephalus) leporis* Pomerantsev, 1946
37. *Rhipicephalus (Rhipicephalus) turanicus* Pomerantsev, 1946
38. *Boophilus microplus* (Canestrini, 1888)
39. *Boophilus annulatus* (Say, 1821) /or *B. calcaratus* (Birula, 1895) /

also shelter *Ornithodoros coniceps*. Domestic chickens are usual hosts of *Argas persicus*, but this tick also infests wild birds in Afghanistan, as it does in Central Asia. Notably, where *A. persicus* has been introduced elsewhere in the world it rarely if ever parasitizes wild birds. Wild birds are hosts of immature *Hyalomma marginatum* subsp. and adult and immature *Ixodes arboricola* and *I. redikorzevi*. Much more study should be devoted to tick, virus, and bird interrelationships in Afghanistan.

Carnivores and hedgehogs are commonly and often heavily infested by adult *Haemaphysalis erinacei turanica* and *Rhipicephalus turanicus;* immatures of these ticks feed chiefly on insectivores, rodents, and pikas *(Ochotona)*. The ubitquitous red fox is also host of immatures of these species, as well as of adult and immature *R. leporis* and *Ixodes crenulatus*. These ticks are all Central Asian, though *R. turanicus* and *I. crenulatus* also occur far westward in southern areas of the Palearctic.

The role of mountain-dwelling wild sheep and goats, such as the ibex, markhour, and argali *(Capra* and *Ovis)* in maintaining populations of distinctive *Haemaphysalis* and *Dermacentor* species is especially interesting. The pika *(Ochotona)* also supports these tick genera and *Ixodes* in high mountains. All tick species found chiefly at or above 7,000 feet altitude in Afghanistan also occur in western Himalayan foothills and mountains. Two species unknown elsewhere are *Ixodes himalayensis*, a parasite of pikas *(Ochotona)*, and *Haemaphysalis cornupunctata*, a parasite of sheep and goats. Other high mountain species also range westward into Irano-turanian and Central Asian highlands. Examples are *H. kashmirensis* (adults on sheep and goats, immatures on *Agama* lizards) and *Dermacentor niveus* (adults on artiodactyls, immatures on small mammals). Some species occur chiefly at high altitudes but also occasionally in foothills and valleys; examples are *D. marginatus* and *H. sulcata*.

Small-size mammals (hedgehogs and shrews (Insectivora), hares and pikas (Lagomorpha), and ground squirrels, mice, rats, hamsters, gerbiis, and voles (Rodentia)) support a variety of tick species. The most numerous and universal are larvae and nymphs of the many *Hyalomma* species. Adult *Hyalomma* parasitize domestic cattle, camels, horses, sheep, and goats. However, immature stages of the important *H. anatolicum anatolicum* are unique in that they usually feed on the same hosts as adults. Most of the omnipresent *Hyalomma* species are widely distributed in the southern Palearctic; some extend into arid northern areas of the Ethiopian Region. *H. asiaticum* ranges westward only to the eastern Mediterranean and *H. schulzei* only to the Egyptian coastal desert. Two species *H. kumari* and *H. marginatum isaaci*, are essentially Indian. The general and widespread interplay between numerous *Hyalomma* (and other tick species), small and large vertebrates, and wild and domestic vertebrates causes these ticks to be exceptionally important in the epidemiology of human and animal infections. Among these are viruses such as Crimean-Congo hemorrhagic fever, Wad Medani, Dera Ghazi Khan, Dhori, and Kyasanur Forest disease; rickettsiae such as Siberian tick typhus

(Rickettsia siberica) and Q fever *(Coxiella burneti)*; and numerous veterinary infections and animal husbandry problems such as tick paralysis, damaged hides, blood and weight loss, etc.

Various small size mammals and reptiles inhabiting burrows in numerous semiarid biotopes and irrigated deserts are hosts of *Ornithodoros tartakovskyi* from lowlands to 8,000 feet altitude in Afghanistan and in Irano-turanian and highland Central Asian Subregions. In USSR, this tick has been found infected with relapsing fever spirochetes *(Borrelia latychevi)*, plague *(Pasteurella pestis)*, Q fever *(C. burneti)*, and *Leptospira*. *O. tartakovskyi* also transmits a gerbil-infecting filaria *(Dipetalonema witei)*. Numerous small and large mammals inhabiting burrows, caves, or stables are hosts of the more widely distributed *Ornithodoros tholozani*, which is a common vector of relapsing fever *(Borrelia persica)* in Afghanistan. The stable-infesting *Ornithodoros laborensis* has seldom been collected in Afghanistan.

Proceedings of the 3rd International Congress of Acarology, Prague, 1971

THE BIOLOGY AND ECOLOGY OF THE QUILL MITE *SYRINGOPHILUS BIPECTINATUS* HELLER, 1880

E. F. GRITSENKO
Poultry Breeding Farm, Krasnodar, USSR

Mites inhabiting the quills of poultry and cage birds, the syringophilids, have been known to science for about 90 years. Despite its considerable history the quill mite has been studied insufficiently. Only the morphology of adults has been well studied and described. The morphology of the immature stages as well as the biology and ecology have not been elucidated at all in professional literature, or these subjects have been dealt with inaccurately.

The present paper gives some data on the morphology and ecology of *S. bipectinatus*. Our studies, based on considerable factual material, proves that the development of syringophilids takes place in a simple bi-nymphal cycle: egg – larva – protonymph – teleonymph – adult. Larvae have no secondary sexual characters and all are homotypic. The protonymph, hatched from the larva, is not morphologically differentiated; only towards the end of its development before moulting into the teleonymph are some changes in the body configuration observed. In the teleonymph stage, male and female specimens are easily distinguished. The male teleonymphs are short and broad, while the opisthosoma in female specimens is considerably shorter. The adult male are longer than male teleonymphs, but shorter than the females. The adult females are shorter than female teleonymphs, their opisthosoma is fairly long, but somewhat narrow. The morphological structures in the ontogeny of syringophilids have been studied and the results published by us.

Entry into the calamus of a feather is gained, by females only, by way of the superior umbilicus. After 2 – 4 days they lay eggs, from which larvae are hatched after 12 – 13 days; larvae moult after 6 days and undergo metamorphosis into protonymphs. The development of protonymphs lasts 6 days, and after moulting they became male or female teleonymphs, from which adult males or females finally emerge. 12 days after their emergence the males and females copulate. The calamus populated by syringophilids usually contains 4 – 5 times fewer males than females.

Our observations and experiments showed that the calamus may be penetrated by unfertilized females or female teleonymphs. In both cases females lay eggs which undergo normal development. From unfertilized eggs, however, only males develop. Thus a compensatory parthenogenesis occurs in syringophilids. This phenomenon has been observed in syringophilids for the first time.

Until recently the problem of how the syringophilids feed inside the calamus was unsolved. The assumption that syringophilids are predators does not hold good any more, because other groups of mites which could serve as prey, have not been found in the calamus. After carefully studying the feeding habits of syringophilids it became clear that during feeding the needle-like chelicerae are buried in the wall of calamus and penetrate through it into the tissue cells of the feather follicle. This applies to all mobile developmental stages. However, larvae and protonymphs are unable to pierce the wall of the calamus and they make use of the channels made in the calamus by a female. The cause of infestation with and routes of distribution of quill mites are of great interest. After studying the fauna of syringophilids it became clear that certain bird species are parasitized by specific mites.

The feeding habits of *S. bipectinatus*, like those of *S. columbae*, *S. minor minor*, cause intense itching. As a result of incessant irritation, under the influence of a nerve impulse the muscle tonus relaxes and a newly developing feather pushes out the old one from the feather follicle. Also, due to this incessant irritation of the nerve fascicles of the feather follicle, chickens pick their feathers containing mites in the quills. These feathers are subsequently replaced by new ones, but during a certain loss of wing feathers a development of small feathers in chickens takes place. The development of new feathers requires a certain amount of nutritive substance from the organism of the bird and this fact may be responsible for a certain decrease in the productivity of poultry.

POPULATION DYNAMICS OF A PARASITIC MITE *DEMODEX CAPRAE* (TROMBIDIFORMES: DEMODICIDAE)

R. R. LEBEL, W. B. NUTTING
Zoology Department, University of Massachusetts,
Amherst, Massachusetts, USA

Demodex caprae populations develop within papules in the skin of the dairy goat, *Capra hircus*. Nutting and Lebel (1969) described the sequence of events in growth and decline of these papules. Invasion of the hair follicle by one to several adult mites is followed by reproduction and harvesting of follicular epithelial cells, which undergo metaplasia. Eventually the papule ruptures; a granulomatous reaction ensues, followed by hyperkeratinization. The mites are extruded and/or phagocytosed by giant cells. Finally, a craterous scar is formed. Notable economic losses due to infestations of this mite have been reported from several African nations, where it was found to be the most common skin disease of goats: 41 % incidence in one report (Bwangamoi, 1969). Except in extreme cases, infestation may go unnoticed for a long time, as papules may be hidden from sight by the host hair. We have observed cases in which a single goat had over 100 papules, with a total population of some 5 million mites!

The follicle opening is closed early in papule development, isolating the mite population from the outside world. Thus, a number of population growth parameters are obviated: emigration, immigration, predation and (usually) bacterial invasions. Under these unique conditions virtually the only important ecological factors are food availability, waste removal, host physiological condition and mite reproductive potential. Space per se does not seem to be limiting, since papules apparently continue to grow until rupture and/or population senescence.

To ascertain rates of population growth, weekly observations were made over a six month period. Papule diameters were recorded, and occasionally papules were punctured for collection of the mites. The cheeselike contents, consisting of mites and a lipid-soluble substance produced by the host (Stromberg, unpublished), were suspended in benzene. Aliquot samples were mounted in Hoyer's medium on microscope slides. Extensive systematic counts were made by stage of the life cycle: ovum (E), larva (L), protonymph (P), nymph (N), female, male, nymph with adult inside and ovigerous female (Figure 1: A–H). From 10 to 100 % of the total papule population were counted. Earlier experiences in counting only the first hundred mites encountered on a slide proved unreliable. The "least to 20" method (Nutting and Desch, pers. comm.) seems sufficient, provided care is taken to suspend the papule contents uniformly and spread the aliquot evenly on the slide; all stages are tabulated until the least abundant has been observed 20 times.

In all, ninety thousand mites were counted for this report, representing papules from 2 to 12 mm in diameter. Some observations were also made on sections of papules, prepared with the paraffin method and stained with hematoxylin and eosin.

Timing of papule diameter increase is highly erratic; Polyakov (1959) made a similar observation for the nodules due to *D. bovis*. Neither the suspected host hormonal cue for mite increase nor the factors involved in the arrest of population growth have been determined to date. Papules most frequently cease growth at about 8 mm diameter, but have been observed by one of us (W.B.N.) to reach 32 mm. Durant (1944) reported a 50 mm papule, which developed from a smaller one which had been punctured and partially emptied of its contents. Some of our observations suggest that such mechanical interference, when unaccompanied by chemical treatments, may stimulate rapid population growth. In general, however, papules grow at a rate of approximately 2 mm each 3 weeks (estimated average of our most reliable data.)

A simple model of population growth applies to *D. caprae*. Examination of papule sections and of papules *in situ* indicate that they are approximately hemispherical in shape. Treatment of the papule as a hemisphere, and of the mite as a cylinder 400 μm long and 60 μm in diameter (volume .00113 mm^3) provides an accurate basis for prediction of population size in a papule of known diameter: e.g. 119,000 for an 8 mm papule, with an aliquot count estimate of 118,000. This cylinder size accounts for average mite dimensions as well as the lipid matrix and mite exuviae. The dome of the hemisphere tends to flatten when diameter exceeds 10 mm, reducing the accuracy of the model (400,000 predicted for 12 mm; aliquot sample count estimate 327,000.)

There is a clear shift in proportions of mature and immature mites as the papule grows (Figure 2). Mites which seem to have died prior to removal from the host were quite rare even in large populations, suggesting a long adult life span (ca. 15 weeks). While the immature component decreases with time, the proportions of the various stadia within that component remain relatively stable: E 31.1 ± 10.2, L 36.6 ± 9.9, P 12.2 ± 3.1, N 20.1 ± 6.5 (percentages and standard deviations; 8 papules.) Ova become progressively less abundant, indicating gradual decline in average female fecundity.

During papule growth up to 7 mm diameter, population increases rapidly; for each millimeter increment of papule diameter two generations of mites must occur (Figure 3). Based on our estimate of a 2 mm increase for each 3 week period, the ovum maturation time must be about 5 days in young populations. The relative constancy of proportion among immature stages suggests applicability of the formula proposed by Spickett (1961) for *D. folliculorum:* $\frac{N\,1}{N\,1} = \frac{D\,1}{D\,2}$ (N = number of mites in a given stage; D = duration of that stage.) Thus we estimate: E 37 hours, L 42, P 15, N 25, during rapid population growth.

Our data suggest, however, that the rapidity of mite development declines as

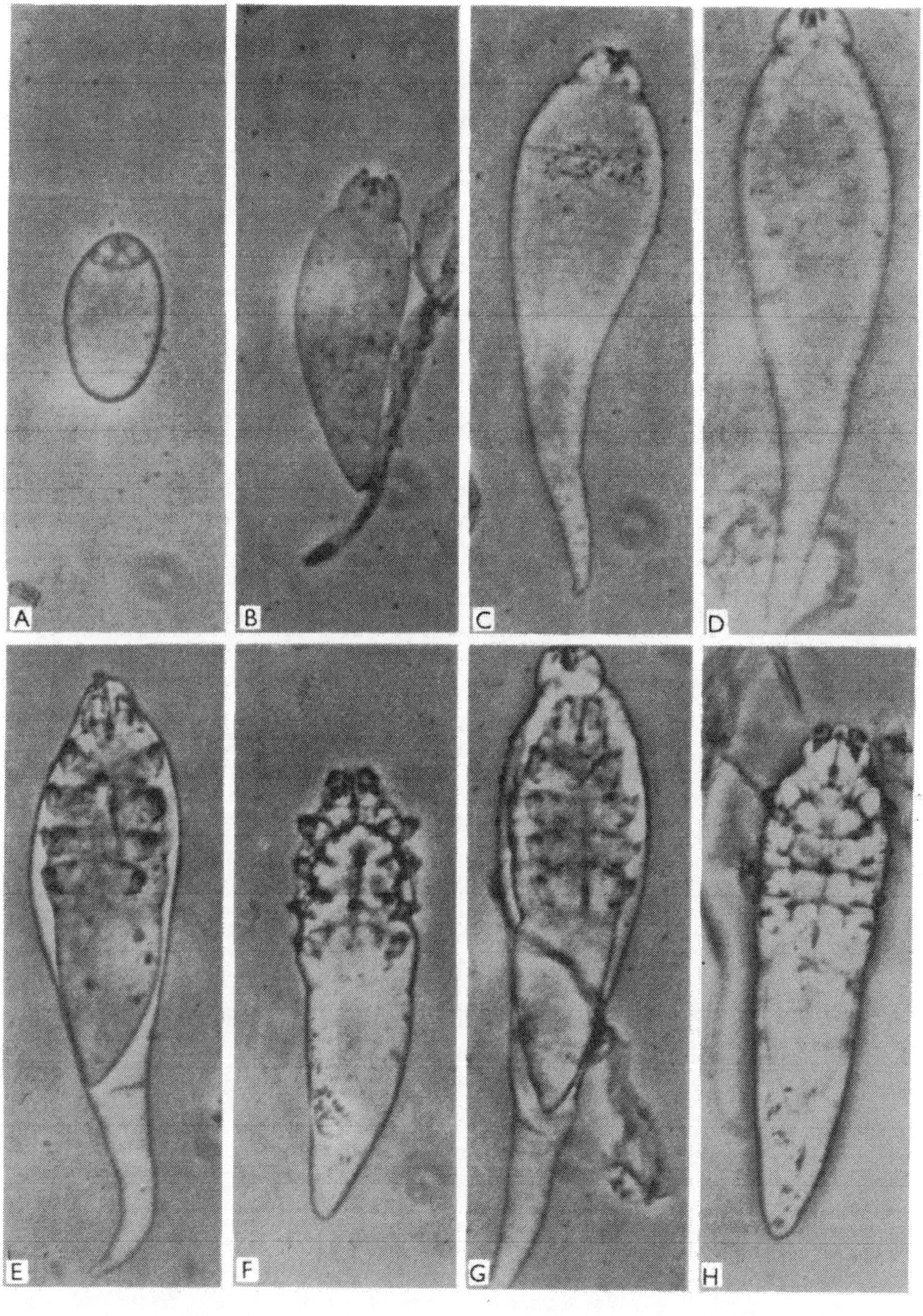

Fig. 1. *Demodex caprae*, life cycle: A) ovum; B) larva; C) protonymph; D) nymph; E) nymph with male inside; F) male; G) nymph with female inside; H) female.

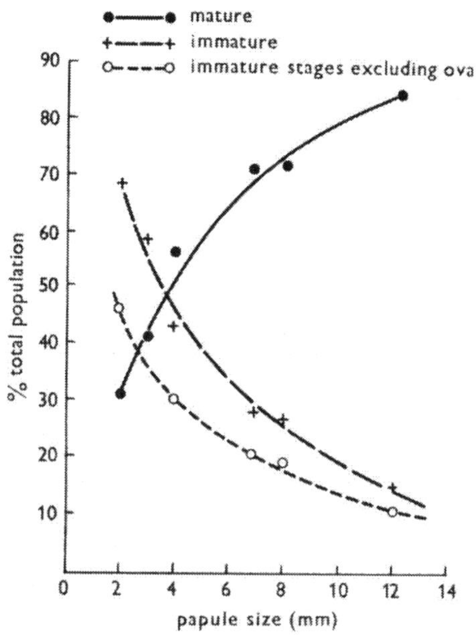

Fig. 2. *Demodex caprae,* population structure with respect to papule diameter.

populations age. Average duration of each stage may even double in populations over 100,000. The reason for this is likely a combination of gradual reduction of food availability and increasing contamination by accumulating waste metabolites. As the hemispherical cavity increases from 4 to 8 mm in diameter, the average cellular surface area (index of food availability) per mite is halved (254 to $127 \times 10^{-5} mm^2$). It is suspected that metabolic processes are depressed, especially in mites near the center of the papule, where both food and oxygen are in short supply. This has already been suggested for *D. bovis* (Nemeseri and Szeky, 1961). Thus the close correspondence of our extended developmental rates with those reported by Spickett (1961) for *D. folliculorum* (E 70 hours, L 64, P 36, N 72) may corroborate the opinion that sebum, the nutrient source in Spickett's cultures, is insufficient for normal demodicid development (Nutting, 1964). Undernourishment could have acted to extend development time as well as shorted adult life spans, which Spickett estimated to be only 120 hours. Unsworth (1946) however, calculated 19 days as the time required for ova of *D. canis* to develop to adults.

The combination of apparent postreproductive longevity and depressed metabolism, which may reduce oviposition rates, could account for the observed accumulation of adults and the simultaneous increasing rarity of ova. It may be that some females undergo an extended period of quiescence under these unfavorable conditions, depositing few ova. They may be held in reserve until papule rupture affords elevated oxygen tension; a burst of activity then would facilitate invasion of new fol-

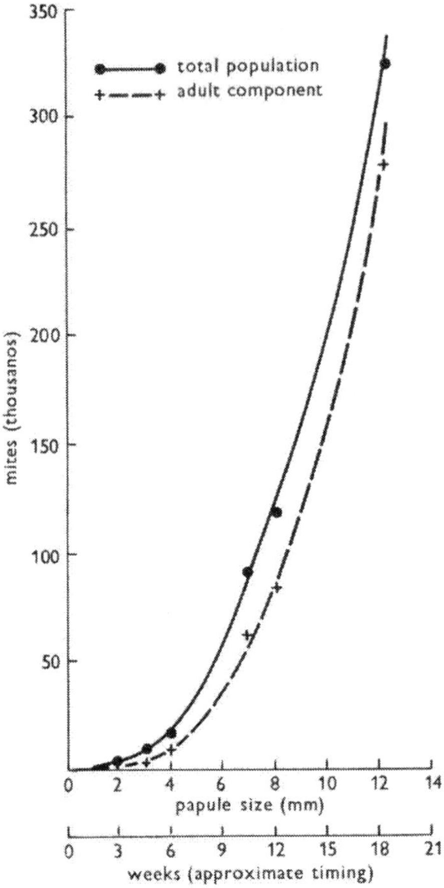

Fig. 3. *Demodex caprae*, population size with respect to papule diameter and time.

licles, where a fresh food supply could support rapid oviposition and development of the population in an incipient papule.

Among adult mites, sex ratio changes somewhat as populations grow. Five small papules (2 to 4 mm) had male/female ratios of 1:10 (7 to 16), while in three large populations (7 to 12 mm) they were 1:35 (28 to 40). Desch and Nutting (1971) observed 1:38 in *D. phylloides*. But Nutting *et al.* (1971) reported 1:2 for *D. gapperi*, and Desch *et al.* (1971) found 1:1 in *D. carolliae*. Exoskeletons of defunct males were exceedingly rare in all our counts, leading us to reject the hypothesis that imbalanced sex ratios reflect differential longevity in this species. Polyakov (1956) reported that male *D. bovis* live only 3 to 5 days while impregnating females before the latter invade follicles to initiate new nodules. He did not explain on what basis he made this estimate. Our data for *D. caprae* support the conclusion that demodicids with small sex ratios may be thelytokous. In thelytoky, scarceness of males may be expected to lead to their further rarification, since unimpregnated females deposit ova which develop exclusively to females; this notwithstanding the possibility that single impregnation may be sufficient for fertilization of all the singly produced ova.

Until culturing techniques for demodicids are perfected, resolution of many of these problems remains difficult if not impossible. To date only limited success has been met in maintenance of these mites away from the host environment to which they are so intimately adapted.

References

Bwangamoi, O.: A survey of skin diseases of domestic animals and defects which downgrade hides and skins in East Africa II. Goats. Bull. Epizoot. Dis. Afr. 17:197–203, 1969.

Desch, C.E., Lebel, R. R., Nutting, W. B., Lukoschus, F.: Parasitic mites of Surinam I. *Demodex carolliae* sp. nov. (Acari: Demodicidae) from the fruit bat *Carollia perspicillata.* Parasitology 62:303–308, 1971.

Desch, C. E., Nutting, W. B.: Demodicids (Trombidiformes: Demodicidae) of medical and veterinary importance. Third Int. Congress Acarol., Prague, 1971.

Durant, A.J.: Demodectic mange in the milk goat. Vet. Med. 39:268–270, 1944.

Nemeseri, L., Szeky, A.: Demodicosis in cattle. Acta Vet. Acad. Sci. Hung. 11:209–221, 1961.

Nutting, W.B.: Demodicidae – status and prognostics. Acarologia 6:442–454, 1964.

Nutting, W. B., Emejuaiwe, S. O. Tisdel, M. O.: *Demodex gapperi* sp. n. (Acari: Demodicidae) from the red-backed vole, *Clethrionomys gapperi.* J. Parasitol. 57:660–665, 1971.

Nutting, W. B., Lebel, R. R.: Lesions due to *Demodex caprae:* origin, development and degradation. Prog. 44th Ann. Meet. Am. Soc. Parasitol. Washington, 1969.

Polyakov, D.K.: Scheme of biological development of the agent of demodicidosis in cattle. Vsesn. N. Issled. Vet. Sanit. Trud. 1:14–15, 1956. (In Russian.)

Polyakov, D.K.: Morphological changes in the skin of cattle under the effect of *Demodex bovis* as a characteristic of the many processes in the biology of the parasite. Vsesn. N. Issled. Vet. Sanit. Trud. 14:34–35, 1959,(In Russian.)

Spickett, S.G.: Studies on *Demodex folliculorum* Simon (1842) I. Life history. Parasitology 51:181-192, 1961.

Unsworth, K.: Studies on the clinical and parasitological aspects of canine demodectic mange. J. Comp. Pathol. 56:114–127, 1946.

EPIDEMIOLOGICAL ANALYSIS OF THE INCIDENCE OF SCABIES IN ONE OF THE DISTRICTS OF CZECHOSLOVAKIA

V. MĚRKA, P. PALIČKA
Military Medical Research Institut, Hradec Králové; Regional Hygiene Station, Karviná, Czechoslovakia

During the 1960's an increase in the incidence of scabies was reported in practically every country in Europe (Měrka, 1971) and many authors have sought its cause: Scabies is not properly diagnosed in time. Untreated, or incorrectly treated, patients are a source of the further spread of scabies for a long time. The number of atypical cases with a mild clinical course and absence of some of the signs is increasing. The predominance of mild forms, which is attributable to better observance of hygienic principles, is a great obstacle to the rapid liquidation of scabies, as the diagnosis of such cases is difficult even for the experienced dermatologist and as the inconveniences are negligible the patient either does not see a doctor at all, or not for a long time. Increasing sexual promiscuity may be one of the factors promoting the spread of scabies. Shrank and Alexander (1967) attribute the wave-like incidence of scabies, with roughly 15-year intervals between the individual epidemics, to immunobiological processes – hypersensitization during the epidemic, followed by the development of a given degree of resistance to the parasite, in young subjects. The next epidemic does not occur until these hypersensitized subjects have been replaced by non-sensitized population, i.e. after about 15 years. Some authors claim that the incidence of scabies has risen an all strata of society, despite improved living standards. There are some problems concerning etiology of scabies.

No concrete data on the incidence of this parasitosis in Czechoslovakia have so far been published with the exception of our preliminary report (Palička et al. 1971).

From 1961 to 1969 we carried out an epidemiological analysis of the incidence of scabies in the district of Karviná, which has a population of about 300,000. The results are given in Fig. 1, which shows a steady increase in the number of notified cases of scabies in the Karviná district since 1966 – in some years by as much as 100 %.

The absolute figures for the incidence of scabies in the individual months of 1967, 1968 and 1969 are given in Fig. 2. Like other authors (Franken and Elste, 1969, Gołębiowska and Szarmach, 1968, Herrmann and Humann, 1969, Obrtel, 1970), we recorded seasonal variation of the incidence of scabies. In all three years, its incidence was highest in the autumn (September – November), i.e. after the summer

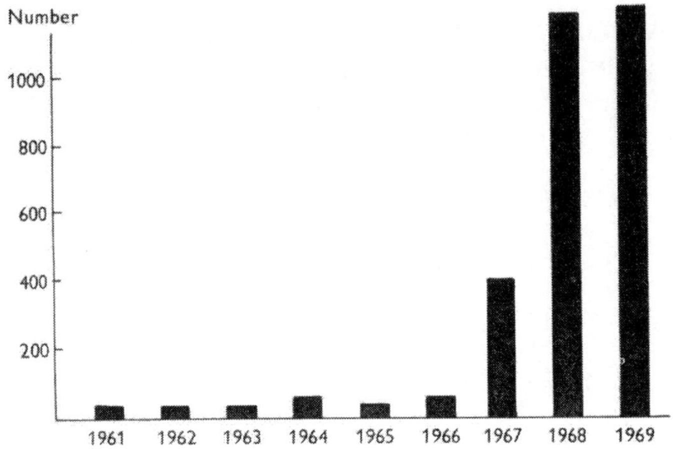

Fig. 1. Absolute number of notified cases of scabies in Karviná district in 1961 to 1969.

holidays. During this period of raised incidence compared with the mean for the whole year, no differences were found in its distribution according to occupation. In an age group analysis we found a relative increase in the under 15 category.

Epidemiological analysis of cases of scabies notified in 1967 to 1969 showed that there was usually a familial incidence. In 1967 the proportion of patients with a familial incidence was 28.6 % of the total number, in 1968 it was 51.5 % and 57.5 % in 1969. With the autumn increase, the number of cases with a familial incidence also rose by 12.3 % to 21.2 % in the individual years compared with the other months. This increase is statistically highly significant in all three years.

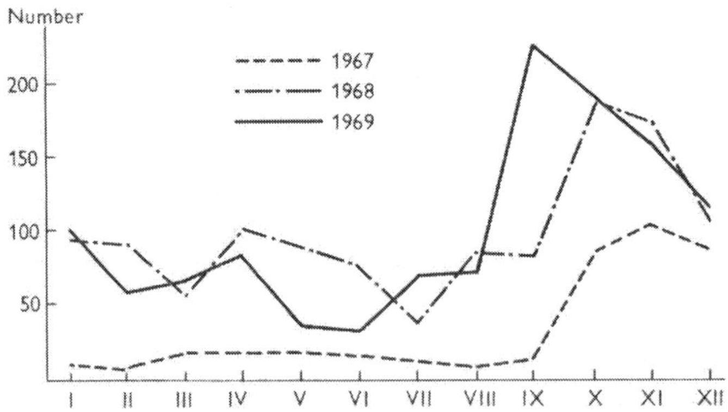

Fig. 2. Seasonal incidence of scabies in individual months.

Fig. 3 gives the specific morbidity rate (per 100,000 population) in the Karviná district in relation to age. The total morbidity rate was 144 in 1967, 411 in 1968 and 418 in 1969. The figure shows that the highest specific morbidity rate during these years was recorded in children of school age (6–15 years).

With reference to this finding, from March to May, 1970, we did dermatological examination in 4,374 children attending five elementary schools in Karviná with the following results: The 21 cases of scabies diagnosed from the clinical signs and confirmed "ex iuvantibus" by the use of an antiscabietic, represent 0.48 % of the normal, i.e. healthy, population. None of these 21 children had been to see a doctor on account of scabies and they can therefore all be regarded as reservoirs and long-term sources of infestation. If we add the number of cases detected by active investigation and the number of notified cases, we obtain, for this age group, the theoretical morbidity rate of about 1,300 cases per 100,000 child population.

We also examined the families of the 21 school children in whom scabies was detected by active investigation. Among the total of 69 subjects, scabies was confirmed in 15. This gives a familial incidence of 22 % not counting the original cases and of 40 % including the original cases. These data seem to confirm the view that the contemporary spread of scabies in Czechoslovakia occurs mainly in families and that these should therefore receive the greatest attention in anti-epidemic work.

During the past three years an epidemic spread of scabies in normal child communities such as elementary schools was found only in exceptional cases, and only where the children were also housed together (in dormitories, hostels, etc.).

It can be concluded that scabies is starting to become a serious problem in

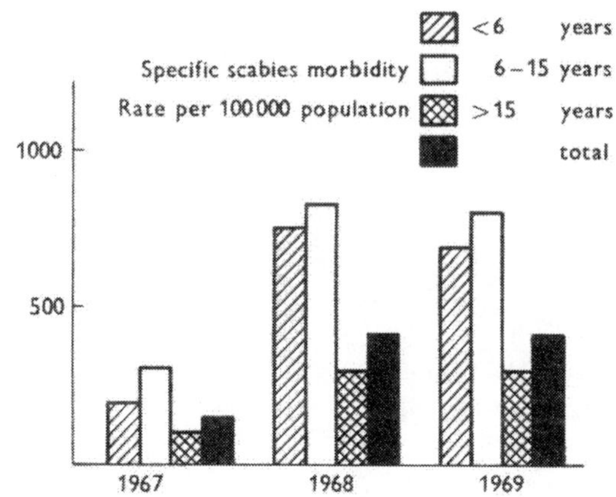

Fig. 3. Specific scabies morbidity rate per 100,000 population in main age groups.

Czechoslovakia. We shall have to pay far more attention than formerly to this parasitic disease.

References

Franken, E., Elste, G.: Beobachtungen über vermehrtes Vorkommen von Scabies-Infektionen. Deutsch. Ges. Wes. 24, 1750–1753, 1969.

Gołębiowska-Podgórczyk, I., Szarmach, H.: Czy epidemia swierzbu? Polski Tygodn. Lek. 23, 1931–1933, 1968.

Herrmann, W.P., Humann, B.: Jahreszeitliche Morbiditätsschwankungen bei der Scabies. Hautarzt 20, 467, 1969.

Měrka, V.: Možné příčiny zvyšující se incidence svrabu. Voj. zdravotn. Listy 11, 150–153, 1971.

Obrtel, J.: Zvýšený výskyt svrabu a jeho příčiny. Čs. Dermatol. 45, 131–132, 1970.

Palička, P., Malíš, L. Zwyrtek, K.: K problematice výskytu scabies a pedikulózy na Karvínsku. Čs. Epidemiol. 20, 116–120, 1971.

Shrank, A. B., Alexander, S. L.: Scabies – another epidemic? Brit. Med. J. 1, 5541: 669–671, 1967.

ON THE MORPHOGENESIS OF THE ATTACHMENT CEMENT OF SOME IXODID TICKS

D. E. MOORHOUSE
Department of Parasitology, University of Queensland, Brisbane, Australia

The patterns of attachment of ticks to their hosts for the purpose of feeding, have been described for seven genera in the family Ixodidae, *(Ixodes, Aponomma, Amblyomma, Dermacentor, Rhipicephalus, Boophilus,* and *Haemaphysalis).* Within the limits of the species examined, the pattern is constant for the genus, except with *Ixodes* where there are two distinct patterns (Moorhouse, 1969). Attachment varies from the simple penetration of the host's integument by the full lengths of the chelicerae and hypostome, to situations where only the distal extremities of the mouthparts penetrate the skin. In these latter cases support is provided by a cement substance produced from the salivary glands of the ticks. This hardens after enveloping the portions of the mouthparts external to the host, and firmly adheres to the host's epidermis, thus forming an 'organ' of attachment.

Consideration of the patterns of attachment shown by the various genera suggests evolutionary relationships which closely accord with the classification of the Ixodidae proposed by Warburton (1907). Here, Warburton divided the Ixodidae on anatomical grounds, into two sections, Prostriata which comprised the genus *Ixodes* alone, and Metastriata which included all other genera. The Metastriata he in turn divided into Longirostrata *(Amblyomma, Aponomma* and *Hyalomma)* and Brevirostrata *(Dermacentor, Rhipicephalus, Boophilus* and *Haemaphysalis).*

The most primitive form of attachment in the Ixodidae appears to be represented by those members of the Prostriata which merely insert their mouthparts into the host tissues without the formation of cement (Group 1 of the *Ixodes,* as described by Moorhouse, 1969). *(I. holocyclus, I. pseudorasus,* and the unidentified species). Subsequent work has shown that *I. ricinus* also belongs to this group. Here, the histological picture of the attachment is similar to that found with the larvae of the argasid tick, *Argas persicus.* (Moorhouse, unpublished). The argasids are generally accepted to represent the primitive types of ticks (Nuttal et al, 1911). The striking feature of this type of attachment is the extensive necrosis of the host tissues around the mouthparts, along with accumulations of polymorphonuclear leucocytes in the tissues both around the mouthparts as well as beneath their tips. The accumulation of these leucocytes must result in destruction of host tissues as was shown to occur during the feeding of the dog tick *Rhipicephalus sanguineus* (Tatchell and Moorhouse, 1970). The final result must be a weakening of the site of attachment and loss of mechanical stability for the tick. At intervals during feeding, *I. holocyclus* frequently burrows more deeply into the host, presumably to overcome loss of stability. The author has found female *I. holocyclus* naturally attached to wild hosts where even the posterior margin of the scutum was burried beneath the host's epidermis. This same tick can readily become detached from the host without suffering damage to the mouthparts and it will reattach to either the same, or a different host, and continue with its engorgement.

During the feeding of *I. holocyclus*, a free-flowing liquid can periodically be seen to spread over the skin from the site of attachment. This appears to be tick saliva which passes between the components of the mouthparts. Such a lateral escape of saliva is not surprising in view of the absence of ducts, limiting membranes, or other devices in the tick's feeding apparatus. Histological studies show that destruction of the host's epidermis extends over a greater distance than the dermal destruction and it is presumed to result from the saliva, which in these ticks must have a histolytic activity. This same pattern of destruction is also found with *I. ricinus* larvae and the larvae of *Argas persicus*. As the lesions caused by these members of the Prostriata open to the surface they are readily subject to bacterial contamination and hence secondary infection.

The members of the Metastriata all produce cement which encases the mouthparts and effectively prevents the lateral escape of saliva. A result is that the lesion forms principally within the dermis and the cement provides a seal which reduces the possibility of secondary infection. Cement appears to be inert insofar as the host is concerned and does not provoke an inflammatory response. However, with the genus *Dermacentor* there is a marked hyperplasia of the epidermis in contact with the cement, but no inflammation.

With the Longirostrata (*Amblyomma*, and *Aponomma* of Moorhouse, 1969, and a single unidentified species of *Hyalomma*), the cement forms a simple tube closely applied to the length of the mouthparts, and irrespective of the instar the full lengths of the mouthparts are inserted into the host tissues. The feeding lesions develop beneath the orifice of the cement tube. There is no destruction of the host's epidermis, nor of the dermal tissues in contact with the cement. Insofar as the stability of the tick is concerned, this situation provides a considerable advance over that previously described. However, there appear to be disadvantages in that the very length of the mouthparts, particularly those of the adults, may place the feeding lesion well below the highly vascularised outer layers of the dermis, and hence the principal source of the tick's nutrient. Furthermore, with species such as *Amblyomma triguttatum* which are mostly found attached to the ears of their hosts, the mouthparts, with their surrounding cement layer, may pass completely through the ear. When this happens with *A. triguttatum* an asymetrical cap of cement may be laid down over the tips of the mouthparts, that is on the opposite surface of the ear to which the tick is attached, and the lesion develops to the side of the mouthparts. Although the situation found with the attachment of the Longirostrata is an advance insofar as mechanical stability is concerned there would seem to be doubtful advantage in regard to feeding.

With the Brevirostrata *(Dermacentor, Rhipicephalus, Boophilus* and *Haemaphysalis)* penetration of the host tissues by the mouthparts is restricted to entry into the epidermis and here the cement assumes the role of providing the mechanical support as well as effectively completing the mouthparts. Unlike the Longirostrata, where the depth of penetration is a direct function of the lengths of the mouthparts of the different instars, the depth of penetration in the Brevirostrata is constant and is independent of the instar. In all species investigated the cement assumes the form of a cone with the base firmly affixed to host's tegument. The ticks are well-placed to take advantage of the capilliary network lying in the highly reactive upper levels of the dermis immediately below the mouthparts, and furthermore are provided with firm mechanical support.

The second group of *Ixodes,* i.e. those which secrete cement *(I. tasmani* and *I. japonensis,* Moorhouse, 1969) is of considerable interest, for with these ticks the mouthparts penetrate partly into the dermis, but they also produce a cone of supporting cement. The depth of penetration by the mouthparts as well as the height of the cement cone increases from larvae to nymphs and from nymphs to adults. This suggests that they constitute a link, not between the Prostriata and the Longirostrata of the Metastriata as could be expected, but between the Longirostrata and the Brevirostrata. However, further speculation on their position must await the collection and processing of much more *Ixodes* material.

References

Moorhouse, D.E.: Proc. 2nd Int. Congress Acarol., 319-327, 1969.

Nuttall, G.H.F., Warburton, C., Cooper, W.F. and Robinson, L.F.: Ticks: A monograph of the Ixodoidea. Part II, Ixodidae. Cambridge University Press, London, 1911.

Tatchell, R.J. and Moorhouse, D.E.: Science 167, 1002–1003, 1970.

Warburton, C.: J. Econ. Biol., 2, 89–95, 1907.

ADAPTATIONS IN DEMODICIDS (TROMBIDIFORMES: DEMODICIDAE) FOR UTILIZATION OF THE MAMMALIAN SKIN COMPLEX

W.B. NUTTING
Department of Zoology, University of Massachusetts, Amherst, Massachusetts, USA

Demodicids have often been cited (e.g., Kaestner, 1968) as remarkably adapted for survival in a unique habitat, the hair follicle. Such statements are, however, largely inferential since few measurements and considerations of the adaptive features of these mites are available in the literature.

The following report will consider this topic under the headings of anatomical, physiological and behavioral adaptations. A brief section on adaptive host-parasite interactions and the evolutionary rate of demodicid adaptations will conclude the account.

Anatomical adaptations: All known members of the genus *Demodex* are obligate symbiotes, in all stages of the life cycle (ovum, larva, protonymph, nymph, adult male and female), on mammals. Most reside in hair follicles, although some burrow epidermal pits or into (modified) sebaceous glands, others enter the ducts of Meibomian glands, and one is even found aberrantly in the blood vessels and lymph nodes. All initial invasions except for the epidermal dwelling species are thought to occur either through these ducts or the openings of the hair follicles. Species which burrow into the epidermis (e.g., *D. criceti)* cut pits to fit both length and breadth of the mite.

Mite length, usually highly variable, seems to bear no relationship to the length of the duct or follicle; in fact several species including *D. molossi* and *D. folliculorum* are often found with opisthosomae extending from the habitat. The greatest width of the adult mites seem to conform closely to the opening of the invasion site (Table I). It should be noted that species such as *D. canis* and *D. caprae*, which have to contend with large hairs in the follicular orifice, are approximately one-half the follicular diameter, whereas *D. gapperi* very closely approximates the opening of the Meibomian gland duct.

Other anatomical modifications include:

1. Ova – configurations range from oval *(D. criceti)*, spindle-shaped *(D. aurati)*, boomerang-shaped (Lukoschus et al.; in press), Y–shaped (Nutting et al., 1968), snake-like and T–shaped (Desch et al.; in press) (see Fig. 1). In the Y–shaped ovum of *D. gapperi* the prongs are laid upstream in the flow from the Meibomian glands; their spread dimensions (c. 50 μm) not only approximate the cross-sectional dimension of the duct (50 μm) but the prongs are pliable, which could increase the frictional drag under flow pressures from the gland, and furthermore, the egg pos-

Table I. Relationships between *Demodex* spp. and their habitats. Presence or absence of a proctodeum is also noted. (All values in micrometers)

Demodicid (Host)	Habitat	Mite Width* (range in life cycle)	Adult Mite* (width male/female)	Proctodeum+ male/female	Habitat Width (max. at skin surface)
Demodex criceti (*Mesocricetus auratus*)	epidermal pits	mite cuts body-size pits; eggs laid therein		present present	
Demodex aurati (*Mesocricetus auratus*)	hair follicle	16.6–25.0	21.1 25.0	present present	30.8
Demodex canis (*Canis familiaris*)	hair follicle	26.6–36.9	32.9 36.9	absent present	66
Demodex phylloides (*Sus scrofa*)	hair follicle	27.6–54.2	49.7 54.2	absent present	44
Demodex caprae (*Capra hircus*)	hair follicle	46.5–58.5	58.5 56.7	absent present	121
Demodex gapperi (*Clethrionomys gapperi*)	Meibomian duct	29.9–46.9	46.9 43.7	absent absent	50
Demodex longissimus (*Carollia perspicillata*)	Meibomian duct total	48.4–80.1 claws – 434.9	73.8 80.1	absent absent	120**
Demodex molossi (*Molossus molossus*)	Meibomian duct total total	64.8–91.0 ova – 163.5 claws – 338.0	82.5 91.0	absent absent	220**

* Average of 20 specimens.
** Data from Desch et al. (In press).
+ A proctodeum is present in one undescribed demodicid resident in the Meibomian glands.

terior is pointed. Despite this, our high egg to adult female ratio (to 3:1) vs. low larval to adult ratios (to 1:3) indicate some egg loss from the population to the outside environment (Nutting et al., 1971).

In three demodicids, the eggs are operculate, and with egg teeth in the developing larvae (Fig. 1, C). Egg-teeth are located closely apposed to the opercular sulci, and seem of utility in forcing the opercular lid ajar for rapid hatching.

2. Immature stages – most species possess epimeral scutes, posteriorly pointed shelves, which seem to aid in anchoring these usually weak-legged stages to the walls of the habitat. These are of varying sizes and shapes: mere vestiges in *D. gapperi,* simple oval shelves in *D. criceti,* and sharp prongs *(D. aurati).* As added holdfasts, *D. gapperi* and *D. molossi* carry over the egg prongs to the larval stages as one or two dorsal finger-like extensions. In the larvae, protonymphs, and nymphs of *D. longissimus* and *D. molossi* legs III have been modified as massive claws which are found deeply embedded in the duct wall for the former, superficially so for the latter (Table I). The larval stage of *D. molossi* (Fig. 2, A) shows the most remarkable set of adaptations: it has, other than the structures mentioned above, large and ventro-laterally reflexed palps, legs I and II with solenidia, and large epimeral scutes. *D. molossi* has, as legs IV of the nymphal stage, a small prominence, with three finger-like protuberances.

Demodex gapperi is also highly modified in that the protonymphal stage is wanting and legs III of the larvae are long and latero-ventrally situated.

3. Adults – marked sexual dimorphism is present in *D. antechini* (Nutting and Sweatman, 1970): the male is long, streamlined and broad in podosoma but the female short, slender in podosoma with a swollen and rounded opisthosoma. Longer, stronger legs are common for males as in *D. gapperi* (Fig. 2, B).

In most reported instances the males are smaller than females (e.g., *D. caprae, D. aurati)* although in a few instances the reverse is true (e.g. *D. bovis, D. antechini).* Males are apparently usually more active than females.

Physiological adaptations: All stages in the life cycle of *D. caprae* are very vulnerable to dessication; death will occur in approximately 1 ½ hours at 20°C and R.H. of 30 to 48 %. Despite this, demodicids survive 4 days in mineral oil *(D. canis)* and 4 to 5 hours in 2% gluteraldehyde *(D. caprae).* Adult exoskeletons distend in 0 to 0.5% NaCl, but show little change in 24 hours in 0.7 to 1.2% NaCl. No dimensional changes in the proctodeum were found at any of these concentrations.

The adaptive significance of the proctodeum (see Desch et al., 1970) is not explained as yet. If it is a hydrostatic organ it is difficult to correlate this function with the patterns of difference noted in species which are adapted to similar habitats, such as the Meibomian complex (Table I). The proctodeum is often of unusual shape and sexually dimorphic, as shown in Fig. 2, D.

Stromberg and Nutting (in MS) suggest that the crystals commonly found in demodicids represent adaptations for waste removal and serve as inherited codes for the elimination of the nitrogenous wastes. In females of *D. antechini* these crystals segregate into two discrete batches much as chromosomes in cell division; one batch remains with the female, the other is incorporated into the ovum (Fig. 2, C).

Behavioral Adaptations: Differential adaptive behavior within the life cycles is implicit in the discovery that nymphs and larvae of *D. aurati* are found below the opening of the sebaceous gland whereas adults and ova lie above (Nutting and

Rauch, 1963). Adult mites do move from the hair follicle and ascend hairs, presumably oriented to host transference. Adult males and females were also noted to wander across the skin surface.

Operculate ova and egg-teeth also seem behaviorally related to survival in Meibomian ducts; rapid hatching to the larval stage would seem necessary. In *D. gapperi* egg laying is a rapid process and undoubtedly a protective behavioral mechanism.

Spickett (1961) reported that *D. folliculorum, sensu latu,* is stereotactic and that during certain periods in the life cycle (nymph, adult) this is reversed. He coupled these observations with the prominent leg development in these stages to speculate that these were the stages of transference between host individuals.

Behavior is remarkably adaptive in terms of success in transference. This minute, weak-legged, easily-dried, single-egg-at-a-time organism transfers between active mammals 100 % effectively in both the golden hamster (*D. aurati*) and man (*D. folliculorum*) - see Nutting, 1964 and 1965.

Host parasite interactions: Complex interactions between the host organism and *Demodex* spp. have been discovered, as for example between *Antechinus stuartii*, a marsupial mouse, and *D. antechini* (Table II). Within this pattern we can also note the delicate balance of the mite reproduction and the increase of host follicular epithelium (Fig. 2, E), the mite food supply. In cases in which this relationship breaks

Table 2. Host–parasite interactions for *Antechinus stuartii* and *Demodex antechini*. Adaptive responses are probably de novo and specific for the mite but ancient and non-specific for the host

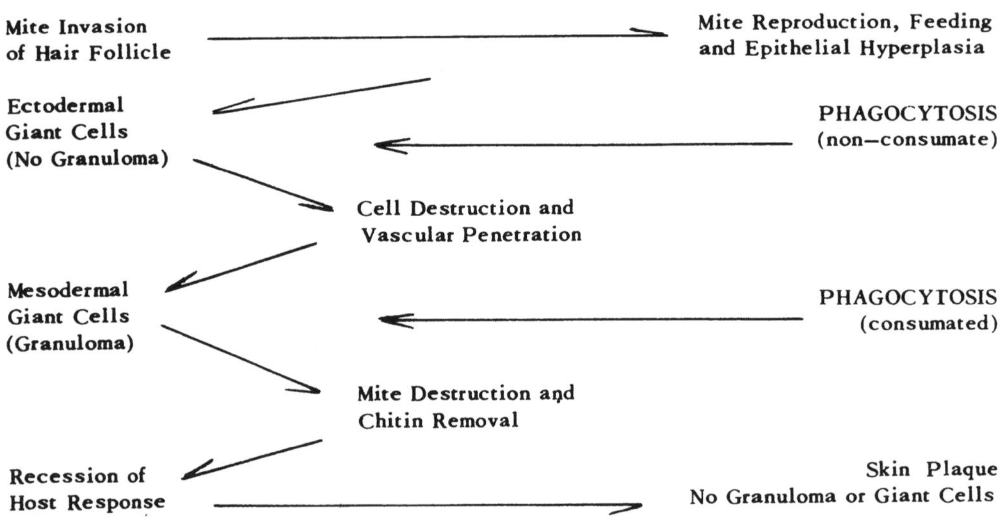

Diagram from Nutting and Beerman, 1965.

down and the mites penetrate to the dermis no response occurs in the host tissue–a unique situation.

Such unusual patterns of interaction would seem to indicate either a long history of host-parasite adaptation or very rapid adaptation of the parasite to host physiological mechanisms which have a wide range of response to any physical, chemical or biotic agent invading the skin.

In conclusion, we should also mention that these anatomical, physiological, behavioral and host-parasite adaptations could have a rapid rate of evolutionary adaptation. The following may apply: In species with small populations (c. 50 mites per host in *D. gapperi)* we would expect acceleration of genetic drift. Helle and van Zon (1967) have shown that mutation rates in mites are unusually high. Up to five species of *Demodex* are found on one host individual (Lukoschus, pers. comm.) – lending credence to synhospitalic species derivation.

Further studies will undoubtedly disclose new, unusual and interesting adaptations in this genus *Demodex.*

References

Desch, C., Nutting, W.B. and Lukoschus, F.: Parasitic mites of Surinam VII: *Demodex longissimus* n. sp. from *Carollia perspicillata* and *D. molossi* n. sp. from *Molossus molossus* (Demodicidae: Trombidiformes): Meibomian complex inhabitants of neotropical bats (Chiroptera). Acarologia. (In press.)

Desch, C., O'Dea, J., and Nutting, W.B.: The proctodeum – a new key character for demodicids (Demodicidae). Acarologia 7:522–526, 1970.

Helle, W. and van Zon, A.Q.: Rates of spontaneous mutation in certain genes of an arrhenotokous mite, *Tetranychus pacificus*. Entomol. Exptl. Appl. 10: 189–193, 1967.

Kaestner, A.: Invertebrate Zoology. Vol. 2, Interscience Publishers, N.Y., 1968.

Lukoschus, F.S., Jongman, R.H.G. and Nutting, W.B.: Parasitic mites of Surinam XII. *Demodex melanopteri* sp. n. (Demodicidae: Trombidiformes) from Meibomian glands of the neotropical bat *Eptesicus melanopterus*. Acarologia. (In press.)

Nutting, W.B.: Demodicidae – status and prognostics. Acarologia 6:441–454, 1964.

Nutting, W.B.: Host-parasite relations: Demodicidae. Acarologia 7:301–317, 1965.

Nutting, W.B., Emejuaiwe, S.O. and Tisdel, M.O.: *Demodex gapperi* sp. n. (Acari: Demodicidae) from the red-backed vole, *Clethrionomys gapperi*. J. Parasitol. 57:660–665, 1971.

Nutting, W.B. and Rauch, H.: Distribution of *Demodex aurati* in the host *(Mesocricetus auratus)* skin complex. J. Parasitol. 49:323–329, 1963.

Nutting, W.B. and Sweatman, G.K.: *Demodex antechini* sp. nov. (Acari, Demodicidae) parasitic on *Antechinus stuartii*. Parasitology 60:425–429, 1970.

Nutting, W.B. and Beerman, H.: Atypical giant cells in *Antechinus stuartii* due to demodicid mites. J. Invest. Dermatol. 45:504–509, 1965.

Nutting, W.B., Tisdel, M.O. and Emejuaiwe: Y- shaped ovum of a demodicid from *Clethrionomys gapperi*. J. Parasitol. 54:393–394, 1971.

Spickett, S.G.: Studies on *Demodex folliculorum* Simon (1842) I. Life history. Parasitology 51:181–192, 1961.

Stromberg, B. and Nutting, W.: Adaptive features of the exoskeleton and "pigment" deposits in *Demodex* spp. (Demodicidae). (In manuscript.)

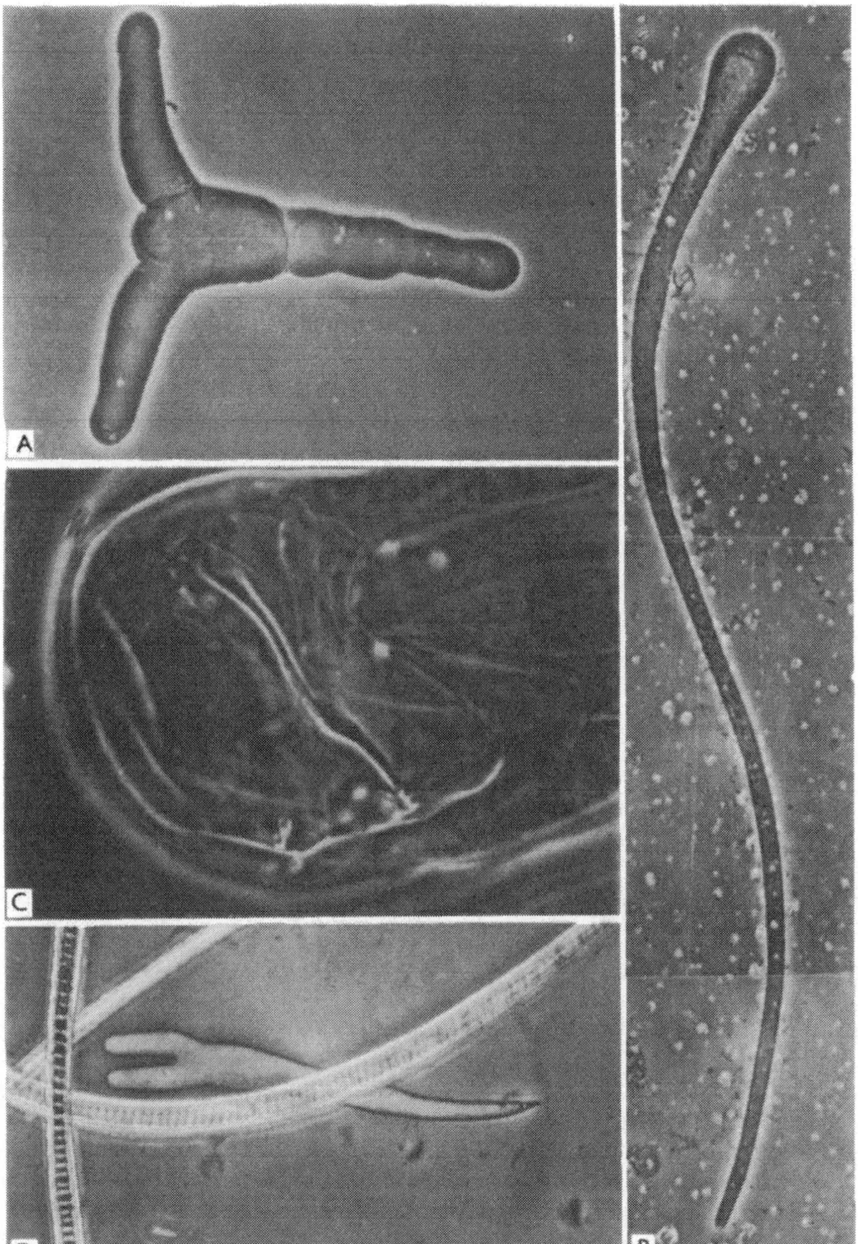

Fig. 1. A)—D): Adaptations in ova of demodicids. A) T-shaped ovum of *D. molossi* (from Desch et al.; in press); B) Snake-like ovum of *D. longissimus* (ibid); C) Anterior end of B to show eggteeth (arrow) near opercular sulcus; D) Y-shaped ovum of *D. gapperi* (Nutting et al., 1968).

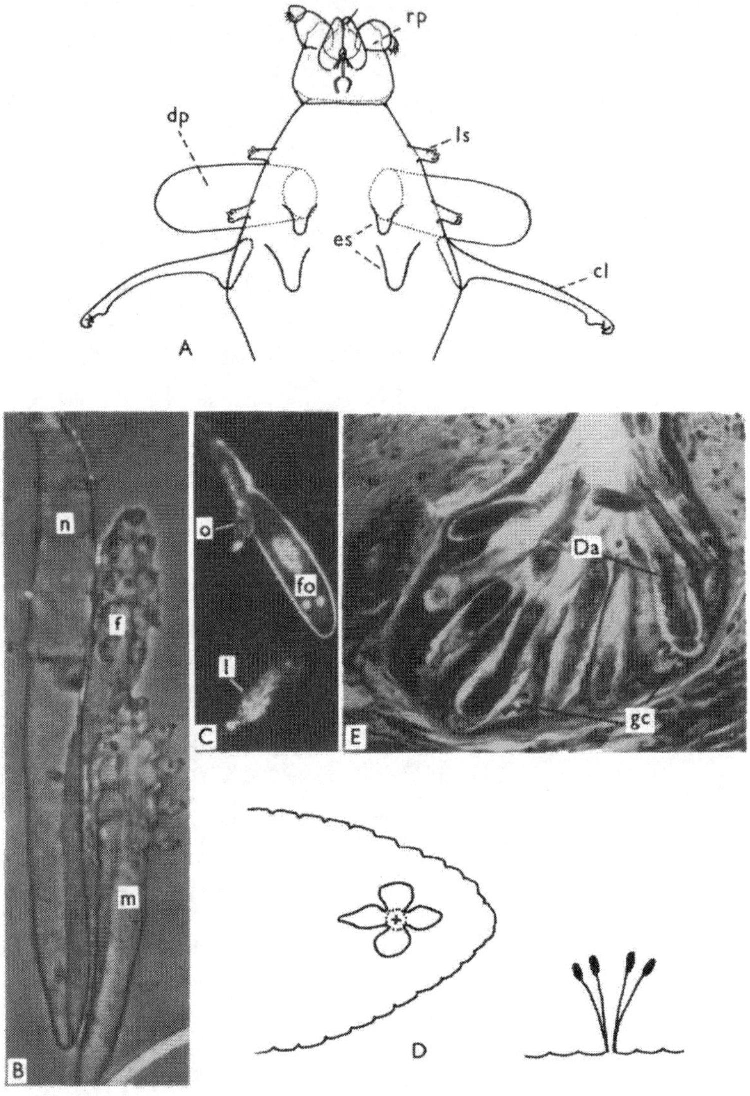

Fig. 2. A)—E): **Adaptations in immature and adult demodicids.** A) Ventral view of anterior portion of larva of *D. molossi* — note epimeral scutes (es), dorsal prominences (dp), claw-like legs III (cl), reflexed palpi (rp), and legs I, II reduced but with sensoria (ls) (from Desch et al.; in press); B) Stages in the life cycle of *D. gapperi* — adult male (m), adult female (f), and nymph (n) (Nutting et al., 1971); C) Female opisthosoma (fo), ovum (o), and larva (l) of *D. antechini* to show crystals of guanine-like waste products (from Stromberg and Nutting, in manuscript); D) Sexual dimorphism in the proctodea of *D. antechini* — female left, male right (Nutting and Sweatman, 1970); E) Host-parasite interactions. Increase in adult and immature *D. antechini* (Da) provoke metaplasia and epidermal giant cell formation (gc) (Nutting and Beerman, 1965).

A SUSCEPTIBILITY OF ANIMALS TO PSOROPTES INVASION

S. NYKOLSKY, O. TCHABLIN
Stavropol Agricultural Institut, Stavropol, USSR

The question about the specificity of the organism causing psoroptes invasion is not clear. According to the results of studies by Dubinin (1950), Kemper and Peterson (1953) the causal agent of psoroptosis is one species – *Psoroptes equi* causing a disease in non-specific hosts. Palimpsestov (1956), Nykolsky and Orlov (1958) did not see any disease in animals during the invasion of psoroptes from non-specific hosts. We are of the opinion that *Psoroptes* mites are host-specific.

The following experiments were carried out by us:

Experiment I. The auditory canal of each of four rabbits, aged 3.5 – 5 month, was artificially infested with 150 *Psoroptes* mites of cattle. The internal surface of the auditory canal was scarified and closed with wet cotton. For 45 days the authors did not observe the development of psoroptosis though the rabbits were excited (itching) and there arose a brief inflammation in the auditory canal. 28 days after insertion 25 mites were taken off rabbit No. 2 and passed into the auditory canals of rabbits 5 and 6. No development of the disease was observed.

Experiment 2. Three rabbits, 3 month of age, were artificially infested with *Psoroptes* taken off sheep. The mites perished in 7 – 10 days.

Experiment 3. Three rabbits, (1, 2, 3) were inoculated subcutaneously with 0,1 ml of allergen prepared from rabbit mite *Psoroptes*; two drops of such allergen were instilled on the conjunctiva of the left eye. Rabbits 1 and 2 had a distinct *Psoroptes* invasion, but not rabbit 3. Three hours after the introduction of the allergen we observed sore oedema with hyperaemia in rabbits 1 and 2, the size of which was 11.0 – 13.5 and 11.0 – 13.5 mm. In 6 hours the oedema enlarged to 20.0 – 30.0 and 15.0 – 19.0 mm with a gradual disappearance in 12 hours. Rabbit 3 remained well. The conjunctiva of rabbits 1 and 2 was hyperaemised in the first 3 hours after allergen instillation.

Experiment 4. The allergen preparation of mite *Psoroptes* from sheep was introduced subcutaneously under the tail in the dose of 0.2 ml to 4 psoroptosis sheep, 4 drops of allergen were instilled on conjunctiva. The authors observed a big hot and sore oedema with light hyperaemia. Eight hours after the injection of allergen the oedema enlarged to 30.0 – 45.0 and 60.0 – 70.0 mm (Fig. 1) and disappeared in 28 to 32 hours. The investigation of blood smears revealed the increase of eosinophil cells.

Experiment 5. (Cross-use of allergens.) The allergen prepared from rabbit mite *Psoroptes* was tested on psoroptosis sheep. The allergen prepared from sheep mite *Psoroptes* was tested on psoroptosis rabbits. The doses and method of use were analogous to exp. 3 and 4. The allergic response proved to be absent. These results show the strong specificity of the casual organisms of psoroptosis.

The allergic situation due to psoroptosis is related to specific *Psoroptes* species and this circumstance may be used as a criterion of specificity.

The morphological and biological characteristics of *Psoroptes* mites and also the allergic responses of animals is an indication of the different specificity of the causal organisms of cattle, sheep and rabbit psoroptosis.

References

Dubinin, V.V.: *Psoroptes* mites, Moscow, 1954. (In Russian.)

Nykolsky, S.N.: Trudy selkhoz. inst., Stavropol, v. 8; 471-479, 1958. (In Russian.)

Kemper, H.E., Peterson, H.O.: Transmission of Psoroptic sheep scab to cattle. (Mimeographed). Animal Disease and Parasite Research Division, APS, USDA, Albuquerque, N.M., CNOK; 1 – 3, 1953.

Palimpsestov, M.A.: Trudy parazitologov Ukrainy, v.1: 144–155, 1961. (In Russian.)

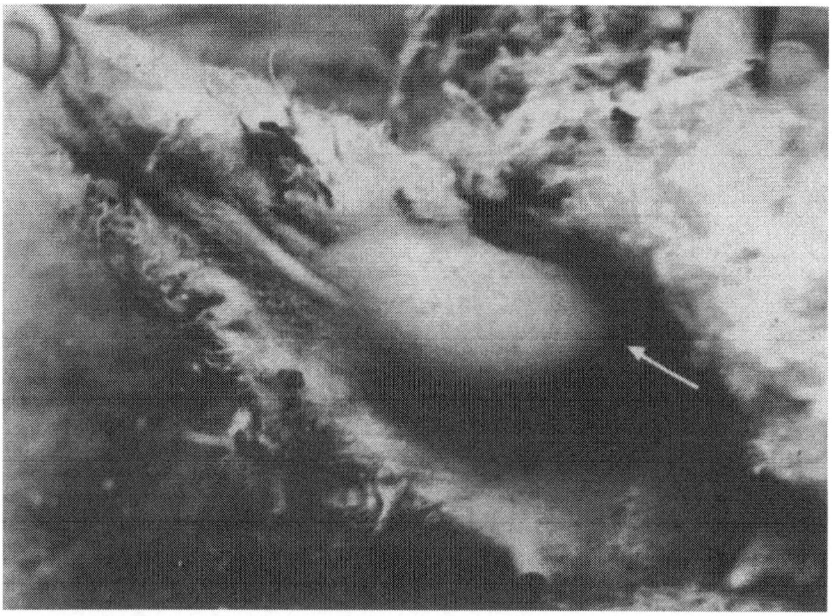

Fig. 1. Dermal allergic response of sheep.

Proceedings of the 3rd International Congress of Acarology, Prague, 1971

CONTROL OF ACARIASES AND PIROPLASMOSES OF FARM ANIMALS IN THE USSR

D. K. POLYAKOV
All-Union Research Institute of Veterinary Sanitation, Moscow, USSR

Apart from the systematic direct treatment of farm animals, the control of acariases as well as the control of free-living parasitic mites, and especially ticks, is carried out in the USSR by treatment of tick biotopes in pastureland and rural areas with acaricides. The biotopes of parasitic mites may be permanent or temporary. For example, the permanent tick biotopes are those places which are visited annually by farm animals and which are populated by rodents serving as maintenance hosts of larvae and nymphs of ticks. In some farms such permanent biotopes may be farm buildings where the animals are kept. The temporary tick biotopes are cultivated areas where farm animals are pastured and where parasitic ticks occur and drop off the pastured animals.

A wide range of control measures are applied, including: 1. chemical methods, based on specific acaricides, preparations of universal effects, disinfectants; 2. physical methods, including fire treatment of premises or burning down of weeds, freezing, flooding of tick biotopes; 3. biological methods, based on the use of predacious mites, "breakdown" of biocenoses and changes inflicted in microrelief and microclimate of an area covered by tick biotopes.

The following acaricides are widely used: chlorophos (0.0 dimethyl – 2.2.2 – trichlorhydroxyethylphosphonate), 1 % dilution is applied to cattle at a dosage of 1 – 3 l per animal. After application the preparation remains acaricidal for 7 days. In some regions especially in the hot season, the period of residual acaricidal action is shorter. The causes of this phenomenon are being studied. The chlorophos preparation is also commonly used against poultry mites *Dermanyssus gallinae* in 0.5 – 2 % dilution at a dosage of 100 – 200 ml per 1 m^2 of hen-houses and poultry yards; trichlormetaphos–3 (–0–methyl–0 ethyl –0–2.4.5– trichlorphenylthiophosphate) is the trolen analog, synthesized in the USSR and is produced in the form of a concentrated emulsion (active ingredients 50 %). However, this preparation is only permitted for the control of ticks parasitizing young animals and cattle bred for meat. A 1% emulsion is applied at a dosage of 1 – 3 l per animal. Acaricidal action lasts 7 days. In the control of poultry mites a 0.5 –1 % emulsion of trichlormetaphos–3 at a dosage of 100 – 200 ml per 1 m^2 of hen-houses and poultry yards is used; sevin (1–naftyl–N–methylcarbamate) is a 85 % soluble powder applied in the form of 0.85 % emulsion to all age-groups of cattle, except lactating animals because sevin may penetrate into the milk. After application of the sevin suspension at a dosage of 1 – 3 1 per animal, its residual acaricidal action lasts 7 days. Sevin is used for the control of overwintering immature stages of ticks in the form of 7.5 % dust at a dosage of 100 – 200 g per animal.

The application of sevin in the control of poultry mites and argasid ticks in poultry farms and enterprises has proved to be highly effective. The poultry mites are controlled in hen-houses and poultry yards before mechanical cleaning by the application of sevin (0.1 % suspension at a dosage of 200 ml per 1 m^2). After the premises have been cleaned, a similar dosage of sevin (0.25 % suspension) is sprayed from special USSR-made apparatuses. In exceptional cases hen-houses are treated with the birds present. Techniques of treatment with sevin (0.25 % suspension) by means of an aerosol nozzle "TAN" with compressor, producing 500 ml of suspension per minute, have been worked out. The aerosols used (at a pressure capacity of 4.5 – 5 atm.) with dispersion of 40 microns enabled the quantities of the preparation to be reduced 4 – 8 times, with identical effectiveness. The larvae of *Argas persicus* are eradicated by 7.5 % sevin dust at a dosage of 5 – 15 g per 1 m^2. Control measures in the tick-infested premises are carried out by 1.5 % chlorophos solution or a 1 % sevin suspension at a dosage of 200 ml per 1 m^2.

In certain regions veterinarians prefer to apply sodium arsenite (Na_3AsO_3). Methods of its application to all kinds of farm animals have been worked out. The cost of sodium arsenite is 10 times lower than that of the above mentioned acaricides, with similar acaricidal action. The preparation is used as a water solution containing 0.16 % of the substance.

In the control of poultry mites 1 % carbophos emulsion and 0.02 % DDVP emulsion are also used; 3 % carbophos and 0.5 % DDVP emulsion at equal dosages are applied against *Argas persicus* ticks. Control measures against psoroptic mange of sheep consist in dipping the animals in 0.03 % emulsion of lindan. These measures are mainly taken as a protection against mites.

Special attention is directed to biological methods of controlling parasitic mites. In integrated systems of control predacious Cheyletid mites have been found to be useful against poultry mites.

Melioriation and agrotechnical measures involving fundamental changes of natural and economic conditions in certain tick-infested areas are taken effectively against ticks. An inventory has been made of tick-infested biotopes, primarily those adjoining the settlements, giving an account of reasons for and accomplishment of fundamental changes inflicted in the area of tick biotopes, thus eliminating their further existence there.

The above-mentioned tasks undertaken basically by the All-Union Research Institute of Veterinary Sanitation, in co-ordination with other veterinary research centres, have considerably decreased the occurrence of acariases and piroplasmoses in the USSR.

ON THE DIAGNOSTICS OF HUMAN SCABIES

M. PŘÍVORA, M. ŠEVCOVÁ
Institut of Epidemiology and Microbiology, Prague, Czechoslovakia

In recent years, it has been reported in the literature that the number of cases of scabies among the human population is rising. Attention to this phenomenon, if we limit ourselves to Europe, has been called, for instance, by Busvine (1966) and Morgan (1968) in Britain, Gołębiowska-Podgórczyk and Szarmach (1968) in Poland, Herrmann and Steigleder (1967) in Germany, Orru et al. (1968) in Sardinia, and Palička et al. (1971) in Czechoslovakia. According to Busvine, the incidence of scabies in man is subject to a certain variation in individual years; in England an increase was noted during both world wars and subsequently in recent years.

If we go into the details of the reported data, we find that they are based solely on clinical diagnoses. It is not entirely without interest to listen to the opinion of especially the older dermatologists (for example Trýb, 1941) that the clinical diagnosis of scabies is not really a simple matter, especially in atypical cases. In the latest scabies epidemic, under progress at present, one often hears dermatologists say that the scabies takes an atypical course and its localization is different from what text-books state.

However, there is another point of essential importance: scabies is a disease, probably the only disease, whose causative agent and methods of its demonstration are known, but where hardly any doctor bothers to identify the causative organism, thereby confirming the clinical diagnosis. Thus scabies has acquired a unique and somewhat bizarre position among all infectious diseases.

In our opinion, the cause lies in the difficulties of the laboratory diagnosis and demonstration of the itch mite and the insufficient co-operation between acarologists and clinical dermatologists. As regards the method, it is quite surprising that most of the existing text-books and monographs (for example Busvine, 1966; Craig and Faust, 1945; Gradwohl and Kourí, 1948; Hunter et al., 1966; Chmel et al., 1954; Korting, 1970; Obrtel, 1965; Pavlovsky, 1951) state that it consists in the search for itch mites in their burrows and extracting them with a needle or lancet. This method, which has been known for a long time but is quite inconvenient for practical purposes because of the time required, has surely greatly contributed to the mistrust of clinicians for the laboratory diagnosis of scabies. Fortunately, there are also other possible methods. Alexander (1968, 1969) reports that a suitable method is to scrape off the surface of a burrow with a sharp scalpel held perpendicular to the skin. This method is rapid and easy and, especially, requires little time. However, it is obvious that some practice is needed if it is to be performed really quickly and especially without damaging the corium. We ourselves have employed a different method, making use of the keratolytic effect of lye: suspect spots of efflorescence are slightly moistened with 5 – 10 % potassium lye and after about one minute the softened and partially dissolved keratin is scraped off with a blunt instrument; the material obtained is then transferred into a drop of either potassium lye or Swann's solution on a slide. Specimens are examined at a low magnification – we ourselves have been using a 100x magnification (Ševcová, 1971). The diagnosis rate of the method is relatively good: out of 36 patients examined in this manner this year itch mites were demonstrated in 24, five preparations at the most being examined per patient. This method is apparently essentially identical with that which we subsequently found in a paper by Rook et al. (1968) who say: "The burrows may be moistened with liquor potassae and gently scraped with a blunt scalpel".

From a technical point of view the potassium lye method seems to be quite acceptable: although somewhat slower than the method of scraping off dry burrows, as suggested by Alexander, it is technically simpler. The whole procedure, moreover, need not be performed in the surgery, but can carried out in the laboratory, as is usual in the diagnosis of other, particularly parasitic diseases or in collecting material samples for examination for special purposes. The result can be known within an hour at most, in any case in the same day if patients are examined serially. The diagnosis rate is fair, especially considering that the negative results in our cases were obtained in patients in whom the clinician was not quite certain about the diagnosis.

Naturally, objections have also been raised against the method, namely that it is annoying to the patient or is even detrimental to him because scars may be produced. As for the first objection, it is obvious that the examination really does bring some inconvenience to the patient – about as much as any other laboratory examination procedures does, e.g. the collection of a blood sample or a swab for bacteriological or other purposes. The places where potassium lye has been applied may itch and it is advisable to warn the patient about this in advance, wash the sites with water immediately after the examination and perhaps apply some acid ointment (for instance one containing boric acid). As for the detriment to the patient, we have not encountered even a trace of anything of the sort so far.

The two last methods, i.e. the one suggested by Alexander and that tested by ourselves and mentioned also by Rook et al., could substantially add to the accuracy of the diagnosis of scabies. One should fully realize that our contemporary knowledge of the agent of scabies in man is mininal as far as other species than *Sarcoptes scabie* are concerned. I have in mind the animal species that can also elicit the disease in man. We cannot obtain the relevant data without identifying the agents and hence without active co-operation between acarologists and dermatologists. Clinicians themselves will hardly find enough time to devote it to the laboratory diagnosis of the condition; frequently they are convinced that they can safely recognize it clinically. It is clearly desirable that scabies be grouped among those diseases whose diagnosis is complete only if it includes laboratory examination and identification of the parasite species. The epidemic now in progress and the enhanced interest in scabies create favourable prerequisites for this. But the iniciative should this time come from acarologists.

Of course, the chief motive of this demand is the benefit of the patient – the point being that only accurate diagnosis can lead to a more accurate and more effective therapy.

References

Alexander, J.O'D.: Scabies and Pediculosis. The Practitioner 200: 632–644, 1968.

Alexander, J.O'D.: Scabies in Children. Clin. Pediat. 8: 73–85, 1969.

Busvine, J.R.: Insects and Hygiene. 2nd Ed., London, Methuen, pp. 254–266, 1966.

Chmel, L., Řehák, A., Hegyi, E.: Laboratórne vyšetrovacie metódy v dermatologii a venerologii. Bratislava, VSAV, 1954.

Craig, C.F., Faust, E.C.: Clinical Parasitology. 4th Ed., London, H. Kimpton, pp. 582–586, 1945.

Fain, A.: Étude de la variabilité de *Sarcoptes scabiei* avec un revision des Sarcoptidae. Acta Zool. Pathol. Antverp. 47: 41–138, 1968.

Gołębiowska-Podgórczyk, I., Szarmach, H.: Czy epidemia świerzbu? Polski tyg. lek. 23: 1931–1933, 1968.

Gradwohl, R.B.H., Kourí, P.: Clinical and Laboratory Methods and Diagnosis. Vol. III, 4th Ed., London, H. Kimpton, 1948.

Herrmann, W.P., Steigleder, G.K.: Scabies. Deutsch. med. Wschr. 92: 1557–1561, 1967.

Hunter, G.W., William, W., Swartzwelder, J.C.: A Manual of Tropical Medicine. 4th Ed., Philadelphia and London, W.B. Saunders, 1966.

Jírovec, O.: Parasitologie für Aerzte. Jena, G. Fischer, pp. 513–516, 1960.

Korting, G.W.: Scabies (Krätze). Therapie der Hautkrankheiten. Stuttgart-New York, Schattauer Verl., 1970.

Morgan, J.K.: More cases of Scabies. Brit. Med. J. 1: 641, 1968.

Obrtel, J.: Dermatovenerologie. Praha, SZdN, pp. 169–174, 1965.

Orru, A., Pirastu, A., Pinetti, P.: La scabbia ricompare. Rass. Med. Sarda 71: 365–375, 1968.

Palička, P., Mališ, L., Zwyrtek, K.: K problematice výskytu scabies a pedikulózy na Karvinsku. Čs. epidemiol., mikrobiol., imunol. 20: 116–120; 1971.

Pavlovsky, E.N.: Učebnik parazitologii čeloveka. 6 vyd., Moskva, Medgiz, pp. 275–278, 1951.

Rook, A., Wilkinson, D.S., Ebling, F.J.G.: Textbook of Dermatology. 2nd Ed., Oxford-Edinburgh, Blackwell Sci. Publ., pp. 1001–1008, 1968.

Schirren, J.M.: Zur Kasuistik der Scabies norvegica sive crustosa. Hautarzt 21: 407–413, 1970.

Ševcová, M.: Prispevek k diagnostice scabies. Cs. epidemiol., mikrobiol., imunol. 20: 113–115, 1971.

Szarmach,H.,Gołębiowska-Podgórczyk, L.: Aktualne problemy rozpoznania swierzbu. Polski tyg. lek. 23: 1402–1404, 1968.

Trýb, A.: Praktická dermatologie. 2nd Ed., Praha, Melantrich, pp. 110–115, 1941.

TICK DISTRIBUTION IN INUNDATED AND NON-INUNDATED BIOTOPES

B. ROSICKÝ, V. ČERNÝ, S. HEJNÝ
Institute of Parasitology, Czechoslovak Academy of Sciences, Prague
and Botanical Institute, Czechoslovak Academy of Sciences,
Průhonice near Prague, Czechoslovakia

Between 1961 and 1963 we studied the occurrence of ticks along 3 routes intersecting an area in southern Moravia, known as a natural focus of tularemia. One of the routes was situated south of Břeclav town, in an annually inundated territory. A total of 44 sites, each $100 m^2$ large, were selected in it, of which 14 were situated in sections that remained dry and 30 in flooded sections. During the period of tick activity 535 collections were made, comprising 3 species: *Ixodes ricinus* (L.), *Haemaphysalis concinna* Koch and *Dermacentor reticulatus* (Fabr.). The first species was the most abundant.

From the geobotanical aspect the area studied may be characterized as follows: elevated sections above the inundated littoral line, overgrown with *Potentilla albae – Quercetum* s.l. association including a large concentration of thermophilic species, are diffusely dispersed over the vast area of inundated forests composed of *Salici-Populetum*, *Ulmo-Fraxinetum* and *Querco-Carpinetum* associations. Open areas of meadows are overgrown with *Caricion gracilis* and *Cnidion venosi* associations, while in the swamps *Phragmition* is common.

In comparison with other non-inundated localities near Břeclav, the frequency of occurrence of *I. ricinus* in the area studied was very low. Of 308 collections made by the flagging method in 1961 only 54 (17.5 %) were positive, of 96 collections made by the same method in 1962 only 8 (8.3 %), and of 131 collections made in 1963 only 17 (13.0 %) yielded positive results. After assessing separate results obtained both in non-inundated and inundated biotopes, distinct differences became apparent. During the three years 57 (32.4 %) positive collections were made out of 176 collections in non-inundated biotopes and a total of 327 ticks (17 ♂, 16 ♀, 116 N, 178 L) was collected. The average number of ticks per one flagging was 1.86 specimens. In inundated biotopes there were 22 (6.1 %) positive collections out of 359 collections made and a total of 42 ticks (4 ♂, 3 ♀, 33 N, 2 L) was collected. The average number of ticks per one flagging was 0.12 specimens. The differences observed were apparent from a comparison of two tick-infested sites 10 m apart, of which one was situated in a non-inundated forest mound and the second just below. Out of 15 collections made, 13 in the first site and 2 in the second site were positive, totalling 75 ticks and 3 ticks respectively. In the non-inundated section of the route *I. ricinus* occurred primarily in tree stands and to a lesser degree in open terrain. The maximum single collection constitute 14 nymphs. Moreover, in the inundated area the majority of ticks was collected in growths of deciduous trees, the maximum of ticks collected being 1 male and 4 nymphs. In the open terrain a sporadic occurrence of *I. ricinus* was ascertained four times.

During 1961 and 1963 the other two tick species were found in small numbers and in non-inundated biotopes only. *H. concinna* was obtained during 5 collections, totalling 1 male, 3 nymphs and 6 larvae, while *D. reticulatus* was found in 4 instances, totalling 6 females and 1 male.

All results indicate that the floods have a negative influence on the survival of ticks, consequently on the numerosity of their populations, as has also been demonstrated under laboratory conditions. (Honzáková, 1971). As regards their abundance as well as their frequency the ticks occur primarily in non-inundated biotopes where their hosts concentrate during the floods and from where they are disseminated into the environs, after the waters subside.

The mentioned distribution excludes any contact of some important reservoir animals of tularemia (*Arvicola terrestris, Ondatra zibethica*) with ticks as potential vectors of this disease. The low numbers of ticks or their absolute absence in the non-inundated biotopes also limits to the minimum the possibility of human infection with other diseases transmitted by ticks.

References

Honzáková, E.: Survival of some ixodid tick species submerged in water in laboratory experiments. Folia Parasitol. (Praha), 18: 155–159, 1971.

Proceedings of the 3rd International Congress of Acarology, Prague, 1971

THE HISTOLOGICAL IDENTIFICATION OF ACARINA IN SKIN BIOPSIES

J. ŠLAIS
Institute of Parasitology, Czechoslovak Academy of Sciences,
Prague and Šikl's Department of Pathology, Medical Faculty,
Charles University, Plzeň, Czechoslovakia

In biopsies of the skin of man, formations of doubtful origin may be found. For diagnostic reasons it should be decided whether these are foreign bodies or a section through the body of a skin parasite. This communication deals with findings of *Demodex* mites, itch mites, ticks and chiggers. The pattern of changes caused by these parasites in the skin may, sometimes, be atypical and, therefore, the correct identification of the parasite can be obtained only from histological sections.

The finding of a *Demodex* mite (*Demodex folliculorum*) in the sebaceous duct is mostly incidental and not associated with the skin affection. It is quite exceptional to find a distinct clinical affection of the skin which had been caused by *Demodex* mites, e.g. blepharitis (Morgan and Coston, 1964). Important for the identification are sections through the thoracic parts of the mature form of the mite formed by the basal segments of the four pairs of legs with internal muscle fibres. In sections through the posterior portion of the body, the cells of the sexual organs of the parasite are a marked feature.

In scabies, the origin of atypical skin changes may be disclosed only if females of the itch mite (*Sarcoptes scabiei*) are found. We succeeded to disclose the etiology of a doubtful skin affection in a seven month-old baby by finding itch mites in the cup-like skin defects in histological sections (Šlais, 1964) (Fig. 1). In sections through the body of the parasite, the glandular corpuscles and the dorsoventral muscles are very marked. In older parts of the skin, eggs covered with a thin chitinous sheath, embryonic cells and yolk cells may be encountered. (Fig. 2).

The changes in the skin caused by blood-sucking ticks (Ixodoidea) are considerably more distinct. When the tick has completed its blood feed and detached itself from the skin, there remains a crater-like defect with a necrosis spreading to the subcutis. It is then possible to encounter a section of the "trunk" and to study the body of the parasite (idiosoma). The microscopic anatomy of the tick enables its identification with even standard histological methods. Giemsa staining demonstrates the layered structure of the soft body covering, the branches of the Malpighian gland and, particularly, sections through the branches of the gut caeca. The tracheae and the bundles of their branches with their spiral chitinous support stain violet with metachromacy and even the layered anorganic corpuscles in the rectum can be demonstrated.

Frequently, skin is excised from the site of tick attachment, but mostly, the major part of the tick's body has been torn off by the patient. The parts remaining in the wound are mostly only remnants of the soft body covering and the highly chitinized anterior portion of the body (gnathosoma). The defect lies at the level of the skin surface and, in a series of histological sections, it was possible to find the proper mouth organ (proboscis) anchored in the subcutis. In sections through the cephalic end of the tick, we found characteristic porous plates of an as yet obscure functional importance. The proboscis itself is armed with coarser teeth on the mandibular sheath (chelicerae) as compared with the minute teeth at the bottom of the anterior end of the hypostome.

While the tick pierces the skin with its proboscis and merges this into the wound, the activity of the chigger larva is responsible for the origin of a special sucking tube which, in the past, had been considered to be an anatomical organ of the parasite. Trombiculid larvae *(Neotrombicula autumnalis)* cause skin lesions in the epidermis of man with a marked reaction in the subcutis. Exceptionally, these may be studied in histological section (Šlais, 1960). The sucking tube (stylostome) originates through a coagulation of tissue fluid under the influence of salivary excretions and the sucking of the parasite in the skin lesion. The larva buries its chelicerae into the horny layer of the skin and excretes the secretion of its salivary glands into the wound. Tissue colliquation is responsible for the origin of a lesion which becomes permeated with tissue fluid. This fluid coagulates under the presence of the salivary excretion and the larva maintains a capillary lumen by sucking in this coagulated substance.

Our schematic illustration (Fig. 1) indicates the clear relationship of the larva to the attacked skin. Even when the larva has detached herself from the skin, the coagulated stylostome surrounded with cell exudate, remains in the disturbed epi-

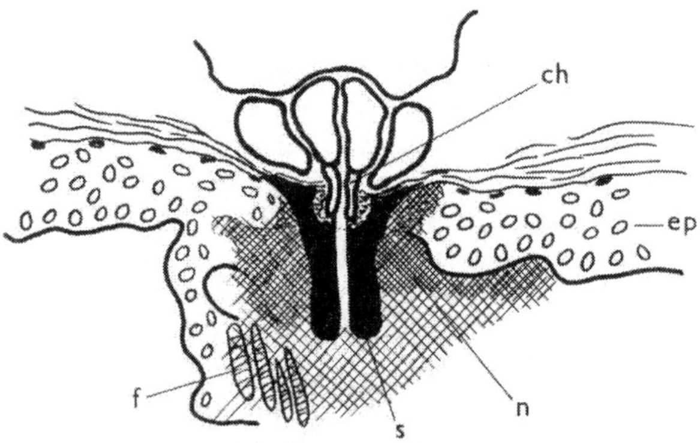

Fig. 1. Schematic illustration showing the relationship between the oral organ of the larva of *Neotrombicula autumnalis* and the affected skin of a white mouse. ch – chelicerae, ep – epidermis, f – hair follicles, n – necrosis, s – stylostome.

dermis. Its structure consisting of acid mucosubstances, can be demonstrated even during the phase of its gradual limitation by regenerating skin.

The stylostome represents the mode of feeding in the skin of these trombiculid species which remain with their host for a longer period and are responsible for the origin of ulcerous skin defects. Even a larva completely covered by hypertrophic and inflamed skin, forms always a sucking hollow with a stylostome at the bottom of the lesion. (Plate II, Figs. 1, 2). The incidence of larvae of *Cheladonta costulata* was discovered in eastern Slovakia. The larva of this species having highly developed salivary glands caused deep ulcers in the skin of *Microtus arvalis* Pallas (Daniel and Šlais, 1957). It has been found that also other cases of the so-called endoparasitism of larvae of the family Trombiculidae occurring mostly in tropical areas, are the consequence of protracted parasitation and its intensity and of the prolonged inflammatory reaction of the skin (Hoeppli and Schumacher, 1962).

References

Daniel, M., Šlais, J.: K otázce intradermálního parasitismu a jeho morfologie u larev sametek *Euscböngastia ulcerofaciens* (Acari: Trombiculidae). Čs. biologie 6: 365–371, 1957.

Hoeppli, T., Schumacher, H.H.: Histological reactions to trombiculid mites, with special reference to "natural" and "unnatural" hosts. Z. Tropenmed. Parasit. 13: 419-428, 1962.

Morgan, R.J., Coston, T.O.: *Demodex* blepharitis. Sth. med. J. (Bgham, Ala.) 57: 694–699, 1964.

Šlais, J.: Histopatologie trombidiosy u hlodavců a člověka. Plzeň. lék. sbor. 13: 45–54, 1960.

Šlais, J.: Příspěvek k morfologii svrabu. Čs. parasit. 11: 257–262, 1964.

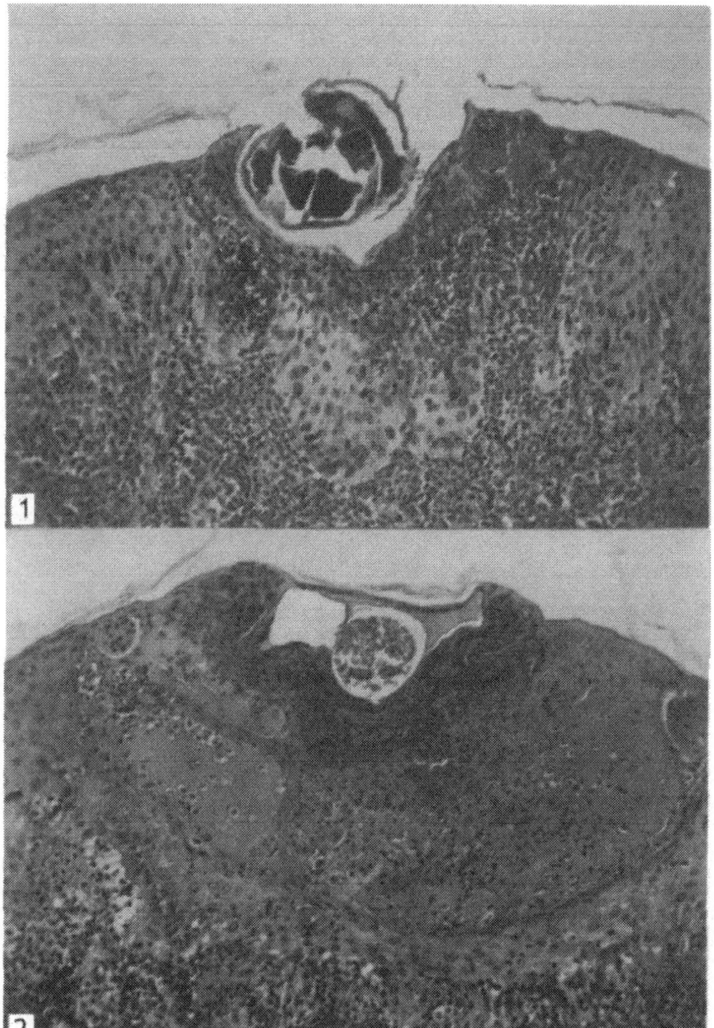

Plate I.

Fig. 1. Cross section through a female *Sarcoptes scabiei* in the cup-like skin defect lined by a thin horny layer. Cell infiltrate surrounding the defect permeates also the papillae of the corium. The structure of the papillae is completely obscured by the infiltrate. Also the corium underlying the defect is greatly infiltrated. Weigert-van Gieson (135×).

Fig. 2. The cup-like defect raised above the surface of the epidermis contains a section through an egg. Under the wide border of necrotic cells of the epidermis and the infiltrate, there are intraepidermic vesicles filled mostly with homogeneous exudate. Strips of preserved epidermic cells separate the vesicles from one another. These strips extend from the bottom of the defect to the basal layers. Spongiosis of the epidermic cells is visible. On the left — developing subepidermic vesicles. Weigert-van Gieson (135×).

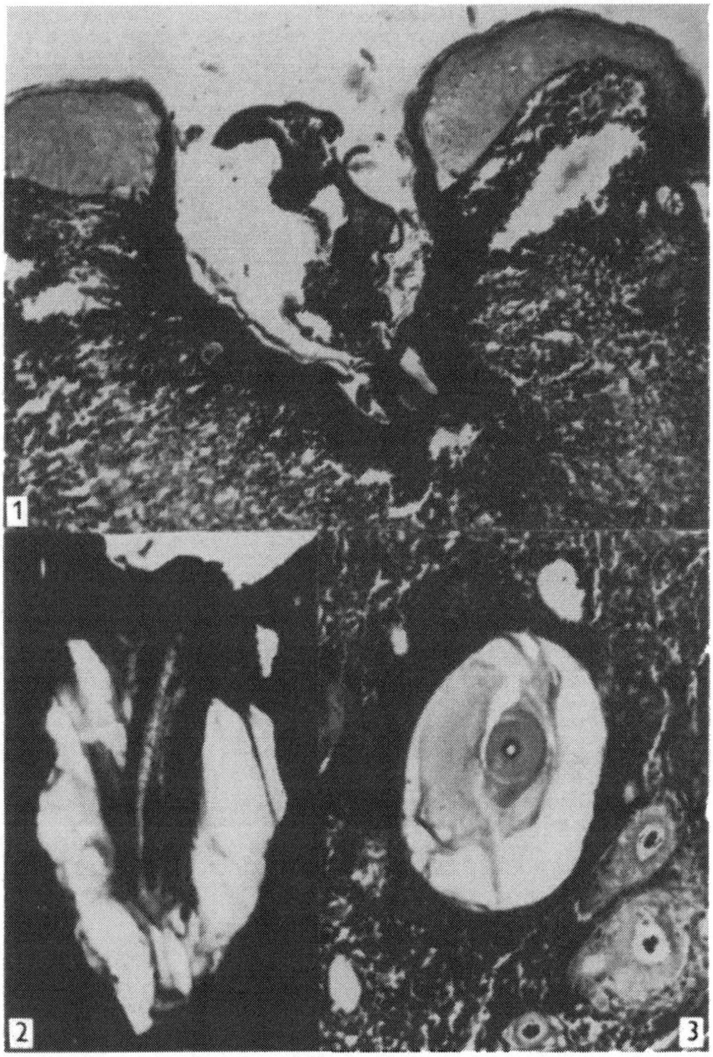

Plate II.

Fig. 1. Larva of ***Cheladonta costulata*** **in a deep skin ulcer of** ***Microtus arvalis*****. A wide dark necrotic zone of the edematous and infiltrated subcutis can be seen under the hypertrophic, interrupted skin surface. The sucking tube (stylostome) in the sucking hollow at the base of the lesion is raised above it; it originated from the sucking of the parasite on tissue fluid. The remaining light space of the sucking hollow is occupied by the secretion of the larval salivary glands. Molybden haematoxylin (90 ×).**

Fig. 2. Longitudinal section (left) and cross section (right) through the stylostome in the sucking hollow at the base of an ulcer in the skin of *M. arvalis,* formerly occupied by a larva of *Ch. costulata.* The central canal is visible in the stylostome; the liquid content of the sucking hollow, precipitated through fixation into irregular formations.

ECOLOGICAL DISTRIBUTION OF HOUSE-DUST MITES IN EUROPE

F. TH. M. SPIEKSMA
Department of Allergology, University Hospital, Leiden, The Netherlands

I. Introduction

From medical investigations by Storm van Leeuwen (1930, 1932) it was already known that house-dust allergic asthma patients were most frequently living in low-lying countries, dwelling in old, damp houses, who suffered their most severe asthma attacks in the autumn. These data suggested a biological source of the house-dust allergen.

When in 1964 we found that house dust – i.e. dust that is collected from the floor of rooms in houses – was a typical habitat for one or more species of *Dermatophagoides*, the above-mentioned epidemiological data became the basis for our subsequent ecological field observations (Spieksma et al., 1967).

Later, it was stressed on, especially by British authors (Maunsell et al., 1968), that also mattresses and other padded also furniture provide a favourable habitat for *Dermatophagoides*.

II. 'Field' observations

1. Season

Initially, we studied the seasonal influences on the size of the populations of house-dust mites. The not surprising result was that the highest number of mites occurred at the end of the summer (Spieksma et al., 1967). Not only *Dermatophagoides*, but the numbers of all other genera of Acaridiae found in dust had their maxima in September.

2. Dampness of houses

Secondly, we examined the number of mites in dust from houses with different degrees of dampness.

We divided 150 houses into five groups with different degrees of dampness and found that, as a group, there was a significant correlation with the numbers of mites in dust (Spieksma et al., 1967).

3. Climate

Thirdly, we compared the numbers of mites in dust from houses in Leiden, situated in the low-lying western part of The Netherlands, with those in dust from houses in two places of Switzerland, namely Basle, at a medium altitude of about 500 meters above sea-level, and Davos at about 1600 meters with a typically cold high-mountain climate.

We found many mites in the Dutch houses, and very few in Davos, with Basle in an intermediate position (Fig. 1) (Spieksma et al., 1971).

All these findings showed that mites, especially the genus *Dermatophagoides*, occur most frequently and abundantly in those places and under those conditions where attacks of asthma in allergic people are seen. And vice versa.

Through a great number of medical experiments we, and many other authors could prove that *Dermatophagoides* produces a substance to which many people become allergic (Voorhorst et al., 1964; 1969; 1970; Mitchell et al., 1969; Wharton, 1970; May, 1970).

III. Laboratory studies

It was the task of the ecologist to investigate the conditions controlling the development of populations of *Dermatophagoides* in dust (Spieksma, 1967).

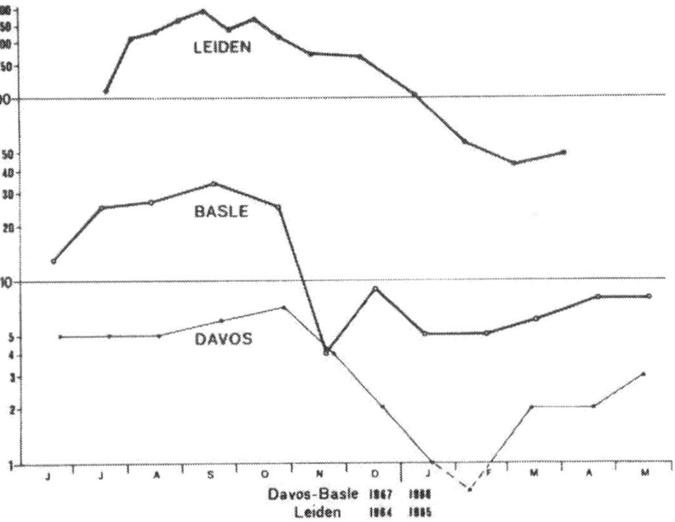

Fig. 1. Influence of climate at different altitudes upon the occurrence of *Dermatophagoides pteronyssinus* in house-dust(Leiden at the sea level, Basle - about 500 m, Davos - 1,600 m above sea level).

Three a-biotic factors in particular have been studied: viz. food, temperature, and humidity.

1. Food

Initially we tried to set up cultures of *Dermatophagoides pteronyssinus* using fishmeal as food. The results were not too bad, but it did not seem to be the best food for them. Then, I found in "A Manual of Parasitic Mites of Medical and Economic Importance" (Baker et al., 1956) that some specimens of *Dermatophagoides* had been kept alive for more than a year on scrapings of cuticle from around human fingernails.

Biological and medical investigators in our institute found that mites are very well satisfied with human skin scales (Spieksma, 1967).

From the ecological point of view this is very important, because the production of skin scales per person per day is 500 – 1000 mg. This is more than enough for the highest numbers of mites found in dust, even taking into account the fact that people spend part of the day outside their homes.

Food is no limiting factor for the occurrence of *D. pteronyssinus* in dust in houses where people are living, or have been so recently. The same holds for mattresses.

Also the skin scales from other animals than man can be used as food.

There is the interesting observation of our Japanese colleagues (Miyamoto et al., 1968), that *Dermatophagoides* also grows very well on dried water-fleas, but I do not think this is their natural food.

2. Temperature

Concerning the temperature (Fig. 2) we found that the optimum temperature for *D. pteronyssinus* was about 25°C (Spieksma, 1967), but they can still grow and multiply between 17 and 32°C. Probably the limits of survival are still broader, depending on the period they are exposed to higher or lower temperatures. The conditions of temperature as they occur in houses of between 17 and 23°, do permit the development of population of mites in dust and mattresses and are also not a serious limiting factor.

3. Humidity

To establish the influence of humidity on the growth of populations of mites is not quite so simple. The humidity of materials and of air is rather hard to determine on a microscale, and the evalution of its influence on these small organisms.

It is this humidity, which, we think, is the most important factor in controlling the presence and the abundance of house-dust mites in the different topographical regions, different houses, and different seasons (Spieksma, 1967, et al., 1971; Voorhorst et al., 1969).

In one of my experiments (Spieksma, 1967) on the influence of the relative hu-

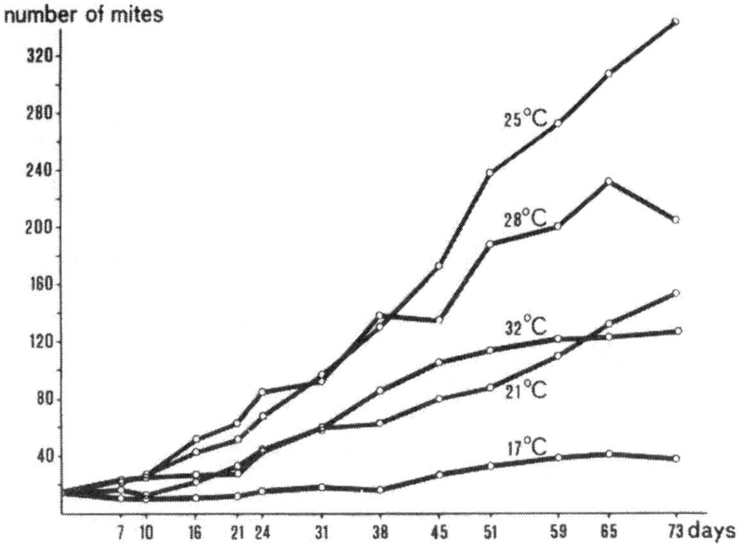

Fig. 2. Influence of temperature upon the population density of *Dermatophagoides pteronyssinus.*

midity of the air, a number of cultures containing approx. 150 living specimens of *D. pteronyssinus* were placed at 25°C temperature in four different degrees of relative humidities (Table 1).

From the result it seems to be clear that 80 % is optimal, and that 75 % is already too dry for an increase of the population. So, we have an optimal value of relative air humidity, which is remarkably sharply limited at the lower side. The influence of the growth of moulds at higher humidities is quite another story, as they obscure the real influence of the humidity over 80 %.

This experiment was executed in a series of climate cells with an open ventilation system. Fresh air together with water vapour, controlling the humidity, was blown into the cells at the top. Through an opening at the bottom of the cell air was allowed to escape.

We have tried to illustrate and summarize these data about humidity in a model. (Fig. 3). In this model, a schematic and simplified representation is depicted of the situation in dust, with respect to the humidity degrees under stable conditions of temperature and air currents.

At the top, you see the situation on wet material as it occurs in houses, when water penetrates through walls, floors and other constructional materials, for instance water in the superficial soil layers or rain. This is not exceptional in the greater part of The Netherlands.

At the bottom is given the situation on dry material as it is found in well constructed houses on dry or rocky soils, for instance in mountain areas, like Davos, or the High Tatras in Czechoslovakia.

Table 1. Numerical increase in laboratory populations of *Dermatophagoides pteronyssinus* at various humidities

relative humidity of the air (in %)	*number of mites after 20 days*
70	42
75	122
80	775
85	-- (moulds)

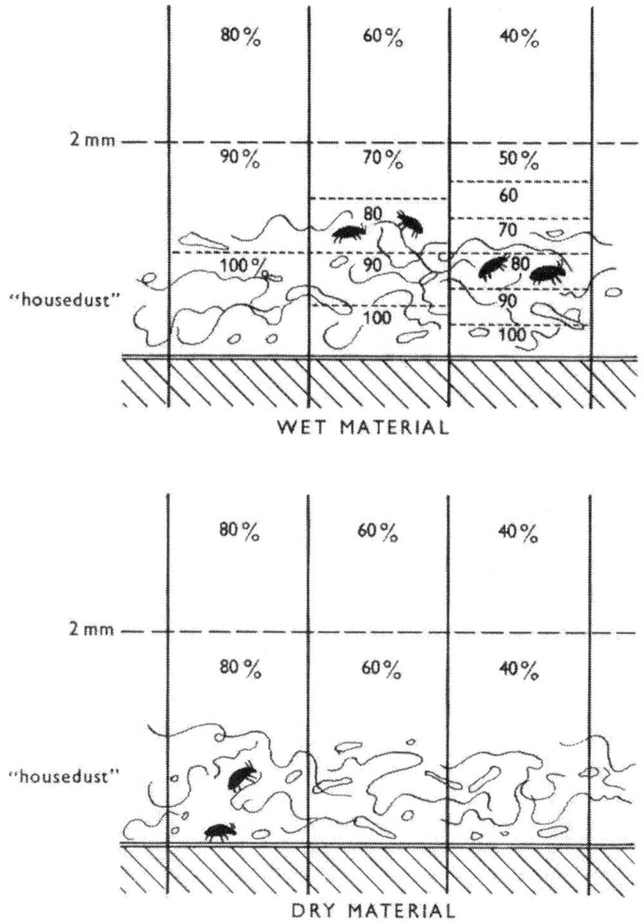

Fig. 3. Diagram depicting the influence of relative air humidity and substrate upon the occurrence and dispersal of *Dermatophagoides pteronyssinus* in house-dust.

The left represents an extremely high air humidity, which might occur in late summer and early autumn in unheated houses.

The right shows a low degree of humidity, which can be found in winter and in cold high-mountain climates, when, by heating the house, the relative humidity drops to these low values.

In the middle is the more or less average situation found during most of the year. We must, of course, realize that the actual situation in a house is always a mixture of these elements. In a damp house not all materials are wet; and in a dry house occasional wet places might occur. The humidity of the indoor air fluctuates daily, and may differ from room to room.

Nevertheless, with this model one can understand that wet places in houses are the most important cause for the presence of mites; that a cold climate, which makes people light their stoves, causes a low air humidity indoors, which hampers the growth of mite populations.

References

Baker, E.W. and 4 co-workers: A manual of parasitic mites of medical or economic importance. Nat. Post Control Ass., New York, 1956.

Maunsell, K., Wraith, D.G. and Cunnington, A.M.: Mites and house-dust allergy in bronchial asthma. Lancet, 1267, 1968.

May, K.L.: Roztocze, kurz domowy i dychawica oskrzelowa. Pol. Tyg. Lek., 25: 622, 1970.

Mitchell, W.F., Wharton, G.W., Larson, D.G., and Modic, R.: House dust, mites and insects. Ann. Allergy 27: 93, 1969.

Miyamoto, T., Oshima, S., Ishizaki, T. and Sato, S.: Allergenic identity between the common floor mite (*Dermatophagoides farinae* Hughes, 1961) and house dust as a causative antigen in bronchial asthma. J. Allergy 42: 14, 1968.

Spieksma, F. Th. M.: The house-dust mite *Dermatophagoides pteronyssinus* (Trouessart, 1897), the producer of the house-dust allergen. Thesis, Leiden, 1967.

Spieksma, F. Th. M. and Spieksma-Boezeman, M. I. A.: The mite fauna of house dust with particular reference to the house-dust mite *Dermatophagoides pteronyssinus* (Trouessart, 1897) (Psoroptidae: Sarcoptiformes). Acarologia 9: 226, 1967.

Spieksma, F.Th. M., Zuidema, P. and Leupen, M.J.: High altitude and house-dust mites. Brit. Med. J. 9: 82, 1971.

Storm van Leeuwen, W: Asthmafürsorge. Med. Welt 4, 1930.

Storm van Leeuwen, W., de Lind van Wijngaarden, C.: Asthma, Bronchitis und Schnupfen in Zusammenhang mit der Jahreszeit. Münch. med. Wschr. 15: 583, 1932.

Voorhorst, R. és Osvath, P.: A házi por atka ereditü allergéneiröl. Orvosi Hetilap 111: 1456, 1970.

Voorhorst, R., Spieksma-Boezeman, M.I.A. and Spieksma, F.Th.M.: Is a mite (*Dermatophagoides* sp.) the producer of the house-dust allergen? Allergie u. Asthma 10: 329, 1964.

Voorhorst, R., Spieksma, F.Th.M., and Varekamp, H.: House-dust atopy and the house-dust mite *Dermatophagoides pteronyssinus*. Stafleu, Leiden, pp. 1–159; Figs. 1–73, 1969.

Wharton, G.W.: Mites and commercial extracts of house dust. Science 167: 1382, 1970.

SPATIAL RELATIONS OF HOUSE DUST MITES[1]

G.W. WHARTON
Acarology Laboratory College of Biological Sciences,
The Ohio State University,
Columbus, Ohio, USA

Since 1964, when Voorhorst et al. published data that demonstrated a relationship between *Dermatophagoides pteronyssinus* Trouessart and house dust allergy, much related work has been done in many countries (Bronswijk and Sinha, 1971). In surveys of household surfaces for mites many different species have been found but the most frequent are *D. pteronyssinus* and *D. farinae*, (Wharton, 1970). These two species can be cultured in the laboratory by a number of methods. In this paper differences in behavior of these two species will be reported.

Methods

Mites that were collected from samples of house dust were placed in cultures each of which consisted of a 15 ml glass bottle, a cap made of cigarette paper held in place by a plastic rim, hair or cotton fibers as a substrate, and dog food or yeast as food. These cultures were kept at room temperature and 75 % relative humidity. The relative humidity was maintained by enclosing the cultures in a container in which they could come into equilibrium with a saturated solution of NaCl. Subcultures were made at intervals until thousands of cultures were available for study and the production of mites. The containers used were glass desiccators or large plastic trash cans with tight fitting tops. Cultures in the trash cans were placed in small plastic bags, 4 or 5 at a time and suspended over the salt solution that covered the bottom of the containers. When the effect of temperature on the cultures was studied desiccators were placed in refrigerated incubators.

These cultures were observed with a dissecting microscope when mites were to be collected or subcultures made. The movements of individual mites were studied with a compound microscope. The mites were placed in Plexiglas chambers and at times TV tape recordings of their movements were made. The TV tapes were in turn used to produce moving pictures. Measurements of the mites were made with the aid of a phase contrast microscope from cleared specimens mounted

[1] This work has been supported by NIH–69–65 a contract with the Division of Biologics Standards.

on glass slides. Stereoscan electron microscopy provided three dimensional views at high magnification.

Results

Cultures of North American house dust mites *(D. farinae)* were characterized by the migration of large numbers of mites from the substrate. When examined these cultures contained many mites on the lids and upper rims of the culture jars. Cultures of the European house dust mite *(D. pteronyssinus)* were characterized by many mites on the bottom of the jars. These differences in culture characteristics were used in making collections of the mites from cultures. When *D. farinae* was harvested, mites were removed from the lids and rims of the jars (Larson et al. 1969). Harvesting of *D. pteronyssinus* was most rapid if the substrate and food were removed from the jar and the mites collected from the bottom. In cultures very heavily populated with *D. pteronyssinus* collections could also be made from the rims. This technique had the advantage of producing mites with a minimum of contamination.

When both European and North American house dust mites were reared in the same culture jar and at different temperatures both species survived and reproduced for over 200 days in most cases. At 15 °C cultures contained 59 % *D. pteronyssinus* and 41 % *D. farinae* and all cultures contained both species. At higher temperatures *D. farinae* did better than *D. pteronyssinus* and comprised more than 80 % of the mites at both 20 and 25 °C. At 20°C one of ten cultures became negative for *D. farinae* and 2 of 10 negative for *D. pteronyssinus*. At 25°C 4 of 10 cultures became negative for *D. pteronyssinus*. All cultures contained many mites even those that were negative for one or the other species.

Observation of these species as they walked on smooth glass or plastic surfaces revealed that they were attached to the surface by the pretarsi (Figure 1) and the long posterior setae d–5 and 1–5. The mites held their bodies well above the substrate (Figure 2) but both d–5 and 1–5 dragged along the substrate behind them. In the case of *D. farinae* great difficulty was experienced in moving backwards and when a mite came to an obstacle it most usually responded by climbing over it and thus moving away from the bottom of the culture. Similar observations were made on *D. pteronyssinus*. They also were characterized by d–5 and 1–5 trailing behind them on the bottom but they tended to avoid and get around obstacles by turning rather than climbing. Occasionally specimens were observed to move backwards.

The length of d–5 in the two species was about the same for each. (ca. 185 μm). The lateral seta however, that is 1–5, was much longer (310 μm) in *D. farinae* than it was in *D. pteronyssinus* (250 μm). The differences observed between right and left sides of the specimens and between the d–5 setae of the two species were not significant. Right and left 1–5 seate within a species were not significantly different but between species the difference was highly significant (Table I).

Table I. Measurements of d–5 for and 1-5 for *Dermatophagoides farinae* and *Dermatophagoides pteronyssinus* in μm

Species	d–5		1–5		d–5	1–5
	right	left	right	left	mean	mean
D. pteronyssinus	190	177	249	250	184	250
D. farinae	187	184	312	309	186	310
t	0.5	1.4	10.0	8.7	$t_{.05} = 2.1$	

The simple difference in length of 1–5 between *D. farinae* and *D. pteronyssinus* may explain in large part the findings that in small cultures *D. farinae* are found concentrated on the lids while *D. pteronyssinus* is found on the bottom. Because of such differences the ecological niche of the two species is different and they tend to exploit different environments. The fact that the two species survived together for 200 days in the same 15 ml jar demonstrates that competition between the two species even in such restricted an environment was not too severe.

References

Van Bronswijk, J. E. M. H., Sinha, R. N.: Pyroglyphid mites (Acari) and house dust allergy. J. Allergy 47: 31–52, 1971.

Larson, D. G., Mitchell, W. F., Wharton, G.W.: Preliminary studies on *Dermatophagoides farinae* Hughes, 1961, (Acari) and house dust allergy. J. Med. Entomol. 6 (3): 295–299, 1969.

Wharton, G.W.: Mites and commercial extract of house dust. Science 167: 1382–1383, 1970.

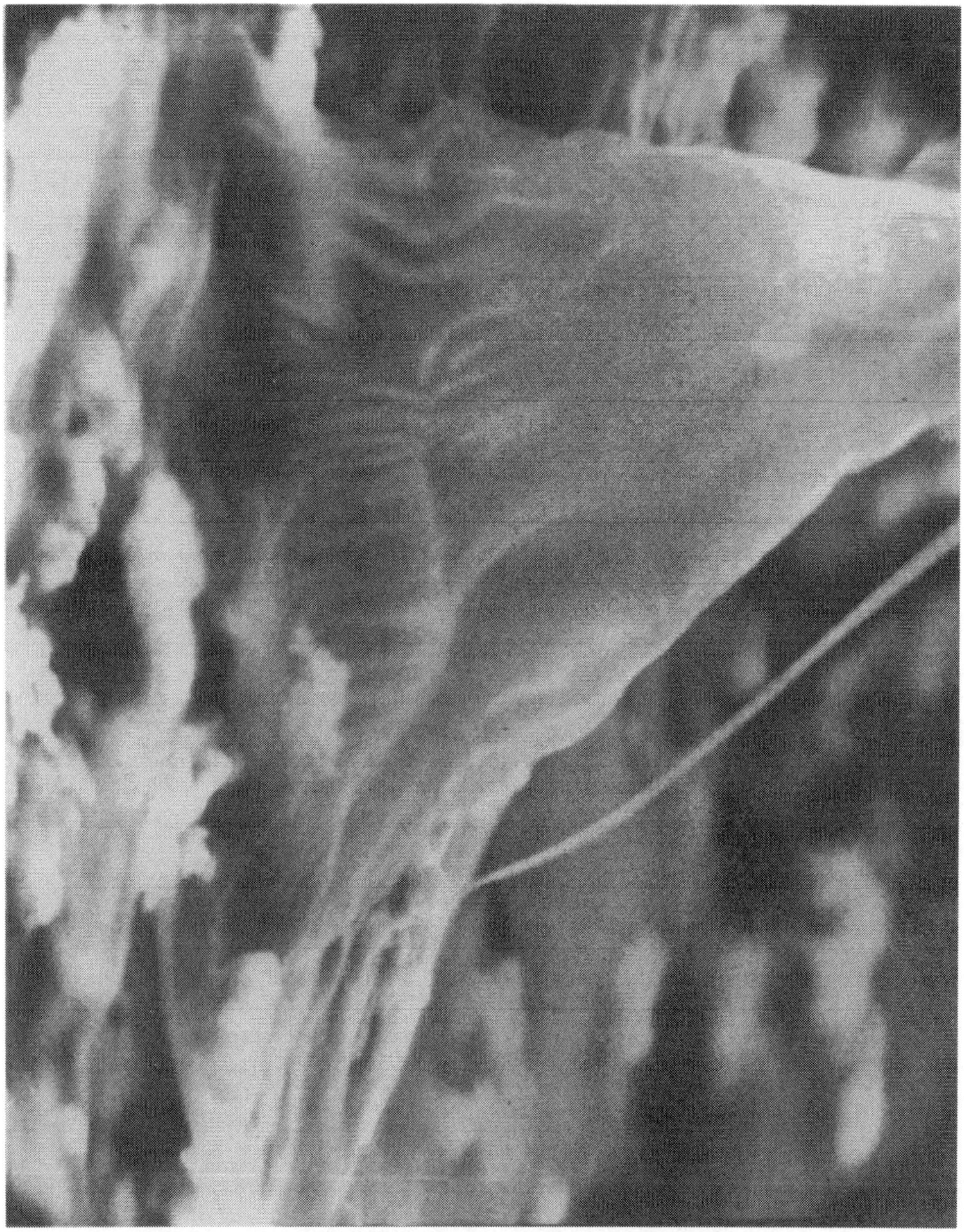

Fig. 1. An SEM micrograph of a pretarsus of *Dermatophagoides farinae* greatly magnified.

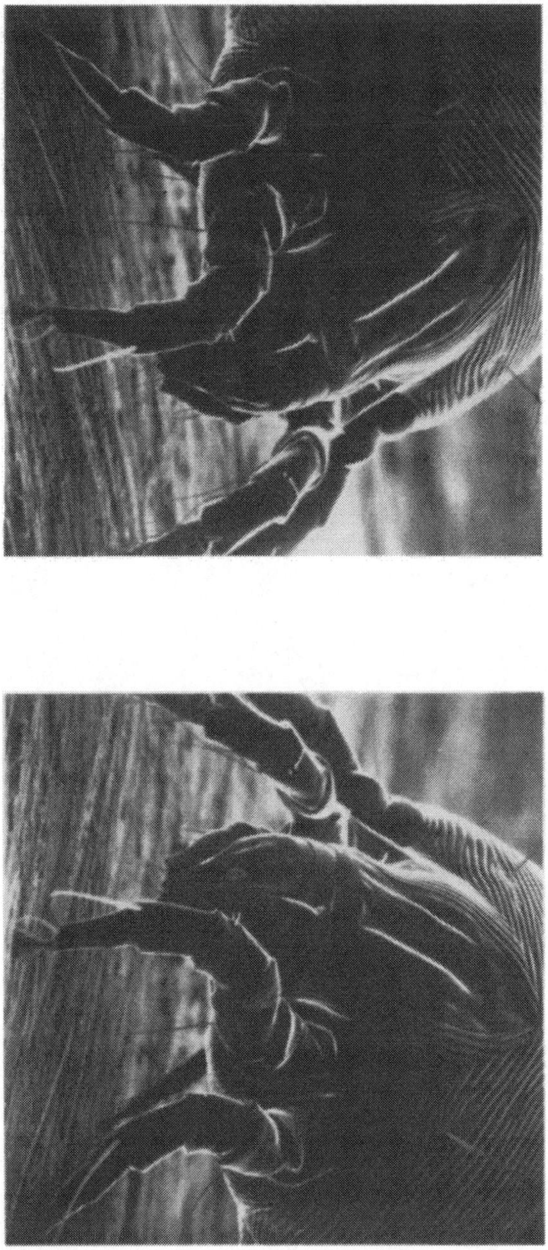

Fig. 2. A three dimensional picture of the protonymph of *Dermatophagoides farinae*. The two pictures can be resolved into a 3 D presentation if a hand mirror is held on the right side of the nose and the image on the right side of the page is viewed in the mirror with the right eye while the left eye views the picture on the left side of the page. Manipulation of the mirror can cause right and left images to be superimposed. When this occurs the protonymph will appear in 3 D.

Section IX A

MITES AND TICKS IN NATURAL FOCI OF TICK-BORNE ENCEPHALITIS

Chairman
V. V. Kucheruk (USSR)

Assistant Chairman
J. M. Kolman (Czechoslovakia)

TICK-BORNE ENCEPHALITIS AS A HAZARD FACTOR IN WORKING ENVIRONMENT

J. AŠMERA, F. HEINZ
Regional hygienic station, Ostrava, Czechoslovakia

In the present study we are dealing with the problems of occupational tick-borne encephalitis (TBE) cases in north-eastern Moravia involving employees of the State Woods and Forests.

We set up the following aims:

a) to map in detail the natural foci of tick-borne encephalitis in the region studied;
b) to ascertain separate natural foci valencies on the basis of the prevalence of tick-borne encephalitis virus in *Ixodes ricinus* ticks;
c) to introduce preventive measures, in some cases in connection with the declaration of certain selected wooded areas as hazardous working environment;
d) to specify the criterion for the acknowledgement of occupational disease.

Material and methods

Our research was concentrated in those areas where human cases of illness exhibiting their varied clinical symptoms of tick-borne encephalitis, were reported from. We shall not mention the laboratory methods; primarily we used methods worked out and described by Libíková et al.

Results

Beginning with the year 1956 we started to trace the epidemiological regularity and clinical progress of tick-borne encephalitis cases in humans reported from all districts of Ostrava and from all districts of northern Moravia. On the basis of the incidence of the disease and subsequently of the results of parasitological and virological investigations we marked the limits of those areas where infection had been actually detected and which were inhabited by ticks. A survey of known natural foci of tick-borne encephalitis in northern Moravia is given in Fig. 1.

The dynamics and epidemiological importance of natural foci of tick-borne encephalitis in northern Moravia were thoroughly studied up to the end of 1970. We isolated 211 strains of the tick-borne encephalitis virus, of which 176 strains were isolated from the tick *Ixodes ricinus* (Linné, 1758), 2 strains from the brains of small

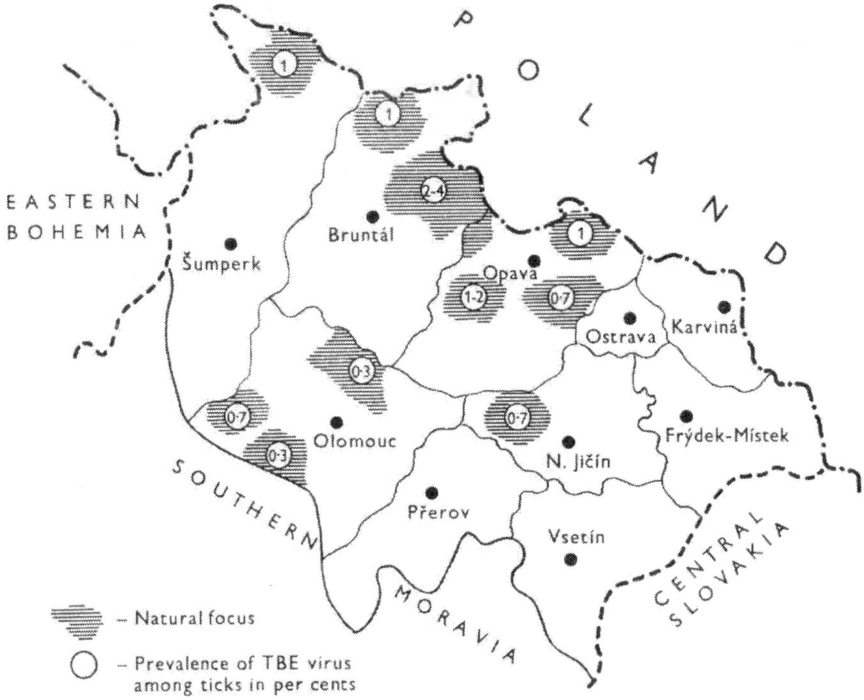

Fig. 1. Survey of natural foci of TBE in northern Moravia.

mammals (one from *Apodemus flavicollis* Melchior 1834, one from *Clethrionomys glareolus*, Schreber 1780) and 33 strains from humans (Nedvídek et al. 1958, Ašmera et al. 1962, 1968, Heinz et al. 1968).

At the Department of occupational diseases in Ostrava and Olomouc 65 cases of occupational tick-borne encephalitis were recorded in the years 1960–1970. From the total number of patients 34 men and 31 women were aged between 17 and 56 (average age 36.4 years). According to professions (Fig. 3) the group of forest workers prevailed (a total of 45, 69.2 %), then 16 foresters (24.6 %) and 4 of other professions (6.6 %). 29 persons were exposed in a hazardous environment for less than 5 years, 36 persons for 6–10 years and 13 persons for more than 21 years. The average exposure in a hazardous environment was 9 years.

We were able to verify the total length of work disability caused by tick-borne encephalitis in 60 patients who were ill for 21–365 days. On an average they were ill for 105 days. As a consequence of the disease according to the objective neurological examination secondary neurasthenic syndrom was diagnosed in 35 cases, small organic complaints in 9 cases, amyotrophy of muscles in 8 cases, strabismus and defective hearing in 3 cases, myalgy in 2 cases, cervicobrachial syndrom in 2 cases, light damage of muscle groups, paresis of upper limbs and light extrapyramidal syndrom in 1 case respectively. Four patients, at the time of examination in the Department of occupational diseases lacked objectively ascertainable changes. Three workers became invalids and were given a full pension.

On the basis of laboratory tests, we declared the three most exposed wooded areas to be a hazardous working environment, in close co-operation with the Forestry administration in northern Moravia centered in Krnov. Considering that we had no vaccination material of high activity (Smorodintsev, Ilyenko 1962, Libíková 1963), from 1960 onwards we emphasized, as preventive measures, the use of protective clothes tight in the lower part and impregnated with 3 % Nera emulsion or with some other repellent preparation -dimethyl-phalate, diethyltoluamid (Rosický, Weiser 1952, Rosický, Havlík 1954). As protective clothes we used normal linen overalls with shirt-fronts. The overalls were provided by the forest enterprise in charge of the wooded area in question. Impregnation with 3 % emulsion of DDT was regularly renewed in the wooded area, in most cases after washing. The employees were made economically interested in strictly observing the preventive measures. Following these preventive measures, we were concerned with paying more attention to the employees at the time of the peak occurrence of *Ixodes ricinus* ticks, i.e. from April 15 to October 15. The employees were encouraged to examine themselves after work and to remove any ticks found. We took care to inform the employees about the risk of tick-borne encephalitis infection. The next measure was the dusting-sanitation of the most dangerous foci in spring by means of 10 % DDT in a quantity of 75 kg/ha.

Discussion

By the complex comparison of virus-carrying ticks and of immunity gained by forest employees in separate natural foci, we placed the examined foci in the group of foci of middle valency (Fig. 2).

The natural foci of tick-borne encephalitis in northern Moravia are characterized as foci of a theriodic type, where the source of blood for ticks *Ixodes ricinus* are small mammals and larger wild animals. Partly, in the Bruntal district, there are the so-called theriodic-bosquematic mixed foci, where the source of blood for ticks are wild as well as domestic animals. The correlation between irruptions of mice and the increase of cases of tick-borne encephalitis was pointed out by Havlík (1954). On the basis of evaluation of clinical cases, hospitalized in northern Moravia in the years 1957–1964, we see that the year of irruption of *Microtus arvalis*, in most cases coinciding with an irruption of *Clethrionomys glareolus*, is followed by the increase of tick-borne encephalitis cases; the epidemic occurrence can be observed 2 years after the irruption year. These results should be taken into consideration when practical measures are introduced to the natural foci of tick-borne encephalitis.

We found out that strict observance of protective measures, especially impregnation of protective clothing with 3 % Nera emulsion and sanitary education, could influence the incidence of tick-borne encephalitis of forest workers. During the years 1960–1970, in areas where preventive measures were adhered to, only four cases were recorded among forest workers and two cases in team-workers (student and house-

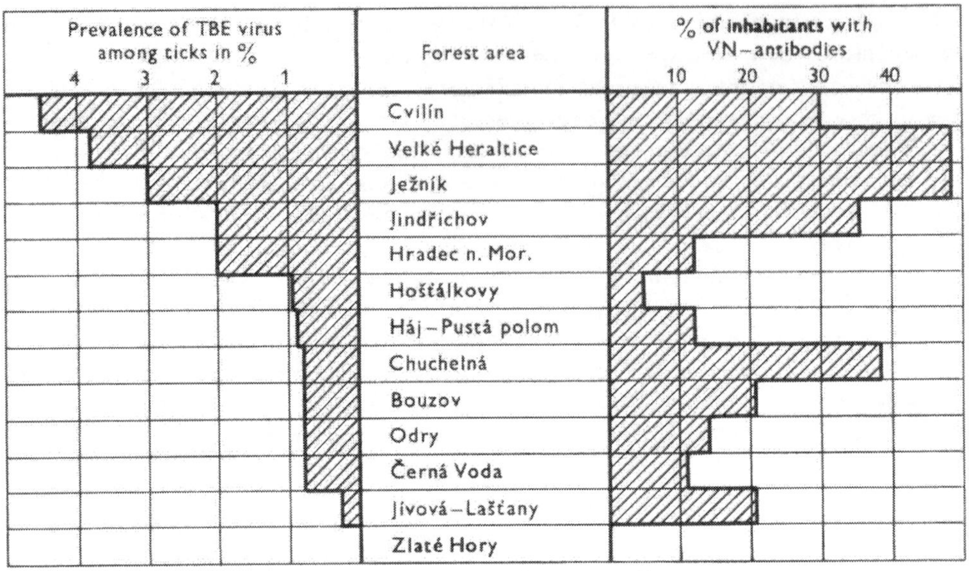

Fig. 2. The prevalence of TBE virus in ticks and the percentage of inhabitants with virus-neutralizing antibodies against the TBE virus in some natural foci in Northern Moravia.

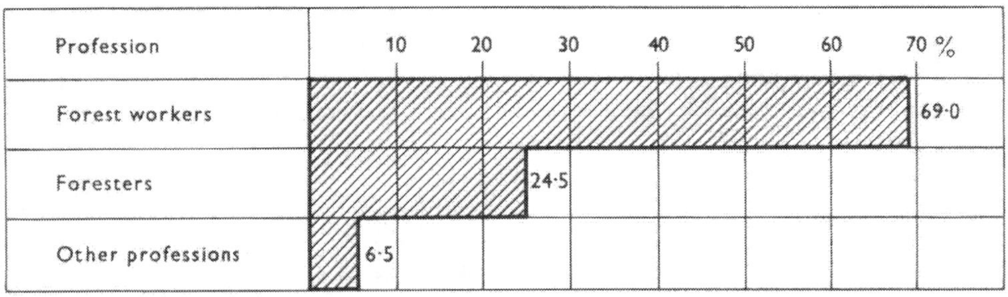

Fig. 3. Survey of occupational TBE cases according to professions.

wife) in comparison with 31 cases in the other wooded areas where no such measures were introduced.

Our research studies showed that the natural foci in wooded areas declared to be hazardous working environments and where preventive measures were observed, proved to be according to laboratory tests, of greatest valency in northern Moravia in the years 1964–1966. They studies confirmed the fact that all preventive measures were of great importance.. It should be noted that other diseases of the inhabitants of the above mentioned areas showed no change.

As to sanitation by means of acaricides, there is the possibility to use, for short-

time sanitation (three weeks), HCH in the quantity of 40 kg/ha, applied from ground spraying machines (Černý, Hodinářová, Kalinovská 1959). For a long-term sanitation DDT should be applied in early spring, during the final melting of snow, in the quantity of 50–75 kg/ha. Dusting is possible only when dealing with small areas; during the time of growth the biocenosis of forests may be adversely affected.

References

Ašmera, J., Šeděnka, B., Nedvídek, J.: Výsledky parazitologického průzkumu v přírodních ohniscích klíšťového zánětu mozku v bývalém Ostravském kraji. Čs. Parazitol. 9 : 5, 1962.

Ašmera, J., Heinz, F., Mlýnková, H., Opravil, E., Šeděnka, B., Kupec, V: Dosavadní výsledky průzkumu přírodních ohnisek klíšťové encephalitidy v Severomoravském kraji. Čs. epidemiol., mikrobiol., imunol. 18 : 4, 1969.

Blaškovič, D., Libíková, H.: Arbovirusinfektionen. In: Gsell O., Mohr W.: Infektionskrankheiten. Band I, Berlin, Heidelberg, New York, Springer-Verlag, pp. 151, 1967.

Černý, V., Hodinářová, D., Kalinovská, M.: Pokusy s hubením klíštěte obecného (*Ixodes ricinus* L.) v přírodě popraškem HCH. Čs. epidemiol., mikrobiol., imunol. 8 : 61, 1969.

Havlík, O.: Význam myších kalamit pro epidemiologii československé klíšťové encefalitidy. Čs. hyg., epidemiol., mikrobiol. 3, 300. 1954.

Henner, K.: Specialní neurologie. SZdN, pp. 491, Praha, 1952.

Nedvídek, J., Boschetty, V., Harudová, L.: Epidemiologie a klinika klíšťové encefalitidy v jednom přírodním ohnisku Ostravského kraje. Čs. epidemiol., mikrobiol., imunol. 7 : 9, 1958.

Rosický, B., Weiser, J.: Škůdci lidského zdraví. Praha. Přírodovědecké vydavatelství, 1952.

Rosický, B., Havlík, O.: Přírodní ohniskovost československé klíšťové encefalitidy. Sborník československá klíšťová encefalitis. SZdN, p. 23, Praha, 1954.

Rosický, B.: Poznámky k ekologii klíštěte *Ixodes ricinus* L. se zvláštním zřetelem k přírodním ohniskům nákaz. Věst. Čs. spol. zool. 18 : 41, 1954.

Rosický, B.:Notes on the Classification of Natural Foci of Tick-borne Encephalitis in Central and Southeast Europe. J. Hyg. Epidemiol., Microbiol., Imunol. (Praha), 2 : 431, 1959.

Rosický, B.: On the Structure and Development of Elementary Foci of Infections in Cultivated Areas. Advances in Biological Sciences. Academia, Praha, pp. 134, 1962.

Smorodintsev, A. A., Ilyenko, V. I.: Results of laboratory and epidemiological study of vaccination against tick-borne encephalitis. In: Biology of Viruses of the Tick-borne Encephalitis Complex. Academia, Praha, 332 pp. 1962.

INFECTIVITY OF *IXODES RICINUS* L. TICKS IN THE MIXED NATURAL FOCUS OF TICK-BORNE ENCEPHALITIS (TBE) AND UUKUNIEMI (UK) VIRUS

J. M. KOLMAN
Institute of Parasitology, Czechoslovak Academy of Sciences, Prague, Czechoslovakia

The primary isolation of numerous strains of the Uukuniemi virus – from *Ixodes ricinus* ticks – in 1963 in an area which was hitherto known as a natural focus of tick-borne encephalitis virus, initiated both the study of biological, physical, and chemical properties of the isolated strains of the Uukuniemi virus and the start of a systematic survey of the infectivity of ticks which are specific vectors of both viruses occurring together in this area. The results acquired by investigation of this area during the past four years are summarized here (Table 1).

Table 1. Results of Virus Isolation Experiments from *Ixodes ricinus* Ticks in the "Potepli" Region in the Years 1966–1969

Locality	Year	Number of ticks tested	Number of strains isolated			
			Strain UK	M.I.R.	Strain TBE	M.I.R.
Large experimental area	1967	1881	6	3.19	1	0.53
	1968	1858	3	1.61	1	0.53
	1969	2082	6	2.88	0	–
13 small control areas	1966	645	1	1.46	2	2.93
	1967	501	4	7.59	0	–
	1968	271	1	3.07	0	–
	1969	416	3	6.55	0	–
Trapping lines I–V	1967	1329	0	–	2	1.50
	1968	1446	4	2.76	0	–
Total		10.429	28	2.70	6	0.57

M.I.R. = minimal infection rate (Meyers et al., 1960).

The size and situation of the investigated area were chosen in such a manner as to characterize the whole area by its phytocoenology and fauna, its macro-relief, degree of soil humidity, microclimate, composition of geological setting, orientation to the cardinal points and by the degree of its isolation.

The ticks were collected from the defined area only by flag-dragging, *i.e.* from the so called large experimental area (extent 25,600 sq. metres) divided by marks into 256 squares (having 100 sq. metres each), from 13 small control areas (20 × 20 metres) and from 5 trapping lines (about 1,400 metres long) divided into sections 25 metres long and leading through the experimental area in such a manner, in order to cross nearly all important biotopes of the investigated area.

Collections of the ticks were carried out in spring and autumn of each year as follows: from each complex of 4 squares (400 sq. metres) of the large experimental area, from each of the 13 small areas, and from each of the 25 metres long sections of the trapping lines, 30 ticks (nymphs for the most part) were collected and tested by isolation tests in suckling 1–2 day old mice. In the case of positive isolation the viral strains were either immediately lyophilized or antigens were prepared of them for immunization and serologic identification; tests for re-isolation of the virus from the initial material – detritus of the ticks – kept on dry ice, were carried out.

As can be seen from the above survey, we isolated and identified in the mentioned localities of the investigated area 28 strains of the UK and 6 strains of TBE virus from 1966 until 1969. The minimal infection rate of ticks by UK virus was 79 % higher than by TBE virus in the assayed area.

In the large experimental area we succeeded by chance to follow the start and development of the dynamic process of the horizontal diffusion of infected ticks into the area. The start of this process was observed as early as 1966 outside the western border of the area, in the small control area designated as "I". In 1967 two further strains of the UK virus were isolated in the same small area and at the same time 6 strains were isolated at the western and northern border inside the large experimental area. In the following years the front of the diffusion wave distinctly moved along the diagonal of the area leading from its north-western to south-eastern corner. In 1969 the front of the diffusion wave spread over the half of the whole experimental area.

On the basis of evaluation of all hitherto known conditions established for survival and securing of circulation of both assayed arboviruses in this relatively thoroughly studied ecosystem during many years, it follows that these conditions ought to be entirely identical both for the TBE and the UK virus. However, the results of our study showed that this was not the case and that contemporarily the existence and circulation of the UK virus is far more favourably stimulated in the biotic and spacial structure of the investigated natural focus than TBE virus (Fig. 1, 2).

It is conceivable that our investigations – though extended over some years, but in relation to the above findings still temporarily limited – do not enable us at the present stage to submit even an approximate explanation of this phenomenon, nevertheless the acquired findings suggest further directions of our working effort and selection of methodical processes for further study of these very interesting problems.

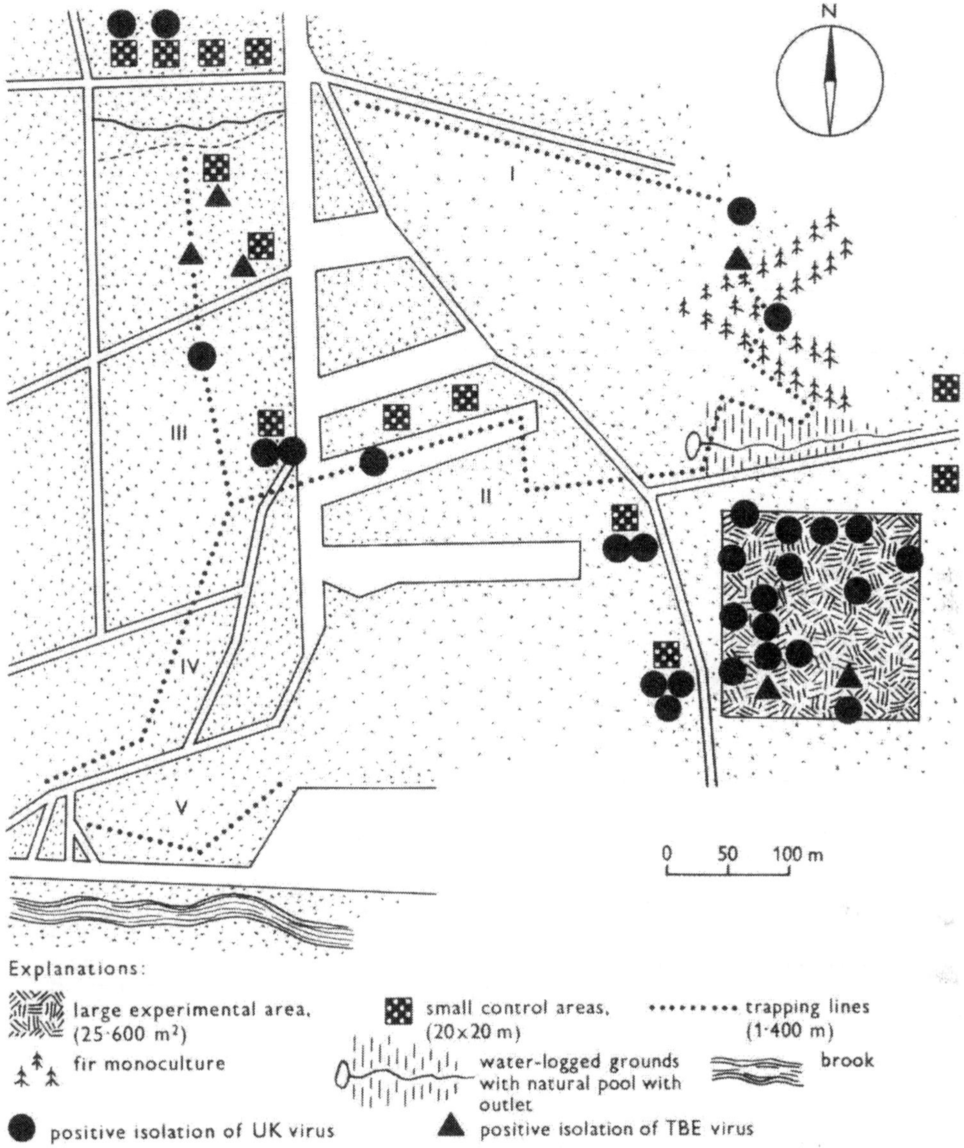

Fig. 1. Schematic map of the examined locality of Potepli; distribution of positive isolations of the UK and TBE virus.

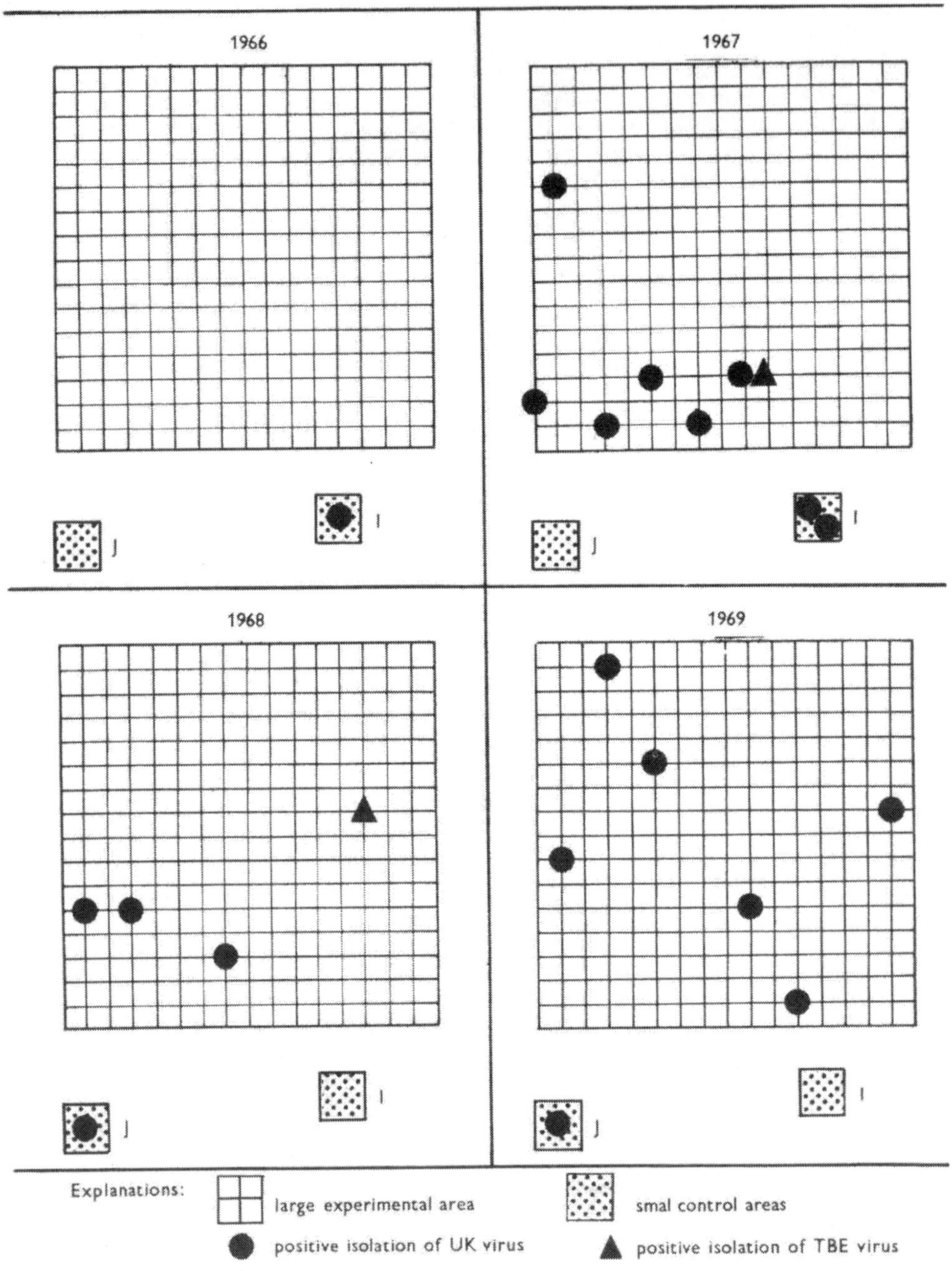

Fig. 2. Dynamics of horizontal diffusion of infected *Ixodes ricinus* L. ticks in the large experimental area in the years 1966–69.

References

Meyers, E.G., Loomis, E.C., Fujimato, F.Y., Ota, M.J., Lennette, E.H.: California encephalitis surveillance program mosquitovirus relationship. Am. J. Hyg. 71 : 368 – 377, 1960.

METHODOLOGICAL PRINCIPLES OF MAPPING THE OCCURRENCE OF IXODID TICKS

E. I. KORENBERG
The Gamaleya Institute of Academy of Medical Sciences,
Moscow, USSR

Without precise cartographic methods it is impossible to solve the problem of studying the occurrence of Ixodid ticks and the factors governing their distribution within a certain area. In our opinion, however, mapping the occurrence of ticks represents one of the divisions of zoological cartography – one of the newest trends in the special cartography of nature. The great practical significance of maps showing the distribution of Ixodid ticks has led, in particular, to the recent organization of a number of WHO Regional Reference Centres for studying the biology and occurrence of Ixodid ticks. The methodological principles of compiling different scale maps, showing the spatial structure of tick populations, are now being intensively elaborated and discussed. The aim of this paper is to sum up briefly certain results of this trend of work.

There are two main approaches:

1. The presence of ticks or the average level of their numerosity is indicated on the contours of landscape, geobotanical or other natural maps. It is usually considered that contours of the same type have a similar number of ticks. The compiled maps, in this case, reflect a very generalized parasitological (tick) characteristic of the territory.

2. An effort is made to illustrate the actual spatial structure of the tick population, which is the direct object of study and mapping. For different species in different parts of the range, and for differently scaled maps, they possibly may or may not correspond to the division of the territory in respect to other natural features.

In our opinion, the second trend is of particularly great interest to parasitologists, and that is why we shall examine it in greater detail. Special attention has been devoted to the mapping of the occurrence of *Ixodes persulcatus* P. Sch. – one of the most widespread species of Ixodid ticks in Eurasia, since data concerning other species are practically non-existent. An analysis of material was carried out beginning with large-scale work, gradually going over to investigations on a smaller scale, since this makes it possible simultaneously to touch upon certain questions of generalized phenomena depicted on the maps.

Large scale maps reflect the direct results of studying the characteristics of the distribution in certain parts of a comparatively small area. Data for maps on a detailed large scale (1 : 5,000–1 : 10,000) may be obtained by counting adult unfed ticks using any generally accepted method on the routes, whose dense network is comparatively evenly spread over the given area. The subsequent fixing of sites of one or several individuals on the same basis presents a good picture of the degree of tick distribution at the moment of mapping. This method may be employed to practically all meadow species of ticks.

Maps on a scale of 1 : 10,000–1 : 100,000 may reflect the actual indices of the abundance of arthropods obtained for one standard unit of counting, for example, the abundance of ticks for 1 flag-hour. A field survey is conducted by evenly spacing the counting points over the given territory. As regards *Ixodes persulcatus*, this method has been comparatively well described (Kucheruk et al., 1965; Zemskaya et al., 1965). In 1967–1968 we surveyed a territory of more than 25 km^2 on the border of the southern taiga and coniferous – broad leaved forests in the east of the East European plain. The field survey of the distribution of *Ixodes persulcatus* was conducted with count workload equal to an average of 5 flag-hours per 1 km^2. A comparison of its results with a large scale geobotanical map showed that the tick distribution did not coincide with the "mozaic" geobotanical cycles. This has been confirmed by the conclusions of a number of other investigators (Birulya, Zalutskaya, 1961; Zemskaya et al., 1965) who pointed out that the population of *Ixodes persulcatus* has its own spatial structure which does not coincide with the chorology of vegetative groups of the lower level.

In order to compile maps of a medium scale (1 : 300,000–1 : 1,000,000), in accordance with the generally acknowledged principles of special natural cartography, field surveys of tick distribution must be carried out on a large scale basis. However, it is practically impossible to carry out such a survey over the entire territory under study, and hence it is conducted on a random basis at so-called key points. The data gathered here must be projected over a considerably larger area. Thus, the key problem, the most complicated problem in medium scale mapping, is to elaborate the principles and practical methods of extrapolation. Moreover, the specific features of a medium scale map are such that the results of each actual count cannot express the level of grouping the tick populations, correlated with the scale of the map. Because of this, it is necessary to elaborate ways of transforming the factual material of a large scale field survey into data that can be depicted on medium scale maps.

As our investigations have shown and described in detail in special publications (Korenberg et al., 1967, 1969), elementary statistical methods allow us to group various variants of tick distribution on the territory at key points into several types. Moreover, each one corresponds to a definite level of average number of arthropods. Therefore, in our opinion, medium scale maps can be used for showing a territory characterized by one type of distribution, and as a result of this, a similar abundance of arthropods. The experience in compiling such a map has been described in detail in a separate paper (Korenberg, Karpenko, 1972 in litt.).

Small scale maps are compiled by summing up and generalizing the accumulated material in laboratory conditions. They can reflect the largest units of spatial differentiations in the population of a certain species. As in the case with the medium scale maps, success in this case depends on how well-founded such units are. So far, however, no investigations have been carried out in this direction. Such investigations must be considered extremely necessary, since it will be possible to divide the ranges of Ixodid ticks according to the level of their number and analyse the causes of their uneven distribution within the range as a whole.

Small scale mapping of Ixodid ticks has actually not gone beyond the stage of compiling the most primitive maps of the distribution of certain species; moreover, only the first steps have been taken in this direction. However, our experience in compiling general reference maps of the range of *Ixodes persulcatus* and *Ixodes ricinus* (Korenberg et al., 1969, 1971) and cadastral reference maps of the distribution of *Ixodes trianguliceps* (Korenberg and Lebedeva, 1969) convinces us that for the most common species of Ixodid ticks, such maps can already be compiled on the basis of the published information and on data available elsewhere.

The elaboration of different approaches to conducting field surveys and principles of compiling maps for nidicolous and certain other biological groups of arthropods remains an important and urgent problem in the cartography of Ixodid ticks.

References

Birulya, N.B., Zalutskaya, L.I.: On problem about interlandscape foci of spring-summer encephalitis. Tezisy Dokl. Soveshch. Vopr. Zool. Issled. Prirod. Ochag. Kleshch. Entsef., Obshch. Ispyt. Prirod., Moskva: 4–7 , 1961 (In Russian.)

Korenberg, E.I., Dzyuba, M.I. and Zhukov, V.I.: Range of *Ixodes ricinus* in the USSR. Zool. Zh., 50, 1 : 41–50. 1971 (In Russian.)

–, Karpenko, A.S.: Principles of mean scale charting of tick distribution. Zool. Zh., 51, 1972 (In Russian.)

–, Kucheruk, V.V., Pogorelenko, L.I., Suvorova, L.G., Kovalevsky, Yu.V.: Types of distribution of taiga ticks as the basis for division into districts of the territory of tick-borne encephalitis. In "Aktual. Probl. Spetsifich. Profilakt. Virus. Zabolev," Moskva: 143–144, 1967 (In Russian.)

–, –, –, –, –, et al.: Main features of tick distribution on vast territories and principles of its investigation (example of Udmurtsk forests). In "Tick-borne encephalitis in Udmurtia and Adjacent Areas", Izhevsk: 142–161, 1969 (In Russian.)

–, and Lebedeva, N.N.: Distribution and some general features of the ecology of *Ixodes trianguliceps* Bir. in the Soviet Union, Folia Parasitol. 16 : 143–152, 1969.

–, Zhukov, V.I., Shatkauskas, A.V. and Bushueva, L.K.: The distribution of *Ixodes persulcatus* in the USSR. Zool. Zh., 48, 7 : 1003–1014, 1969 (In Russian.)

Kucheruk, V.V., Zemskaya, A.A., Kovalevsky, Yu.V., Korenberg, E.I., Pchelkina, A.A., Suvorova, L.G., Tupikova, N.V.: Methods of study of large scale maps of the area structure of human natural foci diseases. In: "Methods of investigations in medical geography", Moscow: 169–177, 1965 (In Russian.)

Zemskaya, A.A., Suvorova, L.G., Tupikova, N.V.: Use of cartographical method in studying the distribution of *Ixodes persulcatus* in natural foci of tick-borne encephalitis. In: "Methods of investigations in medical geography", Moscow, 196–208, 1965 (In Russian.)

TICKS IN RELATION TO MAINTENANCE HOSTS OF ARBOVIRUSES IN CENTRAL EUROPE

O. KOŽUCH, J. NOSEK, E. ERNEK, M. LABUDA, J. CHMELA
Institute of Virology, Slovak Academy of Sciences, Bratislava
and Regional Hygiene Station in Olomouc, Czechoslovakia

Only those animal species, in which a viral infection induces a relatively long-lasting and high-titre viraemia, adequate to infect ticks, can be considered as maintenance hosts (reservoirs) of arboviruses in nature. To be effective, the reservoirs must occur in a biome frequently and in abundance, have a good population renewal, be highly susceptible to the virus and have a high tick load. The tick load of small mammals varies during the year with vector-activity and correlates indirectly with the density of hosts.

In animals with a short population renewal, only such populations are of importance as coincide with the maximum of seasonal activity of the vector. Periodical fluctuation of maintenance hosts may occur, e.g., small rodents which, during overcrowding, can essentially support the circulation of virus in many biotopes.

In the whole distribution area of natural foci of tick-borne encephalitis, small insectivores are of basic importance as very frequent maintenance hosts (reservoirs) of the virus, these are: the common shrew *Sorex araneus*, the European mole *Talpa europaea* and small rodents such as the bank vole *Clethrionomys glareolus* which are in some places associated with the yellow-necked mouse *Apodemus flavicollis* and the long-tailed fieldmouse *Apodemus sylvaticus* although the range of hosts acting as vectors of TBE virus was found to extend to more than 150 species of vertebrates.

On the basis of our lengthy investigations in the elementary foci of tick-borne encephalitis (TBE) in the Tribeč region and Northern Moravia it may be stated that 1. several viruses can circulate in an elementary focus (TBE, Tribeč, Uukuniemi), and 2. the average per cent of viruliferous ticks varies from year to year. Similarly, neutralizing antibodies against TBE virus in populations of small mammals vary during the year as well as during successive years. The average proportion of positive mammals amounts to 8.6 per cent and that of viruliferous ticks to 0.16 per cent (about 1 per cent in "microfoci") (see Table 1).

In the foci examined, hedgehogs, shrews, moles, bank voles, yellow-necked and field mice are very frequent hosts of ticks and are responsible for the maintenance of TBE virus in nature. These maintenance hosts were confirmed in our transmission experiments with ticks, by frequent isolations of virus from their blood, and in experimental infection.

Table 1. Neutralizing antibodies against TBE virus and isolated strains of TBE, Uukuniemi (U), Tribeč (T) viruses from small mammals and percentage of viruliferous ticks in an elementary focus Topolčianky

Year	Rodentia	Insectivora	% posit.	Number of isolated strains	Ticks	% posit.	Density of small mammals per hectare	Average infestation L	Average infestation N	Density of ticks per hectare (max. activity) L	N	A
1956[1]	0/92	0/1	–	1	–	–	–(65)x	4	< 1	–	–	–
1959	4/61	–	6,5	0	–	–	–(55)	17	< 1	–	–	–
1963[2]	depression	0/2	0	0	0/ 266	0	10(33)	< 1	< 1	–	–	–
1964[2]	0/30	1/11	2,4	0	2/ 527	0,4	25(30)	2	< 1	–	–	–
1965	4/28	11/46	20,2	2	2/1753	0,1	20(60)	20,3	1,2	–	–	–
1966	17/196	7/32	10,5	0	0/335	0	45(75)	5	< 1	–	–	–
1967	33/216	5/16	16,3	4	U1 2/667 T1	0,3	50(90)	2	< 1	300.000	48.000	16.000
1968	1/35	0/2	2,5	0	5/2083 T1	0,2	20(42)	1,5	< 1	380.000	36.800	16.000
1969	0/28	0/4	0	0	3/909	0,3	15(55)	2	< 1	380.000	31.000	26.500
1970	14/72	0	19,4	0	U1 2/2847	0,07	–(86)	< 1	< 1	265.000	80.000	3.300

x – Autumn; [1] and [2] – see references.

Frequence of infestation and infestation rates are most important for continual transmission of virus. On the other hand the varying population density of small mammals, mainly small rodents, influence infestation rate especially in the main periods of reproduction of small mammals – at the beginning of summer and in autumn.

The composition of a tick population varies from year to year in the same locality depending upon abiotic and biotic conditions. The relationship of a tick to the enviromental conditions is expressed in the length of its life cycle. The life cycle of the tick *Ixodes ricinus* in the examined regions lasts mainly three years, exceptionally two years. The population density of ticks was established by means of a carbon dioxide method. Although at present we do not know the critical densities of ticks and hosts and frequence of mutual relationships for the possibility of continual circulation of TBE virus in an elementary focus, the results from Topoľčianky (Tribeč) and Jívová (Northern Moravia) presented here indicate that the data given in table 1 do not seem to be the critical for the circulation of a virus, because the TBE virus was isolated either from ticks or mammals or its circulation confirmed serologically (with the exception of the year 1963). The population densities of small mammals and ticks in the Tribeč region are more stable than in Northern Moravia where the average proportion of positive mammals (calculated during 5 years) amounts to 14 per cent and that of viruliferous ticks to 0.2 per cent.

The differences in positivity of small mammals is caused probably by a different composition of the mammalian population (higher densities of the common shrew and the common vole).

Our observations on the frequence of infestation by *Ixodes ricinus* tick of marked small rodents agree with previous data as well. The infestation of individuals during the maximum of seasonal incidence of ticks increased during one day twice or more. It was further observed that the ticks collected from some places more often contained a virus. There are places where the home-ranges of different host species overlap in space and time. Isolation experiments from ticks and from the blood of marked mammals confirm the existence of the "microfoci" as well.

References

Ernek, E., Škoda, R.: Virologické poznatky z výskumu z oblasti Topoľčianky. Biológia 13, 481, 1958.

Grešíková, M., Nosek, J.: Isolation of tick-borne encephalitis virus from *Ixodes ricinus* ticks in the Tribeč region. Bull. Wl. Health Org. 36, Suppl. 1, pp. 67–71., 1967.

Kožuch, O., Grulich, I., Nosek, J.: Serological survey and isolation of tick-borne encephalitis virus from the blood of the mole (*Talpa europaea*) in a natural focus. Acta virol. 10, 557–560, 1966.

Nosek, J., Kožuch, O., Lichard, M., Ernek, E.: Synecology of the tick-borne encephalitis virus in the control area. (In Slovak). In Blaškovič et al.: Experimental study of immunization of domestic milk-giving animals with live attenuated tick-borne encephalitis virus. Biolog. práce SAV XV/5, p. 45, 1970.

CAUSES AND DYNAMICS OF CONTACTS BETWEEN RURAL INHABITANTS AND *IXODES PERSULCATUS* IN NATURAL FOCI OF TICK-BORNE ENCEPHALITIS

V. V. KUCHERUK
The Gamaleya Institute of Academy of Medical Sciences, Moscow, USSR

The majority of tick-borne encephalitis cases are rural inhabitants. A study of their contacts with natural foci of tick-borne encephalitis is carried out by making house to house enquiries among the inhabitants in certain communities. The questioning is carried out either periodically, at equal intervals (10–15 days), during the entire period of the activity of the ticks, or only once, namely after the termination of this period. A file with the results of the enquiry is compiled for every inhabitant. A certain number of questions are given in a simple form, requiring a simple answer. Besides general demographic data, the information for each person should include: dates, reasons and duration of each visit to the forest, location of the visited area and its distance to the nearest settlement, the number of ticks that attached themselves to the person during the visit. Usually, such enquiries are combined with random investigations of persons for immunity to tick-borne encephalitis.

On an average, 70–90 % of the rural inhabitants living within the range of tick-borne encephalitis visit the forest during the course of the year (Table 1). The frequency of visiting the forest is not determined by the geographical position of a given community, but by the distance from it to the nearest forest. In regions where large areas of land are cultivated, the percentage of inhabitants visiting the forest differs considerably for different communities (from 23 to 96 % in the Tatar ASSR, from 46 to 96 % in the Kirov region). And on the contrary, in those areas where the forests have been well preserved, the percentage of inhabitants visiting the forest varies very little in communities (from 85 to 97 % in Komi ASSR, from 80 to 95 % in the Chita region). During the summer season one rural inhabitant visits the forest in the European part of the USSR (Kirov region) from 3 to 12 times, while in central Siberia (Krasnoyarsk area) from 23 to 43 times (Popov, 1963; Fastovskaya and Prisyagina, 1968).

The frequency of contacts between rural inhabitants and ticks depends, to a great extent, on the geographical conditions affecting the community. In the mid-taiga the frequency of contacts is not high, while in the southern taiga and the coniferous broad-leaved forests it reaches a maximum level, while in the broad-leaved forests it drops again somewhat (Table 1). In the southern taiga every 8th–4th rural inhabitant notices that he or she has been bitten by a tick during the summer season.

Sub-zone	Type of forest	Area or region district	No. of communities investigated	No. of inhabitants questioned	Percentage of inhabitants having visited the forest			having had contact with ticks			having been bitten by ticks			Authors
					avg.	min.	max.	avg.	min.	max.	avg.	min.	max.	
Mid–taiga	Mid–taiga European spruce forests	Komi ASSR – several districts	12	1280	92	85	97	1.2	0	4.4	0.6	0	2.4	Vorontsova, Yegorova, 1969
	Mid–taiga central Siberian larch-pine and larch forests	Chita region, Tungukochensky district	6	2080	85	80	96	–	–	–	2.8	0	4.4	Gorin et al., 1967
	Mid–taiga sparse East Siberian larch forests	Krasnoyarsk area, Boguchansky district	3	1090	–	–	–	3.6	1.6	13.4	–	–	–	Grasis & Prisyagina, 1968
South taiga	South taiga European spruce forests	Kirov region, Malmyzhsky district	13	3200	63	46	96	11	7	21	6	4	12	Popov, 1963
	South taiga European spruce forests	Udmurt ASSR, several districts	35	7220	84	44	100	23	0.5	60	14	0	32	Kucheruk et al., 1969
	Mountain taiga West Siberian spruce-fir forests	Novosibirsk region, Toguchinsky district	–	more 2000	–	–	–	–	48	67	–	29	40	Prigorodov, 1966
	South taiga spruce-cedar-fir forests	Krasnoyarsk area, Birilyussky district	4	1920	–	–	–	73	54	100	25	16	74	Yarotsky, 1960
	South taiga birch and birch-pine forests	Krasnoyarsk area, Kozulsky district	5	1230	–	76	100	–	39	83	–	28	69	Fastovskaya & Prisyagina, 1968
	Mountain larch East Siberian forests	Chita region	–	–	81	–	–	–	–	–	27	–	–	Gorin et al., 1967
	Mountain taiga Far East spruce forests	Khabarovsk area, Obluchensky district	–	–	74	–	–	–	–	–	44	–	–	Volkov & Reznik, 1967
	Sub–taiga West Siberian aspen-birch forests	Omsk region, Sidelnikovsky district	2	440	–	–	–	35	–	–	20	–	–	Busygin et al., 1962
Coniferous-broad-leaved forests	European broad-leaved-spruce forests	Udmurt ASSR, several districts	10 10	2550 2550	71 71	49 49	93 93	30 30	10 10	58 58	21 21	6 6	50 50	Kucheruk et al., 1969
	Far East coniferous-broad-leaved forests	Khabarovsk area, Vyazemsky district	2	900	67	–	–	24	–	–	10	–	–	Levantovskaya et al., 1966
Broad leaved forests	Pri-Ural oak-lime tree forests and secondary birch forests in their place	Tatar ASSR, Naberezhno-Chelinsky district	16	6200	–	23	96	–	–	–	–	4.5	11	Lapshina & Gilmanova, 1967
	Pri-Ural oak-lime tree forests	Chelyabinsk region Ashinsky district	–	–	94	–	–	27	–	–	–	–	–	Voinov & Fyodorova, 1969
	Far East broad-leaved forests	Khabarovsk area, Vyazemsky district	2	900	74	–	–	55			21			Levantovskaya et al., 1966
	Far East broad-lived forests	Khabarovsky area, Bikinsky district	–	–	82	–	–	–	–	–	32	–	–	Volkov & Reznik, 1967

Table 1. Intensity of forest visits and the degree of contact of rural inhabitants with *Ixodes persulcatus* in various regions of the USSR

It is possible to designate a limited number of main reasons (7–9) which. when taken altogether, account for more than 90 % of the visits to the forest during the season of tick activity. Most of the reasons (cutting wood, looking after grazing livestock, making hay, berry-picking, picnics) are quite similar for the entire forest zone of the USSR. Another group of reasons (logging or doing field work, gathering wild garlic, fishing) are determined by actual economic or landscape conditions of certain regions. Some of the reasons are more or less constant for the entire season of tick activity, others (for example making hay, berry-picking or gathering wild garlic) coincide with a comparatively short period of time. The share of this or that main reason in the overall balance of visits to the forest by inhabitants of one or several neighbouring communities changes from year to year. This is determined by the actual ecological and economic situation for this or that year.

The degree of danger from visiting the forest for different reasons is not the same. In the Kirov region contact with ticks and bites by the latter were noticed most frequently during berry-picking, picnics and cutting wood. The infection of people with tick-borne encephalitis was usually linked with the same reasons. Contact with ticks was quite high during the grazing of livestock but their bites were comparatively rare. During the time of hay-making contacts with ticks are rare, and their bites are even more rare (Fig. 1). In the Udmurt ASSR stroll through the forest ended in tick bites in 15 % of the cases, cutting wood – 10 %, tending grazing livestock – 3 % and hay-making only in 2 % of the cases (Kucheruk et al., 1969). In the Krasnoyarsk area, during the gathering of wild garlic, 32–55 % of the visits to the forest ended with tick contacts and 13–19 % – tick bites. Corresponding indices for cutting wood are 13–26 % and 3–5 %, while for hay-making the figures are 0.5–2 % and 0.2–1 % (Fastovskaya and Prisyagina, 1968).

There are a number of factors determining the probability of contacts with ticks due to various reasons of visiting the forest. The most important factors are: main season for visits to the forest, mode of moving through the forest and actual sites visited, the intensity of work and the degree of attention given to the tick attacks.

The frequency of rural inhabitants' visits to a forest during the period of tick activity fluctuates considerably. The intensity of visits to the forest begins to increase immediately after the melting of the snow, drying of the soil and the beginning of the growing season. During the whole of the month of May and the first half of June, the intensity of forest-visits changes very little. In the second half of June there is a sharp increase in the number of visits to the forest due to the ripening of berries and the beginning of the hay-making period. The highest level of visits to the forest occurs in the middle and at the end of July. Because of this, and taking into account the declining tick activity, the high index of contact with ticks is maintained throughout the whole of June and the first of half of July (Fig. 2). The weather of a certain season determines, to a considerable degree, the overall volume of contacts of the rural inhabitants with the natural foci and the vectors of tick-borne encephalitis.

The information obtained from questioning the inhabitants makes it possible to evaluate the main parameters determining the intensity of contacts of rural inhabi-

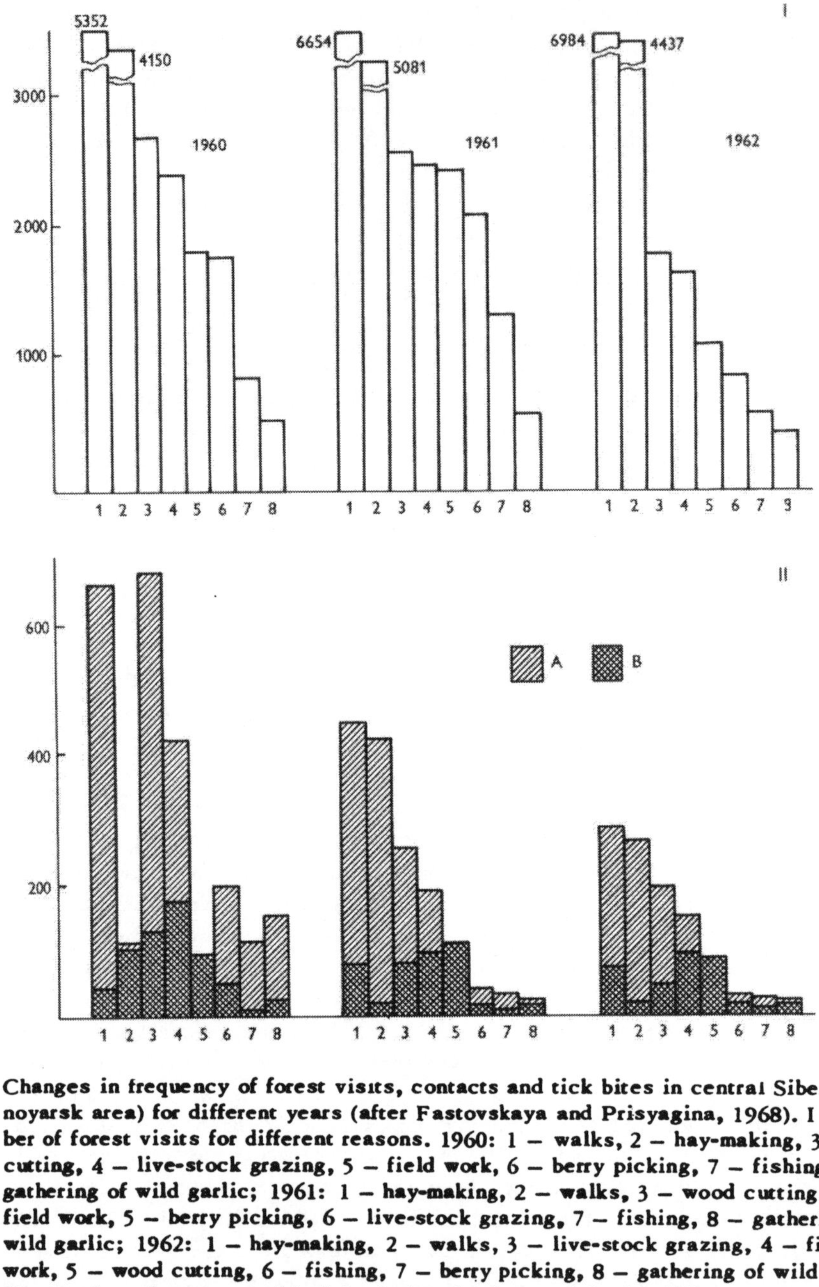

Fig. 1. Changes in frequency of forest visits, contacts and tick bites in central Siberia (Krasnoyarsk area) for different years (after Fastovskaya and Prisyagina, 1968). I – Number of forest visits for different reasons. 1960: 1 – walks, 2 – hay-making, 3 – wood cutting, 4 – live-stock grazing, 5 – field work, 6 – berry picking, 7 – fishing, 8 – gathering of wild garlic; 1961: 1 – hay-making, 2 – walks, 3 – wood cutting, 4 – field work, 5 – berry picking, 6 – live-stock grazing, 7 – fishing, 8 – gathering of wild garlic; 1962: 1 – hay-making, 2 – walks, 3 – live-stock grazing, 4 – field work, 5 – wood cutting, 6 – fishing, 7 – berry picking, 8 – gathering of wild garlic. II – Number of contacts with ticks during the forest visits for different reasons. A – Number of persons who came into contact with ticks, B – number of persons to whom ticks became attached.

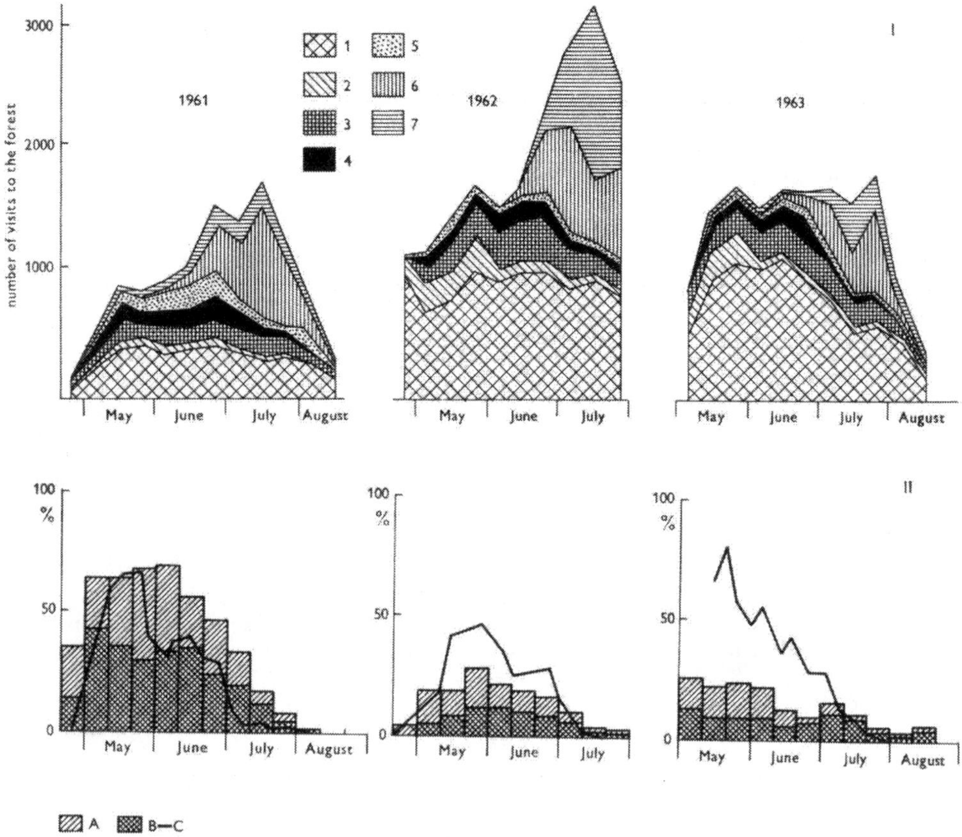

Fig. 2. I – Seasonal dynamics and reasons for the visits to the forest in the Kirov region. 1 – timber cutting and forest planting, 2 – walks, 3 – live-stock grazing, 4 – wood cutting, 5 – wood-cultivating, 6 – berry picking, 7 – hay-making. II – Seasonal changes in the percentage of persons noticing tick attack and tick attachment during their visits to the forest. A – % of persons in contact with ticks; B – % of persons bitten by ticks; C – seasonal activity of ticks in continuous route.

tants with natural foci and the vectors of tick-borne encephalitis. These parameters may be used as the basis for building mathematical models to evaluate the risk of people becoming infected with tick-borne encephalitis (Lvov et al., 1968; Bolshov Goldfarb, 1970).

References

Bolshov, L.N., Goldfarb, L.G.: A study of the epidemic process in tick-borne encephalitis. 2. Mathematical model. In: "Questions of epidemiology and prevention of tick-borne encephalitis", Moscow, 154–171, 1970. (In Russian.)

Busygin, F.F., Prigorodov, V.I., Shaiman, M.S.: Contact of the population in rural settlements in the endemic areas of Novosibirsk and Omsk regions. Abridged report to the enlarged session of the Committee for control of tick-borne encephalitis, Omsk 18–21, 1962. (In Russian.)

Fastovskaya, E.L., Prisyagina, L.A.: Contacts of the population with natural foci and vectors of tick-borne encephalitis. In: "Questions of epidemiology of tick-borne encephalitis and biological laws in its natural focus, Moscow, "Meditsina", 344–360, 1968. (In Russian.)

Gorin, O.Z., Lvov, D.K., Vasenin, A.A.: Estimating the risk of infecting the population of East Trans-Baikal with tick-borne encephalitis on the basis of the data about immunological structure. In: "Materials of the Problem Commission of the AMS USSR. Poliomyelitis and viral encephalitis." 2nd Ed., Arboviruses, Moscow, 34–45, 1967. (In Russian.)

Grasis, V.K., Prisyagina, L.A.: Immunological state of the population in certain regions of the Krasnoyarsk area. In: "Questions of epidemiology of tick-borne encephalitis and biological laws in its natural focus", Moscow, "Meditsina", 329–343, 1968. (In Russian.)

Kucheruk, V.V., Korenberg, E.I., Shulepova, T.G., Lebedev, I.P., Smirnov, A.V.: The basic results from studying the natural focality, epidemiology and preservation of tick-borne encephalitis in the Udmurt ASSR during recent years. In: "Tick-borne encephalitis in Udmurtiya and neighbouring areas", Izhevsk, 229–235, 1969. (In Russian.)

Lapshina, G.N., Gilmanova, G.H.: On the method of determining the epidemiological potential of natural foci of tick-borne encephalitis. Materials of scientific conference at the Kazan Inst. Epodemiol. Microbiol., Kazan, 60–62, 1967. (In Russian.)

Levantovskaya, S.I., Bereta, L.A., Trop, I.E., Feyerstein, M.S., Okunzova, E.V.: Material from studying population's contacts with ticks in two types of foci of tick-borne encephalitis. Scientific notes Khabarovsk Sci. Res. Inst. Epidemiol. Microbiol. 8 : 51–55, 1966. (In Russian.)

Lvov, D.K., Bolshov. L.N., Rudik, A.P., Goldfarb, L.G., Kruopis, Yu.I.: Experience in estimating the intensity of infection with tick-borne encephalitis. Med. Parazitol. Parazitar. Bol. XXXVII, 3 : 274–279, 1968. (In Russian.)

Popov, V.F.: Epidemiology of tick-borne encephalitis in the southern areas of Kirov region Abridged Cand. Diss., Moscow, 1963. (In Russian.)

Prigorodov, V.I.: Tick-borne encephalitis in rural localities. Abridged Cand. Diss., Moscow., 1967. (In Russian.)

Voinov, I.N., Fyodorova, N.V.: A study of the contacts of the population of Chelyabinsk region with the vector of tick-borne encephalitis. In: "Natural focal diseases of the Urals, Siberia and the Far East", Sverdlovsk, 63–65, 1959.

Volkov, V.I., Reznik, B.I.: Immunological structure of the population in foci of tick-borne encephalitis of the Khabarovsk area. In: "Materials of the Problems Commission of the AMS USSR. Poliomyelitis and viral encephalitis". 2nd Ed., Arboviruses, Moscow, 24–26, 1967. (In Russian.)

Vorontsova, T.A., Yegorova. A.D.: Investigating the risk of infection by tick-borne encephalitis in Komi ASSR. In: "Tick-borne encephalitis in Udmurtiya and neighbouring areas", Izhevsk. 229–235, 1969. (In Russian.)

Yarotsky, L.S.: Material on epidemiology of tick-borne encephalitis in its endemic focus in south-east of Prichulimye. Med. Parazitol. Parazitar. Bol. XXIX. 1 : 15–27, 1960. (In Russian.)

TICKS AS VECTORS OF TICK-BORNE ENCEPHALITIS (TBE) VIRUS IN EUROPE

J. NOSEK and D. BLAŠKOVIČ
Institute of Virology, Slovak Academy of Sciences,
Bratislava, Czechoslovakia

In the northern coniferous and temperate deciduous forest which spreads from west of the Urals to the Atlantic Ocean, the common tick is *Ixodes ricinus* (Fig. 1). This tick is the main vector of TBE virus western subtype in the European part of the USSR, Scandinavian countries, Western, Central and Southern Europe and possibly also in Greece and Turkey. We found that the other tick-species – *Haemaphysalis concinna, H. inermis, H. punctata, Ixodes hexagonus, Dermacentor marginatus* and *D. reticulatus* – only acted exceptionally as vectors in certain biomes or areas.

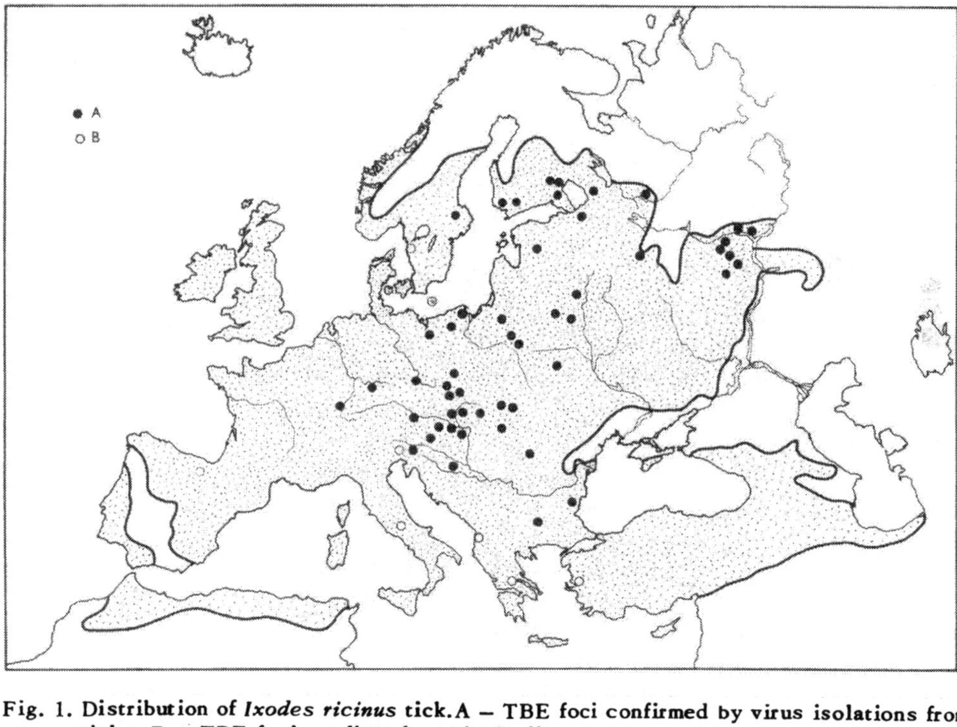

Fig. 1. Distribution of *Ixodes ricinus* tick. A – TBE foci confirmed by virus isolations from ticks, B – TBE foci confirmed serologically.

Numerous virus isolations have been made from the nymphs and/or adults (males and females) out of six of about 50 species occurring in Europe, the bulk of the isolations being from *Ixodes ricinus*.

In Central Europe the tick *Ixodes ricinus* shows a characteristic curve of seasonal dynamics. This is a two-peaked curve with the higher peak in April and the beginning of May and a lower one in September and October. The three- or two-year developmental cycle at lower elevations may be extended to four, five or six years, especially in higher altitudes. This complicated and long lasting development is resposible for so-called stockpile of ticks in nature. Population density and also the proportion between the number of nymphs and adults in particular regions is considerably different. These observations have demonstrated again the fluctuation of the numbers of *I. ricinus* and the proportion between the nymphs and adults in the same locality during the year and during successive years. Dependence on biotic factors is well expressed by the association of ticks with certain plant communities.

According to our observations in Central Europe the plant communities of oak tier, beech-oak tier, oak-beech tier, beech tier, beech-spruce tier and of alluvial plains are characterized by the highest population density of ticks. The composition of plant communities is as follows: oak tier: *Tili-Querceta* and *Ulmi-Fraxineta carpini, Corni-Querceta pubescentis* and *Corni-Querceta carpini, Carpini-Querceta* and *Ligustri-Querceta, Aceri-Campestris Querceta, Querceta petrae*, locust-tree-coopice *Robinieta*; beech-oak tier: *Tili-Querceta, Betuli-Querceta, Corni-Querceta carpini superiora, Fagi-Querceta Tiliae, Querceta-Fagina inferiora*; oak-beech tier: *Corni-Querceta carpini superiora, Fageta-Quercina, Querci-Fageta tiliae*; beech tier: *Fageta typica* and *Tili-Fageta*, Alluvial plains: *Fraxino-Alneta, Fraxini-Querceta roboris, Ulmi-Fraxineta pop·li, Ulmi-Fraxineta carpini.*

The associations of particular tick species with relevant plant communities enables the scientific prognosis for the occurrence, density and stockpile of viruliferous ticks as a potential source of human infection and disease. The main purpose for indicating these parameters is the mapping of these data.

Considering our knowledge of the ecology of TBE virus and its vectors the medical surveillance program evaluating ecological studies should be based upon:

- the mapping of population densities of the main vector and its seasonal incidence,
- the mapping of repartition of tick synusies and main host distribution,
- the establishment of critical density of vector populations, which is significant for the exposure of man to the infection.

The possible geographical distribution of TBE can be inferred on the basis of the established distribution of its known vectors. Plotting the population density, seasonal incidence and composition of tick synusies is also an important aspect of both disease transmission and control studies. When geographical distribution is correlated with the known seasonal activity of the tick vector, prediction can be made as to when tick-borne encephalitis can be expected to occur as well as where they may occur.

The transstadial development of TBE virus in its specific vectors depends on a variety of intrinsic and extrinsic factors. During the feeding of a tick on a viraemic host the intense cell activity associated with digestive processes may favour virus multiplication and rapid infection of salivary glands. The proportion of ticks becoming infected depends upon the height of viraemia in the host at the time of tick feeding (in the tick's previous stage). The infection threshold necessary to infect a certain percentage of ticks has been determined for various tick species During the premoulting period the virus multiplies in ticks. However, after moulting, an ecliptic phase occurs which may last for several weeks depending on the tick species (it is relatively short in *Dermacentor reticulatus*, 10 days, longer in species with a long life cycle, e.g., *Ixodes ricinus* – a few weeks) and temperature. The favourable conditions for infection in "short-feeding" ticks as, e.g. larvae and nymphs of *Haemaphysalis inermis* (for 30' –120') exists if the feeding coincides with the high level viraemia. Conversely, in "slow-feeding" ticks this possibility is much higher.

The active transmission of the TBE virus may be either mechanical or biological (involving development in vectors), The virus multiplies in the vector and is subsequently transmitted to the vertebrate host by means of infectious saliva during feeding. After the tick-bite, the virus multiplies in the local susceptible cells, and is further spread to the regional lymphatic glands from where it goes into the blood stream via the lymphatic routes thus producing viraemia.

The TBE virus survives in an active state in tick instars for long periods, for instance we obtained in adults of *I. ricinus* the survival of TBE virus for at least nine months but only for 4–6 weeks in *D. reticulatus* and *D. marginatus* nymphs. This persistence of virus in vectors is advantageous for the successful survival of the virus in nature during unfavourable climatic conditions, e.g., during the winter period in Europe.

The average proportion of viruliferous ticks in Central European TBE foci varies a great deal, from 0.1 to 3 per cent, about 1 per cent in microfoci. The high population density of *Ixodes ricinus* in many regions is associated with the grazing of farm animals or intense game management. Dispersal of ticks has a mosaic-like pattern, it is rarely uniform. Ecotons may play an important role in the epidemiology of TBE, also the formation of new biotopes by man's activities is very important in the case of pastures situated on hills in a cultivated area. Pastures represent disclimax communities. Although *Ixodes ricinus* tick has almost been eliminated by man's activities in some biotopes, their population density remains high in others.

Experimental transmission of TBE virus has been obtained through the above mentioned species including *Ixodes arboricola*. The evidence of transovarial transmission occurring in nature is lacking but such transmission has been obtained under some unusual circumstances and may play a minor role.

DERMACENTOR MARGINATUS AND *D. RETICULATUS* TICKS, POSSIBLE VECTORS OF TICK-BORNE ENCEPHALITIS VIRUS IN CENTRAL EUROPE

J. NOSEK, O. KOŽUCH and M. LABUDA
Institute of Virology, Slovak Academy of Sciences,
Bratislava, Czechoslovakia

Dermacentor marginatus and *D. reticulatus* ticks have an ecologically limited and mosaic-like pattern of distribution, but the ecological conditions for each species are quite different. The biotopes of both species are under direct and continuous human influence. Both species occur in tick-borne encephalitis foci of boskematic and therio-boskematic type, where they are members of different tick synusies.

Dermacentor marginatus and *D. reticulatus* ticks were confirmed as vectors of tick-borne encephalitis virus in our previous transmission experiments (Kožuch and Nosek, 1971). The habitats of *D. reticulatus* tick are often inundated. The survival of *D. reticulatus* ticks under water was reported by Mačička, Nosek and Rosický (1956), recently by Honzáková (1971). In order to find out whether viruliferous ticks can also survive a certain period in water, the following experiments were carried out.

Laboratory bred *D. reticulatus* larvae, in special capsules, were allowed to feed on white mice. The mice, weighing 6–8 g, were infected intraperitoneally with 0.1 ml of 10 % brain suspension of TBE virus, hedgehog strain J 13 from the 9th mouse passage. The virus in 10 % suspension was 10^8 mouse intracerebral LD_{50} per 0.03 ml. The larvae of *D. reticulatus* fed on simultaneously infected white mice for 4 days. The larvae were infected on the 3rd May. After feeding they were immediately tested for the presence of the virus and were confirmed as viruliferous (100 per cent). They were kept in a thermostat at a temperature of 22 °C – 24 °C and 90 %– 95 % relative humidity. The engorged larvae were periodically exposed to light.

The premoulting period in *D. reticulatus* nymphs lasted 12 days. Ten days after hatching – 25th May – the nymphs, 100 in number, were placed on a *Carex hirta* plant in two glass tubes fixed on each side with fine nylon net. These containers were placed under water at a temperature ranging from 14 °C–17 °C between meadow vegetation on the sandy soil. After 3 days half of the nymphs had survived. These nymphs were tested immediately for the presence of virus parallel with nymphs kept in a thermostat and the rest of the submerged nymphs were used for transmission experiments on the white mice. Adults which hatched from these nymphs were used in the next similar experiment. The premoulting period in adults lasts 20 days and the examination for presence of virus was carried out on July, 9.

The nymphs and adults were tested individually and the suspensions were prepared in Earl solution containing 10 per cent heated calf serum with added penicillin and streptomycin. After clarification, each pool was inoculated into one litter of suckling mice 1–4 days old. The positive samples were titrated by the intracerebral inoculation of white mice, weighing 6–8 g, with 0.03 ml volumes. The results of the individually tested nymphs and adults are summarized in Table 1.

Table 1. Titer of the virus in nymphs and adults after 3 days immersion and in ticks kept in a thermostat

Water		Thermostat	
Number of ticks	Titer per 0.03 ml ic mouse LD_{50}	Titer per 0.03 ml ic mouse LD_{50}	Number of ticks
1 n	3.5	1.0	1 n
2 n	4.0	0	2 n
3 n	4.0	0	3 n
4 n	4.0	0	4 n
5 n	3.0	0	5 n
6 n	2.0	4.0	6 n
7 n	4.0	< 1	7 n
1 ♀	3.7	4.0	1 ♂
2 ♀	4.0	4.0	2 ♂
3 ♀	3.6	4.0	3 ♂
4 ♀	3.6	5.0	4 ♂
5 ♀	4.5	4.5	5 ♀
6 ♀	4.0	4.0	6 ♀
7 ♀	4.5	4.5	7 ♀
8 ♂	4.0	1.5	8 ♀
9 ♂	< 1	< 1	9 ♀
10 ♂	3.0	4.0	10 ♀
11 ♂	4.0	3.7	11 ♀
12 ♂	4.0	4.5	12 ♀
13 ♂	3.6		
14 ♂	4.5	Number of examined adults	24
15 ♂	4.5	Number of positive	12
16 ♂	3.0		
17 ♀	4.0	Examined submerged nymphs	7
18 ♀	4.5	Positive nymphs	7
19 ♀	4.0		
20 ♂	3.6	Examined nymphs	7
21 ♂	3.0	Positive nymphs	3
Examined submerged adults	44		
Positive adults	21		

The transmission rate in nymphs was 75 per cent. Conversely, the transmission rate in nymphs which were kept in the thermostat was 60 per cent only. The survival of adults was 100 per cent. The infection rate in submerged adults and in adults kept in the thermostat was 48 or 50 per cent respectively.

This experiment indicates that the nymphs and adults can survive the brief summer flood without loss of infectivity. Adults can survive in cool and clean water for more than 100 days; conversely, in water with organic residue (leafy litter) they survive for about only one month (Honzáková, 1970).

The very high infection rate in submerged nymphs may be explained by the influence of carbon dioxide which can accelerate their metabolism and virus multiplication in ticks. The second explication of this phenomenon may be in the selection of less viable nymphs in a water environment and the survival of individuals with high metabolism.

On the basis of our experience we support the view that nymphs are most important for the maintenance cycle of TBE virus in nature and they can essentially support the circulation of the virus at the beginning of summer in some localities. Biological transmission of TBE virus with *D. marginatus* and *D. reticulatus* was demonstrated in our experiments but other aspects of transmission such as the dynamics of virus multiplication in ticks, the persistence of virus in ticks have still to be investigated.

D. marginatus and *D. reticulatus* ticks, as members of different vector synusies, can support the maintenance of TBE virus in nature in certain areas, but their role as vectors in Central Europe is relatively small. They can transmit the TBE virus to grazing milk-producing animals, e.g., goats, which may result in an alimentary infection in man.

References

Honzáková, E.: Survival of some Ixodid tick species submerged in water in laboratory experiment. Folia parasitol. (Praha) 18 : 155–159, 1971.

Kožuch, O., Nosek, J.: Transmission of tick-borne encephalitis (TBE) virus with *Dermacentor marginatus* and *D. reticulatus* ticks. Acta virol. (In press.)

Mačička, O., Nosek, J., Rosický, B.: Observations on the bionomics, life-cycles, and economic importance of the tick *Dermacentor pictus* Herm. in Central Europe. (In Slovak, Russian and German summaries). Biologické práce II/12, 49. Bratislava: Publishing House of Slovak Academy of Sciences, 1956.

THE OCCURRENCE OF THE TICK *HAEMAPHYSALIS CONCINNA* KOCH IN RELATION OF TICKBORNE ENCEPHALITIS

W. SIXL
Zoological Institute, University of Graz, Austria

In Austria *H. concinna* is present in the provinces: in Burgenland, in Niederösterreich, in Oberösterreich. This species prefers such localities as moist, light, scattered woods, with a well developed undergrowth, furthermore woodclearings, edges of woods, the biotope of reed bents in the lowlands (around the Neusiedler Lake) and the water-meadows of the Danube. The preferred habitats of these ticks are related to specific plant eco-systems, namely: *Phragmites communis, Cariceto-remotae-Fraxinetum, Querco Carpinetum, Querco luculetum, Quercetum cerris* and the moist steppe biotopes in the environment of the Neusiedler Lake and the neighbouring Zick lakes.

The adults of *H. concinna* are active from the end of March to the end of July. The nymphs are present from the middle of April to October; however, the larvae are living from the end or middle of May to the middle of October. The adults show the highest activity from June to the beginning of July.

According to the investigations in Austria and Czechoslovakia it can be said that *H. concinna* parasitizes mammals and reptiles but clearly prefers birds. The pattern of the distribution is linked with this special preference (Fig. 1). Therefore *H. concinna* has a scattered distribution. Only in the Leitha Mountains and in the east of the Neusiedler Lake the distribution is general. The development from egg to adult takes three years.

In Central-Europe, Austria, Czechoslovakia, and Hungary 87 hosts have been found to be parasitized by all three stages of the tick (Nosek, 1971). I have observed the following hosts in Austria: Adults and nymphs prefer *Cervus elaphus, Capreolus capreolus, Sus scrofa, Vulpes vulpes, Canis familiaris, Mustela erminea, Mustela nivalis, Erinaceus roumanicus, Lepus europaeus, Oryctolagus cuniculus, Apodemus flavicollis, Apodemus sylvaticus, Clethrionomys glareolus, Acrocephalus melanopogon, Locustela fluviatilis.*

Larvae infested: *Acrocephalus melanopogon, Locustela fluviatilis, Apodemus flavicollis, A. sylvaticus, Mustela erminea.* Larvae and nymphs have also been collected from the burrows of *Citellus citellus* by means of the CO_2-method. In other provinces of Austria all stages have been found on deer (*Cervus elaphus*) roe (*Capreolus capreolus*) and even on the Alpine goat (*Rupicapra rupicapra*).

The larvae wait for the host underneath the broad leaves of grasses, shrubs and on deciduous saplings. The nymphs sit on the tips of the grasses and on the surface

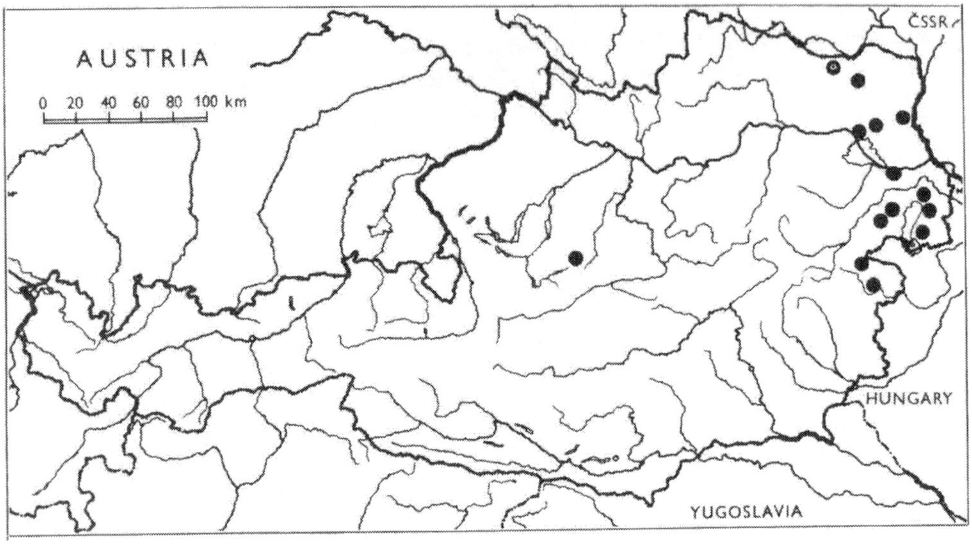

Fig. 1. The distribution of *Haemaphysalis concinna* in Austria.

of leaves up to a height of 60 cm, forming groups with the adults. The adults wait for the host in the same place as the nymphs, congragating to a maximal height of about 100 cm. An exception is that nymphs and adults can be found on the tips of reeds reaching a height of two meters.

In direct sunlight they gather on the dark shaded underside of leaves and grasses. In the environment of the Neusiedler Lake and in the Leitha Mountains field investigations were carried out in 1969, 1970, 1971. They showed that the activity of *H. concinna* is very dependent on relative humidity. Active stages were observed on plants at ecoclimatic temperatures from +20 °C to +40 °C and a high relative humidity (90–100 %). When the relative humidity is lover, the stages are inactive on the plants while they also remain immobile at 40 °C. In particular the activity increases significantly before and after thunderstorms. Optimal activity occurs between 15 °C and 20 °C, and at a high relative humidity; this we confirmed by experimentation. All stages are active throughout the day and the whole night, when the relative humidity is sufficiently high. According to our investigation at the Leitha Mountains site, the phases of highest activity during the day occur between 6–9 a.m. and 5–9 p.m. which coincides with the periods of highest relative humidity (Fig. 2).

The preferred parasitopes or locations on the birds are the regions around the rostrum, the eyelids and the ears. Larvae and nymphs infest the extremities of small mammals, the penis, the abdomen and the head. In Artiodactyla these stages parasitize the phalanges and the head region. The adults prefer the neck, the legs and the abdomen of roe, Alpine goat, deer and wild boar.

In all stages *Hunterellus hookeri* has been found as a parasite. In Europe larvae and nymphs of *H. concinna* often parasitize animals infected with TBE-virus, They

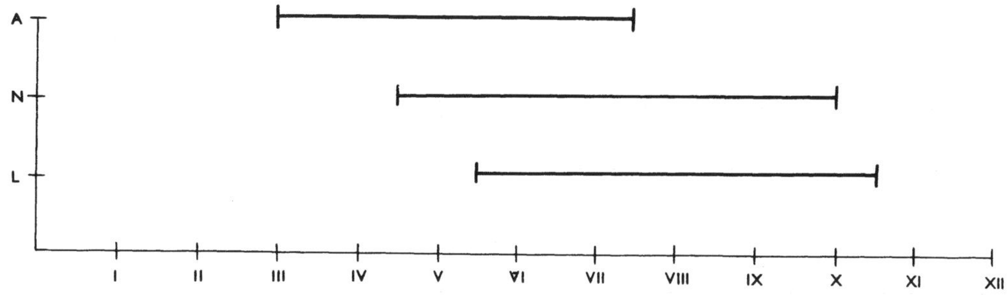

Fig. 2. Seasonal occurrence of *Haemaphysalis concinna*: L = larvae, N = nymphs, A = adults.

are present in the plant-ecosystems which represent typical foci. *H. concinna* forms synusies with *Dermacentor reticulatus*, *Ixodes trianguliceps* and *Ixodes ricinus* in the region of Neusiedler Lake and Sieggraben, and with *I. ricinus*, *I. hexagonus* and *I. redikorzevi* at Parndorf. It is known that some of these species are vectors of the TBE-virus. During our investigations in the Leitha Mountains (Schützen and Donnerkirchen) three strains of tick-borne encephalitis-virus were isolated in collaboration with the Virological Institute of the Slovak Academy of Sciences in Bratislava. This is the first time that *H. cocinna* has been established as a vector of TBE-virus in Central Europe. In the region of the Neusiedler Lake *H. concinna* is probably the main organism resposible for maintaining the TBE-virus in nature. This is particularly interesting in relation to the high density of this tick found at the Leitha Mountains (Schützen) and the Neusiedler Lake sites (Fig. 3). *Quercetum cerris* is the ecological focal point. It is found in warm localities and forms a park-like wood with humid areas and water-shed which drains at Schützen. The focus represents a theriodic type and the most important animals for the maintenance of the population density of ticks are: fallow deer, deer, roe, wild boar, hare and fox as well as small mammals.

Our results confirm that *H. concinna* is a possible vector of TBE-virus in certain regions of Europe. Therefore, accurate investigations of the distribution and of the host-spectrum of this particular tick species are required.

References

Nosek, J.: The Ecology, Bionomics and Behavior of *Haemaphysalis (Haemaphysalis) concinna* Tick. Z. Parasitenk. 36, 233–241, 1971.

Riedl et al.: Isolierung des Zeckenencephalitisvirus (TBE-virus) aus der Zecke *Haemaphysalis concinna* Koch. Archiv Hyg. (In press.)

Sixl, W. et al.: Heimische Zecken I. Mitt. naturw. Ver., Steiermark, 99 : 218–219, 1969.

Fig. 3. Biotope *Haemaphysalis concinna* near the Neusiedler lake.

Section IX B

MITES AND TICKS IN NATURAL FOCI OF OTHER DISEASES THAN TBE

Chairman
B. Rosický (Czechoslovakia)

Assistant Chairman
J. Nosek (Czechoslovakia)

Proceedings of the 3rd International Congress of Acarology, Prague, 1971

BURROW TICK *IXODES TRIANGULICEPS* IN NATURAL FOCI OF HAEMORRHAGIC FEVER WITH RENAL SYNDROME IN BROAD-LEAVED FORESTS OF EASTERN EUROPE

A. F. KATELINA, Yu. A. MYASNIKOV, T. V. PANINA
Regional Sanitary-Epidemiological Station, Tula
The USSR AMS Institute of Poliomyelitis and Virus Encephalitides,
Moscow, USSR

Natural foci of haemorrhagic fever with renal syndrome (HFRS) occur widely in mixed and broad-leaved forests of Europe, from Scandinavia, Czechoslovakia and Yugoslavia in the west to the Urals in the east. The commonly recognized source of infection in these foci are voles of the genus *Clethrionomys*; however, the problem of the modes of circulation of the causative agent in natural foci remains obscure since up to now there are no methods of cultivation in the laboratory of the causative agent of this disease. The greatest importance in the circulation of the agent among rodents was assigned to gamasid mites (Preobrazhensky et al., 1970). Japanese and Soviet workers succeeded in demonstrating infection of these mites with HRFS virus (Chumakov et al., 1965). Traub et al. (1954) believed that trombiculid mites could also transmit the agent among rodents. It has been recently suggested that this function may be performed by burrow ticks *Ixodes trianguliceps* which are quite abundant in natural foci of infection (Bashkirev et al., 1965; Katelina, Panina, 1966).

For 20 years (1951–1970) we have been studying the biology of *Ixodes trianguliceps* ticks in the broad-leaved forests of the Tula region where active natural foci of HFRS are localized. During the observation period over 45 thousand small mammals of 15 species have been examined and 2305 *I. triangulic.eps* ticks have been picked up from them. This species made up 69.8 % of all ixodid ticks collected from the animals i.e. dominated absolutely over other species. From 31 774 bank voles *(Clethrionomys glareolus)* 1560 ticks *I. triangulicebs* were collected. Both the number of bank voles examined and the number of *I. triangulicebs* ticks collected from them constituted about 70 % of the total number of examined animals of all species and burrow ticks collected from them.

Other animals of 7 species (harvest mouse, birch mouse, mole, field vole, house mouse, lesser bicolored white-toothed shrew, water vole) were caught in small numbers (less than 2 thousand (3 %)) and no *I. triangulicebs* ticks were picked up from them.

The index of occurrence (number of animals with ticks per 100 animals caught) of *I. triangulicebs* ticks on bank voles varied in some years from 1.2 (1969) to 10.0

Table 1. The rate of infestation of some species of small mammals with *I. triangulicefs* ticks

Species of animals	Number of animals examined	Index of species domination	Index of occurrence	Index of abudance (per 100 animals)
Clethrionomys glareolus	31 774	70.1	2.8	4.8
Microtus arvalis	20 044	6.7	1.8	5.0
Apodemus sylvaticus	3 503	7.7	2.5	4.3
Apodemus flavicollis	1 459	3.2	3.1	5.3
Apodemus agrarius	1 165	2.6	4.0	7.2
Sorex araneus	3 229	7.1	3.1	6.3
Sorex minutus	158	0.3	7.6	35.0
Neomys fodiens	25	0.05	8.0	28.0

(1953), and the index of abundance (per 100 animals) varieu from 1.7 to 14.0 in the same years. The ticks were particularly numerous in the years of depression in the bank vole population density (1962, 1964, 1967), whereas in the years of high population densities of the bank vole (according to registrations in August–September) the coefficient of occurrence and the index of abundance of ticks were not high (Fig. 1). This inverse relationship may apparently be explained by the lag in the pace of tick reproduction from that of bank vole reproduction.

Ixodes trianguliceps ticks were found on animals from April to December. From January to March only 646 animals were examined which is quite insufficient for the detection of ticks occurring singly. However, in February when only 87 animals were examined, one larva was found. In April, all the developmental phases of ticks were found in almost equal numbers which facilitated the transmission of the infection (Fig. 2). The first numan infections with HRFS are known to occur in April. The index of the abundance of the imago rapidly decreased by June, and later adult ticks were found occasionally till October. Male ticks were found (7 specimens only) in April–June, and one male tick was found in September. It should be mentioned that 5 males were found on animals, in the absence of female ticks, indicating the possibility of blood-sucking by adult male ticks. Only on one bank vole in June did we find simultaneously 2♂, 3♀ and one nymph of the tick *I. trianguliceps*.

Nymphs on animals were quite numerous from April until August, with a maximum in April and July, from September on only single specimens were found and by November they disappeared completely. The dynamics of larva density were of a two-peak nature. From April until June the number of larvae on animals increased owing

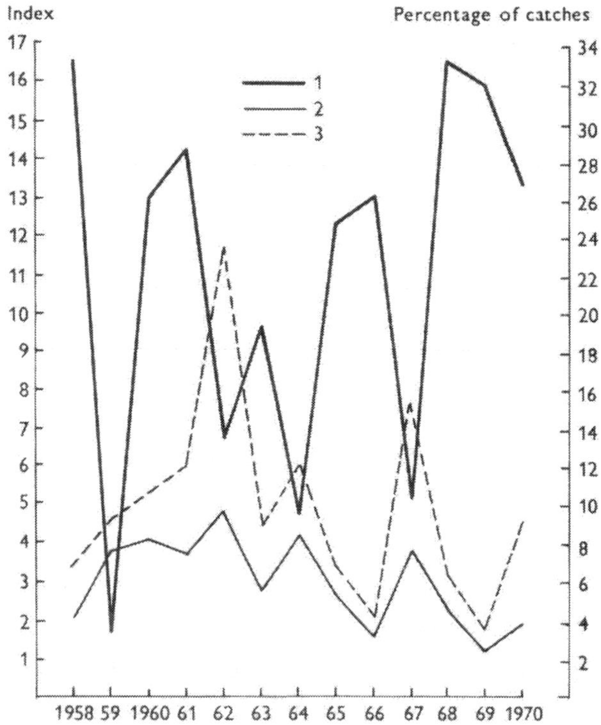

Fig. 1. Relationship between the population density of *Ixodes trianguliceps* ticks and the population density of *Clethrionomys glareolus*: 1 – Per cent of catching of *C. glareolus* in August–September; 2 – Index of occurrence of *I. trianguliceps* on *C. glareolus*; 3 – Index of abundance (per 100 animals) of *I. trianguliceps* on *C. glareolus*.

to the activation of overwintering larvae or the feeding of larvae which emerged from eggs in spring. In July there was a sharp reduction (8-fold as compared with that in June) in the density of larvae, apparently due to the reduced potential of dissemination of infection among bank voles in this month. During August–September, there is a second, greater rise in the density of larvae (emerging from the eggs laid in spring), and from October to December their numbers decline again. It is interesting that in October there is the greatest number of human infections with HRFS. This may serve as an indirect conformation of transovarial transmission of the agent of HFRS by *I. trianguliceps* ticks. If this transmission exists, the emerging larvae may be donors of the agent; however, the further alimentary spread of infection among bank voles in autumn which reach at this time high population densities cannot be excluded.

With regard to the animal-to-animal transmission of infection, of particular interest is group-parasitizing by ticks. In April, among animals on which *I. trianguliceps* ticks were found, 23.8 % of the animals had 2 or more ticks and nearly every animal had at least one nymph or imago. If transovarial transmission of infection is

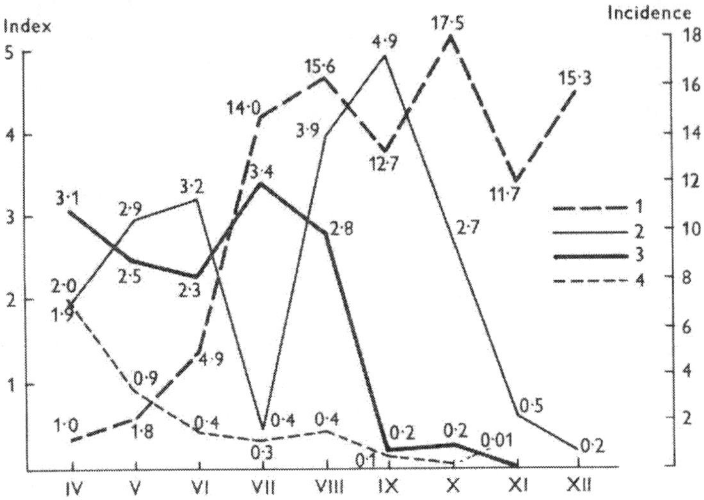

Fig. 2. Indices of abundance of individual phases of development of *I. triangulicebs ticks* on *C. glareolus* (per 100 animals) during the epidemic season of HFRS and incidence of this infection among the human population: 1 – Incidence of HFRS by months (per cent to the annual index); 2 – Index of abundance of larvae of *I. trianguliceps*; 3 – Index of abundance of nymphs of *I. trianguliceps*; 4 – Index of abundance of imago *I. trianguliceps*.

excluded, *i.e.*, if it is assumed that only nymphs and imago can be donors of infection, then the percentage of animals on which two or more ticks feed simultaneously, including at least one nymph or imago, gradually decreases by August to 11.9 %, and in September–October this value sharply decreases to 1.8–0.9 %. The average number of tick-recipients per tick donor in this case was 3.7. If the possibility of transovarial transmission is admitted. in which case larvae may also be donors of infection, the percentage of animal-group hosts increases considerably. It persists at high levels (18.3–27.4 %) with some variations from April to October, and only from November onwards group parasitizing of ticks on bank voles ceases. With admittance of transovarial transmission, there are 2.8 tick-recipients per one tick donor.

Of interest is the maximum number of ticks parasitizing one animal. In August to September, it was not rare to find 20–30 larvae on *C. glareolus, M.. arvalis* and *A. sylvaticus*, and sometimes together with larvae one nymph could be found. In October 1964, 20 *I. tringuliceps* larvae were collected from a small animal such as the *Sorex minutus*. In June 1962, from one bank vole 69 larvae, 8 nymphs and two female *I. trianguliceps* were collected. On 29 occasions this tick was found on animals together with *I. ricinus* ticks (from May till November), on 15 occasions together with *Dermacentor pictus* ticks (from June till September), and on two animals all the three species of ticks were found simultaneously, a fact of unequivocal significance in transmission of tularemia infection.

References

Bashkirev, T. A., Boiko, V. A., Charanova, T. I., Gancharova, M. I., Kuznetsova, N. A., Andronnikov, V. A., Danovsky, L. V.: Features of epidemiology and means of prophylaxis of haemorrhagic fever with renal syndrome in the Middle Povolzhie. In: "Haemorrhagic Fever with Renal Syndrome in Preduralie and Middle Volga", Ufa, 18 – 22, 1965. (In Russian.)

Chumakov, M. P., Leshchinskaya, E. V., Belyaeva, A. P., Povalishina, T. P.: Production of experimental fever in psychiatric patients with the purpose of pyrogenic therapy by means of inoculation with the blood serum from patients suffering from hemorrhagic fever with renal syndrome (HFRS) or with a suspension of ground gamasid mites collected in a focus of HFRS in the Upper Volga Basin. In: "Endemic virus infections (haemorrhagic fevers)", Moscow, Publ. House "Meditsina", 12–22, 1965. (In Russian.)

Kalinkovsky. I. S., Izotov, P. V.: Gamasid mites on rodents during epidemic outbreaks of HINN in Primorsk region". In: Abstracts of Papers presented at sci. conf. Vladivostok IEMH, Vladivostok, 24–25, 1962. (In Russian.)

Katelina, A. F., Panina, T. V.: Ixodid ticks – parasites of bank voles in natural foci of haemorrhagic fever with renal syndrome. In: "Problems of natural-focality infections and medical geography" Tula, 69–72, 1966. (In Russian.)

Preobrazhensky, A. A., Izotov, P. V., Preobrazhenskaya, N. K., Soldatov, G. M.: On the problem of the role of gamasid mites in transmission of haemorrhagic fever with renal syndrome. In: "Tick-borne encephalitis and other natural-focality infections." Sverdlovsk, 188–195, 1970. (In Russian.)

Traub, R., Hertig, M., Lawrence, W. H., Harris, T. T.: Potential vectors and reservoirs of haemorrhagic fever in Korea. Am. J. Hyg. 59, 3 : 291–305, 1954.

DISTRIBUTION AND SEASONAL ACTIVITY IN THE PREIMAGO PHASES OF THE IXODID TICKS IN A FOCUS OF HAEMORRHAGIC FEVER (CRIMEAN TYPE)

V. LEVY
Research Institute of Epidemiology and Microbiology, Sofia, Bulgaria

The haemorrhagic fever (Crimean type), with one of its most active foci being in the region of Pazardjik, is a very important problem of the public health of the country. In 1969 from three sets of *Hyalomma plumbeum plumbeum* Panz. ticks, collected from cattle and horses in this region, virus strains of haemorrhagic fever were isolated from ticks for the first time in our country, and the presence of antibodies to the virus was established in the wild rabbit *Lepus europaeus* (Vassilenko et al., 1970).

Parallelly with the virological and serological study, investigations on the species composition, the seasonal activity and the distribution of the ixodid ticks were carried out. Studies on the imago phases were made by Levy (1971).

In the present paper we communicate data pertinent to the pre-imago phases. We set ourselves the task, on the basis of the predominance and abundance of the various tick species, to establish the hosts preferred by the larvae and nymphs of *H. plumbeum*, with the view of directing the virological investigations and establishing the sources of the infection.

Wild mammals and birds were shot for that purpose, while the smaller mammals were caught by means of the "HERO" type killing traps. The studies were carried out in 1968–1970 in the Pazardjik region from March to November.

As a result of these studies the number of wild animal specimens collected amounted to 513 and were: *Testudo graeca* L. – 2 specimens; Aves – 219 specimens of 44 species; Insectivora – 23 specimens of 5 species; *Lepus europaeus* Pall. – 92 sp.; *Citellus citellus* L. – 1 sp.; Muridae – 174 sp. of 6 species and *Vulpes vulpes* L. – 2 specimens.

The number of larvae and nymphs collected from the above animals amounted to 2582, their species were as follows: *Ixodes ricinus*, L. – 7; *Ixodes crenulatus* Koch – 7; *Haemaphysalis punctata* Can et Fanz. – 793; *Dermacentor marginatus* Sulz – 434; *Rhipicephalus sanguineus* Latr. – 8; *Rhipicephalus turanicus* B. Pom. – 9; *Hyalomma aegyptium* L. – 80; *Hyalomma plumbeum plumbeum* Panz. – 1243 and *Hyalomma plumbeum impressum* Koch – 1 specimen.

Table 1 illustrates the predominant ticks by groups and host species, the more

Table 1. Collections of pre-imago phases of ixodid ticks from wild animals

Host species	No. of hosts examined	No. of ticks	Tick species											
			Hyalomma plumbeum			*Haemaphysalis punctata*			*Dermacentor marginatus*			Other species		
			No.	%	inci-dence	No.	%	inci-dence	No.	%	inci-dence	No.	%	inci-dence
Testudo graeca	2	70										70		
Crocidura suaveolens	14	2										2		
Lepus europaeus	92	1836	1138	62.0	12.4	316	17.2	3.4	365	19.9	3.9	17	0.9	0.2
mouse rodents	174	73	3	4.1	0.02	1	1.4	0.01	61	83.6	0.4	8	10.9	0.05
Vulpes vulpes	2	7										7		
Pica pica	5	16	2	12.5	0.4	14	87.5	2.8						
Garrulus glandarius	9	8	2	25.0	0.2	6	75.0	0.7						
Passer montanus	13	1				1								
Alauda arvensis	3	1				1								
Motacilla flava feldegg	1	1										1		
Turdus merula	18	83	49	59.0	2.7	34	40.9	1.9						
Upupa epops	8	10	1	10.0	0.1	1	10.0	0.1	1	10.0	0.1	7	70.0	0.9
Caprimulgus europaeus	4	2	2											
Athene noctua	5	31	25	80.6	5.0				6	19.4	1.2			
Streptopelia turtur	11	1				1		0.1						
Charadrius dubius	1	1	1											
Perdix perdix	60	439	20	4.6	0.3	418	95.6	6.9	1	0.2	0.01			
Other species	91													
Total	513	2582	1243	48.1		793	30.7		434	16.8		112	4.3	

rarely met species being listed together, while parasite-free host species are given only as a total.

From the calculated indices for the predominance it is seen that the larvae and nymphs of *H. plumbeum* are prevalent on *Athene noctua, L. europaeus* and *Turdus merula*; the larvae and nymphs of *D. marginatus* on Muridae, while the larvae and nymphs of *H. punctata* on birds (*Perdix perdix, Pica pica* and *Garrulus glandarius*).

From the abundance indices, it becomes clear that the *H. plumbeum* is most prolific on *L. europaeus*, followed by *A. noctua, T. merula* and other species with indices of abundance lower than 1; *D. marginatus* on *L. europaeus* followed by *A. noctua*, Muridae and some other, and *H. punctata* on *P. perdix* followed by *L. europaeus, P. pica, T. merula* and others.

Zimina et al. (1965) record that the most important part in the feeding of the pre-imago phases of *H. plumbeum* in the Astrakhan region (USSR) is played by the ravens. Studies of this kind are still insufficient. Larvae and nymphs of other tick species not presented in the Table are: *I. ricinus* – from Muridae, *C. suaveolens* and *L.*

europaeus; I. crenulatus – from *V. vulpes*; *R. sanguineus* and *R. turanicus* from *L. europaeus* and Muridae; *H. aegyptium* – from *T. graeca, L. europaeus* and *Upupa epops* and 1 specimen of *H. pl. impressum*, established for the first time in Bulgaria in the transmigratory species *Motacila flava feldegg*, most likely brought from Africa during the passage.

The following animal species were parasite-free: *Corvus cornix* L. – 2 specimens, *Oriolus oriolus* L. – 2 specimens, *Coccothraustes coccothraustes* L. – 4 sp., *Chloris chloris* L. – 3 sp., *Carduelis cannabina* L. – 1 sp., *Passer domesticus* L. – 7 sp., *Passer hispaniolensis* (Temminck) – 4 sp., *Emberiza melanogaster* (Scopoli) – 1 sp., *Emberiza hortulana* L. – 1 sp., *Galerida cristata* L. – 20 sp., *Motacilla flava feldegg* L. – 1 sp., *Motacilla alba* L. – 1 sp., *Lanius collurio* L. – 2 sp., *Hirundo rustica* L. – 5 sp., *Sylvia nisoria* (Bechstein) – 1 sp., *Sylvia atricapilla* L. – 1 sp., *Turdus ericetorum* (Turton) – 2 sp., *Monticola saxatilis* L. – 1 sp., *Oenanthe oenanthe* L. – 2 sp., *Oenanthe hispanica* L. – 1 sp., *Saxicola torquata* L. – 2 sp., *Erithacus rubecula* L., – 1. sp., *Picus viridis* L. – 1 sp., *Dryobates syriacus* (Hemprich et Ehrenberg) – 1 sp., *Asio otus* L. – 2 sp., *Cuculus canorus* L. – 1 sp., *Falco subbuteo* L. – 1 sp., *Accipiter nisus* L. – 1 sp., *Streptopelia decaocto* (Frivalszki) – 3 sp., *Charadrius dubius* (Scopoli) – 1 sp., *Alectoris graeca* (Meisner) – 2 sp., *Erinaceus europaeus* L. – 1 sp., *Talpa europaea* L. – 1 sp., *Sorex araneus* L. – 1 sp., *Crocidura leucodon* Herm. – 6 sp., and *Citellus citellus* L. – 1 sp.

From the pre-imago phases dominating in the region under study were the species *H. plumbeum, H. puctata* and *D. marginatus*, which complies with the data pertinent to the imago (Levy, 1971). In our country *H. plumbeum* is a two-host species. The larvae and the nymphs parasitize mostly rabbits and birds, while the imago cattle, horses, and sheep.

The seasonal activity of the larvae and nymphs of *H. plumbeum*, in comparison with the seasonal activity of the imago on cattle in the same region with the view to elucidate the annual cycle of the development of this species, is presented in Fig. 1.

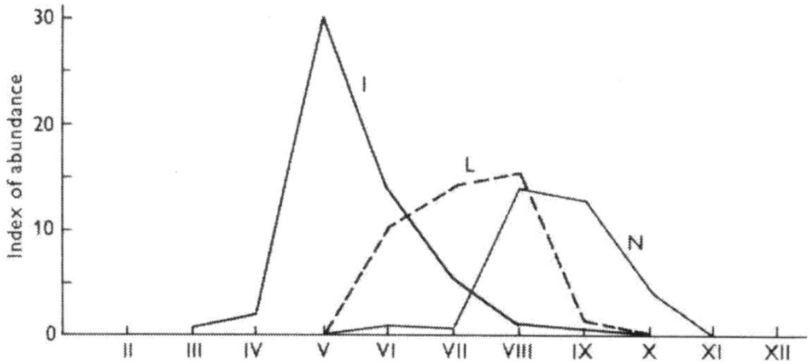

Fig. 1. Seasonal activity of the ticks *Hyalomma pl. plumbeum*; larvae (L) and nymphs (N) on hares, imago (I) on cattle.

The imago of *H. plumbeum* is parasitic from March to September with maximal abundance during the month of May, the larvae from July to September with maximal numbers in July–August, and the nymphs from July to October, with a maximum in August and September. The fed nymphs, at the beginning of August, due to the climatic peculiarities of our country (hot autumn), most likely change into imago, and those fed later hibernate as nymphs.

Figure 2. is illustrative of the seasonal activity curves of the larvae and the nymphs of the *D. marginatus* on *L. europeus*. The larvae of *D. marginatus* are parasitic during June and July, the nymphs from July to September with the maximal abundance in July. The larvae of *H. puctata* are parasitic from June to August with a maximum during July, and the nymphs from July to September. The imago of these two species, in the conditions of our country have two waves of parasitation – a spring and an autumn wave (Serbova, 1955, 1956).

Conclusion

The studies carried out in the Pazardjik region, on the pre-imago phases of the ixodid ticks, reveal that the predominant species in this region are *H. plumbeum, H. punctata* and *D. marginatus*. The host preferred by the larvae and nymphs of *H. plumbeum* are *L. europaeus, A. noctua* and *T. merula*. The part played by the other species of birds with parasites is small, while that of Muridae insignificant. The great number of nymphs of this species during September and October shows that a great part of them hibernate as fed nymphs.

From the studies we carried so far, it is of interest to note that the *L. europaeus, A. noctua* and *T. merula* are preferred hosts for the larvae and nymphs of *H. plum-*

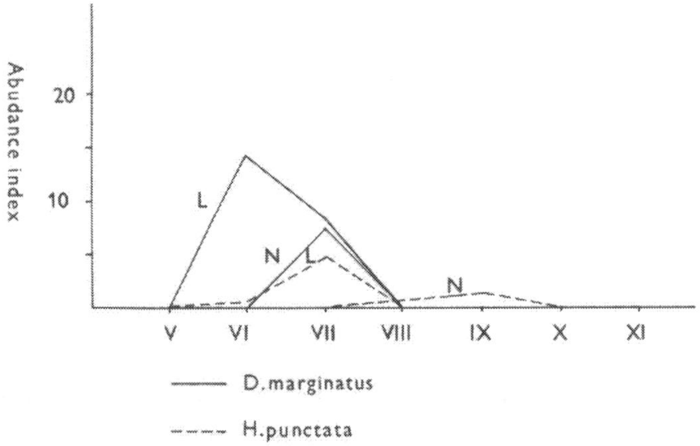

Fig. 2. Seasonal distribution of larvae (L) and nymphs (N) on rabbits.

beum. It may be presumed that they participate in the circulation of the haemorrhagic fever (Crimean type) virus in the outside environment.

References

Levy, V.: The seasonal activity of the ixodid ticks in a focus of haemorrhagic fever (Crimean type) in the Region of Pazardjik. (In print – in Bulgarian).

Serbova, S.: Data on the distribution, biology and ecology of the ixodid ticks, in the Baltchik District, with the view of their epizoologic and epidemiologic significance. Savremena meditzina, V. 2, p. 13–27, 1955.

Serbova, S.: Species composition and seasonal activity of the ixodid ticks in the the Iskra village (Plovdiv region). – Works of RIEM, v. III, p. 209–216, 1956. (In Bulgarian.)

Vassilenko, S., Kirov, I., Katzaro, G., Mihailov, A., Radev, M., Kebedjiev, G., Levy, V., Levy, S.: Studies on the Crimean type haemorrhagic fever in Bulgaria. Letopisi of HEI, V. XXII, p. 153–166, 1970. (In Bulgarian.)

Zimina, J.V., Birula, N.B., Berezin, V.V., Zalutzkaya, L.I., Povalishina, T.P. Stolbov, D.H.: Materials on the zoo-parasitological characteristics of the Crimean type haemorrhagic fever in the Astrakhan region (USSR). – Trudy Instituta Poliomielita i virusnykh encefalitov, T. VII, p. 288–295, 1965. (In Russian.)

Proceedings of the 3rd International Congress of Acarology, Prague, 1971

APPLICATION OF FLUORESCENT ANTIBODY METHOD (FAM) IN THE ISOLATION OF CRIMEAN HAEMORRHAGIC FEVER VIRUS FROM TICKS, THE VECTORS OF THE DISEASE

G. V. POPOV, V. D. LEVY, S. M. VASILENKO, M. P. CHUMAKOV
Research Institute of epidemiology and microbiology, Sofia, Bulgaria
and Institute of poliomyelitis and viral encephalitides, Academy of Medical Sciences, Moscow, USSR

M. P. Chumakov (1945, 1946) was the first to prove the importance of the tick *Hyalomma plumbeum* as the vector of Crimean haemorrhagic fever (CHF) in the Crimea. In Bulgaria as well as in the USSR, a high percentage of seropositive animals was found during serological investigations of farm animals for the presence of antibodies to the CHF virus (Vasilenko et al., 1968). These and other data (Butenko, 1971) indicate a wide distribution of the CHF virus in endemic regions. However, isolation tests of the CHF virus from ticks, conducted on new-born white mice, have not always been successful in obtaining and fixing "tick" strains (Vasilenko and Smirnova, unpublished data). No information was available on the FAM application in the detection of the CHF virus antigen in ticks, the vectors. Consequently few techniques in such studies of ticks have been worked out. Thus the FAM application in the detection of the CHF virus antigen in ticks has become necessary.

Our task consisted in developing a technique of immunofluorescent demonstration of antigen by means of direct FAM in the investigations of ticks for the presence of antigen.

The principle of the technique consists in exposing the salivary gland of the tick and in obtaining a preparation of it stained by FAM. Apart from the salivary glands, haemolymph and uterus were also investigated during our studies. Our experiments showed that it was quite sufficient for our purpose to study the salivary glands (Popov, 1971).

The glands were removed after the dissection of ticks. A direct FAM was applied for the detection of the virus in ticks (Coons and Kaplan, 1950; Popov, 1971). The results obtained in our investigation of ticks are presented in the Table 1. and Figures 1–8.

The Table 1. shows that specific fluorescence of antigen was traced in the tick *H. plumbeum* collected in Bulgaria. In two collections of *H. plumbeum* from Simferopol (Crimea) and Azerbaijan the antigen was not detected; neither was it revealed in a pool of *I. ricinus* ticks, taken as a control.

Figs. 1–8 represent photomicrographs of salivary gland, in which specific fluorescence of the CHF virus antigen was detected.

Table 1. Investigation of ticks by FAM for the presence of CHF virus antigen

Ticks					FAM specific fluorescence serum No. 150	
Species	Date of investigation	Locality	Host	Number and sex	Salivary glands	Uterus
Hyalomma plumbeum	Sept. 1969	Bulgaria	cattle horses	12 ♀	1 ♀	1 ♀
Hyalomma plumbeum	Apr. 1970	Simferopol	cattle horses	29 ♀	–	–
Hyalomma plumbeum	Apr. 1970	Azerbaijan	cattle horses	6 ♀	–	–
Ixodes ricinus	May 1970	Bulgaria	cattle horses	14 ♀	–	–

The preliminary results obtained by us gave an impetus to seek a direct detection of the CHF virus antigen in salivary glands of ticks, the vectors of the disease, and they were confirmed by the studies of Zgurskaya et al. (1971) concerning ticks infected by feeding on infected animals, as well as ticks from natural habitats.

A special point in the investigations of ticks by FAM is the problem of checking. The control tests of immunofluorescence reaction which are required at present (Goldman, 1968) cannot be applied in the investigation of salivary glands of ticks by FAM. This is due to the fact that only one authentic preparation may be made from one salivary gland and the application of the blocking test or variations of staining sera proves to be physically impossible because of lack of homogeneous material. Therefore we propose to use the following criteria as a control: a) application of labelled gamma globulins against CHF, obtained from persons who have had the disease; these globulins should be preliminarily tested on tissue cultures and cryostat brain sections from new-born white mice infected with the CHF virus, i.e. the standard FAM control tests should be observed; b) staining of salivary glands of control ticks free from the CHF virus (*Ixodes, Dermacentor* or *Haemaphysalis* collected in healthy regions); c) negative results in the staining of salivary glands from the pool of vectors investigated.

Our results and their confirmation, as well as the localization of the Powassan virus in the salivary glands of the ticks *D. andersoni* (Chernesky and McLean, 1969) suggest the use of the techniques used by us for more profound studies on the determination of infectivity of ticks, the vectors of viral infections.

References

Butenko, A.M.: A contribution to the studies on etiology, laboratory diagnosis and immunology of Crimean haemorrhagic fever; ecology of the virus – the causative agent. Diss. Dokt., 1971. (In Russian.)

Coons, A.H., Kaplan, M.H.: Localization of antigen in tissue cells II. Improvements in a method for the detection of antigen by means of fluorescent antibody. J. Exptl. Med., 91 : 1–13, 1950.

Chernesky, M.A., McLean, D.M.: Localization of Powassan virus in *Dermacentor andersoni* ticks by immunofluorescence. Canad. J. Microbiol., 15,: 1399–1408, 1969.

Chumakov, M.P.: A new viral tick borne disease – haemorrhagic fever in the Crimea (acute infections capillaro-toxicosis). Simferopol, 13–43, 1945. (In Russian.)

–, Crimean haemorrhagic fever (acute infections capillarotoxicosis). Krymizdat, 1946. (In Russian.)

Goldman, M.: Fluorescent antibody methods. New York–London, 1968.

Popov, G.V.: Immunofluorescent and electron-microscopic studies on the Crimean haemorrhagic fever virus with the application of comparative studies on model viruses. Diss. Dokt., 1971. (In Russian.)

Vasilenko, S.M., Chumakov, M.P., Butenko, A.M., Smirnova, S.E., Popov, G.V.: To the problem of etiology of Crimean haemorrhagic fever (CHF) in Bulgaria. Mater. XV. n. sess. in-ta poliomyelita i virusnykh enceph., No. 3, 90–92, 1968. (In Russian.)

G. V. Popov, V. D. Levy, S. M. Vasilenko, M. P. Chumakov

Zgurskaya, G. N., Popov, G. V., Beresin, V. V., Smirnova, S. E., Chumakov, M. P.: Application of the fluorescent antibody method (FAM) in the isolation of the CHF virus from ticks, the vectors. Mater. n. sess. in-ta virusologii im. D. I. Ivanovskogo (In press – in Russian.)

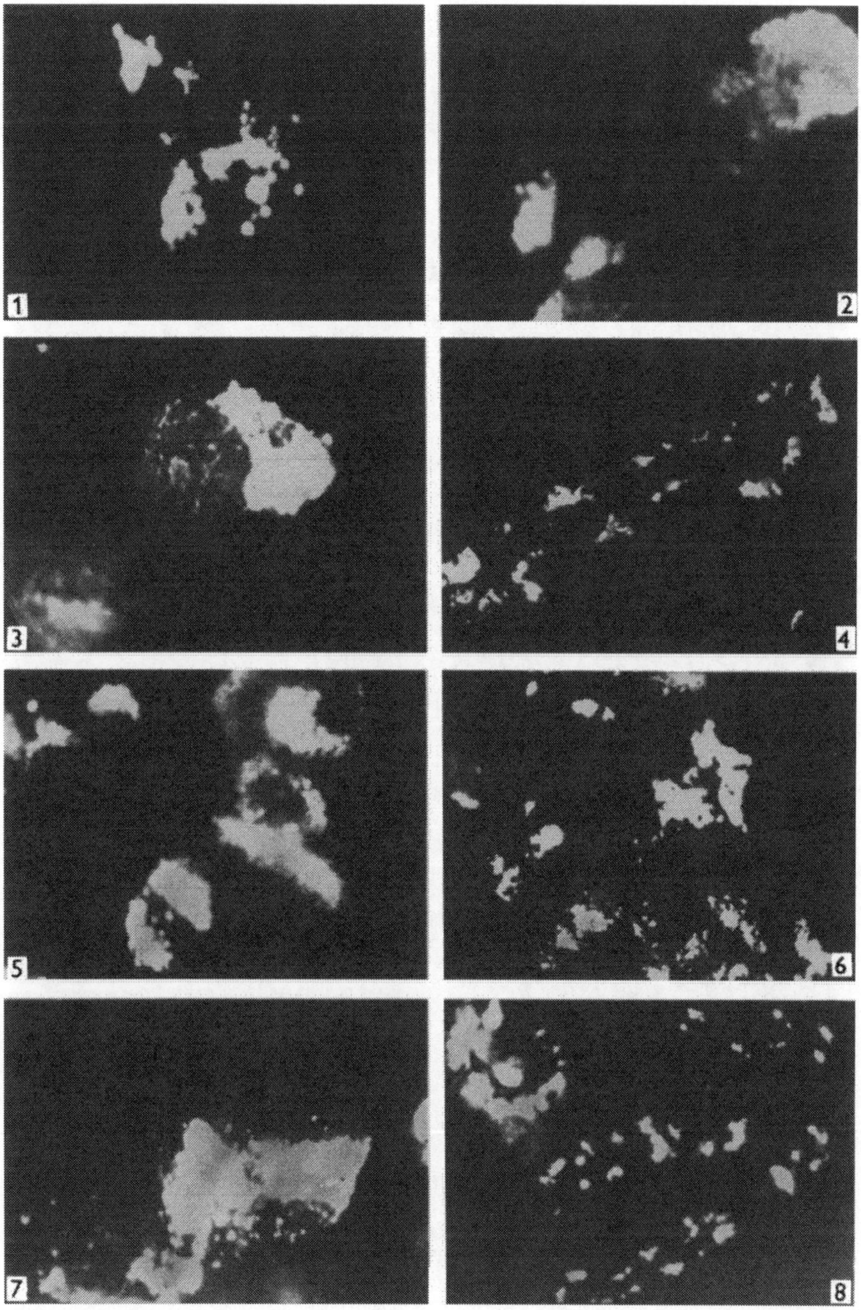

Plate I.

Direct FAM. Virus CHF — tick *H. plumbeum* (♀). Salivary gland labelled γ-globulin from man with a history of CHF.

Figs. 1—3. General appearance of salivary gland (objective 10 ×, 20 ×, eyepiece 7 ×); Figs. 4—8. Separate gland-parts (5) and separate cells (objective 100 ×, eyepiece 7 ×).

Proceedings of the 3rd International Congress of Acarology, Prague, 1971

POTENTIALITIES OF ACCUMULATION OF OMSK HAEMORRHAGIC FEVER VIRUS IN THE POPULATION OF IXODID TICKS OF THE NATURAL FOCUS DURING THE EPIZOOTIC SEASON

O. V. RAVDONIKAS, M. P. CHUMAKOV, D. I. IVANOV, P. V. KORSH, Yu. A. MYASNIKOV
The USSR AMS Institute of Poliomyelitis and Virus Encephalitides, Moscow;
RSFSR MPH Omsk Institute of Natural Focality Infections, Omsk, USSR

In an epizootological investigation of the natural focus, the main task is to elucidate the mechanism of turnover of the causative agent and its quantitative characteristics at different stages of the epizootic process. This is based on the features of biocenotic relations between blood-sucking arthropods and susceptible warm-blooded animals determining the composition of the parasitic system of the natural focus of human infectious disease. The population densities of both and different possibilities of simultaneous combined parasitizing of different vectors determine the greater or lesser degree of the probability of mediatory dissemination of the causative agent among susceptible animals in the territory of a natural focus (Ravdonikas et al., 1969a).

In the present work we used the material which we collected in 1967 in Novokarasuk the natural focus of Omsk haemorrhagic fever (OHF) in the vicinity of Novokarasuk state farm, Omsk region (Ravdonikas et al., 1968; Ivanov et al., 1968). Since ixodid ticks are long-term reservoirs of OHF virus and its main vectors, we confine ourselves to discussing only this group of blood-sucking arthropods determining dissemination of the virus in the focus.

The population density of active ticks feeding on warm-blooded animals was measured by the index of feeding (number of tick-infested animals per 100 trap ditches) without making corrections for natural losses in the course of subsequent metamorphosis and further existence. The 10-day indices of feeding reflect the renewal of the population at the next instar level and their total for the period of activity permits judgement of the trend of accumulation of the overwintering phases of development.

In order to be able to judge potentialities of mediatory transmission of the agent in the course of parasitizing, the index of feeding was calculated not by species and phases of development of ticks but by their combinations including all species and their developmental phases feeding simultaneously on one animal. We divided all combinations into 4 main groups:

1. Single ticks of any species and developmental phase which may infect the animal but are not important in mediatory dissemination because of the absence of other ticks on the host.

2. Monophase combinations consisting of specimens of one developmental phase of one species. The presence among them of an infected tick provides for the spread of the agent within the given developmental phase of the given species.

3. Heterophase combinations including ticks of two or three developmental stages of one species. Mediatory transmission of the agent in this case promotes its intraspecies dissemination.

4. Interspecies combinations are characterized by parasitizing on one host simultaneously of representatives of different developmental phases of two or more species of ixodid ticks. These combinations are the only means of interspecies exchange of the agent in ixodid ticks (Ravdonikas et al., 1969b). (Fig. 1.)

For the analysis we used the data of tick infestation rates of animals caught in trap ditches. The stability of the latter throughout the entire season is contributory to the catching of only migratory animals picking up ticks from a relatively large territory due to which their tick infestation rates are higher. Of importance is also the fact that in trap ditches we examine predominantly animals caught alive which eliminates losses of ticks rapidly deserting a dead animal.

Table 1. presents data characterizing a renewal of the populations of ixodid ticks by the index of feeding of different combinations of parasitizing. Data for the first 10-day period of June when the trap ditches did not work have been taken from materials of trap catching, since the indices for these ways of catching hosts of *Ixodes apronophorus* ticks are quite similar throughout the epizootic season.

Two peaks are typically sharply differentiated. In May–June, the parasitizing of *Ixodes apronophorus* ticks is observed mainly in mono- and heterophase combinations. At this time, wide dissemination of the virus occurs amongst all developmental phases of this tick. In July this species is substituted by *Dermacentor pictus* tick observed mainly in mono- and heterophase combinations as well as together with *I. apronophorus* and *I. persulcatus* ticks. During this period intensive dissemination of the virus occurs owing to the activity of larvae and nymphs of *D. pictus* with a one-year cycle. Most marked interspecies exchange of the virus is also characteristic. In August, parasitisation is at a low level and is characterized by the appearance of large numbers of *I. apronophorus* larvae of the generation of the given year. (Fig. 2.)

Table 2. presents data reflecting the trend of accumulation in nature during the epizootic season of overwintering developmental phases of ixodid ticks capable of preserving the virus until next spring. The main factor in the preservation of the virus in winter is *I. apronophorus* which constitutes about three quarters of the entire overwintering population. *Ixodes persulcatus* (7.1 %) is not very important because its imago parasitizes large mammals and because of that they cannot transmit the virus mediatorily to their larvae and nymphs or to ticks of other species. Of even lesser importance is *D. pictus* owing to the owerwintering of only adult forms which in spring feed mainly on large animals and are incapable of the mediatory transmission of the preserved virus to other blood-sucking arthropods including their own larvae and nymphs. At the same time, their role may be important because of the existence of transovarial transmission which requires further study.

Fig. 1. Indices of feeding of different combinations of ixodid ticks parasitizing small mammals in the Novokarasuk natural focus of OHF in Omsk region according to data of catches in catching ditches (in first 10 days of June – into traps), 1967 . Sums of 10-day indices of different combinations: 1. Single parasitizing. No mediatory transmission: a) *Ixodes apronophorus* and *I. persulcatus* (L, N, I); b) *Dermacentor pictus*. 2. Monophase parasitizing. Mediatory transmission of the agent within one phase of development: a) *Ixodes apronophorus*: b) *Dermacentor pictus*; c) *Ixodes persulcatus*. 3. Heterophase parasitizing. Mediatory transmission of the agent within different phases of development of one species: a) *Ixodes apronophorus*; b) *Dermacentor pictus*. 4. Multi-species parasitizing. Mediatory transmission of the agent between different species: a) *Ixodes apronophorus* and *Dermacentor pictus*; b) *Ixodes persulcatus* and *Dermacentor pictus*.

Table 1. Index of feeding of ixodid ticks in Novokarasuk natural focus of OHF, Omsk region, 1967, according to registration by catching ditches[x]

10–days periods	Combination patterns	Single		Monophase					Heterophase		Interspecies		Total
		ticks with 3–year cycle	*Dermacentor pictus*	*Ixodes persulcatus*	*Ixodes apronophorus*		*Dermacentor pictus*		*Ixodes apronophorus* all phases	*Dermacentor pictus* larva and nymph	*Ixodes apronophorus* and *Dermacentor pictus*	*Ixodes persulcatus* and *Dermacentor pictus*	
				larva	larva	nymph	larva	nymph					
May	2	1.19	–	2.36	0.68	–	–	–	–	–	–		4.23
June	1[x]	3.8	–	–	20.0	1.54	–	–	103.02	–	–	–	128.36
	2	4.44	–	–	4.44	1.37	–	–	6.8	–	–	–	17.05
	3	1.59	0.83	2.57	1.29	0.71	2.45	–	1.29	–	0.71	2.3	12.74
July	1	3.56	4.44	–	0.55	–	31.94	3.23	1.36	2.44	7.1	–	54.62
	2	3.25	4.75	10.13	1.07	–	45.01	1.75	4.69	23.49	17.27	18.1	129.21
	3	1.98	4.62	–	–	–	14.19	6.41	–	15.8	1.57	–	44.57
August	1	0.64	3.94	0.32	1.13	–	0.64	5.26	–	6.45	2.0	–	20.38
	2	0.6	1.9	–	11.04	0.29	0.45	2.13	0.45	0.29	2.13	–	19.28
	3	0.27	1.38	–	0.75	–	0.34	1.14	–	0.14	1.02	0.14	5.18

[x] In the first 10 days of June trap registrations were used because ditches did not work for catching.

The above data on the features of feeding of ixodid ticks and the formation of the overwintering portion of their populations form the basis on which materials of virologic studies may be projected for obtaining quantitative indices of the intensity of the virus circulation and its accumulation during the epizootic season.

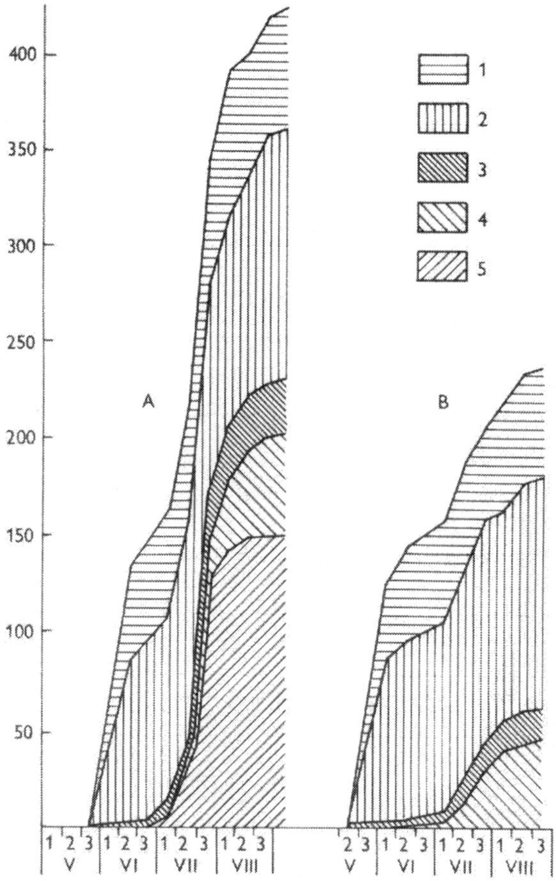

Fig. 2. The sum of 10-day indices of feeding of ixodid ticks on small mammals according to catches into catching ditches (first 10 days of June – into traps) in Novokarasuk natural focus of OHF, Omsk region, 1967 (Increasing sum). Ixodid ticks with potential infection: 1. Nymphs of *I. apronophorus* moulting into imago; 2. Larvae of *I. apronophorus* moulting into nymphs; 3. Larvae and nymphs of *I. persulcatus*; 4. Nymphs of *D. pictus* moulting into imago; 5. Larvae of *D. pictus* moulting into nymphs: A) Trend of population renewal of ixodid ticks in 1967; B) Formation of the overwintering portion of the population of ixodid ticks with potential mediatory infection.

T a b l e 2. Formation of the overwintering portion of the population of ixodid ticks potentially mediatorily infected as a result of group parasitizing. Index of feeding according to collections in Novokarasuk focus of OHF, 1967

10–days periods	Species and phases	*Ixodes apronophorus*		*Ixodes persulcatus*		*Dermacentor pictus*	Total
		Nymph	Imago	Nymph	Imago	Imago	
May	2	0.99	0.51	2.36	–	–	3.86
June	1	82.05	43.25	–	–	–	125.3
	2	10.94	5.06	–	–	–	16.0
	3	2.38	2.38	2.74	–	–	7.5
July	1	3.57	1.92	–	–	1.37	6.86
	2	5.06	3.6	10.48	–	8.66	27.8
	3	1.35	0.21	0.225	0.226	17.04	19.05
August	1	1.31	0.16	0.63	–	11.9	14.0
	2	11.32	0.45	0.16	–	4.22	16.08
	3	0.9	0.07	0.07	–	2.45	3.49
Total		119.87	57.61	16.665	0.226	45.64	239.78

References

Ivanov, D.I., Ravdonikas, O.V., Alifanov, V.I.: Ixodid ticks of mammals of a natural focus of OHF (1967). In: "Materials of the XVth scientific conference of the USSR AMS Institute of Poliomyelitis and Virus Encephalitides", V. 3, 140–142, 1968. (In Russian.)

Ravdonikas, O.V., Chumakov, M.P., Ivanov, D.I., Korsh, P.V.: On mediatory transmission of the agent in blood-sucking arthropods. In "Arboviruses. Materials of XVIth scientific conference of the USSR AMS Institute of Poliomyelitis and Virus Encephalitides", Moscow, V. 2, 188–189, 1969a. (In Russian.)

–, –, –, –: The modes of OHF virus dissemination with participation of ixodid ticks in the natural focus of this infection. Ibid, 186–187, 1969b. (In Russian.)

Ravdonikas, O.V., Korsh, P.V., Ravdonikas, I.O.: The status of the fauna of small mammals as a factor determining the intensity of virus circulation in natural foci of OHF. Ibid, 147–150, 1968. (In Russian.)

RELATIONSHIP BETWEEN TICKS AND RICKETTSIAE OF THE ROCKY MOUNTAIN SPOTTED FEVER GROUP (RMSF) ISOLATED IN SLOVAKIA

J. ŘEHÁČEK, R. BREZINA, M. ŽUPANČIČOVÁ and E. KOVÁČOVÁ
Institute of Virology, Slovak Academy of Sciences, Bratislava, Czechoslovakia

In 1968–1970, thirty-six strains of rickettsiae belonging to the RMSF group were isolated from ticks *Dermacentor marginatus* and *Ixodes ricinus* collected mostly in localities in the Krupina Hills in Central Slovakia (Brezina et al., 1969, Řeháček et al., 1972) (Table 1). These strains were isolated from ticks shown to be positive in the haemocyte test by the direct inoculation of tick suspension into yolk sacs of chick embryos. Most of the strains were isolated from suspensions prepared from single ticks.

All isolated strains multiply very poorly in yolk sacs of chick embryos killing them 3–4 days after infection. So far, a low virulence of these strains is suggested. Guinea pigs react only exceptionally upon the intraperitoneal infection with a short and weak febrile response. Scrotal reaction occurs irregularly.

The infestation with rickettsiae studied by the haemocyte test in ticks, mainly of *D. marignatus*, was very high, especially in arthropods collected in the Krupina Hills. We would like to demonstrate to you the rate of infestation of *D. marginatus* in two selected localities in this region which were studied during a three year period (Table 2). We found also *I. ricinus* and *H. punctata* to be infested with rickettsiae at the same habitat but lower than *D. marginatus*.

Very high infestation of *D. marginatus* adults in nature with rickettsiae offered an excellent model for studies on the distribution of rickettsiae in the haemolymph and organs of these arthropods. The results obtained demonstrated a generalized infection with these rickettsiae in the organisms of both sexes. The highest amount of rickettsiae were observed in the haemolymph (Fig. 1), gut (Fig. 2), Malpighian tube and neural ganglion in males and in the haemolymph, connective tissue (Fig. 3) and Malpighian tube in females. The organs of half-engorged females contained higher amounts of rickettsiae than those of unengorged ticks. These studies performed by the Gimenez staining method and the immunofluorescence technique gave almost the same results.

Rickettsiae showed great polymorphism in different organs of ticks. In haemocytes were found fine rods, in the gut and connective tissue thick rods and dumb-bell like forms. It was interesting to observe that in the ovaries there was in addition to the typical rods, a mass of minute formations displaying diffused fluorescence.

Table 1. Strains of rickettsiae belonging to the RMSF group recovered from ticks collected in 1968–1970 in Slovakia

Number of strains	Locality	Ticks	Isolated in
7	Krupina Hills	*Dermacentor marginatus*, adults of both sexes	1968
13	Krupina Hills	*Dermacentor marginatus*, adults of both sexes	1969
11	Krupina Hills	*Dermacentor marginatus*, adults of both sexes	1970
3	Krupina Hills	*Ixodes ricinus*, females	1970
1	Štiavnica Mountains	*Dermacentor maginatus*, female	1970
1	South Slovakian Karst	*Dermacentor marginatus*, female	1970

To explain more clearly the relationship between rickettsiae and ticks, the propagation of this agent in tick organisms and tick cells was studied in vitro. The results of that agent in tick organs and haemolymph of ticks infected by parenteral injection will be reported in a further contribution presented by us at this Congress. Let us now briefly summarize the results on cultivation of these rickettsiae in tick cells in vitro.

The cultures established from developing adults of *Hyalomma dromedarii* were employed in these experiments. The cells were cultivated on cover slips and placed in test tubes. Before inoculations with rickettsiae, the medium in which the culture was grown, was replaced with an antibiotic-free medium. The results obtained have shown a type of multiplication of this agent in tick cells typical for rickettsiae of the RMSF group (Fig. 4). The rickettsiae eastablished a homogeneous population and were found in cytoplasm as well as in nuclei. They did not occur in cytoplasm vacuoli. The titration experiments carried out for the comparison of the detection of small amounts of rickettsiae demonstrated almost the same results in yolk sacs and tick cells in vitro. Titres varied between 10^2–10^4, however, tick cells have shown markedly higher multiplication.

It was found that laboratory mice bitten by *D. marginatus* females previously ascertained by haemocyte test as infectious with rickettsiae, developed antibodies against the agent. Also larvae and nymphs which originated from these females transmitted rickettsiae to mice. Smears prepared from these ticks showed rickettsiae to be numerous. This demonstrated that the rickettsiae are transmitted transovarially by ticks and by tick bites, probably via saliva. The presence of rickettsiae in ticks collected in early spring (February) suggested the hibernation of the pathogen in arthropods.

Table 2. The infestation of ticks *Dermacentor marginatus* (adults) in the vicinity of the villages V.L. and S. in the Krupina Hills region with rickettsiae of the RMSF group

Locality		Date	%
S. (Krupina Hills)	1968	March 20	15
		April 2	18
		April 25	7
		May 28, June 20	0
		Sept. 12, 24	25
	1969	April 5	18
	1970	May 21	16
		June 16	20
		Sept. 17	32
		Oct. 8	31
V.L. (Krupina Hills)	1968	Febr. 29	43
		March 21	9
		April 5, May 28, June 20	14*
		Oct. 11	38
		Oct. 25	21
	1969	April 5	19
		April 30	8
	1970	May 21	45
		Sept. 17	15
		Oct. 8, 18	26
	1971	Febr. 20	32

* Same result of separate dates.

According to these results as well as those previously mentioned, we consider ticks, especially *D. marginatus*, to be the primary vectors and probably also the reservoir of the rickettsiae of the RMSF group occurring in Slovakia.

References

Brezina, R., Řeháček, J., Áč, P., and Majerská, M.: Two strains of rickettsiae of Rocky Mountain Spotted Fever group recovered from *Dermacentor marginatus* ticks in Czechoslovakia. Results of preliminary serological identification. Acta virol. 13 : 142–145, 1969.

Řeháček, J., Brezina, R., Áč, P., Župančičová, M., and Kováčová, E.: Contribution to the natural focality of rickettsiae belongig to the Rocky Mountain Spotted Fever (RMSF) group in Slovakia. Folia Parasitol. (In press.)

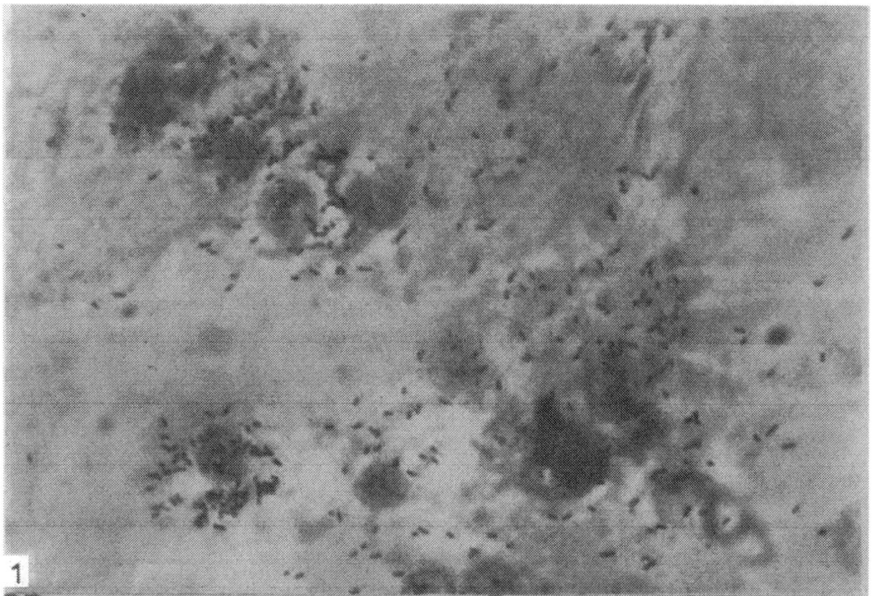

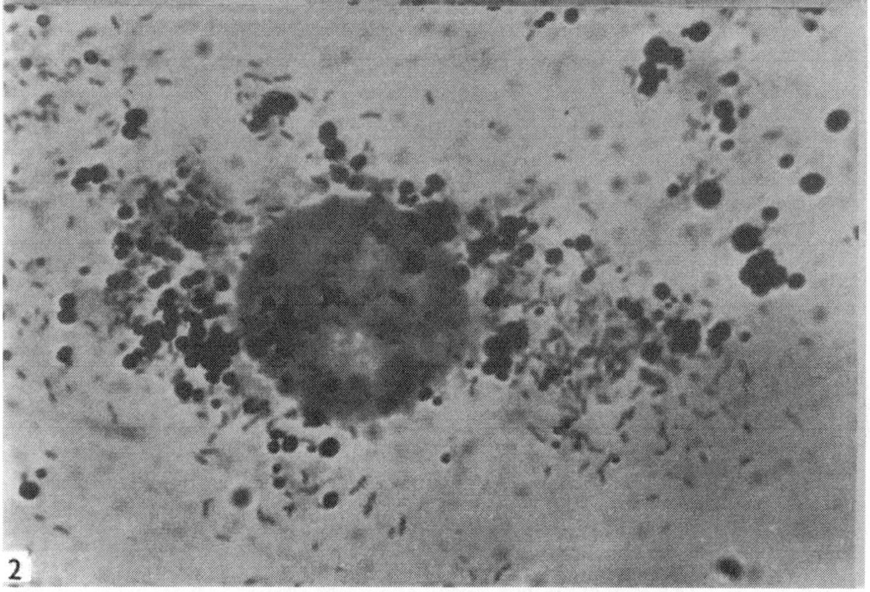

Fig. 1. Rickettsiae of the RMSF group in haemolymph of female *Dermacentor marginatus*.
Fig. 2. Rickettsiae of the RMSF group in gut of female *Dermacentor marginatus*.

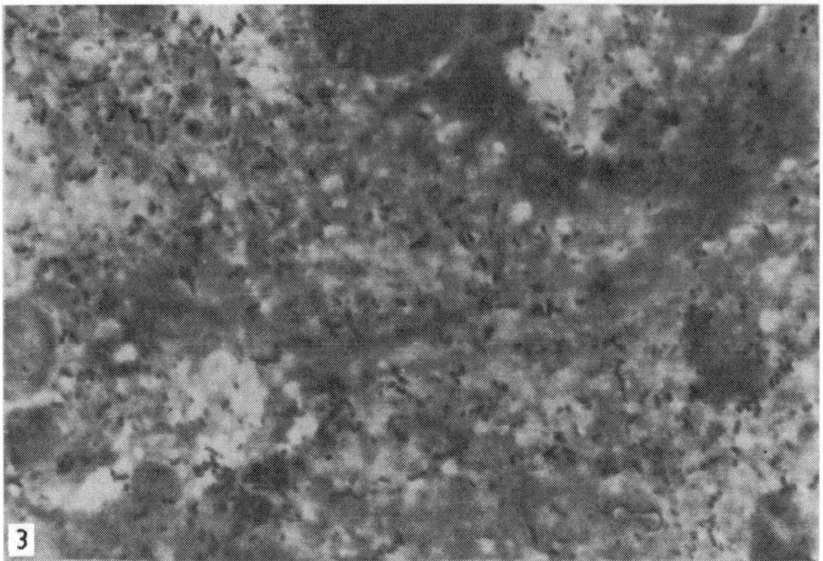

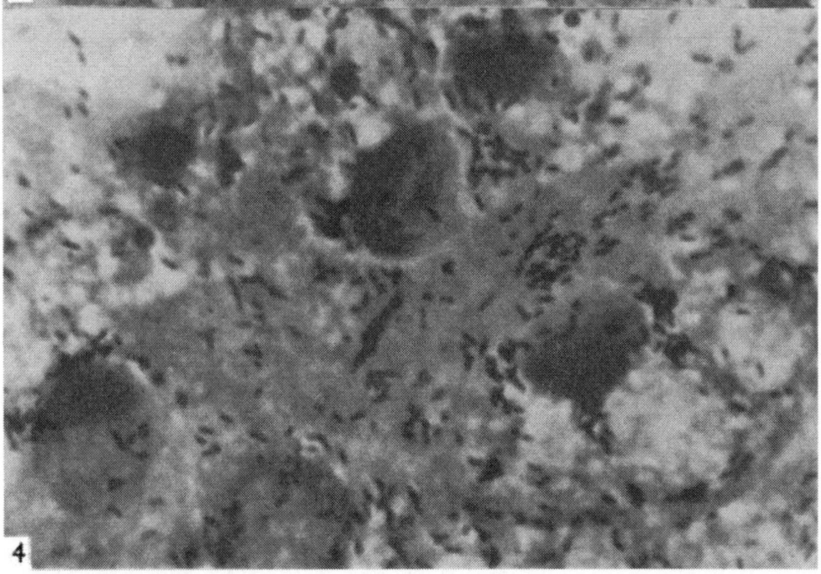

Fig. 3. **Rickettsiae of the RMSF group in connective tissue of female *Dermacentor marginatus*.**
Fig. 4. Rickettsiae of the RMSF group in cells of *Hyalomma dromedarii* in vitro, 8 days after the infection.

Proceedings of the 3rd International Congress of Acarology, Prague, 1971

ATTEMPTS TO USE HALF-ENGORGED TICKS FOR THE TITRATION OF RICKETTSIAE

J. ŘEHÁČEK, M. ŽUPANČIČOVÁ and E. KOVÁČOVÁ
Institute of Virology, Slovak Academy of Sciences, Bratislava, Czechoslovakia

A fundamental problem in experimental work with rickettsiae consists still in the establishment of a reliable medium for their cultivation. Most of rickettsiae multiply readily either in vertebrate and invertebrate tissue culture or embryonated hen eggs or in laboratory animals. This is not always satisfactory and use of laboratory animals such as guinea pigs, Syrian hamsters and mice brings technical problems and has more disadvantages from the economical point of view.

The present paper deals with the observations concerning use of half-engorged ticks as a substrate for the titration of several species of rickettsiae.

Material and Methods

Rickettsiae: The following strains of rickettsiae were used: *Rickettsia prowazeki*, strains Breinl and W 2; *R. mooseri*, a strain originated from Mexico and kindly supplied by Prof. Mooser, Zürich; *R. conori*, strain M_1; *R. akari*, strains Toger and Kaplan; rickettsiae of the Rocky Mountain Spotted Fever (RMSF) group isolated in Slovakia, strains B, 4 and 14, and *Coxiella burneti*, strains Florian, Nine Mile and 27. All strains originate from the collection of the Department of Rickettsiae of our Institute.

Ticks: *Hyalomma dromedarii* Koch from our laboratory stock, *Ixodes ricinus* L. and *Dermacentor pictus* Herm. collected in the field where no evidence of the rickettsiae presence was obtained by systematic investigations lasting several years, *Dermacentor marginatus* Sulz. ticks were collected in Central Slovakia in focus of Q fever and rickettsiae of the RMSF group. These ticks were, prior to experimental use, systematically examined for infestation with rickettsiae by haemocyte test before and after a blood meal. Only negative ticks were used in the experiments.

Arrangement of the experiments: Half-engorged ticks on healthy guinea pigs were carefully washed in running tap water and distilled water. Following that they were intracoelomically injected in the central part of the alloscutum by a tuberculine syringe using No. 1. needle with 0.01–0.05 ml with various doses of rickettsiae from an infected yolk sac. After injection they were stored at 30°C in test tubes in which the humidity was regulated by a periodical moistening with a few drops of distilled water on filter paper. Weekly, for a period of 4 weeks, they were removed, thoroughly washed in distilled water and bled on cover slips by the amputation of the frontal limbs. Dried haemolymph smears were flame fixed and stained by the Gimenez method (Gimenez, 1964) and in a few experiments by indirect methods of fluorescent antibody techniques (Weller and Coons, 1954). (Figs. 1–5.)

Results

First, engorged nymphs of *Hyalomma dromedarii* were tested as a substrate for titration of *Coxiella burneti.* Nymphs inoculated immediately, after dropping off healthy guinea pigs, with various doses of infected yolk sacs were allowed to hatch to adults which were tested for rickettsiae by the haemocyte test. Positive results were achieved in dilution of 10^{-2} and 10^{-3}. Because of a higher sensitivity of half-engorged ticks to *C. burneti* (see later), the application of larger volumes of inocula and the decrease in technical problems when working with adults, the method of involving nymphs in titration experiments was abandoned.

Most of the experiments were performed with half-engorged females. The titers of rickettsiae in ticks were first ascertained weekly for a period of 6 weeks following the inoculation but because of no change in titers between 4 and 6 weeks, we decided to establish ID_{50} at 4 weeks.

Use of half-engorged females of *I. ricinus* ticks for titration of various rickettsiae offered unreliable results. The sensitivity of ticks did not differ too much from that of other ticks but difficulties over tick breeding, injecting and the supply of haemolymph eliminated this tick species from further observations.

The results on using *Dermacentor marginatus, D. pictus* and *H. dromedarii* for the same purpose are summarized in Tables 1 and 2. These ticks demonstrated almost the same sensitivity to all rickettsiae species tested. The detection of rickettsiae in haemolymph by the Gimenez method as well as by the immunofluorescence technique showed almost the same results (Figs 1–5). The sensitivity of ticks to rickettsiae of the typhus group (*R. prowazeki, R. mooseri*) was low. *R. akari* multiplied in ticks unsatisfactorily. *R. conori, C. burneti* and rickettsiae of the RMSF group isolated in Slovakia multiplied in ticks very well. In the case of the last two mentioned organisms the ticks demonstrated a higher sensitivity to very small doses of rickettsiae than did the yolk sacs.

Rickettsiae which multiplied in ticks (*Coxiella burneti* and rickettsiae belonging to the RMSF group isolated in Slovakia) occurred in haemocytes of ticks in various amounts depending on the quantity of the agent inoculated and the time following the infection. The highest number of rickettsiae was always found in haemocytes in ticks inoculated with the largest quantity of agents but on testing at 4 weeks many haemocytes of ticks injected with the smallest amounts of rickettsiae demonstrated a massive infestation too.

Table 1. Attempts to use half–engorged ticks for the titration of rickettsiae

Rickettsiae	Strain	EID_{50} /0.2 ml	*Dermacentor marginatus* weeks				*Hyalomma dromedarii* weeks			
			1	2	3	4	1	2	3	4
Coxiella burneti	Florián	6.0*					5.0	6.0	7.0	7.5[1]
		4.3	6.5**	5.0	6.5	<6.0	2.0	4.0	6.0	5.5
		4.3					3.0	5.0	5.0	ND
		4.3					2.0	3.0	4.0	5.0
	Nine Mile						3.0	4.0	5.0	5.0
							4.5	5.5	6.5	7.0[2]
	27	8.0					3.0	4.5	5.5	6.7
Rickettsia akari	Toger	4.5	2.5	2.3	2.3	2.0	2.0	3.5	4.0	3.0
	Kaplan	4.0					1.5	2.0	2.2	ND
		4.0					ND	2.0	2.5	3.0[3]
Rickettsia conori	M_1	4.5	4.5	<2.0	<4.0	<4.0				
		3.0					2.0	2.5	3.5	ND
Rickettsia prowazeki	W II	4.0					3.5	3.0	4.0	ND
	Breinl	3.5	<2.0	<2.0	<2.0	<2.0				
Rickettsia mooseri		4.3	3.0	3.0	3.0	3.0				
		4.3					1.0	1.5	2.7	2.7
		<3.0					1.5	1.7	3.0	3.0
Rickettsiae of RMSF group isolated in Slovakia	B	3.0	4.5	4.5	4.5	4.0				
		4.0					3.5	5.0	5.0	ND
		3.2					ND	3.0	3.5	4.2
	14	3.2					ND	2.5	3.0	3.5

* log EID_{50} /0.2 ml,
** log ID_{50} /0.001–0.01 ml (drop of haemolymph),
[1] positive fluorescence ID_{50} – 6.5,
[2] ,, – 7.0,
[3] ,, – 3.0,
ND – not done

Table 2. Attempts to use half–engorged females *Dermacentor pictus* for the titration of rickettsiae

Rickettsiae	Strain	EID_{50} /0.2 ml	Ticks							
			Gimenez weeks				Immunofluorescence weeks			
			1	2	3	4	1	2	3	4
Coxiella burneti	Nine Mile		<4.0**	5.7	>8.0	>8.0	<4.0	4.3	ND	8.0
	27	8.0*	<4.0	5.7	>8.0	> 8.0	3.7	5.7	ND	7.5
Rickettsia akari	Kaplan	4.0	1.0	ND	1.0	ND	ND	1.0	ND	3.7
Rickettsia conori	M_1	4.5	2.0	ND	3.5	4.5	1.0	1.3	ND	4.5
Rickettsia prowazeki	W II	4.0	2.0	ND	1.0	1.0	1.0	1.0	ND	1.0
	Breinl	3.5	2.3	ND	<1.0	<1.0	1.0	1.0	ND	<1.0
Rickettsia mooseri		3.0	<1.0	<1.0	<1.0	1.0	<1.0	1.0	ND	1.0
Rickettsiae of RMSF group isolated in Slovakia	4	3.2	2.7	2.3	ND	5.5	2.0	1.0	ND	5.0
	14	3.2	2.0	ND	3.5	5.0	3.0	3.0	ND	5.0

* – log EID_{50} /0.2 ml, ** – log ID_{50} /0.001–0.01 ml (drop haemolymph), ND – not done.

Discussion

To make a final decision on which species of tick to use as a substrate for titration of rickettsiae, several peculiarities of tick behaviour should be taken into consideration. We tried to summarize a few data thought to be important from this point of view concerning ticks examined in our experiments (Table 3).

According to the present results it is possible to use various tick species as laboratory animals for titration of rickettsiae. The high sensitivity of half-engorged tick females to multiply often almost negligible amounts of *C. burneti* and rickettsiae of the RMSF group isolated in Slovakia offers further application of this substrate for experimental work with rickettsiae. We suppose this phenomenon to be probably useful, for example, for the cultivation of rickettsiae which would be concentrated for the preparation of antigen and in rickettsiae isolation experiments from material collected in the field. The isolation experiments would be performed directly in the field when no suitable conditions for work with laboratory animals and tissue cultures are available and when the material should be kept for a longer period.

Materials collected in nature are usually contaminated with undesirable micro-organisms which are not harmful to ticks. This is an advantage in the use of ticks in isolation experiments compared with that of vertebrate tissue culture and chick embryos.

Further application is to use half-engorged ticks for the transportation of rickettsiae for long distances, as Parker suggested already in 1942 with *Ornithodoros* ticks.

Ticks are naturally infested by symbionts of *Wohlbachia* genus. Because of simi-

Table 3. Data on tick capability to be used as a substrate for titration of rickettsiae

Species	*Ixodes ricinus*	*Dermacentor marginatus*	*Dermacentor pictus*	*Hyalomma dromedarii*
Possibility to collect ticks free of undesirable rickettsiae	easy	uneasy	very easy	no
Breeding of ticks in laboratory	difficult (long time period of development very sensitive to humidity changes	easy only in spring and summer time	easy only in spring and summer time	very easy for all year
Sensitivity of ticks to injuring (inoculation, bleeding)	very sensitive	insensitive	insensitive	insensitive
Volumes of inocula applied	low <0,1 ml	high >0,1 ml	high >0,1 ml	very high >0.5 ml
Supply of haemolymph	low	high	high	high
Recommendation for use	do not recommend	do not recommend	recommend as second	recommend as first

lar morphology to pathogenic rickettsiae, attention must be paid to their differentiation. Using the immunofluorescence method for this purpose will be very helpful.

References

Gimenez, D. F.: Staining rickettsiae in yolk sac cultures. Stain Technol. 39 : 135–140, 1964.

Parker, R. R.: *Ornithodoros* ticks as a medium for the transportation of diseae agents. Publ. Health Rep. (Wash.) 57 : 1963–1966, 1942.

Weller, T. H., and Coons, A. J.: Fluorescent antibody studies with agents of varicella and herpes zoster propagated in vitro. Proc. Soc. Exptl. Biol. (N. Y.) 86 : 789–794, 1954.

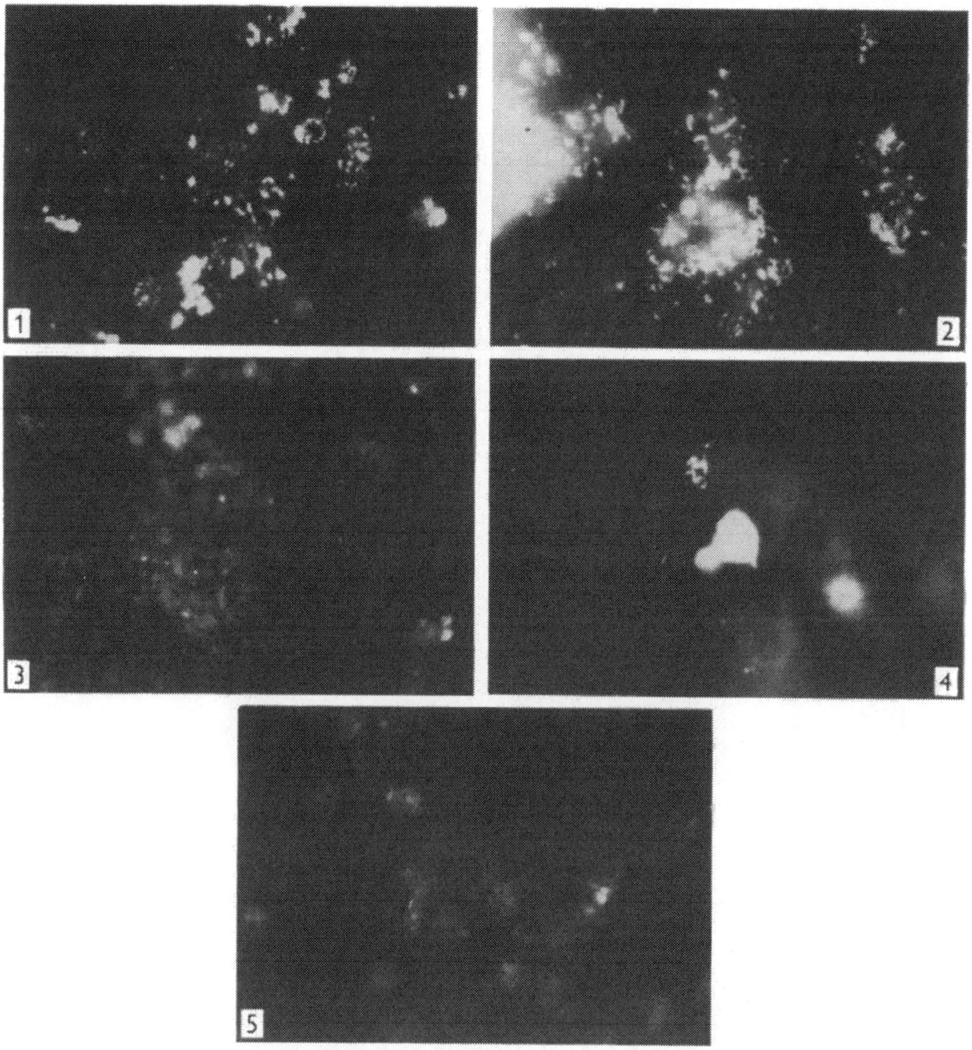

Fig. 1. Haemocytes of *Hyalomma dromedarii* infected with *Coxiella burneti*, Florián strain, dilution 10^{-2}, 4 weeks following the inoculation.

Fig. 2. Haemocytes of *Hyalomma dromedarii* infected with *Coxiella burneti*, Florián strain, dilution 10^{-4}, 4 weeks following the inoculation.

Fig. 3. Haemocytes of *Hyalomma dromedarii* infected with *Coxiella burneti*, Florián strain, dilution 10^{-6}, 4 weeks following the inoculation.

Fig. 4. Haemocytes of *Hyalomma dromedarii* infected with *Rickettsia akari*, Kaplan strain, dilution 10^{-2}, 4 weeks following the inoculation.

Fig. 5. Haemocytes of *Hyalomma dromedarii* infected with *Rickettsia akari*, Kaplan strain, dilution 10^{-3}, 4 weeks following the inoculation.

Section X

BIOLOGICAL, CHEMICAL AND PHYSICAL CONTROL

Chairman
C. A. Hall (USA)

Assistant Chairman
M. Přívora (Czechoslovakia)

CHLOROBENZILATE, AN ACARICIDE OF WIDE SPECTRUM AGAINST CITRUS MITES

H.H. ATTIAH, M.L. WAHBA and S.M. KODIRAH
Acarology Investigations, Crop Protection Department, Ministry of Agriculture, Dokki, Egypt

Among pests of economic importance infesting citrus trees in Egypt are citrus mites. Not only do they affect the yield owing to depletion of nourishment from infested trees, but they also cause injuries to the fruit and hence minimize their market value. Symptoms caused by rust mites, flat mites and bud mites are well known among citrus growers wherever citrus trees are grown. Although leprosis was reported as a result of flat mite infestations in Florida (Knorr et al., 1957) this disease has not yet been recorded in Egypt.

Three different species are considered of great economic importance in Egypt, namely the citrus rust mite *Phyllocoptruta oleivora* Ashmead, the citrus flat mite *Brevipalpus californicus* and the citrus brown mite *Eutetranychus orientalis* Klein. The citrus bud mite *Aceria sheldoni* Ewing is of importance only on lemon trees, since deformities are rarely found on orange or mandarine fruits. Attiah, 1969 recorded the spider mite *Tetranychus neocalidonicus* Marc Andre on citrus trees in a grove located at Abo Hommos Lower Egypt; specimens of the mite are now difficult to collect after withdrawal of infested trees. It is not known whether the previous species exists elsewhere since it is an omnivorous species.

Some different trials were carried on against the citrus mites (excluding *T. neocalidonicus)* to find out the best acaricide to be used against each species. Concerning the flat mite *B. californicus,* Attiah and Wahba concluded that kelthane 42% and chlorobenzilate 50% at rates of 125 and 100 cc per 100 litres water respectively are very efficient. The following trials were conducted against the rust mite, the brown mite and the bud mite.

Methods

A different number of orange or lemon trees was chosen in citrus groves for each of three different experiments according to different conditions, however the method of experimenting was the same. The complete randomized block design was followed; spraying was applied only once by means of a motor sprayer, i.e. in high volume spraying; samples were collected at different intervals and were taken to the laboratory for counting the mites under a dissecting microscope. Analysis of variance was done for the total population of mites existing on samples of each treatment on all sampling dates. All experiments were carried out at Sabbahia Experiment Station, near Alexandria.

The first experiment was conducted against both the rust mite and the citrus brown mite on orange trees. Six different compounds were tried on 22.7.1969, each treatment comprised 6 trees replicated 4 times, besides 24 trees left as control totalling 168 trees. Treatments are indicated in Tables 1 and 2.

Table 1. Populations of *Phyllocoptruta oleivora* following treatment

Treatment[a)]	Number of mites existing on 96 leaves[b)] after days indicated compared to the check						% reduction from check
	7	14	21	28	37	89	
Dithane Z.78 150 grams	0	0	8	11	65	357	87.9
Dithane M.45 100 grams	0	0	0	0	34	247	92.3
Torak 90 % 250 cc	14	90	301	373	417	1686	20.5
Kelthane 45 % 125 cc	14	34	57	202	159	1503	45.7
Chlorobenzilate 25 %, 150 cc	9	42	16	34	23	364	86.6
Roger 40 % 125 cc	48	419	421	503	679	707	23.4
Control	172	357	667	761	1020	645	

a) sprayed on 22.7.69, b) 24 leaves x 4 replicates

The second experiment was done on 11.7.1971 against the citrus rust mite on 144 orange trees, using five different compounds while the third experiment was conducted on the same date against the citrus bud mite infesting lemon trees. Eighty trees were used, i.e. 4 trees replicated 4 times for each of five different treatments including the control. Treatments of both experiments are shown in Tables 3 and 4 respectively.

Sampling for the bud mite was done by collecting 10 twigs from each replicate. Bud mites existing in 3 buds from each twig was counted under the microscope, totalling 120 buds on each sampling date for each treatment.

Table 2. Populations of *Eutetranychus orientalis* following treatment

Treatment [a)]	Number of mites existing on 96 leaves [b)] after days indicated compared to the check						% reduction from check
	7	14	21	28	37	89	
Dithane Z.78 150 grams	81	92	142	107	96	6	63.7
Dithane M.45 100 grams	115	71	79	97	165	9	62.9
Torak 90 % 250 cc	45	17	61	54	43	26	83.0
Kelthane 45 % 125 cc	13	2	6	8	2	0	97.9
Chlorobenzilate 25 %, 150 cc	58	27	46	17	166	0	78.3
Roger 40 % 125 cc	50	40	284	249	293	17	35.3
Control	632	120	261	257	161	11	

a) sprayed on 22.7.69, b) 24 leaves x 4 replicates

Results and Discussion

Compounds used against the citrus rust mite are shown in Tables1 and 3. It is evident from figures that dithane M45 at 100 and 120 grams, dithane Z78 at 150 grams, chlorobenzilate 25 % at 150 cc and chlorobenzilate 50 % at 100 cc all per 100 litres water were effective means of control against rust mites. Torak 90 % 250 cc, kelthane 45 % 125 cc, roger 40 % 125 cc, kelthane 18.5 % 250 cc, omite 57 % 75 cc and pacol 1000 cc per 100 litres water could be considered unsuccessful against the previous mite.

Considering the citrus brown mite, as shown in Table 2, kelthane 45 % 125 cc almost eradicated the mite for around three months. Torak 90 % 250 cc and kelthane 25 % 150 cc were not markedly different giving satisfactory control, while dithane Z78, 150 grams, dithane, M45, 100 grams and roger 40 % 125 cc were unsuccessful.

The citrus bud mite was successfully controlled by chlorobenzilate 50 % 100 cc, while omite 57 % 75 cc, pacol 1000 cc and malathion 300 cc were all inadequate (Table 3).

It is concluded that dithane M45, 100 and 120 grams and dithane Z78, 150 grams

Table 3. Effect of acaricides against the citrus rust mite *Phyllocoptruta oleivora* Ashmead on orange trees

Treatment[a)]	Average number of mites per leaf[b)] after days:					% reduction from check
	7	14	21	28	35	
Dithane M.45 120 grams	0	0	0.1	0.1	0.2	99.7
Chlorobenzilate 50 %, 100 cc	0.3	1.0	0.8	0.9	3.3	94.6
Kelthane 18.5 %, 250 cc	0.8	2.0	5.1	7.5	17.7	71.9
Omite 57 %, 75 cc	0.5	5.3	14.1	16.4	25.1	48.0
Pacol 1000 cc	2.6	7,2	17.6	20.6	32.5	31.7
Control	2.8	17.6	22.5	32.3	42.6	

a) applied 11.7.71, per 100 litres water, b) based on 80 leaves, 20 x 4 replicates

Table 4. Effect of acaricides against the citrus bud mite *Aceria sheldoni* Ewing on lemon trees

Treatment[a)]	Average number of mites per bud[b)] after days:					% reduction from check
	7	14	21	28	35	
Chlorobenzilate 50 %, 100 cc	2.3	3.2	2.5	2.2	0.5	88.9
Omite 57 %, 75 cc	6.1	7.6	10.6	8.6	2.8	63.1
Pacol, 1000 cc	3.4	7.9	9.0	6.2	2.6	69.9
Malathion, 300 cc	15.5	13.9	17.8	9.2	4.4	37.0
Control	22.3	23.6	24.5	16.6	9.6	

a) applied 11.7.71, per 100 litres water; b) based on 120 buds, 30 x 4 replicates

are the best treatments against the citrus rust mite, but are not effective against the brown and the bud mites . Kelthane 45 % 125 cc is extremely effective against the brown mite, but that compound is not satisfactory against the rust mite (not tried against the bud mite). Chlorobenzilate 50 % 100 cc is considered the best treatment against the citrus bud mite, in the meantime it is effective against both the rust mite and the brown mite (since 150 cc of the formulation 25 % was satisfactory against the latter mite). It is of interest to mention that the same compound at the same dosage was effective against the citrus flat mite (Attiah and Wahba, a previous paper). All results were confirmed by statistical analysis.

References

Attiah, H. H., Wahba, M. L.: Phosphorus compounds as a cause of flat mite increase. III. Int. Congr. Acarol., 1971.

Knorr, L. C., Smit, R. F., Du Charme, E. P.: Handbook of citrus diseases in Florida. Bull. 587, U. Florida Exptl. Stations, Gainesville, Florida, 1957.

Proceedings of the 3rd International Congress of Acarology, Prague, 1971

INFLUENCE OF CERTAIN PHYTOCIDES AND ANTIBIOTICS ON THE REPRODUCTION AND DEVELOPMENT OF *TETRANYCHUS* MITES

K. S. CHHABRA
Department of Zoology and Entomology, Punjab Agricultural University, Ludhiana, India

One of the serious problems in the control of plant pests is the development of resistance by them towards pesticides. This problem is of great theoretical and practical importance. At this time we have already registered more than 112 species of insect pests and mites resistant to different pesticides. To control them is becoming difficult year after year. At present, species of spider mites *Tetranychus telarius, T. cinnabarinus, T. viennensis, T. mcdanieli, T. atlanticus, Panonychus ulmi, P. citri*, are known to be resistant to phosphoorganic and other specific acaricides. Thus, the mass reproduction of *Tetranychus* mites under the conditions of the adapted system of chemical control, may be the reason for mite population becoming resistant to the acaricides.

World assortment of insecticides including acaricides is vast, but in spite of that there exists a large number of unsolved problems. In view of this, the present studies were carried out to screen certain new biological active preparations having acaricidal properties. The studies were conducted in the greenhouse, under controlled ecological conditions. The object for the test was *Tetranychus telarius* L. and the food plant was the bean *(Phaseolus vulgaris* L.).

In the course of these studies, it was recorded that a preparation Kansatin-I, extracted from the dried leaves of the hemp plant is toxic to all the development stages of this mite. Thus eggs of different age groups treated with 0.8, 0.9, and 1.0 % of Kansatin-I gave mortality from 24.9 - 79.1 %. In this respect it was noted that mortality of three days old eggs sharply increased in comparison to that of one and two days old eggs. When this preparation was tested against different postembryonic stages of this mite the mortality of larvae ranged between 75.1 - 86.0 % and that of the inactive larval stage 65.4 - 75.8 %. Susceptibility of protonymphs and inactive protonymphal stages to this acaricide fluctuated from 69.7 - 82.5 % and 59.9 - 68.2 %, respectively. Mortality of deutonymphs and its inactive stage ranged from 78.3 - -90.0 % and 70.0 - 83.3 %. Adult stages of the mite were highly susceptible to the Kansatin-I in the concentrations of 0.8, 0.9 and 1.0 %. Their mortality varied in the case of males between 95.0 - 100 % and for female mites 90.0 - 98.3 %.

Another preparation Dinonylphthalate (DNNP) available in the wild vine plant (*Parthenocissus quinquefolia* Michx). was proved to be acaricidal to this mite in the concentrations of 0.4, 0.5 and 0.6 %. Thus, the eggs of the different age groups treated with DNNP killed from 59.0 - 79.0 %. It was interesting to note that the two days old eggs were somewhat resistant to this acaricide in comparison to that of one and three days old. At the same time it was observed that mortality of eggs occurs at the late embryonic stage and the larvae hatched out from the treated eggs were deformed (not fully developed). Mortality of treated larvae with DNNP ranged from 74.3 - 85.5 %. Susceptibility of protonymphs and its inactive stage to the DNNP fluctuated from 71.0 - 82.0 %. Treatment of deutonymphs and its inactive stages with this acaricide gave mortality of 63.6 - 83.3 % and 43.3 - 80.0 %, respectively. Adult stages were somewhat tolerant to this preparation and their mortality fluctuated from 30.0 - 51.6 % in the case of female mites and that of males 45.0 - 78.5 %.

Fusariotoxin, a preparation from the microfungus, *Fusarium sporotrichilla* Bill. proved to be highly toxic to all the development stages of this mite at the concentrations of 0.01, 0.02 and 0.03 % and also induced sterility in the treated female mites. Thus, treatment of eggs of different age groups with this toxin gave mortality of 10.0 - 25.0 % and the mortality of larvae hatched out from the treated eggs ranged from 80.0 - 93.0 % 48 hrs after hatching. Susceptibility of larvae to this toxin 48 hrs after treatment fluctuated from 76.1 - 95.5 % and that of the inactive larval stage from 75.7 - 84.2 %. Mortality of treated proto- and deutonymphs 48 hrs after treatment ranged from 70.0 - 88.7 % and 71.6 - 91.6 % and for their inactive stages 73.0 - 93.0 % and 63.3 - 88.8 %, respectively. An altogether different picture was observed in the case of treated adult stages of this mite. Thus, males treated with this toxin at the concentrations 0.01, 0.02 and 0.03 % 48 hrs after treatment killed 28.3 - 41.6 %, and the treated females 16.6 - 30.0 %. During the course of these studies it was observed that the residual toxic effect of this toxin remained for 15 days on the treated plants.

At the time of carrying out these studies on the acaricidal properties of the new preparations, it was observed that fusariotoxin possessed a sterilizing effect and reduced the egg-laying of the treated female mites. Thus, females treated with 0.01, 0.02 and 0.03 % of toxin after 4 days of the treatment laid 46.9, 26.5 and 20.5 % of eggs respectively. After 8 days of the treatment under the same ecological conditions the reduction in the egg-laying remained 53.1, 73.5 and 79.5 %, respectively. After 12 days of the treatment in the case of females treated with 0.01 % of toxin this effect lessened and the reduction in the egg-laying remained 25.3 %. In the case of 0.02 and 0.03 % of toxin treated females the reduction in the egg-laying was recorded 66.0 and 75.9 %, respectively.

As fusariotoxin possesses an acaricidal and sterilizing effect, studies were carried out to observe the effect of the toxin in the population dynamics of this mites. Under the concentrations of 0.01, 0.02 and 0.03 % of toxin the population of mites 10 days after treatment on the treated plants remained at 28.8, 19.2 and 15.9 %, respectively, whereas, on the untreated plants it was 73.4 %. 15 days after treatment with 0.01 % of toxin, on the treated plants the population of mites increased and it remained at 59.6 %, whereas, in the case of 0.02 and 0.03 % of toxin it was further reduced and remained at 18.0 and 9.1 %, respectively, whilst at the same time on the untreated plants it was increased (104.4 %).

In addition to this, studies were also carried out to observe the effect of fusariotoxin on the development of the different stages of this mite. It has been recorded that the development of the various stages of this mite treated with 0.01, 0.02 and 0.03 % of toxin is depressed to the range of 3 to 4 days in comparison to that of untreated ones. It was interesting to note that the incubation period of the treated eggs remained the same as of the untreated ones. At the same time it was observed that the treated post-embryonic stages of this mite stopped feeding and their body coulour became transparent.

RESISTANCE TO ACARICIDES IN FRUIT TREE RED SPIDER MITE, *PANONYCHUS ULMI* (KOCH), IN SOUTH-EAST ENGLAND

J. E. CRANHAM
East Malling Research Station, Maidstone, Kent, England

Spider mites have a marked capacity for evolving multi-resistant strains and this development seems to be very largely - as A. W. A. Brown (1961) has called it - a "one-way street". This implies the evolution of resistant genotypes which are equally fit and competitive with the susceptible genotype, so that resistance is stable under relaxation of the specific selection pressure. Much of the published work in this field of study relates to *Tetranychus urticae*, *T. cinnabarinus* and *T. pacificus* and there has been relatively little detailed study of *Panonychus ulmi*.

In the present paper, I shall briefly review the development of resistance in *P. ulmi* on top fruit in south-east England over the last 20 years, and our findings on the current status of resistance in work over the last four years. In testing for resistance in field populations, this work was greatly facilitated by the development of a simple assay technique; this involves the use of newly-hatched larvae, hatched from winter eggs on small fragments of bark directly on to leaf disks of Myrobalan plum. Details and advantages of this technique have been described (Cranham, 1969).

The historical perspective

Over the last 20 years, as resistance to one chemical or group of chemicals has developed and become widespread, growers have switched to using chemicals of other groups. Fig. 1 attempts to summarize the general trend of usage and development of resistance. In general terms, the marked development of resistance to a chemical group, i.e. widespread failure to obtain adequate control, has taken about five years of regular use (about 20 generations of the mite). This was reasonably true, successively, for the chlorbenside group, for the organophosphates (O.P's) and for tetradifon. With binapacryl, resistance was often apparent within 3-4 years, but resistance to dinocap has not been found even where that material had been used for ten years. Dicofol has also been used for nearly ten years but only locally and occasionally. Usage has increased following the failure of tetradifon, but it has seldom been used regularly on the same orchard year after year, and usually then only as a single summer spray. With this pattern of usage, resistance has been detected on only two sites and is definitely not of general occurrence as yet.

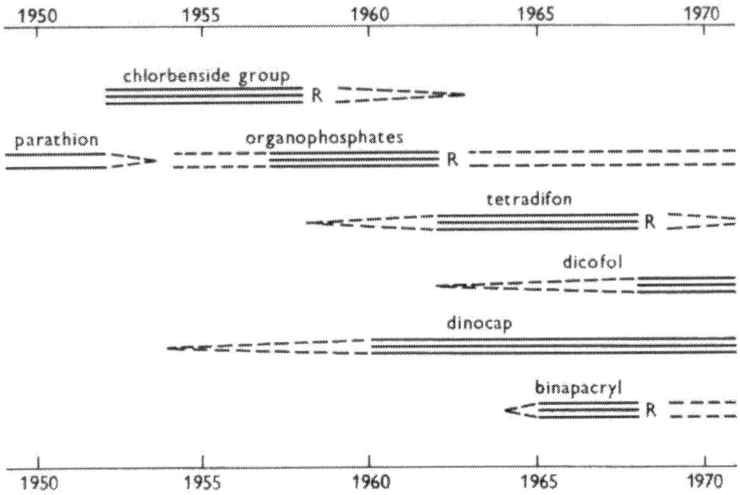

Fig. 1. The development of resistance to acaricides in *Panonychus ulmi* in south-east England. The solid lines indicate the years when materials were widely used as effective acaricides; R indicates the development of widespread resistance.

Each new type of resistance has been specific to a high degree. Thus, the resistance to the chlorbenside group (including fenson and chlorfenson) did not involve tetradifon (and tetrasul); the latter are also ovo-larvicides but clearly have a different mode of action. With the O.P's, it is possible that some degree of resistance to parathion developed in the early 1950's but was then unrecognized. Later, with the more extensive use of a wider range of O.P.'s - including demeton-S-methyl, dimethoate, vamidothion and azinphosmethyl - resistance developed to all those then in common use; but there was no cross-resistance to unrelated chemical groups.

The present status of resistance

Testing for restistance in field populations necessitated establishing the range of response of susceptible mites; and desirably this had to be done for a number of different populations which had not been exposed to the test acaricide. Moreover, when an acaricide had been very widely used, there was evidence that dispersal had carried some resistant mites to unsprayed orchards. Some susceptible samples were obtained from West-of-England orchards where chemical usage had been minimal.

Comparison of resistance to chlorbenside and to the O.P.'s is interesting. The chlorbenside group have not been used in most orchards for over ten years. However, in 1968-70, it was difficult to find samples which were clearly susceptible. It would seem that resistance is quite stable and that dispersal had spread some resistant mites to unsprayed orchards. The accumulated results on a wide range of samples indicate an LC_{50} for susceptible mites of 20 p.p.m. and for resistant mites, 1,000 p.p.m. Using a discriminating dose of 125 p.p.m., all samples included a resistant fraction, usually 10-30 %, but in samples from two orchards it was 90 %.

With O.P.'s, on the other hand, there has been no relaxation of selection following development of resistance in the early 1960's, since O.P.'s have continued in common use for post-blossom insect control. It was possible to establish the susceptible response on samples from four different orchards in isolated situations with no history of post-blossom O.P.'s. However, such orchards are rare. Tests with discriminating doses of dimethoate (31 p.p.m.) and vamidothion (15 p.p.m.) showed that resistant mites formed virtually 100 % of the population in many Essex orchards and 50-100 % in most Kent orchards. O.P.-resistant mites are, in fact, now the commonest type in S.E. England. Examination of the spectrum of resistance to 20 different O.P's has shown that the resistance factor (in larval tests) varies considerably, being 8x to 32x for most chemicals but 4x or less for a few, such as azinphos-ethyl, "Torak" [5-(2-chloro-1-phthalimidoethyl) 0,0-diethyl phosphorodithioate], and "HOE 2960" [1-phenyl-3-(0,0-diethyl-thionophosphoryl)-1, 2, 4-triazol]. These latter can provide good field control. As yet, we have no evidence of more than one mechanism and spectrum of resistance to O.P's, whatever the selecting O.P. Cross-resistance to carbamates, including formetanate, is of a low order, 2x to 4x.

Tetradifon. Resistance to tetradifon, with cross-resistance to tetrasul, occurs virtually only in those orchards with a marked history of usage for 4-8 years, probably under half the number of commercial orchards. In these a proportion of the population is clearly multi-resistant to O.P's and tetradifon. Since resistance was found in 1967-68 to be widespread, usage has declined considerably. In 1970, checks after two years non-use, on orchards where resistance had been found, showed a minority fraction of resistant mites. This suggests stability, but in laboratory work these multi-resistant mites showed markedly reduced fecundity and were more susceptible to some unrelated acaricides, e.g. binapacryl, chlorphenamidine. Further work is indicated.

Binapacryl and dinocap. These dinitro funigicides are in common use, primarily for the control of apple mildew *(Podosphaera leucotricha)*, but also providing a "bonus" acaricidal side-effect. This reduces the frequency with which other specific acaricides are needed (e.g. to one summer spray of dicofol in alternate years), and consequently delays development of resistance to the latter. Against this background, the rather rapid development of mite resistance to binapacryl (8x) is a serious loss, especially since extensive larval tests on a range of field samples revealed a low degree of cross-resistance (< 2x) to dinocap. On the other hand, dinocap remains the outstanding exception - no resistance to it has been detected in orchards where it has been used for up to ten years or more.

Mites resistant to binapacryl show a similar level of cross-resistance to the related material dinobuton. Indeed, it is unexpected that binapacryl and dinocap do not both produce mutual cross-resistance of a similar order. The toxic symptoms look the same, but the facts suggest a difference in the toxic pathway if not the site of action.

New materials. Despite the development of multi-resistant strains, mite control presents no serious difficulties at present because of the commercial availability of new acaricides, each apparently differing in mode of action and outside the existing spectrum of resistance. These include fenazaflor, "Plictran" (tricyclohexyltin hydroxide) and formetanate. Together with the older material dicofol they form a considerable armoury of acaricides, at least for the 1970's. The best practical advice would seem to be to use one material on a farm for not more than three years running, then to switch to another material for three years, and so on. The available materials permit this.

References

Brown, A.W.A.: Insecticide resistance - genetic implications and applications. Wld Rev. Pest Control, 6: 104-14, 1967.

Cranham, J.E.: Monitoring for resistance of fruit tree red spider mite *(Panonychus ulmi)* to acaricides. Proc. 5th Br. Insect. Fungi. Conf. (1969), 104-10, 1969.

THE MITES ASSOCIATED WITH ARMORED SCALE INSECTS

U. GERSON
The Hebrew University, Faculty of Agriculture
Rehovot, Israel

Mites are often reported to be associated with injurious armored scale insects (Homoptera: Diaspididae), but no review on this association is available. The relevant mites belong to the Astigmata, Mesostigmata and Prostigmata, and may be predaceous, mycophagous or saprophagous.

The astigmatic *Hemisarcoptes* is the best known acarine predator of diaspidids. Four species are currently recognized, but it is very difficult to separate them and a comparative systematic study is needed. *Hemisarcoptes* occurs in many parts of the world, from southern Quebec (Samarasinghe and Leroux, 1966) to Mauritius (Williams, 1970). It has also become a pest of scale insects reared in the laboratory as hosts for beneficial coccinellid predators (Sellers and Robinson, 1950).

Hemisarcoptes is disseminated in the field by the lady beetle *Chilocorus;* three beetle species are known to be specifically associated with three species of the mite (Gerson, 1967b). This association was utilized by Bedford (1949) to introduce *Hemisarcoptes* from California into Bermuda. In Israel a close synchronization exists between the mite and *Chilocorus*, as both become most abundant in July. On citrus bark *Hemisarcoptes* attacks the chaff scale in larger numbers than do either of two hymenopterous parasites of the pest (Gerson, 1967a). In one case the mite destroyed about 70 % of a California red scale population on citrus. *Hemisarcoptes* thus has a certain potential for biological control, but systematic and biological studies are needed before it can be put to good use.

Another astigmatic mite, *Thyreophagus*, is often found under the shields of dead armored scale insects. In the laboratory this mite has been continuously reared on fungi.

Mites of the mesostigmatic family Phytoseiidae at times devour armored scale insect larvae (crawlers) (McMurtry, 1963), but feeding on such prey appears to be rather fortuitous.

The Prostigmata have the largest number of species which feed on armored scales, the Raphignathoidea and the Cheyletidae being especially noteworthy. *Saniosulus*, a voracious predator of crawlers (Gerson and Blumberg, 1969), *Eupalopsis* (Gerson, 1967a) and *Neophyllobius* (Richards, 1962) belong to the Raphignathoidea. *Cheletogenes, Cheletomimus* and *Hemicheyletia* are the better known cheyletid predators of armored scales.

Cheletogenes, the most abundant acarine predator on citrus bark in Israel, completes a generation in about 90 days at 28°C while feeding on one crawler a day (Avidov et al., 1969). The mite has a maximum of 13 progeny per female and is fairly tolerant to starvation, desiccation and some pesticides. *Cheletomimus* feeds on citrus scale insects in Israel (Gerson, 1967a) and on the latania scale on avocado in California (Ebeling, 1959). *Hemicheyletia* has been studied in Florida by Kanavel and Selhime (1967).

Balaustium and *Cunaxoides* are amongst other prostigmatic mites reported to feed on armored scales (Ebeling, 1959; Baker and Wharton, 1952).

Another type of association exists between soft scales (Coccidae) and mites of the family Tydeidae. The latter feed on honeydew excreted by the former (Inserra, 1967).

To sum up, a better understanding of the diverse associations existing between mites and scale insects, besides adding to general knowledge of the biology of the mites, may also contribute to the biological control of armored scale insect pests.

References

Avidov, Z., Blumberg, D., Gerson, U.: *Cheletogenes ornatus*, (Acarina: Cheyletidae), a predator of the chaff scale on citrus in Israel. Israel J. Entomol. 3: 77--93, 1968.

Baker, E. W., Wharton, G. W.: An Introduction to Acarology. Macmillan, New York, 1952.

Bedford, E. C. G.: Report of the plant pathologist for the year 1949. Department of Agriculture, Bermuda, 1949.

Ebeling, W.: Subtropical Fruit Pests. University of California, Division of Agricultural Sciences, 1959.

Gerson, U.: The natural enemies of the chaff scale, *Parlatoria pergandii* Comstock, in Israel. Entomophaga 12: 97-109, 1967a.

Gerson, U.: Observations on *Hemisarcoptes coccophagus* Meyer (Astigmata: Hemisarcoptidae), with a new synonym. Acarologia 9: 632-638, 1967b.

Gerson, U., Blumerg, D.: Biological notes on the mite *Saniosulus nudus*. J. Econ. Entomol. 62: 729-730, 1969.

Inserra, R.: Precisazioni morfobiologiche su *Lorryia formosa* Cooreman (Acarina, Tydeidae). Boll. Lab. Entomol. Agr. "Filippo Silvestri" Portici 25: 295-316, 1967.

Kanavel, R.F., and Selhime, A.G.: Biological studies on *Paracheyletia bakeri* (Acarina: Cheyletidae). Flo. Entomolo. 50: 107-113, 1967.

McMurtry, J. A.: Diaspidine scale insects as prey for certain phytoseiid mites. Adv. Acarology 1: 151-154, 1963.

Richards, A. M.: The oyster-shell scale, *Quadraspidiotus ostreaeformis* (Curtis), in the Christchurch district of New Zealand. N.Z.J. Agr. Res. 5: 95-100, 1962.

Samarasinghe, S., LeRoux, E.J.: The biology and dynamics of the oystershell scale, *Lepidosaphes ulmi* (L.) (Homoptera: Coccidae) on apple in Quebec. Ann. Entomol. Soc. Quebec 11: 206-292, 1966.

Sellers, W.F., Robinson, G.G.: The effect of the miticide Neotran upon the laboratory production of *Aspidiotus lataniae* Signoret as a coccinelled food. Can. Entomol. 82: 170-173, 1950.

Williams, J.R.: Studies on the biology, ecology and economic importance of the sugarcane scale insect, *Aulacaspis tegalensis* (Zhnt.) (Diaspididae) in Mauritius. Bull. Entomol. Res. 60 : 61-95, 1970.

MANY-YEAR REACTION OF THE POPULATION OF *IXODES PERSULCATUS* P. SCH. TICKS TO A SINGLE INTRODUCTION OF STABLE RESIDUAL ACARICIDE INTO THEIR HABITATS

N. N. GORCHAKOVSKAYA, YU. S. KOROTKOV
Institute of Poliomyelitis, Moscow, USSR

The objective of the work was (1) to study the conditions for the formation of the population of ticks after its extermination on a relatively large territory (over 12 thousand hectars); (2) to establish a relationship between the duration of the effect of extermination of *I. persulcatus*, the duration of residual action of DDT, and rate of introduction of ticks, when there is a certain abundance of ticks and of their hosts in the adjacent untreated territory.

The material was collected in spring-summer seasons of 1955-1970 in foci of tick-borne encephalitis in the south of Western Siberia (Salair). The station is located in a large tract of forest, where the indigenous chernevaya taiga in the major portion of the territory, has been substituted by secondary aspen-fir forests and partly by aspen forests. Mixed and coniferous forests occupy 75-80 % of the area, and aspen forests and overgrown clearings 10-15 %. The forests are very greatly cluttered up, and vigorous development of underbrush and high grass is characteristic. The population density of small mammals - hosts of preimago stages - is high and stable: the rate of catching of the animals *(Clethrionomys, Microtus, Sorex)* annually varies; in May within 3.5 to 31.2 %, in July-August within 25.7 - 66.5 % (Koshkina, 1970). Alpine hare (0.5 - 5 per kilometer of the route) and cattle (200 - 10 head per km^2 as one goes deeper in the taiga), play an important role in providing food for ticks.

The population density of the ticks was determined by the number of imagoes on the average per one hour and 1 km of the route, and that for larvae and nymphs - by infestation of rodents and insectivora, by % of lines of snap-traps in which animals with larvae and nymphs were found (lines of 25 traps). When the forests were examined 1-3 seasons before treatment during the period of mass activation of ticks, from 0.1 - 1 to 10 - 15 preimagoes per animal and from 5-10 to 40-70 imagoes per person plus flag per one hour of registration were found. Distribution of imagoes in the area of the station is shown in Fig. 1.

The forest has been treated once with 10% DDT dust from an airplane (0.3 - 0.5 g of technical substance per 1 m^2) during 1959 - 1962 according to the tactics of consecutive attack of the focus (Gorchakovskaya, 1962, 1965). The treated territory measures 25 - 30 km along the front and 10 - 12 km deep. As a result of this treatment, a high effect of tick extermination has been obtained and has persisted for many years in a large area.

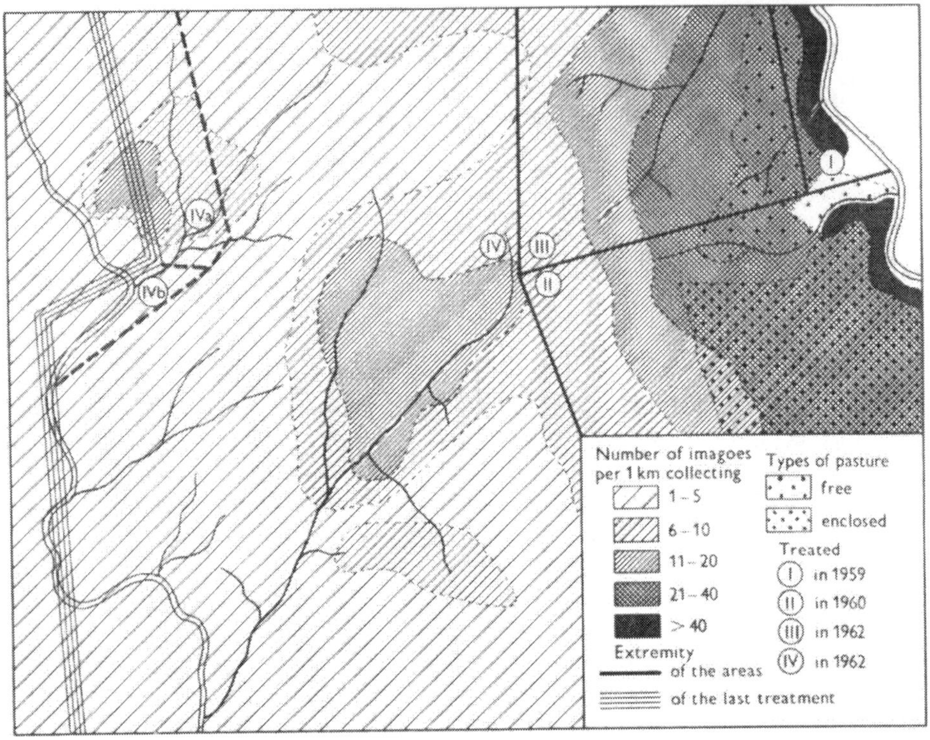

Fig. 1. Scheme of the distribution of imagoes *I. persulcatus* before treatment in June and stages of aviation treatment of the stationary with DDT dust (0.3-0.5 g/ m^2).

In the year of treatment, the population density of preimago was reduced by 41.2-93.6 % in comparison with the initial level (Table 1), and that of imago by 92-100 % (Table 2). The residual density of ticks on the average per season is shown in the Tables. The maximum values in the season were 0.21 tick per 1 animal with the infestation rate of 6.9 % of animals and 2.6 imago per 1 hour of registration. In some cases, the residual population density of ticks in the year of treatment was comparable to the abundance of ticks in a low-density population. Therefore, the Tables also show the calculation of the extermination effect with regard to such controls.

In the 3rd post-treatment year in almost all areas the values of tick abundance are reduced to 0, and the effect of extermination increases to 100 %. Previously, persistence of acaricidal properties in the forest floor in the 2nd and partially in the 3rd seasons (Gorchakovskaya, 1957, 1960, 1965; Gorchakovskaya et al., 1959) had been established. It appears that the contact of ticks with poison for 2-3 seasons is necessary for extermination of all ticks of the population of a species with a 3-year cycle of development.

Recovery of the density of ticks after treatment of their habitats is possible due to: (1) residual density of ticks in the area; (2) flaws of treatment, and (3)

Table 1. INDICES OF ABUNDANCE AND THE EFFECT OF EXTERMINATION OF PREIMAGO PHASES IN AREAS TREATED WITH DDT

Years after treatment	Area I				Areas II – III				Area IV				Areas along the border		
	No. of animals examined	Index of abundance	Effect of extermination to the average initial	Effect of extermination to the minimum in control	No. of animals examined	Index of abundance	Effect of extermination to the average initial	Effect of extermination to the minimum in control	No. of animals examined	Index of abundance	Effect of extermination to the average initial	Effect of extermination to the minimum in control	No. of animals examined	Index of abundance	Effect of extermination to the minimum in control
1	188	0.40	93.6	0.0	1278	0.23	64.1	0.0	140	0.1	41.2	40.0	150	0.06	64.7
2	14	0.08	98.6	47.0	2225	0.05	92.2	70.0	500	0.02	89.0	89.0	171	0.13	0.0
3	475	0.0	100.	100.	1295	0.005	99.9	97.0	538	0.0	100.	100.	264	0.09	47.0
4	105	0.0	100.	100.	2682	0.05	92.0	70.0	521	0.019	88.9	88.9	33	0.08	53.0
5	170	0.0	100.	100.	1668	0.01	98.5	94.0	413	0.005	97.1	97.1	117	0.19	0.0
6	217	0.09	98.6	47.0	1757	0.04	93.7	76.5	104	0.0	100.	100.	95	0.08	53.0
7	31	0.09	98.6	47.0	1575	0.09	98.5	86.0	750	0.01	94.1	94.1	302	0.24	0.0
8	103	0.04	99.4	76.5	531	0.06	91.0	65.0	224	0.04	76.6	76.6	483	0.17	0.0
9	–	–	–	–	1284	0.09	85.9	46.5	300	0.21	0.0	0.0	565	0.19	0.0
10	23	0.04	99.4	76.5	623	0.033	94.9	80.0							
11	216	0.81	87.0	0.0	879	0.24	62.3	0.0							
12	347	0.38	93.9	0.0											

The average initial index of abundance in areas: No I – 6.2, Nos. II - III - 0.64, No. IV - 0.17.
In control areas: abundance 0.06 – 8.2,
rate of occurrence 15.2 – 72.0 %.

Table 2. THE EFFECT OF EXTERMINATION OF ADULT TICKS

	Area I				Areas II – III				Area IV				Areas along the border of treatment			
Years after treatment	No. of km count	No. of ticks per km	Effect to the average initial	Effect to the minimum in control	No. of km count	No. of ticks per km	Effect to the average initial	Effect to the minimum in control	No. of km count	No. of ticks per km	Effect to the average initial	Effect to the minimum in control	No. of km count	No. of ticks per km	Effect to the average initial	Effect to the minimum in control
1	75	0.53	99.0	47.0	90	0.26	98.0	74	64	0.0	100.	100.	130	0.4	92.0	60
2	59	0.02	99.9	98.0	200	0.14	99.0	86	198	0.08	99.8	92.0	85	0.1	98.0	90
3	60	0.0	100.	100.	85	0.0	100.0	100.	52	0.0	100.	100.	95	0.01	99.8	99
4	82	0.0	100.	100.	261	0.02	99.8	97	107	0.0	100.	100.	80	0.2	96.0	80
5	72	0.0	100.	100.	232	0.0	100.	100.	107	0.0	100.	100.	40	0.5	90.0	50
6	15	0.08	99.8	92.0	193	0.0	100.	100.	20	0.0	100.	100.	30	0.7	86.0	30
7	59	0.0	100.	100.	185	0.02	99.9	98	68	0.04	99.7	96.0	60	0.8	84.0	20
8	26	0.0	100.	100.	103	0.03	99.8	97	20	0.05	99.6	95.0	110	1.1	78.0	0
9	101	0.1	99.8	90.0	85	0.05	99.7	95	70	0.5	96.7	50.0	190	1.6	68.0	0
10	56	0.35	99.3	65.0	40	0.2	98.7	80								
11	30	0.8	98.4	20.0												
12	60	1.5	97.0													

The average initial index of abundance in the end of May, June in areas:
No. I – 50 ticks per km of the rout, Nos. II-IV – 15 ticks per km of the route, minimum in the control – 1 tick per km of the route.

introduction of ticks by their hosts. In the territory under study there were no flaws in treatment of the forest; the residual density declined to zero by the 3rd season. In the territory completely freed from ticks, restoration of the population might occur only due to introduction. Thus, we were able to observe the fact of death of the entire population of *I. persulcatus* under condition of treatment of a vast area and subsequently to study the importance of introduction of ticks from the adjacent untreated areas of a forest tract. Only large animals and birds can bring ticks (mostly engorged females and nymphs) to areas distant from the border. Along the

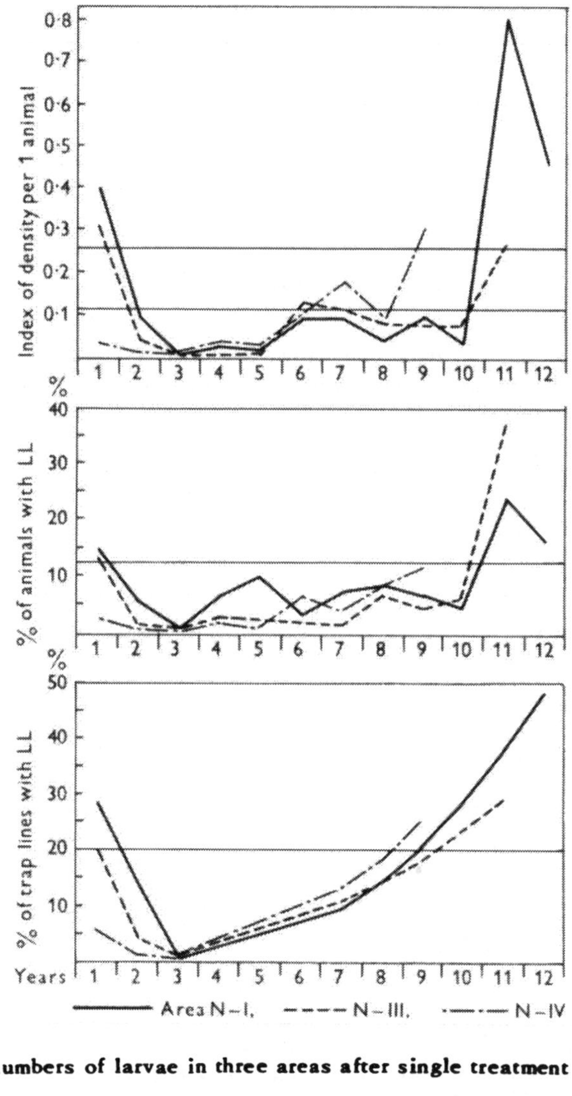

Fig. 2. Changes in numbers of larvae in three areas after single treatment with DDT dust in next years.

border, small and moderate mammals constantly bring in larvae and nymphs.

The high effect of extermination at I - III areas of the station persisted for 10 - 12 years, in the IVth for 9 years (observation period). The effect of extermination of larvae and nymphs in the I - III areas began to decline only in the 11th season, in the IVth area in the 8 - 9th season. Even in the border one-kilometer band the effect of extermination of adult ticks began to decline slowly only in the 5 - 6th season; in the 9th season it was still 68 %. Extermination of preimaginal stages there did not exceed 65 % in the first years and beginning from the 7th season onwards was zero.

It is most interesting to observe the formation of a new population in areas remote from the untreated parts of the forest. Under these conditions of greatest importance is introduction of engorged females capable of giving large progeny upon subsequent development. Fig. 2 shows variations in the density of larvae in three indices during all the years of observations in 3 areas. The thin horizontal line in the figure indicates the level of abundance of ticks in extremely small populations which we conditionally take for the threshold density for a given species under the conditions of our station.

Zero levels of abundance of larvae in the 3rd post-treatment season indicate death of progenies of all egg batches from introduced females in this period. From the 4th season and on there is a slow variable growth of the density of larvae but up to 8 - 11th season their density remains at the subthreshold level. Abundance of larvae in the area closer to the border of treatment (No. IV) approaches the threshold level earliest - in the 8th - 9th season. The rate of occurrence of ticks per line seems to be the most accurate reflection of the increase of abundance of larvae, since it levels the annual variations in the density of animals. These

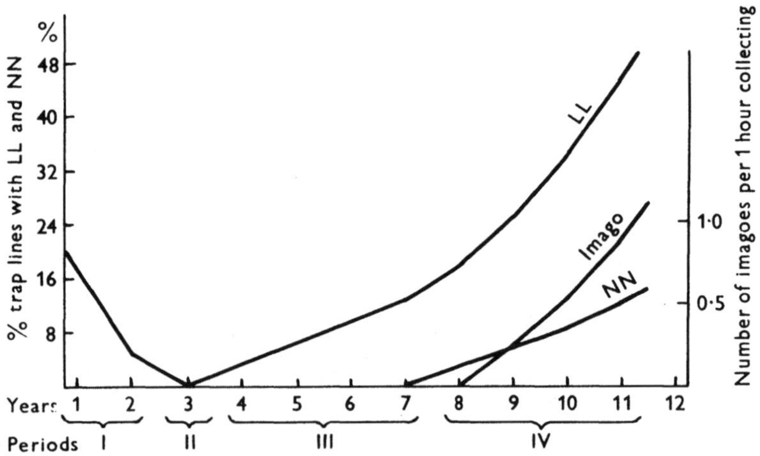

Fig. 3. Changes in numbers of ticks (*I. persulcatus*, LL, NN, imagoes) after single aviation treatment of the forest with DDT (0.5 g/ m²) next years.

curves show slow growth of the density of larvae up to the 7th season, after which the growth changes qualitatively and goes on exponentially.

Despite gradual increase of abundance of larvae which in some years reaches minimum in control by the index of abundance, this is not followed for a number of years by increase of abundance of nymph and imago (Table 2, Fig. 3). First nymphs in these areas appear in the 8th season and imagoes in the 9th season. Their abundance begins to increase only in subsequent years. This phenomenon may be explained by the importance of peculiar features of biology of the species with three-year cycle of development (Gorchakovskaya, Chigirik et al., 1970) as well as by an assumption that DDT retains its acaricidal properties in the forest floor not for 2 - 3 years but for 6 - 7 years with gradual reduction of the effective action. It is possible that the pressure of the residual effect of DDT causes death of larvae and introduced nymphs in the next stage when they begin to move in the thickness of the floor. This assumption is confirmed by the situation in the border-line area (IV b) where there occurs increased introduction of all stages of development of ticks. But even here adult ticks appear en masse only in the 8th season.

Studies of many-year reaction of *I. persulcatus* population to DDT reveal four qualitatively different periods (Fig. 3):

1st period (1 - 2 seasons after treatment) - period of mass contact of ticks with the acaricide; death of practically all specimens (89 - 100 %) living in the area. Characteristic increase of the effect of tick extermination in time. The density of the population is at the subthreshold level.

2nd period (3rd season) - the period of continuing effective action of DDT. The local population has died out. Introduced ticks die because of contact with the forest floor.

3rd period (4 - 7th season) - the period of gradual decline in the DDT effect in the forest floor. Survival rate of the introduced ticks gradually increases, some ticks manage to engorge. But because of characteristics of biology and the action of pressure of the residual properties of DDT a considerable portion of them die. The increase in abundance of larvae is not accompanied by appearance of nymphs and imagoes.

4th period (8th and subsequent seasons). The pressure of the residual action of DDT is apparently lost. Nymphs appear in the 8th season, imagoes in the 9th. The introduced ticks can complete the metamorphosis cycle. Formation of a new population begins. Growth of the abundance of ticks turns from linear into the exponential. The duration of this period depends upon a complex of conditions in the treated and adjacent areas; in considerably isolated areas it may be indefinite (up to 16 years of observation) without tendency for reestablishment which here depends only upon introduction and biological requirements of the species (island forest, ticks are introduced only by birds).

References

Gorchakovskaya, N. N.: On the duration of the effect of extermination of *Ixodes persulcatus* P. Sch. ticks in foci treated with acaricides, and the possibilities of using aerial spraying. Voprosy Virusologii 5: 297-301, 1957.

Gorchakovskaya, N. N.: Control of the main vector and reservoir of tick-borne encephalitis virus in nature. Voprosy medicinskoy virusologii, Issue VI - Neuroinfections, 121-125, 1960.

Gorchakovskaya, N. N.: The tactics of direct extermination of ticks in control of tick-borne encephalitis incidence. Medicinskaya parazitologia i parazitarnye bolezni 1: 67-72, 1962.

Gorchakovskaya, N. N.: Biocenotic changes in a natural focus of tick-borne encephalitis after its treatment with stable acaricides. Theoretical questions of natural foci of diseases. Czechoslovak Academy of Sciences, 325-336, Prague, 1965.

Gorchakovskaya, N. N., Preobrazhenskaya, N. K., Dobrynina, L. I.: On the duration of the effect of extermination of ticks in areas singly treated with acaricides. Zoologichesky Zhurnal, 38: 1353-1359, 1959.

Gorchakovskaya, N. N., Chigirik, E. D., Chumakov, M. P., Naidich, G. N., Levkovich, E. N., Chumak, N. F., Aristova, V. A., Goldfarb, L. G., Arkatovsky, P. A., Istratkina, S. V.: Factors determining many-year reaction of the tick-borne encephalitis vector to the stable and unstable acaricides. Problems of epidemiology and prophylaxis of tick-borne encephalitis. Transactions of the Institute of Poliomyelitis and Virus Encephalitides, 18: 221-251, 1970.

Koshkina, T. V.: The method of prognosing the population densities of small rodents in taiga foci of tick-borne encephalitis (on the example of voles in the foothill taiga of Salair) - Ibid, 55-65, 1970.

METHODOLOGY FOR ASSESSING RESISTANCE IN THE EUROPEAN RED MITE

D. H. C. HERNE
Research Station, Canada Department of Agriculture
Vineland Station, Ontario, Canada

The failure of dicofol sprays to control the European red mite *Panonychus ulmi* Koch in many Ontario fruit orchards showed the need for a satisfactory method for determining whether such failures were due to the development of resistance. The ideal method should be accurate enough to detect resistance in its early stages and yet be simple enough to allow extensive surveys for resistance based on response to a single concentration of pesticide (Herne,1969).

In tests at Vineland, both the topical application method of Harrison (1961) and the microslide method of Voss (1961) proved unsatisfactory for extensive tests with *P. ulmi* adult females. Methods in which leaves are dipped in toxicants are considered suitable for tests of both eggs and adults and offer an alternative to the slide-dip method (Saba,1970).

Several leaf-disc methods have been proposed. Hoyt and Harries (1961) used such a method to detect resistance to dicofol in adult females of *Tetranychus telarius* (L.) (= *urticae)* and *T. mcdanieli* McG. Foott and Boyce (1966) reported their leaf-disc method for adult female *T. urticae* was unsatisfactory for *P. ulmi.*Cranham (1968) used a leaf-disc method for detecting resistance to dicofol and other acaricides in larval stages of *P. ulmi* but stated that it did not give "a precise indication of the magnitude of resistance."

This paper reports on the development of a leaf-disc method for measuring dicofol resistance of *P. ulmi* adult females; its evaluation using dicofol-resistant and susceptible laboratory reference strains as well as several field strains; and its use in estimating, from a single dose of dicofol, the resistance level of *P. ulmi* in 35 Ontario fruit orchards.

Methods

Two reference strains of *P. ulmi* were used: a susceptible (S) strain obtained from apple trees unsprayed for 10 years at Kentville, Nova Scotia; a dicofol-resistant (R) Ontario strain selected periodically in the laboratory for a year and a half with dicofol concentrations ranging from 0.003 % to 0.03 %. The partially-selected R strain, termed intermediate resistant (Int. R), was used in evaluating the test method. The strains were reared on peach seedlings at constant temperature (24°C ± 1°C) and continuous light (12 lm/m²). The 3 to 5-day-old adult females used in all tests were reared on excised peach leaves in dishes containing cotton saturated with water.

To establish the log-dosage probit (ld-p) mortality lines peach discs were sprayed in a Potter tower with 5 ml of a solution of dicofol in acetone. Five leaf discs (2 cm²) were placed, topsurface down, on a dish of wet cotton; 10 mites were transferred to each disc. Tests were repeated 6 times for each concentration. Mortality was assessed after a 48-hr exposure at 27°C ± 3°C and RH 40 % ± 5 %. Mites that did not move when prodded or moved slightly but did not walk were considered dead.

Results and discussion

Evaluation of Test Method and Monitoring Dose: - The ld-p lines for S and R

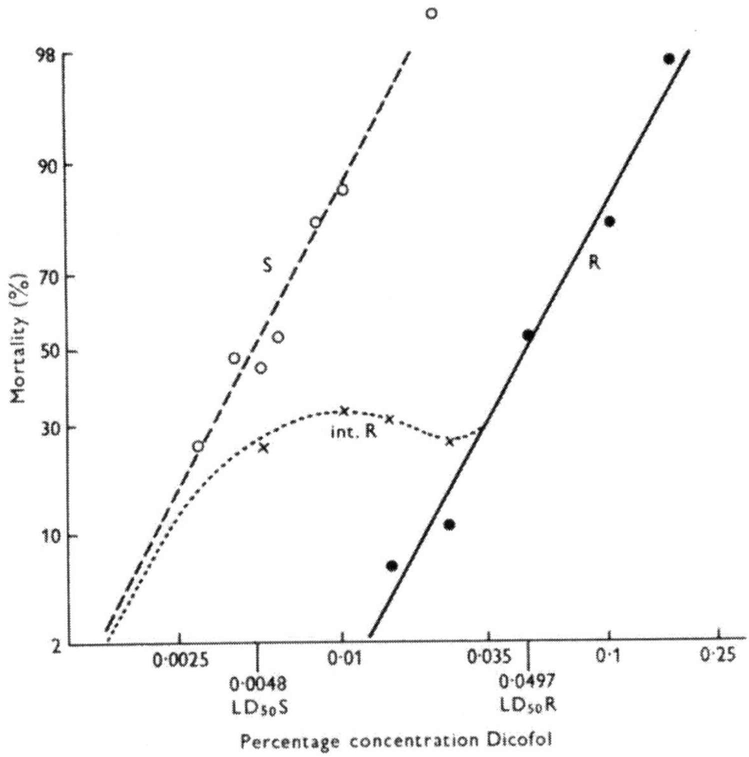

Fig. 1. Dosage-mortality lines for females of the S strain (bars) and R strain (solid), and for the partially-selected R strain (Int. R) (dots) of *Panonychus ulmi.*

strains, and the Int. R strain showed that a monitoring dose of 0.035 % distinguished between susceptible and dicofol-resistant mites (Fig. 1). Because this dose lay at the outer limit of the plateau region of the stepped curve it also gave an estimate of the resistance level in the Int. R strain. After 8 months of selection the strain apparently consisted of 70 % resistant individuals.

The accuracy and reproducibility of the test method was sufficient to distinguish S and R mites. Dicofol LD_{50} concentrations of 0.0048 ± 0.0004 and 0.049 ± 0.0048 % were obtained for the S and R strains respectively. The corresponding LD_{95}'s were 0.014 ± 0.002 % for the S and 0.154 ± 0.019 % for the R strains. The slopes were similar, 3.50 for the S and 3.34 for the R strain. The resistance ratio $\frac{LD_{50}\,R}{LD_{50}\,S}$ for dicofol was a factor of 10, which is relatively small compared to the 1000-fold increase in resistance to tetradifon for *T. urticae* (Overmeer 1966).

A test of mites from two orchards resulted in one straight ld-p line (Orchard A, untreated with dicofol), and one stepped curve similar to that for Int. R mites (Orchard B, dicofol-resistance suspected) (Fig. 2). The population in Orchard B apparently consisted of 50 % resistant individuals. These results were confirmed with

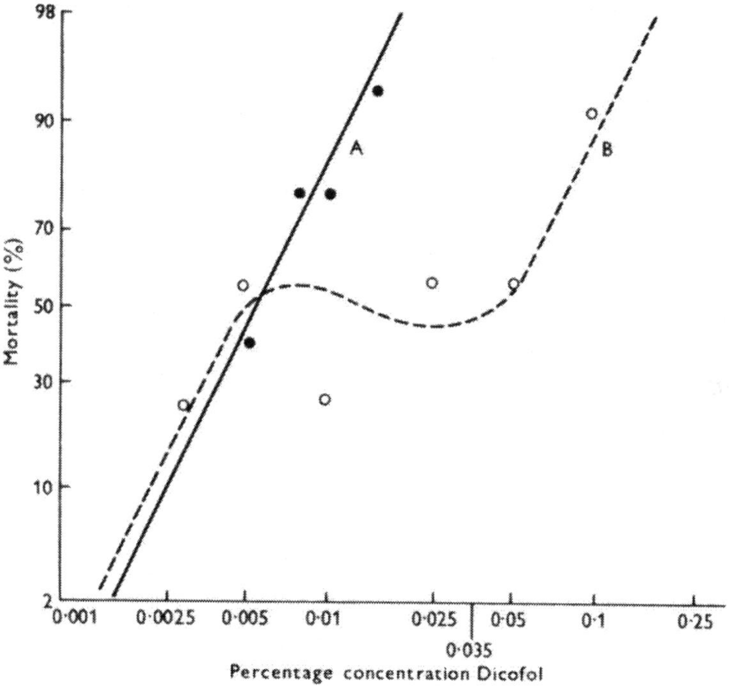

Fig. 2. Dosage-mortality lines for *Panonychus ulmi* females from orchard A where dicofol controlled mites and from orchard B where dicofol failed to control mites. Stepped curve for orchard B mites is extrapolated from only 6 points.

ld-p curves obtained with mites from 6 other orchards where resistance to dicofol was indicated. The mortalities from the monitoring dose alone would have indicated the resistance levels in the populations.

Survey of Dicofol Resistance: – About 200 adult female mites were sampled on leaves from 5 or more trees in each of 35 Ontario orchards where resistance was suspected. Adult females of the F_1 or F_2 generations of sampled mites were tested at the 0.035 % dose of dicofol. Each orchard was assessed with 3 consecutive tests of 50 mites each. The S and R reference strain mites and unsprayed controls were included in each test.

The test results in the survey with the single dose (0.035 %) of dicofol also showed good accuracy. The standard errors of the average percentage mortalities per orchard ranged from 1.8 to 8.5. Resistance ranged from low to high in 19 of the 35 Ontario orchards sampled including pear, peach and apple (Fig. 3).

Selection pressure with dicofol in the different orchards varied somewhat between regions. The general relation between selection pressure and resistance level is also shown in Fig. 3. Six or more applications of dicofol over a 4-year

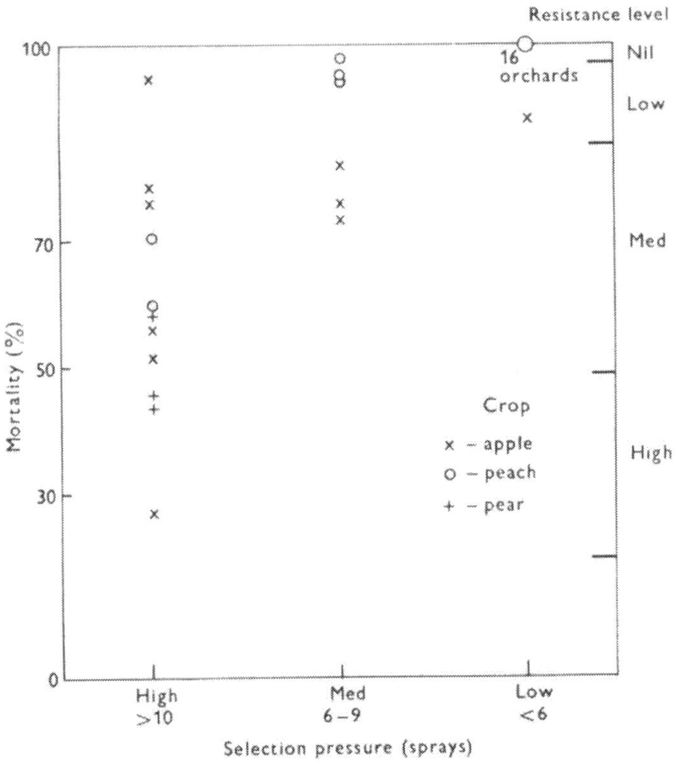

Fig. 3. Resistance levels and the relation to selection pressure to dicofol in *Panonychus ulmi* from 35 Ontario orchards.

period appear necessary before resistance can be detected. The possibility existed of some cross-resistance from other compounds to dicofol. The survey confirmed that resistance to dicofol contributed to inadequate control in more than 50 % of the sampled orchards. In the remaining orchards the lack of control was apparently due to causes other than resistance such as poor spray coverage.

References

Cranham, J. E.: Laboratory determination of resistance to tetradifon in the fruit tree spider mite, *Panonychus ulmi* (Koch). Rep. E. Malling Res. Sta. for 1967 pp. 165-168, 1968.

Foott, W. H. and Boyce, H. R.: A modification of the leaf-disc technique for acaricide tests. Proc. Entomol. Soc. Ont. (1965) pp. 117-119, 1966.

Harrison, R. A.: Topical application of insecticide solutions to mites and small insects. N. Z. J. Sci. 4: 534-539, 1961.

Herne, D.C.: A method for determining resistance in phytophagous mites. Pesticide Progress C. D. A. Research Report, Ottawa 7: 81-86, 1969.

Hoyt, S. C., Harries, F. H.: Laboratory and field studies on orchard-mite resistance to Kelthane. J. Econ. Entomol. 54 (1): 12-16, 1961.

Saba, F.: A simple test method for evaluating response to toxicants in mite populations. J. Econ. Entomol. 64 : 321, 1970.

Voss, G.: Ein neues Akarizid-Austestungsverfahren für Spinnmilben. Anz. Schädlingsk. 34: 76-77, 1961.

SUSCEPTIBILITY OF *IXODES RICINUS* L. TO DDT AND OTHER ACARICIDES IN WESTERN SLOVAKIA

M. KIEFER and D. CYPRICH
Chair of Systematic and Ecological Zoology, Natural Sciences Faculty,
Comenius University, Bratislava, Czechoslovakia

From the epidemiological and epizootic aspects, *Ixodes ricinus* is a tick of great importance. Therefore, its prevalence in a large area of this country necessitates effective control measures be developed in order to provide protection against it or to eradicate it. The first experiments in Czechoslovakia aimed at destroying ticks in the environment were conducted against *Dermacentor marginatus* and were performed in Central Slovakia by Mačička, Rosický and Černý (1955), and in Central Bohemia, an area known as a focus of tick-borne encephalitis, by Černý, Hodinářová and Kalinovská (1959). The results obtained were not satisfactory from the point of view of permanent eradication, only a short-term reduction in the number of ticks was achieved. Objective and comparable data on the initial susceptibility of *Ixodes ricinus* to DDT and other acaricides have not conclusively been obtained in Czechoslovakia to date. Assessment of the level of tick susceptibility to the acaricides in use is important for determinating the causes of physiological resistance development.

The first exact data on the susceptibility of *Dermacentor marginatus* from Central Slovakia to DDT and other acaricides, under laboratory conditions, were obtained by Přívora, Rupeš and Černý (1967). Of the six insecticides used, Imidan was the most effective and DDT the least.

Uspensky (1967) tested *Ixodes persulcatus* from the Western Sayana region (USSR) against the World Health Organization suggested standard insecticides. The most effective preparation was DDT (LD_{50} 0.000334), the weakest Fenthion (LD_{50} 0.000851). This tick is the vector of tick-borne encephalitis in the taiga zone.

Material and methods

During 1971, adult *Ixodes ricinus* ticks were collected in two Western Slovakia areas (Rohožník and Jur near Bratislava) and their sensitivity to DDT and other acaricides was tested. In the Rohožník area, there had been cases of tick-borne encephalitis and a nervous-system disease of obscure etiology among the human population. No insecticides had previously been used in the study areas. Ticks were collected by flagging early in the morning and evening. The tests were performed two days after the collection.

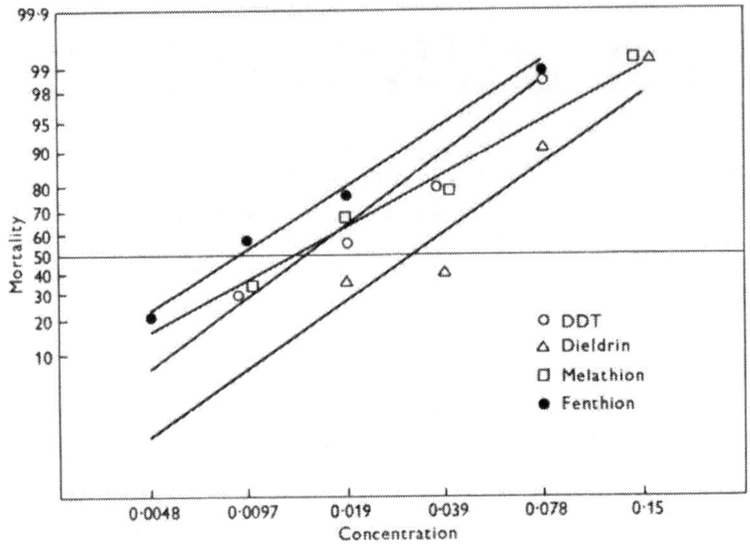

Fig. 1. Sensitivity of *Ixodes ricinus* males and females to different acaricides in the Rohožník area.

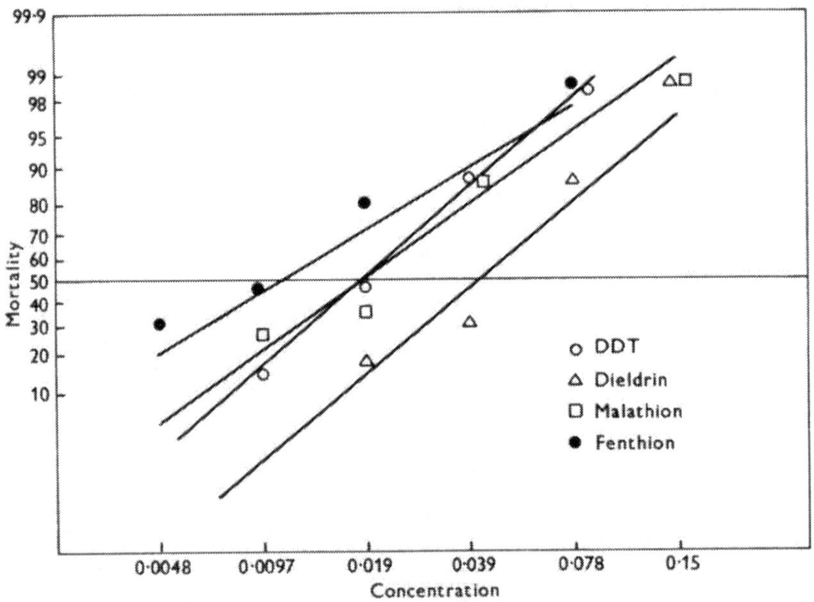

Fig. 2. Sensitivity of *Ixodes ricinus* males to different acaricides in the area of Jur near Bratislava.

Acaricide concentrations were selected after preliminary tests. These concentrations were applied to the dorsum of each individual using a microcapillary tube. After the application, the ticks were placed in groups of 10 into containers adapted to the purpose as described by Nelzina (1951). Results were read 24 h later. Ticks incapable of movement were considered dead. Each test series was repeated four times. The results were evaluated by the probit method.

Results

Figures 1, 2 and 3 present regression lines with mortality probits plotted against the logarithms of the concentrations of the individual acaricides used. The statistical evaluation of the results is presented in Tables 1, 2 and 3.

The experiments revealed that the ticks were susceptible to the acaricides employed. Female and male susceptibilities to DDT and Fenthion were practically identical in the area of Jur near Bratislava. There were minor differences in relation to Malathion and Dieldrin.

Comparison of the individual acaricides in terms of LD_{50} shows that the most effective preparation was Fenthion (LD_{50} ranged from 0.0009477 to 0.001120). The lowest effect was exhibited by Dieldrin (LD_{50} 0.04823 - 0.03262). It was found by means of the χ^2 test that the experimental deviations of the probits from the straight

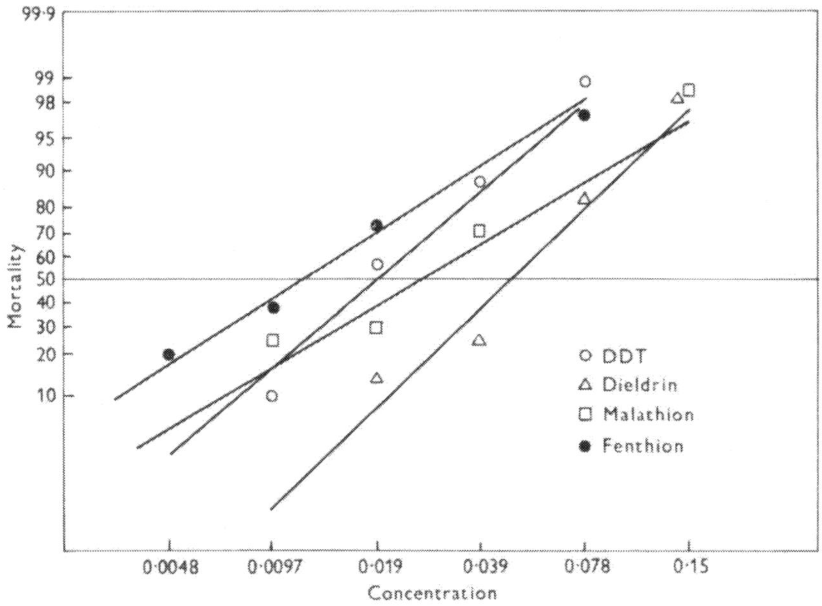

Fig. 3. Sensitivity of *Ixodes ricinus* females to different acaricides in the area of Jur near Bratislava.

Table 1. Characteristics of regression line for *Ixodes ricinus* males and females in the Rohožník area

Acaricide	LD_{50} % of working concentration	LD_{50} max	min	χ^2 calculated	tabular
DDT	0.01522	0.0193	0.0111	0.7072	5.991
Dieldrin	0.03262	0.0977	0.0079	6.5603	5.991
Malathion	0.01479	0.0182	0.0102	1.3203	5.991
Fenthion	0.00947	0.0119	0.0072	3.6332	5.991

Table 2. Characteristics of regression line for *Ixodes ricinus* males in area of Jur near Bratislava

Acaricide	LD_{80} % of working concentration	LD_{50} max	min	χ^2 calculated	tabular
DDT	0.01906	0.0216	0.0165	0.3634	5.991
Dieldrin	0.04140	0.0814	0.0210	4.4377	5.991
Malathion	0.01994	0.0271	0.0181	5.3654	5.991
Fenthion	0.01084	0.0116	0.0070	1.2730	5.991

Table 3. Characteristics of regression line for *Ixodes ricinus* females in the area of Jur near Bratislava

Acaricide	LD_{50} % of working concentration	LD_{50} max	min	χ^2 calculated	tabular
DDT	0.02005	0.0235	0.0167	1.3283	5.991
Dieldrin	0.04823	0.0573	0.0409	4.4096	5.991
Malathion	0.02364	0.0304	0.0186	3.9704	5.991
Fenthion	0.01120	0.01423	0.0687	0.2039	5.991

line were not significant, except for the case of Dieldrin used in the Rohožník area, where the χ^2 value was 6.5603 at two degrees of freedom. The tabular value is 5.991.

References

Černý, V., Hodinářová, D., Kalinovská, M.: Pokusy s hubením klíštěte obecného (*Ixodes ricinus* Linn.) v přírodě popraškem HCH. Čsl. epid. mikrobiol. imunologie, VIII, 1, 1959.

Mačička, O., Rosický, B., Černý, V.: Poznámky k bionómii a vývoju, zdravotnickému a hospodárskému významu pijaka stepného (*Dermacentor marginatus* Sulz.) v strednej Európe. Práce II. sekcie SAV, Bratislava, 1955.

Nelzina, E. N.: The rat tick. Izdatel'stvo AMN SSSR, 1-100. 1951.

Přívora, M., Rupeš, V., Černý, V.: Toxicity of certain Insecticides for the Tick *Dermacentor marginatus* Sulzer, 1776. Folia parasitologica (Praha) 17: 81-84, 1970.

Uspensky, I. V.; On the Susceptibility of *Ixodes persulcatus* P. Sch. Tick to DDT and other Pesticides in West Sayan, USSR. World Health Organization (VBC) 67, 41, 1967.

SUSCEPTIBILITY OF *SANCASSANIA PHYLLOGNATHI* TO SOME INSECTICIDE AEROSOLS

V. MĚRKA, V. RUPEŠ[x], E. VOBRÁZKOVÁ[xx], M. PŘÍVORA[x]
VLVDÚ, Hradec Králové,[x] Institute of Epidemiology and Microbiology Prague,
and[xx] Institute of Parasitology, Czechoslovak Academy of Sciences, Prague, Czechoslovakia

Few data can be found in the literature about the susceptibility of mange mites to insecticides (Burckhardt, Spühler,1955, Měrka,1970, Mielke,1970). This can be explained by the fact that *Sarcoptes scabiei,* though being the causative agent of unpleasant parasitosis, is not the vector of an infectious disease that would stand in the foreground of interest to the World Health Organization. For this reason, papers concerned with this topic are very rare in the epidemiological-microbiological literature. Dermatologists consider scabies to be a closed chapter and do not give much attention to these questions. The testing of susceptibility of mange mites meets with difficulties because a method for their laboratory rearing is not yet known.

To obtain an idea about the applicability of some insecticides for the control of scabies, the present authors have used the mite *Sancassania phyllognathi* as the model mite for their experiments. The mites, regardless of sex or stage of development, in groups of about 100 specimens, are put into dry Huffaker's chambers (Dosse,1957). Open Huffaker's chambers with the mites are exposed to the effect of the aerosol tested in a static aerosol chamber for a period of 30 minutes. After exposure and ventilation, the chambers are closed by means of cover glasses. Huffaker's chambers are then moistened and stored in humid condition at room temperature. On the fourth day the number of live and dead mites are determined under a stereoscopic microscope. Only those mites are considered to be dead which are completely motionless and do not react on the touch with a pin. The mites are not fed during the three days after exposure. *Sancassania* tolerate very well this period of starvation at a suitable relative air humidity. Each experiment included a control group of *Sancassania* which were not exposed to the effect of the insecticid aerosol.

The insecticid aerosol of lindane (gamma isomer of hexachlorocyclohexane), Dichlorvos and Trichlorphon is prepared by means of an indirect glass nebulizer. The median count diameter (MCD) of the aerosol particles is $0.5\mu m$, the median mass diameter (MMD) being $1.6\mu m$. The performance of the nebulizer at a flow rate of 8.7 litres per min. is 160 mg/min.

The nebulizer is a part of a static prism-shaped aerosol chamber. It was switched on for 30 seconds at a time by means of an electromagnetic valve. The concentration of the insecticide tested in the aerosol chamber is adjusted by the concentration of the nebulized insecticide solution. Acetone and dioxane solutions were nebulized. It has been proved in previous experiments that the organic solvents used alone have no acaricide effect.

The results of the experiments were processed by the probit method according to Lichfield and Wilcoxon and the following LC_{50} values were arrived at:

On nebulizing lindane dissolved in acetone 50 % of the experimental mites died at a concentration in the aerosol chamber of 2.02 mg/100 litres. If dioxane solution is nebulized a concentration of lindane 7.36 mg/ 100 litres is necessary to achieve 50% death of the mites (Fig. 1).

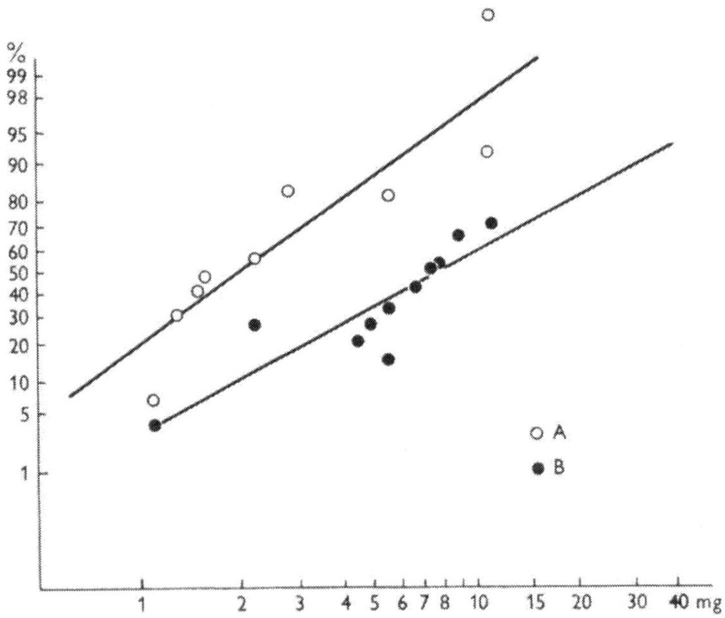

Fig. 1. Effect of lindane aerosol on *Sancassania phyllognathi:* A - aerosol generated from acetone solution, B - aerosol generated from dioxane solution.

On nebulizing acetone solution of Dichlorvos 50 % of the mites died at a concentration of 38.8μg/ 100 litres. If the aerosol is prepared from dioxane solution of Dichlorvos a concentration of 194 μg/ 100 litres is necessary to achieve 50% death of the mites (Fig. 2).

On nebulizing acetone solution of Trichlorphon 50 % of the mites died at a concentration of 0.92 mg / 100 litres. LC_{50} for an aerosol generated from dioxane solution is 7.6 mg / 100 litres (Fig. 3).

The differences caused by the solvent are accounted for by different dispersity of the aerosol. Dioxane is less volatile and consequently, the aerosol particles are probably larger than in the acetone solution.

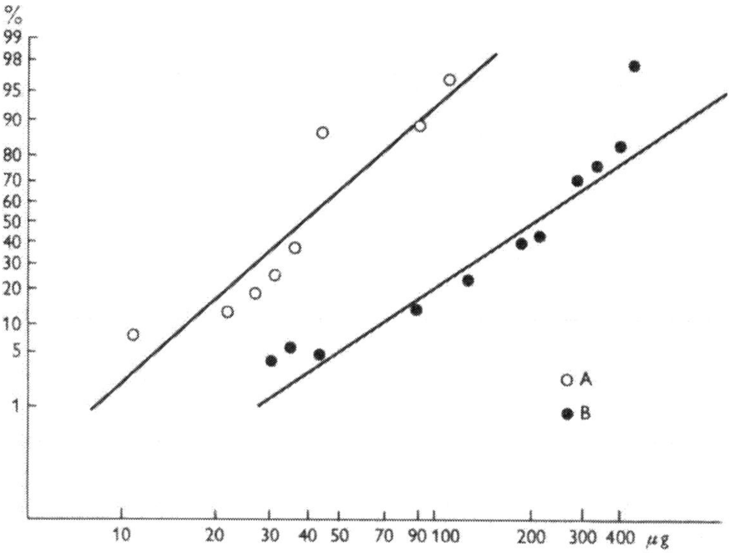

Fig. 2. Effect of Dichlorvos aerosol on *Sancassania phyllognathi:* A - aerosol generated from acetone solution, B - aerosol generated from dioxane solution.

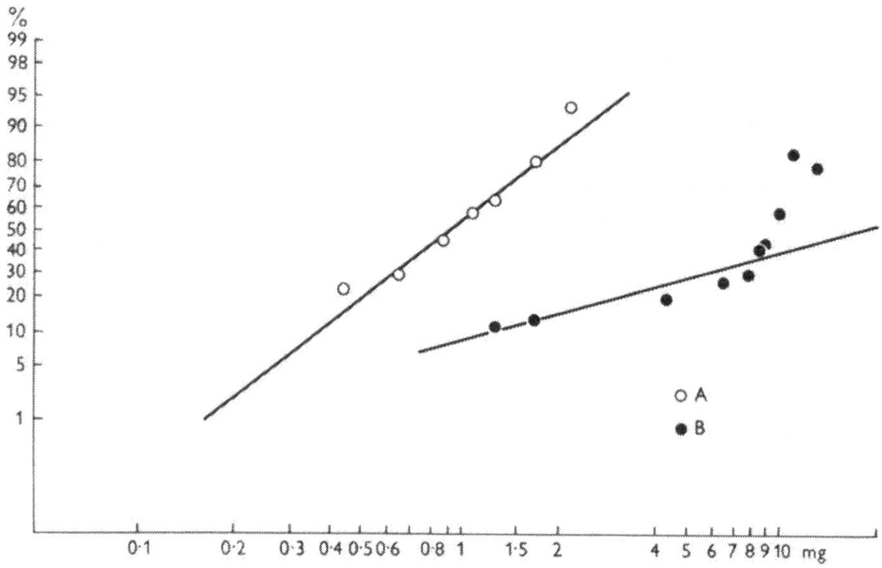

Fig. 3. Effect of Trichlorphon aerosol on *Sancassania phyllognathi:* A - aerosol generated from acetone solution, B - aerosol generated from dioxane solution.

In comparison with lindane, the advantage of Dichlorvos lies in its substantially better acaricide effect, the toxicity for man being almost the same, and also in the possibility of using its vapours for insecticide and acaricide purposes. With regard to the high tension of the vapours, a closed space is rapidly filled by the vapours of Dichlorvos. This property is utilized by "Vapona strips" produced by Shell International Co. An advantage of Trichlorphon in comparison with lindane is its lower toxicity for man.

The acaricide effect of the aerosol depends on the solvent used. The aerosols prepared from acetone solutions are more effective than those prepared with dioxane solutions, which is probably due to different dispersity of the aerosol. On the evaluation according to the results obtained with acetone solutions, the best effect on *Sancassania* is achieved by Dichlorvos which is about 50 times more effective in comparison with lindane and about 25 times more effective as compared with Trichlorphon. Among the aerosols prepared from dioxane solutions, Dichlorvos gave the best results while the effect of lindane and Trichlorphon is approximately the same.

For desinsection in scabies, aerosol or vapours of Dichlorvos would be the most suitable means.

References

Burckhardt, W., Spühler, H.: Die Behandlung der Scabies und Pedikulosen mit Gamma-Hexachlorcyclohexan. Dermatologica 110, 39-41, 1955.

Dosse, G.: Arbeitsmethoden zu den morphologischen und biologischen Untersuchungen von räuberischen Milben. Z. angew. Entomol. 40, 155-160, 1957.

Měrka, V.: Úloha desinsekce u scabies. Čs. dermatol. 45, 135-136, 1970.

Mielke, D.: Ektoparasitenbefall bei Rindern, Schweinen und Schafen und dessen Bekämpfung. Angew. Parasitol. 11, 167-170, 1970.

Proceedings of the 3rd International Congress of Acarology, Prague, 1971

EFFECT OF TRICHLORPHON (DTHP)-BASED PREPARATIONS ON TICKS AND MITES

J. MINÁŘ, V. ČERNÝ, M. PŘÍVORA+ and A. KALINEC+

Institute of Parasitology, Czechoslovak Academy of Sciences, Prague;
+Institute of Epidemiology and Microbiology, Prague, Czechoslovakia

At the present time, organophosphate-based insecticides are employed on an ever growing scale in applied arachnoentomology, because they offer a number of advantages over the chlorinated hydrocarbons which have been used so far. Apart from a high effectiveness, organophosphate insecticides are particularly expedient in that they can be used systemically and, do not usually accumulate in mammalian tissues, which makes them practicable for veterinary purposes.

The preparation Hypocid, developed at the Institute of Parasitology, Czechoslovak Academy of Sciences, displays a high systemic effectiveness against hypodermatosis of cattle (Minář and Dorž, 1969 a, b; Minář et al., 1970) and against lice of domestic animals. This, together with the published reports on the contact employment of organophosphates against parasitic mites (Nepoklonov and Talanov, 1968; etc.), stimulated the present experiments. The aim was to ascertain the degree of effectiveness and the most suitable type of application for these preparations in relation to ticks and other parasitic mites.

The experiments were performed with Hypocid (containing 40% trichlorphon in special solvents), and a newly developed preparation, DTHP-spray (containing 20% trichlorphon).

A tentative laboratory test performed with 1600 *Ixodes ricinus* (L.) larvae, engorged on white mice that had been treated externally with Hypocid clearly showed that the preparation was systemically effective against ticks. In 1970-1971, further laboratory tests and some field trials were carried out to demonstrate the effect of these preparations on ticks *Ixodes ricinus* (L.) and *Dermacentor marginatus* (Sulz.) and mites *Dermanyssus gallinae* (De Geer).

The laboratory tests with DTHP, produced by Spolana, were performed on hungry and engorged females, and hungry males, and nymphs of *I. ricinus* and on *D. gallinae*. The method employed was that of tarsal contact of the arthropods with an insecticide film on the internal wall of glass tubes (Rupeš et al., 1968, 1970) or, in the case of engorged females on Petri dishes. The exposure lasted 24 hr, and for the next 24 hr the arthropods were transferred into a holding tube (Petri dish). Results were read after the 24-hr exposure period and after the 24-hr holding period. The amount of solution used to impregnate the test tubes or Petri dishes was chosen so that the concentration corresponded to the amount of insecticide per $1 m^2$.

Table 1. Sensitivity of unfed ticks *Ixodes ricinus* to DTHP. The numerator gives the number of ticks in experiment, denominator gives the number of deaths

Date of the experiment	Concentration of DTHP (g/m^2)	Checking after exposure			Checking after recovery		
		Males	Females	Nymphs	Males	Females	Nymphs
28.4 – 30.4.	0.02	10/8	10/6	10/8	10/10	10/10	10/10
	0.002	10/8	10/1	10/6	10/10	10/10	10/10
	0.0002	10/5	10/0	10/5	10/10	10/10	10/10
	Control	10/0	10/0	10/0	10/0	10/0	10/0
1.6. – 3.6.	0,002	10/10	10/10	10/10	10/10	10/10	10/10
	0.0002	10/10	10/10	10/10	10/10	10/10	10/10
	0.00002	10/7	10/5	10/7	10/9	10/8	10/10
31.5.– 1.6.	0.000002	10/7	10/1	10/7	10/10	10/7	10/10
	0.0000002	10/5	10/0	10/5	10/9	10/8	10/10
	Control	10/0	10/0	10/0	10/0	10/0	10/0
23.6. – 25.6.	0,0002	10/10	10/8	–	10/10	10/10	–
	0,00002	10/10	10/8	–	10/9	10/9	–
	0,000002	10/7	10/6	–	10/8	10/9	–
	0,0000002	10/5	10/2	–	10/7	10/6	–

Table 2. Sensitivity of mites *Dermanyssus gallinae* to DTHP. The numerator gives the number of mites in experiment, denominator gives the number of deaths

Date of the experiment	Concentration of DTHP (g/m^2)	Checking after exposure				Checking after recovery			
		Repetition				Repetition			
		1.	2.	3.	Total	1.	2.	3.	Total
3.5. – 5.5.	0.02	25/25	25/25	25/25	75/75	25/25	25/25	25/25	75/75
	0,002	25/25	25/25	25/25	75/75	25/25	25/25	25/25	75/75
	0,0002	25/15	25/14	25/12	75/41	25/25	25/24	25/25	75/74
	Control	25/0							
10.5 – 12.5.	0,00002	25/25	25/25	25/25	75/75	25/25	25/25	25/25	75/75
	0,000002	25/24	25/24	25/25	75/73	25/25	25/25	25/25	75/75
	0,0000002	25/25	25/24	25/24	75/73	25/25	25/25	25/25	75/75
	Control	25/0							
18.5. – 21.5.	0,00002	25/24			25/24	25/24			25/24
	0,000002	25/23			25/23	25/24			25/24
	0,0000002	25/23			25/23	25/24			25/24

The results, presented in Tables 1 and 2, show an extremely high effectiveness of DTHP for the hungry stages of *I. ricinus* and for *D. gallinae*. Repetition of the experiment with some of the concentrations gave similar results. The effectiveness for engorged *I. ricinus* females was substantially lower.

In the first field trial, *D. marginatus* imagoes in small linen bags were engorged on the predilection attachment sites, i.e. the distal end of the tail of heifers. Four groups of three animals each, were included in the experiment. The first group received a dose of 10 ml of Hypocid per 100 kg live weight applied to the back, the second group a therapeutic dose against hypodermatosis, i.e. 20 ml per 100 kg, the third group was treated with DTHP-spray applied to the distal part of the tail for 4 sec, which corresponds to a dose of 0.7 g of active substance, and the fourth group was the untreated control. Ten *D. marginatus* females and five males were planted per animal. The effect was checked on the next day and after one week. Seven and 18 days after the first insecticide treatment, ten females and five males per animal were again planted on two animals of each group and the progress of engorgement was checked a week later.

The check performed one day after the insecticidal application, showed that a great majority of the ticks were dead or in tremor in all the groups; a week after the application only two live, partially engorged females were found on one animal treated with Hypocid at the dose rate of 10 ml per 100 kg. The check on ticks planted on the animals one week after insecticide application showed that there were 40 % and 13 % of live imagoes on heifers that had received doses of 10 and 20 ml per 100 kg, respectively, and no live specimens on heifers treated with the DTHP-spray. The check on ticks planted 18 days after insecticidal application showed that, in the same groups of animals, there were 83 % and 27 % of live imagoes, respectively, with the females partially engorged. All ticks were dead on heifers treated with the spray. *Dermacentor marginatus* females on control, untreated animals engorged fully in all experiments.

The results show that the systematic effect of Hypocid on ticks is highest during the period immediately after application but can still be observed in some of the parasites after a week. The spray proved to be more suitable, exhibiting 100 % effectiveness up to 18 days.

The DTHP-spray was subjected to a further field trial in an area where *D. marginatus* is prevalent. In the early part of May, a day prior to turning cattle out to graze, four groups of cows consisting of 10 animals each were treated with the spray on the predilection attachment sites for this species, i.e. the dewlap and tail. The application interval was 10, 20, 30 or 40 seconds, which corresponds to doses of 1.75, 3.5, 7.0 and 15 g of active substance per animal. The treated parts were inspected for the presence of ticks six times in the course of the next four weeks. During this period, one tick was found four weeks after treatment on an animal that had been treated with the lowest dose. The control group of 10 untreated animals displayed, during the observation period, 9 to 44 ticks per animal, with the highest tick counts being found one week, and the lowest four weeks, after turning the cows to grass.

This experiment again shows the high and prolonged effectiveness of DTHP-spray against ticks.

DTHP-spray was also tested against the mite *D. gallinae* in hen-coops on poultry farms. Wooden hen-coops were sprayed with two dilutions of the preparation, containing 20 % and 2.5 % of active substance, applied in amounts of approximately 2.7 and 0.25 g per 1 m^2. In the case of each of the concentrations employed, all the mites were found dead. The first individual mites began to reappear after two weeks, when the spray was repeated. Mites were not detected in subsequent checks during the following month.

It is concluded from these experiments that DTHP-based sprays can be very effective agents in the control of ticks and other parasitic mites, which is of great practical value.

References

Minář, J., Dorž, C.: Tlumení střečkovitosti skotu v Mongolské lidové republice. Veterinární medicína, (Praha), 14, 12 : 695 - 701, 1969a.

Minář, J., Dorž, C.: Iz praktiki lecheniya gipodermatoza krupnogo rogatogo skota. Mezhdunarod. selskochozyaystv. zhurnal,SEV,Moskva, 13, 3 : 100 - 101, 1969b.

Minář, J., Musil, J., Tesařík, L., Zbořil, V.: Tlumení střečkovitosti skotu v šumavské pastevní oblasti Hypocidem. Veterinární medicína (Praha), 15, 5 : 287 - 296, 1970.

Nepoklonov, A. A., Talanov, G. A.: Sovremennye insekticidy, primenjaemye v zhivotnovodstve. Rosselchozizdat, Moskva, 1 - 60, 1968.

PRELIMINARY LABORATORY TESTS WITH TICK REPELLENTS

D. NOVÁK
Laboratory of Entomology OHS, Hodonín, Czechoslovakia

In this communication we shall describe preliminary laboratory tests with several perfumes which may be added to the standard tick repellents to improve their repellency. The World Health Organisation (1963) recommended as the best tick repellents Indalone, diethyltoluamide, dimethylcarbate, dimethylphthalate and benzyl benzoate for general use. The repellents should provide complete protection against ticks, but several provide no more than 90 % protection through various periods of ageing and wear. Tick repellents are most effective when applied to the clothing. Novák (1969, 1970) has described several experiments with perfumes as tick repellents. Enti (1969) has indicated that leaves from the plants *Clausena anisata, Hyptis suaveolens* and *Ocimum gratissimum* are used as repellents in Ghana.

In our laboratory, we tested new perfumes as tick repellents. A ring 7 cm in diameter was drawn on a filter paper, and a 1 cm wide circular "barrier" of the substances to be tested was painted on the filter paper outside this ring. Adult males, females and nymphs of the test ticks *Ixodes ricinus* were placed inside the barrier and the time taken to cross it was recorded. The repellent effects were studied (for a period of 5 minutes) commencing 30 minutes after application. Twenty six perfumes were used: *Clausena anisata, Ocimum gratissimum, Hyptis suaveolens,* fialka 97233 (viola), jasmín Š 97385 (jasmine), cologne, fougere, květina (flower), skořice GK (cinnamon), mateřídouška KS (wild thyme) lesana, juchta GK, mimosa M, azur M, žloutkové M (yolked), hellada M, jedle P/9 (fir), růže M (rose), citronella M, muškát (nutmeg), puškvorec (sweet calamus), břízodehtová bílá (birchtar white), šalvěj albánská (sage), sandela, pačulová (pathouli), technický parfum (technical perfume) and two repellents Dieltamid (diethyltoluamide) and Retox acted as standards. All perfumes were prepared by Aroma, Děčín, Czechoslovakia. During the preliminary laboratory tests the air temperature was 22–24°C and the air humidity 56–87 %.

The results of the preliminary laboratory tests of the best perfumes as tick repellents are shown in the Table 1.

From these results it is concluded that against *Ixodes ricinus* the repellent properties of the perfumes fialka 97233 (viola), jasmín Š 97385 (jasmine) and *Clausena anisata* are more effective than the standard repellents Dieltamid (diethyltoluamide) and Retox. It is suggested that these perfumes may be added to the standard tick repellents to improve their effect.

Table 1. Results of preliminary laboratory tests with the best perfumes as *Ixodes ricinus* repellents

Repellent	Effectiveness in percentage		
	males	females	nymphs
Perfumes			
Clausena anisata	86	67	100
fialka 97233 (viola)	85	70	95
jasmín Š 97385 (jasmine)	85	65	85
Standards			
Dieltamid (diethyltoluamide)	70	45	80
Retox	70	45	80

References

Enti, A.A.: Personal information, 1969.
Novák, D.: Report on some work of the Entomological Laboratory, OHS Hodonín, Czechoslovakia, PANS 15/3 : 362–364, 1969.
Novák, D.: New laboratory tests with perfumes as tick repellents. Arch. Roum. Pathol. Exptl. Microbiol. 29/3 : 539–540, 1970.
Recommended methods for vector control. WHO Technical Report Series, No. 265, 1963.

A NEW CHEMICAL GROUP OF CATTLE ACARICIDES

B.H. PALMER, J.F. MCCARTHY, A. KOZLIK, I.R. HARRISON
Boots Pure Drug Co. Ltd., Nottingham, England

Recent work by Wharton and Roulston, 1970, and by O'Sullivan and Green, 1971, has focussed attention upon the growing significance of organophosphorus (OP), and carbamate, resistant strains of *Boophilus microplus*. As more and more resistant strains have emerged, so the need for a different chemical type of acaricide has become increasingly apparent.

We have recently made a number of new triazapenta diene compounds, which have shown a high level of acaricidal activity against both OP-susceptible and OP-resistant strains of *B. microplus*. The first of these compounds was 1,5-di- (4-chloro-2-methylphenyl)-3-methyl- 1, 3, 5- triazapenta- 1, 4-diene (BTS 23 376). This compound had reached a fairly advanced stage of evaluation when other, more active, members of the series were found, including 1,5-di- (2, 4-dimethylphenyl)-3-methyl- 1, 3, 5-triazapenta- 1, 4- diene (BTS 27 419).

This paper deals with preliminary laboratory and calf experiments carried out with these two compounds.

Methods

(i) Maintenance of *Boophilus microplus* – The parasitic part of the life cycle of *B. microplus* is maintained on 2–6 month-old calves, housed in a quarantine calf unit. Each individual animal is yoked in a crate, on a slatted floor, raised about two feet off the ground (Harrison and Mundy, 1968).

Normally, gravid female ticks begin to drop off a calf 20 days after infestation with tick larvae.

Gravid female ticks, on incubation at 25°C and 90 % R.H., commence laying eggs within 3 days, and normally complete their ovulation by 12 days. Infective larvae hatch from the eggs after a further 4 weeks' incubation.

Three strains of *B. microplus* are maintained.

The Yeerongpilly Strain. This is the original OP-susceptible strain.

The Biarra Strain. Most of the experimental work here reported was carried out using the OP-resistant Biarra strain that has been maintained continuously in this laboratory for 3 years.

The Mackay Strain. This strain has only recently been obtained from Australia.

(ii) Laboratory Tests. The laboratory testing techniques have been described in detail elsewhere (Harrison, 1967).

(a) Spraying Tower Test. One-week-old tick larvae are used as test organisms. Aqueous solutions or suspensions of test compound, at a series of known concentrations, are prepared. One ml of test solution is placed in the reservoir of a Tattersfield spraying tower and evenly sprayed over a 4.25 cm circular dish containing approximately 60–70 tick larvae. As soon as the tick larvae have dried, about 40 of them are transferred to a clean filter paper in the lid of a petri dish.

In each test, a series of concentrations of test compound is used and, from the mortality responses, the median lethal concentration is calculated by a method of probit analysis. Fiducial limits, with 95 % confidence, are also calculated.

A feature of triazapenta diene compounds is the delayed response shown by the affected tick larvae in this test. This is in marked contrast to the rapid "knock down" shown by the OP compounds.

(b) Gravid Female Dipping Test. Groups of 20 fully-fed gravid female ticks are totally immersed for 3 minutes in aqueous solutions (or suspensions) of test compounds. The solutions are drained off and the ticks are air dried before being stuck down on their backs on strips of adhesive tape.

Observations are made after 6 days' incubation at 25°C and 90 % R.H., when the percentage of gravid females that have failed to lay eggs in each group is recorded.

ED_{50}s and fiducial limits are calculated from the dose response, employing the same probit analysis method as used for the spraying tower test.

(iii) Calf Spraying Tests. Calves, individually housed in crates in the calf unit, are infested three times per week for 3½ weeks, when all tick parasitic stages, from larvae to replete gravid females, are present.

Compared with the infestations used by Australian workers, ours are relatively light, since experience has shown that significantly heavier tick burdens result in serious debilitation or even death of infested calves.

Test compounds are applied in aqueous suspensions (or emulsions) by means of a pressurised hand spray.

Challenge inocula, in a known amount, of infective larvae is applied at intervals also after treatment to assess the persistence of acaricidal effect.

Daily recoveries of dropped gravid females are recorded throughout the experiment, commencing 3 days before treatment.

Results

The activity of BTS 23 376 and BTS 27 419 was compared with some established acaricides (see Table 1). These results clearly show that BTS 27 419 is equally outstanding in its acaricidal activity against all three strains of *B. microplus* in the spraying tower.

A comparison was made of BTS 23 376, BTS 27 419 and Dursban in calf spraying tests.

Table 1.

Acaricide	Tick Strain	% Mortality at Following Concns a.i.						LD$_{50}$	Fiducial Limits	
		250 ppm	125 ppm	62.5 ppm	31.3 ppm	15.6 ppm	7.8 ppm		Upper	Lower
Bromophos Ethyl (E.C.)	Yeerongpilly	98.7	91.3	53.8	31.6	23.1	–	42.8	57.4	32.0
	Mackay	59.7	12.4	0.0	0.9	0.0	–	222.2	–	–
	Biarra	44.2	10.8	11.1	3.1	6,4	–	> 250.0	–	–
Carbaryl (W.P.)	Yeerongpilly	–	100.0	94.7	54.6	6.9	–	29.9	33.2	27.0
	Mackay	29.0	0.0	0.0	0.0	–	–	> 250.0	–	–
	Biarra	36.5	1.0	1.0	0.0	–	–	> 250.0	–	–
Chlorfenvinphos (E.C.)	Yeerongpilly	100.0	98.7	77.0	16.1	14.3	–	40.8	59.1	28.2
	Mackay	3.0	4.0	4.0	4.0	3.0	–	> 250.0	–	–
	Biarra	0.0	5.8	4.0	0.0	0.0	–	> 250.0	–	–
Delnav (E.C.)	Yeerongpilly	–	99.0	86.2	29.6	1.6	–	39.7	42.0	37.5
	Mackay	4.0	3.0	2.0	0.0	–	–	> 250.0	–	–
	Biarra	9.9	0.0	0.0	0.0	–	–	> 250.0	–	–
Dursban (E.C.)	Yeerongpilly	–	100.0	81.7	34.1	5.6	–	38.1	40.8	35.6
	Mackay	100.0	84.8	9.1	4.0	–	–	87.8	115.9	66.6
	Biarra	87.4	34.7	1.3	0.0	–	–	150.8	157.8	144.0
BTS 23 376	Yeerongpilly	100.0	97.3	77.8	36.5	9.0	–	37.9	40.7	35.4
	Mackay	99.6	96.1	64.2	17.2	10.0	–	47.4	58.2	38.5
	Biarra	99.7	92.3	75.9	28.4	10.0	–	42.4	47.1	38.1
BTS 27 419	Yeerongpilly	–	100.0	100.0	93.1	70.0	47.5	9.2	11.0	7.8
	Mackay	–	100.0	100.0	84.5	58.3	28.8	12.8	14.8	11.1
	Biarra	–	99.3	99.3	82.9	53.9	39.1	12.3	15.7	9.7

(a.i. = Active Ingredient.)

Table 2.

Days After Treatment	Mean Numbers Gravid Females Collected from Pen Floors			
	Untreated Controls Group	BTS 23 376 @ 0.075 % Group	BTS 27 419 @ 0.025 % Group	Dursban @ 0.075 %
−2	38.3	34.0	59.3	80.3
−1	147.6	144.0	171.3	156.3
0	196.6	203.6	212.3	214.3
1	196.6	413.0	334.0	337.0
2	255.6	83.6	84.0	246.0
3	288.0	24.0	38.6	136.0
4	221.3	7.6	139.0	59.0
5	315.6	36.3	93.0	50.6
6	235.0	8.3	42.3	43.3
7	213.0	7.0	10.6	62.3
8	208.0	3.3	14.0	31.6
9	268.6	3.3	11.0	28.0
10	205.0	2.3	7.0	15.0
11	185.6	5.0	7.0	14.3
12	134.0	5.0	6.0	11.0
13	105.0	2.0	8.0	22.3
14	69.3	1.0	13.0	16.6
15	85.0	10.3	15.0	18.3
16	170.3	1.6	3.3	20.3
17	146.3	1.6	2.0	19.3
18	209.6	14.0	13.3	9.6
19	222.3	12.3	14.0	7.6
20	233.3	10.6	7.3	8.3
21	219.3	8.3	17.3	8.3
22	151.3	6.6	8.0	5.0
23	101.0	1.6	2.3	0.6
24	109.0	1.3	1.3	0.0
25	60.3	0.3	0.0	0.0
26	54.3	1.6	0.0	0.0
27	31.3	0.0	0.0	0.0

Twelve calves were inoculated with infective larvae of the Biarra strain of *B. microplus* on 11 occasions before treatment. The calves were divided into 4 dose groups of 3 calves each, and were treated as follows: –

Group 1: Treated with BTS 23 376 at 0.075 % a.i.

Group 2: Treated with BTS 27 419 at 0.025 % a.i.

Group 3: Treated with Dursban E.C. at 0.075 % a.i.

The remaining 3 calves served as untreated controls.

Each treated animal was sprayed with 8 litres of aqueous suspension of acaricide, delivered at a pressure of 80 p.s.i., through a coarse spray nozzle. Daily collections of dropped gravid female ticks were made for the 3 days preceding treatment, and then for 27 days after treatment (see Table 2).

These results indicate a good control of the tick infestations in all 9 treated calves, when compared with the untreated animals. Both experimental acaricides gave as good control as did Dursban at the recommended concentration of 0.075 % (for treatment of Biarra strain). Preliminary results so far received from Australia and South Africa have been encouraging, and further field trials are in progress.

Acknowledgments. The authors wish to thank their colleagues for their technical assistance with this work.

Thanks are also due to the Directors of Boots Pure Drug Co. Ltd., for permission to publish these results.

References

Wharton, R.H., Roulston, W.J.: Resistance of Ticks to Chemicals. Ann. Rev. Entomol. 15: 381–404, 1970.

O'Sullivan, P.J., Green, P.E.: New Types of Organophosphorus-Resistant Cattle Ticks *(Boophilus microplus).* Letter to the Editor in Australian Vet. J. 47: 71, 1971.

Harrison, I.R., Mundy, A.E.: The Housing of Calves for Experiments with Cattle Tick *Boophilus microplus.* J. Inst. An. Tech. 19: 132–135, 1968.

Harrison, I.R.: The Testing of Compounds Against the Cattle Tick *Boophilus microplus,* with Particular Reference to the Development of 2-cyclohexyl-phenyl N-methylcarbamate. Proc. 2nd Int. Congr. of Acarology, 1967.

PROSPECTS FOR CHEMICAL CONTROL OF *BOOPHILUS MICROPLUS* IN AUSTRALIA

W.J. ROULSTON
Division of Entomology, CSIRO, Long Pocket Laboratories,
Indooroopilly, Brisbane, Australia

Resistance of ticks to chemicals was recently reviewed by Wharton and Roulston (1970) who at that time recognized three strains of *B. microplus* resistant to organophosphorus (OP) compounds and carbamates. These were the Ridgelands strain with a moderately insensitive cholinesterase, the Biarra strain with a highly insensitive cholinesterase and the Mackay strain with sensitive cholinesterase but with greater detoxifying ability than either the Biarra or the OP-susceptible Yeerongpilly reference strain.

At the 2nd International Congress of Acarology, Roulston (1967) reported on the use of Dursban, bromophos-ethyl and phosmet to control Biarra ticks and suggested that it was only a matter of time before strains of ticks resistant to these chemicals appeared. Subsequently two strains have been recognized with increased resistance to at least one of these acaricides. These strains designated Mt. Alford and Gracemere, were reported in 1970 by O'Sullivan and Green (1971) who showed that both had increased resistance to Dursban and diazinon, but, that apart from this the resistance factors in the Mt. Alford and Gracemere strains were similar to those of the Biarra and Ridgelands strains respectively.

Determination of bimolecular rate constants for the *in vitro* reaction between cholinesterases and inhibitors indicated that the cholinesterases of the Mt. Alford and Gracemere strains were indistinguishable from those in the Biarra and Ridgelands strains respectively (H.J. Schnitzerling, unpublished information). The biological effectiveness of phosphorothionates, such as Dursban, is largely dependent on the amount of the parent compound converted to the strongly inhibiting oxon. On this basis Dursban would be less likely to be effective against ticks which contained lower levels of Dursbanoxon. Studies on the penetration and metabolism of radioactive Dursban showed that it penetrated Mt. Alford, Gracemere and Biarra ticks at approximately the same rate but there was relatively little Dursbanoxon present in the Mt. Alford strain compared to the Biarra strain, while the amount in the Gracemere strain was intermediate (C.A. Schuntner, unpublished information). This finding, together with evidence that there were greater amounts of hydrolytic metabolites in the Mt. Alford and Gracemere strains indicated that these strains were able to metabolize Dursban and Dursbanoxon at a greater rate than the Biarra strain. Thus while all these resistant strains had insensitive cholinesterase as a resistance mechanism the Gracemere and Mt. Alford strains in addition had enhanced detoxification.

Chemical control of resistant strains is becoming increasingly difficult and few chemicals with a mode of action other than that involving cholinesterase inhibition are showing any promise. One exception chlorphenamidine, the best known of the formamidine group, was reported by Roulston and Wharton (1967) to be equally effective against OP-susceptible Yeerongpilly and OP-resistant Biarra ticks. In spite of its effectiveness against Biarra ticks, chlorphenamidine was not developed commercially at that time because of some reported toxicity at high concentrations and chemical instability in aqueous suspension.

In the belief that some improvement in the level of tick control could result from mixtures if the two chemicals had a different mode of action, cattle infested with Biarra ticks were sprayed with mixtures of OP compounds and chlorphenamidine, the Biarra strain being chosen as at this time it was the most difficult to control. These mixtures were generally more effective in controlling Biarra ticks than either component used alone. Extensive tests have been made on mixtures containing the OP compound at the concentration registered for use, plus 0.01% chlorphenamidine which is 1/10th the concentration of chlorphenamidine required for effective control if it had been used alone. Whereas a commercial preparation based on 0.05% bromophos-ethyl gave only 78 % control of Biarra ticks, the addition of 0.01% chlorphenamidine increased this level to 99 % (Roulston et al., 1971). Likewise the control produced by 0.025% coumaphos was increased from 41 % to 99 % and by diazinon from 64 % to 99 %.

Control of the Biarra strain under field conditions had been difficult and it was obvious that the Mt. Alford strain, which could be regarded as a Biarra strain with a detoxifying mechanism, was going to be even more difficult to control. Using the same test method (Stone and Haydock, 1962) resistance levels in Mt. Alford larvae determined by O'Sullivan and Green (1971) and CSIRO (unpublished information) were in general agreement and indicated high resistance in larvae to Dursban (123X) diazinon (208X) and to such compounds as pyrimithate (204X) and pirazinon (56X) which have a chemical structure similar to diazinon. However the resistances to bromophosethyl (4.3X) and phosmet (1.5X) differed little from 3.7X and 2.9X of the Biarra strain. On this basis bromophos-ethyl and phosmet were expected to be as effective in controlling Mt. Alford as Biarra ticks.

On a property, where the ticks were estimated to be approximately 70 % Mt. Alford and 30 % Biarra types, spraying trials showed that ten days after treatment both 0.025 % Dursban and the commercial preparation of 0.05% bromophos-ethyl had given 80 % control of ticks when used alone nad 94 % control when each was used in a mixture containing 0.01% chlorphenamidine (Roulston et al., 1971); corresponding figures for 0.075% phosmet were 97 % and 98 %. The Boots experimental compound, BTS 27419 when used alone gave 100% control.

Spraying trials on stalled cattle infested with Mt. Alford ticks have confirmed the unexpected result that Dursban and bromophos-ethyl are equally efficient (or inefficient) against Mt. Alford ticks (Roulston et al., 1971). On the basis of resistance factors obtained by exposing larvae in impregnated paper packets (see above), there was a striking difference in Mt. Alford and Biarra ticks with respect to Dursban but not to bromophos-ethyl. However a more recent test in which larvae were immersed in suspensions of bromophos-ethyl has demonstrated greater differences in the resistance factors to bromophos-ethyl of the Biarra (8X) and Mt. Alford (23X) strains of ticks.

Because of the upsurge of resistant strains chemical control of ticks is becoming increasingly difficult and while no doubt new acaricides such as Bayer 6896 and Boots BTS 27419 do offer considerable promise the history of resistance in cattle tick to chemicals does not allow any optimism that they will provide long term control unless there is some change in the way acaricides are used.

References

O'Sullivan, P.J., Green, P.E.: New types of organophosphorus-resistant cattle ticks *(Boophilus microplus)*. Australian Vet. J. 47 : 71, 1971.

Roulston, W.J.: Acaricide resistance in the cattle tick *Boophilus microplus* in Australia. Proc. 2nd Int. Congr. Acarol. Nottingham, 1967.

Roulston, W.J., Wharton, R.H.: Acaricide tests on the Biarra strain of organophosphorus resistant cattle tick *Boophilus microplus* from southern Queensland. Australian Vet. J. 43: 129-34, 1967.

Roulston, W.J., Wharton, R.H., Schnitzerling, H.J., Sutherst, R.W., Sullivan, N.D.: Mixtures of chlorphenamidine with other acaricides for the control of organophosphorus resistant strains of cattle tick *Boophilus microplus*. Australian Vet. J. (in press).

Stone, B.F., Haydock, K.P.: A method for measuring the acaricide susceptibility of the cattle tick *Boophilus microplus* (Can.). Bull. Entomol. Res. 53: 563-78, 1962.

Wharton, R.H., Roulston, W.J.: Resistance of ticks to chemicals. Ann. Rev. Entomol. 15: 381-404, 1970.

TICK RESISTANT CATTLE FOR THE CONTROL OF *BOOPHILUS MICROPLUS*

R.H. WHARTON, K.B.W. UTECH and R.W. SUTHERST
Division of Entomology, CSIRO, Long Pocket Laboratories,
Indooroopilly, Brisbane, Australia

The one-host cattle tick *Boophilus microplus* has constituted a threat to Australia's cattle industry ever since it was introduced almost a century ago from Java (Seddon,1968). Its origin is uncertain, but undoubtedly the tick and Zebu *(Bos indicus)* cattle have been associated for many thousands of years in Asia and a state of equilibrium between host and parasite has evolved. In Malaysia, Thailand and the Philippines, *Bos indicus* cattle are infested with small numbers (usually less than ten engorged females per animal) of *B. microplus* (R.H. Wharton, unpublished). This level of infestation is evidently sufficient to maintain transmission of *Babesia argentina* and *B. bigemina* since these blood parasites are present in the indigenous cattle and cause serious clinical disease in imported non-immune *Bos taurus* dairy cattle in Asia (Legg, 1959).

In Australia, until comparatively recently, the cattle industry was based almost exclusively on cattle of European *(Bos taurus)* breeds which had no prior contact with *B. microplus*. The initial contacts were disastrous, mainly because of the effects of *Babesia*, but also because very large numbers of ticks engorged on the cattle causing serious production losses and mortalities. In an effort to limit these losses, chemicals were applied to the cattle, first arsenic, then chlorinated hydrocarbons and in more recent years, organophosphorus compounds. Although these chemicals have been successful in limiting production losses they are expensive and the ability of the tick to develop resistance to them has led to a series of crises in the cattle industry. The present situation wherein at least six distinct strains of organophosphorus resistant *B. microplus* have developed since these chemicals were introduced some 15 years ago indicates the futility of expecting stable control based on chemicals (Wharton and Roulston,1970, W.J. Roulston unpublished).

It is obviously desirable that an alternative approach based on sound ecological principles, but without reliance on chemicals should be developed. The use of cattle which are resistant to tick infestation, expressed by their ability to prevent the maturing of large numbers of female *B. microplus*, provides this alternative approach. The improved control which results from the use of resistant cattle has been demonstrated experimentally by Wilkinson (1962) using selected Shorthorn *(B. taurus)* cattle, by Riek (1962), Francis and Little (1964) and Wharton et al.

(1969) using Zebu x British cattle and by cattle owners who have introduced Zebu *(B. indicus)* cattle into their herds.

Concept of the use of resistant cattle

The concept envisages the management of tick populations by utilising host resistance as a major mortality factor in the tick's life system and the restoration, in a relatively sophisticated cattle industry, of the equilibrium between host and parasite which evolved under natural conditions. It requires an understanding of the population dynamics of *B. microplus* and in particular the relationship between the larval productivity of the engorged female, contact between host and parasite in the pastures and mortality of parasitic ticks on the host.

Implications in relation to control

The principles on which the control of *B. microplus* are based are (a) prevention of live-body weight losses and death from the direct effects of ticks and (b) prevention of clinical disease from infection of cattle with *Babesia argentina* and *B. bigemina*. To achieve satisfactory control, therefore, tick populations should not rise above levels which cause an appreciable effect on cattle but not fall below the level where *Babesia* transmission becomes intermittent. There is evidence to indicate that infestations of 20–30 engorged females per day have little effect on production and infestations of this order have been recorded on cattle properties where tick fever is not a problem (R.H. Wharton unpublished). No information is available to indicate the minimum tick population needed to maintain tick fever immunity in cattle but there is a distinct possibility that under some circumstances the use of resistant cattle could reduce populations below the level where *Babesia* transmission is interrupted (see below).

Resistance is associated primarily with Zebu *(B. indicus)* cattle (Riek,1962) and cattle of these breeds (Brahman, Africander, Sahiwal, Sindhi) and their crosses with European *(B. taurus)* cattle will be required. Cattle can be rated for tick resistance on the basis of field infestations by counting semi-engorged females (Wilkinson,1962) and under these conditions *B. taurus* cattle have been shown to carry 2–4 times as many ticks as *B. indicus* x *B. taurus* cattle (Wharton et al.,1969, Seifert,1971). However a measure of the level of resistance is essential for an understanding of how tick populations may be controlled by resistance. This may be determined by infesting cattle with known numbers of tick larvae and estimating the yield of engorged females by counting semi-engorged females 4.5–8mm in length on days 17–30 after infestation (Wharton and Utech,1970). The resistance level, based on a minimum of two infestations of 20,000 larvae, 14 days apart, is expressed as the percentage of female ticks which fail to mature assuming a 1 : 1 sex ratio. Thus cattle on which 100 or 1000 ticks mature have resistance levels of 99 or 90 % respectively.

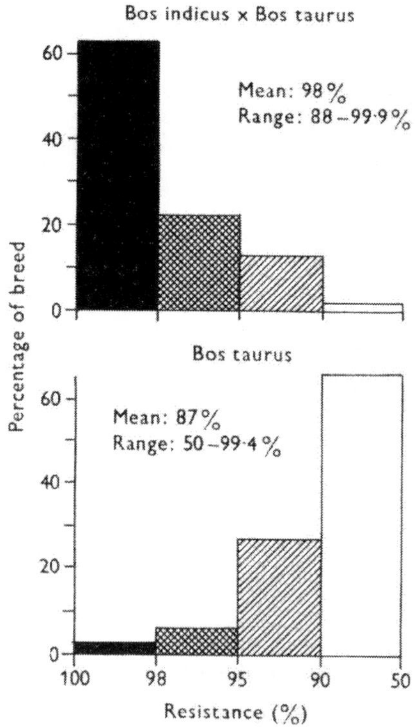

Fig. 1. Frequency distribution of resistance to *Boophilus microplus* in *Bos indicus* x *Bos taurus* and *Bos taurus* cattle.

Figure 1 shows the distribution, range and levels of resistance that have been determined for herds of *B. indicus* x *B. taurus* and *B. taurus* breeds. A wide range of resistance occurs in both breeds, but the mean for *B. indicus* x *B. taurus* was 98 % compared with 87 % for *B. taurus* and whereas over 80 % of the Zebu cross cattle could be regarded as resistant (level > 95 %) less than 10 % of the European cattle were in this category.

Different levels of resistance will be required for the various climatic conditions under which cattle are maintained. Thus Zebu cross cattle with a resistance level of 98.5 % just failed to produce satisfactory control in the wet tropics rainfall 203cm. mean temperature 23°C, but almost eliminated ticks in one year in the subtropics rainfall 76 cm. mean temperature 21.5°C (R.H. Wharton unpublished).

Although it has been shown that tick resistance in *Bos taurus* cattle is heritable (Wharton, Utech and Turner,1970) the rate of progress towards the development of resistant herds will be slow. In a herd in which selection is based only on tick resistance recent estimates suggest that an improvement of 1 % resistance per generation may be possible. Thus it may require at least 10 generations to achieve a resistance level of 97–98 % which can be achieved with a single mating between

B. indicus and *B. taurus* cattle (R.H. Wharton and K.B.W. Utech unpublished). Selection for resistance is also possible in cross-bred cattle but the culling of the relatively small percentage of less resistant cattle would be more practical. In Figure 1 the 15 % *B. indicus* x *B. taurus* cattle with resistance levels of less than 95 % would produce almost as many engorged females as the remaining 85 % and culling would not only be desirable for breeding but also an important factor in limiting tick populations.

References

Francis, J., Little, D.A.: Resistance of Droughtmaster cattle to tick infestation and babesiosis. Australian Vet. J. 40 : 247–253, 1964.

Legg, J.: Report to the Government of the Federation of Malaya on animal health problems with special reference to protozoan blood parasites. FAO Report No. 1090, Rome, 26 p., 1959.

Riek, R.F.: Studies on the reactions of animals to infestation with ticks. VI. Resistance of cattle to infestation with the tick *Boophilus microplus* (Canestrini). Australian J. Agric. Res. 7 : 227–232, 1962.

Seddon, H.R.: Diseases of domestic animals in Australia. Part 3. Arthropod infestations (Ticks and Mites). Serv. Publs. Dep. Hlth Australian Ind. Hyg. 7, 170 p., 1968.

Seifert, G.W.: Variations between and within breeds of cattle in resistance to field infestations of the cattle tick *(Boophilus microplus)*. Australian J. Agric. Res. 22 : 159–168, 1971.

Wharton, R.H., Harley, K.L.S., Wilkinson, P.R., Utech, K.B. and Kelley, B.M.: A comparison of cattle tick control by pasture spelling, planned dipping, and tick resistant cattle. Australian J. Agric. Res. 20 : 783–797, 1969.

Wharton, R.H., Roulston, W.J.: Resistance of ticks to chemicals. Ann. Rev. Entomol. 15 : 381–404, 1970.

Wharton, R.H., Utech, K.B.W.: The relation between engorgement and dropping of *Boophilus microplus* (Canestrini) (Ixodidae) to the assessment of tick numbers on cattle. J. Australian Entomol. Soc. 9 : 171–182, 1970.

Wharton, R.H., Utech, K.B.W., Turner, H.G.: Resistance to the cattle tick, *Boophilus microplus* in a herd of Australian Illawarra Shorthorn cattle: its assessment and heritability. Australian J. Agric. Res. 21 : 168–181, 1970.

Wilkinson, P.R.: Selection of cattle for tick resistance and the effect of herds of different susceptibility on *Boophilus* populations. Australian J. Agric. Res. 13 : 974–983, 1962.

Section XI

PHYSIOLOGY, GENETICS AND BEHAVIOUR

Chairman

W. Knülle (West Berlin)

Assistant Chairman

F. Dusbábek (Czechoslovakia)

ON THE MORPHOLOGICAL AND PHYSIOLOGICAL CHARACTERISTICS OF THE ALIMENTARY CANAL OF THE BULB MITE *RHIZOGLYPHUS ECHINOPUS* (FUMOUZE ET RABIN)

I. A. AKIMOV
Institute of Zoology, Ukrainian Academy of Sciences,
Kiev, USSR

The morphological and ecological characteristics of the bulb mite *Rhizoglyphus echinopus* are rather different from those in other representatives of the family Acaridae (Nesbitt, 1945). In fact it was shown that these mites have a capacity for living under water and feeding on rotten substrate.

The first anatomical data of the bulb mite were described by Haller (1880). Much attention to the structure of the alimentary system of this species was given by Bekker (1959). But functional characteristics of intestinal sections of *Rhizoglyphus echinopus* have still been quite insufficiently investigated.

Material and methods

Total micropreparations of mites stained with haematoxylin according to Heidenhain, Mallory and with carmine according to Best (Romeis, 1948, Pearse, 1960). were investigated. The pH in different parts of the gut was determined visually under the microscope after adding of colour indicators to the food. Digestion of proteine in the intestine was observed with the same method, using as food the complex of protein-congo-red which desintegrated during the proteine hydrolysis with the excretion of this stain (Bressler et al., 1969). Extraintestinal action of amylase was determined quantitatively by the modified method of Somogyi-Nelson and Smith and Roy (Asatiani, 1957) after a 10-hour exposure of 100 samples of mites in 1 ml of starch gel.

Results

The digestive system of the bulb mite consists of the buccal apparatus, pharynx, oesophagus, ventriculus, colon, postcolon and rectum. Their structures are similar to those of *Caloglyphus berlesei, C. michaeli, Tyrophagus infestans* and *Acarus siro* (Prasse, 1967). The buccal apparatus is of a chewing type (Fig. 1). Chelicerae are large, massive with denticles on digitus mobilis and digitus fixus. Medial surface of chelicerae with mono- and double thorns directed forward (Fig. 1). The hypostome shields the buccal apparatus not only from below and from the sides but partly from the front. The anterior margin of the hypostome is broad, medially incurved and with soft chitinous parts laterally of this curvature (Fig. 1.). The dorsal surface of the hypostome has maxillar denticles, three on each side. Epipharynx (labrum) with four pairs of pectinate outgrowths. The oesophagus does not reach the ventral surface of the ventriculus but gradually enlarges to reach the anterior part of the latter, according to Haller (1880). Small dispersed fractions of Indian ink, and flakes of neutral-red and Congo-red settled on microvilli of glandural cells. Food lumps are formed from larger fractions of mucus and fragments of digestive cells. The peritrophical membrane of lumps in the colon stained blue, using Mallory's method. Salivary glands are similar to those in *Acarus siro* (Hughes, 1950).

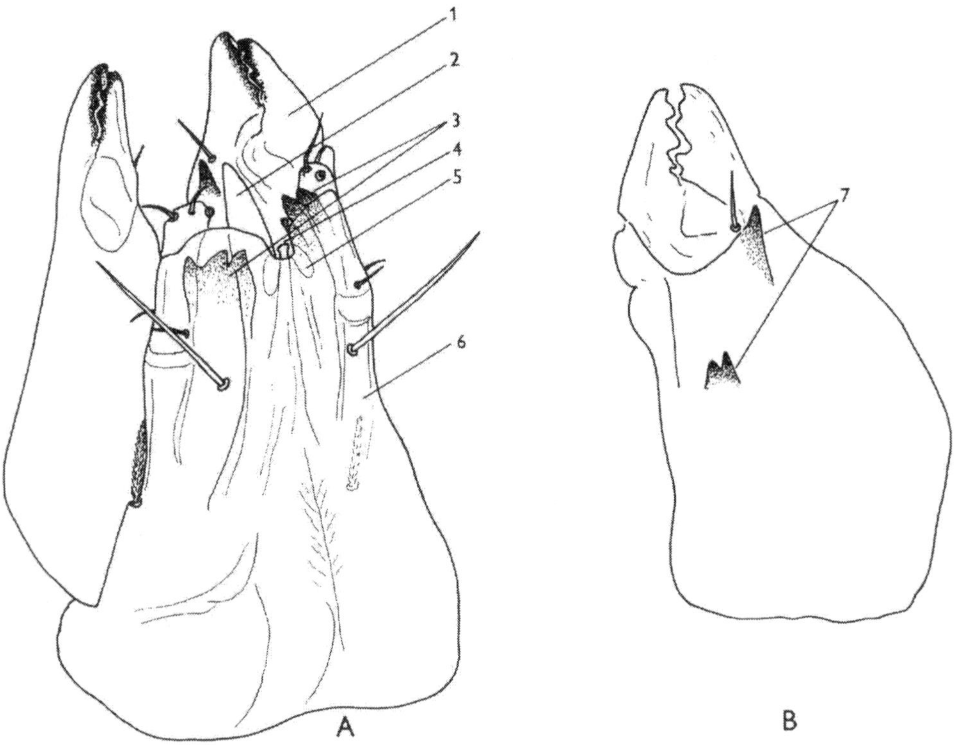

Fig. 1. The buccal apparatus of the bulb mite: A) The gnathosoma (ventral side): 1. Chelicerae; 2. Labrum; 3. Maxillar denticles; 4. Hypostomal curvature; 5. Soft chitinous part; 6. Hypostome; B) The chelicerae: 7. Mono- and double thorns.

In the starch paste on which the mites were feeding, reduced sugars could not be detected but only dextrines. The results of these determinations show that in crowded places the bulb mite substrate is subjected to the action of α-amylase due to the excretion of the mites.

During the alternate movements of chelicerae the maxillae and the medial spears of chelicerae chew the food. From the preoral cavity the food moves into the periodically contracted pharynx and then into the peristalting oesophagus. The ventriculus and diverticula of starving mites are filled with fluid food in 3–5 minutes. The frequency of peristaltic movements of the ventriculus and diverticula is 10–20 per minute. The formation time of the first food bolus in the ventriculus and its stay in the colon intercommunicates and depends upon peculiarities of the food substrate in particular on the concentration of solid indigested fragments. When mites were fed on IM solution of glucose while increasing the concentration of Indian ink (from 0.24 to 59 mg/ml) this time increased from 14 to 20 minutes but the time of stay of the food bolus in the colon decreased from 33 to 29 minutes. The first food bolus remained in the colon 2–2.5 times longer than in the ventriculus in spite of rather considerable variations of glucose concentration in the food. Covered with the peritrophic membrane the bolus staying in the colon is condensed and fluid is filtered through this membrane and released again into the ventriculus. If the mites feed on stained solutions or on proteins stained with (Congo-red), to the end of presence of the bolus in the colon the stained fluid is filtered through the membrane. This shows a hydrolysis of protein. Before passing into the postcolon the size of the bolus decreased 2–2.5 times. In the postcolon the stain rapidly filters from the bolus when its concentration is rather decreased. In the ventriculus the pH is 7.2–7.4, in colon 7.2 to 7.6, in the postcolon 7.6–8.0. This level remained unchanged even if pH of food substrate deviated from 2.2 to 9.0. When fluid passed from the colon to the ventriculus, a change in indicator coloration was noticed. It shows a definite regulating significance of the colon in buffering the system.

Discussion

The preparation of food before consumption by the bulb mite begins just before its ingestion by means of extraintestinal starch hydrolysis. Excretion of α-amylase on the substrate promotes the reproduction of mould readily eaten by mites and this probably takes place not only in connection with the secretion of salivary glands but also with the excrements. The presence of the medial curvature of the hypostome with lateral areas of soft chitin and with maxillar denticles and medial thorns of chelicerae indicates a definite though imperfect mechanical treatment of the food. At the same time the pectinate outgrowths of the epipharynx enclosing the buccal cavity from above may be similar to those in other mites sucking fluid (Belozerov, 1957). Mouth apparatus, pharynx and oesophagus are adapted to swallowing and conducting fluid as well as solid food. In the ventriculus, besides intracellular digestion typical for Chelicerata (Snodgrass, 1948), the mucusation of food, formation of a food bolus from solid substances and fragments of digestive cells, absorption of nutrient solutions is taking place. The presence of microvilli on glandular and digestive cells of the ventriculus probably points to contact digestion (Ugolev, 1963). The formation of a food bolus in the ventriculus should be regarded not only as a method of faecal excretion but also as preparation of food to luminal digestion in precedent stages. The colon digests not only fragments of food but also those substances which absorbed to indigestible substances in solution. The importance of luminal digestion was proved by the prolongated presence of food in the colon and also by the presence of a buffering system which supported the definite pH level in the intestine and this was very similar to the pH optimum of the digestive enzyme of mites (Barabanova, 1971). The surplus of food and its extra intestinal treatment compensates for the insufficiency of its digestion in mites.

In the postcolon, together with very short Malpighian tubules, the excretion of metabolic products is probably taking place. The epithelium of the postcolon is very similar to the excretory organ of spider mites (Bekker, 1957).

Thus, the characteristics of bulb mite ecology and especially methods of feeding are closely linked with the morpho-functional adaptations of all parts of the digestive system.

References

Asatiani, V.S.: Biochemical photometry. Part III, Akademia Nauk USSR, 1957. (In Russian.)

Barabanova, V.V.: A study of the activity of some nutritive enzymes of the bulb mite – *Rhizoglyphus echinopus*. III. Int. Congress Acarol., Abstracts, Prague, pp. 25, 1971. (In Russian.)

Bekker, E.G.: To the problem of the order Acarina. Trud. Inst. morph. zhivotn. 27 : 151 to 162, 1959. (In Russian.)

Belozerov, V.N.: Some data on the biology and anatomy of the mite *Poecilochirus necrophori* Vitzt. (Parasitiformes, Parasitidae). Zool. Zhurnal 36 (12) : 1802–1812, 1957. (In Russian.)

Bressler, S.E., Blinkova-Loukianets, A.A., Popov, A.G.: Colorimetrical method for measuring the proteolytical activity in different biological media. Laborat. delo, n. 7, pp. 406–409, 1969. (In Russian.)

Hughes, T.E.: The physiology of the alimentary canal of the *Tyroglyphus farinae*. Quart. J. Microscop. Sci. 91 (1) : 45–61, 1950.

Nesbitt, H.J.: Revision of the family Acaridae (Tyroglyphidae) (Order Acari), based on comparative morphological studies. Part. I. Historical, morphological and general taxonomic studies. Canad. J. Res. Sect. D, 23 : 139–188, 1945.

Prasse, J.: Zur Anatomie und Histologie der Acaridae mit besonderer Berücksichtung von *Caloglyphus berlesei* (Michael, 1903) und *C. michaeli* (Oudemans, 1924). I. Das Darmsystem. Wiss. Z. Univ., Halle, 5 : 789–812, 1967.

Pearse, A.G.E.: Histochemistry, theoretical and applied. London, 1960.

Romeis, B.: Mikroskopische Technik. Müchen, 1948.

Snodgrass, R.E.: The feeding organs of Arachnida including the mites and ticks. Smith. Misc. Coll. 110 (10) : 1–93, 1948.

Ugolev, A.M.: Contact digestion. Acad. Sci. USSR, Moscow–Leningrad, 1963. (In Russian.)

AN ELECTRON-MICROSCOPE INVESTIGATION OF THE MIDGUT EPITHELIUM OF THE ADULT FEMALE OF *DERMACENTOR MARGINATUS* SULZ.

V. N. BELOZEROV and V. A. TYMOPHEEV
Biological Research Institute of Leningrad University,
Leningrad, USSR

The midgut epithelial cells of starved mosquitoes are characterized by the presence of whorls of granular endoplasmic reticulum which unfold following ingestion of a blood meal (Bertram et Bird, 1961; Stäubli et al., 1966). In mosquitoes and other blood sucking insects digestion takes place in the gut lumen. It would therefore be of interest to examine the fine structure of the midgut epithelium and its changes in blood-sucking ticks in which intracellular digestion occurs.

We have undertaken an electron-microscope investigation of the midgut epithelial cells in adult females of *Dermacentor marginatus* (unfed 7-months old specimens, ticks after attachment, stages of slow and rapid feeding, and also after engorgement). The midgut diverticula were dissected in cold saline and fixed in osmium tetroxide with subsequent treatment with uranyl acetate. The dehydrated material was embedded in araldite. The sections cut with a LKB-3 ultra nicrotome and stained with lead citrate were examined with UEM-B-100 electron microscope.

Observations of the midgut epithelium of *D. marginatus* revealed that its cells lie on the basement membrane of a thin laminate appearance. It is attached to the haemocoele by connective tissue cells and also by muscles and tracheoles. The basement membrane of unfed ticks (just after attachment) forms numerous folds. Progress in ingestion of a blood meal results in unfolding and stretching of the membrane (its thickness decreases 3–4 times). Basal parts of the epithelial cells are distinguished by the powerful system of septa (e.g. invaginations of the outer cell membranes) which is characteristic of tissue with an intense transport function, for instance of coxal glands playing an important role in the water and ion exchange in argasid ticks (Groepler, 1969; Hecker et al., 1969) and of salivary glands possessing the same function in ixodid ticks (Dzhafarov, 1964, 1965; Balashov et Dzhafarov, 1966). The apical and side surfaces of the cells are covered by microvilli which form the so-called striated border. The microvilli are spaced at intervals which equal their diameter (in this respect ticks are different from insects and other animals with luminal digestion in which the microvilli are placed very close together). Apart from some common cell organelles (nucleus, mitochondria, Golgi complex, elements of endoplasmic reticulum) there are in the epithelial cells a number of different vacuole- and granula-like inclusions. The cellular fine structure reveals distinct changes associated with ingestion and digestion of a blood-meal.

The unfed ticks (Fig. 1.) are characterized by a loose cytoplasm with a number of optically empty places. The endoplasmic reticulum is represented by single vesicles and short vesiculate strands scattered in the cytoplasm (the strands are situated usually close to the cell walls). Mitochondria are scarce, they have a sausage-like form and are located mainly toward the apical and basal ends of the cell. There are numerous typical granula of a laminate structure which appear to be reserve food inclusions for late stages of usage. There occur also oval or egg-shaped structures with a concentrically layered content and a dense cover. The gut lumen appears to be empty.

In attached ticks the midgut cells become active and more highly organized. The ingestion of the first portions of a meal induces pinocytosis and the transportation of the food substances into the cell. Mitochondria increase in number. In the perinuclear zone an activation of the Golgi complexes occur and also the formation of a highly organized granular endoplasmic reticulum. There appear vacuoles of different size and optical density in the cytoplasm. But a significant number of inclusions of both types remain which are characteristic for unfed ticks. The nuclei are rich in chromatin and have 1–2 nucleoli.

The next stage (of growth and slow feeding, Figs. 2 and 3A) is characterized by an occurrence of a highly organized system of rough endoplasmic reticulum within the cells, in the perinuclear zone. Between the ergastoplasm and the cell surface the cytoplasm is filled with large optically empty vacuoles and small dense vacuoles. The outer layer is rich in small pinocytic vacuoles. Numerous mitochondria with well-developed cristae are situated in both vacuolated layers. At this stage a differentiation takes place in the nuclear structure as determined by the physiological state of the ticks. In diapausing ticks (which were maintained at long-day conditions before feeding) the nuclei are more vacuolized and short of chromatin. In non-diapausing ticks (which were re-activated by cold treatment before feeding) the nuclei retain their active structure. Except for these peculiarities connected with the nuclear appearance, the epithelial cells in ticks of both states (with developing and non-developing gonads) are similar.

During the stage of rapid feeding (Fig. 3B) the epithelial cells are being filled with large vacuoles of homogeneous content. The cytoplasm with organelles (mitochondria and endoplasmic reticulum) remain between these vacuoles and in narrow perinuclear and peripheral outer layers. The same picture of cells filled with large vacuoles is characteristic for engorged and dropped-off ticks. This is accompanied by poorer staining of the cells.

The revealed peculiarities of the ultrastructural organization of the midgut epithelial cells in ixodid ticks and its changes (especially in respect to the endoplasmic reticulum) reflect some functional differences between ticks and blood-sucking insects. The peculiarities are connected, undoubtedly, with the prevalence of intracellular digestion and with an accumulation of food reserves in epithelial midgut cells of ixodid ticks. The ultrastructural organization of digestive cells of ixodid ticks warrants a more thorough study to clarify some aspects of physiology of these parasitic arthropods and their interrelations with agents of tick-borne diseases.

References

Balashov, Y.S., Dzhafarov, T.E.: The structure of salivary glands of the tick *Hyalomma asiaticum*. Zool. Zhurn. 45 : 1134–1143, 1966. (In Russian.)

Bertram, D.S., Bird, R.G.: Studies on mosquito-borne viruses in their relation vectors. I. The normal fine structure of the midgut epithelium of the adult female *Aedes aegypti*. Trans. Roy. Trop. Med. Hyg. 55 : 404–423, 1961.

Dzhafarov, T.E.: An electron microscope investigation of pyramidal alveoli of salivary glands in some ticks. In: Elektronomikr. i fluorescent. mikr. kletki (Leningrad): 31–37, 1964. (In Russian.)

–: An electron microscope investigation of secretory alveoli of salivary glands of the tick *Ornithodorus moubata*. Cytology (Leningrad) 7 : 233–236, 1965. (In Russian.)

Groepler, W.: Feinstruktur der Coxalorgane bei der Gattung *Ornithodorus*. Z. Wiss. Zool. 178 : 235–275, 1969.

Hecker, H., Diehl, P.A. et Aeschlimann, A.: Recherches sur l'ultrastructure et l'histochimie de l'organe coxal d'*Ornithodorus moubata*. Acta Trop. 26 : 346–360, 1969.

Stäubli, W., Freyvogel, T.A.: Structural modification of the endoplasmic reticulum of midgut epithelium cells of mosquitoes in relation to blood-intake. J. Micr. 5 : 189–204, 1966.

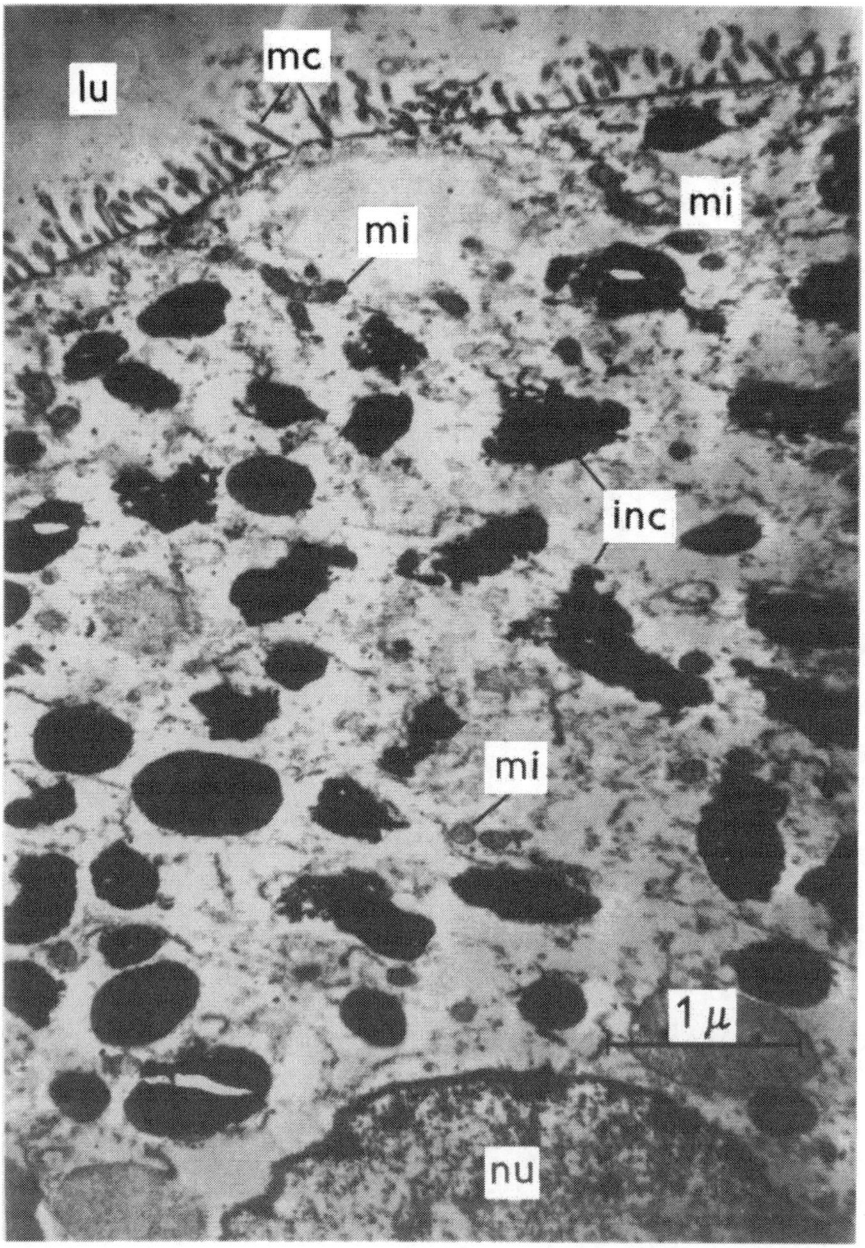

Fig. 1A) The epithelial cells in an unfed tick: The apical part of the epithelial cell.

Abbreviations: bm — basement membrane; cpz — endoplasmic reticulum; inc — cellular inclusions; lu — lumen of the gut; mc — muscular cells; mi — mitochondria; mv — microvilli; nu — nucleus; pi — pinotic vacuoles; s — septae at basal part of the cell; h — haemocoele; w — cell walls.

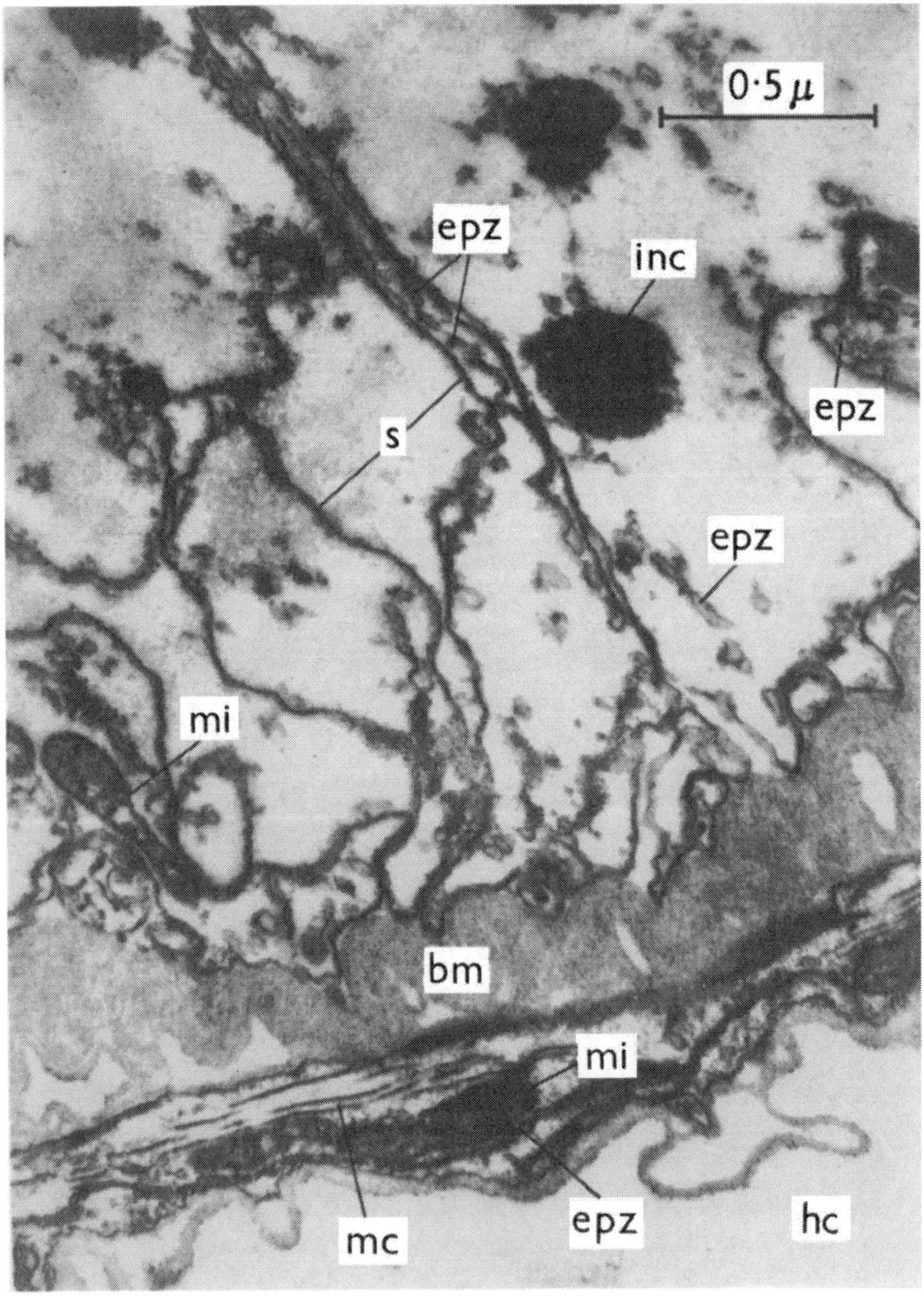

Fig. 1B) The epithelial cells in an unfed tick: The basal part of the epithelial cell.

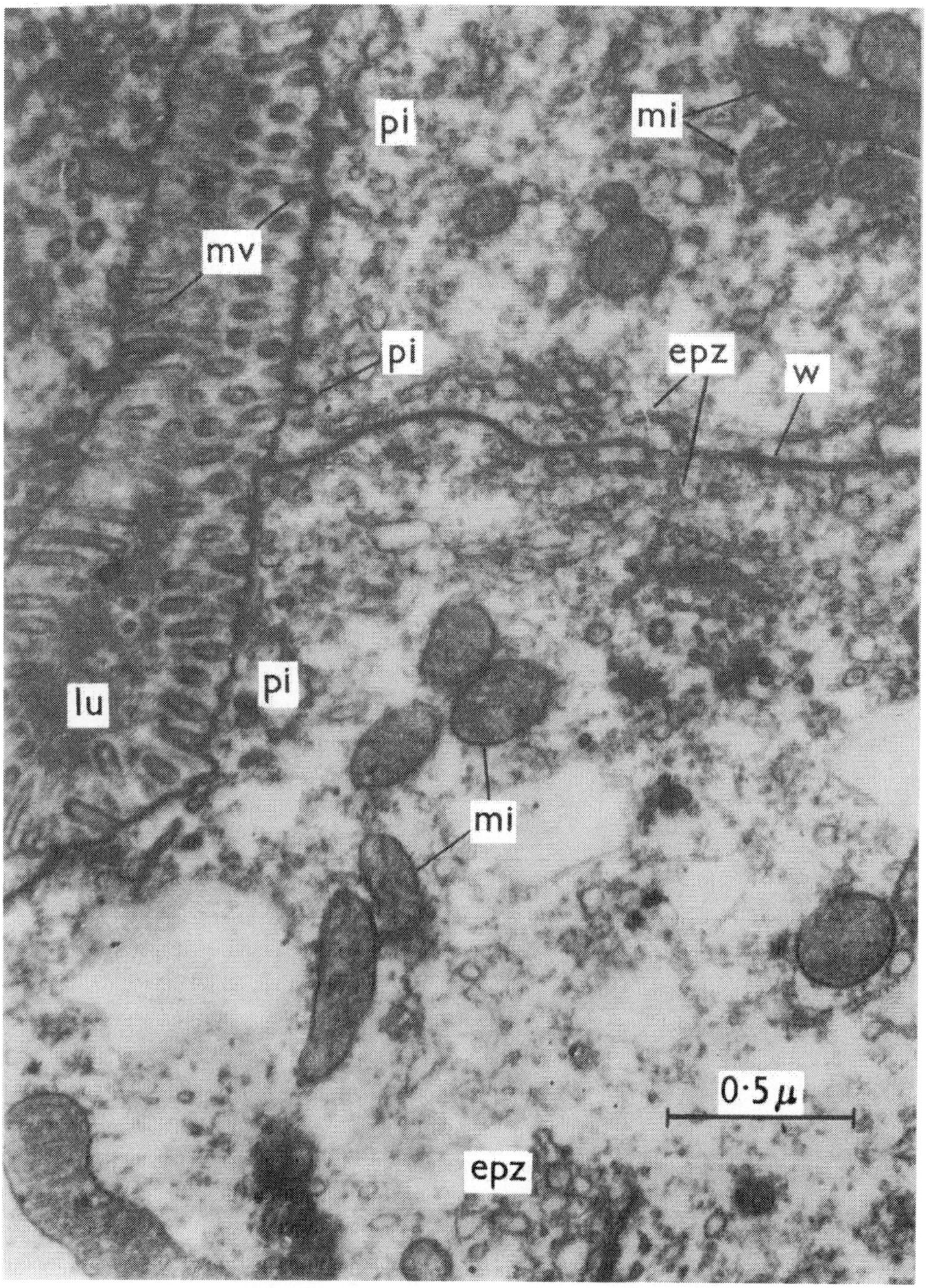

Fig. 2A) The epithelial cells in a tick at the stage of slow feeding: The outer parts of the cells along the pocket-like invaginations of the midgut epithelium.

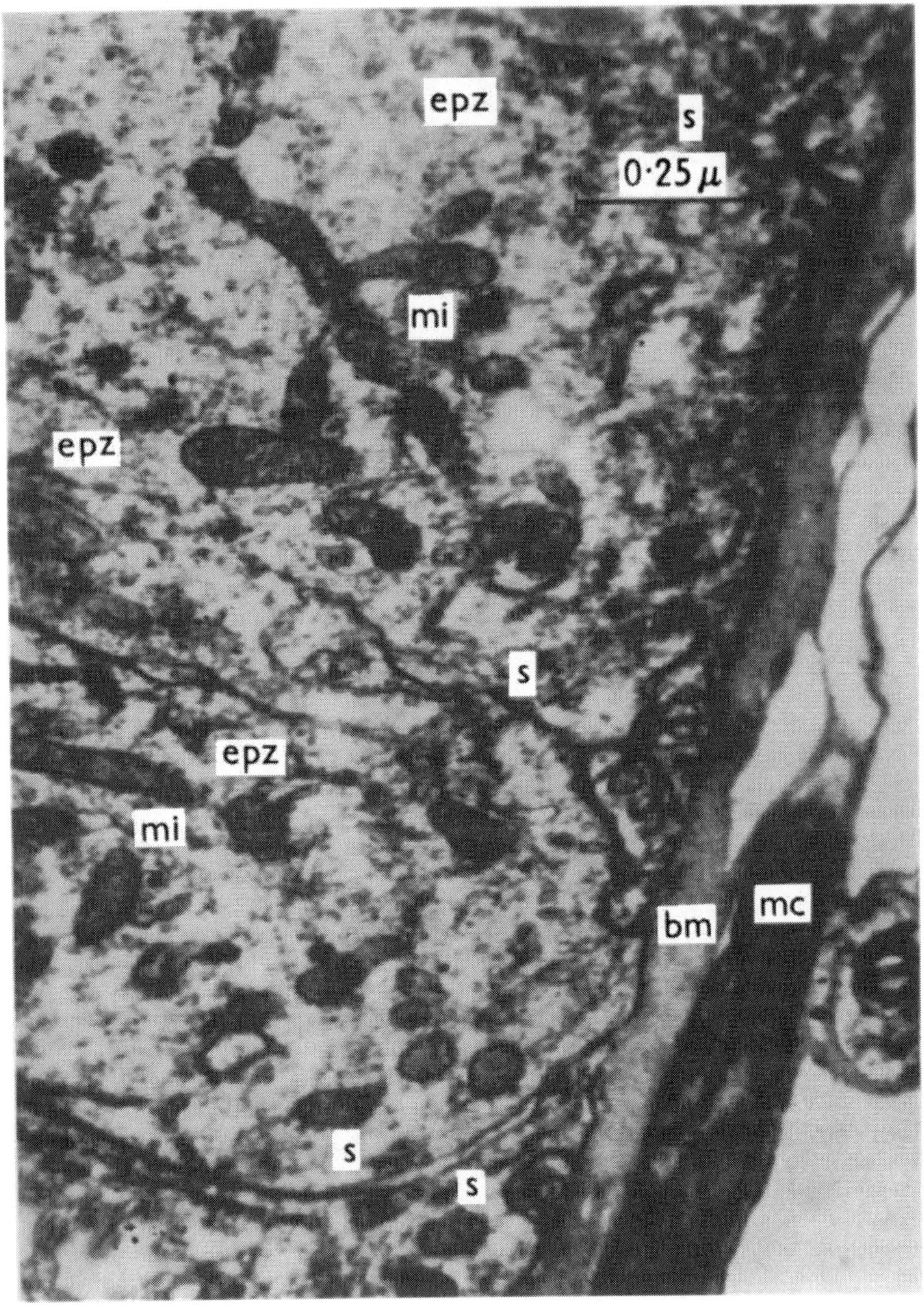

Fig. 2B) The epithelial cells in a tick at the stage of slow feeding: The basal part of the cells.

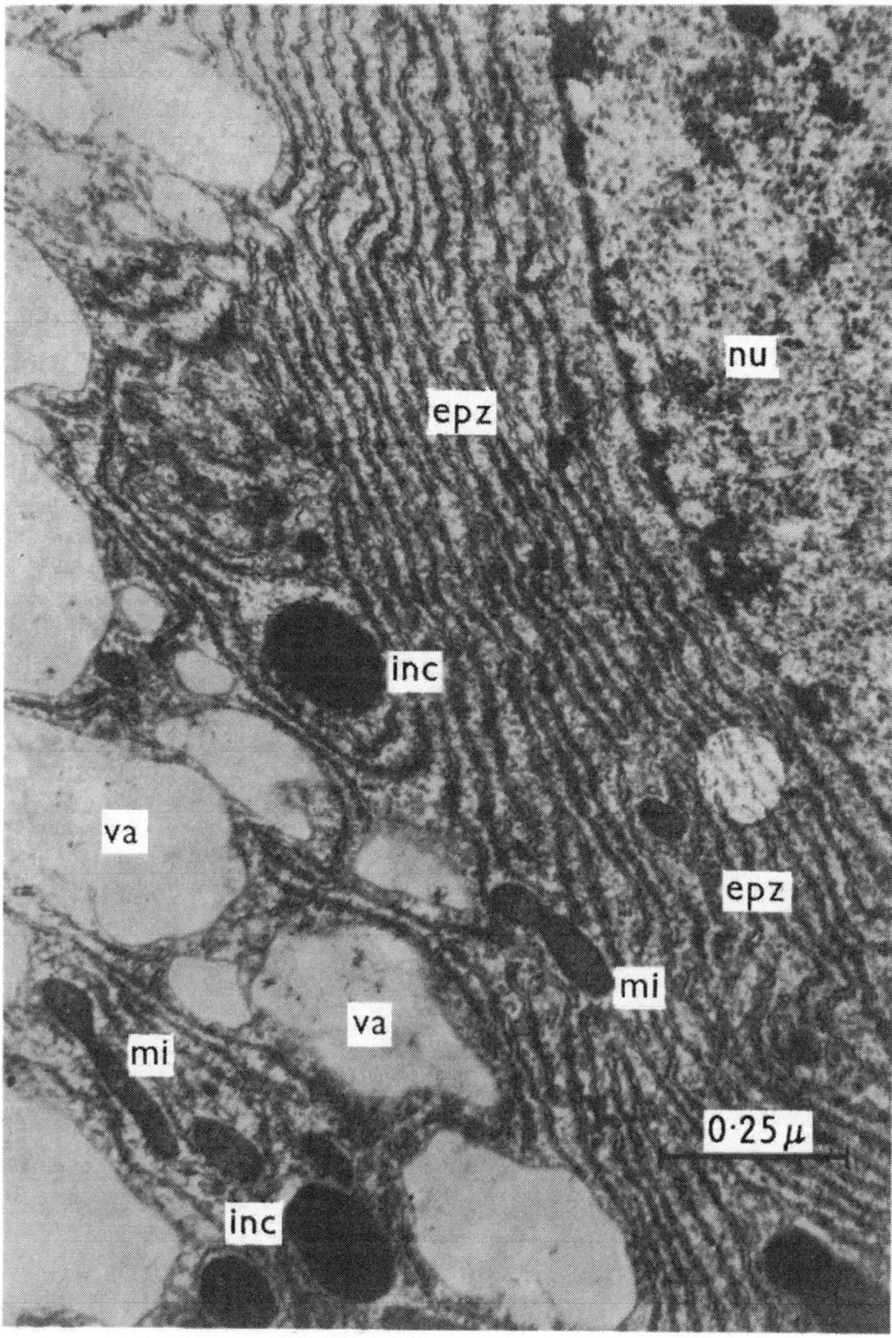

Fig. 3A) The perinuclear zone of the epithelial cell in a tick at the stage of slow feeding, with the development of rough endoplasmic reticulum.

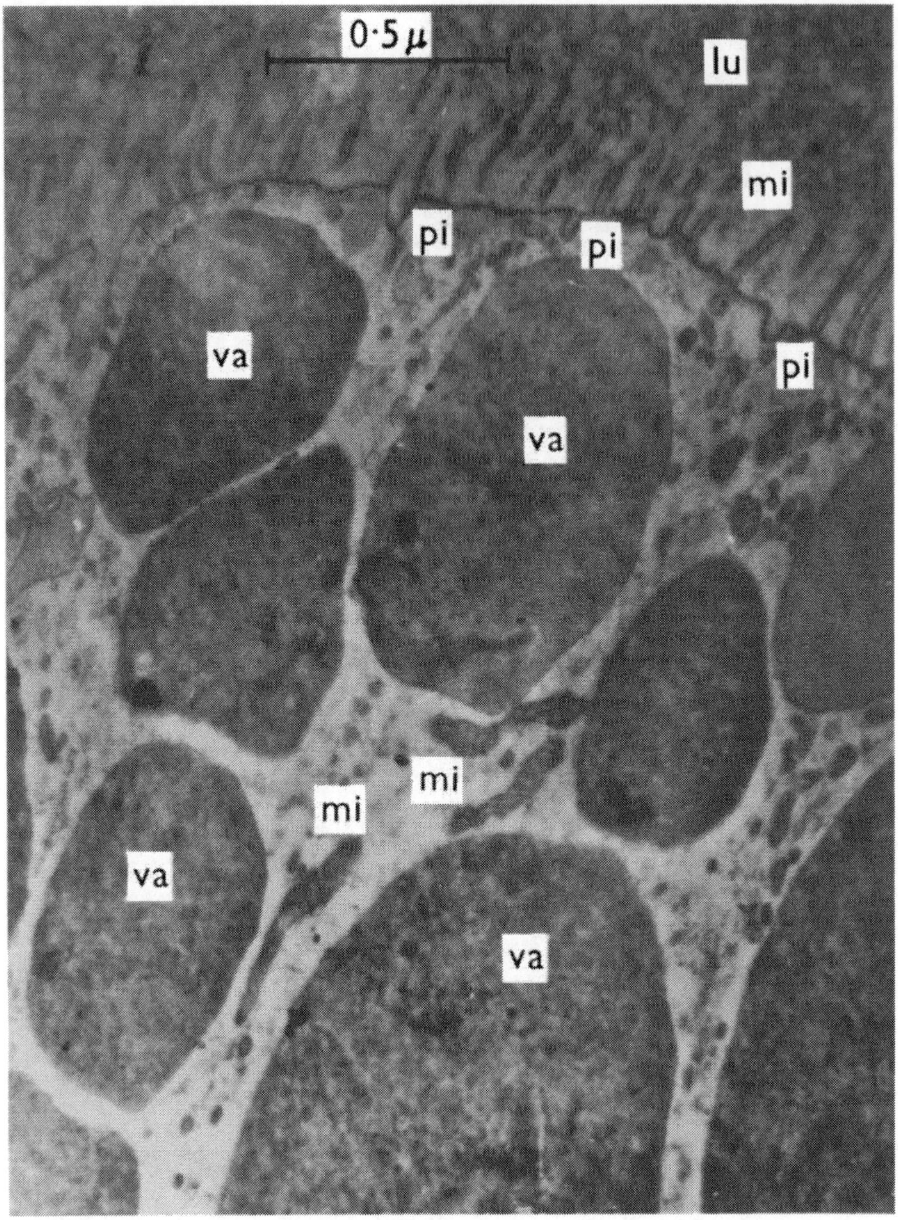

Fig. 3B) **The apical part of the cell in a tick at the stage of rapid feeding (before engorgement); large vacuoles fill the cell.**

THE SALIVARY GLANDS OF ARGAS PERSICUS (OKEN, 1818)

W. A. CHINERY
Department of Microbiology, University of Ghana Medical School,
Accra, Ghana

From studies on the salivary glands of some ixodid ticks (see Till 1961, Chinery 1963, 1965, 1970) it is evident that they differ histologically from the salivary glands of *Argas persicus*. In view of the difference in the feeding behaviour and the rate of engorgement of blood between most genera of argasid and ixodid ticks, the present study was undertaken to investigate what differences might exist histochemically between the alveoli of the salivary glands of ixodid and argasid ticks and also the possibility that this might elucidate further the functions in ixodid ticks of the other types of cells and alveoli which so far are not clear.

Materials and methods

The salivary glands of ticks at various stages of attachment (viz. unfed, engorging, fully engorged, post ovipositing and post fed ticks) were fixed in various histological and histochemical fixatives, embedded in paraffin wax and sectioned and stained by various histochemical methods given in Pearce (1968) and Barka and Anderson (1963).

Results

Histology. The salivary glands of *A. persicus* comprise a pair of compact elongated organs resembling a bunch of grapes and lying on their side in the body from the base of the first pair of legs to the base of the fourth pair of legs (Fig. 1).

They are about a fifth the length of the tick and are of the same length in males and females of similar size.

Histologically the salivary glands which are similar in both sexes consist of two types of alveoli, namely types I and II' (see Fig. 2). The type I alveoli correspond to the alveoli of the second type of Robinson and Davidson (1913), and the type II' alveoli correspond to the alveoli of the first type of these authors.

Because the type II' alveoli of *A. persicus* are comparable to the type II alveoli of ixodids but are not identical to the latter, they will be referred to as type II' alveoli.

The type I alveoli occupy the anterior third of the main salivary duct (m.s.d. Figs. 2 and 3).

During engorgement, there is an increase in the secretory activity of the peripheral cytoplasm indicated by increase in R.N.A., and slightly larger granules accummulate at the periphery of the central cavity, and these stain pink to red colour in haematoxylin-eosin. There is no appreciable increase in the size of these alveoli during engorgement.

The type II' alveoli form the rest of the gland. They are more numerous than the type I alveoli and occupy the posterior two thirds of the glands (see Fig. 2). They consist of 8 to 14 cells. These cells are of two types (a) and (i) (see Fig. 4). The (a) cells are few, being one or two in number and are always located at the base of the alveolar duct. The (a) cells contain small compact granules which stain bright red in haematoxylin-eosin and take up a wine colour in Mallory's triple stain after formal sublimate fixation. The (i) cells contain granules of various sizes (2 μ to 8 μ) and appearance as well as colloid-like masses (c.m. see Fig. 4). Some of these granules like those of the (a) cells are small (2 μ in size) and compact (s.g. Fig. 4), some are of medium size (4 μ) and compact (m.g. Fig. 4). Others are large (8 μ in size) compact granules (l.g. Fig. 4) while others are small shiny granules (sh.g. Fig. 5) of various sizes ranging from 1 μ to 3 μ. These granules stain red and bright red in haematoxylin-eosin and cherry red to deep red in Mallory's triple stain after formal sublimate fixation. The centre of the alveolus contains a lumen (lu. Fig. 5) into which the cells empty their secretion and from which arises the alveolar salivary duct.

The appearance and intensity of staining of the granules remains the same in the unfed, engorging, fully fed as well as post fed male ticks and ovipositing female ticks, the morphological and histological appearance of these alveoli in both sexes being the same.

Histochemistry. The histochemical results on the salivary glands are given in Table 1. From histochemical observations it can be concluded that the peripheral cytoplasm of the type I alveolus contains a glycogen-protein complex.

The granules in the (a) cells of the type II' alveoli contain proteins while the granules in the (i) cells lying adjacent to the (a) cells contain a glycogen-protein complex. The granules and the colloid-like masses in the rest of the cells contain a carbohydrate-protein complex which could be a neutral mucopolysaccharide, a mucoprotein or a glycoprotein.

Discussion

Morphologically there is a marked difference between the salivary glands of *Argas persicus* and those of ixodid ticks, those of *A. persicus* being very compact and limited to the anterior half of the body and consist of alveoli of only two types while those of the ixodid ticks are loose racemose and occupy a considerable amount of space in the body cavity reaching the posterior one quarter of the body and contain alveoli of three types in the unfed ticks and four types in attached males of some ixodid ticks. In *A. persicus* there are only two types of alveoli while there are three types of alveoli in the female ixodid ticks studied: Till (1961), Chinery (1962, 1965,

Table I. Summary of Histochemical Results

	Type I alveolus
Peripheral Cytoplasm	P.A.S. ++ Glycogen ++ Methylene Blue Extinction > pH. 4 Metachromazia – Glycogen ++ Non-Specific Proteins ++ Tyrosine: Significant Amino groups: Moderate Tryptophane: Moderate Lipids: Appreciable R.N.A. Moderate
	Type II' alveolus
Secretory 1. Granules in (a) Cells	P.A.S. – Non-Specific Proteins +++ Tyrosine: Significant Amino groups: Appreciable Tryptophane: Appreciable Sulphydryl groups: Significant
Secretory 2. Granules in (i) Cells	P.A.S. ++++ Glycogen –, ++ Metachromazia –, + Induction of Metachromazia ++ Methylen Blue Extinction > pH. 4 Non-Specific Proteins +++ Tyrosine: Significant Tryptophane: Traces to moderate Amino groups: Appreciable to significant Sulphydryl groups: Traces to significant
3. Cytoplasm of Type II Cells	P.A.S. ++ Non-Specific Proteins ++ R.N.A. + Alkaline Phosphatase +, +++

1970) and four in the attached males of some ixodid ticks Till (1961), Chinery (1963, 1965). These differences in morphological appearance, number and types of alveoli suggest differences in the functions of the salivary glands of *A. persicus* and ixodid ticks as will be discussed below.

Since most ixodid ticks attach for long periods (viz. 5–14 days or more) during engorgement, while the adults of *A. persicus* engorge very rapidly (viz. 1 hour or less), ixodid ticks must therefore produce more salivary secretion. Whereas ixodid ticks imbibe tissue juice and blood slowly for several days before a final rapid phase of engorgement of blood and therefore require more salivary secretion for the dissolution of host tissues, *Argas persicus* feeds almost exclusively on blood. This is reflected in the appearance of the salivary glands of ixodid ticks during engorgement. In ixodid ticks, the types II and III alveoli enlarge considerably due to enlargement of certain particular cells.

These are the (b) cells, the predominant type of cell in the type II alveoli and the (e) cells which, though inconspicuous, in the type III alveoli of unfed ixodid ticks occupy more than half the size of these alveoli in fully engorged females (Chinery, 1965). Granules believed to be precursors of an anti-coagulant (Chinery, 1965), accumulate in the (b) cells which are comparable to the (i) cells of the type II' alveoli of *Argas persicus*. In *A. persicus*, the type II' alveoli comparable to the type II alveoli of ixodids, do not show any appreciable increase in size and the (i) cells already contain the granules believed to be the precursor of an anticoagulant which in *Argas persicus* should be immediately available in the salivary secretion at the beginning of engorgement since the adult ticks engorge rapidly (less than one hour) on blood.

The salivary glands of most ixodid ticks are also known to produce a "cement" which attaches the mouthparts of the tick to the host, Gregson (1960), Arthur (1969). Recently Chinery (in press) has shown that this "cement" is the product of the secretion of particular cells in the types II and III alveoli of the ixodid tick *Haemaphysalis spinigera*. In *A. persicus* which does not produce a cement substance during engorgement, the type III alveoli of ixodid salivary gland is absent and the type II' alveoli of *A. persicus* contain very few of these cement producing cells comparable to those in the type II alveoli of ixodids. In some of the ixodid ticks so far studied (Till, 1961, Chinery, 1965), another type of cell is differentiated within the type II alveoli and another type of the salivary alveolus (type IV) becomes distinguishable in the salivary glands during attachement of males. In *Argas persicus*, the morphological, histological and histochemical appearance of the type II alveoli remains the same in both sexes during their short period of engorgement, and neither additional cell types are differentiated in the type II' alveoli of the male nor other types of alveoli formed in the salivary glands of the engorging male. It is interesting to note that unlike ixodid ticks, both male and female ticks of *A. persicus* engorge at the same rate. Although cytolisins have not been demonstrated directly in the salivary secretion of *A. persicus*, the anticoagulant properties of the saliva of *A. persicus* have been demonstrated, (see Nuttall and Strickland, 1908).

Anticoagulants are normally known to be acid mucopolysaccharides. Although in this study the presence of acid mucopolysaccharides was not demonstrated in the granules of the (i) cells, the granules of these cells contained a carbohydrate-protein complex which shows B metachromazia in azure A after treatment with concentrated sulphuric acid. This could be a neutral mucopolysaccharide a mucoprotein or a glycoprotein.

From observations on the appearance of the secretory granules of the (i) cells in unfed ticks and ovipositing female ticks, it is possible to conclude that these granules represent different stages in the elaboration of salivary secretion. The small granules observed in some of these cells representing the initial stage, the compact granules representing the intermediate stage and the colloid-like masses representing the final stage in the elaboration of the secretion, the latter being formed presumably by the merging together of the large compact granules.

However, the function of the type I alveoli of *Argas persicus* which also occurs in the salivary glands of *Ornithodorus moubata* (Chinery, unpublished observation) remains undetermined. Further studies are likely to shed more light on the function of these alveoli which are common to both argasid and ixodid ticks.

I wish to thank Dr. E.N.W. Oppong of the Faculty of Agriculture, University of Ghana, Legon for the supply of ticks. I am also thankful to Prof. W.N. Laing and Dr. E.C. Christian of the Department of Pathology, Ghana Medical School, for providing facilities for photomicrography. I am also grateful to Mr. S.W. Briandt and his staff of the Medical Illustrations Unit, Ghana Medical School for photographic proceeding.

References

Arthur, D.R.: Tick feeding and its implications. In Advances in Parasitology (Ed. Ben. Dawes), Vol. 7, pp. 275–292. Acad. Press, London and New York, 1969.

Barka, T., Anderson, P.T.: Histochemistry. Theory, practice and bibliography. Harper and Row, Publishers Inc., New York, Evanson and London, 1963.

Chinery, W.A.: Cytochemical observations on the salivary glands of an ixodid tick. Demonstrations, Lab. meeting of the Roy Soc. Trop. Med. Hyg., Trans. Roy Soc. Trop. Med. Hyg. 56 : 268, 1962.

–: The histology of the salivary glands, gut and genital organs of the ixodid tick *Haemaphysalis spinigera* Neumann. Ph. D. Thesis, University of London, 1963.

–: Studies on the various glands of the tick *Haemaphysalis spinigera* Neumann 1897. Part III. The salivary glands. Acta Tropica 22 : 321–349, 1965.

–: Studies on the salivary glands of the tick *Amblyomma variegatum*. Abstract, Biennial Conference of the West African Science Association. April, 1970. University of Ibadan, Nigeria, 1970.

–: The origin and nature of the "cement" substance observed in skin lesions on rabbits resulting from the attachment and feeding of adult ticks of *Haemaphysalis spinigera*. J. Med. Entomol. (In press.)

Gregson, J.D.: Morphology and functioning of the mouthparts of *Dermacentor andersoni* Stiles. Acta Trop. 17 : 48–79, 1960.

Nuttall, G. H, F., Strickland, G.: On the presence of an antigoagulin in the salivary glands and intestines of *Argas persicus*. Parasitology 1 : 302 – 310, 1908.

Pearce, A.G.E.: Histochemistry, Theoretical and applied. Third edition, Vol. 1, pp. 759, 1968.

Robinson, L.E., Davidson, J.: The anatomy of *Argas persicus* (Oken 1818). Part II, Parasitology 6 : 217–256, 1913.

Tatchell, R.J.: Salivary secretion in the cattle tick as a means of water elimination. Nature, London, 213 : 940–941, 1967.

Till, W.M.: A contribution to the anatomy and histology of the brown ear tick *Rhipicephalus appendiculatus* Neumann. Mem. Entomol. Soc. South Africa, No 6, 1961.

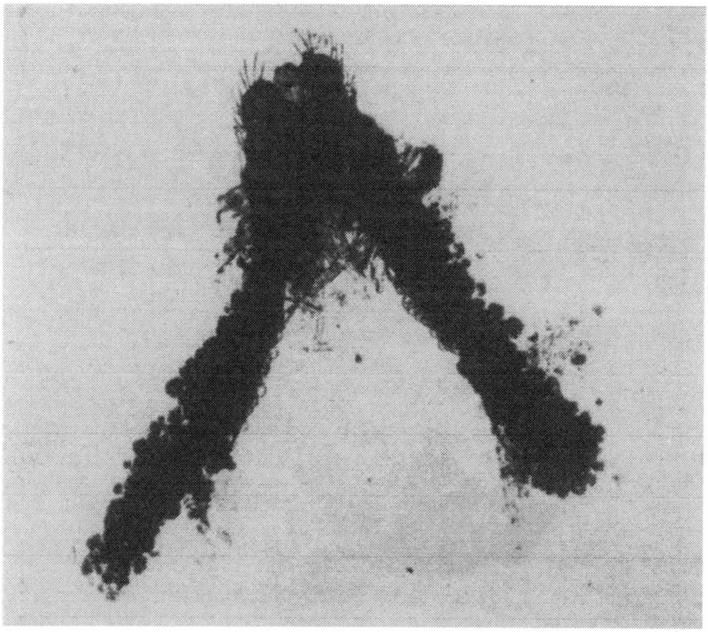

Fig. 1. Whole mount of the mouthparts and salivary glands of an adult female *Argas persicus*. (70 ×)

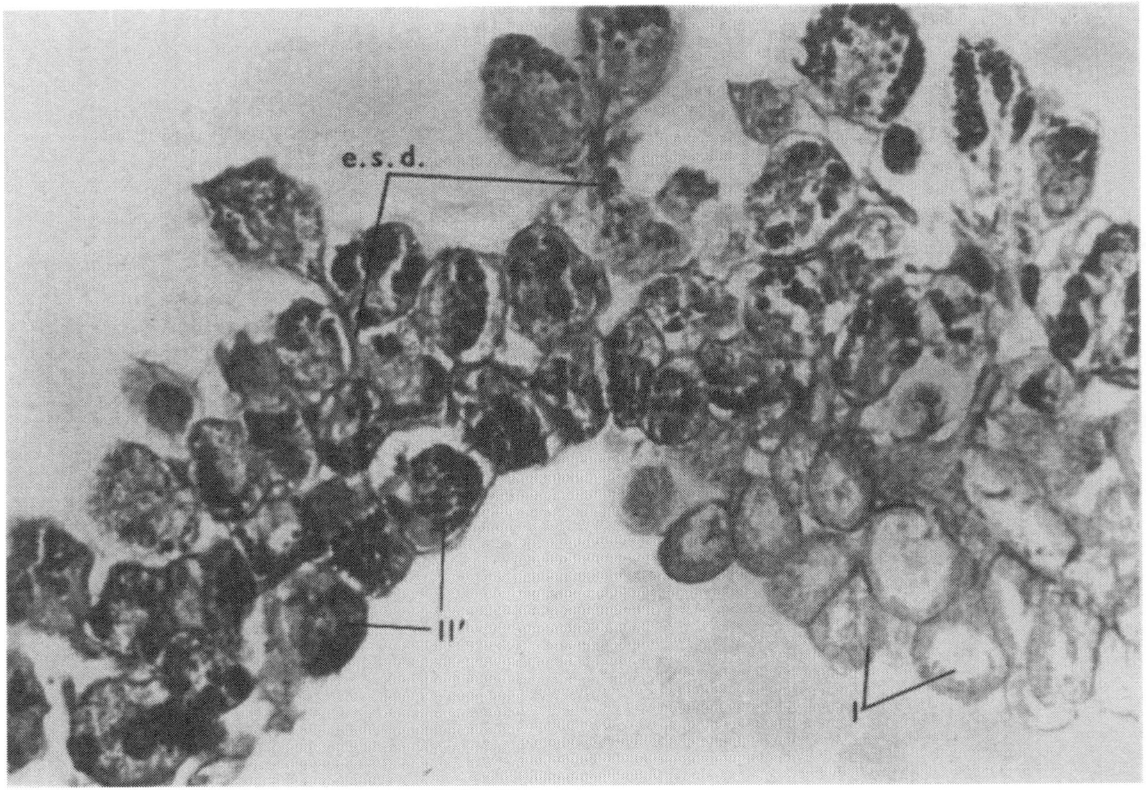

Fig. 2. A longitudinal section through the salivary gland showing type I and II alveoli. Fixed in formal sublimate and stained with triple stain (Himes and Moriber). (256 ×).

Explanation of abbreviations: c.m. — colloid-like mass; e.s.d. — efferent salivary duct; l.g. — large granules; lu. — lumen; m.g. — granules of medium size; m.s.d. — main salivary duct; s.g. — small granules; sh.g. — shiny granules.

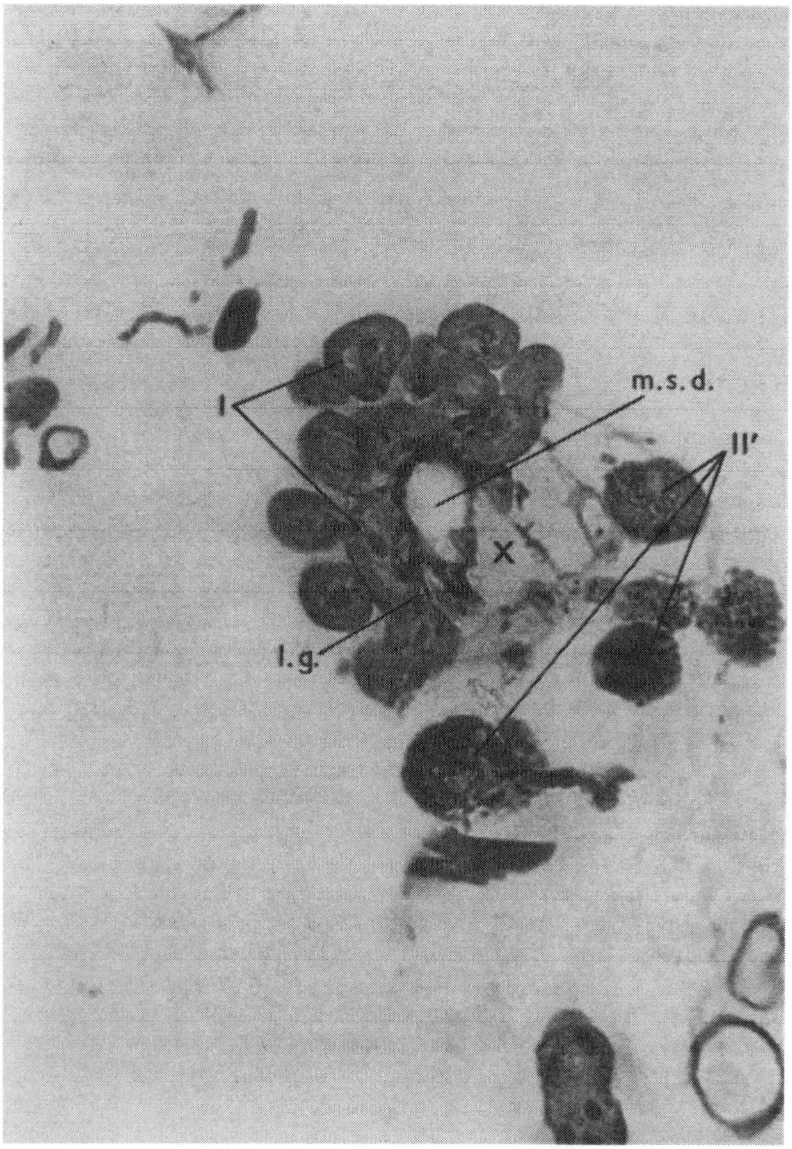

Fig. 3. A transverse section through the salivary gland showing the type I alveoli restricted to the dorsal, mesial and ventral sides. Fixed in alcoholic Bouin's fixative and stained with haematoxylin-eosin. (256 ×).

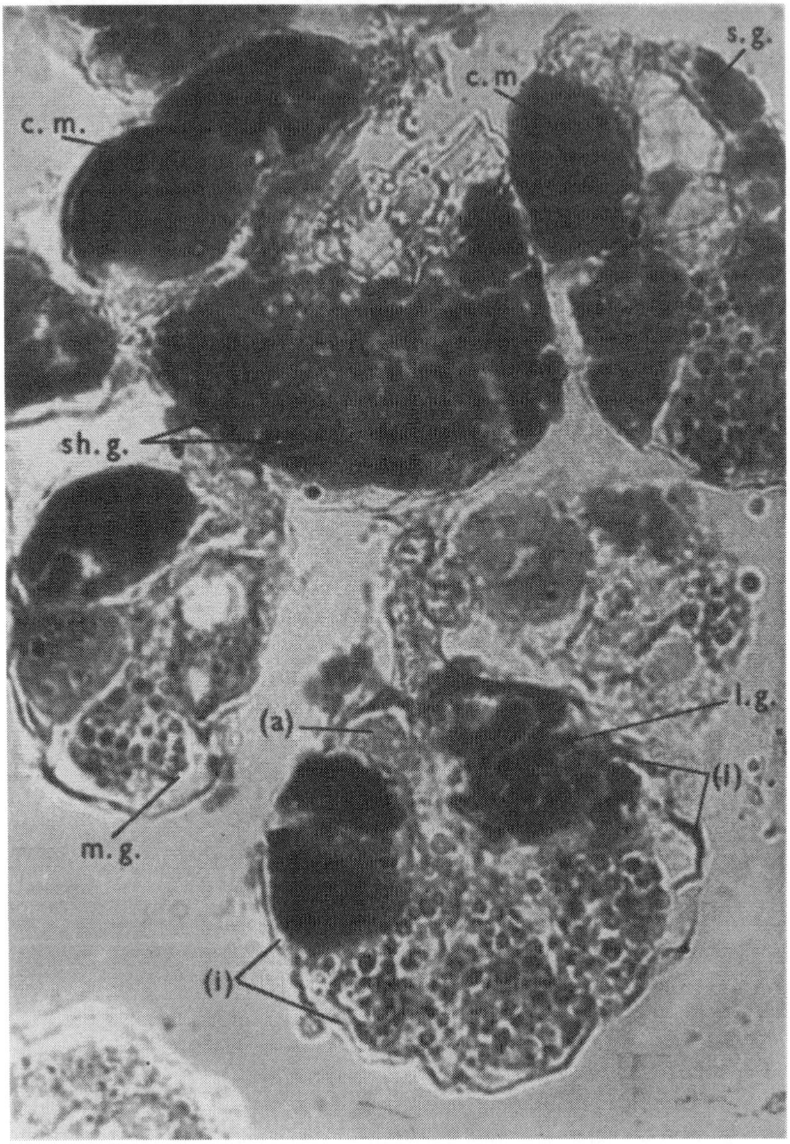

Fig. 4. A transverse section through the salivary glands showing the (a) and (i) cells and the small granules (s.g.) the granules of medium size (m.g.) and the large granules (l.g.). Fixed in formal sublimate and stained with triple stain (Himes and Moriber). (464 ×).

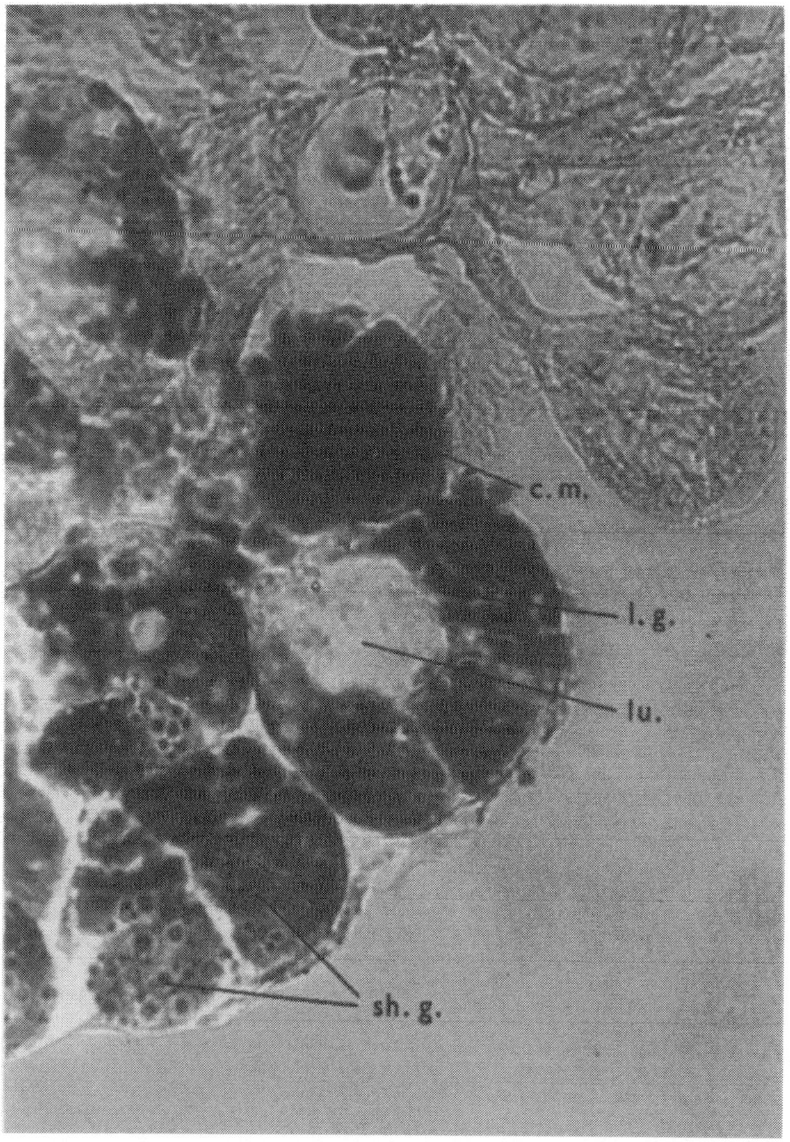

Fig. 5. A transverse section trough the salivary glands showing the lumen in the type II alveolus. Fixed in formal sublimate and stained with triple stain (Himes and Moriber). (464 ×).

KINETICS OF WATER EXCHANGE BETWEEN A MITE, *LAELAPS ECHIDNINA*, AND THE SURROUNDING AIR

T. L. DEVINE and G. W. WHARTON
Institut für Angewandte Zoologie, Freie Universität Berlin, 1 Berlin 41, West-Berlin
The Acarology Laboratory, College of Biological Sciences, The Ohio State University, Columbus, Ohio, USA

Abstract

The temporal dependence of water exchange between an arthropod and its surroundings fits a mathematical model based on diffusion theory. The model requires that the mass change rate of *L. echidnina* results from a zero order component consisting primarily of water sorption and a first order component made up largely of water loss. Tentatively, transpiration of water is associated with primarily the tracheal system, and sorption of water primarily with other surfaces. Placing mites in a CO_2 atmosphere greatly increased their water loss, but had little effect on sorption of water. The effect of increase in temperature on the sorption rate (m_S) and transpiration rate constant (k_T) was to increase them. The influence of the activity of the vapor phase in the ambient air (a_V) was primarily on sorption. The seven-fold increase in transpiration that occurs when mites die is in part explained by loss of the life process that inhibits transpiration.

Results of this study will be published in another journal under the title, "Kinetics of water exchange between a mite, *Laelaps echidnina*, and the surrounding air".

COPULATION AND SPERMATOPHORE FORMATION IN SOFT AND HARD TICKS*

B. FELDMAN-MUHSAM
Laboratory of Medical Entomology, The Hebrew University – Hadassah Medical School, Jerusalem, Israel

The reproduction i.e., copulation, fertilization and oviposition aroused the interest of acarologists already at the beginning of this century (Samson, 1909; Nuttall and Merriman, 1911). However, these processes in ticks are so unusual and different from those in other groups of Arthropods, that even continued research by Robinson (1942), Wagner-Jevseenko (1958), Tatchell (1962) and others, carried on in the last decades, could not solve most of the major problems involved and we are still facing many question marks. Today, I shall confine myself to presenting a few points which we succeeded in elucidating concerning one of the three phases of reproduction, i.e., copulation including spermatophore formation and sperm transfer from the male to the female.

We all know that ticks are divided into two families, the Argasidae and the Ixodidae; the latter subdivided again into prostriata and metastriata that is to say Ixodinae and Amblyomminae. I would here suggest to raise the two subfamilies to the status of families, to wit; Ixodidae and Amblyommidae, and in the following I shall refer to them accordingly. I presume that many of you have felt before the need to separate the *Ixodes* from the other genera of Ixodidae, as *Ixodes* differs in many features of external and internal morphology, in several biological characteristics and – as we shall see – also in some notable details of its ways of copulation.

We shall consider here merely the different features of copulation in the three families. While in general lines copulation is similar in all three groups, details differ; Ixodidae, in the sense we have suggested, including only the prostriata, are in many respects an intermediate group between the Argasidae and the Amblyommidae, fam. nov.

Let us summarize the main features of copulation in the three groups.

In Argasidae copulation begins when the male introduces the whole capitulum including the palps into the female genital aperture. After several minutes of intermittent movement into and out of the female genital aperture, the male extrudes a transparent baloon (Fig. 1), the ectospermatophore, into which after 1–2 seconds, he injects his sperm. He finally closes the full spermatophore with a "stopper" we called

* Supported by N.I.H. grant AI-07711, U.S. Public Health Service.

endospermatophore, a very important structure which – as I shall demonstrate – changes its form in adaptation to its function.

On its full formation, the male attaches the hypostome to the spermatophore's neck and grasping its lower part with its palps, deposits it on the female genital aperture. However, to avoid its sticking to the rough parts of the mouth, just before grasping, the male starts secreting saliva (Figs. 2 & 3), which lubricates all parts. He continues to secrete the saliva until he leaves the female (Feldman-Muhsam et al., 1970).

The spermiophores which up to now have been in the ectospermatophore, have evidently to proceed to the female genital tract. Soon after the spermatophore has been deposited on the female genital aperture (Fig. 4), the endospermatophore which hitherto served as a "stopper" starts to change its form and function by everting and evaginating its two lobes, after which the spermiophores flow through the neck into the two lobes of the endospermatophore which is by now well located in the female genital tract. The process of sperm transfer from the ecto- into the endospermatophore is accomplished without any intervention by the male and can be achieved in vitro and observed under the stereoscopic microscope (Figs. 5 & 6). This proves that the male mouth parts are not being used to cut the tip of the neck of the spermatophore to let the sperm flow into the female genital tract, as suggested by some authors (Tatchell, 1962; Moorhouse, 1966).

The sperm transfer is triggered by a gas, which we have shown to be CO_2, probably originating from an enzymic reaction at the periphery of the ectospermatophore, producing a pressure whereby the spermiophores are being extruded.

In *Ornithodorus savignyi* (the species of Argasidae here under consideration) the two lobes of the endospermatophore full of sperm are still connected with each other and with the neck of the spermatophore by two tiny tubes. After a few minutes these tubes become disconnected from the neck of the spermatophore, leaving the empty ectospermatophore on the exterior of the female genital aperture for some hours or even days. The spermfilled capsules ensuing from the everted lobes of the endospermatophore remain in the so-called "uterus" of the tick, a structure which should rather be called "spermatophorotheca" (Feldman-Muhsam, 1967), since its function is to keep the sperm capsules.

In the genus *Ixodes* copulation resembles on the whole that of the Argasidae. In relevant studies Arthur (1962) and Moorhouse (1966) state that during copulation the male introduces only the hypostome and chelicerae into the female genital aperture, whereas the palps do not enter it and only splay out adhering to the female cuticula. No further details have been given for the genus *Ixodes* by these authors.

In other genera of the Ixodidae, the metastriata – which we suggest to establish as a new third family, the Amblyommidae, the male continues, before as well as after the formation of the spermatophore to apply his still joined mouth parts to the female genital aperture (Feldman-Muhsam and Borut, 1971). If, at this moment, male and female are separated one can observe that the chelicerae are extruded from their shafts to be introduced into the female vagina. These different modes of copulation are correlated with the structure of the female genital aperture. In the Argasidae where the whole capitulum with the palps is introduced, the female genital aperture is large and elastic; in the Amblyommidae, its diameter is small and enclosed by chitinous flaps which prevent its expanding; whereas in *Ixodes* the genital opening while small, lacks these chitinous flaps, whereby the introduction of the male hypostome and chelicerae is made possible.

These then are the differences in the pattern of copulation between the three families under consideration.

The ectospermatophore is similar in all three families, but the endospermatophore is different. In the Argasidae, as we have described it in the course of the previous Congress of Acarology (Feldman-Muhsam, 1969), the endospermatophore consists of two lobes each containing a sac of proteinaceous "plates". The anatomy of the endospermatophore of the *Ixodes* has not been studied by the authors who described its copulation, and neither had we the opportunity to do so since *Ixodes* ticks are extremely rare in the Middle East.

In Amblyommidae the endospermatophore, though containing components similar to those in the Argasidae, differs in its morphology. It should be stressed that – though we tried scores of times – we never succeeded in obtaining evagination of the spermatophore in vitro, thus we do not know which part of the endospermatophore forms the capsule of spermiophores found upon copulation in the receptaculum seminis of the female. We do know, however, that it contains symbionts – which incidentally will never attach to the spermiophores as they do in most species of Argasidae – as well as all other components present in the endospermatophore of Argasidae, though their arrangement varies.

Copulation in the Argasidae results in two spermfilled capsules enclosed in the spermatophorotheca. Each of these capsules is provided with a tubelike extension which forms the opening for the spermiophores exit.

In Amblyommidae copulation produces one capsule filled with sperm and having two tubelike extensions located in the receptaculum seminis and not in the spermatophorotheca as it is in the Argasidae. The number of capsules found in the female depends on the structure of the endospermatophore.

In Ixodidae the receptaculum seminis appears to be formed by a constriction of the posterior part of the spermatophorotheca, thereby presenting an intermediate stage between the two other families, the Argasidae and the Amblyommidae.

To conclude, we should like to state that, though limited by time we could not go into the details of other morphological and biological variations, we consider that – in view of all the demonstrated differences in the pattern of copulation as also in the morphological details of parts of reproductive organs discussed here – our suggestion to raise the two subfamilies to the status of families seems to be fully justified.

References

Arthur, D.R.: Ticks and Disease. Pergamon Press, 1962.

Feldman-Muhsam, B.: Spermatophore formation and sperm transfer in *Ornithodoros* ticks. Science 156 : 1252–1253 1967.

Feldman-Muhsam, B.: The components of the spermatophore, their formation and their biological functions in argasid ticks. Proc. 2nd Int. Congress Acarol. (Sutton, Bonington, 1967) : 357–362, 1969.

Feldman-Muhsam, Borut, S., Saliternik-Givant, S.: Salivary secretion of the male tick during copulation. J. Insect Physiol. 16 : 1945–1949, 1970.

Feldman-Muhsam, B., Borut, S.: Copulation in ixodid ticks. J. Parasitol. 57 : 630–634, 1971.

Moorhouse, D.E: Observations on copulation in *Ixodes holocyclus* Neumann and the feeding of the male. J. Med. Entomol. 3 : 168–171, 1966.

Nuttall, G.H.F., Merriman, G.: The process of copulation in *Ornithodorus moubata.* Parasitology 4 : 39–44, 1911.

Robinson, G.G.: The mechanism of insemination in the argasid tick *Ornithodorus moubata.* Parasitology 34 : 195–198, 1942.

Samson, K.: Zur Spermiohistogenese der Zecken. S. B. Ges. Naturf. Fr., Berlin, 8 : 486–499, 1909.

Tatchell, R.I.: Studies on the male accessory reproductive glands and the spermatophore of the tick *Argas persicus* Oken. Parasitology 52 : 133–142, 1962.

Wagner-Jevseenko, O.: Fortpflanzung bei *Ornithodorus moubata* und Übertragung von *Borrelia duttoni.* Acta Trop. 15 : 118–168, 1958.

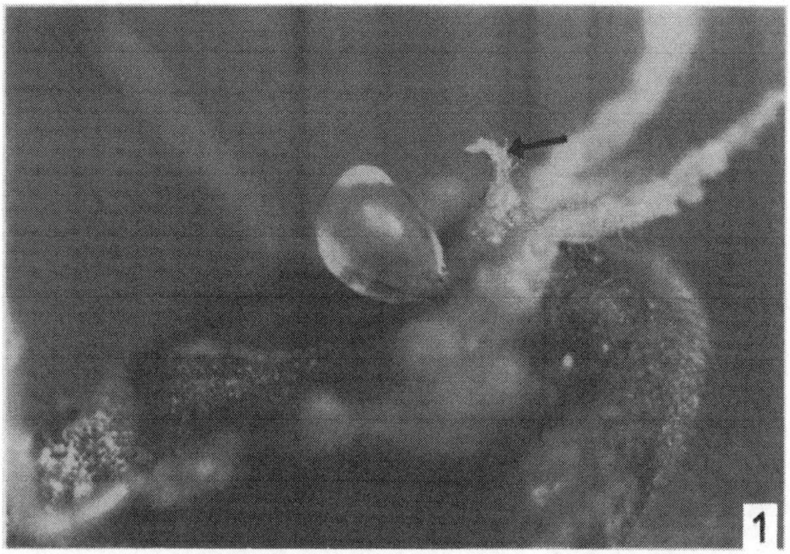

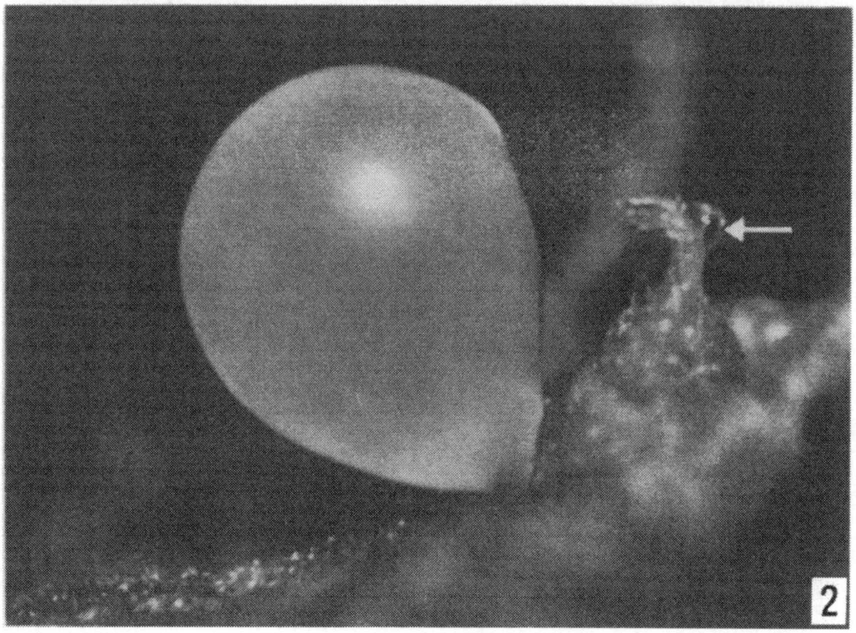

Plate I.

Fig. 1. The ectospermatophore extruded from the male genital aperture: the male *Ornithodoros savignyi* has been separated from the female.

Fig. 2. The same male as in Fig. 1., a second later. Droplet of saliva (arrow) at the tip of the chelicerae (Side view).

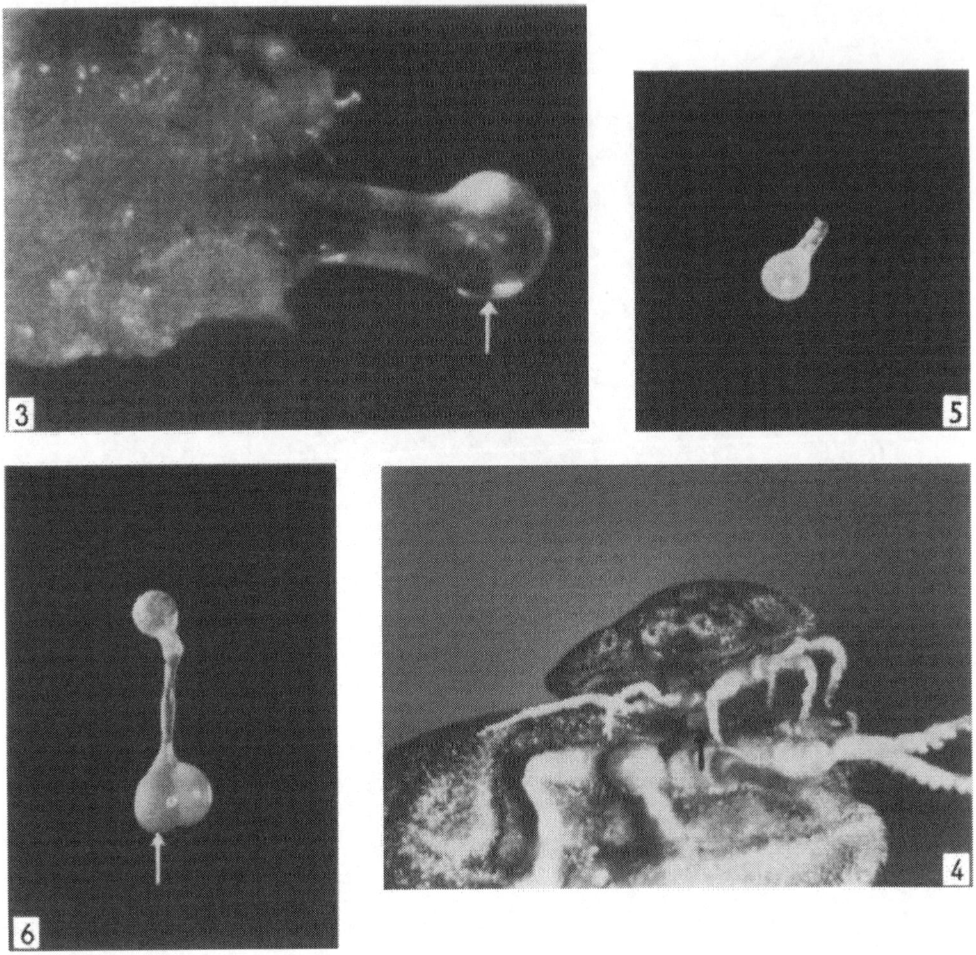

Plate II.

Fig. 3. ***Ornithodoros savignyi* male.** Droplet of saliva (arrow) at the tip of the chelicerae. (Dorsal view).

Fig. 4. End of copulation of *Ornithodoros savignyi.* The spermatophore (arrow) has just been deposited on the female genital aperture; the male leaves the female.

Figs. 5., 6. Evagination of spermatophore of *Ornithodoros savignyi* in vitro: 5. Beginning of evagination; 6. The fully evaginated endospermatophore. Note the two lobes (arrow).

CONTINUOUS INSEMINATION IN *EUTROMBICULA SPLENDENS* (ACARINA: TROMBICULIDAE)

L. J. HOLMAN
Department of Natural Sciences, Longwood College, Virginia, U.S.A.

Regardless of the fact that we now know how sperm are transferred from male to female in trombiculids, and regardless of the existence of information on longevity and productivity of chigger cultures and even the existence of records of the number of eggs produced by females over a period of time, there appears to be no evidence to indicate whether a female is receptive only once or whether she may be inseminated repeatedly. Most of the information relating to reproductive potential of chigger cultures and of single females has accrued incidental to studies on culture methods or life histories (Wharton, aided by Fuller, 1952; Neal and Barnett, 1961; Holman, 1961; and others). Lipovsky, Byers and Kardos (1957) described the mode of insemination of chiggers as the deposition of stalked spermatophores by males and the subsequent acquisition of sperm sacs by the females. Whatever the factors may be that induce the female to pick up spermatophores, i.e., that make her receptive, the physical presence of the male is not requisite. However, from the observations of Holman (1961) it does appear that the acquisition of one or more spermatophotores is prerequisite to oviposition. Furthermore, she found no significant difference between the number of eggs produced by single females isolated with five spermatophores and the number of eggs produced by females kept constantly with males.

The opportunity for the present study arose when several white larvae appeared in subcultures of *Eutrombicula splendens* derived from a strain which had been maintained in culture since 1957 by Dr. Hansell F. Cross of Georgia State College, U.S.A. Four of the white chiggers were successfully reared; and from a single pair of adults nine successive generations of white chiggers have been obtained. The normal red stock cultures have since bred true. Reciprocal matings between white and red mites invariably produced only red larvae, suggesting that *red* behaves as a simple dominant. If this is indeed true, then ovipositing females from a pure white culture should ultimately yield red larvae should they acquire additional spermatophores derived from red males.

Materials and Methods

Culture vessels used in this study contained a layer of plaster of paris/charcoal mixture as a humidifier. Larvae were allowed to feed on a caged mouse and when engorged were collected and placed in four-ounce wide mouth jars for rearing. Collembolan eggs were provided as food for active post-larval stages. By isolating the mites in one-dram vials during the imagochrysalid stage some 200 virgin females were secured; and from these, 20 females were chosen by a random method of selection for use in the experiment. Ten of the white virgin females were paired with white males and ten were paired with red males. Eggs resulting from these pairings were removed to separate containers for development. Subsequent matings between white and red strains were made as follows: twenty four days after the initial pairings were made, ovipositing females of white pairs were separated from the original males. Five of them were placed in a culture jar with five red males and five were placed in a second jar with five white males. Similarly, ovipositing females of the mixed pairs were separated from the red males into two groups of four each. Five white males were placed with the first group and five red males with the second group. These sequential matings resulted in the following potential crosses: (1) white females first with white males then with red males; (2) white females first with white males then again with white males as a control for culture (1); (3) white females first with red males then with white males; (4) white females first with red males then again with red males as a control for culture (3). Eggs were collected from each culture and kept separately as in the first matings. Observations were discontinued on the experimental cultures after five months. At this time, however, the males present in the control cultures were exchanged. These two cultures were maintained for another three months and eggs were collected as before. Larvae emerging from the respective egg cultures were classified as to color.

Table 1. Color of larvae obtained from sequential matings between white and red strains of *Eutrombicula splendens*

Matings	No. of Larvae after 1st exchange of males		No. of Larvae after 2nd exchange of males**	
	Red	White	Red	White
1. White females X (white males)* X red males	67	367		
2. White females X (white males)* X white males	0	225	894	341
3. White females X (red males)* X white males	200	27		
4. White females X (red males)* X red males	189	0	555	235

*Males removed after 24 days.
**Males of cultures 2 and 4 exchanged after 5 months.

Results

In the original matings all larvae obtained from the white pairs were white, whereas all larvae from mixed pairings were red. Counts of larvae from subsequent cultures are shown in Table 1. In every case where males of a different color were substituted for the original males, both red and white larvae were later recovered from the jars of eggs obtained from that culture.

Discussion

It became apparent during the progress of the experiment that eggs laid by white females were white regardless of the color of the males. Pigmentation in known hybrid eggs did not become evident before the deutoval stage but was well developed in newly emerged larvae. This is as might have been anticipated since maternal tissues are responsible for formation of the outer egg membranes. The growth of mold was a constantly recurring problem in jars containing developing eggs and apparently contributed to the death of a large number of embryos. It is, therefore, not feasible to attribute statistical significance to the data obtained. Nevertheless, the fact remains that white females mated to white males and producing white larvae exclusively began to produce red larvae after being placed with red males. Conversely, white females mated to red males and producing red larvae exclusively began to produce white larvae after being placed with white males. It is clear from this the females in each case picked up spermatophores subsequent to first insemination. In addition, the times at which males were exchanged during the experiment demonstrate that the females were obtaining additional spermatophores approximately at one month and also at six months following original mating.

Acknowledgments. The author is indebted to Dr. Hansell F. Cross for providing a starter culture of *Eutrombicula splendens*, and to Miss Dian Turner who set up the experiment as part of a special projects course in Biology at Lonwood College.

References

Holman, L. J.: Reproductive biology of *Trombicula (Eutrombicula) splendens* Ewing, 1913, with emphasis on the role of the spermatophore. Doctoral Dissertation, University of Maryland, College Park, Maryland, 54 pp., 1961.

Lipovsky, L. J., Byers, G. W. Kardos, E. H.: Spermatophores – the mode of insemination of chiggers (Acarina: Trombiculidae) J. Parasitol. 43 : 256–262, 1957.

Neal, T. J., Barnett, H. C.: The life cycle of the scrub typhus chigger mite, *Trombicula akamushi*. Ann. Entomol. Soc. Am. 55 : 295–303, 1961.

Wharton, G. W., aided by Fuller, H. S.: A manual of the chiggers. Mem. Entomol. Soc. Wash., No. 4, Washington, D.C. 185 pp., 1952.

Proceedings of the 3rd International Congress of Acarology, Prague, 1971

WATER VAPOR SORPTION FROM THE ATMOSPHERE: EVIDENCE FOR ACTIVE AND PASSIVE COMPONENTS IN LARVAE OF *DERMACENTOR VARIABILIS* (SAY)

W. KNÜLLE and T.L. DEVINE
Institut für Angewandte Zoologie, Freie Universität Berlin, West–Berlin

Abstract

Ticks absorb water vapor from the atmosphere. Gross transpiration, gross sorption, and net movement of water between the atmosphere and larvae of *Dermacentor variabilis* (Say) were measured by means of tritium-labelled water, and roles played by active and passive processes in these water movements were examined.

The zero-order rate constant for the active transfer of water from the atmosphere into larvae of *D. variabilis* is 0.044 μg/h at 92.5 % R.H. and 25 °C.

The experiments favor the view that active sorption of water vapor from the atmosphere by the larva does not take place over the entire humidity range; it seems that this inward transport of vapor by hydrated larvae becomes ineffective below a relative humidity of 85 %. This limit coincides with the relative humidity above which partially dehydrated larvae can resorb lost water from the air, previously designated as the critical equilibrium humidity. If the critical equilibrium humidity specific to an arthropod (range from 45 % to 95 % R.H.) generally corresponds to the lower limit of humidity for active sorption of water from the atmosphere into the organism, then the latter limit is also specific for species or closely related groups.

Results of this study will be published in another journal under the title, "Evidence for active and passive components of sorption of atmospheric water vapour by larvae of the tick *Dermacentor variabilis*."

FEEDING OF TICKS ON SENSIBILIZED ANIMALS

V. A. MUSATOV
Institute of Agriculture, Velikie Luki, USSR

Our previous studies showed that during the feeding of the tick species studied the immune state of the host changed, i. e. the host became sensibilized. The antigenic properties of salivary glands secretion play a major role among the factors influencing this phenomenon. The research on the peculiarities of feeding, growth and development of ticks on sensibilized animals constitutes great scientific interest and practical importance.

Methods

The feeding of adult ticks of three species: *Rhipicephalus bursa, Hyalomma plumbeum plumbeum* and *Hyalomma asiaticum* was studied on sensibilized animals, on sheep under laboratory conditions and on naturally infested cows. The animals became sensibilized as a rule, after the first feeding of the mentioned tick species, and the sensibilization was confirmed by the results of immunobiological tests. For the purpose of control two feedings were studied: the first was the feeding of tick species under study on a given animal and the second was the feeding of ticks on nonsensibilized animals. The ticks were allowed to feed on a fresh patch of skin on the dorsum in numbers not exceeding those of the first tick pool.

The process of feeding, growth and development was observed daily and the results were evaluated according to data obtained by biometric assessment of values characterizing these indicators for females after their dropping off (under laboratory conditions). Nine main indicators (Table 1) have been studied during 15 years from a comparative aspect. Characteristics of the external morphology and of the structure of the internal organs in females were studied by dissection as well as histologically.

Results

The ticks that were about to feed on sensibilized animals did not attach themselves readily, they often changed the site of attachment before selecting a definitive one. Even with the naked eye an irregular development of females could be observed . Individual females became detached from the animal after a few days

Table 1. Main indicators of feeding, growth and development studied in three species of ixodid ticks

Species		n	% of engorded ticks	Average length of feeding (in days)	Average weight (in mg)	Length of body (in mm)	Thickness (in mm)	Width (in mm)	Contents of dry substance in body of engorded females (%)	Potential fertility (number of eggs per 10 mg of female weight)	Actual fertility, percentage of live larvae after 30 days from number of eggs
Rhipicephalus bursa	Control	447	95.6	8.0	677	13.6	7.7	9.2	39.2	100	92
	Experiment	318	84.3	10.3	492	12.2	6.5	8.9	37.2	73	41
Hyalomma plumbeum pl.	Control	340	93.0	10.7	1030	17.5	9.0	11.2	38.5	110	87.1
	Experiment	196	64.0	12.8	700	15.2	6.6	10.4	33.0	79	40.6
Hyalomma asiaticum as.	Control	312	91.2	10.9	1573	20.7	9.6	12.5	37.6	116	72.5
	Experiment	156	72.5	12.9	843	17.0	6.8	10.9	31.6	113	36.6

of feeding and did not attempt to attach themselves again to this animal. In vitro they died relatively soon, while the females fixed to the non-sensibilized animals often also terminated their feeding.

The histological investigations of skin showed that the attachment and normal feeding of ticks on such animals in the initial period were prevented by changes in the fixative mechanism of parasites. The capsule round the hypostome was not formed deep within the skin (as is usual), but on its surface or in the epithelium of the cuticula. Due to this fact the females did not reach with their hypostomes the skin layers rich in large blood vessels and did therefore not receive sufficient quantities of blood. The dropping-off period for females which had finished feeding extended for many days, and single females remained attached to the skin for unlimited lengths of time without showing any change in their size. This fact may be explained by their inability to extract the hypostome from the pathologically changed skin. The degree of such an impeded feeding process and development was higher in females of the genus *Hyalomma* than in those of the other genera.

Table 1. shows quantitative indicators of growth and development of females which fed on sensibilized and control animals. These data indicate that on sensibilized animals, during a long period of feeding, the females gained less weight and size, accumulated less dry substance in their bodies and suffered a reduction in their reproductive ability. The reliability of the difference between the values in females from sensibilized animals on one hand, and in females from control animals on the other, was demonstrated by mathematical analysis and does not leave any doubts. Hence, these values reflect objective natural laws. During our experiments we observed cases when not a single female *H. plumbeum* feeding on separate sensibilized animals, under similar conditions, achieved full engorgement, while the viability of larvae was reduced 7–13 times, in comparison with the control animals.

The contents of the gut of females from the sensibilized animals was fairly jelly-like, often colourless, with dark red flakes. This was due to the fact that the females imbibed blood which was not pure and contained some other components (skin lysates, inflammatory exudate, etc.).

Histological investigation of the mid-gut sections showed that the size of secretory cells was considerably reduced, their vacuolation lessened and the cycle of changes in the gut epithelium was disturbed. In our opinion these disturbances reflected also the reduction of physiological activity of gut epithelium.

Discussion

The phenomenon of parasitism is one of the forms of inter-relationships among animals. The studies on the host-parasite relationship in the system of "tick - - mammal" represent a considerable contribution to the deciphering of complicated processes of inter-relationships among live organisms. Our data show that in this system, under influence of many factors accompanying the feeding of the first pool of ticks, the immune state of the host underwent a change and sensibilization set in. In such animals primarily the conditions for attachment of ticks were disturbed; to a certain degree the biological adaptation to parasitism became therefore disturbed. Considerable reactive changes of the host's skin tissues, clearly of pathologic character, disturbed the conditions for the parasite's growth and development. The presence of the pathologically changed skin surface and the consumption of food with a changed composition were the causes of the above mentioned serious disturbances in the process of feeding, growth and development of tick females.

SPERM MATURATION IN TICKS: AN EXAMPLE OF CAPACITATION IN INVERTEBRATES?[1]

J. H. OLIVER, JR. AND L. P. BRINTON
Dept. of Biology, Georgia Southern College, Statesboro, Georgia, U. S. A.
Rocky Mountain Laboratory, Hamilton, Montana, U. S. A.

Capacitation (physiologic changes of spermatozoa occurring between insemination and fertilization) in certain mammals was first reported by Austin (1951) and Chang (1951). Since then many papers have dealt with capacitation in several species of mammals and one with it in frogs (Shivers and James, 1970). Spermatozoa of many invertebrates may capacitate in the female before penetrating ova, although this possibility has apparently been ignored by most biologists. Capacitation has not been reported for a single invertebrate. Perhaps the reason for this is that most workers researching on capacitation are unfamiliar with invertebrate reproduction and/or a rather narrow concept of capacitation held by some persons. Moreover, many students of reproduction in invertebrates are unaware of capacitation in mammals. Clearly, some dramatic changes occur in spermatozoa of certain invertebrates between insemination and fertilization. Our purpose here is to cite a few examples of these changes in arthropods and to emphasize and further explain some of these changes in tick spermatozoa.

Development of spermatozoa in male and female *Dermacentor occidentalis* Marx was studied in Shen's saline solution with phase contrast optics, squash preparations stained with aceto-orcein, the usual histological sections stained with Mallory's triple stain (using Delafield's hematoxylin) and the electron microscope (tissue fixed in glutaraldehyde, post-fixed in osmium tetroxide, embedded in Araldite and stained with uranyl acetate and Reynolds lead citrate). Sections were examined in Siemans I and IA electron microscopes at 80 Kv.

Several scientists (Samson, 1909; Casteel, 1917; Oppermann, 1935) realized long ago that germinal cells of ticks were not fully mature in the male and that elongation of "prospermia" occurred upon transfer to females. These authors and others including Rothschild (1961) and Reger (1962, 1963) did not understand fully the process of elongation and were unaware of certain nonmorphological attributes of the spermatozoa.

Growth and differentiation of male germinal cells in ticks terminating in

[1] Supported in part by U.S. Public Health Service research grants AI–06169 and AI–09556 from the National Institutes of Health.

aflagellate spermatozoa have many interesting aspects including capacitation. Oliver and Brinton (in manuscript) present a detailed account of this development which includes unusually large spermatocytes, formation of chromosomal vesicles or karyomeres, presence of intercellular bridges during spermatogenesis and part of spermiogenesis, halting of spermatid development in males and necessity of transference to female for final growth and differentiation (capacitation), unusual type of motile processes, backward locomotion if nucleus is in anterior part of cell, and comparatively large size of mature spermatozoa. The average spermatozoon, such as that of man or sea urchin, is approximately 60 μ long, whereas tick spermatozoa are 1000 μ in some argasid species (Rothschild, 1961) and 200 μ in the ixodid tick *Dermacentor occidentalis*.

In *D. occidentalis*, attachment to a host is necessary for spermatogenesis and spermiogenesis to proceed, although the exact nature of the developmental stimulus remains unknown. Nevertheless, after feeding, cells undergo reductional and equational divisions, grow and differentiate, and finally become elongate spermatids approximately 100 μ long. No additional development of spermatids occurs unless the germinal cells are transferred to a female. If they reach the reproductive tract of the female, additional differentiation and enlargement occur within a few hours. Approximately five days minimum are required from transfer of spermatophores to female until fertilization occurs. Capacitation involves sub-terminal rupturing of the peripheral-most membranes at the pointed tip of the elongated spermatid, an extension of the inner core out of the newly open end of the outer sheath with simultaneous sliding of outer sheath back over the inner core and eventually turning in and contributing to the base of inner core at the opposite end of the cell. Finally, the entire outer sheath turns in and carries the long fusiform nucleus with it. At this stage of development the cell processes (probably function in locomotion) are exposed on the exterior surface of the spermatozoon and extend the length of the cell. It is unclear whether they exist as long parallel ridges (Rothschild, 1961) or are attached at only the head end (Reger, 1962). Our observations support the former because parallel ridges were seen at all levels of the spermatozoon in cross sections with the electron microscope. The portion of inner core which was broadened laterally in the elongated spermatid becomes the biconcave, paddle-like region in the mature spermatozoon. The definitive position of the nucleus and acrosome is in the opposite half of the gamete, i. e. in the long, round, narrow region. Mitochondria are scattered throughout the cell, but are consistently more abundant near the tip of the elongating inner core during development and in the mature spermatozoon.

In addition to morphological changes including doubling of length, capacitation in tick spermatozoa also results in gamete movement and development of penetrating ability. In saline, three types of movement were observed in capacitating spermatids and spermatozoa: writhing movements of the anterior end, serpentine contortions beginning anteriorly just behind the enlarged head (paddle-like region) and traveling posteriorly and a steady gliding movement resembling motion of some ciliated protozoa. The first type, in contrast to the latter two, does not result in progressive movement and the last type was seen only in mature spermatozoa. Although spermatozoa can move, it is doubtful if motility is of major importance in transfer of spermatids in the male reproductive tract and subsequent migration up the oviducts into the ovary. Pressure in testes from enlargement of cells and contraction of dorsoventral muscles attached to the respective body walls, probably force the spermatids from the testes and down the vasa deferentia. Hinton (1964) believes that once insect and vertebrate spermatozoa leave the follicle, their motility, if any, plays no effective role in their passage through the male reproductive tract. Contractions of oviducts in *D. occidentalis* appear responsible for the major relocation of spermatozoa in the female. Large clumps or masses of spermatids in the male genital tract and masses of spermatozoa in the oviducts would seem to preclude sperm motility as a significant factor in gamete relocation. The gametes are so crowded and entangled it would be virtually impossible for them to move en masse by their own efforts. Moreover, the ovary and oviducts of mated females exhibit peristalsis in saline. Seminal fluid probably enhances rhytmic contractions of the oviducts in ticks as has been reported for other animals, i. e. the bug *Rhodnius prolixus* (Davey, 1958).

Pharmacologically active compounds in the seminal fluid are present in the accessory glands of *Rhodnius prolixus*, the roach *Periplaneta americana* (Davey, 1960) and the tick *Argas persicus* (Tatchell, 1962). Catecholamines (DOPA, dopamine, epinephrine and norepinephrine) were detected in accessory gland extract of *Periplaneta* and the tick *Ornithodoros tholozani* (Galun and Warburg, 1967). Catecholamines have been shown to also accelerate movements of the heart and malpighian tubules in *Periplaneta* (Davey, 1964) and the foregut of *Locusta migratoria* (Freeman, 1966). Some evidence suggests that DOPA also initiates development of oocytes in *O. tholozani* (Galun and Warburg, 1967).

The importance of spermatozoan mobility in the penetration of tick ova has not been determined unequivocally. The role of mobility is especially difficult to ascertain since the definitive location of the nucleus and acrosome in the spermatozoon is in the posterior part of the gamete. Nevertheless, it is becoming increasingly evident that sperm motility may be necessary for fertilization, but that it plays little part in relocation in the female tract of certain other animals. Dead spermatozoa are moved to the upper levels of the oviducts of cows as rapidly as are living ones (VanDenmark and Hays, 1954), and the same happens in the bug *Rhodnius prolixus* (Davey, 1958).

After capacitation spermatozoa are in the oviduct wall as well as in the lumen. It is unclear whether gametes penetrated these cells or were phagocytized by them. Nevertheless, many cells are found with spermatozoa inside them and it is tempting to suggest that capacitation renders spermatozoa able to penetrate various types of cells including ova. Unpublished data on the intracellular presence of spermatozoa in the ovarian luminal epithelium of the closely related Rocky Mountain Spotted Fever tick, *Dermacentor andersoni*, suggest that entry is due to penetration and not phagocytosis. Capacitated rabbit spermatozoa become penetrating cells, not only penetrating ova but also other cells (Bedford, 1965), and Soupart, considering capacitation in mammalian spermatozoa, offered the same hypothesis regarding the initiation of penetrating ability (Williams et al., 1967).

Capacitation has been demonstrated conclusively in several mammals (hamsters, rats, sheep and rabbits), and there is some evidence of its presence in man (Williams et al., 1967). A different, albeit definite capacitation occurs in tick sperm. There is a strong suggestion of its occurrence in other arthropods, i. e. sperm activation in bedbugs (Davis, 1965) and gross morphological changes in sperm of the fly, *Sciara coprophila* (Phillips, 1966). It is still uncertain whether some type of capacitation occurs in most animals, but this possibility is favored because it occurs in widely related taxa. Moreover, it would be biologically favorable for most male animals to possess uncapacitated or inactive spermatozoa and for these cells to be "switched on" only when conditions were favorable for fertilization.

References

Austin, C. R.: Observations on the penetration of the sperm into the mammalian egg. Australian J. Sci. Res., B. 4 : 581–589, 1951.

Bedford, J. M.: Effect of environment on phagocytosis of rabbit spermatozoa. J. Reprod. Fert. 9 : 249–256, 1965.

Casteel, D. B.: Cytoplasmic inclusions in male germ cells of the fowl tick, *Argas miniatus*, and histogenesis of the spermatozoon. J. Morphol. 28 : 643–683, 1917.

Chang, M. C.: Fertilizing capacity of spermatozoa deposited into the Fallopian tubes. Nature (London) 168 : 697, 1951.

Davey, K. G.: The migration of spermatozoa in the female of *Rhodnius prolixus* Stahl. J. Exptl. Biol. 35 : 694–701, 1958.

Davey, K. G.: A pharmacologically active agent in the reproductive system of insects. Can. J. Zool. 38 : 39–45, 1960.

Davey, K. G.: The control of visceral muscles in insects. Adv. Insect Physiol. 2 : 219–245, 1964.

Davis, N. T.: Studies of the reproductive physiology of Cimicidae (Hemiptera) –– II. Artificial insemination and the function of the seminal fluid. J. Insect Physiol. 11 : 355–366, 1965.

Freeman, M. A.: The effect of drugs on the alimentary canal of the African migratory locust *Locusta migratoria*. Comp. Biochem. Physiol. 17 : 755–764, 1966.

Galun, R., Warburg, M.: Studies on the reproductive physiology of the tick *Ornithodoros tholozani* (Laboulbène and Mégnin): the effect of mating on oogenesis. Věstník Čs. spol. zool. (Acta Soc. Zool. Bohemoslov.) 31 : 329–334, 1967.

Hinton, H. E.: Sperm transfer in insects and the evolution of haemocoelic insemination, p. 95–107. In: K. C. Highnam (ed.) Insect Reproduction. Symp. Roy. Entomol. Soc. London, No. 2, 1964.

Oppermann, E.: Die Entstehung der Riesenspermien von *Argas columbarum* (Shaw) (*reflexus* F.). Z. mikr. anat. Forsch. 37 : 538–560, 1935.

Phillips, D. M.: Observations on spermiogenesis in the fungus gnat *Sciara coprophila*. J. Cell.Biol. 30 : 499–517, 1966.

Reger, J. F.: A fine-structure study on spermiogenesis in the tick *Amblyomma dissimile* with special reference to the development of motile processes. J. Ultrastr. Res. 7 : 550–565, 1962.

Reger, J. F.: Spermiogenesis in the tick *Amblyomma dissimile*, as revealed by electron microscopy. J. Ultrastr. Res. 8 : 607–621, 1963.

Rothschild, L.: Structure and movements of tick spermatozoa (Arachnida: Acari). Quart. J. Microsc. Sci. 102 : 239–247, 1961.
Samson, K.: Zur Spermiohistiogenese der Zecken. S. ber. Ges. Naturf. Fr. (Berlin) 8 : 486–499, 1909.
Shivers, C. A., James J. M.: Capacitation of frog sperm. Nature (London) 227 : 183–184, 1970.
Tatchell, R. J.: Studies on the male accessory reproductive glands and the spermatophore of the tick, *Argas persicus* Oken. Parasitol. 52 : 133–142, 1962.
VanDenmark, N. L., Hays, R. L.: Rapid sperm transport in the cow. Fertility and Sterility 5: 131, 1954.
Williams, W. L., Abney, T. O., Chernoff, U. W., Dukelow, W. R., Pinsker, M. C.: Biochemistry and physiology of decapacitation factor. J. Reprod. Fert. Suppl. 2 : 11–23, 1967.

FEEDING BEHAVIOR AND NUTRITIONAL REQUIREMENTS OF SOME ACARI

J.G. RODRIGUEZ
University of Kentucky, Lexington, Kentucky, U.S.A.

Phytophagous Mites

Progress has been deliberate in the development of a chemically defined for the plant-feeding mite, *Tetranychus urticae* Koch. The complexity of this problem stems with the highly specialized mode of feeding and the precise feeding requirements that characterize this mite. There are other complications. For example, we have recently discovered that this species and *T. turkestani* Ugarov and Nikolski reacted, favorably or unfavorably, to nonnutritive chemicals (Dabrowski and Rodriguez, 1971). These species were tested for their reaction to essential oils of strawberry foliage from a cultivar relatively susceptible to injury from these phytophagous mites. Generally the essential oils showed a repellent action; however, 0.01 % and 0.1 % *trans*–2–hexen–1–ol, mixtures of 0.1 % and 1.0 % octyl alcohol + fufural, and 0.1 % solution of α –terpineol + methyl salicylate attracted *T. urticae*. Six oils at low concentrations were also significantly attractive to *T. turkestani*.

We have also shown that water-extractable components form part of the biochemical milieu responsible for resistance of certain strawberry clones to *T. urticae* and *T. turkestani*. Gustatory studies with ^{32}P-labeled water-soluble extracts showed that these spider mites were also quite responsive to taste. The rate and amount of ingestion over a 24-hr test period related to the relative resistance or susceptibility of the strawberry clones to the mites (Rodriguez et al., 1971). To be sure, water extractable components are as yet undetermined but the evidence is that phenolic compounds are involved (Dabrowski and Rodriguez, 1971). A number of phenolic compounds that occur in strawberry foliage were bioassayed for gustatory acceptance by *T. urticae* and we found that most of the compounds inhibited feeding. Ingestion measurements using a sugar solution tagged with ^{32}P showed that chlorogenic acid, p–hydroxyphenylacetic acid, 3, 4, 5–trihydroxybenzoic acid, 2, 5–dihydroxybenzoic acid, p–hydroxybenzoic acid, p–hydroxycinnamic acid, and vanillic acid inhibited food intake significantly. After 24 hr feeding, kaempferol, cyanidin chloride, morin dihydrate, and ferrulic acid at 10^{-6}M were the only compounds that did not significantly reduce feeding.

These signal chemicals, then, play an extremely important role in nature. In artificial culture the investigator is hard put to duplicate them. In the artificial culture of mites, we may be faced with a situation to having to investigate signal chemicals as these may ultimately be the key to initiating normal feeding behavior.

We can, nevertheless, report on some recent advances in artificial culture. A new diet had been developed for *T. urticae* that reflects the findings of Walling et al.(1968) with respect to fatty acid metabolism and recent findings by Ekka et

Table 1. Composition of Basic Synthetic Diet per 100 ml Diet (1X concentration) for *Tetranychus urticae*

L-amino acids		Water soluble vitamins	
Alanine	150 mg	Folic acid	0.34 mg
Arginine	200	Inositol	6.00
Aspartic acid	200	P-aminobenzoic acid	0.50
Cysteine	50	Pyridoxine HCl	0.10
Glutamic acid	180	Riboflavin	0.20
Glycine	200	Thiamine HCl	0.20
Histidine	100	Nicotinic acid	0.30
Isoleucine	100	Biotin	0.0001
Leucine	160	Ca-pantothenate	0.50
Lysine	120	Choline chloride	10.00
Methionine	60	B_{12}	0.002
Phenylalanine	60	Ascorbic Acid	100.00
Proline	60		
Serine	80	**Lipids**	
Threonine	120		
Tryptophan	40	B-sitosterol	0.50 mg
Tyrosine	50	Stigmasterol	0.44
Valine	120	Stearic acid	1.00
RNA	100	Palmitic acid	0.50
		Oleic acid	0.50
Sugars		Linoleic acid	0.50
		Linolenic acid	3.00
Sucrose	3.0 g		
Levulose	0.5	**Mineral salts**	
Dextrose	0.5		
		K_2HPO_4	37.50 mg
Fat soluble vitamins		$Na_2HPO_4 \cdot 12H_2O$	10.60
		$MgSO_4 \cdot 7H_2O$	15.60
Vitamin A	3.0 mg	$CaCl_2 \cdot 2H_2O$	3.62
Vitamin E	5.0	$CoCl_2 \cdot 6H_2O$	0.08
		NaFe	1.00
		NaMn	0.50
		NaCu	0.08
		NaZn	0.12

al.(1971) involving sterols in the bean plant and mite carcasses, (Table 1). The former oviposition rate of 0.34 egg/female/ day, was improved to 0.65 egg/female day and the viability of eggs was improved from 3.6 to 19.2 percent. Other gains were in the lengthening of the oviposition period from 25 to 36 days and the shortening of the pre-oviposition period from 3.5 to 1.5 days. These improvements were largely attributed to the elimination of cholesterol and substituting plant sterols and the addition of increased icrements of alpha tocopherol. Work is continuing in improving dietetics, including feeding stimulants.

Stored Food Mites

The acarid mites, *Tyrophagus putrescentiae* (Schrank) and *Caloglyphus berlesei* (Mich.) are important mites because they attack stored foods, especially the soft

Table 2. Composition of the axenic diet used in feeding *Tyrophagus putrescentiae* and *Caloglyphus berlesei*.

Compound	g/100ml	Compound	g/100ml
L–alanine	0.26	Sucrose	8.0
L–arginine	.34		
L–aspartic acid	.60	Aphacel	1.0
L–cysteine	.09		
L–glutamic acid	.94	Wesson's Salts (Mix W)	2.5
Glycine	.28		
L–histidine	.18	Myristic acid	.056
L–isoleucine	.23	Palmitic acid	.494
L–leucine	.40	Stearic acid	.268
L–lysine	.31	Oleic acid	.494
L–methionine	.07	Linoleic acid	.070
L–phenylalanine	.25	Linolenic acid	.028
L–proline	.33	Cholesterol	.500
L–serine	.26		
L–threonine	.20	Tween 80	.01
L–tryptophan	.08		
L–tyrosine	.15	Vitamin Diet Fortification	
L–valine	.25	Mixture*	.40
RNA	.10	Streptomycin	.05
Agar	3.0	Methyl–p–hydroxybenzoate (15% in 95% ethanol)	1.0ml
		Formaline (10%)	0.1ml
		KOH (22.5% in H_2O)	0.5ml

*Nutritional Biochemicals Corporation. Mixture total includes 0.339 g dextrose.

moist variety. *T. putrescentiae* is found in much drier situations than *C. berlesei* and so is much more common. The critical water activity for the first species is a_w 0.70 ($a_w = \frac{\text{relative humidity}}{100}$) while a_w for *C. berlesei* is estimated at 0.95, hence the food substrate or microenvironment has to reflect those requirements. The dietetic problems involved with feeding acarids artificially are relatively simple compared to those concerning the plant-sucking tetranychids, for the acarids have chewing mouth parts. Furthermore, the air-organism interface is primarily superficial and complications in culturing associated with a tracheal system, such as tetranychids posses, are not present. We have reported on a chemically defined diet that is very adequate to culture *T. putrescentiae* and this has been done axenically (Rodriguez and Lasheen, 1971). Table 2 gives the formulation of this diet. We now report that this same diet appears to meet the requirements of *C. berlesei*. The difference in the two diet requirements is in water content of the finished diet prior to being inoculated with sterile larvae of these two species. The diet tubes for *T. putrescentiae* are followed to dry overnight with a 1.8 % loss of moisture prior to inoculation while the diet tubes for *C. berlesei* are ready for use immediately after the agar based diet cools.

Table 3. Results of deletion experiments to determine essentiality of amino acids in *Caloglyphus berlesei*. Counts were made at the time interval indicated

Diet	3 Day Larvae	14 Days Adults	14 Days Larvae	29 Days Adults	29 Days Larvae
Expt. 1 & 2					
Control	124	34	38	81	434
Deleted					
Tyrosine	147	33	126	106	489
Phenylalanine	68	0	42	0	19
Phenyl alanine and Tyrosine	85	1	47	0	16
Threonine	132	0	66	0	25
Tryptophan	90	0	63	0	21
Valine	99	0	50	0	19
Expt. 3 & 4					
Control	186	60	258	276	639
Deleted					
Arginine	182	0	114	0	41
Isoleucine	173	0	104	0	24
Leucine	140	0	64	0	15
Lysine	168	0	116	0	20
Histidine	167	1	98	0	11
Cystine	158	10	75	13	26

Studies to characterize essentiality of amino acids were conducted for *T. putrescentiae* (Rodriguez and Lasheen, 1971), and the amino acids shown to be essential were those same ones essential for the rat: threonine, valine, methionine, isoleucine, leucine, tyrosine, phenylalanine, lysine, histidine, arginine and tryptophan. The mite is capable of synthezing cysteic acid, hydroxyproline, aspartic acid, serine, glutamic acid, proline, glycine, alanine, and cystine. Although our studies employing radiometric procedures for determining synthesis of amino acids in *C. berlesei* are not yet complete, we have nonetheless determined their essentiality with the use of the classic deletion method (Table 3). The data show that the same situation prevails with this mite as with *T. putrescentiae*, and that both of these species have apparently similar nutrition requirements.

Research in the signal chemicals that surely must be involved in acarid feeding requirements is apparently lacking. This area is most assuredly one that is fruitful for study.

References

Dabrowski, Z.T., Rodriguez, J.G.: Studies on resistance of strawberries to mites. 3. Preference and nonpreference responses of *Tetranychus urticae* and *T. turkestani* to essential oils of foliage. J. Econ. Entomol. 64 : 387–391, 1971.

Dabrowski, Z.T., Rodriguez, J.G.: Gustatory responses of *Tetranychus urticae* Koch to phenolic compounds of strawberry foliage. Postepy Nauk Rolniczych (Progress in Agricultural Science) Progress in Acarology, part II., 1971. (In press.)

Ekka, Irma, Rodriguez, J.G., Davis, D.L.: Influence of dietary improvement on oviposition and egg viability of the mite *Tetranychus urticae*. J. Insect Physiol., 1971. (In press.)

Rodriguez, J.G., Dabrowski, Z.T., Stoltz, L.P., Chaplin, C.E., Smith, W.O., Jr.: Studies on resistance of strawberries to mites. 2. Preference and nonpreference responses of *Tetranychus urticae* and *T. turkestani* to water-soluble extracts of foliage. J. Econ. Entomol. 64 : 383–387, 1971.

Rodriguez, J.G., Lasheen, A.M.: Axenic culture of *Tyrophagus putrescentiae* in a chemically defined diet and determination of essential amino acids. J. Insect Physiol. 17 : 979–985, 1971.

Walling, M.V., White, D.C., Rodriguez, J.G.: Characterization, distribution, catabolism, and synthesis of the fatty acids of the twospotted spider mite, *Tetranychus urticae*. J. Insect Physiol. 14 : 1445–1458, 1968.

AN ACTIVE CONSTITUENT OF THE SALIVA OF THE CATTLE TICK, *BOOPHILUS MICROPLUS*

R. J. TATCHELL* AND K. C. BINNINGTON
CSIRO, Division of Entomology, Long Pocket Laboratory, Indooroopilly, Brisbane, Australia

A variety of toxic, lytic and anticoagulant functions have been ascribed to the salivary secretions of ticks. Frequently the evidence supporting these claims has fallen short of complete proof. Recently however, Neitz et al. (1969) have purified a toxic component of the saliva of *Ornithodoros savignyi* and Hellman and Hawkins (1967) have described anticoagulins and fibrinolysins from *Ornithodoros moubata*.

The work of Hellman and Hawkins was particularly valuable in that it diferentiated between anticoagulants of salivary and gut origin and thereby demonstrated a flaw in studies based on whole tick homogenates (Markwardt, 1963).

The saliva of *Boophilus microplus* was stated to lack anticoagulants and cytolytic agents (Tatchell, 1969) and host tissue destruction during the feeding of *Rhipicephalus sanguineus* was shown to be caused by the intense infiltration of host polymorphonuclear neutrophil leucocytes (Tatchell et Moorhouse, 1970). It was postulated that pharmacological activity was present in tick saliva; this paper describes a method of purification and assay of an active constituent of *B. microplus* oral secretion.

Materials and Methods

Boophilus microplus of the "Yeerongpilly" strain were used within 4 h of natural detachment for the collection of saliva after stimulation by topical pilocarpine application (Clarke and Hewetson, 1971). Care was taken to exclude samples from ticks showing evidence of excessive drug trauma i. e. leakage of gut contents across the cuticle in the region of topical application; this could be minimised by reducing the concentration of pilocarpine to 25%.

Saliva was pooled and kept at −30°C until 10 ml batches had been accumulated. Preliminary separation and concentration was accomplished by ultrafiltration across "Diaflo" membranes excluding molecular weights (MW) greater that 15,000 followed by washing and reduction in volume to 1 ml over the 500 MW excluding membrane. This concentrate was then applied to a 40 cm × 1 cm Sephadex LH20 column, eluted

* Present address: NAMRU–3, American Interests Section, Spanish Ambassy, Cairo, Egypt.

with Tyrode's solution and collected in 2 ml fractions for UV spectral analysis and bioassay.

Standard pharmacological bioassay techniques using the guinea-pig ileum and duodenum, rat duodenum and vagina, rabbit duodenum, ileum and vagina were attempted. Predictable and reproducible results were only obtained with a well-oxygenated rat fundus preparation in 5 ml Tyrode's medium using a water-jacketed organ bath at 37°C. Natural saliva (Gregson, 1960) and homogenates of salivary glands dissected under ice-cold 1.3% NaCl from < 4.5 mm ticks were also tested.

Results and Discussion

The rat fundus preparation responds to the addition of *B. microplus* "natural" saliva, salivary homogenate, or purified active component of pilocarpine-saliva by an immediate sustained contraction. The effect is removed by washing the preparation with a flow of fresh medium whereupon relaxation occurs.

The active component emerged from the Sephadex column as a compact peak; at a flow rate of 12 ml/h activity was located within 3 consecutive 2 ml fractions in the region of fraction 15. In samples containing protein the activity left the column shortly after the protein peak. As protein would travel at the solvent front this suggests a MW approaching the exclusion limit for this gel. A feature of the procedure was an unexplained degradation of the gel. The flow rate steadily diminished and more than 3 runs/column were rare. It is possible that the gel beads were being attacked or that the pores were being blocked, perhaps by tick cement precursors.

Preliminary experiments showed that pilocarpine could be removed from the saliva by washing over the 500 MW "Diaflo" membrane through which the pilocarpine readily passed. It was also found that the non-pilocarpine pharmacological activity passed through the 1000 MW "Diaflo" membrane. "Diaflo" membranes are not intended for use in absolute MW determinations but these results indicated that the active material probably has a MW > 500 and < 1000 - 1500.

On the basis of these results and the lack of response from all the bio-assay preparation except the rat fundus it was unlikely that the material was acetyl choline, an acetyl cholinesterase, 5–hydroxytryptamine (serotonin), adrenalin or nor–adrenalin. These conclusions were confirmed using specific antagonists or inhibitors for these substances in concentrations which eliminated the bioassay response to the known substance. However, sodium meclofenamate was found to be a very powerful antagonist. Under some conditions sodium meclofenamate is a more or less specific antagonist of bradykinin (Collier and James, 1967); this proved to be so with the rat fundus preparation tested against standard bradykinin (gift of Sandoz Ltd., Basle). Despite these results the *B. microplus* material is resistant to the effects of incubation with chymotrypsin or pepsin under conditions which destroyed the activity of standard bradykinin; it is therefore unlikely to be a polypeptide. Further proof that it was not bradykinin was provided by the lack of response of rat duodenum or guinea-pig ileum after addition of chymotrypsin (Edery and Grunfeld, 1969).

The biological significance of this active substance was studied, with the help of Dr. D. H. Kemp, by intradermal injection into calves which had not been exposed to infestations of *B. microplus*. There was a clear difference between the saliva constituent and a saline control in the extent to which capillary permeability increased after the injection. The results were similar whether based on the leakage of $^{125}I_2$ -labelled albumin or Evans Blue-labelled albumin. Measurements were made on wheal size and radioactivity or dye amounts in biopsies taken after intradermal injection (Table 1).

Table 1. Increased capillary permeability following intradermal injections of the pharmacologically active fraction of *B. microplus* oral secretion. Measured by leakage of $^{125}I_2$ – and Evans Blue–labelled plasma into the tissues and their subsequent estimation by liquid scintillation counting, visual estimation and spectrophotometry with the assistance of Dr. D. H. Kemp

$^{125}I_2$ –Labelled albumin

Injection	Counts–background/300 sec	Counts/100 mg biopsy
Saliva fraction	3757	1288
Saliva fraction	3158	1068
Saline	1985	719
Saline	1701	575

Evans Blue–Labelled albumin

Injection	Visual wheal estimate	Wheal diameter (mm)	Spectrophotometric estimate of weight of Evans Blue/biopsy (μg)
Saliva fraction	++	8	16.2
Saliva fraction	++++(+)	12	22.2
Saliva fraction	++++	11	20.6
Saline	+	6	9.9
Saline	+(+)	6	11.1
Saline	+(+)	6	6.7

The conclusion from these and previous results is that *B. microplus* does not have a salivary anticoagulin and, unlike *Ornithodoros* (Howell, personal communication) and *Argas* (Tatchell, unpublished), does not contain proteolytic enzymes in its salivary secretion. The saliva does contain a pharmacologically active material which causes contraction of a rat fundus preparation and an increase in capillary permeability after intradermal injection into the bovine host. The latter response would obviously aid in tick feeding. Tatchell and Moorhouse (1969) found that dilatation of adjacent superficial skin capillaries occurred very soon after the attachment of larval ticks suggesting that pharmacologically active material was being injected by the tick. The material apparently cannot be a polypeptide kinin. The antagonism by sodium meclofenamate is therefore particularly interesting in that the tick may have developed a chemical which can mimic some of the effects of bradykinin without being susceptible to natural degradation by kinases.

References

Clarke, R. H., Hewetson, R. W.: A modification to the collection of saliva from *Boophilus microplus*. J. Parasitol. 57 : 194–195, 1971.

Collier, H. O. J., James, G. W. L.: Humoral factors affecting pulmonary inflation during acute anaphylaxis in the guinea-pig in vivo. Brit. J. Pharmacol. 30 : 283–301, 1967.

Edery, H., Grunfeld, Y.: Sensitization of smooth muscle to plasma kinins: effects of enzymes and peptides on various preparations. Brit. J. Pharmacol. 35 : 51–61, 1969.

Gregson, J. D.: Morphology and functioning of the mouthparts of *Dermacentor andersoni* Stiles. Acta Trop. 17 : 48–79, 1960.

Hellmann, K., Hawkins, R. I.: The action of tick extracts on blood coagulation and fibrinolysis. Thrombos. Diathes. Haem. 18 : 617–625, 1967.

Markwardt, F.: Blutgerinnungshemmende Wirkstoffe aus Blutsaugenden Tieren. VEB, Jena, 1963.

Neitz, A. W. H., Howell, C. J., Potgieter, D. J. J.: 18. Purification of a toxic component in the oral secretion of the sand tampan *Ornithodoros savignyi* Audouin (1827). J. South Africa Chem. Inst. 22 : 142–149, 1969.

Tatchell, R. J.: Host-parasite interactions and the feeding of blood-sucking arthropods. Parasitology 59 : 93–104, 1969.

Tatchell, R. J., Moorhouse, D. E.: The feeding processes of the cattle tick *Boophilus microplus* (Canestrini) Part II. The sequence of host-tissue changes. Parasitology 58, 441–459, 1968.

Tatchell, R. J., Moorhouse, D. E.: Neutrophils: their role in the formation of a tick feeding lesion. Science 167: 1002–1003, 1970.

BLATTISOCIUS TARSALIS: BEHAVIOR IN ASSOCIATION WITH THE MOTH *EPIZEUXIS AEMULA*

A. E. TREAT
The American Museum of Natural History, Dumont, New Jersey, U. S. A.

Until recently the ascid mite *Blattisocius tarsalis* (Berlese, 1918) was known chiefly if not exclusively from man-made habitats such as hay barns, grain bins, and laboratory insect cultures, where it is a predator upon the eggs and early instars of various insects (Darst and King, 1969). In 1969 I reported the frequent occurrence of this mite on feral adults of the small herminiine noctuid *Epizeuxis aemula* Huebner, which is common throughout most of temperate North America. The larvae of this moth feed upon many kinds of leaves, usually dry and fallen. On rare occasions they have been damaging to dry fodder, and have sometimes been reported as attacking spruce needles. Often double brooded, they hibernate as larvae, pupating in the spring within thin, net-like cocoons covered with frass or leaf fragments. They are strongly and negatively phototaxic, resistant to extremes of temperature and humidity, and immune to molds such as those that occur in damp leaf litter.

Among 314 of these moths collected at light in Tyringham, Massachusetts, during the years 1967 to 1970, a total of 88 carried females of *Blattisocius tarsalis*, the over all incidence amounting to 28 percent. Incidence on female moths was about 40 percent as compared with only 14 percent on males. The number of mites per infested moth ranged from one to 17, averaging three to four. Pinned specimens of *Epizeuxis aemula* in The American Museum of Natural History provided evidence may occur locally in eastern and southern Canada and in the United States as far west as Texas. With the exception of only two host specimens, the mites were found on noctuids of no other species, not even those of the genus *Epizeuxis* whose range and habits are said to be similar to those of *E. aemula*.

Phoresy by females of *B. tarsalis* upon adults of cultured grain and flour moths has been reported by others, but with few behavioral details and with no indication of how or when the mites find and board their hosts. The usual site of occupancy is the dorsum of the metathoracic and first abdominal segments on both sides of the body. The tympanic recesses are occasionally entered, but the ears are not damaged.

I can state unequivocally that adults of *Epizeuxis aemula* are often actually parasitized by *B. tarsalis* females. I have seen the mites with their mouthparts inserted into the bodies of the moths. I have watched the act of feeding and later observed multiple puncture wounds at the occupied sites, but I have not noted any shortening of life or impairment of function in the parasitized insects. As one would expect from the form of their chelicerae, the mites are attached only during feeding, and then only lightly; at other times they can move about freely.

Moths in captivity may retain their mites for several days or may lose them within a few hours. Eggs of the moth are laid in pavements, often on or between the surfaces of dead leaves. Attempts to rear the moths together with the mites, though only partly successful, have yielded interesting observations. Mites in various stages were often seen on or near the egg masses of *Epizeuxis*, occasionally with their mouthparts penetrating the egg surface. Some eggs were visibly damaged, but the number of mites declined, and egg mortality caused by predation was very slight. Casual encounters between mites and young moth larvae were frequent, but such larvae were apparently not attacked or injured. Full grown caterpillars were sometimes boarded, however, and on one occasion I observed two adult mites with their mouthparts deeply imbedded in the ventral abdominal integument of a large moth larva. Engorgement proceeded for more than forty minutes. The host remained passive until the end of this period. Then it swung its head toward the area occupied by the two mites, whereupon they promptly left its body. The caterpillar showed no adverse effects. The fungivorous mite *Tarsonemus confusus* Ewing was abundant in some of the rearing dishes. These mites were occasionally attacked and apparently consumed by an adult *Blattisocius*, and their eggs may have furnished some food for the ascids.

Most interesting was the behavior of the *Blattisocius* in relation to the pupae and emerging adults of the moth. The cocoons are of loose texture. When several cocoons of different ages were present in the same rearing dish, the female mites were somehow able to recognize the one containing a pharate adult. Hours before eclosion, they assembled on or in that cocoon, moving about now and then, and sometimes boarding or palpating the pupal cuticle, but mostly resting motionless near the head end. Male mites wandered about the dish continually, often visiting the assembled females and attempting copulation, though without success, since the females, perhaps already inseminated, regularly evaded their efforts.

The emergence of the moths took place with little obvious warning beyond a gradual color change beginning a few days prior to the event, and a protraction of the pupal abdomen some minutes beforehand. Eclosion was almost explosively rapid, but the mites were evidently ready for it. As soon as the pupal skin opened, and before any part of the moth was free, the mites boarded the thin strip of exposed imaginal surface in the cleft lateral to the antennae. Within four or five seconds after the cleft appeared, the moth darted out of the pupal skin and with wings still unexpanded ran rapidly about the dish, often putting its head down as if trying to bury itself. After a few minutes it came to rest and quickly expanded its wings. The mites were still aboard.

The timing and speed with which the mites mount the emerging moths suggest a highly evolved behavioral pattern perhaps including response to a kairomone produced by the pharate host. In any event, the female mites appear able (1) to distinguish cocoons containing pharate adults from those containing younger pupae, (2) to recognize the anterior end of such cocoons or their occupants, and (3) to respond almost instantly to the first exposure of adult surface. The literature on *B. tarsa-*

lis does not reveal whether similar behavior occurs in relation to its common microlepidopteran hosts. It would be of interest to know whether there exist behavioral and genetic differences between populations associated with these insects and those associated with *Epizeuxis aemula.*

References

Berlese, A.: Centuria quarta di Acari nuovi. Redia 13 : 134, 1918.

Darst, H., King, E.W.: Biology of *Melichares tarsalis* in association with *Plodia interpunctella.* Ann. Entomol. Soc. Am. 62 : 747–749, 1969.

Treat, A.E.: Association of the mite *Blattisocius tarsalis* with the moth *Epizeuxis aemula.* J. New York Entomol. Soc. 77: 171–175, 1969.

Section XII

ECOLOGY OF MITES AND TICKS

Chairman
G. W. Wharton (USA)

Assistant Chairman
M. Daniel (Czechoslovakia)

A STUDY OF THE LIFE CYCLE OF *PSEUDOBRYOBIA DRUMMONDI* (EWING) (ACARINA: TETRANYCHIDAE)

M. J. ABBATIELLO
Department of Biological Sciences, State University A. & T. College Farmingdale, New York, USA

The spider mite, *Pseudobryobia drummondi* (Ewing) is a species that is indigenous to the arid regions of southwestern United States and western and central Mexico. This species is found commonly in that area on the creosote bush, *Larrea tridentata.* In addition to being found on the creosote bush, it has been collected in Arizona from *Cucurbita palmata, Ephedra fasciculata* and *Sorghum halepense* (Tuttle and Baker, 1968).

This article presents, through a series of drawings of the life stages, the characteristic setation and setal additions on the body and legs of the larval, nymphal and adult stages of *P. drummondi.* All the life stages observed and figured were of specimens collected on the creosote bush, during the mite collecting survey taken of western and central Mexico, by Doctors Tuttle, Baker and Abbatiello in July–August 1970.

In the comparative study of the dorsal body setae of the four life stages the illustrative material on the adults represents no new information since the description of the species (Figs. 4 and 5). And surprisingly to me no change occurs in the dorsal setation of the immature forms. The dorsal propodosomal and dorsal hysterosomal setal number and arrangement of the adult are present in the larval stage (Fig. 1), and the same number and arrangement are retained by the protonymph (Fig. 2) and deutonymph (Fig. 3). The 32 dorsal body setae seen in the adult are maintained in the same relative position in all stages of the life cycle.

As was to be expected, a detailed study of the venters and legs, on the other hand, showed a normal progression in setal additions. The venter of the larva (Fig. 6) exhibits specific setae. Present are 2 pairs of ventrocentrals, 3 pairs of anals and 2 pairs of post-anal setae. No gnathosomal or pre-genital setae are present. The ventral setae are few in comparison to the number present in later stages. A significant change occurs in the ventral setal pattern of the protonymph (Fig. 7). Two new pairs of setae are added; a pair of gnathosomal setae and a pair of pre-genitals. All other setae seen in the larval stage remain in the protonymph in the same relative position. The deutonymph (Fig 8) resembles the protonymph in general appearance, with a few specific differences in setation. All ventral setae of the protonymph are maintained in the same relative position. The noticeable characteristics of this stage include the addition of the third pair of ventrocentral setae and a sin-

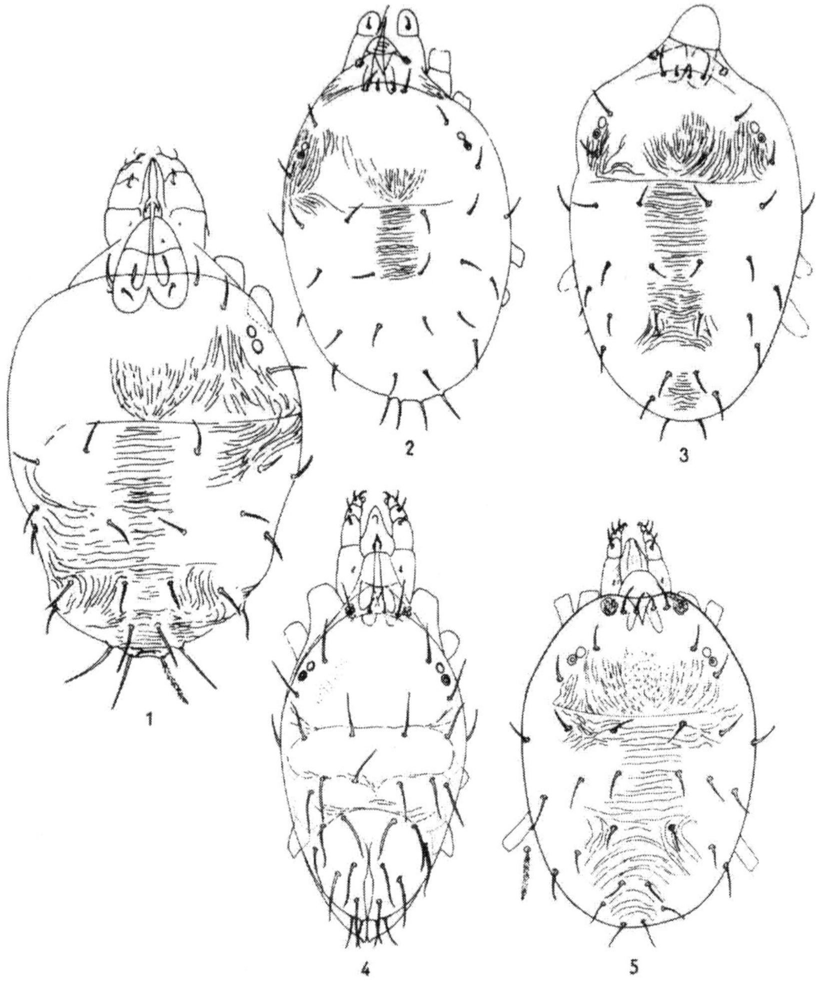

Fig. 1. *Pseudobryobia drummondi.* Dorsum of larva. Fig. 2. *Pseudobryobia drummondi.* Dorsum of protonymph. Fig. 3. *Pseudobryobia drummondi.* Dorsum of deutonymph. Fig. 4. *Pseudobryobia drummondi.* Dorsum of adult male. Fig. 5. *Pseudobryobia drummondi.* Dorsum of adult female.

gle pair of pre-anal setae. Cagle (1946, 1949, 1956) and Reeves (1963) in their investigations of the life histories of tetranychid mites indicate that the deutonymphs destined to become males are readily distinguishable from females by the smaller size of the body and the abdomen being more pointed than those that will become females. Therefore, on the basis of these observations, I am making the assumption that the deutonymph figured in this article is a female deutonymph. And that the pre-anal setae seen in this deutonymph are probably the genital setae of the adult female. The basic setal ventral pattern persists from the deutonymph into the adult and the setae are present in the same location as the preceding stage. In the adult female (Fig. 10) an additional pair of genital setae are added and the adult male (Fig. 9) exhibits the characteristic 5 pairs of genito-anal setae.

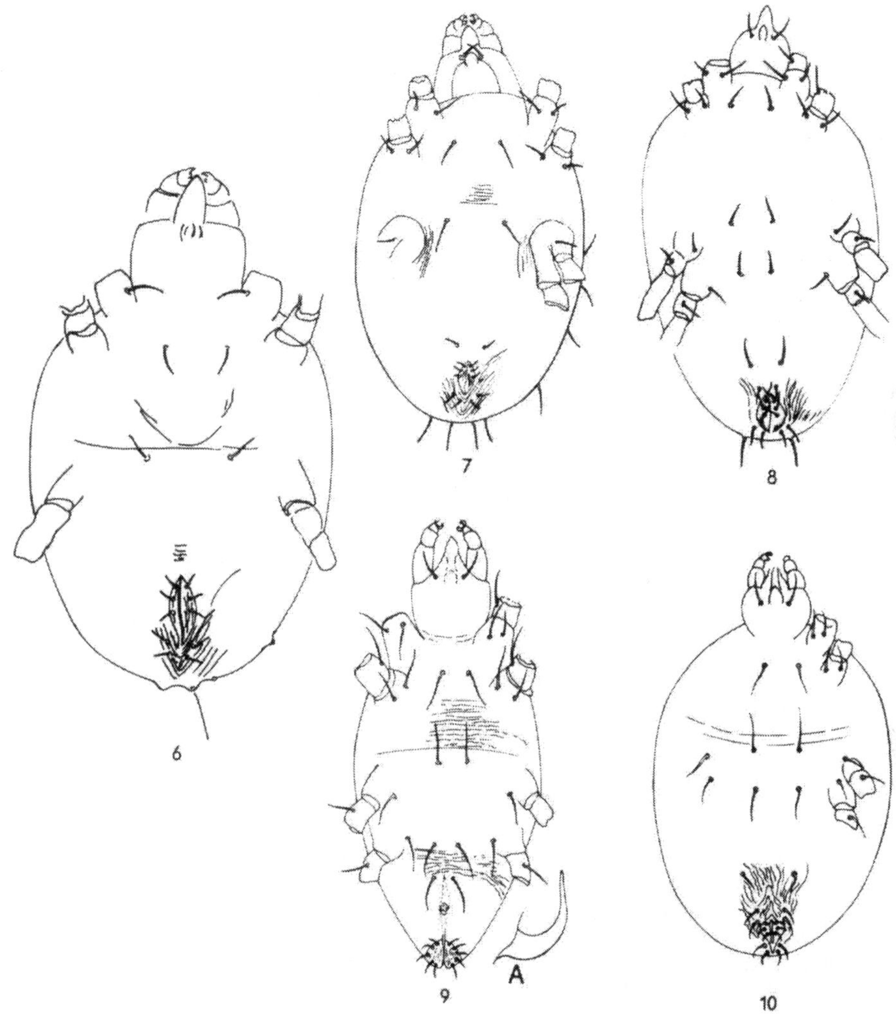

Fig. 6. *Pseudobryobia drummondi.* Venter of larva. Fig. 7. *Pseudobryobia drummondi.* Venter of protonymph. Fig. 8. *Pseudobryobia drummondi.* Venter of deutonymph. Fig. 9. *Pseudobryobia drummondi.* Venter of adult male: A) aedeagus. Fig. 10. *Pseudobryobia drummondi.* Venter of adult female.

Leg I of the larva (Fig. 11A)* carries a total of 18 setae which are distributed by segment from coxa to tarsus in the following order: 1-0-3-4-4-6 (Table 1). Setation of leg I of the protonymph (Fig. 11B) increases to 26 and are distributed by segment from coxa to tarsus as follows: 2-1-4-5-5-9 (Table 1). In the deutonymph (Fig. 11C) the number of leg I setae increases to 32. The setal formula of leg I has changed to 2-1-6-5-5-13 (Table 1). In the adult the number of setae increases on certain segments of leg I yielding the formula 2-1-10-7-9-16 for a total of 45 setae (Table 1). The setal formulae and setae added for legs II, legs III, and for legs IV (Figs. 12A, B) at different stages of development are shown in Table 1. A trichobothrium seta, which is barely differentiated from the other setae is present on the tibia of legs I–IV and is retained in the same relative position in all stages of the life cycle.

Table 1. Comparions of number of setae on segments of legs of *Pseudobryobia drummondi* (Ewing), each formula gives the number in the order of the following leg segments: coxa, trochanter, femur, genu, tibia, tarsus, formulae include trichobothria and exclude solenidia and duplex setae

Stage	Leg I	Leg II	Leg III	Leg IV
Larva	1-0-3-4-4-6	0-0-3-3-5-6	0-0-2-2-4-6	
Protonymph	2-1-4-5-5-9	2-1-3-4-5-8	1-1-2-3-4-8	0-0-2-2-3-6
Deutonymph	2-1-6-5-5-13	2-1-3-4-5-9	1-1-2-3-4-8	1-1-2-3-3-8
Adults	2-1-10-7-9-16	2-1-6-7-7-13	1-1-4-6-6-12	1-1-3-5-5-12

The setal changes that occur are probably not unique to *P. drummondi*. Yet, the general trend of such events as setal progression and addition may be similar for a larger group (tribe, subfamily, family) of mites, and may hold the key to understanding the relationship between groups.

I wish to express my sincere thanks to Dr. Edward W. Baker, of the Entomology Research Division, U.S. Department of Agriculture, for his verification of the illustrations, invaluable guidance and reading the manuscript. My appreciation is extended to my colleague Dr. Donald M. Tuttle, of The University of Arizona for his aid in mounting the specimens. And finally, I wish to thank Miss Maryanne Johnson, of the Audio-Visual Services Department and Mr. Robert Doman, of the Photographic Department for their technical assistance in the preparation of the illustrations.

* Note: O indicate dorsal position and X a ventral position for leg setae.

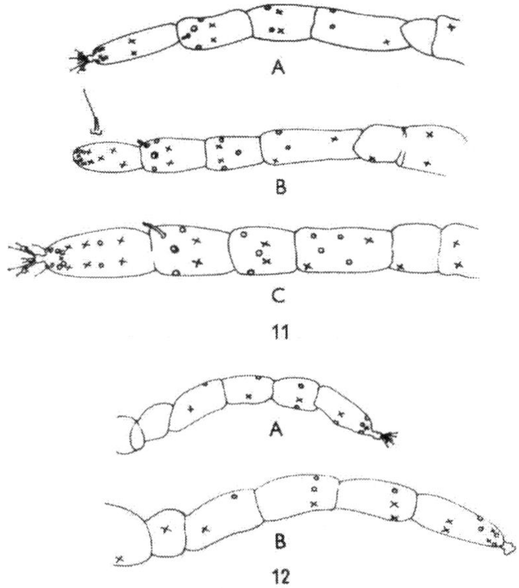

Fig. 11. Legs I of *Pseudobryobia drummondi*: A) larva; B) protonymph; C) deutonymph.
Fig. 12. Legs IV of *Pseudobryobia drummondi*: A) protonymph; B) deutonymph.

References

Cagle, L. R.: Life history of the European red mite. Va. Agr. Exptl. Sta. Tech. Bul. 98 : : 1–20, 1946.

–: Life history of the two-spotted spider mite. Va. Agr. Exptl. Sta. Tech. Bul. 113 : 1–31, 1949.

–: Life history of the spider mite *Tetranychus atlanticus* McG. Va. Agr. Exptl. Sta. Tech. Bul. 124 : 1–22, 1956.

Reeves, R. M.: Tetranychide infesting woody plants and a life history of the elm spider mite *Eotetranychus matthyssei* n. sp. Cornell Univ. Agr. Exptl. Sta. Mem. 380 : 1–99, 1963.

Tuttle, D. M., Baker, E. W.: Spider mites of southwestern United States and a revision of the family Tetranychidae. Univ. of Arizona Press. 1–143, 1968.

PARASITE-HOST RELATIONSHIPS IN FEATHER MITES

V. ČERNÝ
Institute of Parasitology, Czechoslovak Academy of Sciences,
Prague, Czechoslovakia

The feather mites (Analgoidea) make a group characterized by permanent parasitism and consequently closely associated with their hosts. Their occurrence on birds is of ancient date, which fact is confirmed by a considerable morphological and physiological specialization of many species. The beginning of the development and independence of this group from the tyroglyphoid complex dates back to the time between the Secondary and Tertiary Periods and its intensive evolution has been going on since Oligocene (Dubinin,1951).

With respect to hosts the feather mites are characterized by different degrees of specificity, from forms associated with only one host species to forms with a wide host range. According to Dubinin (1950) this host specificity is due, on the one hand, to the influence of environmental conditions under which each species had been developing and, on the other, to the length of the period in which this development took place. Hence it follows that on historically old host groups there exists a large number of closely specialized mite forms and vice versa, in phylogenetically younger host groups there are forms which can transfer even to unspecific hosts. In any case the duration of the biological parasite-host system is of far greater importance than the absolute age of both of them. This is the reason why on various hosts there are species whose specificity is developed to a different degree.

I should like to illustrate examples of the particular degrees of host specificity existing in species and genera of feather mites. The specificity on the level of families is omitted because their systematics is being revised at present. My conclusions are primarily on taxons which have been recently scrutinized or revised. It should be noted that the conclusions concerning supraspecific categories of mites are to a certain extent subjective because they depend on the scope within which e.g. individual genera are considered. To a lesser degree this also applies to some categories of hosts, whose taxonomy is not unanimously understood by all ornithologists.

A) Distribution of feather mite species

1. Monoxenous species are distributed on a single host species. Such an occurrence is well known with many feather mites. For example the latest revision of the genus *Proctophyllodes* has revealed that 47 species out of 123, i.c. 38 %

have been found on a single bird species (Atyeo et Braasch, 1966). The species of scores of recently described monotypic genera have been indicated to be monoxenous as well. Many species of this category also exist among other genera. However advancing research may reveal that many species listed here actually belong to the next group after all.

2. Oligoxenous species are distributed on more than one host species, the lowest degree in this category being bixenous species.

a) Species with monogeneric distribution occur only on members of the same host genus, for example *Diomedacarus gigas* on the genus *Diomedea*, *Avenzoaria totani* on *Tringa*, *Zachvatkinia stercorarii* on *Stercorarius*, *Calcealges yunkeri* on *Zosterops*, *Hemicalcealges schistopygus* on *Lonchura*, *Bregetovia limosae* and *Montchadskiana buchholzi* on *Limosa*, *Ptiloxenus major* on *Podiceps* etc. Sometimes the specificity at the level of subgenus may be observed, such as the occurrence of *Microspalax puffini* on the subgenus *Puffinus*, *M. thyellodromae* on *Thyellodroma* and *M. ardennae* on *Ardenna*, all hosts belonging to the genus *Puffinus*.

b) Species with monofamilial distribution occur on the representatives of host genera of the same family. For example *Ptyctophyllodes trogonis* on Trogonidae, *Freyana anatina* on Anatidae, *Ardeacarus ardeae* on Ardeidae, *Mesalgoides picimajoris* on Picidae, *Psilobrephosceles ortygometrae* on Rallidae, *Proctophyllodes huitzilopochtlii* on Trochilidae, *Zachvatkinia puffini* on Puffinidae, *Gabucinia delibata* on Corvidae. This type of distribution is also quite frequent.

c) Species with monoordinal distribution parasitize the representatives of several families of the same order. This distribution may be observed with *Alloptes tubinarii*, found on Procellariiformes. In many cases this is rather a specific occurrence within the framework of a certain suborder such as *Proctophyllodes cotyledon*, *Pterodectes bilobatus* and *Mesalgoides oscinum* on Passeres, *Diproctophyllodes dielythra* and *Nycteridocaulus tyranni* on Tyranni, or *Uniscutalges elegans* on Pici. This type of distribution on hosts from several families is less frequent.

d) Species with polyordinal distribution parasitizing hosts of various orders, occur very rarely. Older acarological literature contains a number of references to such cases, but these should be taken critically due to possible incorrect data concerning hosts as well as morphological characteristics of the parasite itself. For example in the latest monograph on the genus *Proctophyllodes* (Atyeo and Braasch, 1966) the species *P. anthi* is given as the parasite of hosts of the families Motacillidae, Alaudidae and Jyngidae and *P. polyxenus* of some families of Passeriformes and of Strigidae. It must be noted however that even in recent literature many cases of the occurrence of certain species on hosts from different orders are indicated, but should be regarded as an accidental occurrence or as a result of accidental contamination of the hosts during their collection in the field and examination in the laboratory.

B) Distribution of genera of feather mites

The classification is similar to that of species, the survey does not include monobasic genera.

a) Genera with monogeneric distribution are not so numerous as subsequent categories. For example *Dinalloptes* and *Michaelichus* on the host genus *Phalacrocorax*, *Pseudavenzoaria* on *Tringa*, *Pomeranzevia* on *Numenius*, *Burhinacarus* on *Burhinus*, *Onychalloptes* on *Phaeton*, *Chauliacia* on *Apus*.

b) Genera with monofamilial distribution are quite frequent. For example *Laronyssus* on the host family Laridae; *Geranolichus* on Gruidae; *Tillacarus* on Numididae; *Alloptoides* on Anatidae; *Dichobrephosceles* on Scolopacidae; *Hemicalcealges* on Ploceidae; *Dermonoton* on Strigidae; *Ardeialges* on Ardeidae; *Metanalges* on Rallidae; *Eustathia* and *Neochauliacia* on Apodidae; *Trochilalges* on Trochilidae; *Falculifer* and *Pterophagus* on Columbidae; *Anapodema*, *Calaobia*, *Membranolobus* and *Vexillaria* on Bucerotidae.

c) Genera with monoordinal distribution are also frequent. For example *Kramerella* on the host order Strigiformes, *Grallobia* on Ralliformes, *Hemifreyana* on Charadriiformes. However, in this category the specificity is mostly revealed at the level of suborder, e.g. *Nycteridocaulus* on Tyranni; *Anisodiscus*, *Joubertophyllodes* and *Monojoubertia* on Passeres; *Freyanopterolichus* on Ciconiae; *Avenzoaria*, *Bychovskiata*, *Bregetovia*, *Freyanomorpha* and *Sokoloviana* on Charadrii; *Hyonyssus* on Pici.

d) Genera with polyordinal distribution are also quite frequent on hosts both from 2 orders and from many orders. For example *Zachvatkinia* on Charadriiformes and Procellariiformes; *Homeobrephosceles* on Gruiformes and Charadriiformes; *Uniscutalges* on Piciformes and Coraciadiformes; *Pseudogabucinia* on Ciconiiformes, Falconiformes and Gruiformes; *Coraciacarus* on Coraciadiformes, Cuculiformes, Falconiformes, Apodiformes and Gruiformes; *Brephosceles* on Gaviiformes, Procellariiformes, Anseriformes, Gruiformes and Charadriiformes.

The study of specific and generic specificity may in some cases provide data applicable to ornithosystematics. The occurrence of a certain taxon on certain hosts may indicate their phylogenetic affinities. However, the possibility of their secondary infestation due to ecological and ethological factors should be taken into consideration and individual cases should be considered from all aspects and with regard to other groups of parasites. As a whole the feather mites are a very suitable group for such phyletic conclusions.

References

Atyeo, W. T. and Braasch, N. L.: The feather mite genus *Proctophyllodes* (Sarcoptiformes: Proctophyllodidae). Bull. Univ. Nebraska State Mus. 5 : 1–354, 1966.

Dubinin, V. B.: About the specificity of feather mites in connection with the evolution of their hosts. Usp. sovr. biol. 29 : 442–457, 1950. (In Russian.)
Dubinin, V. B.: Feather mites (Analgesoidea). Part I. Introduction. Fauna SSSR, Paukoobraznye VI, 5. Izd. AN SSSR Moskva–Leningrad, 1951. (In Russian.)

INTERRELATIONSHIPS OF MESOSTIGMATIC MITES IN THE NESTS OF *CLETHRIONOMYS GLAREOLUS*

M. DANIEL
Institute of Parasitology, Czechoslovak Academy of Sciences, Prague, Czechoslovakia

Studies on nest communities arouse a special interest not only as a theoretical problem, but also in view of epidemiological practice, because a number of nidicolous parasites have been proved or supposed to be vectors of natural focus diseases. This fact stresses the necessity of searching for a concise expression of results obtained in order to compare them mutually and to retain the possibility of using the published results directly as source data for eventual calculations.

A marked component of nest communities is represented by mites, especially from the group Gamasoidea, that are a frequent object of research in nests of most diverse hosts. Predominant papers are those listing the species found. The mites in them are classified according to food specialization, and their relationship to the cenosis of the nest is expressed by a percentage of the total number of nidicolous organisms etc. Against both mentioned aspects objections should be raised on principle. In most gamasoid mites a strict food specialisation is out of the question because their character in this respect is very changeable and cannot be therefore taken as a basis for further analyses. Likewise the mere percentual calculation is too static and does not give a true picture of actual interrelationship in particular species.

We have made an attempt to solve the problem of interspecific relationships in mesostigmatid mites from the nests of *Clethrionomys glareolus* by the method of association analysis which aims to assess the relationship of each pair of species in the given group. This relationship is assessed by mutual statistic dependence of both species. A high statistic dependence of these species may be, as a rule, interpreted as a high degree of harmony in ecological requirements, which means, that both species are equally influenced by environment.

To ascertain the statistic dependence usually a contingency table combined with the criterion χ^2 or Fisher's test is often used. For the solution of problems of interspecific relationships we do not consider this method to be quite suitable. Undesirable above all, is a strong influence exerted upon the result by quantity in the fourth cell of contingency table (both species absent). In the evaluation of interspecific relationships this means that the results based on contingency tables reveal a high number of associations, of which many, however, are false. We have therefore chosen a method which does not involve the category "both species absent", and instead used the affinity index originally derived by Fager (1957) for the purposes of phytosociology. Strictly speaking, it is not the association in the original sense. That is why some authors designate the indicator of this type as "index of interspecific overlapping". In this paper we use the term association sensu lato which includes all these indicators. Among various indices from the category "interspecific overlapping" Fager's index is of indisputable priority because it can be statistically evaluated by a t-test.

The data on mites from 150 nests of *Cl. glareolus* collected in all seasons of the year in forest biotopes of Vsetínské Beskydy (northern Moravia) were taken as a basis. A total of 5,609 mesostigmatid mites belonging to 42 species was found. The findings with low frequencies (i.e. occurrence in less than 5 nests) were eliminated from the statistical test and consequently a total of 20 species listed in Table 1. remained to be used in the analysis. The influence of biotope and location of the nest were chosen as main objects of our studies. The nests were classified in 4 combinations: forest – nests situated at the ground level and underground; clearing – nests situated at the ground level and underground.

The results were demonstrated in graphs either by means of matrices or in the form of association graphs empirically derived. In both cases the knowledge that the graph is a set of objects and simultaneously a set of correspondences, was taken as a basis. The objects are species (they form vertices in the graph), the correspondences are statistically significant associations among them (edges connecting in graph).

Table 1. A survey of numerical symbols of gamasoid mites from the nests of *Clethrionomys glareolus* whose interspecific relationships have been statistically analyzed

Symbol	Mite species	Symbol	Mite species
1	*Pergamasus* sp.	11	*Geholaspis* sp.
2	*Eugamasus kraepelini*	12	*Proctolaelaps pygmaeus*
3	*Eugamasus remberti*	13	*Androlaelaps sardous*
4	*Eugamasus* sp.	14	*Hypoaspis heselhausi*
5	*Euryparasitus emarginatus*	15	*Eulaelaps stabularis*
6	*Cyrtolaelaps mucronatus*	16	*Haemogamasus horridus*
7	*Cyrtolaelaps minor*	17	*Haemogamasus nidi*
8	*Gamasellus silvestris*	18	*Haemogamasus hirsutus*
9	*Veigaia nemorensis*	19	*Haemogamasus hirsutosimilis*
10	*Macrocheles pauperior*	20	*Hirstionyssus isabellinus*

The species in them are arranged and designated by numerical symbols, 1–14 indicating carnivores, sapro-, copro- and necrophagous organisms, while 15–20 represent species with differently developed degrees of hematophagia.

If we direct our attention only to species (i.e. in the language of graphs: the vertices) we obtain information only about the species composition which is in particular nest combinations rather uniform and consequently it might have been concluded that there is no difference between individual communities studied. A different picture presents itself if we pay attention to edges, i.e. correspondences (in our case the so-called statistically significant associations between species). The number of edges (correspondences) is a very important indicator because it shows close relationships of mites and a higher degree of organization of nidocenosis. While studying in detail the matrices of particular nest categories we see, that the differences are not only in the number, but also in the quality of vertices and edges. There are relatively few different vertices in particular groups, but there are plenty of different edges. In other words, the identical mite species under different conditions of biotope and nest location differ in their interrelationships.

The nests of the forest (underground) combination reveal three pronounced groups of edges (i.e. positive associations) on the highest level 0.001 (Fig. 1). The first group connects the species 3–6 (i.e. *Eugamasus remberti, Eugamasus* sp., *Euryparasitus emarginatus* and *Cyrtolaelaps mucronatus*). The second group includes the species 15–18 (i.e. *Eulaelaps stabularis, Haemogamasus horridus, H. nidi* and *H. hirsutus*). The third group of association combines the first and second group, especially through the species *E. emarginatus* and *Eugamasus* sp. on one hand and *E. stabularis* on the other.

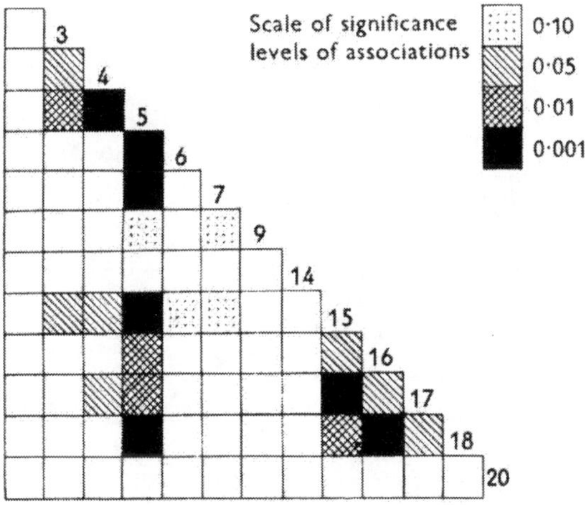

Fig. 1. Matrix of interspecific associations of gamasoid mites in the nests of *Clethrionomys glareolus* (forest – nests situated underground).
Numerical symbols indicate particular mite species in accordance with Table 1.

In nests of the forest (at the ground level) combination there are traces of all three groups although they are very attenuate (Fig. 2.), as for the number of edges and for the significance level of their associations. In the clearing (at the ground level) combination the positive associations are numerous, but on the low level, diffusely dispersed all over the matrix (Fig. 3). No groups come into consideration.

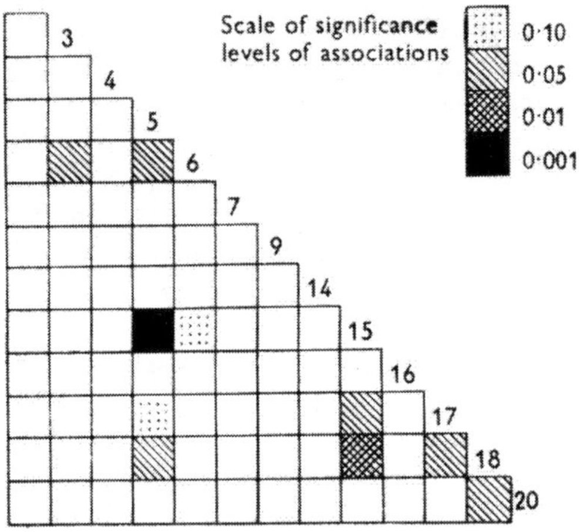

Fig. 2. Matrix of interspecific associations of gamasoid mites in the nests of *Clethrionomys glareolus* (forest – nests situated on the ground).
Numerical symbols indicate particular mite species in accordance with Table 1.

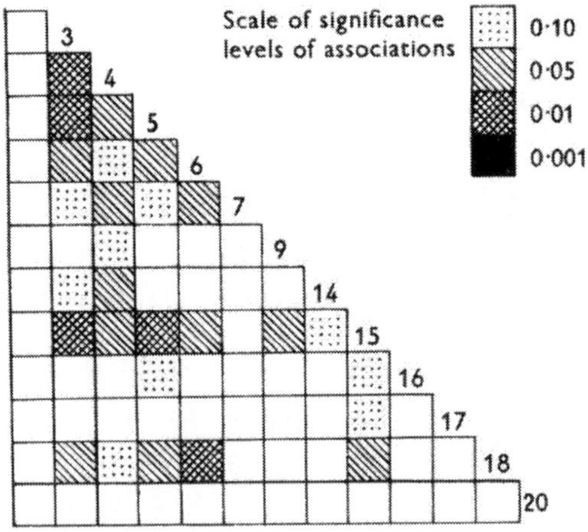

Fig. 3. Matrix of interspecific associations of gamasoid mites in the nests of *Clethrionomys glareolus* (clearing – nests situated on the ground).
Numerical symbols indicate particular mite species in accordance with Table 1.

The species designated by lower numeral symbols are related to the nest association rather than to the nest parasites themselves. The results obtained in both biotopes correspond partly with the fact that *C. glareolus* is primarily a forest animal and partly that the nests situated at the ground level have less definitely formed conditions of environment and are more exposed to the contact with the surrounding mite fauna in the litter.

Another form of demonstrating the groups of associated species is the empirically derived association diagram, where vertices are depicted in the way close to a layman's idea of knots in a net and lines representing edges. These association diagrams may be well used not only for demonstrating results, but also for considerations about the circulation of pathogenic agents in the nest community etc. (Fig. 4, 5.)

If we assess the results from the aspect of particular mite species, the most associated species was *E. stabularis*; according to the number of associations *Euryparasitus emarginatus*, *Eugamasus* sp., *Haemogamasus hirsutus*, *H. nidi* and *H. horridus* followed next.

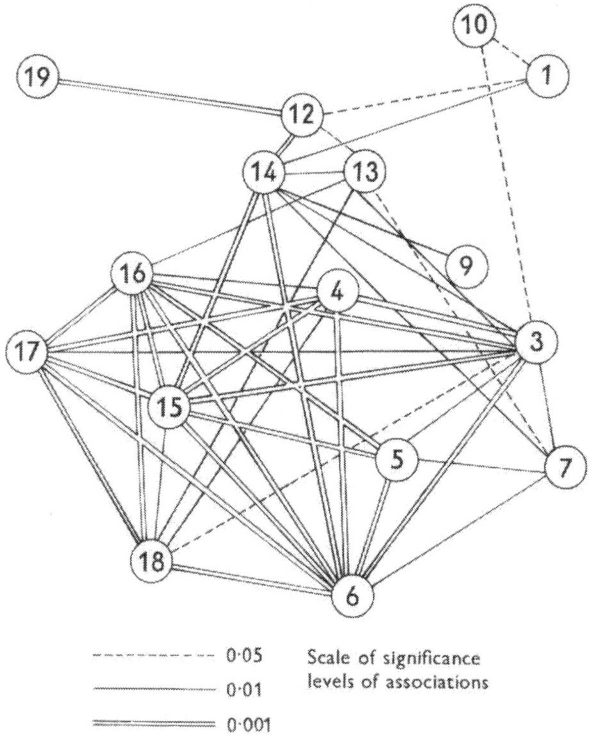

Fig. 4. **Association diagram of relationships of gamasoid mites in the nests of *Clethrionomys glareolus* (total group studied).**
Results based on evaluation of pairs of species by contingency table. Vertices include numerical symbols of particular mite species in accordance with Table 1.

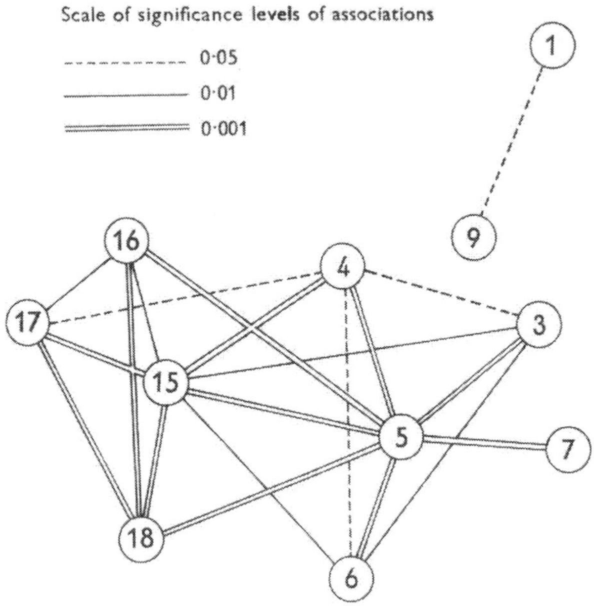

Fig. 5. Association diagram of relationships of gamasoid mites in the nests of *Clethrionomys glareolus* (total group studied).
Results are based on evaluation of pairs of species by Fager's affinity test. Vertices include numerical symbols of particular mite species in accordance with Table 1.

Out of the total number of 42 gamasoid species found only 13 species were associated at a different degree. If we regard the nidocenosis as a group of species directly or indirectly associated, in which causal interrelationships exist, only these 13 species make the nest community of gamasoid mites under given conditions. Although others occurred in the nests, they are virtually a part of different communities. Such an analysis leads to a precise delimitation of the nest community, which may be of practical importance in epidemiological considerations.

We assume that the comparison of results presented, with those obtained by traditional methods of evaluation of nest materials, clearly demonstrates the simplicity, lucidity and easy comparability of results, as well as a wider and more profound assessment, and consequently provides further detailed information. We, therefore, propose to accept this method as a part of standard evaluation of extensive groups of nest organisms.

References

Fager, E. W.: Determination and analysis of recurrent groups. Ecology 38 : 586–595, 1957.

THE ACARINIA ZONE OF BATS UNDER CONDITIONS OF CENTRAL EUROPE

F. DUSBÁBEK
Institute of Parasitology, Czechoslovak Academy of Sciences, Prague, Czechoslovakia

The terms acarinium and zone of acarinia have been created by Rosický (1953) and defined in detail by Mrciak and Rosický (1959) and Rosický and Mrciak (1967). In harmony with their definition the acarinium in the acarinia zone of bats in Central Europe is considered in this paper as a group of mites belonging to the orders Parasitiformes and Acariformes, parasitizing bats of one colony and permanently or temporarily inhabiting their permanent or periodically visited shelters. The group of acarinia of particular bat species make the zone of bat acarinia.

The zone of bat acarinia in Central Europe is characterized by the following features:

1. Isolation of the zone of bat acarinia. The bats are an ecologically and bionomically isolated group of hosts feeding in the air and therefore lacking any terrestrial feeding ground. The contact with members of other acarinia zones is consequently possible only in the bat shelters.

2. Stability of acarinia in this zone. The bats use their shelters permanently or periodically for many years. Due to this fact the succession takes place very slowly in such shelters, but it is accompanied by the seasonal microsuccessions resulting from seasonal changes of shelters (winter shelters, temporary shelters and summer shelters).

3. Passivity of bats in using the shelters. The bats do not build their shelters, neither arrange them as other mammals do; nor do they store supplies. The active functions of the hosts in the shelters are eliminated. The thermoregulatory function of the nest is taken over by congregations of hosts in colonies, and thus their passive influence is increased, e.g. their guano causes changes in the microclimate of the shelter (changes of humidity, concentration of ammonia etc.)

4. Spaciousness and variety of original shelters as well as their microclimatic stability make facultative and obligatory commensalism possible for a wide range of mite species.

5. Transition of ectoparasites originally exophilic (parasites of feeding ground, Beklemishev, 1951) to endophilia (parasitism in shelter) is marked e.g. in the families Ixodidae and Trombiculidae. Trombiculids and ixodids usually attack their hosts outside of the nests or roosts but they parasitize bats in their shelters because they have practically no contact with bats except in the roosts.

6. Transition of many endophilic ectoparasites to epizoic permanent type of parasitism (e.g. in the famil y Spinturnicidae or some Macronyssidae). This is probably an adaptation to offset the difficulty of finding a host in the relatively spacious shelters of bats. Mites of the family Spinturnicidae are actually the only gamasoid mites with a permanent type of epizoic parasitism.

On the basis of their bionomy and ethology the mites occurring in the acarinia of Central-European bats may be divided into several groups (Fig. 1):

1. Cavital muco-hematophagous endoparasites, parasitizing primarily nasal cavities (Gastronyssidae, Ereynetidae, Trombiculidae – *Doloisia*). The data on their occurrence and bionomy are fragmentary and reveal that particular species attack mostly bats of one host family and that their specificity is consequently on the level of mono- up to oligoxenia.

2. Tissue kerato-mucophagous endoparasites of the families Sarcoptidae, Psorergatidae and Demodicidae, parasitizing mainly the skin of bats, subcutaneous layers or mucosa of mouth cavity. Their physiological adaptation is also very high but due to frequent keratophagia their specificity remains mainly at the level of oligoxenia.

3. Epizoic ectoparasites with obligatory hematophagia or lymphophagia, of the families Spinturnicidae and Myobiidae whose specificity is determined primarily by trophic relationships to the host. These phylogenetically old groups whose evolution was concurrent with the evolution of the hosts, are trophically and physiologically very

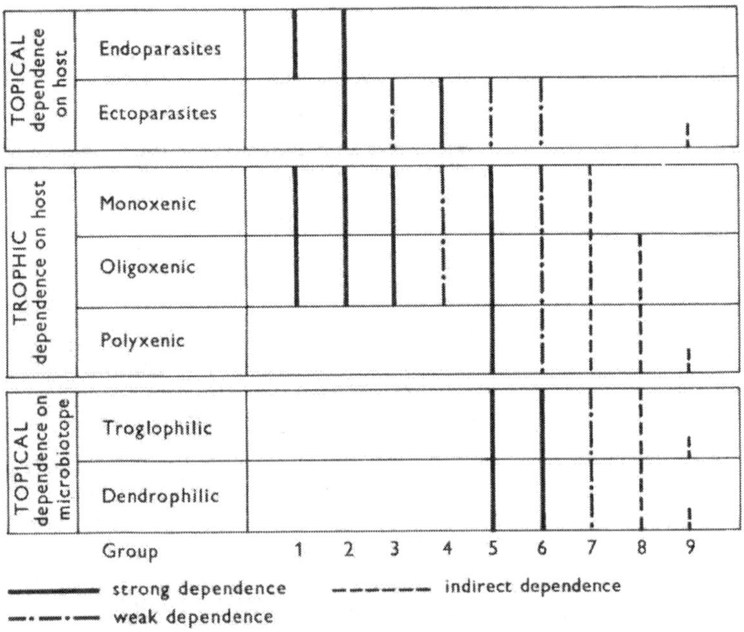

Fig. 1. Dependence of mites of different groups on hosts and outer environment in the zone of bat acarinia.

much specialized. The specificity in them has reached the highest possible degree that is a single parasite species is specific for a single host species (monoxenia), less frequently a single species of mite may parasitize several species of the same genus or family (oligoxenia).

4. Epizoic ectoparasites with obligatory kerato-kolagenophagia, of the family Labidocarpidae, whose specificity is determined primarily by topical relationships to host. The system of attachment to the body of host in them has reached such a high degree of specialization that most species are adapted only to a certain type of hair which is revealed by a strict topical localization on the body of host. The choice of host is determined more by the possibility of attachment on their bodies than by physiological specialization. Specialization for a group of species within the scope of a genus or family (oligoxenia) is the most frequent.

5. Endophilic ectoparasites with transition from periodic to permanent type of parasitism and obligatory hematophagia, whose presence in the acarinium is determined not only by trophic relationships to the host but also by their own requirements for a specific biotope or microbiotope. These are mites of the family Macronyssidae, whose physiological adaptation has reached a unique level. We find here mono-, oligo- and polyxenous species (Fig. 2). Monoxenia is typical for species with a dominant epizoic type of parasitism. Species with an endophilic type of parasitism, for which oligo- and polyxenia is characteristic, are mostly associated with a certain

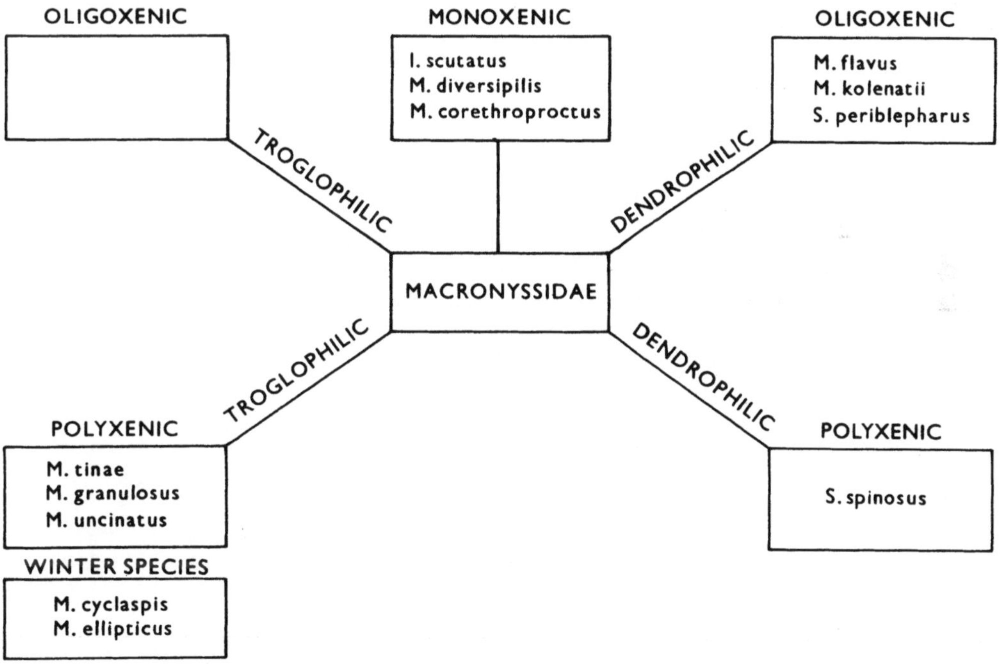

Fig. 2. Relation of mites to hosts and outer environment in the family Macronyssidae (Group 5).

type of microbiotope; therefore we may distinguish here troglophilic and dendrophilic species. Two troglophilic species *Macronyssus ellipticus* and *M. cyclaspis* should be specially mentioned because they may be regarded as the so-called winter species of mites, and in this respect are similar to fleas of the genus *Nycteridopsylla.* The period of their reproductive activity falls in the winter months and this is why they occur on bats only in their winter shelters. In summer colonies of bats these two species have never been met with in Central Europe.

6. Endophilic temporary or periodical ectoparasites with obligatory hematophagia in some stages, whose presence in the acarinium is determined primarily by their own requirements for a specific microbiotope. They are members of the families Ixodidae, Argasidae and Trombiculidae, species predominantly troglophilic, exceptionally also dendrophilic (e.g. *Argas vespertilionis*), characterized by mono-, oligo- and polyxenia.

7. Obligatory commensals of the genus *Nycteriglyphus* of the family Rosensteiniidae, trophically apparently quite dependent on the products of metabolism of host, i.e. guano (indirect trophic dependence) and occurring only in the shelters of bats. Likewise some species of the genus *Oribella (O. cavatica, O. forsslundi)* of the family Thyrisomidae are known only from the shelters of bats.

8. Facultative commensals i.e. free-living inhabitants of bat shelters, occurring usually also outside the zone of bat acarinia.

9. Parasitic mites of other host groups accidentally introduced in the zone of bat acarinia.

References

Beklemishev, V.N.: (Parasitism of arthropods on terrestrial vertebrates; the ways of its origin). Med. Parazitol. 20 : 151–160, 233–241, 1951. (In Russian.)

Mrciak, M., Rosický, B.: O vzťahoch roztočov radu Parasitiformes k ich hostitel'om, najmä drobným cicavcom. Biológia 14 : 241–264, 1959.

Rosický, B.: Klasifikace roztočů v přírodních ohniscích přenosných chorob. Čs. Biol. 2 : 27–32, 1954.

–, Mrciak, M.: The concept of the acarinium and the acarinia zones. Folia Parasitol. (Praha) 14 : 349–360, 1967.

Proceedings of the 3rd International Congress of Acarology, Prague, 1971

INFLUENCE OF TEMPERATURE ON POPULATION INCREASE OF FOUR DUNG-LIVING MACROCHELID MITES WITHIN OPTIMUM RANGE

A. FILIPPONI
Istituto Superiore di Sanita, Roma, Italy

Population growth studies were conducted with 4 dung-living and predatory macrochelid mites at different temperatures on Rodriguez's standard medium, frozen house fly eggs and live rhabditid nematodes being offered as food. The species *Macrocheles penicilliger* (Berl.), *M. peniculatus* Berl., *M. subbadius* (Berl.) and *M. muscaedomesticae* (Scopoli), were selected. They are closely related, but have a different degree of association in nature. The first two, besides, are thelytokous, the others arrhenotokous.

The determination of optimum temperature, thought of as the range within which there is maximum population increase, was the primary concern in gaining information on species niche requirements, but the effects of temperature on components of speed of multiplication were also comparatively investigated in the 4 species.

The "working range", i.e. a more restricted temperature range in which the assembling of requisite data for calculation of population increase parameters can be made possible, was established for each species on the basis of developmental duration and oviposition rate curves, obtained within a wide range of temperatures from 10° to 40°C : 10°, 13°, 15.5°, 20°, 22.5°, 26°, 28.5°, 33°, 36° and 40°. *M. penicilliger* laid eggs from 10° to 28.5°, showed optimum oviposition range (with values not significantly different from maximum) at 20°–26° and minimum development duration at 26°. *M. peniculatus* laid eggs from 13°–33°, showed optimum oviposition range at 20° to 28.5° and minimum development duration at 28.5°. *M. subbadius* laid eggs from 15.5°–36°, showed optimum oviposition range at 26°–28.5° and minimum development duration at 33°. *M. muscaedomesticae* laid eggs from 20° to 36°, showed optimum oviposition range at 28.5°–36° and minimum development duration at 36°. Accordingly, the w.r. for the 4 species was settled respectively at 22°–26°, 24°–28°, 25°–29° and 28°–34°.

Cohort of 16 females, each composed of 2 subsamples (corresponding to 2 laboratory cultures of the same field strain) were bred at every 2°C of species w.r. Females were reared singly in culture tubes, renewed once a day, at 75% R.H. The consequent moisture gradient in culture and the alternative food (house fly eggs and nematodes) aimed at meeting the different requirements of various stages, and offered a "multiple niche" environment. Adult progeny was counted from daily oviposition for a more reliable census and only virgin mothers were used for arrhenotokous forms, in order to avoid intersex competition during progeny development. For all species, strains used in these experiments were different from those used in the previous ones.

In the respective working ranges, temperature proved to affect adult female longevity only in *M. penicilliger* and *M. subbadius*. Total fertility (adult progeny/female) and fertility rate (adult progeny/female/day in a constant reproductive age span) were influenced in all species. Generally, the intensity of the 2 population attributes fell off from the lower to the upper limit of the w.r.; but in *M. penicilliger* the maximum fertility was reached at 24° and in *M. subbadius* maximum fertility and fertility rate were both reached at 27°. Temperature influenced also l_x and m_x curves, but only in *M. penicilliger* and *M. subbadius* and to a different degree. Lastly, subsampling contributed additional variance only in the case of one of the thelytokous forms, *M. peniculatus*, giving evidence in favour of a good degree of standardization of cultures.

Using eggs laid by females of the same laboratory stock as above and bred under optimal conditions in order to avoid the phenomenon of facultative larviparity, we investigated the mean duration of development from egg to adult at the same temperature as the cohorts.

Temperature clearly affected developmental duration within the w.r. of each species, the males, when present, having a shorter duration than the females. Mean female developmental duration in hours was for *M. penicilliger* 152.4 at 22°, 137.5 at 24° and 111.1 at 26°; for *M. peniculatus* 92.6 at 24°, 75.0 at 26° and 71.0 at 28°; for *M. subbadius* 56.9 at 25°, 47.4 at 27° and 44.1 at 29°; and for *M. muscaedomesticae* 59.8 at 28°, 42.3 at 30°, 49.5 at 32° and 49.9 at 34°, the rising of the last 2 values being probably due to the closeness of the upper limit of effective temperature range of the species.

The influence of temperature on sex ratio, expressed as female probability, was investigated in *M. muscaedomesticae*, with 5 experimental populations started with 30 virgin females and males at 28° and 32°. Medium and food were progressively increased. Population samples were drawn each 5 days and adult progeny was counted from eggs laid over the following day by 30 females randomized from the population samples, for an evaluation of s.r. at birth.

Evidence was accumulated which showed that s.r. in the population and to a lesser extent in the progeny varies with real fluctuations of an oscillatory type and that population s.r. is negatively correlated with s.r. at birth. Oscillations appeared more violent at 32° than at 28°, but when total mean sex ratios were compared (pooled data of the first 5 population samples at 28° vs. that at 32°) no significant difference resulted. The s.r. at birth of the species was estimated from the 2 experiments as 0.440, corresponding to a s.r. in population of 0.758. The s.r. at birth of *M. subbadius* was estimated from a single experiment at 27° to be 0.449, this corresponding to a s.r. in population of 0.700.

Capacity for increase data are reported in Table 1. The power to increase rises from the first to the 4th species: the relative intensity of r_m can give basic information on the potentiality of a single species as a predator. It is worth noting that 3 out 4 species are able to exceed the speed of multiplication found in *Musca domestica.*

Table 1. Capacity for increase data in optimum range

Species	Temp. °C	Sex ratio	R_o	T (days)	r_m (per day)
1–*Macrocheles penicilliger* (Berl.)	22°	1	52.25	13.69	0.289
	24°	1	42.84	11.63	0.323
	26°	1	23.24	11.21	0.281
2–*Macrocheles peniculatus* Berl.	24°	1	77.14	8.73	0.498
	26°	1	75.74	7.67	0.564
	28°	1	60.26	7.75	0.529
3–*Macrocheles subbadius* (Berl.)	25°	0.449	50.76	5.80	0.677
	27°	0.449	59.68	5.06	0.808
	29°	0.449	53.32	4.60	0.865
4–*Macrocheles muscaedomesticae* (Scopoli)	28°	0.440	70.45	5.48	0.776
	30°	0.440	58.07	4.48	0.906
	32°	0.440	55.87	4.81	0.837
	34°	0.440	36.17	4.91	0.731

The maximum r_m indicates the middle of the optimum range we are searching for. This can be evaluated for the 4 species as 22°–26°, 24°–28°, 27°–31° and 28°–32°, i.e. nearly overlapping the investigated w.r., except in the case of *M. subbadius* whose range shifted up by 2°C, due probably to genetic differences in strains used in the 2 series of experiments. So every species proved to have its own optimum temperature range with *M. pericilliger* adapted to a lower temperature and *M. muscaedomesticae* to the higher, just overlapping the optimum temperature range of *Musca domestica.*

A comparison (Fig. 1 and 2) between the r_m temperature curve, the curves of 2 main components, R_o and T, and the curves of the minor components, such as total fertility and fertility rate (influencing R_o) and longevity and development duration (influencing T) suggests that the net reproductive rate (R_o) and the mean length of a generation (T) were both affected by temperature, even near the optimum range, so that (since $r_m = \log_e R_o / T$) only r_m or its antilog R_m may be used in looking for the optimum range of a species. Moreover, the trend of R_o and T curves appeared species specific as a consequence of the different influences of the minor components on these statistics.

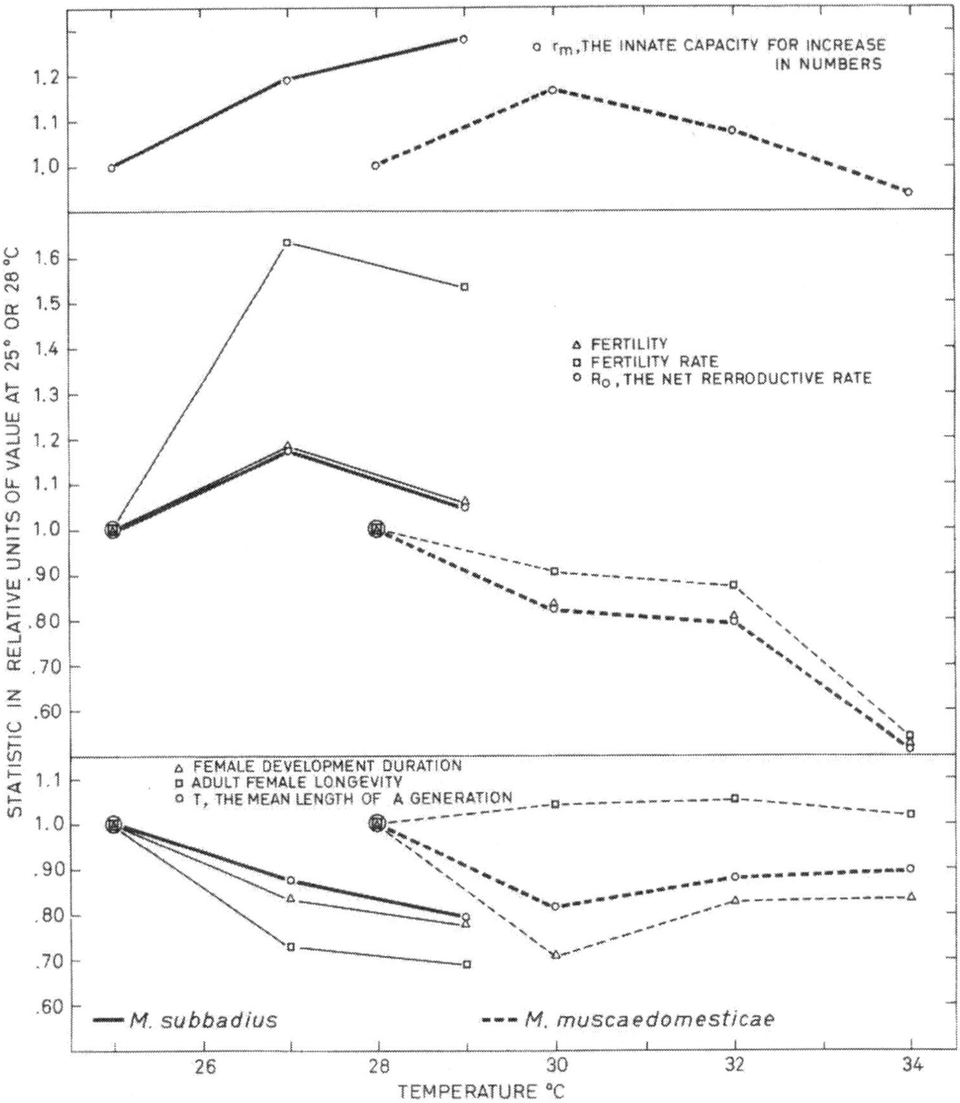

Fig. 1. Relation to temperature of innate capacity for increase and of some of its components at optimum range.

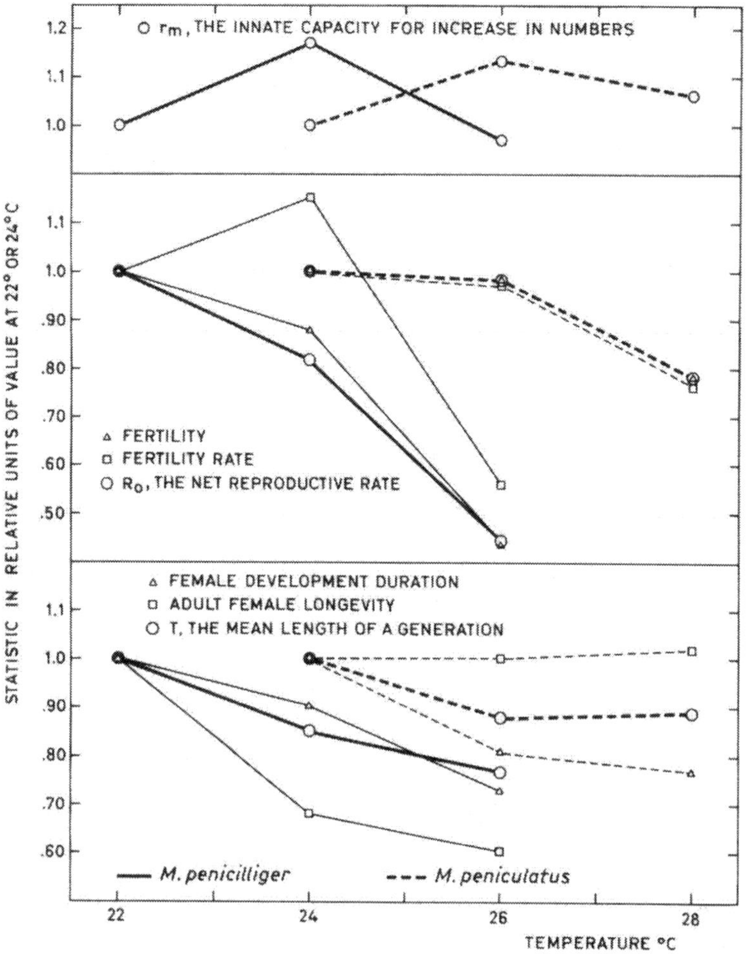

Fig. 2. Relation to temperature of innate capacity for increase and of some of its components at optimum range.

POPULATION REGULATION IN QUILL MITES (ACARINA: SYRINGOPHILIDAE)+

J. KETHLEY
Acarology Laboratory, The Ohio State University, Columbus*, USA

Abstract

The wing quills of the House Sparrow (*Passer domesticus* L.) may host as many as three species of quill mites, suggesting ecological differences in the quill mite habitat. Detailed investigation of a single species of quill mite, *Syringophiloidus minor* (Berl.) reveals adaptations that maximize mite success within the confines of the feather calmus, and suggests a mechanism that allows partitioning of the wing by two or more different sized quill mite species.

Quill mites live inside the feather calmus and feed on host tissue fluids by piercing the quill wall with long stylet-like chelicerae. Mite reproduction and development take place within the calmus. Only adult female mites disperse. Females can move into newly developing quills during host moult in the fall, or into the developing feathers of nestlings in the spring. The quills in which the mites live represent a closed system, with calmus volume limiting the number of mites that may mature and calmus wall thickness limiting the ability of the mites to feed. Female mites cease egg production before the quill contains enough mites to fill it. As a result, the quill does not become filled with immature forms that would be incapable of mating and dispersing. Males of *S. minor* are smaller than the females and have a frequency of 1 : 11, thus over 95% of the limiting resource, calmus volume, is occupied by females, the potential dispersants.

The size of a female is evidently limited by two opposing selective pressures: the smaller they are, the more can disperse from a given quill, but if they become too small the mouthparts will not be able to reach through the quill wall during feeding. Thus each quill size has an optimum mite size, and the mite species occur in a corresponding diversity of sizes.

* Present address: Field Museum of Natural History, Division of Insects, Roosevelt Road at Lake Shore Drive, Chicago, Illinois, 60605.

+ Accepted for publication in Ecology.

DENSITY AND HOST RELATIONS OF *IXODES TRIANGULICEPS* BIR. IN WOODLAND AND GRASSLAND AREAS

T. G. O'DONNELL
Animal Ecology Research Group, Department of Zoology,
Oxford, England

Ixodes trianguliceps, the forest tick, is a three host parasite of small mammals, primarily rodents and insectivores. It has a wide geographic distribution, ranging from Ireland in the west, to the vicinity of Lake Baikal in the eastern U.S.S.R. (Arthur 1963, Korenberg & Lebedeva 1969). The general distribution, together with areas where detailed studies on the biology and ecology of this tick have been made, are indicated in Fig. 1.

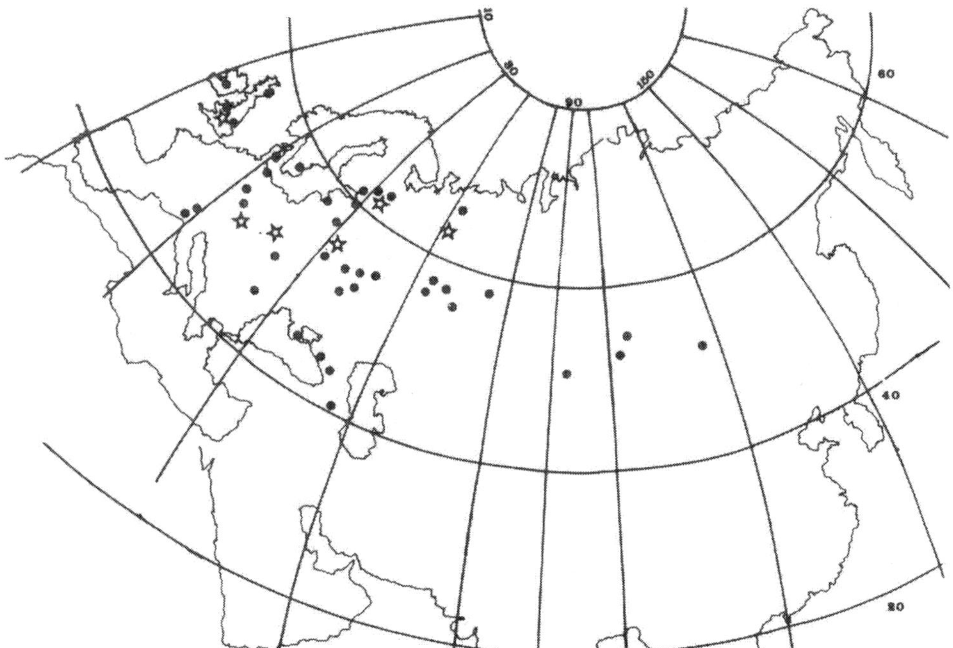

Fig. 1. The distribution of *Ixodes trianguliceps*. Black points-areas where the tick has been recorded; asterisks – areas where detailed observations on its general biology have been made.

This study was carried out as part of a larger investigation into small mammal ectoparasite ecology (O'Donnell 1971) in Wytham Woods near Oxford in central England. Samples from three host species, *Apodemus sylvaticus* (L.), *Clethrionomys glareolus* (Schr.) and *Microtus agrestis* (L.) were collected at approximately five week intervals over a two year period. Rodents were trapped using Longworth live traps baited with oats and containing sterile wood wool in the nest box as insulating material. Animals were killed using chloroform, and then skinned. The skin was dissolved by digesting it in a buffered trypsin solution and was then boiled in 10% KOH (Cook 1954). Ectoparasites were then filtered from the resulting "soup". The insulating wood wool was processed in a woodland litter extractor (Kempson, Lloyd and Ghelardi 1963), and any parasites that had migrated from the host's body to the surrounding material were collected. The seasonal fluctuations of the tick population, its host preferences and the effect of host habitat on infestation were studied. The results, based on a collection of nearly one thousand small rodents, are substantially different from those recorded by previous workers.

In this paper the index used to indicate changes in the density of the parasite population is the mean infestation level, plotted on a logarithmic scale. This was derived for each sample by dividing the number of ticks collected by the total number of potential hosts trapped.

Seasonal Dynamics of Ixodes trianguliceps

Fig. 2 shows the changes in population density of larval, nymphal and adult ticks as collected from all hosts over the study period. Larvae were active throughout the year with population density building up over summer to reach a mid-autumn peak. Nymphs were active from late spring to autumn, reaching peak density at mid-summer. Adults, generally females, were recorded from April to October, but so infrequently that it is impossible to detect a distinct peak. Only larvae were sufficiently numerous to be examined on the basis of each host species (Fig. 3) and on each they showed a similar pattern of activity to that outlined above.

In no case was there any evidence of an annual bi-modal activity-cycle as shown by Vysotskaya (1951) and Lachmajer (1962) for central Europe, or as shown by Cotton and Watts (1967) for the same area in central England.

Effect of Host Habitat on Tick Infestation

All hosts trapped were subjectively categorised as coming from one of three broadly defined habitat types.

I. Dense canopy woodland largely dominated by beech (*Fagus silvatica* L.), hazel (*Corylus avellana* L.) and sycamore (*Acer pseudoplatanus* L.), with open or lightly undergrown ground and field layers.

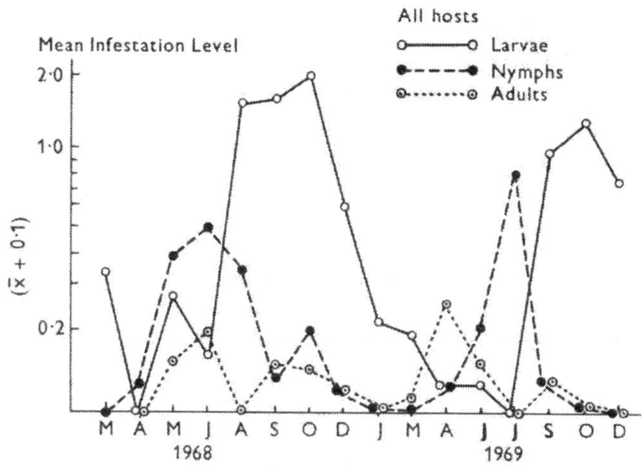

Fig. 2. Changes in the population density of larval, nymphal and adult ticks, over the study period.

II. Open canopy woodland containing an association of oak (*Quercus robur* L.) and birch (*Betula pendula* Roth.) or sycamore, having a ground and field layer densely overgrown to a height of 1 metre with brambles (*Rubus* sp.).

III. A plantation of young deciduous and coniferous trees, in grassland dominated by *Arrhenatherum elatius* (L.) and *Dactylis glomerata* L.

Only *Apodemus* and *Clethrionomys* were trapped in all three habitat types, *Microtus* being confined to the grassland area. In the case of both *Apodemus* and *Clethrionomys* the habitat in which they were trapped appeared to make little difference to their infestation with larval ticks. However, nymphal ticks were less prevalent on hosts trapped in lightly undergrown woodland over the whole study period (p 0.001). The overall results indicate that host habitat is not an important parameter controlling the occurrence of *I. trianguliceps* on small rodents in Wytham Woods. This finding conflicts with the results of both Vysotskaya (1951), Nikitina (1960) and Lachmajer (1962) working in broadly similar habitats, who maintain that host habitat exercises a direct effect on infestation with this tick.

Relative Infestation of the Three Host Species

Fig. 3 shows that infestation with larvae does not appear to be significantly affected by host species. Table I indicates that adult ticks also are similarly unaffected. However the overall mean infestation level of *Clethrionomys* with nymphal *I. trianguliceps* is significantly greater than that of either *Apodemus* or *Microtus* (p 0.01). This result again is at variance with previous results from these, and related host species.

Table 1. Overall infestation of the three host species with larval, nymphal, and adult *Ixodes trianguliceps*

Species	No. of hosts	Larvae		Nymphs		Adults	
		No. of larvae	x̄ infestation	No. of nymphs	x̄ infestation	No. of adults	x̄ infestation
Apodemus sylvaticus	317	203	0.64	27	0.09	10	0.03
Clethrionomys glareolus	375	295	0.79	89	0.24	10	0.03
Microtus agresis	291	155	0.53	8	0.03	20	0.07
Totals	983	653	0.66	124	0.13	40	0.04

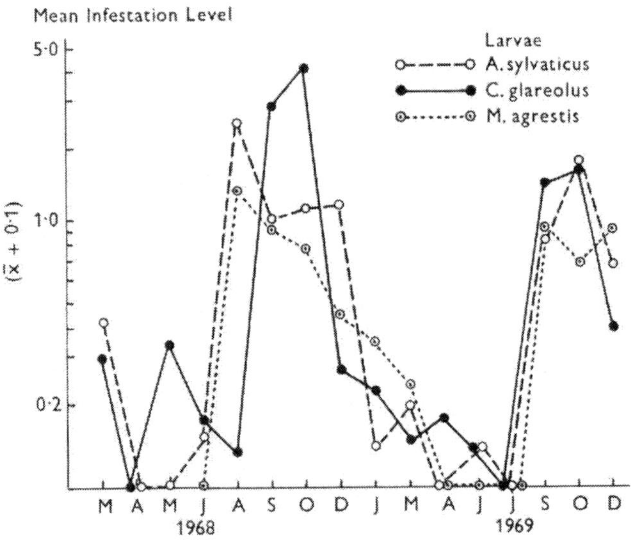

Fig. 3. Mean infestation levels of each host species with larval ticks over the study period.

Thus in summary it appears that in the area where the study was carried out *I. trianguliceps* exhibits pronounced seasonal changes in population density with a unimodal cycle of activity for all stages over the year. Neither host habitat nor host species appeared to have any effect on infestation with larval ticks, though nymphs may by influenced to some extent by both of these factors. The lack of agreement between these results and those of previous workers is inexplicable on the basis of the data available though some suggestions as to the possible causes may be made. Firstly, it is possible that the differences observed between this study and those of Vysotskaya (1951), Nikitina (1960) and Lachmajer (1962) represent real differences in the populations of *I. trianguliceps* studied, these being the result of distinct behavioural or physiological responses to the differing biotic and abiotic conditions prevailing in England vis à vis central Europe. Secondly, they may be due to differences in the collection techniques used. Thus it is possible that previous workers underestimated the numbers of ticks leaving the cooling bodies of snap trapped hosts. Alternatively where hosts were live trapped (as in Cotton & Watts 1967) it is possible that the nest box insulating material (hay or straw) was not examined sufficiently carefully, leading to incomplete samples. The importance of this material as a refuge for ticks is shown by the fact that in this study up to 17 % of a tick sample has been recovered from this source.

I express my thanks to Professor Pringle for making the facilities of the Zoology Department available to me, and to Oxford University for permitting me to carry out my field work in Wytham Woods.

References

Arthur, D. R.: British Ticks. Butterworths, London. 213 pp., 1963.

Cook, E. F.: A modification of Hopkins technique for collecting ectoparasites from animal skins. Entomol. News 5, pp. 35–37, 1954.

Cotton, M., Watts, C. H. S.: The ecology of the tick *Ixodes trianguliceps* Birula (Arachnida; Acarina; Ixodoidea). Parasitology 57, pp. 525–531, 1967.

Kempson, D., Lloyd, M., Ghelardi, R.: A new extractor for woodland liter. Pedobiologia 3, pp. 1–21, 1963.

Korenberg, E. I., Lebedeva, N. N.: Distribution and some general features of the ecology of *Ixodes trianguliceps* Bir. in the Soviet Union. Folia Parasitol. (Praha) 16, pp. 143–152, 1969.

Lachmajer, J.: The ecology of the tick *Ixodes trianguliceps* Bir., 1895. Bull. Inst. Mar. Med. Gdansk 13, pp. 149–160, 1962.

Nikitina, N. A.: The biology of *Ixodes trianguliceps*. Med. Parazitol. 6, 708–712, 1960.

O'Donnell, T. G.: The ecology of ectoparasites of three species of small rodents. Phil. Thesis, Oxford University, 1971.

Vysotskaya, S. O.: On the biology of the ixodid tick *Ixodes trianguliceps* Bir. Parazitol. Sborn. Zool. Inst. Akad. Nauk. S. S. S. R. 13, pp. 105–110, 1951.

THE PROTONYMPH OF *PIMELIAPHILUS* (PTERYGOSOMATIDAE) AND ITS SIGNIFICANCE RELATIVE TO THE CALYPTOSTASES IN THE PARASITENGONA[1]

I. M. NEWELL
Department of Biology and Department of Entomology,
University of California, Riverside, California, U.S.A.

The number of active nymphal instars in the Trombidiformes ranges from none (e.g. Podapolipodidae) to three (e.g. *Isobactrus*, Halacaridae). Other genera have one active nymphal instar (*Copidognathus*, Halacaridae) or two (*Tetranychus*, Tetranychidae). In the last three examples, all of the nymphs are active; there are no calyptostases. Each larva or nymph does, however, go through an obligatory inactive period toward the end of its development, when it is unable to move. This is associated with the production of the cuticle of the next instar, and the establishment of new myocuticular junctions. The quiescent period is ended by the splitting of the old cuticle and the emergence of the next instar. Only *one* cuticle is left behind, that of the preceding larva or nymph.

In the development of typical Parasitengona (e.g. Trombidiidae, Trombiculidae, Erythraeidae, Hydrachnidae, etc.), there is only one active nymphal instar. There are, however, two calyptostases, one between the larva and active nymph or "nymph", the other between "nymph" and adult. These are inactive forms, devoid of appendages. When the "nymph" emerges, it leaves behind two cuticles, one belonging to the larva, the other to the first calyptostase. The adult also leaves behind two cuticles, that of the "nymph" and that of the last calyptostase. In the Parasitengona, an elaborate and confusing terminology has evolved around these calyptostases (nymphophan, nymphoderma, teleiophan, teleioderma, etc.). At the Second International Congress of Acarology, this was reviewed in detail by Cassagne-Méjean (1969), who urged the suppression of the complex system of terms and its replacement by the simple terms protonymph, deutonymph and tritonymph as used in other groups of Acari.

This was suggested as early as 1938 by Grandjean, who pointed out that the one active nymph in the development of typical Parasitengona should be called the deutonymph, with the protonymph and tritonymph being the first and second calyptostases respectively. This would bring the terminology of the nymphal instars in this group into line with that of other Trombidiformes. Some workers have accepted this

[1] This study was supported by Grant No. GB 15840 from the National Science Foundation to the University of California, Riverside.

quite readily, but others have not. Even in some very recent publications the old system of names is repeated in detail.

As far as the writer can tell, largely from verbal communications, much of the hesitation to adopt the simplified terminology comes from a reluctance to equate a motile nymph having eight functional legs and a normal complement of setae with a bag of tissue enclosed in a simple chitinous sac.

During the course of studies on the Pterygosomatidae, the writer has made observations which are significant to this problem. The development of female Pterygosomatidae (not the male) is comparable with that of the Parasitengona in that there is a single active nymphal instar and two calyptostases. The active nymphal instar must be regarded as the deutonymph, while the calyptostases must be regarded as the protonymph and tritonymph.

In most genera of Pterygosomatidae (e.g. *Geckobiella*), the protonymph and tritonymph are sac-like, with small limb-vestiges, devoid of setae as in the Parasitengona. But in *Pimeliaphilus gloriosus* Newell and Ryckman 1966, the protonymph and female tritonymph have greatly elongated (but unsegmented) legs, and in addition they have a well developed chaetotaxy on the legs, the idiosoma, and the gnathosoma. In short, they are fairly complete Acari, lacking only segmentation in the appendages, ambulacra, and the ability to move. They even have eyes, or at least corneae. (The material described here was mounted in Hyrax, after digestion in pepsin, so the eyespots, if present, would have been hydrolyzed.)

Fig. 1 shows a protonymph which has been dissected from a larval cuticle. The legs have a well developed chaetotaxy, but are devoid of ambulacra, which would be functionless in an immotile nymph. The greatest number of long setae (9 or 10) is found on leg I, the next greatest (5) on II (Fig. 2). There are 3 on leg III, and 0 on IV (Fig. 3). These setae are constant in position and number, but the writer is unable to correlate them with specific setae in either the larva or deutonymph.

In addition to the long setae, a number of setal vestiges are also present (Fig. 2). Again, these are relatively constant in form and position from one individual to another, but cannot be correlated with setae in larval or deutonymphal stages. They do, nevertheless, constitute a definable chaetotaxy. They could be named and numbered as precisely as the setae in any of the active stages, but since there is presently no useful purpose in doing so, I have not done this. The leg cuticle is striated, but the striae have been omitted from Figs. 1 and 2, for purposes of simplification.

Fig. 4 shows the gnathosoma, with cheliceral and pedipalpal vestiges. Here, the chaetotaxy is reduced to a single long seta and a terminal vestige on each pedipalp. Peritremal vestiges are also present, but are not shown in the figure.

Calyptostatic nymphs of all species of *Pimeliaphilus* examined by the writer have a chaetotaxy of the type described for *P. gloriosus*. Fig. 5 shows the protonymphal chaetotaxy of *P. plumifer* Newell and Ryckman 1966, plus the dorsal idiosomal chaetotaxy of the deutonymph. The larval cuticle was completely removed from the specimen. The deutonymphal setae are dotted, whether they were concealed by the protonymphal cuticle (as in the dorsocentral portion of the idiosoma), or fully exposed

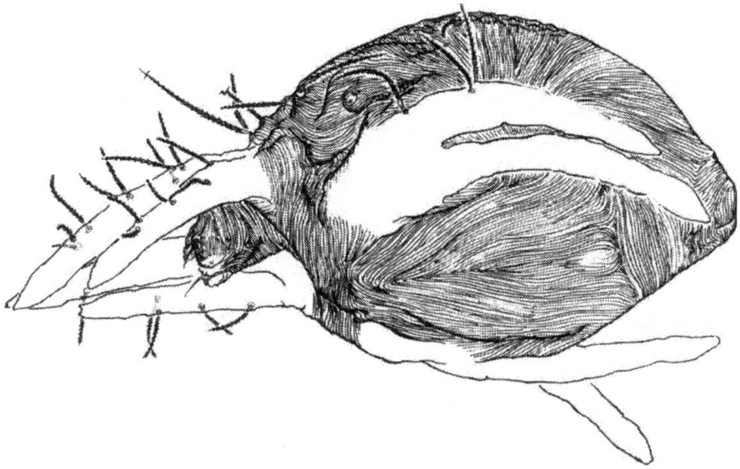

Fig. 1. ***Pimeliaphilus gloriosus***, protonymph, ventrolateral view.

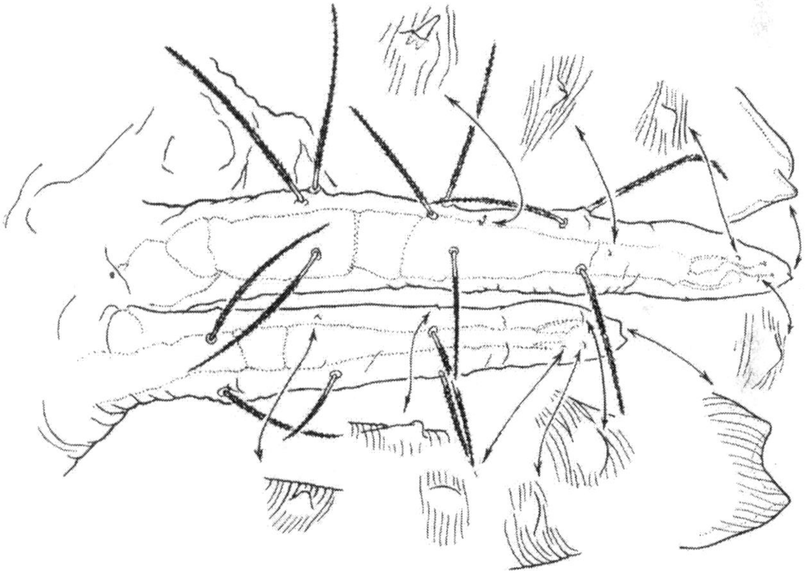

Fig. 2. ***Pimeliaphilus gloriosus***, protonymph, legs I, II, right, dorsolateral.

(as along the anterior and lateral margins), in order to differentiate them from those of the protonymph. The leg setae of the protonymph are shown, but those of the deutonymph, which are still inside the cuticle of the protonymphal legs, are omitted

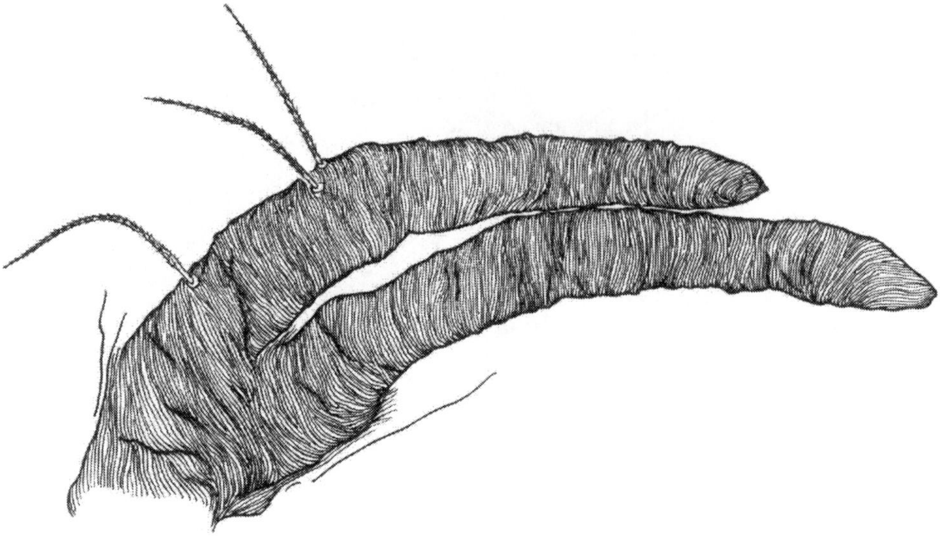

Fig. 3, *Pimeliaphilus gloriosus*, protonymph, legs III, IV, right, ventrolateral.

Fig. 4. *Pimeliaphilus gloriosus*, protonymph, gnathosoma, ventrolateral.

for simplicity. This specimen had been stopped in the process of molting as the deutonymph was preparing to emerge from the protonymphal cuticle. Splitting of the idio-

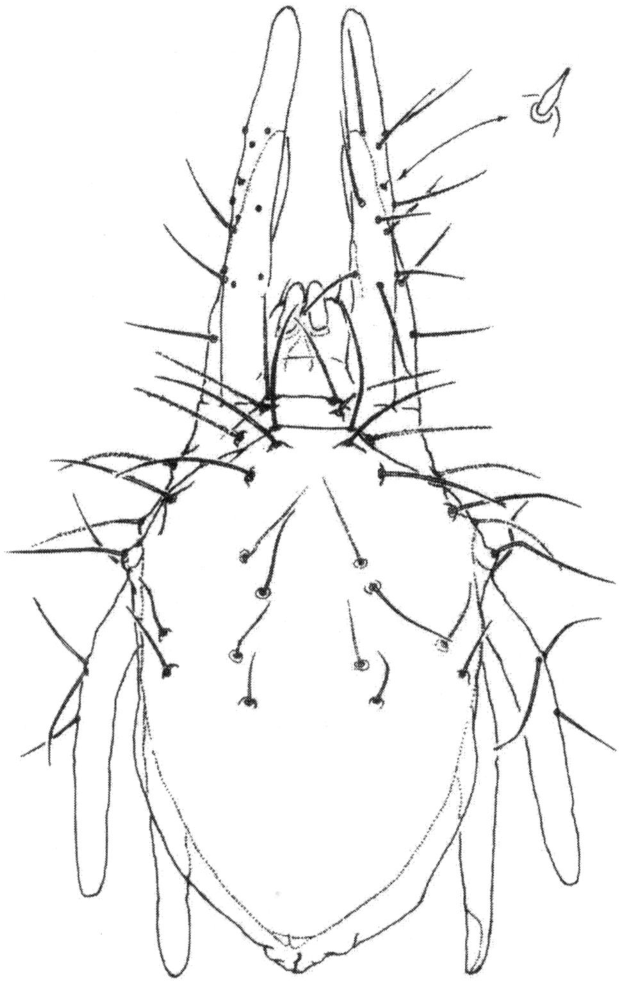

Fig. 5. ***Pimeliaphilus plumifer*, protonymph, showing dorsal chaetotaxy. Dotted setae are deutonymphal setae.**

somal cuticle is horizontal, beginning in the anterior portion and extending around the lateral margins. The legs of the deutonymph are withdrawn from the protonymphal cuticle in the usual manner.

The significance of the calyptostases in the Pterygosomatidae to those in the Parasitengona is quite clear. There can be no reasonable question that the protonymphs described above represent a true nymphal stage, despite the fact they are never motile, and show structural regressions associated with this lack of motility. At the same time, there are other genera in the Pterygosomatidae (*Geckobiella, Pterygosoma*) in which there is no chaetotaxy, but small limb vestiges are present. It is then a very simple step to the condition found in many Parasitengona in which the protonymph and tritonymph have neither chaetotaxy nor appendages. But it is still appropriate to refer to these as protonymph and tritonymph. In doing so, we could eliminate the cumbersome, obsolete terminology still used by some authors in the Parasitengona, and bring the terminology into line with that used for other Trombidiformes.

What is the evolutionary significance of the nymphal calyptostases in Pterygosomatidae and the Parasitengona? Grandjean (1938) considered that they are a manifestation of evolutionary regression. The present writer believes however, that the almost universal persistence of these stages in the groups in which they occur indicates that they represent a type of specialization which is being rigidly retained, rather than a type of evolutionary regression. In other words, structural regression in this case must not be mistaken for evolutionary regression. The Pterygosomatidae and Parasitengona appear to have evolved a functional reliance upon the calyptostase in their ontogeny. Perhaps it is essential to the complex internal reorganizations which occur in the ontogeny of these groups. Certainly, in those groups of the Trombidiformes which lack calyptostases, the dropping (or addition?) of nymphal stages appears to be far more easily accomplished, evolutionarily, than in the Parasitengona. In the writer's experience, the prime example of this is in the Halacaridae, where mono-, di-, and trinymphal groups occur. The number of nymphal instars corresponds very closely, if not precisely, with generic boundaries as presently determined by structural characteristics.

A point of further significance in the calyptostases of *Pimeliaphilus* is in the evolutionary origin of the Parasitengona. Consider the following:

(1) Both Pterygosomatidae and Parasitengona have parasitic larvae.

(2) In the Pterygosomatidae, the host-range is large, including insects and reptiles. In the Parasitengona the host range is even larger, including arthropods and vertebrates.

(3) The Pterygosomatidae include both oligotrichous and polytrichous adults. Polytrichy is the rule in the terrestrial Parasitengona, but oligotrichy in the aquatic forms.

(4) The Pterygosomatidae and the Parasitengona have calyptostatic nymphs. Calyptostatic nymphs have not been reported in other groups of Trombidiformes. The least modified calyptostatic nymphs presently known are in *Pimeliaphilus* (Pterygosomatidae), the most highly modified are in the Parasitengona.

Pterygosomatidae, like the Parasitengona, have parasitic larvae, but the latter group is characterized by a two-host life cycle. In all parasite groups in which one-host and two-host life cycles occur, one-host forms are, for good reasons, invariably considered more primitive. If this reasoning is applied to the Pterygosomatidae, these provide the most logical candidates among extant Acari, for the ancestral groups of the Parasitengona.

References

Cassagne-Méjean, F.: Sur les calyptostases des Hydrachnelles. Proc. 2nd Int. Congr. Acar., pp. 93–97, 1969.

Grandjean, F.: Sur l'ontogénie des Acariens. C. Roy. Acad. Sci. Paris, 206(3) : 146–150, 1938.

Newell, I.M., Ryckman, R.E.: Species of *Pimeliaphilus* (Acari, Pterygosomidae) attacking insects, with particular reference to species parasitizing Triatominae. Hilgardia 37(12) : 403–436, 1966.

PRELIMINARY CONSIDERATIONS OF A POPULATION MODEL FOR *BOOPHILUS MICROPLUS* IN AUSTRALIA

R. W. SUTHERST and R. H. WHARTON
Division of Entomology, CSIRO, Long Pocket Laboratories, Indooroopilly, Brisbane, Australia

Boophilus microplus has been the subject of intensive research over many years, and sufficient information is now available to encourage us to investigate the development of a model to explain the population dynamics of the tick in Australia. The initial model will serve to test the extent of our knowledge of the tick. Later it is hoped to refine the model to the stage at which it can help to assess the severity of the tick challenge to cattle, in habitats of varying favourability for *B. microplus*. As discussed by Wharton, Utech & Sutherst (1971) the use of tick resistant cattle provides a logical approach to the control of *B. microplus*. It is hoped that a simulatory model will enable us to determine the appropriate degrees of tick resistance required to give satisfactory control in different habitats.

We have made the following assumptions:

(1) Zebu *(Bos indicus)* cattle, *B. microplus* and *Babesia* blood parasites form natural host-parasite systems in the conditions under which they coexist in Asia.

(2) In Asia a stable situation exists between the numbers of *B. microplus* and those of its bovine host; the equilibrium being due essentially to the degree of host resistance to ticks which has evolved in Zebu cattle in different habitats.

(3) That the ecological process involved automatically in the numerical limitation of *B. microplus* is intraspecific competition for a feed on the host animals.

(4) That the numbers of *B. microplus* on cattle in Australia can be stabilized satisfactorily by selection of the appropriate degree of host resistance for particular conditions of climate, pasture and soil.

In the development of the model, the life cycle of *B. microplus* can be considered in three separate parts.

1. The larval productivity of engorged female ticks in pastures. 2. The intensity of host-tick contacts. 3. The mortality of ticks on the host.

Larval productivity is defined as the mean number of active larvae produced by ticks which fall into pastures. It is influenced by the mortality factors which act on the engorged female and the eggs. These factors are primarily physical, although localized predation may be severe (Wilkinson, 1970). For the successful production of larvae, temperatures above 38°C and prolonged exposure to desiccating atmospheres of less than 70 % RH must not be experienced either by the ticks or by their eggs

(Hitchcock, 1955). Adequate indicators are therefore required of the microclimate in pastures. To be of practical use on a geographical scale, they must be readily obtainable from standard meteorological data. A quantitative relationship is being sought between a meteorological index and larval productivity.

Early estimates of larval productivity have demonstrated the great significance of climate. Thus, a mean larval productivity of 250 from a potential of about 2000 was recorded over a three month period in the summer of 1968 when there were 37 cm of rain; whereas over the same period in 1969 with 10 cm of rain, the mean was only 3.6. The physical factors determining the larval productivity are now being investigated; preliminary results indicate that pasture cover and near surface soil moisture are important and interrelated with temperature in determining the microclimate in the immediate vicinity of the tick.

There are also two temporal factors to consider. Firstly, oviposition and egg development require at least 3–4 weeks; lethal conditions that occur during this period must be accounted for, as must the cumulative effects of unfavourable conditions. Secondly, the length of time during which conditions remain favourable is important. In tropical and subtropical Australia it is usual to have a wet summer and a dry winter; further south the length of the winter increases and leads to tick development being greatly extended or even arrested and eggs fail to hatch (Snowball, 1957; McCulloch and Lewis, 1968). The degree of favourability of the wet season will determine the magnitude of population changes from one generation to the next; and the length of the favourable periods will determine the number of generations that the ticks will undergo before unfavourable conditions return. The first aspect must be investigated experimentally, the second by examination of long-term meteorological data.

Tick-Host Contacts. The numbers of larvae finding a host and the rate at which they do so must be considered when assessing the tick challenge. Two major factors influence the total proportion of larvae finding a host: the stocking rate, and the longevity of larvae on the pasture. The expected asymptotic increase in the cumulative number of larvae being removed by cattle from pastures has been confirmed.

Mortality on the Host. Wilkinson (1962) and Wharton *et al.* (1969) have already demonstrated the striking effect of host resistance on tick populations in the field; it is therefore ecologically sound to exploit this mortality factor for tick control.

The effects of different degrees of tick resistance, larval productivities, and rates of tick-host contacts on tick populations can be seen in Table 1. Specific degrees of tick resistance refer to the mortality of the parasitic stages on cattle infested with 20,000 larvae. Table 1. shows that the 99 % resistance level is severe but would be necessary where all factors favour a large larval challenge; the 90 % resistance level would be inadequate in most situations.

Host resistance must be considered both as a population and as an individual attribute. With a wide spectrum of resistance within a herd, some animals will be producing many more ticks than others as shown by Wharton, Utech and Sutherst (1971).

Table 1. The interaction between tick resistance of cattle, larval productivity and tick-host contact on populations of *Boophilus microplus*

Degree of host resistance (%)	Larval productivity (% of potential)	Tick-host contact (% larvae finding a host)	Change in population sizes of *Boophilus microplus* Generation N	→	Generation N+1
90	200 (10 %)	10	1	→	1
95	400 (20 %)	10	1	→	1
99	2000 (100 %)	10	1	→	1
90	400 (20 %)	50	1	→	10
95	400 (20 %)	50	1	→	5
99	400 (20 %)	50	1	→	1

The host-parasite system for *B. microplus* is summarized in Figure 1. It can be seen that the maintenance of equilibrium is dependent upon a number of factors:

(i) In a stable situation the degree of host resistance may not be affected or counteracted sufficiently by variations in weather, pasture conditions, etc., to prevent intraspecific competition between ticks from acting as a mechanism which so adjusts their numbers that their deleterious effects on cattle are minimal.

(ii) Unstable situations are due to host resistance being either too high or too low for competition to serve as in (i). Very high resistance could lead to extinction of *B. microplus*, very low resistance either to extinction of both *B. microplus* and the cattle, or to the destruction of susceptible hosts and an increase in the resistance of the herd, leading eventually to a stable situation.

The fluctuating rainfall in Australia may produce situations where either long periods of favourable conditions allow the tick populations to increase to intolerably high numbers, or long unfavourable periods cause the local extinction of the ticks. The former would require supplementary control such as with acaricides, while the latter will tend to be minimized by favourable foci and poor condition of the hosts reducing their resistance. Inoculation for protection against babesiosis may be required as a precautionary measure where tick populations are maintained at low levels.

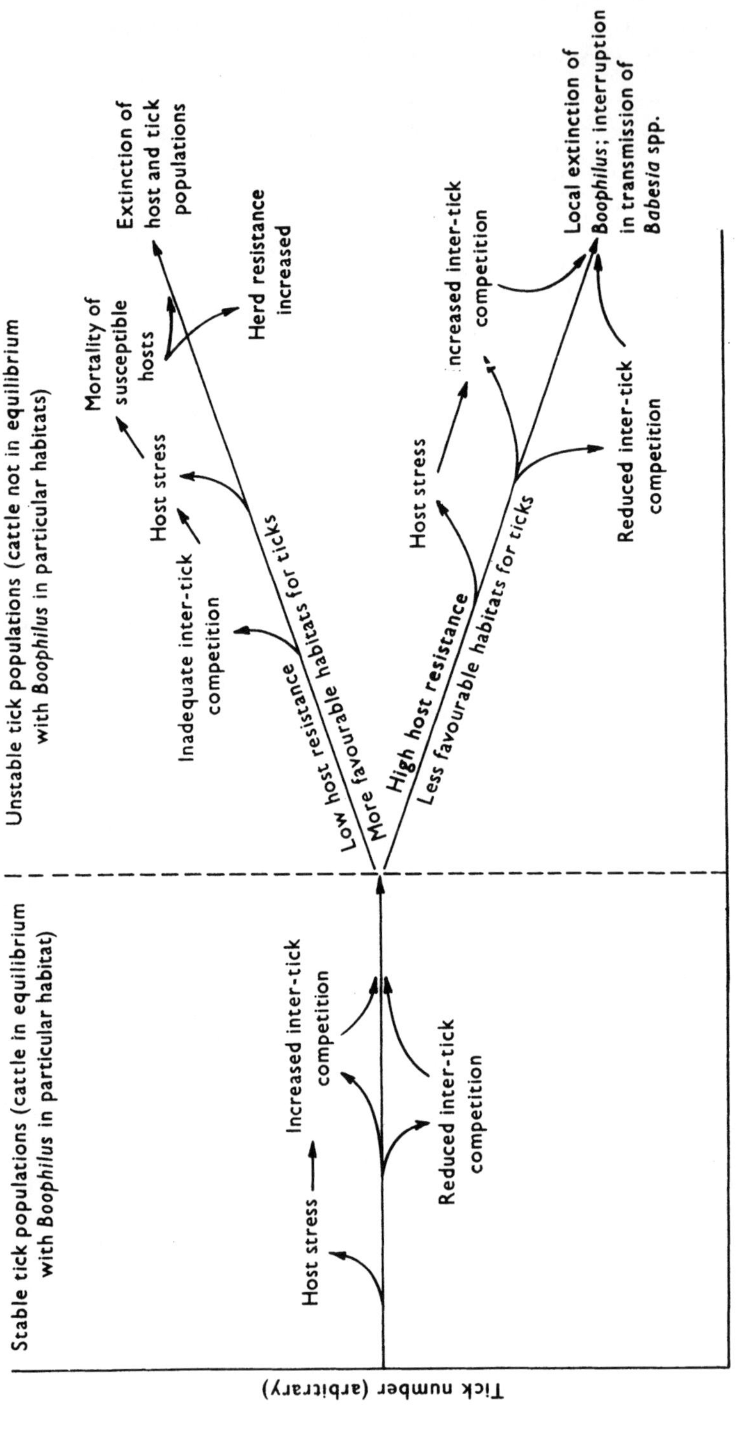

Fig. 1. Hypothetical host-parasite system for *Boophilus microplus* and cattle.

References

Hitchcock, L. F.: Studies on the non-parasitic stages of the cattle tick *Boophilus microplus* (Canestrini) (Acarina: Ixodidae). Australian J. Zool. 3, 295–311, 1955.

McCulloch, R. N., Lewis, I. J.: Ecological studies of the cattle tick, *Boophilus microplus*, in the north coast district of New South Wales. Australian J. Agric. Res. 19, 689 – 710, 1968.

Roberts, J. A.: Resistance of cattle to the tick *Boophilus microplus* (Canestrini). II Stages of the life cycle of the parasite against which resistance is manifest. J. Parasitol. 54, 667–673, 1968.

Snowball, G. J.: Ecological observations on the cattle tick, *Boophilus microplus* (Canestrini). Australian J. Agric. Res. 8, 394–413, 1957.

Wharton, R. H., Harley, K. L. S., Wilkinson, P. R., Utech, K. B., Kelley, B. M.: A comparison of cattle tick control by pasture spelling, planned dipping, and tick-resistant cattle. Australian J. Agric. Res. 20, 783–97, 1969.

Wharton, R. H., Utech, K. B. W, Sutherst, R. W.: Tick resistant cattle for the control of *Boophilus microplus*. Proc. 3rd Int. Congr. Acarol, 1971.

Wilkinson, P. R.: Selection of cattle for tick resistance and the effect of herds of different susceptibility on *Boophilus* populations. Australian J. Agric. Res. 13, 974–83, 1962.

Wilkinson, P. R.: Factors affecting the distribution and abundance of the cattle tick in Australia: observations and hypotheses. Acarologia 12, 492–508, 1970.

TERMINATION OF DIAPAUSE IN LABORATORY-REARED *DERMACENTOR ANDERSONI* ADULTS

P. R. WILKINSON
Research Station Lethbridge, Alberta, Canada

Some investigations on the acaricidal control, physiology and paralyzing ability of *Dermacentor andersoni* can be carried out on wild-caught ticks or ticks from outdoor rodentaria (Wilkinson, 1968), and such ticks can retain their normal spring activity for six or more months if stored in unlit incubators at 5°C. However, a regular supply of laboratory cultured ticks would facilitate year-round investigation and meet needs for ticks of known ancestry.

No difficulty has been experienced in rearing this tick to the unfed adult stage using methods similar to those described by Gregson (1966), although a nymphal diapause or quiescence is seen in the field (Wilkinson, 1968). If the unfed adult ticks are allowed to remain at 25°C, a high proportion will either fail to feed or engorge much more slowly than wild-caught ticks. The poor feeding is still seen after 3–9 months at 25°C, showing that starvation alone is insufficient to stimulate feeding. Gregson (1966) stated that "conditions for breaking the diapause have not yet been perfected."

In response to this challenging problem, a heuristic series of trials of stimuli found successful in terminating diapause in ticks and other arthropods (e.g. Belozerov, 1965; Beck, 1968) was commenced in 1963. Many variations in length and temperatures of chilling periods and photoperiod manipulations were tried in 1963–64, and these led to the series of 41 separate trials involving 195 separate treatments in 1965–70.

Methods

Production of unfed adult

Parent engorged female ticks oviposited at 25°C and 95% R.H., and the larvae and nymphs were fed on rabbits. Engorged larvae and nymphs were kept at 75% R.H., over saturated salt solutions. The incubators were lit only during examination.
In some of the later trials, engorged nymphs were placed in questing arenas, standing in water trays, similar to those depicted by Wilkinson (1964), where they moulted in humid conditions and the resulting adults were free to ascend sticks and 'quest'.

Assessment of termination of diapause

In the usual test, a capsule of 10 pairs of ticks from each treatment were applied to each of two sheep, and each sheep also received a capsule of control wild or outdoor 'rodentarium' ticks. The sheep were normally previously uninfested. The ticks were inspected at seven days and ten days, full engorgement and often detachment by seven days being the norm for wild ticks, and ten days being selected as the period for assessing later engorgement, and removal of ticks not spontaneously detached at seven days.

The ticks were rated according to the following scale and the spontaneously detached ticks were weighed. Weighing is not a direct indication of degree of engorgement, since it is largely dependent on the size of the flat female which varies with strain of tick and the size of the engorged nymphs. Progress of engorgement is followed by color changes of the cuticle and the stages 'red', 'tan', and 'grey' will be familiar to those working with this species. The following rating system was adapted to provide concise comparisons of degree of engorgement following the various treatments: -

No ticks attained tan or larger stages	0
Some ticks attained tan stage but none attained grey stage	1
Some ticks attained grey stage but none detached spontaneously	2
Some ticks attained grey stage and detached spontaneously	3

In one series (40)*, control ticks were deliberately omitted to test whether laboratory reared ticks could cause paralysis. The sheep became paralyzed on the seventh day (highest rated capsule 2), 90 ♂ and 90 ♀ being the original infestation.

Results

The results for 1965–70 have been summarized in Table I and the treatments producing Rating 3 at seven days have been summarized in Table II. Copies of these tables and an appendix will be deposited in the library of the Canada Department of Agriculture, Research Station, Lethbridge, Alberta. Copies may be obtained on request. These tables are too long for publication here, so the following statement will be limited to some preliminary conclusions from the 195 treatments tested.

Rating 3 at seven days was achieved after 16 of the treatments and all of these involved three or more months chilling at 5°C. In 15 out of the 16 treatments the ticks were subjected to an increase in photoperiod and temperature, e.g., from 8L to 16D during chilling at 5° to 13L–11D at 10°C, the last-mentioned regime lasting for one to two weeks before feeding. In the one exception, the ticks were exposed to 12L–12D at 5°C for three months prior to feeding. Incandescent lights were as effective as fluorescent lights. An 8L–16D photoperiod did not counteract the inhibiting effect of a 25°C temperature on diapause termination, but at this photoperiod, more

* Figures in parentheses refer to Series numbers in Table I.

ticks ascended sticks in an arena than in a comparable group kept in darkness. Active questing was not necessarily followed by engorgement within seven days. The more complex naturalistic treatments involving monthly temperature variations and freedom to 'quest' on sticks were often, but not always, more effective than the simpler treatments. The following treatments were ineffective in terminating diapause, contrary to experience with some insects; heat-shock (24 hours at 35°C) (11D) wounding of opisthosoma with a needle (4F), and soaking with water (8B, 40A). The exposure of the ticks over ammonium salts was ineffective, despite apparently favourable indications from another laboratory, and a literature reference (Beck, Shand, and Garland, 1969) to effectiveness of injection of ammonium salts (38C, 39A, 39D). Juvenile hormones remain to be tested.

Rating 3 was obtained on four occasions in the August–December period, but a circaannus rhythm (Brown, 1968) in the tick, favouring rapid engorgement in the February–June period, cannot be excluded as a possibility and needs further investigation.

Suitably stored wild ticks engorged in seven days in any month of the year, so there was no evidence of seasonal variation in host susceptibility (c.f. Rowan and Gregson, 1935, a paper based on few ticks and inadequate controls.)

Inbred strains appeared to lose some vigor after several generations in the laboratory, in comparison to F_1 ticks.

Some geographic variation in diapause requirements can be expected, so that these observations will not necessarily apply in detail to *D. andersoni* from areas other than B. C.

The trials suggest that classical graphs of percentage of individuals terminating diapause, plotted against one or two environmental factors, may sometimes be misleading, and that resumption of full vigor may require complex stimuli. It is much easier to obtain Rating 3 at ten days rather than seven days. Further analytical experiments will be needed to define the conditions for termination diapause more precisely.

References

Beck, S. D.: Insect photoperiodism. Academic Press, New York and London, 1968.

Beck, S. D., Shane, J. L., Garland, J. A.: Ammonium-induced termination of diapause in the European Corn Borer, *Ostrinia nubilalis*. J. Insect. Physiol. 15 : 945–951, 1969.

Belozerov, V. N.: Lectures at the 16th and 17th Annual N. A. Kholodovsky memorial reading. U.S.S.R. Academy of Sciences, Moscow, 1965.

Brown, F. A.: Endogenous rhythmicity reviewed with new evidence. Sciencia 103 : 1–16, 1968.

Gregson, J. D.: Ticks. In: Insect colonization and mass production. Edited by Smith, C. N., Academic Press, New York and London, pp. 49–72, 1966.

Rowan, W., and Gregson, J. D.: Winter feeding of the tick, *Dermacentor andersoni* Stiles. Nature 135 : 652, 1935.

P. R. Wilkinson

Wilkinson, P. R.: A barrier for ticks and crawling organisms. J. Econ. Entomol. 57 : 414, 1964.
Wilkinson, P. R.: Phenology, behaviour and host-relations of *Dermacentor andersoni* Stiles in outdoor "rodentaria" and in nature. Can. J. Zool. 46 : 677–689, 1968.

ELECTROGRAMS OF FEEDING TICKS RECORDED CONTINUOUSLY BY TELEMETRY

G. K. SWEATMAN, G. TOMEY and G. KATUL
School of Medicine, American University of Beirut, Beirut, Lebanon

Abstract

Electrograms recording the feeding behaviour of certain ixodid ticks when attached to a host have been studied by Gregson (Proc. 2nd Int. Congress Acarol., p. 329–339, 1969) and Sweatman and Gregson (J. Med. Entomol. 7 : 575–584, 1970). Temperature was the important factor controlling the total engorgement period, even though those parts of the electrograms interpreted as pharyngeal sucking periods showed no clear relationship with temperature. These recordings, however, were short-term and discontinuous. A new monitoring system has been developed which permits continuous observation of the feeding behaviour. Both the feeding electrogram and host skin-surface temperature at the site of the tick are being recorded by telemetry from a free-ranging host.

CLOSING ADDRESS

B. ROSICKÝ
Vice-President of the Czechoslovak Academy of Sciences, Prague

Dear colleagues and friends,

Our Congress has come to an end. During these last six days we have dealt with and discussed the most important problems of acarology. Let us have a look at what new ideas has the Congress brought us.

In my opinion, the most important one entails the necessity to publicize the importance of acarology more intensively. The public should be informed not only on a national scale, but on a global scale as well, about the significance of acarology to mankind. As a first step, acarologists in individual countries should gradually unite either in independent societies or in sections incorporated in already existing zoological, parasitological and similar societies. This step would create a solid working base for further development of acarology and would provide pre-requisites for the foundation of a World Federation of acarological societies.

I am convinced that you will approve the proposal that a letter be addressed to the World Health Organization (WHO) as well as to the Organization for Food and Agriculture (FAO) by the newly elected International committee of the Congress, underlining the importance of acarology to human health and nutrition, and requesting support for further acarological research from these international governmental organizations of the United Nations. In this connection we shall also propose to the relevant WHO and FAO representatives to assign some of their officers to maintain a permanent contact with the new International Congress of Acarology Executive Committee.

The Third International Congress has given many new impulses for further elucidation of the importance of acarology.

Above all, the economic and medical importance of mites has been underlined. Invaluable information has been obtained from separate sections, e.g. on the research results concerning the active role of mites in the circulation of organic substances in soil; new data concerning the effects of predacious mites of the family Phytoseiidae on Tetranychid mites; data concerning the effects of Tetranychid mites on the physiological condition of plants; the elucidation of the ecology of hypopi and their classification; the import of mites along with transportation of plant products, etc. From the medical aspect, a number of contributions have brought new data on the dissemination of arboviruses and other pathogenous agents by ticks and elucidated the relationship of arboviruses to ticks in general ;

on the role of Trombiculid mites in rickettsioses; the significance of nidicolous mites in the maintenance of natural foci of diseases; the exceptional importance of house dust mites in connection with the incidence of allergies in man; the continuous measurements of microclimatic effects on the development and hibernation of mites and ticks, etc. New suggestions have also been made regarding the problems of natural focality of diseases transmitted by mites and ticks. Our knowledge of the involvement of whole complexes of mites and ticks in the transmission of natural focus diseases, especially arboviroses, has been widened and deepened. Useful knowledge was also obtained concerning the control of mites and ticks and their resistance to acaricides. Our Congress has also shown the perspectives, in both the theoretical research and in practice, of further studies of genetics, physiology, behaviour and ecology of mites and ticks. Of paramount importance has been our symposium on scanning electron microscopy which offers unexpected possibilities for further studies of mite morphology and phylogeny. Invaluable suggestions as regards the classification of higher units, have been made by taxonomic section.

The Czechoslovak Organizing Committee will consider it an honour to prepare for press as soon as possible the Proceedings of this Congress which we believe will further publicize acarology in the world.

In all sections and in the symposium as well lively discussions followed the papers submitted. Very gratifying has been the participation of young acarologists from various countries and this fact holds a promise for the development of acarology in future. The discussions have given much impetus to further development of the various branches of acarology. Acarologists have thus obtained further information on fields related to their own professional interest.

Materials submitted in separate sections have confirmed further advances of acarology achieved since the Congress in Nottingham.

Allow me to thank all participants of the Third International Congress of Acarology, in particular all colleagues from abroad, for a very active and valuable cooperation and for taking part in friendly and cordial discussions on acarological problems so dear to our hearts and our interest.

Allow me to thank the Presidium of the Czechoslovak Academy of Sciences which made this Congress possible by their support and understanding the necessity for a further development of acarology.

We are much obliged to the just resigned International Committee, which has prepared new projects for further organized work of acarologists. We also thank sectional chairmen and assistant chairmen who have made exceptional efforts to maintain a high scientific level of our sessions.

I would also like to thank all staff members of the Institute of Parasitology of the Czechoslovak Academy of Sciences who devoted many months, days and hours in preparing and organizing this Congress.

I would like to express special thanks to the members of the Organizing Committee representing leading workers of Czechoslovak acarological research centres, for their contribution to the success of the Congress, in particular to Dr M. Daniel, Dr V. Černý, Dr K. Šamšiňák, Dr F. Dusbábek and others.

We also appreciate the difficult and very useful work of our interpreters who have helped to overcome the language barrier in our sessions.

I believe that you are satisfied with the Congress and with its scientific standard. I wish you all good health and much energy for the further development of all trends of acarology, and continued success until our next meeting at the Fourth International Congress of Acarology.

I hereby declare the Congress closed. Thank you.

THE RULES OF THE INTERNATIONAL CONGRESS OF ACAROLOGY[1]

A. The name of the organization shall be the International Congress of Acarology herein-after referred to as the Congress.

B. Membership

1. All individuals and organizations engaged in the study of the Acari are eligible for membership.
2. Membership shall be automatic on payment of a fee as defined in Sections F (2) and F (3) unless the person registering shall indicate otherwise.
3. The Congress may confer Honorary Membership upon distinguished Acarologists. Their number shall not exceed seven at any one time.
4. Membership shall lapse if the appropriate fee has not been paid by the commencement of the Congress.

C. Organization of the Congress

1. The Congress shall carry out its work through the following bodies:
   (a) The Officers
   (b) The Executive Committee

D. The Officers

1. The Officers shall consist of the President, the immediate Past-President, the Secretary and the Treasurer.
2. The President shall be a resident of the country in which the Congress will be held.

E. Executive Comittee

1. The Executive Committee shall consist of the Officers and eight other members.
2. The Executive Committee shall conduct the business of the Congress through the Officers and shall be responsible for the continuing activities of the Congress.
3. The major business of the Executive Committee shall be as follows:
   (a) to maintain a list of members of the Congress;
   (b) to decide the membership fee;
   (c) to handle the funds of the Congress and to decide on its expenditure;
   (d) to arrange for the nomination of members as candidates for election as Officers of the Congress and members of the Executive Committee at the Plenary Session of the Congress;
   (e) to decide on the venue of each Congress;
   (f) to supervise the programme of the Congress and the publication of its Proceedings;

---

[1] These Rules were presented by the Executive Committee to the Third International Congress of Acarology and accepted by the Congress with minor emendations as included here.

These Rules were adopted with the explicit understanding that they would be subject to emendation by the Congress. The Executive Committee is preparing a list of proposed modifications to clarify the Rules and make them more workable, and these proposals will be presented to the Fourth Congress. Any Acarologist who wishes to make suggestions regarding the Rules is urged to transmit them to the Secretary.

(g) to sponsor a World List of Acarologists giving addresses and information of the particular interest of each, to be published at regular intervals. This list is to include all active Acarologists and is not to be limited on the basis of participation in the Congress or the payment of any sum.

F. The Congress

1. The organization of each Congress will be delegated to a local committee constituted by the President and hereinafter referred to as the Congress Organizing Committee. The President shall ensure that the Executive Committee is consulted on matters of substance such as the structure of the Congress, the topics represented and the publication of the Proceedings of the Congress.
2. The Congress registration fee for full participants will consist of two parts:
   (a) a sum to cover the cost of membership;
   (b) other Congress costs included under this heading.
3. Members unable to attend the Congress can maintain their membership by payment of that part of the registration fee referred to under 2 (a) of this section. They shall receive the usual Congress circulars and the World List of Acarologists but shall not be entitled to receive the Congress programme or other services provided under 2 (b) of this Section. All Acarologists listed in the world list shall receive a copy of this publication , the manner of distribution to be decided by the Executive Committee.

G. The Plenary Session of the Congress

1. At the Plenary Session the following business shall be transacted:
   (a) the discussion and, if seen fit, the approval of the reports of the Secretary and Treasurer, and any recommendations made by the Executive Committee;
   (b) the discussion of any proposal made in writing by at least two members provided it has been sent to the Secretary 6 months before the Congress meets;
   (c) voting on any proposed changes of the Rules;
   (d) announcement of the venue for the next Congress;
   (e) the election of the Officers and members of the Executive Committee duly nominated in accordance with Section 1 (1).
2. Voting on proposals shall be done by a show of hands. A secret paper ballot shall be made for all contested elections and in these circumstances scrutineers shall be appointed.
3. Decisions shall be passed by a simple majority vote, with the exception of proposals concerning changes of the Rules, where a two-thirds majority of the votes cast shall be required.

H. Change of Rules

1. Proposals for changes in the Rules can be made by the Executive Committee or by at least 20 members. They must be communicated to the Secretary at least six months before the next Congress and be published in a Congress circular. The Secretary shall send voting papers by post to those members not attending the Congress.

I. Other Regulations

1. The elected Officers and other members of the Executive Committee shall serve from the end of one Congress to the end of the following Congress. The Secretary shall seek nominations of members for election to the above positions prior to the Congress. Nominations, proposed and seconded, must be in the Secretary's hands not later than 48 hours before the Plenary Session. No Officer or member of the Executive Committee, other than the Secretary and Treasurer shall serve for more than three Congresses consecutively.
2. The business of the Congress may be conducted by post.
3. The Executive Committee may set up Sub-committees for carrying out specific functions, and they may co-opt other members to serve on them.

4. The Executive Committee may settle any question not covered by the Rules if it sees fit. It may also give decisions in all doubtful cases as regards the interpretation of the Rules.

LIST OF PARTICIPANTS

Abbatiello, M., Agricultural and Technical College, State University, Farmingdale, New York 11735, USA
Akhmetbekova, R.T., Institute of Zoology, the Kazakh Academy of Sciences, Alma-Ata, USSR
Akimov, I.A., Institute of Zoology, Ukrainian Academy of Sciences, Kiev-30, USSR
Anastos, G., Dept. of Zoology, University of Maryland, College Park, Maryland 20740, USA
Aoki, J., National Science Museum, Tokyo, Japan
Ašmera, J., Regional Centre of Hygiene and Epidemiology, Ostrava, Czechoslovakia
Atanasov, N., Plant Protection Institute, Sofia, Bulgaria
Athias, F., INRA, Faune du Sol. 21-Dijon, France
Athias-Henriot, C., INRA, Faune du Sol, 21-Dijon, France
Attiah, H., Plant Protection Dept., Ministry of Agriculture, Dokki, Cairo, Egypt, ARE
Atyeo, W.T., Dept. of Entomology, University of Georgia, Athens, Georgia 30601, USA
Axtell, R.C., Dept. of Entomology, North Carolina State University, Raleigh, North Carolina 27607, USA
Babenko, L.V., The Martsinovsky Institute of Medical Parasitology and Tropical Medicine, Moscow G-435, USSR
Bader, C., Museum of Natural History, CH-4051 Basel, Switzerland
Balashov, Yu.S., Zoological Institute, USSR Academy of Sciences, Leningrad B-164, USSR
Batista Collazo, O., Escuela de Ing. Agrónomos, Universidad de Habana, Cuba
Bazan-Strzelecka, H., Universytet Łodzki, Museum Katedry Ewolucjonizmu, Łodz, Poland
Belozerov, V.N., Dept. of Entomology, University of Leningrad, Leningrad B-164, USSR
Bengston, M., Dept. of Prim. Industries Meiers RD, Indooroopilly, Brisbane, Australia
Bernini, F., Istituto di Zoologia, Università di Siena, Siena, Italy
Boczek, J., Warsaw Agricultural University, SGGW, Ursynów, Warsaw, Poland
Bognar, S., University of Horticulture, Budapest, Hungary
Bohnsack, K.K., Dept. of Zoology, San Diego State College, San Diego, California 92115, USA
Bravenboer, L., Proefstation voor de Groentenen, Fruitteelt onder Glas, Naaldwijk, Netherlands
Bregetova, N.G., Zoological Institute, USSR Academy of Sciences, Leningrad B-164, USSR
Brennan, J.M., Rocky Mountain Laboratory, U.S. Public Health Service, Hamilton, Montana 59840, USA
Bronswijk, J. van, Katholieke Universiteit, Nijmegen, Netherlands
Bulanova-Zakhvatkina, E.M., The Lomonosov State University, Fac. of Biology and Soil Science, Moscow B-234, USSR
Bund, C.F., van de, Phytopathological Service of the Netherlands, Wageningen, Netherlands
Carmona, M.M., Estãçao Agronómica Nacional, Oeiras, Portugal
Černý, V., Institute of Parasitology, Czechoslovak Academy of Sciences, Praha, Czechoslovakia
ChannaBasavanna, G.P., University of Agricultural Sciences, Hebbal, Bangalore-24, India
Chhabra, K., present address: Zukovskov, 6, Kiev, USSR
Chinery, W., Ghana Medical School, Accra, Ghana
Chmielewski, W., Institute of Plant Protection, Poznań, Poland
Coineau, Y., Laboratoire Arago, Faculté des Sciences de Paris, 66 Banyuls-sur-Mer, France
Conroy, J.C., Dept. of Biology, University of Winnipeg, Winnipeg 2, Manitoba, Canada

Costa, M., Kibbutz Mishmar Haemek, Kiryat, Tivon, Israel
Cranham, J.E., East Malling Research Station Maidstone, Kent, England
Cser, Z., Institute of Parasitology, Veterinary College, Budapest, Hungary
Czajkowska, B., Dept. of Applied Entomology, Warsaw Agricultural University, SGGW, Ursynów, Warsaw, Poland
Dąbrowski, Z., Warsaw Agricultural University, SGGW Ursynów, Warsaw, Poland
Daniel, M., Institute of Parasitology, Czechoslovak Academy of Sciences, Praha, Czechoslovakia
Danielová, V., Institute of Parasitology, Czechoslovak Academy of Sciences, Praha, Czechoslovakia
Desch, C.E., Dept. of Zoology, University of Massachussetts, Amherst, Mass. 01002, USA
Devine, T.L., Institute of Applied Zoology, West Berlin
Diehl, P., Dept. of Zoology, Cambridge University, Cambridge, England
Dinsdale, D., New University of Corelaine, Corelaine, Northern Eire
Downing, F., ICI Ltd. Alderlay Park, Macclesfield. Cheshire, England
Dubinina, H.V., Zoological Institute, USSR Academy of Sciences, Leningrad B-164, USSR
Dusbábek, F., Institute of Parasitology, Czechoslovak Academy of Sciences, Praha, Czechoslovakia
Düszgünez, Z., Ankara Universitesi Ziraat Fakültesi, Bitki Koruma Kürsüsü, Ankara, Turkey
Edler, A., Ecological Dept., Zoological Institute, S-223 62 Lund, Sweden
Engelmann, H.D., Staatliches Museum für Naturkunde, 89 Görlitz, DDR
Evans, G.O., Dept. of Agricultural Zoology, University College, Dublin 9, Eire
Eyndhoven, G.L. van, Zoological Museum, Amsterdam, Netherlands
Fain, A., Institut de Médecine Tropicale, 2000 Antwerpen, Belgium
Feider, Z., Laboratorul de Zoologie, Universitatea Al.I. Cuza, Iaşi, Romania
Feldman-Muhsam, B., Hebrew University, Jerusalem, Israel
Filipponi, A., Istituto Superiore di Sanità, Roma, Italy
Filippova, N.A., Zoological Institute, USSR Academy of Sciences, Leningrad B-164, USSR
Fritzsche, R., Institut für Phytopathologie, 432 Aschersleben, DDR
Gadjiev, A.T., Institute of Zoology, Azerbaijan Academy of Sciences, Baku-73, USSR
Gasser, R., Reserch Dept., J.R. Geigy A.S., Basel, Switzerland
Georgiev, B., Veterinary Institute, Plovdiv, Bulgaria
Gerson, U., Faculty of Agriculture, The Hebrew University of Jerusalem, Revohot, Israel
Giboda, M., Institute of Hygiene and Epidemiology, Košice, Czechoslovakia
Gledhill, T., Freshwater Biological Association, River Laboratory, Wareham, Dorset, England
Goncharova, A.A., Dept. of Biology, The Chita Medical Institute, Chita, USSR
Gorchakovskaya, N.N., Institute of Poliomyelitis and Viral Encephalitides, USSR Academy of Medical Sciences, Moscow B-27, USSR
Griffiths, D., Pest Infestation Laboratory, Agricultural Research Council, Slough, Bucks., England
Gritsenko, E.F., Poultry Breeding Farm, Krasnodar 41, USSR
Haarløv, N., Royal Veterinary and Agricultural College, 1870 Copenhagen V, Denmark
Hafez, M., Dept. of Entomology, Faculty of Sciences, Cairo, ARE
Halašková, V., Institute of Zoology, Charles University, Praha, Czechoslovakia
Halík, V., Žitná 36, Praha, Czechoslovakia
Hall, C.A., Lake Jackson, Texas 11566, USA
Hammen, L., van der, Rijksmuseum van Natuurlijke Historie, Leiden, Netherlands
Hammer, M., Roland, Fredensborg, Denmark

Hammer, O., Kastanie Allé, Allerød, Denmark
Hansell, R., Dept. of Zoology, University of Toronto, Ontario, Canada
Harrison, I.R., Thurgarton Research Station, The Priory, Thurgarton, Nottingham NG 14 7GX, England
Hatzinikolis,N.E., Institute Pathologic Benaki, Kiffisia, Athens, Greece
Heinz, F., Regional Centre of Hygiene and Epidemiology, Ostrava, Czechoslovakia
Herne, H.C., Research Station, Canada Dept, of Agriculture, Vineland Station, Ontario, Canada
Hirschmann, W., Am Kavierlein 26, 851 Fürth i.B., BRD
Hoffman, J.R., College of Natural Sciences, Michigan State University, East Lansing, Michigan 48823, USA
Holman, L.J., Dept. of Natural Sciences, Longwood College, Farmville, Virginia 23901, USA
Honzáková, E., Institute of Parasitology, Czechoslovak Academy of Sciences, Praha, Czechoslovakia
Hoogstraal, H., c/o Spanish Embassy, American Interests Section, Cairo, Egypt, ARE
Hudáková, A., Natural History Faculty, Comenius University, Bratislava, Czechoslovakia
Hushcha, G.J., Institute of Zoology, Ukrainian Academy of Sciences, Kiev-30, USSR
Hyland, K., University of Rhode Island, Kingston, Rhode Island, USA
Jedlička, L., Natural History Faculty, Comenius University, Bratislava, Czechoslovakia
Jeleva, M.A., University of Sofia, Sofia, Bulgaria
Kadjaya, G.Sh., Institute of Zoology, Georgian Academy of Sciences, Tbilisi 30, USSR
Kaiser, M.N., c/o Spanish Embassy, American Interests Section, Cairo, Egypt, ARE
Kethley, J., Field Museum of Natural History, Chicago 111, USA
Kevan, D.K., Dept. of Entomology, McGill University, Mac Donald College, P. Quebec, Canada
Khudyakov, I.S., The Kirov Military Medicine Academy, Dept. of Biology, Leningrad K-9, USSR
Kiefer, M., Natural History Faculty, Comenius University, Bratislava, Czechoslovakia
Kitahara, K., Faculty of Medicine, Kagoshima University, Kagoshima, Japan
Kmeť, M., Central State Veterinary Institute, Bratislava, Czechoslovakia
Knülle, W., Institut für Angewandte Zoologie der freien Universität Berlin, 1 Berlin 41, West Berlin
Kolebinova, M., Zoological Institute, Bulgarian Academy of Sciences, Sofia, Bulgaria
Kolman, J.M., Institute of Parasitology, Czechoslovak Academy of Sciences, Praha, Czechoslovakia
Korotkov, Yu.S., Institute of Poliomyelitis and Viral Encephalitides, USSR Academy of Medical Sciences, Moscow B-27, USSR
Kováčová, E., Institute of Virology, Slovak Academy of Sciences, Bratislava, Czechoslovakia
Kožuch, O., Institute of Virology, Slovak Academy of Sciences, Bratislava, Czechoslovakia
Kramárová, L., Natural History Faculty, Comenius University, Bratislava, Czechoslovakia
Kramář, J., Natural History Faculty, Charles University, Praha, Czechoslovakia
Kratochvil, J., Institute of Vertebrate Zoology, Czechoslovak Academy of Sciences, Brno, Czechoslovakia
Krivolutsky, D.A., Institute of Evolutionary Morphology and Ecology, USSR Academy of Sciences, Moscow B-133, USSR
Kropczyńska-Linkiewicz, D., Dept. of Applied Entomology, Warsaw Agricultural University, SGGW, Ursynów, Warsaw, Poland
Kucheruk, V.V., The Gamaleya Institute of Epidemiology and Microbiology, USSR Academy of Medical Sciences, Dept. of Natural Foci Infections, Moscow D-98, USSR

Kudon, L., Zoology Dept., University of Rhode Island, Rhode Island, USA
Kurcheva, G.F., Institute of Evolutionary Morphology and Ecology, USSR Academy of Sciences, Moscow B-133, USSR
K unst, M., Natural History Faculty, Charles University, Praha, Czechoslovakia
Lebel, R., Boston College, Chestnut Hill, Massachusetts, USA
Levy, V., Research Institute of Epedemiology, Sofia, Bulgaria
Lindquist, E., Entomology Research Institute, Ottawa 2, Ontario, Canada
Lom, J., Institute of Parasitology, Czechoslovak Academy of Sciences, Praha, Czechoslovakia
Loomis, R.B., Dept. of Biology, California State University, Long Beach, California 90801, USA
Luxton, M., Molslaboratoriet, Femmoler. Denmark
Měrka, V., VLVDÚ, Hradec Kralové, Czechoslovakia
Minář, J., Institute of Parasitology, Czechoslovak Academy of Sciences, Praha, Czechoslovakia
Mironescu, J., Muzeul de Istorie Naturala, Iaşi, Romania
Mitrofanov, V.I., Botanical Gardens, Jalta, Crimea, USSR
Moorhouse, D.D., Dept. of Parasitology, University of Queensland, St. Lucia, Brisbane, Australia
Moritz, M., Zoologisches Museum, Berlin, DDR
Mrciak, M., Natural History Faculty, Comenius University, Bratislava, Czechoslovakia
Murphy, P.W., University of Nottingham, School of Agriculture, Sutton, Bonington, England
Musatov, V.A., Agricultural Institute, Dept. of Zoology, Pskov Region, Velikie Luki, USSR
Müller, J., VEB.-Fahlberg-List. Magdeburg, DDR
Myasnikov, Yu.A., Institute of Poliomyelitis and Viral Encephalitides, USSR Academy of Medical Sciences, Moscow, USSR
Natcheff, P.D., Higher Agricultural Institute, Sofia, Bulgaria
Naudo, M., National Museum of Natural History, Lab. of Zoology, Paris, France
Neronov, V.N., WHO, Geneva, Switzerland
Newell, I., Dept. of Biology, University of California, Riversides, California, USA
Nosek, J., Institute of Virology, Slovak Academy of Sciences, Bratislava, Czechoslovakia
Novák, D., Entomological Laboratory, Hodonín Hospital, Hodonin, Czechoslovakia
Nutting, W.B., Zoology Dept., University of Massachusetts, Amherst, Mass. 01002, USA
O'Donnell, T., Animal Ecology Research Group, Dept. of Zoology, Oxford, England
Oliver, J.H., Dept. of Biology, Georgia Southern College, Statesboro, Georgia 30458, USA
Országh, I., Natural History Faculty, Comenius University, Bratislava, Czechoslovakia
Parr, W.J., Glasshouse Crops Research Institute, Littlehampton, Sussex, England
Pegazzano, F., Research Institute of Agricultural Zoology, 50125 Firenze, Italy
Peterson, P.C., Dept. of Entomology, Youngstown State University, Youngstown, Ohio, USA
Petrova, A., Zoological Institute, Bulgarian Academy of Sciences, Sofia, Bulgaria
Piffl, E., Zoologisches Institut der Universität Wien, Wien 1, Austria
Polyakov, D.K., All-Union Research Institute of Veterinary Sanitation, Moscow, USSR
Přívora, M., Institute of Hygiene and Epidemiology, Praha, Czechoslovakia
Punčochář, P., Hydrobiological Laboratory, Botanical Institute, Czechoslovak Academy of Sciences, Praha, Czechoslovakia
Rack, G., Zoologisches Institut und Museum, 2000 Hamburg 13, BRD
Radovsky, F.J., Bernice P. Bishop Museum, Honolulu, Hawaii, USA
Rajski, A., Faculty of Zoology W.S.R., Szczeçin, Poland
Řeháček, J., Institute of Virology, Slovak Academy of Sciences, Bratislava, Czechoslovakia
Rehse-Küpper, B., Abt. für Virologie, Univ. Nervenklinik, 5 Köln-Lindenthal, BRD
Reinhard, R., VEB-Fahlberg-List, Magdeburg, DDR
Robaux, P., Laboratoire d'Ecolgie Générale, 91 Brunoy, France

Rodriguez, T.G., Dept. of Entomology, University of Kentucky, Lexington, Kentucky 40506, USA

Rohde, CH.J. (Jr.), Dept. of Biological Sciences, Northern Illinois University, Dekalb, Illinois 60115, USA

Rosický, B., Institute of Parasitology, Czechoslovak Academy of Sciences, Praha, Czechoslovakia

Roulston, W.J., CSIRO, Long Pocket Laboratories, Indooroopilly, Queensland, Australia

Rupeš, V., Institute of Hygiene and Epidemiology, Praha, Czechoslovakia

Ryba, J., Institute of Parasitology, Czechoslovak Academy of Sciences, Praha, Czechoslovakia

Saba, F., Vero Beach Laboratories, Vero Beach, Florida 32690, USA

Samšiňák, K., Institute of Parasitology, Czechoslovak Academy of Sciences, Praha, Czechoslovakia

Schuster, R., Zoologisches Institut der Universität Graz, Graz, Austria

Sepasgozarian, H., Faculty of Agriculture, Karadj, Iran

Shehata, K.K., Acarology Research Unit, National Research Center, Dokki, Cairo, ARE

Shereef, G.M., Dept. of Entomology, Moscow University, Moscow B-234, USSR

Shcherbak, G.J., Institute of Zoology, Ukrainian Academy of Sciences, Kiev-30, USSR

Silvere, A.P., Institute of Experimental Biology, Estonian Academy of Sciences, Tallin-Harju, USSR

Sitnikova, L.G., Zoological Institute, USSR Academy of Sciences, Leningrad B-164, USSR

Şixl, W., Zoologisches Institut der Universität Graz, Graz, Austria

Šlais, J., Institute of Parasitology, Czechoslovak Academy of Sciences, Praha, Czechoslovakia

Spieksma, F.T.M., Akademisch Ziekenhuis, Leiden, Netherlands

Steinhausen, W., Leupoldsdorf, BRD

Strandtmann, R., Texas Techn. University, Lubbock, Texas, USA

Sweatman, G.K., American University of Beirut, Beirut, Lebanon

Szymański, S., Dept. of Parasitology, Polish Academy of Sciences, Warsaw, Poland

Tadros, M.S., Faculty of Agriculture, Alexandria University, Alexandria, ARE

Tagiltsev, A.A., Kazakh Institute of Epidemiology, Microbiology and Hygiene, Alma-Ata, USSR

Tareev, V.N., Institute of Agriculture, Ussurijsk, USSR

Teschner, D., Zoologisches Institut, Braunschweig, BRD

Travé, J., Arago Laboratoires, Banyuls-sur-Mer, France

Treat, A., American Museum of Natural History, Dumont, New York, USA

Ugolini, A., Ministero dell Agriculture e delle Foreste, Bologna, Italy

Uspenskaya, I.G., Institute of Zoology, Moldavian Academy of Sciences, Kishinev, USSR

Vaněk, J., Institute for Shaping and Protection of Countryside, Czechoslovak Academy of Sciences, Říčany, Czechoslovakia

Vaszilková, A., Central State Veterinary Institute, Bratislava, Czechoslovakia

Vávra, J., Institute of Parasitology, Czechoslovak Academy of Sciences, Praha, Czechoslovakia

Vercammen-Grandjean, P., University of California, San Francisco Medical Center, San Francisco, USA

Verner. P., Dept. of Entomology, Research Institute of Food Industry, Praha, Czechoslovakia

Vobrázková, E., Institute of Parasitology, Czechoslovak Academy of Sciences, Praha, Czechoslovakia

Volgin, V., Zoological Institute, USSR Academy of Sciences, Leningrad B-164, USSR

Vrie, M., van de, Fruit Control Station, Wilhelminadorp, Netherlands

Walwork, J., Dept. of Zoology, University of London, London, England

Webb, N., Nature Conservancy, Wargham, Dorset, England
Wharton, G.W., Dept. of Zoology and Entomology, Ohio State University, Columbus, Ohio, USA
Wilkinson, P., Research Station, Lethbridge, Alberta, Canada
Woolley, T.A., Colorado State University, Fort Collins, Colorado, USA
Yankovskaya, A.I., Zoological Institute, USSR Academy of Sciences, Leningrad B-164, USSR
Žďárková, E., Dept. of Entomology, Research Institute of Food Industry, Praha, Czechoslovakia
Zohdy, G., present address: Research Institute of Plant Production and Plant Protection, Praha-Ruzyně, Czechoslovakia
Župančičová, M., Institute of Virology, Slovak Academy of Sciences, Bratislava, Czechoslovakia

INDEX

B

C

G

H

I

J

K

L

M

Q

R

S

U

V

W

X

Z

FSC
www.fsc.org
MIX
Papier | Fördert
gute Waldnutzung
FSC® C083411

Kolibri 360 GmbH
Ferdinand-Jühlke-Straße 7
99095 Erfurt, Deutschland
produktsicherheit@kolibri360.de